HVAC
Level Three

Trainee Guide

Boston   Columbus   Indianapolis   New York   San Francisco   Amsterdam
Cape Town Dubai London Madrid Milan Munich Paris Montreal Toronto Delhi
Mexico City   Sao Paulo   Sydney   Hong Kong   Seoul   Singapore   Taipei   Tokyo

NCCER

President: Don Whyte
Vice President: Steve Greene
Chief Operations Officer: Katrina Kersch
HVAC Project Manager: Chris Wilson
Senior Development Manager: Mark Thomas
Senior Production Manager: Tim Davis

Quality Assurance Coordinator: Karyn Payne
Desktop Publishing Coordinator: James McKay
Permissions Specialists: Kelly Sadler
Production Specialist: Kelly Sadler
Production Assistance: Hannah Payne
Editors: Graham Hack, Debie Hicks

Writing and development services provided by Topaz Publications, Liverpool, NY

Lead Writer/Project Manager: Troy Staton
Desktop Publisher: Joanne Hart
Art Director: Alison Richmond

Permissions Editor: Andrea LaBarge
Writers: Troy Staton, Thomas Burke, Terry Egolf

Pearson

Director of Alliance/Partnership Management: Andrew Taylor
Editorial Assistant: Collin Lamothe
Program Manager: Alexandrina B. Wolf
Assistant Content Producer: Alma Dabral
Digital Content Producer: Jose Carchi
Director of Marketing: Leigh Ann Simms

Senior Marketing Manager: Brian Hoehl
Composition: NCCER
Printer/Binder: LSC Communications
Cover Printer: LSC Communications
Text Fonts: Palatino and Univers

Credits and acknowledgments for content borrowed from other sources and reproduced, with permission, in this textbook appear at the end of each module.

Copyright © 2018, 2013, 2008, 2002, 1996 by NCCER, Alachua, FL 32615, and published by Pearson, New York, NY 10013. All rights reserved. Printed in the United States of America. This publication is protected by Copyright and permission should be obtained from NCCER prior to any prohibited reproduction, storage in a retrieval system, or transmission in any form or by any means, electronic, mechanical, photocopying, recording, or likewise. For information regarding permission(s), write to: NCCER Product Development, 13614 Progress Blvd., Alachua, FL 32615.

41 2024

ISBN-13: 978-0-13-518510-0
ISBN-10: 0-13-518510-6

Preface

To the Trainee

Heating, ventilating, and air-conditioning (HVAC) systems technicians install and repair systems that regulate the temperature, humidity, and the total air quality in residential, commercial, industrial, and other buildings. Refrigeration systems, which are used to transport or store food, medicine, and other perishable items, are also a major part of the HVAC technician's job. Hydronics (water-based heating systems), solar panels, and other specialized heating, cooling, and refrigeration systems are also important facets of the HVAC industry.

As a technician, you must be able to install, maintain, and troubleshoot problems throughout the entire system. You must know how to follow drawings or other specifications to maintain any system. You will have to understand air movement, temperatures, and pressures. You may also need basic working knowledge of other crafts such as sheet metal, welding, basic pipefitting, and electrical practices.

Nearly all buildings and homes in the United States alone use forms of heating, cooling, and ventilation to maintain comfort. The increasing advancement of HVAC technology causes employers to recognize the importance of continuous education and keeping up to speed with the latest equipment and skills. Hence, technical school training or apprenticeship programs often provide an advantage and a higher qualification for employment.

NCCER's HVAC program has been designed by highly qualified subject matter experts. The four levels present an apprentice approach to the HVAC field, including theoretical and practical skills essential to your success as an HVAC technician. The US Department of Labor projects faster than average job growth in the HVAC industry.

We wish you the best as you begin an exciting and promising career. This newly revised HVAC curriculum will help you enter the workforce with the knowledge and skills needed to perform productively in either the residential or commercial market.

New with *HVAC Level Three*

NCCER is proud to release *HVAC Level Three* with updates to the curriculum that will engage you and give you the best training possible.

TIn this edition, you will find a number of important updates to the content. In "Basic Maintenance," new information has been added about technician safety, magnetic bearings, and online work orders. New training on communicating thermostats has been added to "Control Circuit and Motor Troubleshooting." The material from "Zoning, Ductless, and Variable Refrigerant Flow Systems" was updated to include the latest technology, including information from Daiken, one of the industry's top manufacturers of these systems. Coverage of variable speed units and controllers also was added to "Troubleshooting Heat Pumps." The module "Retail Refrigeration Systems" now addresses the use of ozone and water filtration as maintenance and troubleshooting training for these systems.

We wish you success as you progress through this training program. If you have any comments on how NCCER might improve upon this textbook, please complete the User Update form located at the back of each module and send it to us. We will always consider and respond to input from our customers.

We invite you to visit the NCCER website at **www.nccer.org** for information on the latest product releases and training, as well as online versions of the *Cornerstone* magazine and Pearson's NCCER product catalog.

Your feedback is welcome. You may email your comments to **curriculum@nccer.org** or send general comments and inquiries to **info@nccer.org**.

NCCER Standardized Curricula

NCCER is a not-for-profit 501(c)(3) education foundation established in 1996 by the world's largest and most progressive construction companies and national construction associations. It was founded to address the severe workforce shortage facing the industry and to develop a standardized training process and curricula. Today, NCCER is supported by hundreds of leading construction and maintenance companies, manufacturers, and national associations. The NCCER Standardized Curricula was developed by NCCER in partnership with Pearson, the world's largest educational publisher.

Some features of the NCCER Standardized Curricula are as follows:

- An industry-proven record of success
- Curricula developed by the industry, for the industry
- National standardization providing portability of learned job skills and educational credits
- Compliance with the Office of Apprenticeship requirements for related classroom training (*CFR 29:29*)
- Well-illustrated, up-to-date, and practical information

NCCER also maintains the NCCER Registry, which provides transcripts, certificates, and wallet cards to individuals who have successfully completed a level of training within a craft in NCCER's Curricula. *Training programs must be delivered by an NCCER Accredited Training Sponsor in order to receive these credentials.*

Special Features

In an effort to provide a comprehensive and user-friendly training resource, this curriculum showcases several informative features. Whether you are a visual or hands-on learner, these features are intended to enhance your knowledge of the construction industry as you progress in your training. Some of the features you may find in the curriculum are explained below.

Introduction

This introductory page, found at the beginning of each module, lists the module Objectives, Performance Tasks, and Trade Terms. The Objectives list the knowledge you will acquire after successfully completing the module. The Performance Tasks give you an opportunity to apply your knowledge to real-world tasks. The Trade Terms are industry-specific vocabulary that you will learn as you study this module.

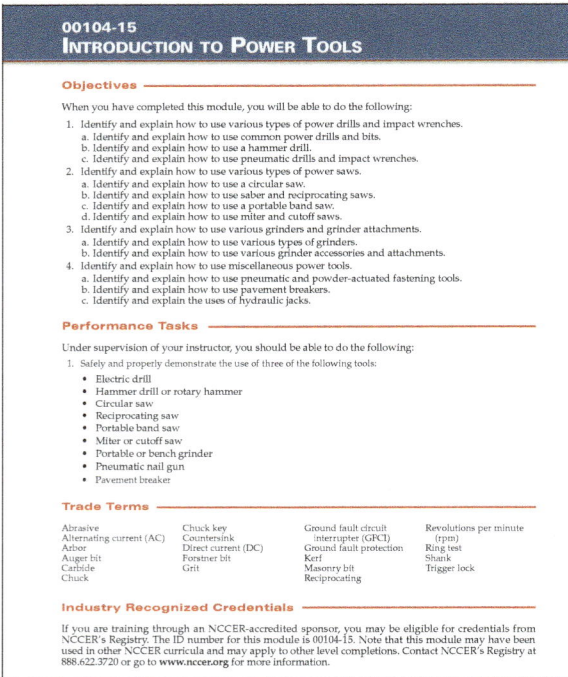

Figures and Tables

Photographs, drawings, diagrams, and tables are used throughout each module to illustrate important concepts and provide clarity for complex instructions. Text references to figures and tables are emphasized with *italic* type.

Notes, Cautions, and Warnings

Safety features are set off from the main text in highlighted boxes and categorized according to the potential danger involved. Notes simply provide additional information. Cautions flag a hazardous issue that could cause damage to materials or equipment. Warnings stress a potentially dangerous situation that could result in injury or death to workers.

Trade Features

Trade features present technical tips and professional practices based on real-life scenarios similar to those you might encounter on the job site.

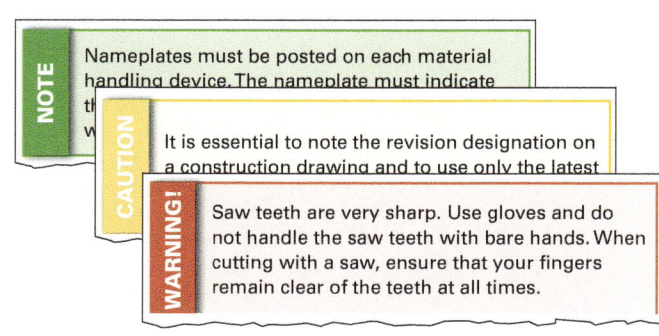

Case History

Case History features emphasize the importance of safety by citing examples of the costly (and often devastating) consequences of ignoring best practices or OSHA regulations.

Going Green

Going Green features present steps being taken within the construction industry to protect the environment and save energy, emphasizing choices that can be made on the job to preserve the health of the planet.

Did You Know

Did You Know features introduce historical tidbits or interesting and sometimes surprising facts about the trade.

Step-by-Step Instructions

Step-by-step instructions are used throughout to guide you through technical procedures and tasks from start to finish. These steps show you how to perform a task safely and efficiently.

> Perform the following steps to erect this system area scaffold:
>
> *Step 1* Gather and inspect all scaffold equipment for the scaffold arrangement.
>
> *Step 2* Place appropriate mudsills in their approximate locations.
>
> *Step 3* Attach the screw jacks to the mudsills.

Trade Terms

Each module presents a list of Trade Terms that are discussed within the text and defined in the Glossary at the end of the module. These terms are presented in the text with **bold, blue** type upon their first occurrence. To make searches for key information easier, a comprehensive Glossary of Trade Terms from all modules is located at the back of this book.

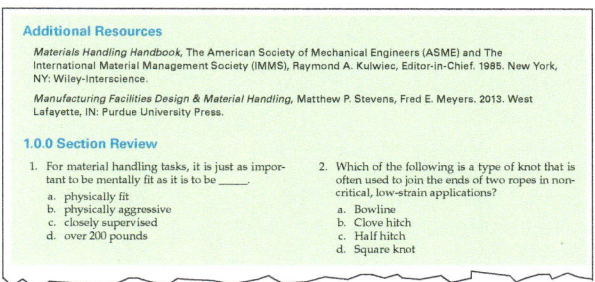

Section Review

Each section of the module wraps up with a list of Additional Resources for further study and Section Review questions designed to test your knowledge of the Objectives for that section.

Review Questions

The end-of-module Review Questions can be used to measure and reinforce your knowledge of the module's content.

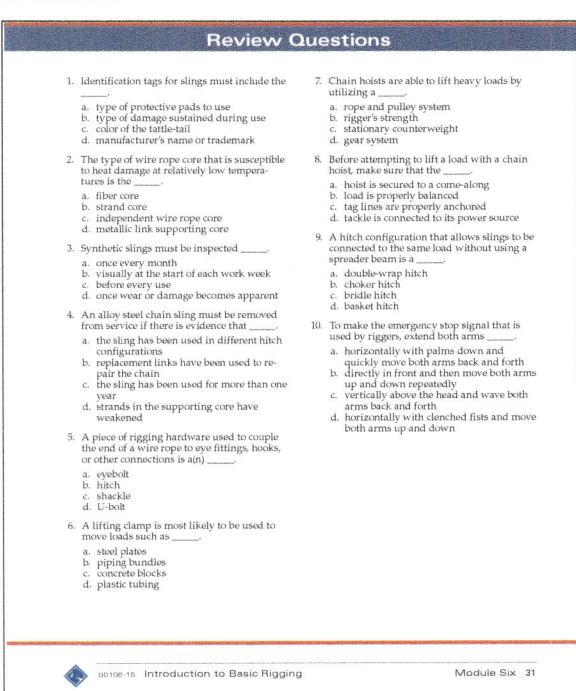

iii

NCCER Standardized Curricula

NCCER's training programs comprise more than 80 construction, maintenance, pipeline, and utility areas and include skills assessments, safety training, and management education.

Boilermaking
Cabinetmaking
Carpentry
Concrete Finishing
Construction Craft Laborer
Construction Technology
Core Curriculum: Introductory Craft Skills
Drywall
Electrical
Electronic Systems Technician
Heating, Ventilating, and Air Conditioning
Heavy Equipment Operations
Heavy Highway Construction
Hydroblasting
Industrial Coating and Lining Application Specialist
Industrial Maintenance Electrical and Instrumentation Technician
Industrial Maintenance Mechanic
Instrumentation
Ironworking
Manufactured Construction Technology
Masonry
Mechanical Insulating
Millwright
Mobile Crane Operations
Painting
Painting, Industrial
Pipefitting
Pipelayer
Plumbing
Reinforcing Ironwork
Rigging
Scaffolding
Sheet Metal
Signal Person
Site Layout
Sprinkler Fitting
Tower Crane Operator
Welding

Maritime

Maritime Industry Fundamentals
Maritime Pipefitting
Maritime Structural Fitter

Green/Sustainable Construction

Building Auditor
Fundamentals of Weatherization
Introduction to Weatherization
Sustainable Construction Supervisor
Weatherization Crew Chief
Weatherization Technician
Your Role in the Green Environment

Energy

Alternative Energy
Introduction to the Power Industry
Introduction to Solar Photovoltaics
Power Generation Maintenance Electrician
Power Generation I&C Maintenance Technician
Power Generation Maintenance Mechanic
Power Line Worker
Power Line Worker: Distribution
Power Line Worker: Substation
Power Line Worker: Transmission
Solar Photovoltaic Systems Installer
Wind Energy
Wind Turbine Maintenance Technician

Pipeline

Abnormal Operating Conditions, Control Center
Abnormal Operating Conditions, Field and Gas
Corrosion Control
Electrical and Instrumentation
Field and Control Center Operations
Introduction to the Pipeline Industry
Maintenance
Mechanical

Safety

Field Safety
Safety Orientation
Safety Technology

Supplemental Titles

Applied Construction Math
Tools for Success

Management

Construction Workforce Development Professional
Fundamentals of Crew Leadership
Mentoring for Craft Professionals
Project Management
Project Supervision

Spanish Titles

Acabado de concreto: nivel uno (*Concrete Finishing Level One*)
Aislamiento: nivel uno (*Insulating Level One*)
Albañilería: nivel uno (*Masonry Level One*)
Andamios (*Scaffolding*)
Carpintería: Formas para carpintería, nivel tres (*Carpentry: Carpentry Forms, Level Three*)
Currículo básico: habilidades introductorias del oficio (*Core Curriculum: Introductory Craft Skills*)
Electricidad: nivel uno (*Electrical Level One*)
Herrería: nivel uno (*Ironworking Level One*)
Herrería de refuerzo: nivel uno (*Reinforcing Ironwork Level One*)
Instalación de rociadores: nivel uno (*Sprinkler Fitting Level One*)
Instalación de tuberías: nivel uno (*Pipefitting Level One*)
Instrumentación: nivel uno, nivel dos, nivel tres, nivel cuatro (*Instrumentation Levels One through Four*)
Orientación de seguridad (*Safety Orientation*)
Paneles de yeso: nivel uno (*Drywall Level One*)
Seguridad de campo (*Field Safety*)

Acknowledgments

This curriculum was revised as a result of the farsightedness and leadership of the following sponsors:

Builders Association of North Central Florida
CareerSafety Center
Center for Employment Training
Duke Energy
Fort Scott Community College
Hubbard Construction
Industrial Management and Training Institute, Inc.
Lee Company
Lincoln Tech
Santa Fe College
Windham School District

This curriculum would not exist were it not for the dedication and unselfish energy of those volunteers who served on the Authoring Team. A sincere thanks is extended to the following:

Corey Driggs
Art Grant
Joseph Pietrzak
Norman Sparks
John Stronkowski
Tony Vazquez
Ted Watts

A sincere thanks is also extended to the dedication and assistance provided by the following technical advisors:

Senobio Aguilera   Lenny Joseph   Chris Sterrett

NCCER Partners

American Council for Construction Education
American Fire Sprinkler Association
Associated Builders and Contractors, Inc.
Associated General Contractors of America
Association for Career and Technical Education
Association for Skilled and Technical Sciences
Construction Industry Institute
Construction Users Roundtable
Design Build Institute of America
GSSC – Gulf States Shipbuilders Consortium
ISN
Manufacturing Institute
Mason Contractors Association of America
Merit Contractors Association of Canada
NACE International
National Association of Women in Construction
National Insulation Association
National Technical Honor Society
National Utility Contractors Association
NAWIC Education Foundation
North American Crane Bureau
North American Technician Excellence
Pearson
Prov
SkillsUSA®
Steel Erectors Association of America
U.S. Army Corps of Engineers
University of Florida, M. E. Rinker Sr., School of Construction Management
Women Construction Owners & Executives, USA

Contents

Module One
Fasteners, Hardware, and Wiring Terminations

Provides trainees with guidance related to working with a variety of fasteners, hardware, and wiring terminations used in HVAC systems. Additionally, direction will be given to the installation of these components in accordance with accepted practices. (Module ID 03313; 10 Hours)

Module Two
Control Circuit and Motor Troubleshooting

Provides trainees with information and skills needed to troubleshoot control circuits and electric motors found in heating and cooling equipment. Developing troubleshooting skills for control circuits and electric motors is vital to the future success of trainees in the HVACR trade. To that end, a portion of this module is devoted to hands-on practice and the successful completion of its required performance tasks. (Module ID 03314; 30 Hours)

Module Three
Troubleshooting Cooling

Reviews the configuration and operation of cooling systems. It discusses the challenge of troubleshooting problems in a refrigeration circuit with the goal of enabling a technician to accurately pinpoint the cause of a malfunction. Solutions for problems that commonly occur in cooling systems are explained in detail. (Module ID 03210; 20 Hours)

Module Four
Troubleshooting Heat Pumps

Provides trainees with guidance related to troubleshooting heat pump systems. Developing the necessary skills to provide these services are vital to the future success of trainees in the HVACR trade. Following a thorough review of the heat pump operating cycle, troubleshooting procedures for components common to heat pumps are presented. (Module ID 03311; 12.5 Hours)

Module Five
Troubleshooting Gas Heating

Reviews the operation of gas-fired furnaces and boilers, and explains the procedures involved in determining the cause of malfunctions. Because the development of troubleshooting skills is vital for success in the HVACR trade, a portion of this module is devoted to hands-on practice and successful completion of the performance tasks. (Module ID 03209; 12.5 Hours)

Module Six
Troubleshooting Oil Heating

Describes the construction and operation of oil-fired heating systems and their components. It contains instructions for servicing and testing of oil furnaces, as well as procedures for isolating and correcting oil furnace malfunctions. (Module ID 03310; 15 Hours)

Module Seven
Troubleshooting Accessories

Provides trainees with information and skills needed to troubleshoot various air treatment accessories used with heating and cooling equipment. Developing troubleshooting skills for accessories is vital to the future success of trainees in the HVACR trade. A portion of this module is devoted to hands-on practice and the successful completion of its required Performance Task. (Module ID 03312; 10 Hours)

Module Eight
Zoning, Ductless, and Variable-Refrigerant Flow Systems

Provides trainees with the information and skills needed to troubleshoot and repair zoned, ductless, and variable refrigerant flow systems. Developing troubleshooting and repair skills for these relatively new heating and cooling products is vital to the future success of trainees in the HVACR trade. To that end, a portion of this module is devoted to hands-on practice and the successful completion of its required Performance Tasks. (Module ID 03315; 15 Hours)

Module Nine
Commercial Hydronic Systems

Reviews some basic properties of water and describes how water pressure is related to the movement of water through piping systems. It also describes various types and components of commercial hot-water heating and chilled-water cooling systems, as well as how they systems function. (Module ID 03305; 12.5 Hours)

Module Ten
Steam Systems

Focuses on the use of steam for storing and moving energy in HVAC systems. It reviews the fundamentals of water that relate to steam and describes the basic steam system cycle. It discusses a steam system's operational components, including steam boilers, their accessories and controls, and steam system loads such as heat exchangers/converters and terminal devices. Steam system valves and piping are covered in detail, including common types of steam system piping arrangements, the components of a condensate return/feedwater system, steam and condensate pipe sizing, and pressure-reducing valves and thermostatic valves. Various types of steam traps and their maintenance requirements are also presented. (Module ID 03306; 10 Hours)

Module Eleven
Retail Refrigerant Systems

Provides trainees with guidance related to retail refrigeration systems. Developing the necessary skills to understand the applications, principles, and troubleshooting of retail refrigeration systems are vital to the future success of trainees in the HVACR trade. (Module ID 03304; 15 Hours)

Module Twelve
Customer Relations

Presents the importance of establishing good relations with customers and provides guidance on how to achieve that goal. The module focuses on the ways that a technician can make a good first impression and describes how to communicate in a positive way with customers. It further covers the elements of a service call and provides guidance for dealing with different types of problem customers. (Module ID 03316; 5 Hours)

Glossary

Index

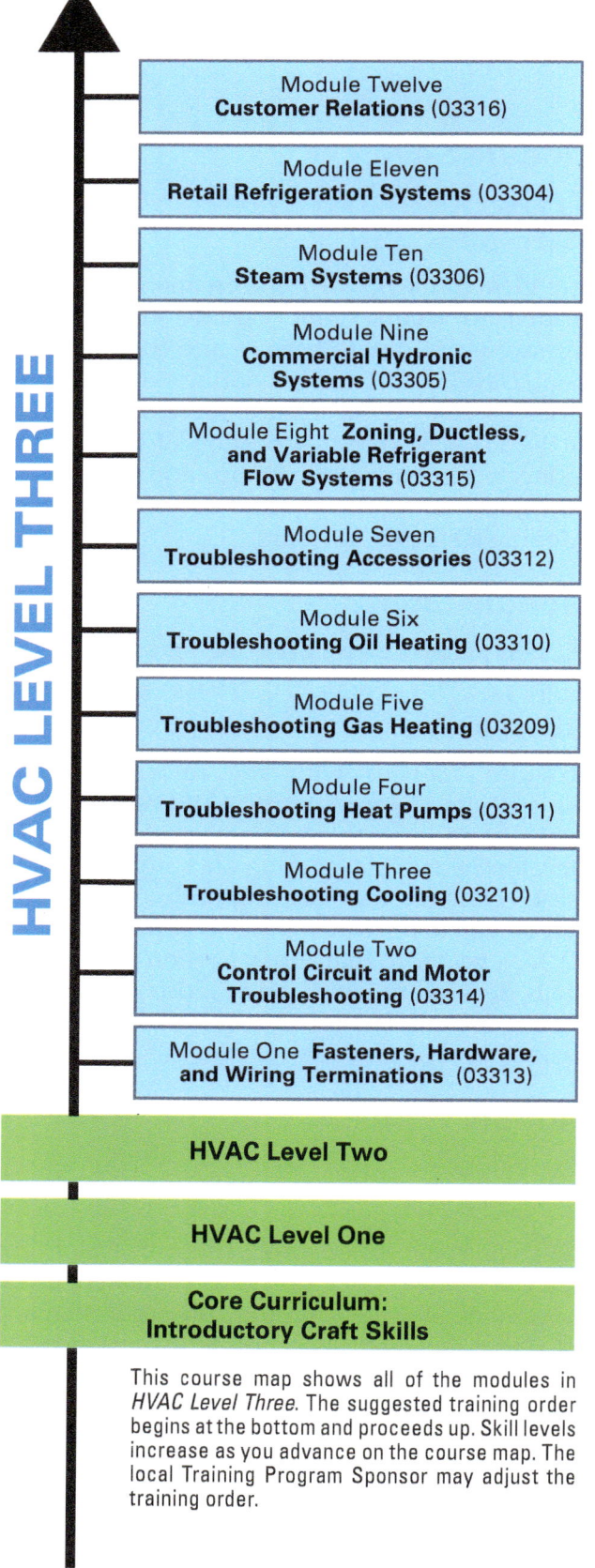

This course map shows all of the modules in *HVAC Level Three*. The suggested training order begins at the bottom and proceeds up. Skill levels increase as you advance on the course map. The local Training Program Sponsor may adjust the training order.

Fasteners, Hardware, and Wiring Terminations

OVERVIEW

During the installation and maintenance of HVAC equipment, technicians work with a variety of fasteners, hardware, and wiring terminations. These components must be carefully selected for the specific application and installed according to accepted practices. Using the correct components ensures that the assembled system will perform properly, and will not fail because of the wrong choice of component or its incorrect installation.

Module 03313

Trainees with successful module completions may be eligible for credentialing through NCCER's National Registry. To learn more, go to **www.nccer.org** or contact us at **1.888.622.3720**. Our website has information on the latest product releases and training, as well as online versions of our *Cornerstone* magazine and Pearson's product catalog.

Your feedback is welcome. You may email your comments to **curriculum@nccer.org**, send general comments and inquiries to **info@nccer.org**, or fill in the User Update form at the back of this module.

This information is general in nature and intended for training purposes only. Actual performance of activities described in this manual requires compliance with all applicable operating, service, maintenance, and safety procedures under the direction of qualified personnel. References in this manual to patented or proprietary devices do not constitute a recommendation of their use.

Copyright © 2018 by NCCER, Alachua, FL 32615, and published by Pearson, New York, NY 10013. All rights reserved. Printed in the United States of America. This publication is protected by Copyright, and permission should be obtained from NCCER prior to any prohibited reproduction, storage in a retrieval system, or transmission in any form or by any means, electronic, mechanical, photocopying, recording, or likewise. To obtain permission(s) to use material from this work, please submit a written request to NCCER Product Development, 13614 Progress Blvd., Alachua, FL 32615.

03313 V5

From *HVAC Level Three, Trainee Guide*. NCCER.
Copyright © 2018 by NCCER. Published by Pearson. All rights reserved.

03313
Fasteners, Hardware, and Wiring Terminations

Objectives

When you have completed this module, you will be able to do the following:

1. Identify the types, uses, and installation methods of fasteners.
   a. Identify various types and the uses of threaded fasteners.
   b. Identify and explain how to use taps, dies, and screw extractors.
   c. Explain how to install and torque fasteners to a specific value.
   d. Identify and explain how to install various types of toggle and anchor bolts.
   e. Identify various types of non-threaded fasteners.
2. Identify and describe the installation of various types of vibration isolators.
   a. Identify and describe vibration isolators used to support and suspend equipment and piping.
   b. Identify and describe methods used to restrain equipment during seismic events.
   c. Explain how to select and install various vibration isolators.
3. Identify common low- and line-voltage electrical termination hardware and explain how to properly terminate wiring connections.
   a. Identify common low- and line-voltage electrical terminating hardware.
   b. Explain how to properly terminate low- and line-voltage wiring connections.

Performance Tasks

Under the supervision of your instructor, you should be able to do the following:

1. Torque threaded hardware to a specific torque value.
2. Select the appropriate drill bit and install an anchor in brick or concrete block.
3. Terminate line- and low-voltage wiring on a compressor contactor.

Trade Terms

AL-CU
Break-away torque
Connector
Lug
Nominal size
Run-down resistance
Seismic activity
Seismic restraint

Set or seizure
Tensile strength
Termination
Thread classes
Thread series
Thread standards
Torque
Vibration isolator

Industry-Recognized Credentials

If you are training through an NCCER-accredited sponsor, you may be eligible for credentials from NCCER's Registry. The ID number for this module is 03313. Note that this module may have been used in other NCCER curricula and may apply to other level completions. Contact NCCER's Registry at 888.622.3720 or go to www.nccer.org for more information.

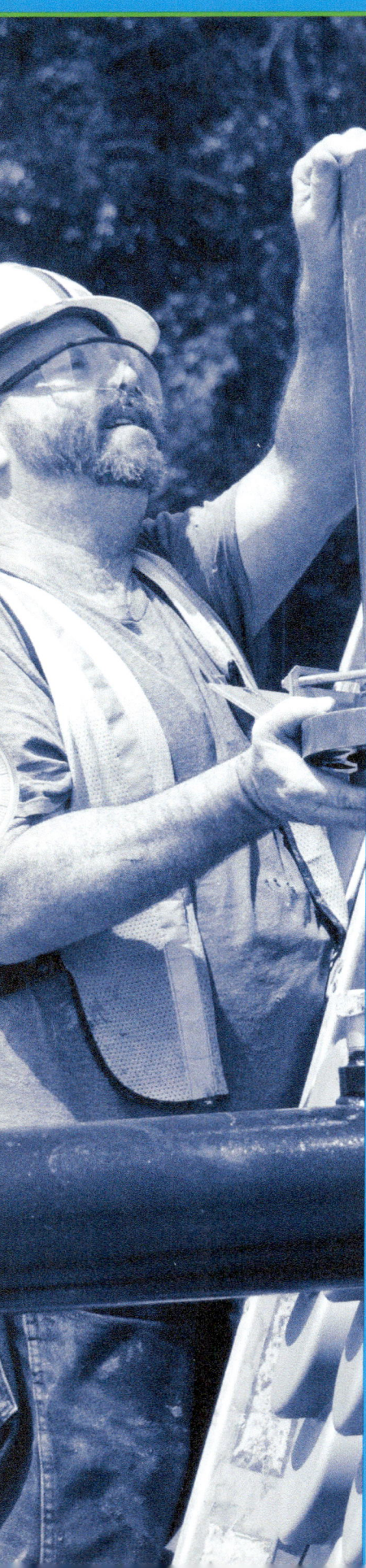

Contents

1.0.0 Fasteners 1
  1.1.0 Threaded Fasteners 1
    1.1.1 Thread Designations 1
    1.1.2 Threaded Fastener Grade Designations 2
    1.1.3 Machine Bolts, Machine Screws, Stud Bolts, and Cap Screws 2
    1.1.4 Nuts 4
    1.1.5 Thread Repair Inserts 5
    1.1.6 Flat and Lock Washers 5
    1.1.7 Set Screws 5
    1.1.8 Screws 6
  1.2.0 Taps, Dies, and Screw Extractors 7
    1.2.1 Taps 7
    1.2.2 Dies 8
    1.2.3 Fastener Extraction 8
  1.3.0 Torquing Threaded Fasteners 9
    1.3.1 Torque Wrenches 11
    1.3.2 Tightening Sequences 12
  1.4.0 Toggle Bolts and Anchor Bolts 13
    1.4.1 Toggle Bolts 14
    1.4.2 Anchor Bolts 14
    1.4.3 Installing Non-expanding Concrete Anchors 16
    1.4.4 Installing Expanding Concrete Anchors 17
    1.4.5 Installing Tapcon® screws 17
  1.5.0 Non-Threaded Fasteners 18
    1.5.1 Retainer and Lock Rings 18
    1.5.2 Pins 19
    1.5.3 Keys 19
    1.5.4 Rivets 19
2.0.0 Vibration Isolators 22
  2.1.0 Types of Vibration Isolators 22
    2.1.1 Compressor, Fan, and Blower Motor Vibration Isolators 22
    2.1.2 Vibration Isolators that Reduce Vibration Between Objects 23
    2.1.3 Vibration Isolators that Reduce Vibration Through Ducts 24
    2.1.4 Vibration Isolators that Reduce Vibration Through Pipes 25
  2.2.0 Seismic Restraints 26
    2.2.1 Seismic Restraints for Residential Equipment 26
    2.2.2 Seismic Restraints for Light Commercial Equipment 26
    2.2.2 Seismic Restraints for Hydronic and Commercial Refrigeration Equipment 26
  2.3.0 Selecting Vibration Isolators and Materials 26
3.0.0 Wiring Terminations 31
  3.1.0 Connectors and Terminations 31
  3.2.0 Installing Terminations 33
    3.2.1 Preparing Conductors for Termination 33
    3.2.2 Installing a Crimp Connector 34

Figures and Tables

Figure 1   Thread designations 2
Figure 2   Grade markings for steel bolts and screws 3
Figure 3   Machine bolts, machine screws, cap screws, and stud bolts 4
Figure 4   Threaded rods supporting ducts 5
Figure 5   Nuts 6–7
Figure 6   Thread repair insert 7
Figure 7   Flat and lock washers 7
Figure 8   Set screws 8
Figure 9   Screw head types and screw drives 9
Figure 10 Lag screw 9
Figure 11 Thread-forming and thread-cutting screws 9
Figure 12 Taps 10
Figure 13 Tap and die set 10
Figure 14 Screw extractor 10
Figure 15 Torque specifications 11
Figure 16 Torque wrenches 12
Figure 17 Proper use of a torque wrench 13
Figure 18 Common fastener tightening sequences 13
Figure 19 Flanged coupling alignment 14
Figure 20 Toggle bolts 14
Figure 21 Toggle bolt installation 15
Figure 22 Concrete inserts 15
Figure 23 Non-expanding concrete anchor 16
Figure 24 Expanding concrete anchor 16
Figure 25 Adhesive anchor 16
Figure 26 Tapcon® screws 16
Figure 27 Installed expanding-type concrete anchor 18
Figure 28 Hanger installed with Tapcon® screws 18
Figure 29 Retainer and lock rings 19
Figure 30 Pin fasteners 20
Figure 31 Keys 20
Figure 32 Rivets 20
Figure 33 Pop rivet tool 21
Figure 34 Hermetic compressor vibration isolators 22
Figure 35 Direct-drive blower vibration isolators 23
Figure 36 Semi-hermetic compressor vibration isolator 23
Figure 37 Vibration-isolating pad 23
Figure 38 Air handler suspended with vibration isolators 24

Figures and Tables

Figure 39 Vibration-isolating hanger 24
Figure 40 Heavy-duty vibration isolator. 24
Figure 41 Ductwork vibration and noise control devices 25
Figure 42 Flexible metal vibration eliminator 25

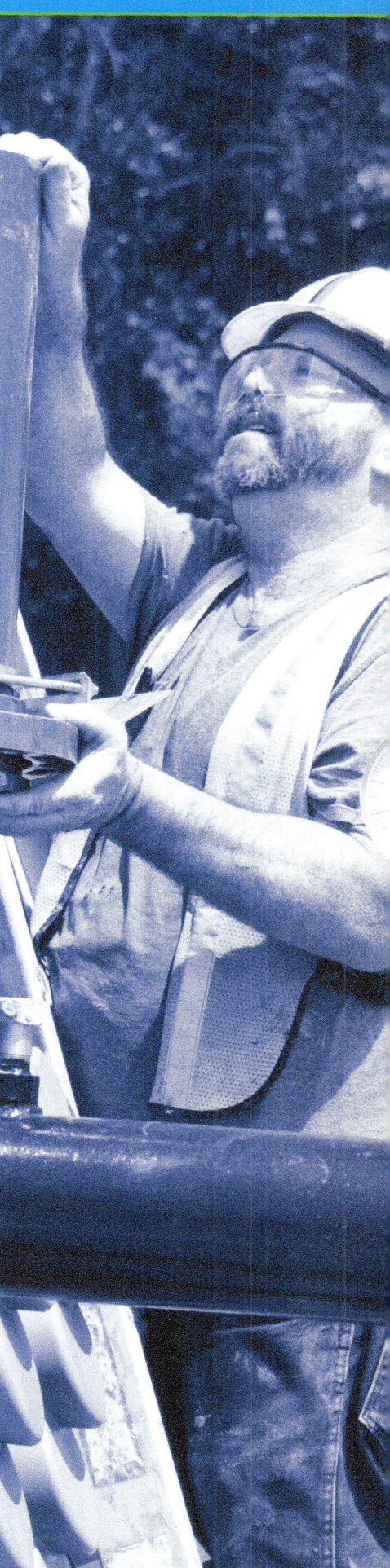

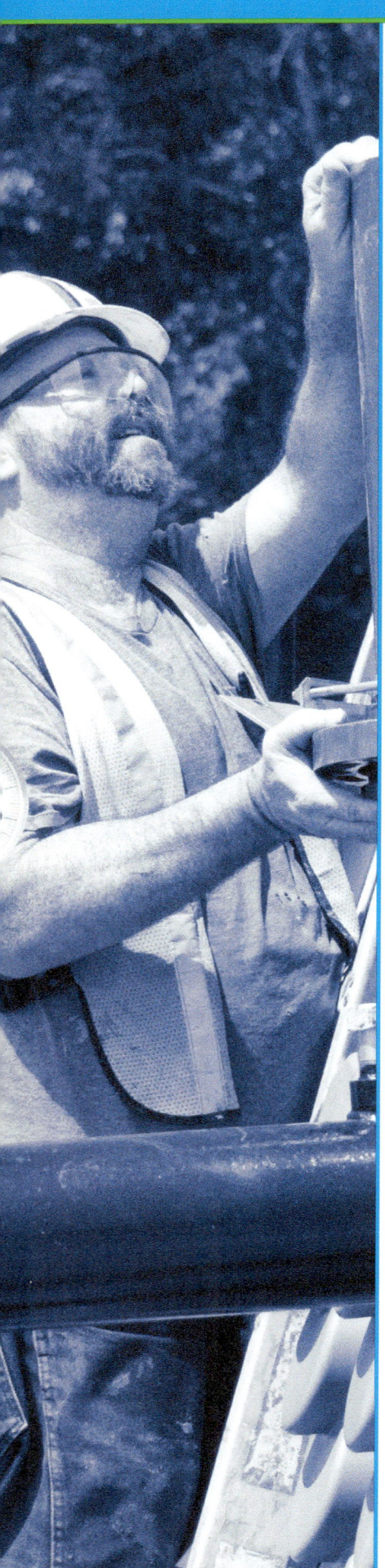

Figure 43 Piping suspended with vibration isolators 25
Figure 44 Bumpers 27
Figure 45 Factory-built seismic curb 27
Figure 46 Suspended equipment seismic restraints 28
Figure 47 Water pump seismic restraints 28
Figure 48 Combination seismic restraint 29
Figure 49 Vibration isolator selection information 29
Figure 50 Crimp-on terminations 32
Figure 51 Basic crimp connector structure 32
Figure 52 Wire stripper/crimper 33
Figure 53 Production-grade wire strippers 33
Figure 54 Proper stripping length 34
Figure 55 Manual crimping tool 34

Table 1 Hole Sizes for Non-Expanding Concrete Anchors 17

SECTION ONE

1.0.0 FASTENERS

Objectives

Identify the types, uses, and installation methods of fasteners.
  a. Identify various types and uses of threaded fasteners.
  b. Identify and explain how to use taps, dies, and screw extractors.
  c. Explain how to install and torque fasteners to a specific value.
  d. Identify and explain how to install various types of anchor bolts.
  e. Identify various types of non-threaded fasteners.

Performance Tasks

1. Torque threaded hardware to a specific torque value.
2. Select the appropriate drill and install an anchor in brick or concrete block.

Trade Terms

Break-away torque: The torque required to loosen a fastener. This is usually lower than the torque to which the fastener has been tightened.

Nominal size: A means of expressing the size of a bolt or screw. It is the approximate diameter of a bolt or screw.

Run-down resistance: The torque required to overcome the resistance of associated hardware, such as locknuts and lock washers, when tightening a fastener.

Set or seizure: In the last stages of rotation in reaching a final torque, the fastener may lock up; this is known as seizing or set. This is usually accompanied by a noticeable popping effect.

Tensile strength: The maximum stress a material can endure before being pulled apart.

Thread classes: Threads are distinguished by three classifications according to the amount of tolerance the threads provide between the bolt and nut. The different classes are designated by a number and letter combination that indicates the degree of fit for a thread. Higher numbers indicate a tighter fit.

Thread series: The combinations of diameter and pitch in threads which are applied to specific thread diameters to identify the coarseness or fineness of a thread.

Thread standards: An established set of standards for machining threads.

Torque: Applying a specific amount of twisting force to a threaded fastener.

Fasteners are used to install many types of parts and equipment. Common fasteners include bolts, screws, pins, and retainers. It is important that the correct fastener be selected. For example, bolts and screws are graded based on strength. Even if the size of a bolt is correct, it may fail if an inferior grade (strength) is installed.

1.1.0 Threaded Fasteners

Threaded fasteners are one of the most common types of fasteners. Many are assembled with nuts and washers while others are installed in threaded holes.

Types of threaded fasteners that will be covered in this section include the following:

- Machine bolts, machine screws, stud bolts, and cap screws
- Nuts
- Toggle bolts
- Inserts
- Flat and lock washers
- Set screws
- Screws

Before reviewing the different types and shapes of threaded fasteners, it is important to understand the various types of threads used. In addition, technicians must understand the various grades of threaded fasteners. The thread type and grade used will depend upon the application for the fastener.

1.1.1 Thread Designations

Fastener threads are made to established **thread standards**. The most common standard is the Unified or American National Standard. There are three series defined by the Unified standard. These series are based on the number of threads per inch for a fastener of a certain diameter. The three series are the following:

- *Unified National Coarse (UNC) Thread* – UNC thread is used for bolts, screws, nuts, and other general applications. Fasteners with UNC threads are used for rapid assembly and disassembly of parts where corrosion or slight damage may occur.

 03313 Fasteners, Hardware, and Wiring Terminations · Module One 1

- *Unified National Fine (UNF) Thread* – UNF thread is used for bolts, screws, nuts, and other uses where a finer thread than UNC is required.
- *Unified National Extra Fine (UNEF) Thread* – UNEF thread is used for thin-walled tubes, nuts, ferrules, and couplings.

Bolt and screw threads are designated by a standard method (*Figure 1*). The standard designations include the following:

- Nominal size (diameter).
- Number of threads per inch.
- Thread series symbol.
- Thread class. The thread class determines how tightly the male (external) and female (internal) threads fit each other. There are three classes, each of which is applied to external and internal threads. Classes 1A, 2A, and 3A apply to external threads, such as a bolt. Classes 1B, 2B, and 3B apply to internal threads, such as a nut. Higher numbers indicate a tighter fit. Therefore, Class 3A and 3B threads provide the tightest fit. Class 1 is the rarest, and is only used when quick disassembly is important. The loose fit provides limited strength and resistance to separation. Class 2 is the most common. Class 3 threads are needed when close tolerances and maximum strength are priorities.
- Left-hand thread symbol. Unless shown, the threads are right hand. This symbol is used only for left-hand fasteners.

Metric screw threads based on the American National Standard are also in common use. Metric M-profile threaded screws are a coarse-thread series of fasteners used for general fastening purposes. Metric MJ-profile threaded screws have slightly different thread characteristics, and are used with aircraft parts and other high-stress applications requiring extra strength.

Coarse pitch threads (fewer threads per inch) are used in routine applications and offer greater stripping strength when used with larger diameter fasteners. Coarse threads are easier to form and nuts are easier to start and run down on the threaded section. Fine pitch threads have a higher tensile strength, and resist loosening when subjected to shock or vibration. Fine threads are useful if a thin nut is needed or where very fine adjustments must be made.

1.1.2 Threaded Fastener Grade Designations

The strength and quality of a fastener can be determined by special grade markings on the head of the fastener. These markings are standardized by the Society of Automotive Engineers (SAE) and ASTM International. Grade markings are sometimes called line markings.

Figure 2 shows the SAE and ASTM markings for steel bolts and screws.

While there are several grades of bolts, the three most commonly used are SAE Grade 2, Grade 5, and Grade 8. Grade 2 is made from low-carbon steel and is used in low-stress applications. For example, a Grade 2 bolt might be adequate for a bracket supporting a sheet metal duct. Grades 5 and 8 bolts use higher-quality steel that has been quenched and annealed. They are used in applications subjected to higher stress or pressure. These higher-grade bolts would be used to secure the cylinder heads of a semi-hermetic compressor, as one example. Higher stress or pressure dictates whether a grade 8 is used rather than a grade 5. Never substitute a lower grade bolt for a higher grade. Always choose bolts that have grade markings when the application has any importance. Bolts without grade markings are likely of poor quality and may fail.

1.1.3 Machine Bolts, Machine Screws, Stud Bolts, and Cap Screws

Machine bolts (*Figure 3*) are used to assemble parts that do not require close tolerances. The tolerance is the amount of variation allowed from a standard. Machine bolts are made with diameters ranging from ¼" to 3" and with lengths from ½" to 30". A machine bolt is tightened and released by turning its mating nut that is usually furnished along with the bolt.

Machine screws are used for general assembly. They have slotted or recessed heads. Machine screws are available in diameters from #6 through #12 (0.060" to 0.2160"). Fractional diameters range from ¼" to ½", and vary in length from 1/8" to 3". They are also made in metric sizes.

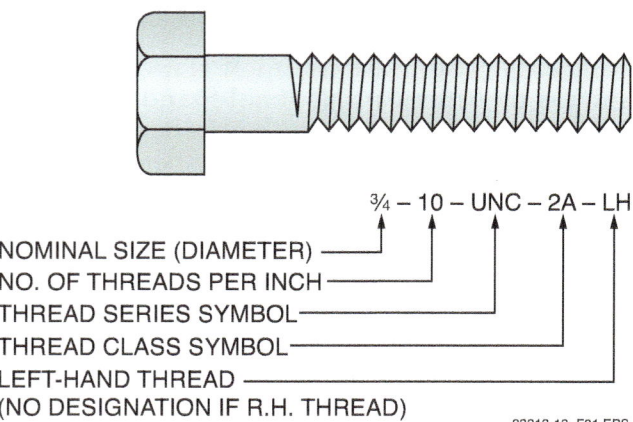

Figure 1 Thread designations.

ASTM AND SAE GRADE MARKINGS FOR STEEL BOLTS & SCREWS

Grade Marking	Specification	Material
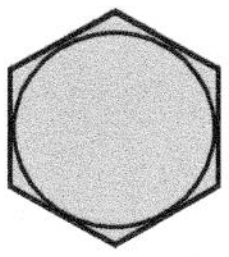	SAE-GRADE 0	STEEL
	SAE-GRADE 1 ASTM-A 307	LOW CARBON STEEL
	SAE-GRADE 2	LOW CARBON STEEL
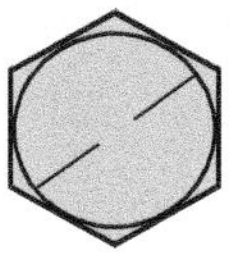	SAE-GRADE 3	MEDIUM CARBON STEEL, COLD WORKED
 A 449	SAE-GRADE 5	MEDIUM CARBON STEEL, QUENCHED AND TEMPERED
	ASTM-A 449	
 A 325	ASTM-A 325	MEDIUM CARBON STEEL, QUENCHED AND TEMPERED
 BB	ASTM-A 354 GRADE BB	LOW ALLOY STEEL, QUENCHED AND TEMPERED
 BC	ASTM-A 354 GRADE BC	LOW ALLOY STEEL, QUENCHED AND TEMPERED
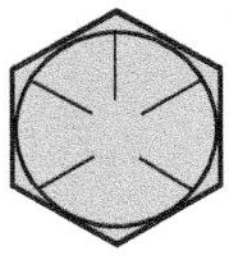	SAE-GRADE 7	MEDIUM CARBON ALLOY STEEL, QUENCHED AND TEMPERED ROLL THREADED AFTER HEAT TREATMENT
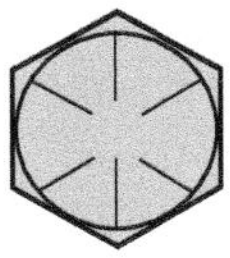	SAE-GRADE 8	MEDIUM CARBON ALLOY STEEL, QUENCHED AND TEMPERED
	ASTM-A 354 GRADE BD	ALLOY STEEL, QUENCHED AND TEMPERED
 A 490	ASTM-A 490	ALLOY STEEL, QUENCHED AND TEMPERED

ASTM SPECIFICATIONS
 A 307 – LOW CARBON STEEL EXTERNALLY AND INTERNALLY THREADED STANDARD FASTENERS.
 A 325 – HIGH STRENGTH STEEL BOLTS FOR STRUCTURAL STEEL JOINTS, INCLUDING SUITABLE
   NUTS AND PLAIN HARDENED WASHERS.
 A 449 – QUENCHED AND TEMPERED STEEL BOLTS AND STUDS.
 A 354 – QUENCHED AND TEMPERED ALLOY STEEL BOLTS AND STUDS WITH SUITABLE NUTS.
 A 490 – HIGH STRENGTH ALLOY STEEL BOLTS FOR STRUCTURAL STEEL JOINTS, INCLUDING
   SUITABLE NUTS AND PLAIN HARDENED WASHERS.
SAE SPECIFICATION
 J 429 – MECHANICAL AND QUALITY REQUIREMENTS FOR THREADED FASTENERS.

Figure 2 Grade markings for steel bolts and screws.

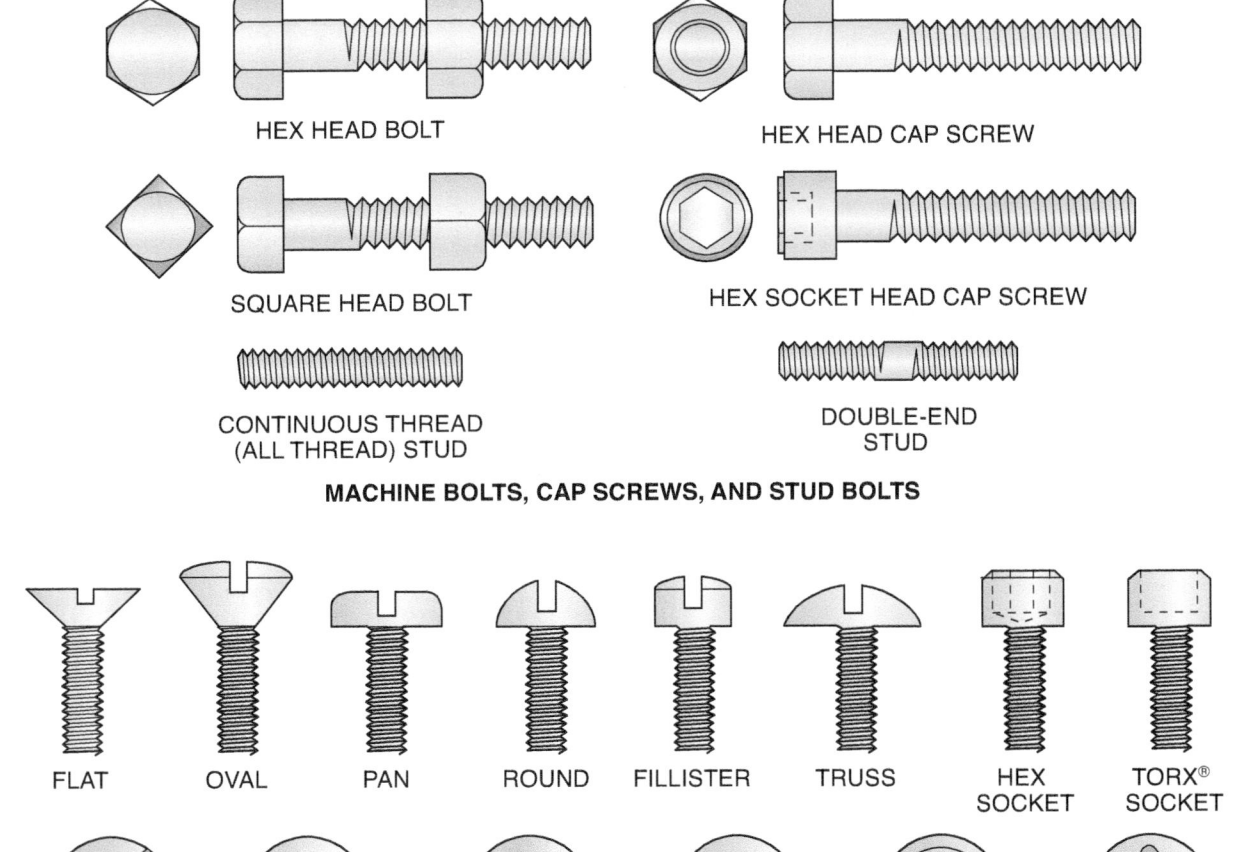

Figure 3 Machine bolts, machine screws, cap screws, and stud bolts.

A machine screw normally mates with an internally threaded hole into which it is tightened or released. They can also be used with matching nuts.

Cap screws are generally used on assemblies that need a finished appearance. They pass through a clearance hole in one part of the assembly and are screwed into a threaded hole in the other part. This clamps the parts together when the cap screw is tightened. They are made to close tolerances with machined or semi-finished bearing surfaces under the head. Cap screws come in coarse and fine threads, and in diameters from ¼" to 2". Lengths from ⅜" to 10" are available. They are also made in metric sizes.

Stud bolts are headless bolts, threaded either along the entire length or on both ends. One end can be screwed into a threaded hole. The part to be clamped is fitted over the other end of the stud, and a nut and washer are screwed on to fasten the two parts together.

Threaded rod is nothing more than a metal rod that has threads on both ends, or is threaded over its entire length. Standard lengths are available, such as 6 foot or 10 foot lengths, but longer rods can be cut to any length. The rod is usually made of low-carbon steel since the rods are often used in low-load applications. Typical HVAC applications include use as a pipe hanger or as part of a trapeze assembly for supporting a duct (*Figure 4*).

1.1.4 Nuts

Nuts used with most threaded fasteners have either hex (hexagonal) or square shapes and are used with bolts having the same shaped head. *Figure 5* shows different types of nuts.

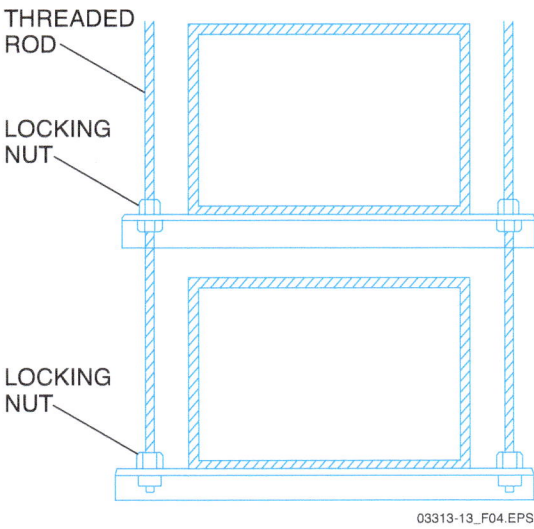

Figure 4 Threaded rods supporting ducts.

Some special-purpose nuts include the following:

- *Acorn nut* – These are used when appearance is important or when there are exposed sharp threads on the fastener.
- *Castellated (or castle) and slotted nuts* – After the nut is tightened, a cotter pin is fitted into one set of slots and through a hole in the bolt. The cotter pin keeps the nut from loosening.
- *Self-locking nut* – This has a nylon insert or is slightly deformed so it cannot work loose. Once a self-locking nut is used and removed, it should be thrown away and replaced with a new one.
- *Wing nut* – These are used where frequent adjustments and service are necessary. Wing nuts allow for loosening and tightening without the use of a wrench.
- *Jam nut* – This is a thin nut used to lock a standard nut in place. Note that jam nuts are installed on the fastener first, leaving it under the standard nut. The jam nut is often found on top, which is incorrect.
- *U-nuts and J-nuts* – These are used with soft, thin panel materials, or with thin sheet metal. The U-nut and J-nut have a threaded hole or stamped opening that accepts threads. They are slipped onto the edge of thin, relatively soft material to provide a quick method of attachment.
- *Cage nut* – These are square nuts enclosed in a stamped sheet metal box. The cage has a spring clip at the bottom which can be clipped into a square hole in a rack. The nut is loose enough inside the cage to allow for some small misalignment when mounting equipment.

1.1.5 Thread Repair Inserts

Thread repair inserts, or heli-coils (*Figure 6*), are a special kind of fastener used to provide high-strength threads in soft metals and plastics. They are also used to replace damaged or stripped threads in a tapped hole. Inserts are made in standard sizes and forms including metric sizes. The insert, which is larger in diameter than the tapped hole, is compressed during installation and allowed to spring back, permanently anchoring the insert in the tapped hole.

1.1.6 Flat and Lock Washers

Flat washers (*Figure 7*) provide an enlarged surface used to distribute the load from bolt heads and nuts. Flat washers are made in light, medium, heavy-duty, and extra heavy-duty series.

Fender washers are similar to flat washers with the difference being they have a larger outside diameter to bridge oversized holes or other wide-clearance requirements.

Lock washers are used to keep bolts or nuts from working loose. They are placed between the flat washers and the bolts or nuts. Some common types of lock washers include the following:

- *Split ring* – Commonly used with bolts and cap screws
- *External* – Provides the greatest resistance
- *Internal* – Used on small screws
- *Internal-external* – Used for oversized mounting holes
- *Countersunk* – Used with flat or oval-head screws

1.1.7 Set Screws

Set screws (*Figure 8*) are usually made of heat-treated steel and are used to secure objects within or against another object. For example, set screws are commonly used to fasten a pulley or gear to a shaft. Set screws are classified by head styles and point styles.

Set screws are usually headless (also known as blind), meaning the screw is fully threaded yet has no head. This type of screw is usually driven using fluted, hex socket, slotted or square heads. Once the set screw is screwed into the threaded hole, it uses a clamping force through the bottom tip that protrudes through the hole using friction as a fastener.

The cup point is the most common type of set screw. It is well suited for use in high vibration applications because it is very durable and shows a high resistance to wear while having strong grip capabilities. The knurled cup point is also designed to lock to prevent the screws from working loose during vibration.

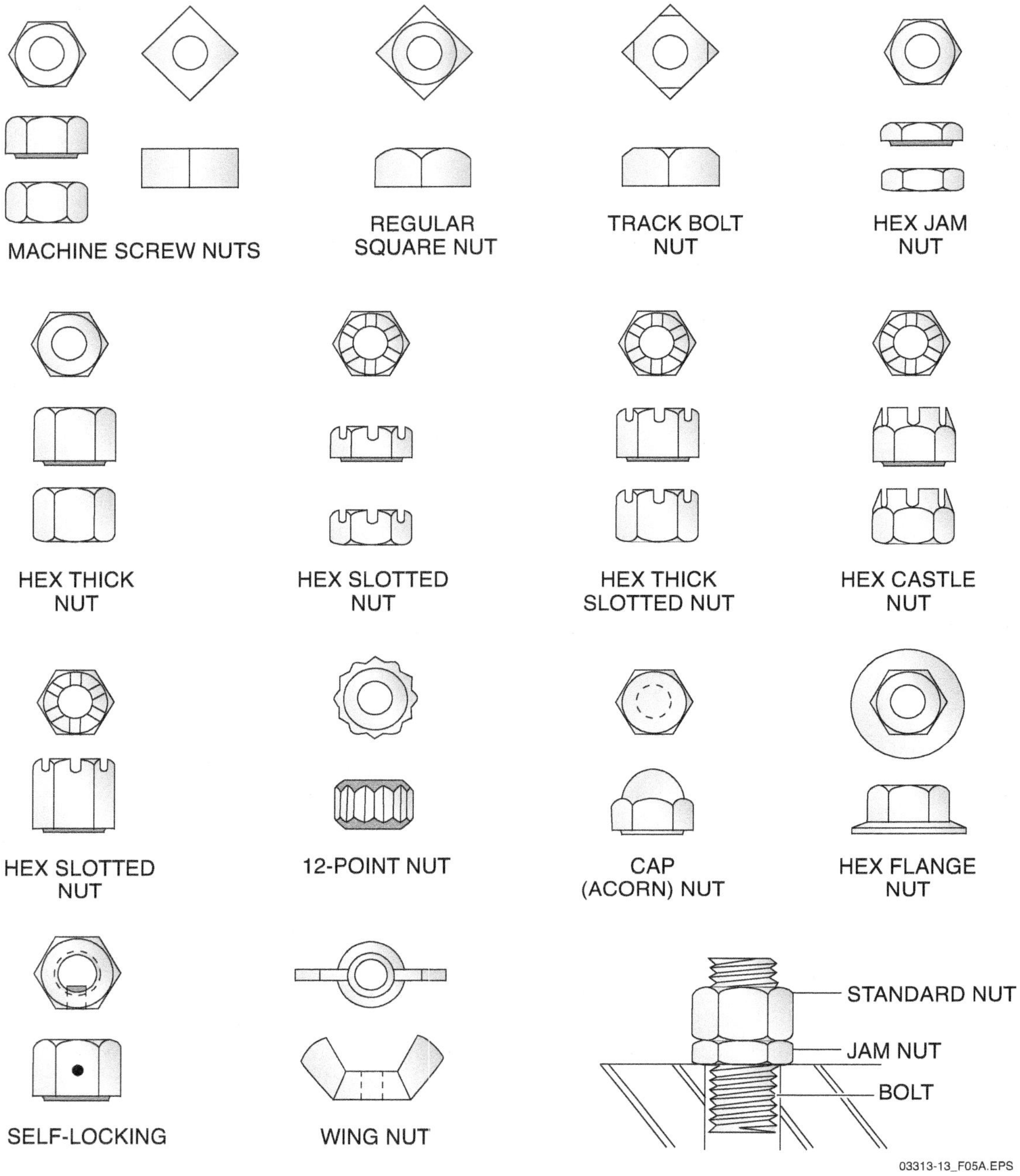

Figure 5 Nuts. (1 of 2)

1.1.8 Screws

Screws are used to fasten materials in a secure manner or used if the fastened materials have to be taken apart. The two most common screw head types are round and flat. Slots types and head types are varied to fit both hand and powered drivers (*Figure 9*). Screw sizes are based on the diameter or body of the screw and are given in gauge numbers from No. 2 to No. 24. Screws larger than ¼" diameter are sized in fractions of an inch. Lengths range from ¼" to 6". Screws can be made of different metals and are often coated or plated to reduce corrosion.

Lag screws are used for heavy-duty fastening of objects to wood structures. When used with a lead lag shield, lag screws can be used to fasten objects to concrete or masonry (*Figure 10*).

Thread-forming screws (*Figure 11*) are used mainly to fasten light-gauge metal parts. They form a thread as they are driven. This eliminates tapping the hole. Some also drill their own hole, using a tip that is shaped like a common high-speed drill bit.

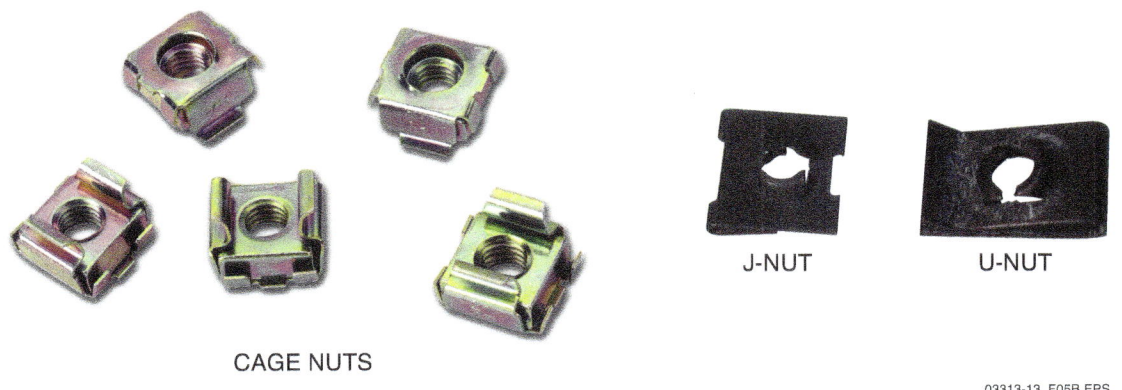

Figure 5 Nuts. (2 of 2)

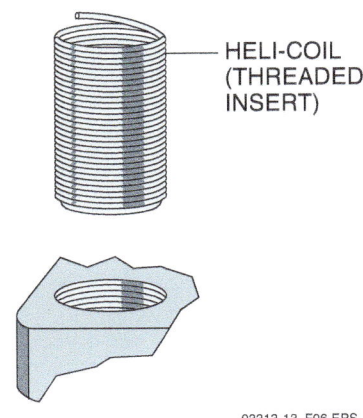

Figure 6 Thread repair insert.

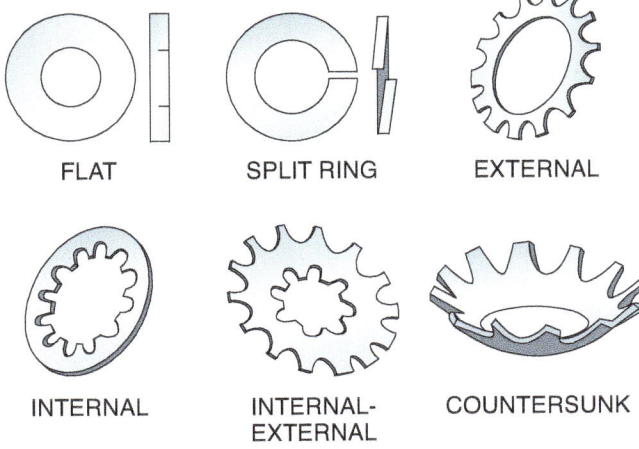

Figure 7 Flat and lock washers.

Thread-cutting screws cut threads into the metal as they are driven into a pilot hole. They are made of hardened steel and are used to join heavy-gauge sheet metal and nonferrous metal parts.

Self-drilling sheet metal screws are commonly used in the installation of sheet metal duct systems. These screws increase productivity by decreasing the number of steps required for the job. However, standard thread-forming screws often have a very sharp point for a tip. When they are used on thinner gauges of sheet metal, pre-drilling is not necessary. The sharp point helps penetrate the metal to get the screw started.

1.2.0 Taps, Dies, and Screw Extractors

Taps and dies are used to cut threads in materials such as metal, plastic, and hard rubber. Screw extractors are used to remove fasteners that have broken off flush to a surface or below it.

1.2.1 Taps

Taps (*Figure 12*) are used to cut internal threads in materials; dies are used to cut external threads on bolts and rods. Taps are used to thread holes that go all the way through a piece of material, or holes that end inside the material (blind holes). Common types of taps include the following:

- *Taper taps* – Taper taps are used when starting a tapping operation or when tapping through holes. They have a chamfer length of 8 to 10 threads.
- *Plug taps* – Plug taps are used following the taper tap. They have a chamfer length of 3 to 5 threads.
- *Bottoming taps* – Bottoming taps are always used after the taper and plug taps have already been used. They are used for threading the bottom of a blind hole and have a chamfer length of only 1 to 1½ threads.

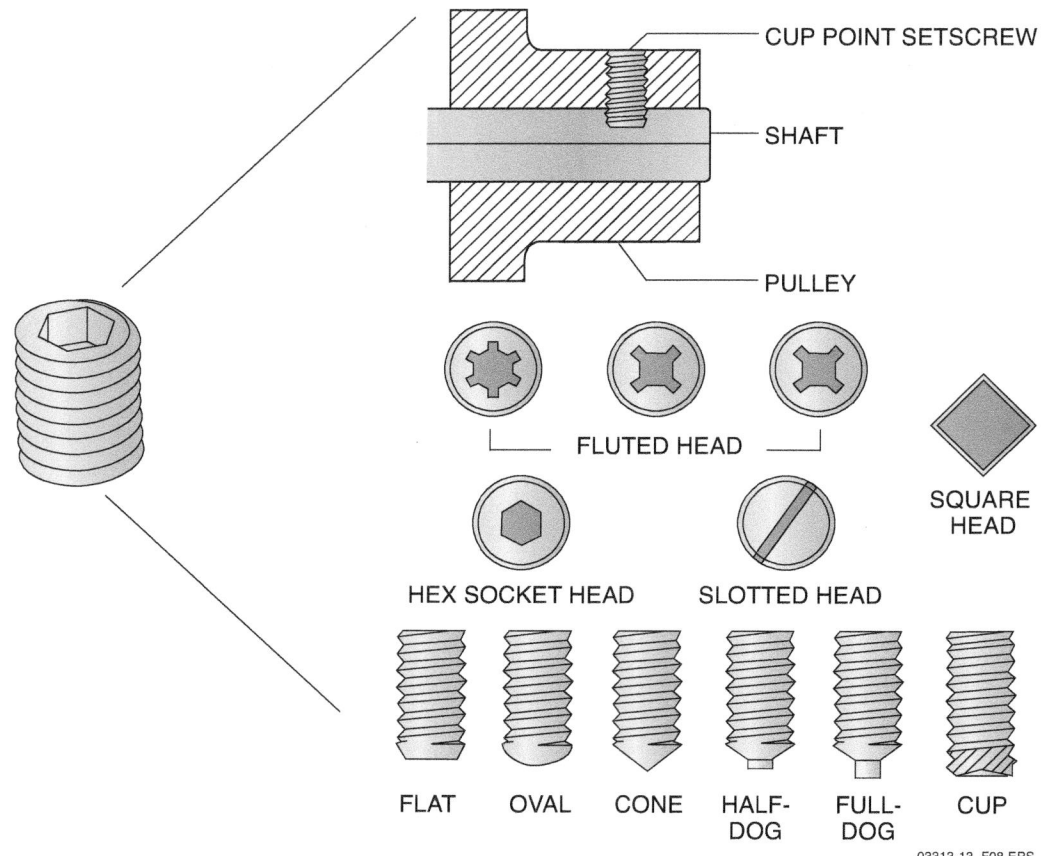

Figure 8 Set screws.

- *Pipe taps* – Pipe taps are used where extremely tight fits are necessary, as with pipe fittings. The diameter of the tap from end to end of the threaded portion is tapered at the rate of ¾ inch per foot.

1.2.2 Dies

Dies are made in a variety of shapes and sizes and are solid or adjustable. Common types of dies include the following:

- Pipe dies will cut American Standard Pipe Threads or other pipe (such as metric) threads.
- Rethreading dies (die nuts) are used for repairing damaged or rusty threads on bolts or screws. They are usually hexagon-shaped and can be turned using any type of wrench that will fit.
- Screw adjusting dies are adjusted by a screw in the die.
- Open adjusting dies are adjusted by three screws in the die holder.
- Two-piece rectangular dies are adjusted by setscrews and held in ordinary or ratchet diestocks.

Taps and dies can be obtained in complete sets that include various taps and dies, as well as diestocks, tap wrenches, guides, and the screwdrivers and wrenches necessary to loosen and tighten adjusting screws. *Figure 13* shows a common tap and die set.

1.2.3 Fastener Extraction

A bolt may become seized if it has been in place for a long time, was forced into the bottom of a hole, or became jammed in place. In these cases, the head may break off or the nut may become stripped. This can leave the bolt or screw locked in place. In order to remove the fastener, a special tool called a screw extractor is used. Screw extractors are commonly referred to as easy-outs (*Figure 14*). To use a screw extractor, a hardened drill of the correct size is used to drill a hole centered and down the axis of the fastener. The screw extractor is a tapered tool with a spiral that is the reverse of the thread of the fastener. The taper is inserted into the fastener and turned in the direction that would be used to unscrew the fastener. The spiral grips the inside of the hole, and breaks loose the fastener, allowing the piece to be removed. Before attempting to break the fastener loose, satu-

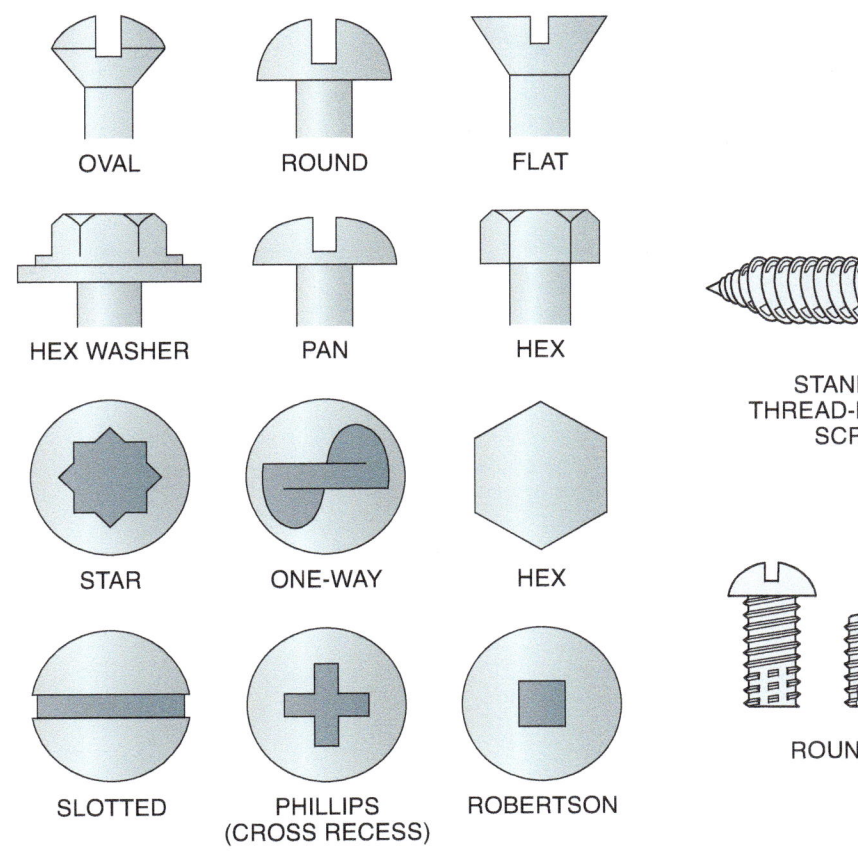

Figure 9 Screw head types and screw drives.

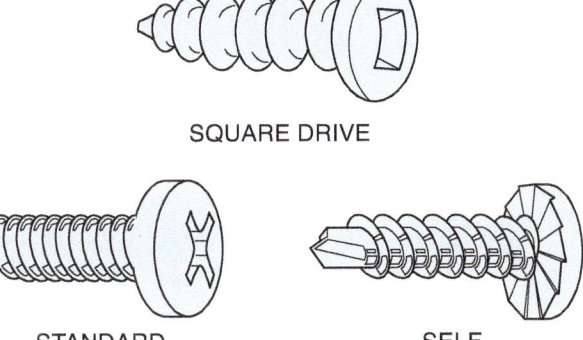

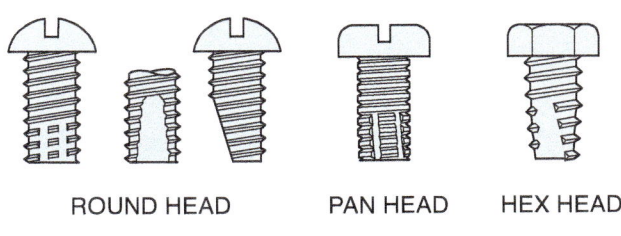

Figure 11 Thread-forming and thread-cutting screws.

rate the threads with penetrating oil and let the oil soak in. A light tap with a hammer may help loosen the fastener. Applying heat with a torch and/or applying dry ice to the joint may also help loosen the fastener. If there is enough of the broken fastener available, a nut may be welded to the stud and a wrench used to back out the broken fastener.

1.3.0 Torquing Threaded Fasteners

To properly torque threaded fasteners, technicians must know the type of fastener in use and the correct torque to be applied. *Figure 15* shows typical torque specifications for fasteners. The

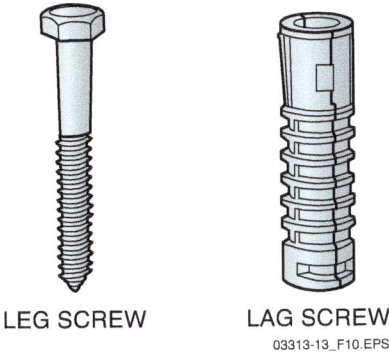

Figure 10 Lag screw.

Fastener Extraction—Art or Science?

Extracting a seized or broken fastener can be a challenge. Special tools such as screw extractors are helpful, but there are many other ways that stubborn fasteners can be removed using readily available tools and materials. Technicians often have to get creative. For example, if the hex head of a bolt is rounded off, a file can be used to flatten two of the sides enough so they can be gripped with a wrench. A hacksaw can be used to cut a slot in the top of a broken bolt so that a screwdriver can be used to remove the stud. Penetrating oil can help break the bond of a rusted connection. The use of heat or cold can cause metal to expand or contract, helping to break the bond holding the bolt. The vibrations caused by light hammer taps can also help break the bond.

(A) TAP

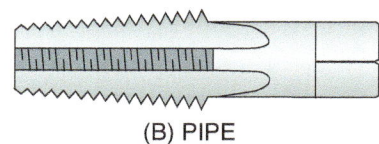
(B) PIPE

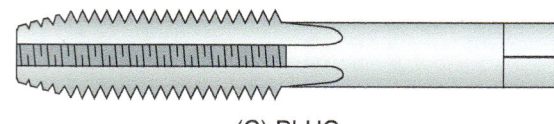
(C) PLUG

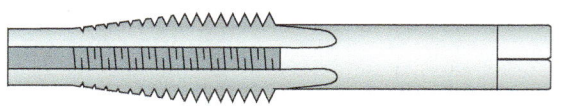
(D) TAPER

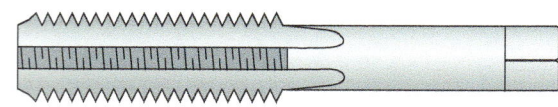
(E) BOTTOMING

Figure 12 Taps.

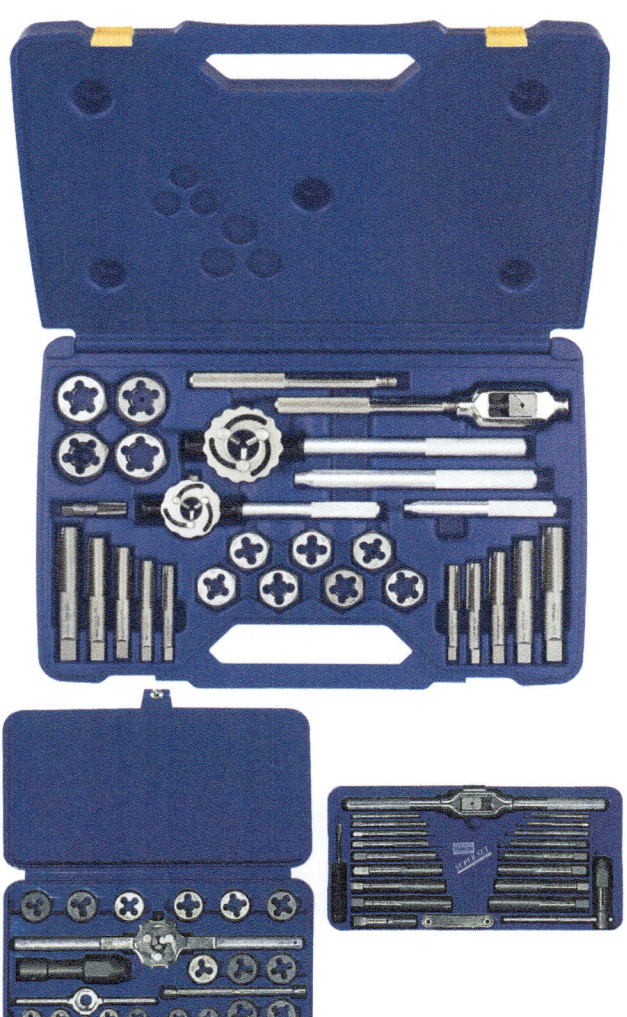

Figure 13 Tap and die set.

Figure 14 Screw extractor.

TORQUE IN FOOT POUNDS

FASTENER DIAMETER	THREADS PER INCH	MILD STEEL	STAINLESS STEEL 18-8	ALLOY STEEL
¼	20	4	6	8
5/16	18	8	11	16
⅜	16	12	18	24
7/16	14	20	32	40
½	13	30	43	60
⅝	11	60	92	120
¾	10	100	128	200
⅞	9	160	180	320
1	8	245	285	490

SUGGESTED TORQUE VALUES FOR GRADED STEEL BOLTS

GRADE	SAE 1 OR 2	SAE 5	SAE 6	SAE 8
TENSILE STRENGTH	64,000 PSI	105,000 PSI	130,000 PSI	150,000 PSI

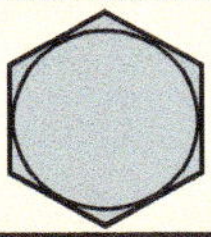

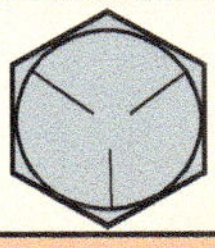

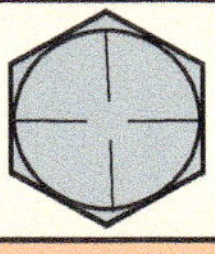

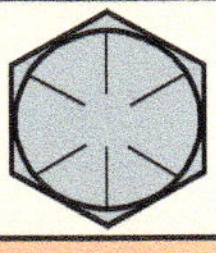

BOLT DIAMETER	THREADS PER INCH	FOOT POUNDS TORQUE			
¼	20	5	7	10	10
5/16	18	9	14	19	22
⅜	16	15	25	34	37
7/16	14	24	40	55	60
½	13	37	60	85	92
9/16	12	53	88	120	132
⅝	11	74	120	169	180
¾	10	120	200	280	296
⅞	9	190	302	440	473
1	8	282	466	660	714

Figure 15 Torque specifications.

specifications are usually stated in inch-pounds for small fasteners or foot-pounds for larger fasteners. However, the proper torque to be applied may be specified in manufacturer' assembly instructions or similar sources of information. Tightening sequences are also important to ensure proper fit and alignment. This section describes the guidelines you should follow when torquing threaded fasteners.

1.3.1 Torque Wrenches

Torque wrenches (*Figure 16*) are used to tighten fasteners to the specified value. The click-type torque wrench is likely the most common type used in the HVAC trade. The correct torque is set by twisting the wrench handle in or out, aligning the index mark with the desired torque value. When tightening a fastener and the torque setting has been reached, a slight break or give can be felt in the torque wrench handle. The beam-type torque wrench does not require an initial setting. It is simply pulled until the needle deflects enough to align with the mark for the desired torque value.

When assembling equipment using a threaded fastener, each fastener should be inspected for damaged threads. If damaged, replace the fastener with a new one. If the threads are dirty or rusty, clean them with a wire brush. In most cases, no type of lubricant, including oil, should be used. This can change the torque value. A fastener to be torqued should only be lubricated if the specification for the toque value instructs the installer to do so. Always follow the manufacturer's instructions and recommended torque values when installing fasteners on equipment.

(A) BEAM-TYPE TORQUE WRENCH

(B) CLICK-TYPE TORQUE WRENCH

(C) DIGITAL TORQUE WRENCH

03313-13_F16.EPS

Figure 16 Torque wrenches.

If the fastener is being installed in a threaded hole, inspect the threads in the hole for damage. If the threads are damaged, the hole should be re-tapped. Be sure to use the correct size and type of tap for the fastener being used. If the hole is blind, a bottom tap will be required.

The correct size torque wrench for a job is one that will read between 25 and 75 percent of its scale when the required torque is applied. This allows for adequate capacity and provides satisfactory accuracy. Avoid using an oversized torque wrench because the scale divisions are too coarse, making it difficult to get an accurate reading. Using a wrench that is too small will not allow for extra capacity in the event of set or seizure or run-down resistance. All threaded fasteners must be clean and undamaged in order to get accurate readings. The calibration of torque wrenches must be checked periodically to guarantee accuracy.

The following terms, sometimes used in manufacturers' service literature, must be understood when using a torque wrench:

- *Break-away torque* – The torque required to loosen a fastener. This is generally lower than the torque to which it has been tightened. For a given size fastener, there is a direct relationship between tightening torque and breakaway torque. This relationship is determined by actual test. Once known, the tightening torque can be checked by loosening and checking breakaway torque.
- *Set or seizure* – In the last stages of rotation in reaching a final torque, seizing or set of the fastener may occur. When this happens, there is usually a noticeable popping sound and vibration. To break the set, back off and then again apply the tightening torque. Accurate torque settings cannot be made if the fastener is seized.
- *Run-down resistance* – The torque required to overcome the resistance of associated hardware, such as locknuts and lock washers, when tightening a fastener. To obtain the proper torque value where tight threads on locknuts produce a run-down resistance, add the resistance to the required torque value. Run-down resistance must be measured on the last rotation or as close to the makeup point as possible.

After determining the correct torque value, set the controls or indicator on the wrench to that value. This is not required if using a beam-type torque wrench. Place the torque wrench on the object to be tightened (nut or bolt) and hold the head of the wrench with one hand to support and align the wrench and object being tightened (*Figure 17*). Secure the other end of the fastener with a second wrench if required. As the object is tightened, watch or listen for the indication that the proper torque has been achieved. Once the torquing is complete, always return the torque wrench setting to zero for storage.

1.3.2 Tightening Sequences

When tightening bolts on flanges and similar surfaces, the bolts must be tightened to the proper torque and in the proper sequence. Tighten each

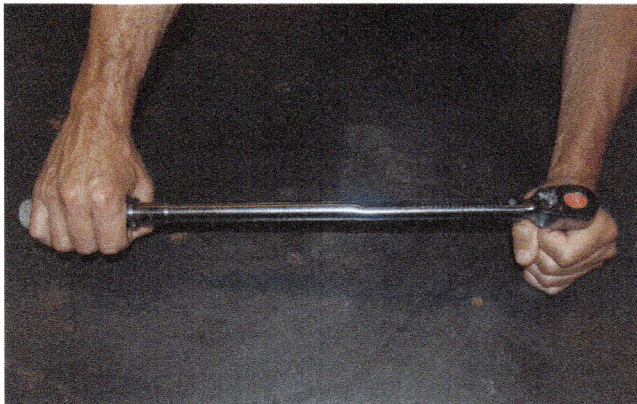

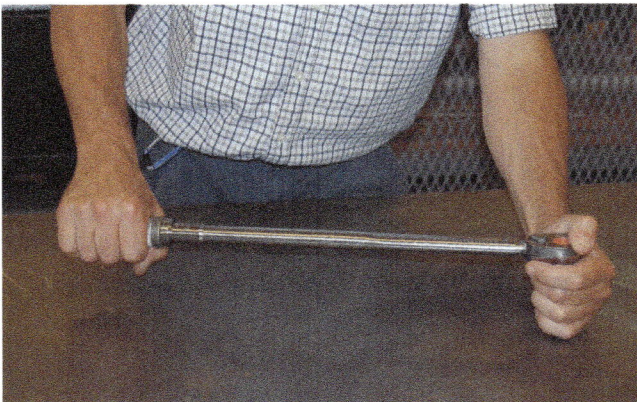

DIRECTION OF ROTATION DEPENDS ON DIRECTION
OF THREADS ON BOLT

Figure 17 Proper use of a torque wrench.

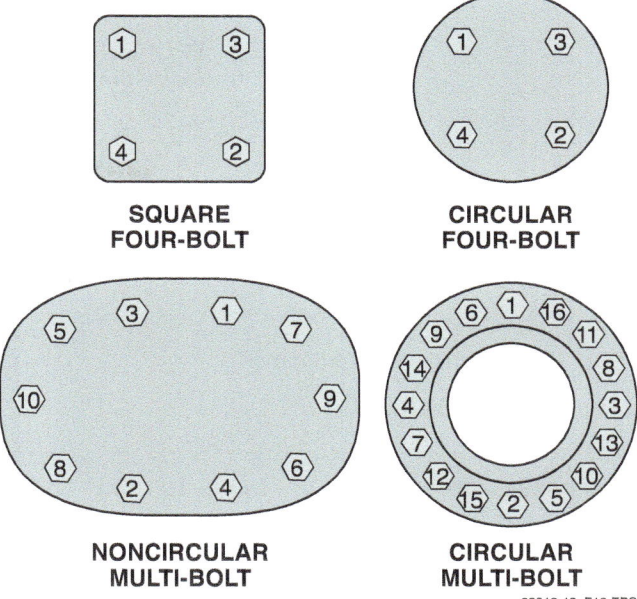

Figure 18 Common fastener-tightening sequences.

fastener only a small amount at a time until snug, following the proper tightening sequence for the bolt pattern and fastener type. After all the fasteners are snug, tighten to the final torque. This prevents warping or damaging the flange or machine part and helps prevent leaks. The numbers in *Figure 18* show the proper tightening sequences for some common bolt patterns.

When flanged ends (*Figure 19*) are bolted together, they must be properly aligned to reduce stress on the flange faces and on the fasteners as they are being torqued.

> **CAUTION**
>
> Tighten fasteners only small amounts at a time, following the proper tightening sequence for the bolt pattern and fastener type before reaching the final torque. Failure to follow the correct tightening sequence or over-tightening can result in damage to the fasteners, any sealing gaskets, or the object being fastened.

1.4.0 Toggle Bolts and Anchor Bolts

Toggle bolts and anchor bolts are specialized fasteners used to secure objects in place. Toggle bolts are used to fasten a part to a hollow wall or panel. Anchor bolts are used to fasten parts and equipment to concrete and/or masonry.

Reusing Nuts and Bolts

When new nuts and bolts are torqued, a certain amount of stretching and/or deforming takes place. If the nut/bolt combination is taken apart and reused, additional stretching and deforming takes place which further weakens the combination. In low-stress applications this may not be a problem. However, in high-stress environments such as the head bolts on a semi-hermetic compressor, it may be wise to replace the bolts with new. Compressor manufacturers sell replacement cylinder head bolts in kit form.

Note that jam nuts should be torqued differently. When a jam nut is used, the thin jam nut is placed on the fastener first, not last. It is then torqued to 25 to 50 percent of the final torque required. The second, standard nut is then placed on top and torqued to the final torque value. While the upper nut is torqued, the jam nut must be held with a spanner wrench or other wrench thin enough to prevent interference with the rotation of the upper nut.

03313 Fasteners, Hardware, and Wiring Terminations   Module One   13

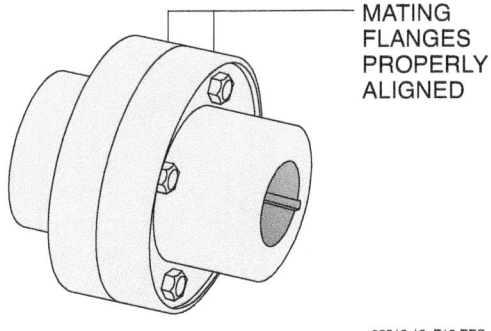

Figure 19 Flanged coupling alignment.

1.4.1 Toggle Bolts

As a general rule, toggle bolts (*Figure 20*) should only be used to support light loads. In many cases, the surface will be sheet rock or a similar wall or siding material. Regardless of the strength of the bolt, the wall material may not be sound enough to reliably support a great deal of weight.

To install a toggle bolt, first drill the correct size hole in the wall or panel. The hole must accommodate the wings of the toggle. Remove the screw from the toggle bolt, insert it in the object to be fastened, and then re-insert the screw into the toggle assembly. Compress the wings of the toggle toward the head of the bolt, and insert it into the hole. Once this is done, the spring-loaded wings will deploy and engage the back side of the wall. Use a screwdriver to draw the wings against the backside of the wall and secure the object (*Figure 21*).

1.4.2 Anchor Bolts

Types of concrete and masonry anchors include the following:

- Concrete inserts
- Non-expanding concrete fasteners
- Expanding concrete fasteners
- Adhesive anchors
- Tapcon® screws

Concrete inserts (*Figure 22*) are imbedded in wet concrete and the hanger extension rod is screwed into the anchor after the concrete has set. The nut is welded to the insert. Other attachments can be added, depending on the requirements of the system.

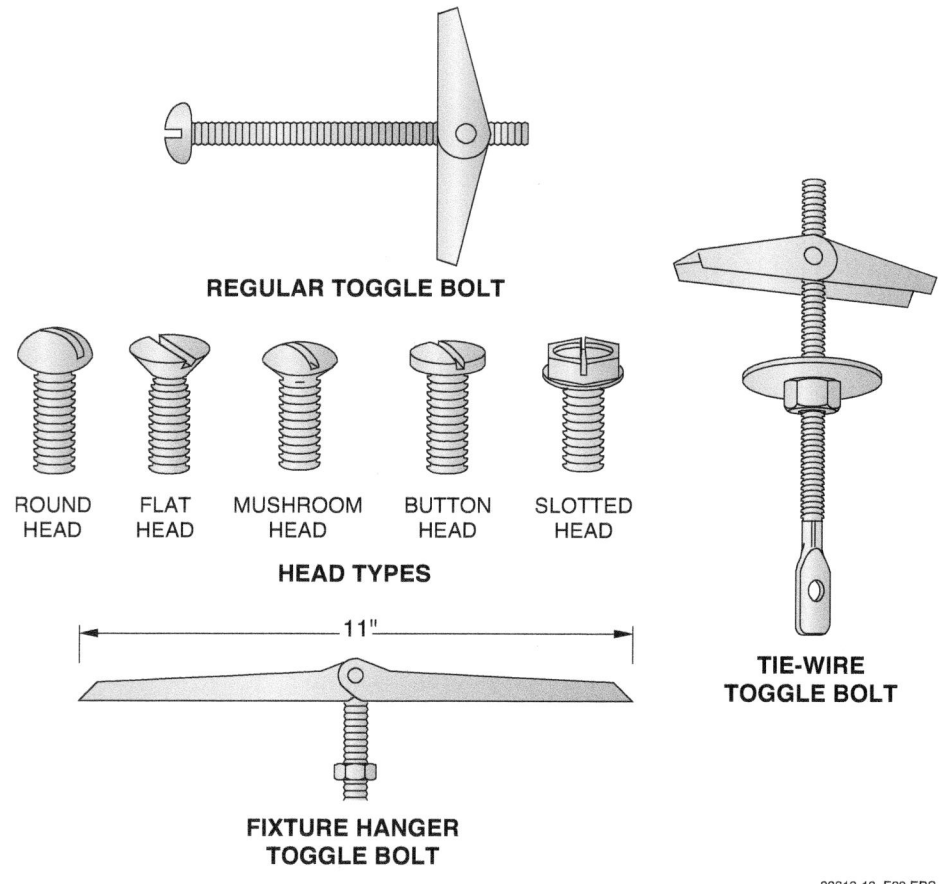

Figure 20 Toggle bolts.

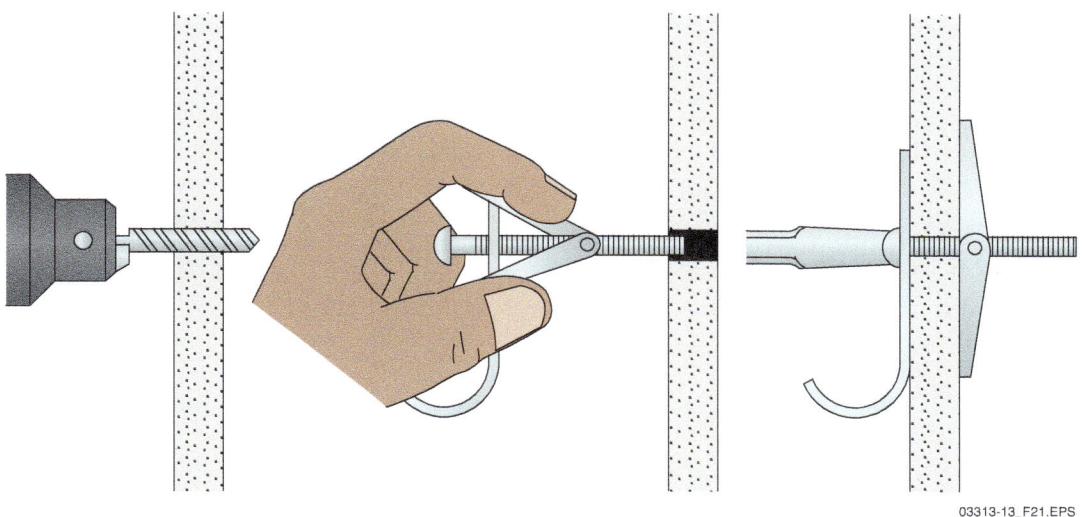

Figure 21 Toggle bolt installation.

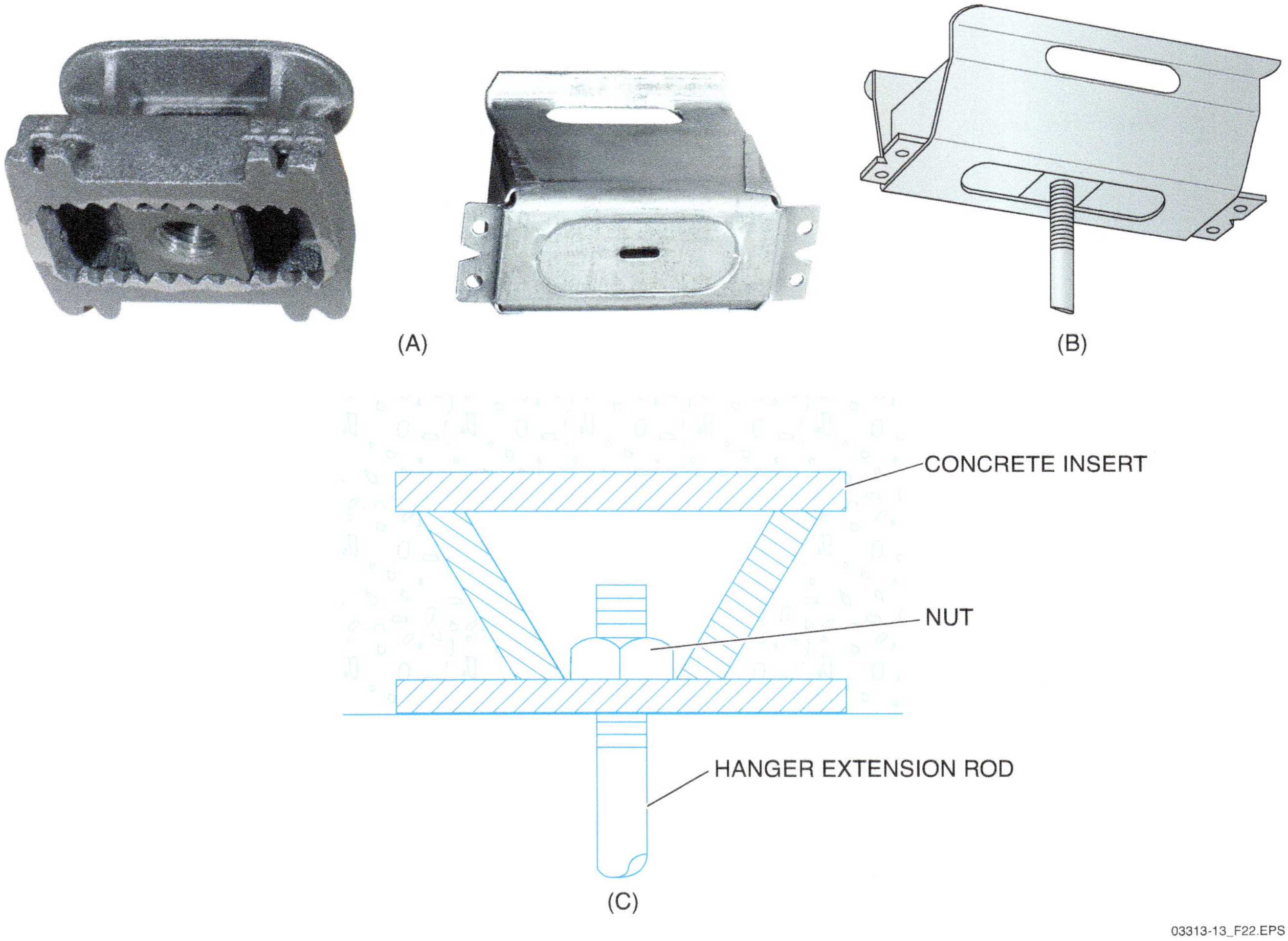

Figure 22 Concrete inserts.

Non-expanding concrete anchors (*Figure 23*) are placed in a hole drilled in hardened concrete and are held in place with grout or epoxy. Expanding concrete anchors (*Figure 24*) however, are placed in snug-fitting holes drilled in hardened concrete. As the nut is tightened, it causes the wedge at the base of the anchor to be drawn up into the sleeve. The sleeve expands as a result, wedging it tightly against the walls of the hole.

Bolts can be secured in holes drilled in concrete using an adhesive, like a non-expanding anchor. It is available as pre-packaged capsules of epoxy and hardener. Both components are packed in a glass or plastic capsule that is inserted into the hole. The bolt is inserted to break the capsule and twisted around to mix the two components. The bolt is then left in place until the epoxy hardens (*Figure 25*). Use care when working with the epoxy as it may contain glass fragments that can cause cuts.

Tapcon® screws (*Figure 26*) can be used for light and medium-duty applications where something must be fastened to concrete or masonry. Introduced in 1976, these screws are driven directly into concrete. Various diameters and lengths are available. After a pilot hole of the correct diameter is drilled in the masonry, the screw is driven in, threading its way into the material.

1.4.3 Installing Non-expanding Concrete Anchors

When installing anchors in hardened concrete, make sure the area where the equipment is to be fastened is smooth so that the equipment will have solid footing. Uneven footing might cause the equipment to twist, warp, not tighten properly, or vibrate during operation.

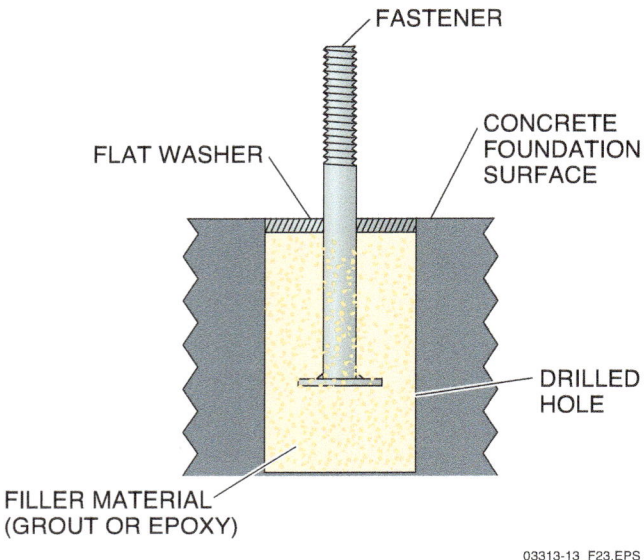

Figure 23 Non-expanding concrete anchor.

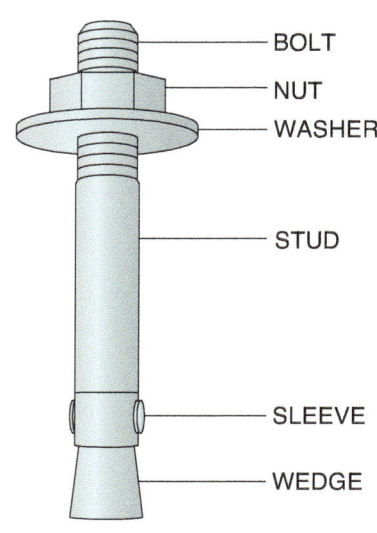

Figure 24 Expanding concrete anchor.

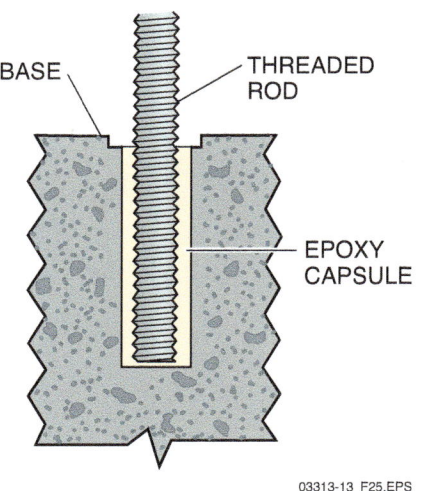

Figure 25 Adhesive anchor.

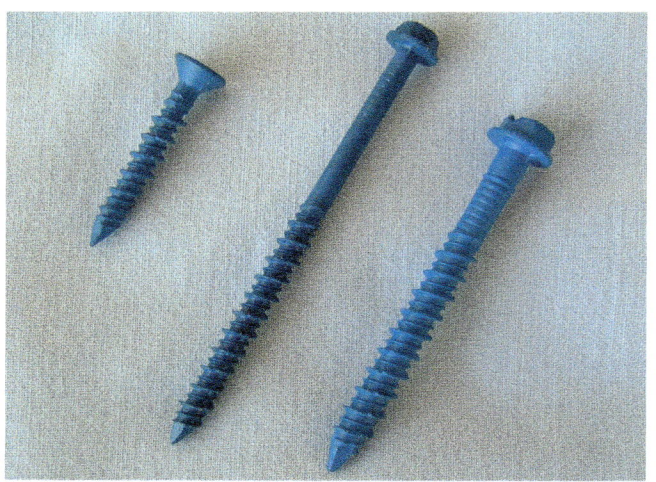

Figure 26 Tapcon® screws.

WARNING!
Drilling in concrete generates noise, dust, and possible flying objects. Always wear safety glasses, ear protectors, respiratory protection, and work gloves. Ensure that others in the area also wear the appropriate PPE.

Carefully inspect the drill and bit to make sure they are in good operating condition. If they are not, you could be injured by parts that chip and fly off during operation.

When installing a non-expansion anchor bolt in hardened concrete, it is installed in a drilled hole filled with a filler material (grout). When installing this type of anchor, the drill bit must be slightly larger in diameter than the head of the fastener and the flat washer used with the fastener. The flat washer must be able to fit down inside the hole, just below surface level.

Also, the anchor bolt should extend out of the hole far enough for the threads and a little of the unthreaded bolt to be above the surface level. Before selecting a bolt length, confirm how much of the bolt needs to protrude from the concrete surface. For example, if a vibration-isolating pad will be placed on the floor under the equipment foot, this will require a longer bolt length. During the drilling process, the drill can be lubricated with water. *Table 1* summarizes the diameter (drill bit size) and depth of the opening to be drilled to install a grouted-in-place non-expanding anchor.

After the hole is drilled, the bolt installed, and the hole filled with grout, make sure the bolt is left straight after working the anchor around in the hole and installing the washer. The washer centers the bolt and holds it until the grout hardens. If the grout hardens and the bolt is not straight, it will be unusable and the job will have to be repeated after the bolt is removed. Allow the grout to fully dry before mounting anything to the anchor bolt.

1.4.4 Installing Expanding Concrete Anchors

When drilling the hole for an expanding-type concrete anchor, the diameter of the hole should be only slightly larger than the diameter of the anchor. The anchor manufacturer will specify the correct diameter drill bit needed for each size anchor. This is the size that should be used. If the hole is too large or out-of-round, the anchor may not be able to wedge itself tightly in the hole. Using the correct size hole, as specified by the bolt manufacturer, ensures that the specifications of strength and hold will remain valid.

The depth of the hole must be such that the anchor does not bottom out prematurely or protrude above the hole. *Figure 27* shows an expanding-type concrete anchor set in a hole of the proper diameter and depth. When drilling the hole, concrete dust may accumulate in the bottom of the hole, preventing complete insertion of the anchor. Use air to blow out any accumulated dust.

1.4.5 Installing Tapcon® screws

Tapcon® screws are driven into masonry or concrete in much the same way wood screws are driven into wood (*Figure 28*). Using this type of fastener requires that a pilot hole of the correct diameter and depth be drilled in the masonry. Failure to use the correct diameter pilot drill can result in two problems. If the pilot hole is oversized, the threads of the screw cannot properly engage the masonry or concrete. Under stress, the screw may pull out. If the pilot hole is undersized, it is dif-

Table 1 Hole Sizes for Non-Expanding Concrete Anchors

DIAMETER OF BOLT TO BE FASTENED	DIAMETER OF OPENING	MINIMUM DEPTH OF OPENING
¼"	¾"	3"
⅜"	1¼"	4"
½"	1½"	4"
⅝"	1¾"	6"
¾"	1¾"	6"
1"	2"	8"
1¾"	2½"	12"

*Concrete must have minimum compression strength of 3000 pounds per square inch.

03313 Fasteners, Hardware, and Wiring Terminations

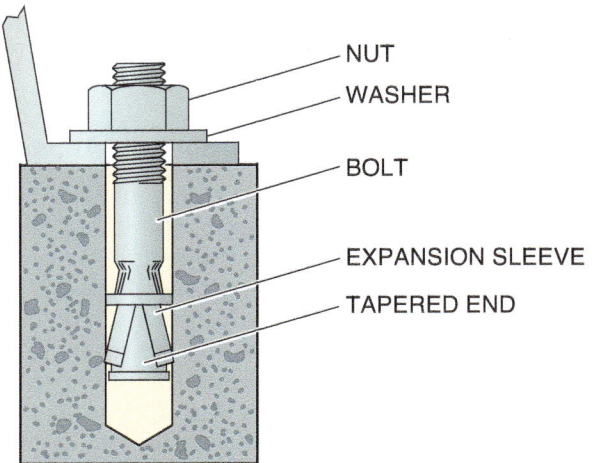

Figure 27 Installed expanding-type concrete anchor.

ficult to drive the screw. Over-torquing the screw often results in the head of the fastener breaking off. The manufacturer will specify the correct diameter masonry drill to use. In many cases, the correct drill is packaged with the screws.

1.5.0 Non-Threaded Fasteners

Non-threaded fasteners have many uses. This section describes the following types of non-threaded fasteners:

- Retainer rings
- Pins
- Keys
- Rivets

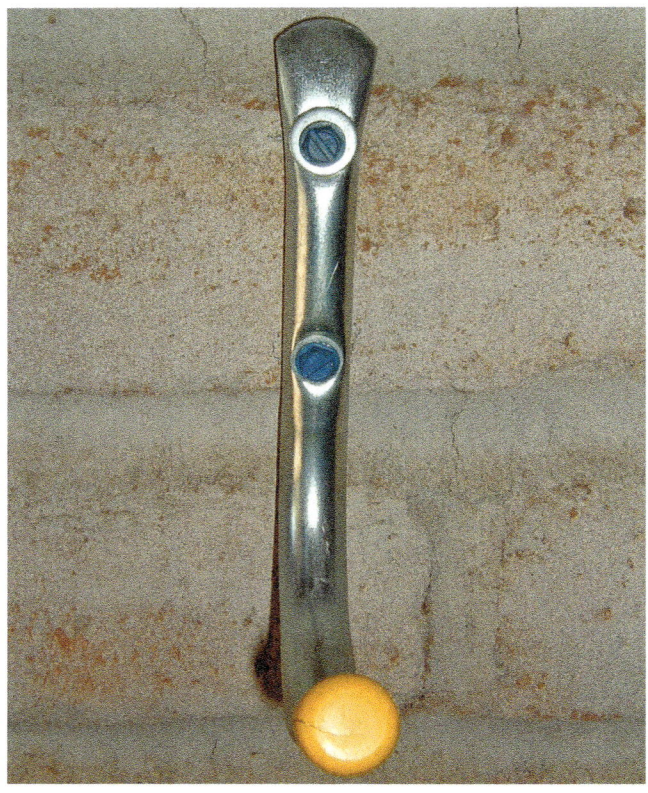

Figure 28 Hanger installed with Tapcon® screws.

1.5.1 Retainer and Lock Rings

Retainer and lock rings (*Figure 29*) are used for both internal and external fastening. Some retainer rings are seated in grooves. Others are self-locking and do not require a groove. Special pliers are used to remove internal and external rings. Self-locking rings snap in place.

Hammer Drills

Hammer drills are commonly used to drill holes in masonry and concrete. The combination of the twisting of the drill bit accompanied by hammer-like blows makes drilling holes in masonry and concrete much easier than is possible with a conventional electric drill.

Figure 29 Retainer and lock rings and lock ring pliers.

1.5.2 Pins

Pin fasteners (*Figure 30*) are used to align mating parts, to hold gears and pulleys on shafts, and to secure slotted nuts.

Some common pins and their uses are as follows:

- *Dowel pins* – These are fitted into reamed holes to position mating parts. They also support a portion of the load placed on the parts.
- *Taper pins* – These are used to fasten gears, pulleys, and collars to a shaft.
- *Cotter pins* – These are fitted into a hole drilled through a shaft. Cotter pins are used to prevent parts from slipping on or off the shaft, and also to keep slotted nuts from working loose.

1.5.3 Keys

Keys (*Figure 31*) are metal parts used to prevent a gear or pulley from rotating on a shaft. One half of the key fits into a seat or groove on the shaft. The other half fits into a keyway in the hub of a gear or pulley.

1.5.4 Rivets

Rivets are used to permanently join two pieces of material. Two common types of rivets are the tinner's rivet and the blind, or pop, rivet (*Figure 32*). To install a tinner's rivet, the rivet is inserted into a hole that has been drilled or punched. The small end of a tinner's rivet is hammered or set in the form of a head. Hollow tinner's rivets are clinched at the small end with a special tool.

03313   Fasteners, Hardware, and Wiring Terminations   Module One   19

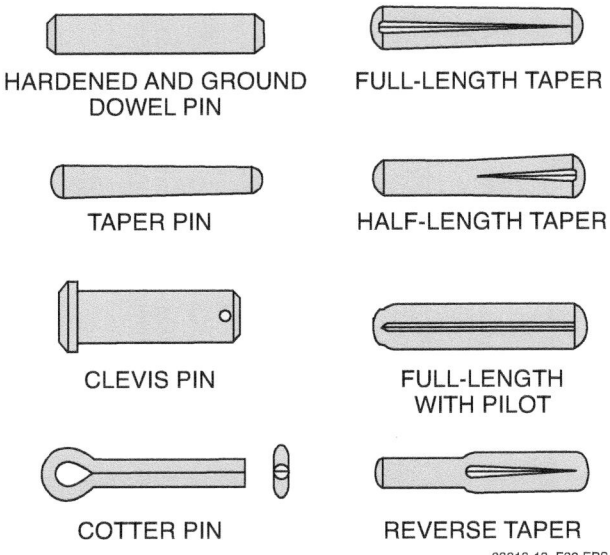

Figure 30 Pin fasteners.

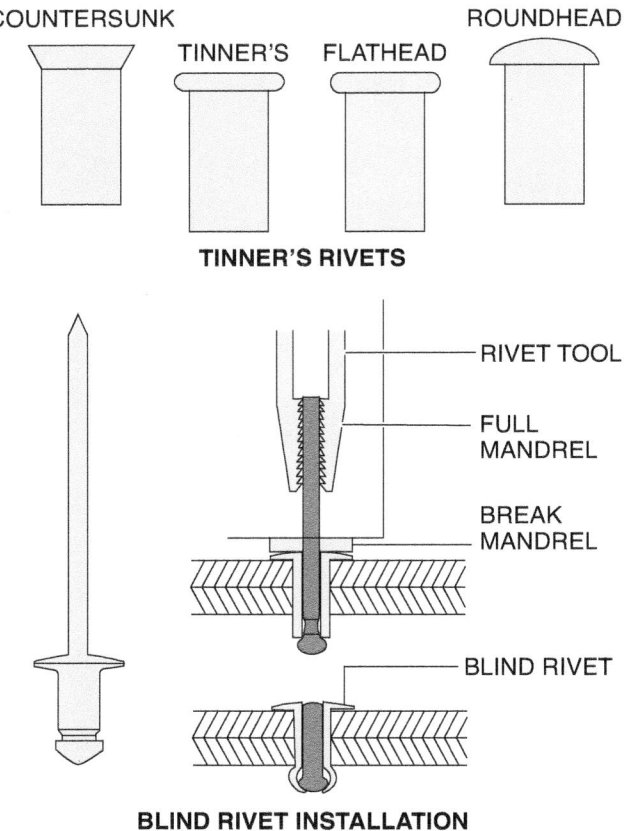

Figure 32 Rivets.

Tinner's rivets are sized in ounces or pounds per 1,000 rivets. For example, a 6-ounce rivet means that 1,000 of these rivets will weigh 6 ounces. As the weight increases, so does the diameter and length of the rivet.

A manual pop rivet tool (*Figure 33*) can be used to install blind rivets in a shop or at the job site. Blind rivets are used to fasten sheet metal, fiberglass, and plastics. They are available in various lengths and diameters. Blind rivets are used when the joint can only be reached from one side. To install a blind rivet, the rivet is inserted into a previously drilled or punched hole and then the rivet is set, or popped, using the tool.

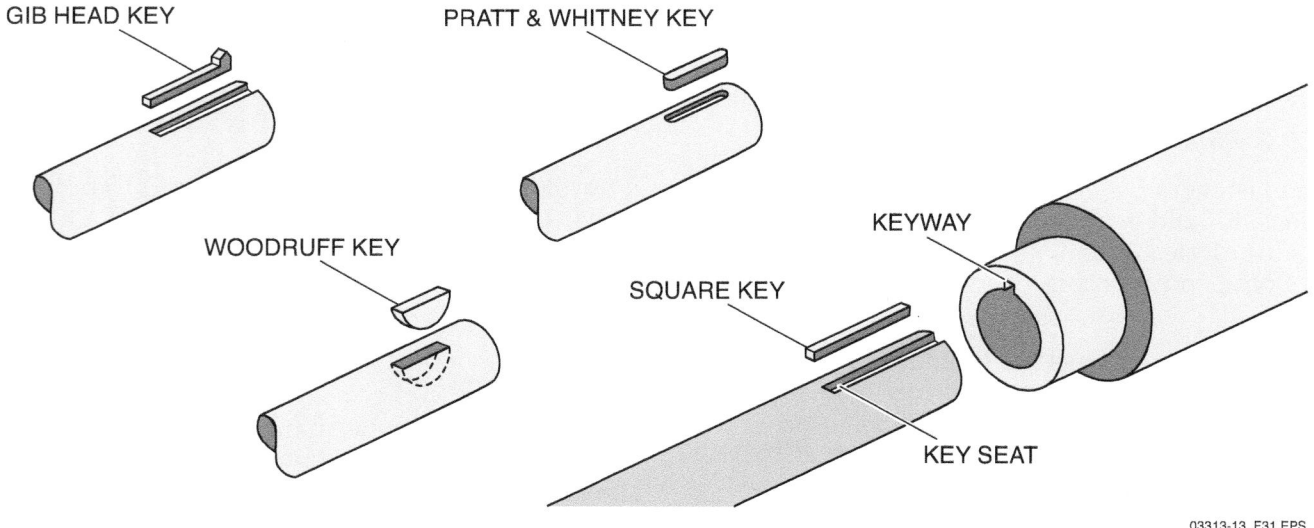

Figure 31 Keys.

Figure 33 Pop rivet tool.

Removing Rivets

Chisel or grind off the heads of tinner's rivets or steel blind rivets; remove the shaft by striking through with a punch. Drill off blind or countersunk rivet heads using a high-speed electric drill, then punch out the shaft.

Additional Resources

Air Conditioning Systems, Principles, Equipment, and Service, Latest Edition. Upper Saddle River, NJ: Prentice Hall.

1.0.0 Section Review

1. Fasteners over 1" in diameter almost always use _____.
    a. coarse threads
    b. fine threads
    c. metric threads
    d. metric fine threads

2. Which type of die is often shaped like a hexagon?
    a. Open adjusting
    b. Screw adjusting
    c. Rethreading
    d. Two-piece

3. When using a correct sized torque wrench, the specified torque reading will show on the indicator in a range between _____.
    a. 40 to 60 percent of full scale
    b. 75 to 100 percent of full scale
    c. 25 to 75 percent of full scale
    d. 50 to 80 percent of full scale

4. Non-expanding concrete anchors are held in place with _____.
    a. folding wings
    b. wedges
    c. adhesive
    d. grout

5. To keep a slotted nut from working loose, use a _____.
    a. key pin
    b. cotter pin
    c. spring pin
    d. dowel pin

Section Two

2.0.0 VIBRATION ISOLATORS

Objectives

Identify and describe the installation of various types of vibration isolators.
a. Identify and describe vibration isolators used to support and suspend equipment and piping.
b. Identify and describe methods used to restrain equipment during seismic events.
c. Explain how to select and install various vibration isolators.

Trade Terms

Seismic activity: Movement of the earth such as during an earthquake.

Seismic restraint: A device or method used to prevent or restrict the movement of equipment during a seismic event.

Vibration isolator: A device that prevents the transmission of vibrations from operating machinery to surrounding objects.

HVAC equipment contains components such as fans, blowers, and compressors, all of which contain electric motors. The electric motors and the devices they drive generate normal noises and vibrations that can be objectionable. If equipment that generates vibrations is in direct contact with other objects, those vibrations can be transferred to the structure, and to pipes and ducts connected to the equipment. For example, an attic-mounted air handler that is installed directly on the attic floor can transmit vibrations to the structural members of the floor, allowing the vibrations to be heard throughout the structure. To reduce the transmission of unwanted noise and vibration, various kinds of vibration isolators are installed on the equipment and on the ducts and pipes connected to the equipment.

2.1.0 Types of Vibration Isolators

Vibration isolators can be grouped based on the type of vibration they are designed to reduce or eliminate. Vibration isolators can be broken into four distinct groups of devices:

- Devices that reduce vibration caused by compressors, fans, and blower motors
- Devices that reduce vibration transmission between objects
- Devices that reduce vibration transmission through ducts
- Devices that reduce vibration transmission through pipes

2.1.1 Compressor, Fan, and Blower Motor Vibration Isolators

Welded hermetic compressors installed in packaged units and split-systems are mounted on grommet-like vibration isolators made of rubber (*Figure 34*). A bolt with an integral washer passes through the center of the grommet and holds the compressor to the unit's base pan. The rubber prevents metal-to-metal contact between the base pan and compressor, keeping any vibrations the compressor generates from being transferred to the rest of the unit.

It is very important to note that isolators used in this way are of little or no value if the hold-down bolt is tightened. The hold-down bolts of compressors are tightened at the factory for shipment of the unit only. However, during the installation or initial startup, the bolts should be loosened so that they are not applying any pressure on the rubber grommets. Crushing the rubber grommet by tightening the bolt eliminates most or all of the isolation characteristics.

A similar arrangement is used to isolate a direct-drive blower motor from the blower housing

Figure 34 Hermetic compressor vibration isolators.

(*Figure 35*). Three grommet-like isolators are attached to the three motor mounts that are located 120 degrees apart around the motor. These bolts must be tight to hold the motor in place, but not tight enough to damage the isolator.

Heavier semi-hermetic compressors are mounted on stronger isolators that use springs instead of rubber (*Figure 36*). These compressors installed in equipment are often equipped with some means to prevent the compressor from moving during shipment. This may take the form of a removable shipping bracket and/or mounting hardware that keeps the mounting springs compressed. Before starting the compressor, the bolts compressing the mounting springs are backed out to allow the mounting springs to decompress and provide isolation. Even though the compressor may be spring-mounted within a unit, equipment manufacturer's sometimes recommend that the entire unit be mounted on spring-type isolators to further reduce the transmission of vibrations.

2.1.2 Vibration Isolators that Reduce Vibration Between Objects

Equipment vibration can be transmitted between the equipment and the object to which it is mounted. For example, a roof-mounted packaged unit can transmit vibrations to the structural members of the roof. Relatively lightweight residential air handlers installed directly on ceiling joists in an attic can be isolated by placing vibration-isolating pads (*Figure 37*) under key locations of the unit. This particular product is constructed of two layers of grooved rubber sheet, separated by a cork center.

When working with all vibration isolators, it is essential to ensure that the isolator is working within its designed load range. The pad shown in *Figure 37* is rated at 50 pounds per square inch. That means that only 4 square inches of pad area is needed for a 200-pound unit. If too much weight is applied, the pad will likely crush down too tightly against its cork center and reduce its effectiveness. If it is under-loaded, it will also fail to provide good isolation. This concept is common to all vibration isolation products. To perform the desired function, they must be properly loaded.

Residential air handlers installed in attics can also be suspended from roof rafters using multiple threaded rods that are coupled through vibration-isolating hangers (*Figure 38*). Each threaded rod contains its own vibration isolator (*Figure 39*). This prevents any vibration from being transmitted through the threaded rods to the roof rafters.

In *Figure 39*, note the spring deflection scale. This provides a quick visual means of determining how much the spring is deflected under the load. Loading a spring isolator to somewhere near mid-range is usually ideal. For example, if a 50-pound load on each corner of a suspended load is expected, vibration isolators with a maxi-

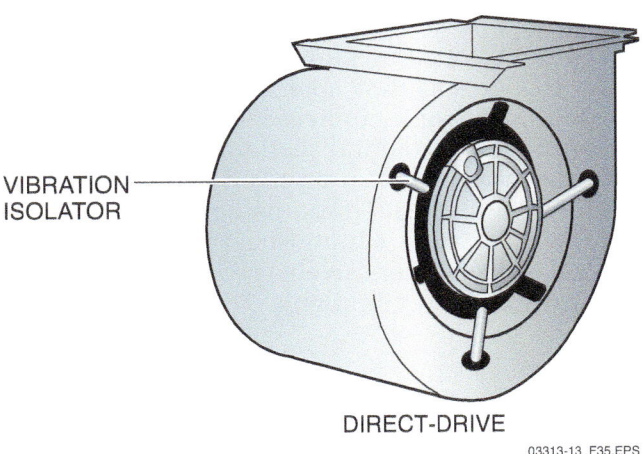

Figure 35 Direct-drive blower vibration isolators.

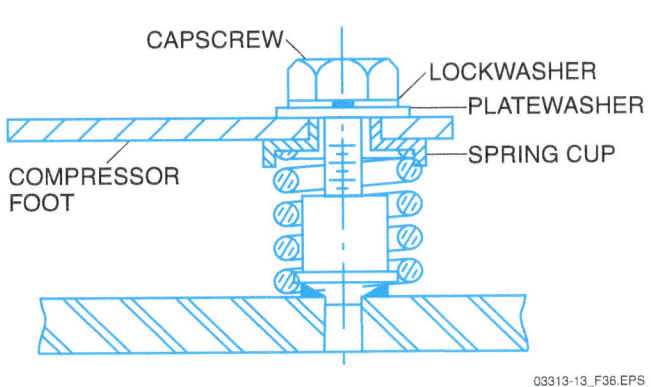

Figure 36 Semi-hermetic compressor vibration isolator.

Figure 37 Vibration-isolating pad.

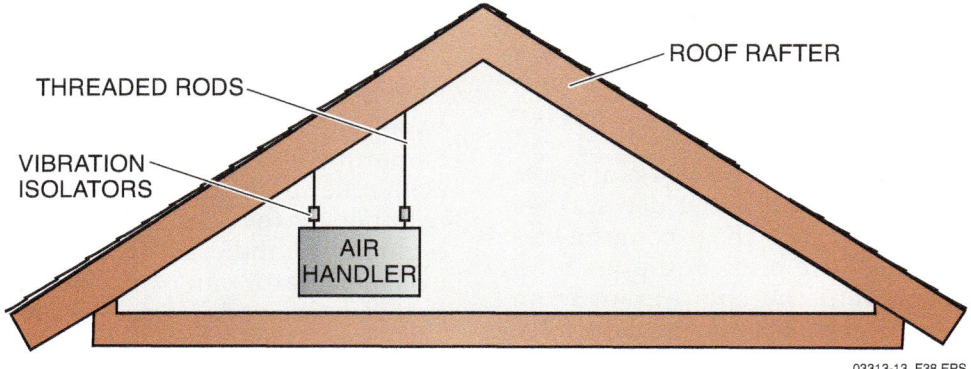

Figure 38 Air handler suspended with vibration isolators.

mum loading of roughly 100 pounds would likely be chosen.

Heavy HVAC equipment, such as a chiller, is mounted on heavy-duty spring-mounted vibration isolators (*Figure 40*). Although it may appear that the upper pad of the isolator is touching the top of the base, it is actually riding on top of the spring. Since the corner and/or side-to-side loads of a large piece of equipment are usually different, it is not unusual to see two, or even three, different sizes of vibration isolators supporting a unit.

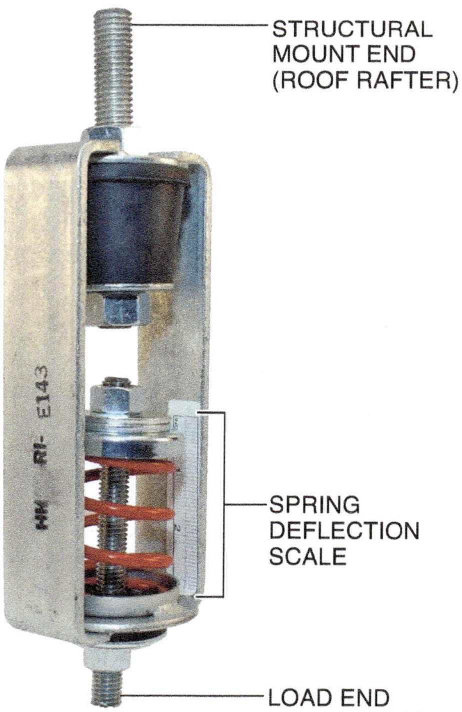

2.1.3 Vibration Isolators that Reduce Vibration Through Ducts

In metal duct systems, any noise or vibration generated by the equipment can be transmitted throughout the structure. Because of the materials used, flexible ducts and ducts made with rigid fiberglass board do not easily transmit vibrations and do not require vibration isolation. The simplest way to prevent vibrations from the air handler into a metal duct system is to install flexible (non-metallic) connectors in the supply and return air ducts as they enter/leave the air handler (*Figure 41*). Flexible connectors can be installed in vertical and horizontal sections of duct. In commercial applications, metal ducts can be suspended with threaded rod using spring-mounted vibration isolators. Lining sheet metal ducts with fiberglass duct liner can also reduce the transmission of noise and vibrations.

Figure 39 Vibration-isolating hanger.

Figure 40 Heavy-duty vibration isolator.

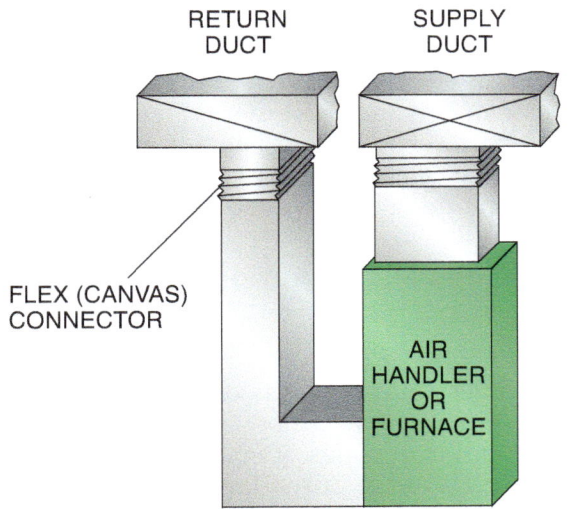

VIBRATION/NOISE CONTROL AT AIR HANDLER

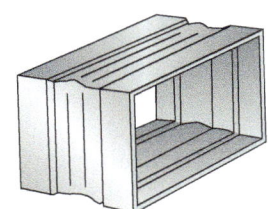

VIBRATION/NOISE/MOVEMENT CONTROL IN DUCT RUNS

Figure 41 Ductwork vibration and noise control devices.

2.1.4 Vibration Isolators that Reduce Vibration Through Pipes

Pipes, like metal duct systems, can transmit any noise or vibration generated by the equipment throughout the structure. Vibrations can be reduced at the source (a pump or compressor) or in the piping system itself. Compressors, especially larger semi-hermetic compressors, can generate noise and vibration at start-up and during operation. The starting torque of the motor can cause the compressor to move as it starts. If rigid pipes are connected to the compressor, they can be stressed by the starting torque and transmit noise and vibrations through the connected piping. To reduce this problem, a flexible metal vibration eliminator (*Figure 42*) can be installed in the suction and/or discharge lines of the compressor. The vibration eliminator is typically made of corrugated bronze with a braided bronze jacket. Stainless steel versions are also available. A wide range of lengths and diameters are available so they can be installed in almost any refrigeration

Figure 42 Flexible metal vibration eliminator.

system. Copper fittings on either end allow the fitting to be brazed into copper refrigerant lines. The corrugations in the body of the device make it flexible and able to absorb equipment vibrations. Note that the isolator in *Figure 42* is designed specifically for refrigerant line use.

Piping suspended by threaded rods (*Figure 43*) can also include vibration isolators. Again, sizing each vibration isolator for the expected load is essential to get the desired performance.

Figure 43 Piping suspended with vibration isolators.

2.2.0 Seismic Restraints

In areas with high seismic activity, local codes often require that HVAC equipment be fitted with some form of seismic restraint to prevent movement if an earthquake occurs. The reason for this is quite simple. If gas-fired equipment moves too much, it can break gas and electrical connections, resulting in a fire. If heavier equipment breaks loose, it can cause additional structural damage and possibly injure bystanders. Equipment can be restrained using angle iron, straps, bumpers, vibration-isolating devices, cables, and other methods.

The variety of seismic restraint requirements for all the different types of HVAC equipment is beyond the scope of this module. For more detailed information on this subject, refer to *FEMA 412, Installing Seismic Restraints for Mechanical Equipment* published by the Federal Emergency Management Agency.

2.2.1 Seismic Restraints for Residential Equipment

There are several important requirements for equipment installed in a residential environment. For example, residential gas furnaces and water heaters with input ratings less than 100,000 Btuh must use flexible connectors between any rigid gas pipe and the appliance. Furnaces and water heaters can be attached to the structure with metal straps bolted to the structure. Furnaces can also be restrained with bumpers. Residential split-system outdoor units should be bolted to a concrete pad using a concrete anchor.

Bumpers can be used to prevent horizontal movement on equipment that is not top-heavy and prone to tipping over. The bumpers shown in *Figure 44* are bolted to concrete.

2.2.2 Seismic Restraints for Light Commercial Equipment

Packaged rooftop units that are curb-mounted can be converted in the field to be resistant to seismic events or the unit can be installed on a factory-built seismic curb like the one shown in *Figure 45*. Equipment such as ceiling-mounted air handlers can be suspended and constrained with a combination of threaded rod hangers with vibration isolators and steel cables (*Figure 46*).

2.2.3 Seismic Restraints for Hydronic and Commercial Refrigeration Equipment

Pumps used in hydronic systems are restrained using a number of devices. In *Figure 47*, vibration isolators and snubbers retain the pump on its pad. Flexible connectors in the pump's supply and return water lines along with vibration-isolated pipe hangers help prevent breakage of the pipes in case of seismic movement. *Figure 48* shows a combination coil spring isolator and seismic restraint that can be used for various kinds of heavy floor-mounted mechanical equipment.

2.3.0 Selecting Vibration Isolators and Materials

The selection of the correct vibration isolator or seismic restraint depends on the type of equipment, its weight, and how and where it is to be installed. Local codes often dictate what devices or materials are required for a particular application. Job specifications also dictate the type of vibration isolation that an application requires.

Vibration isolators and seismic restraints that can be purchased as components include the following:

- Suspension-type vibration isolators
- Spring-type vibration isolators
- Grommet-type vibration isolators
- Vibration-isolating pads
- Flexible metal refrigerant pipe
- Factory-built seismic roof curbs

Manufacturers of these products and devices usually provide printed selection information so that the correct device can be found for any application. In addition to common information such as dimensions, they can provide weight limitations or range of movement limitations. *Figure 49* shows one particular type of vibration isolator, along with its selection data. Note that a single basic assembly can accommodate a number of different springs. The spring design and construction characteristics determine its capacity. The springs are often color-coded, making it easy to visually confirm which spring has been provided in an installation. Note that the isolators at the lower end of the chart, with the highest weight rating, actually are equipped with two springs. A core spring is provided inside the main, outer spring.

Did You Know?

The Great San Francisco Earthquake

In 1906, a devastating earthquake struck the City of San Francisco. Many of the city's water mains were broken, resulting in a loss of water pressure. With little or no water pressure available to fight the many fires that started after the earthquake, fires raged out of control, destroying the city.

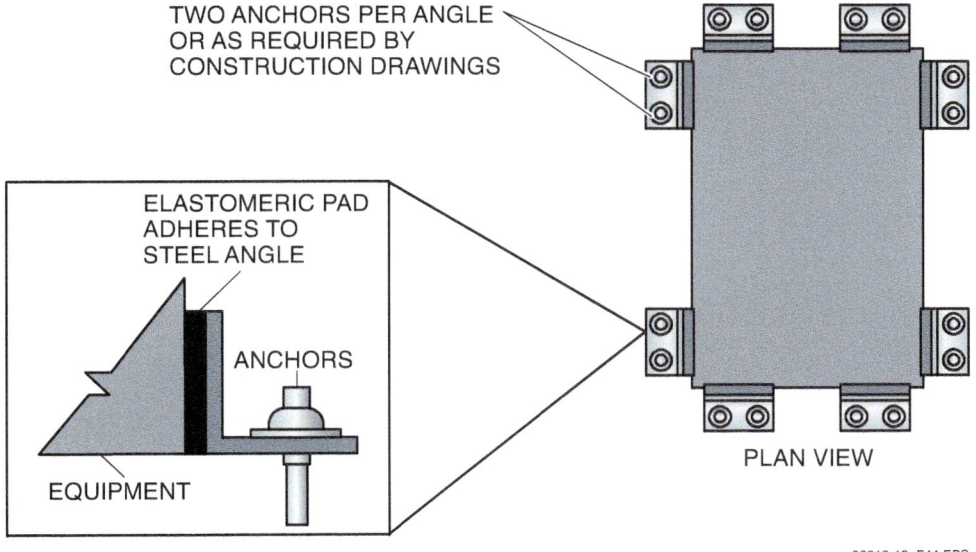

Figure 44 Bumpers.

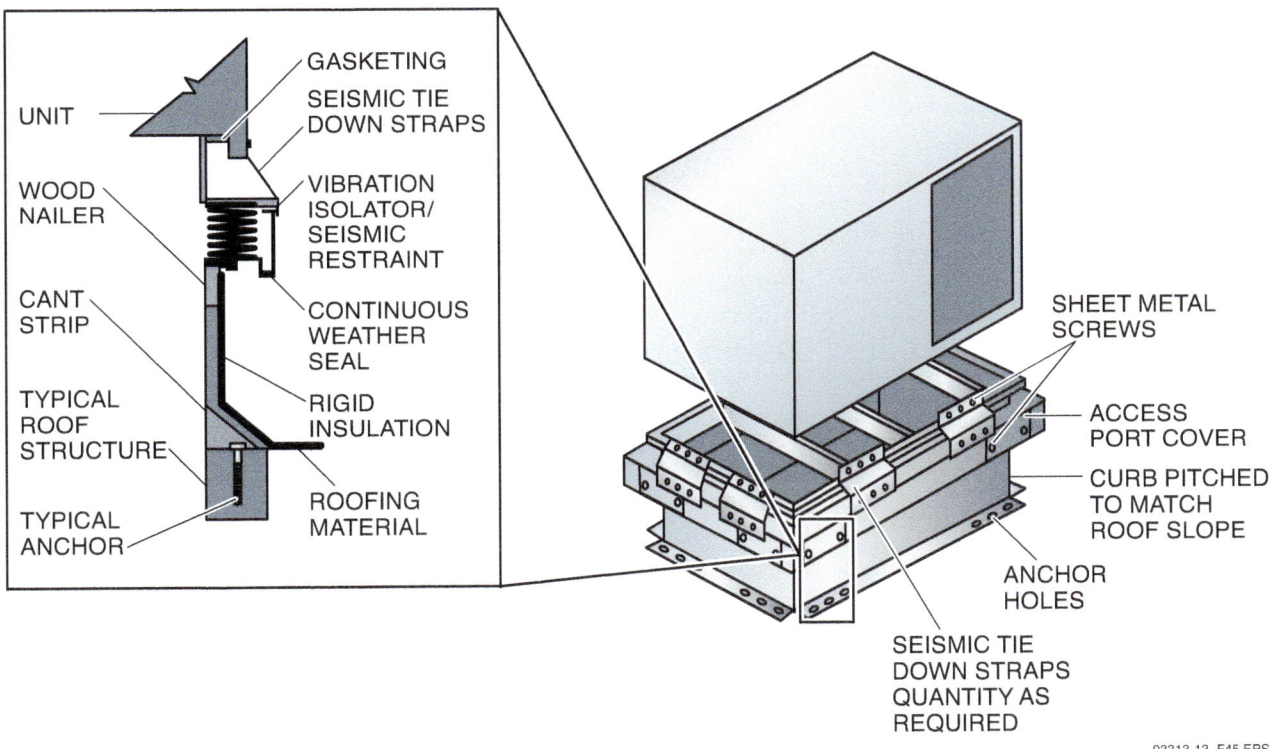

Figure 45 Factory-built seismic curb.

Seismic Gas Shut-Off Valve

The danger of fire when gas lines rupture during an earthquake is obvious. A new gas shut-off valve is now available to help reduce this danger. Installed in the main gas line after the gas meter, this valve shuts off gas flow to the entire structure if an earthquake of magnitude 5.4 or above is detected. If shutoff occurs, the valve must be manually reset.

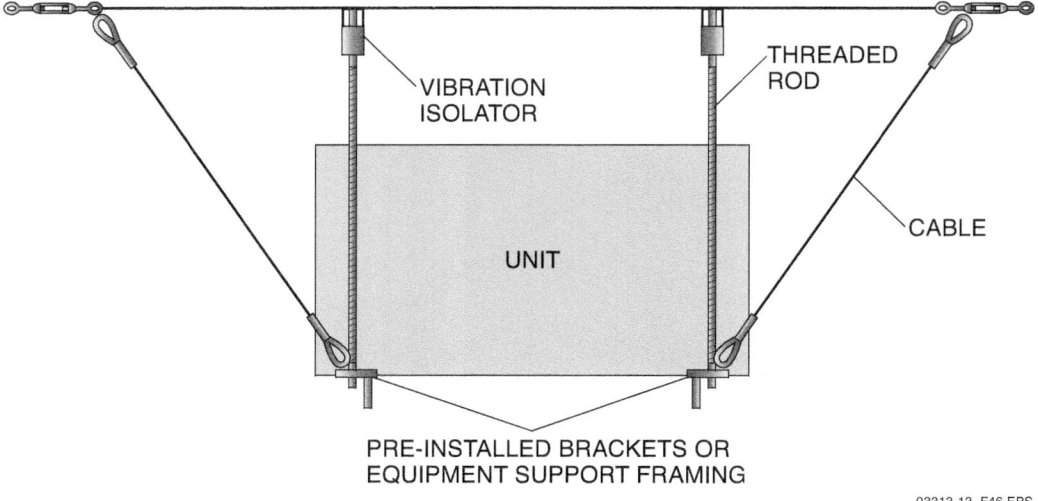

Figure 46 Suspended equipment seismic restraints.

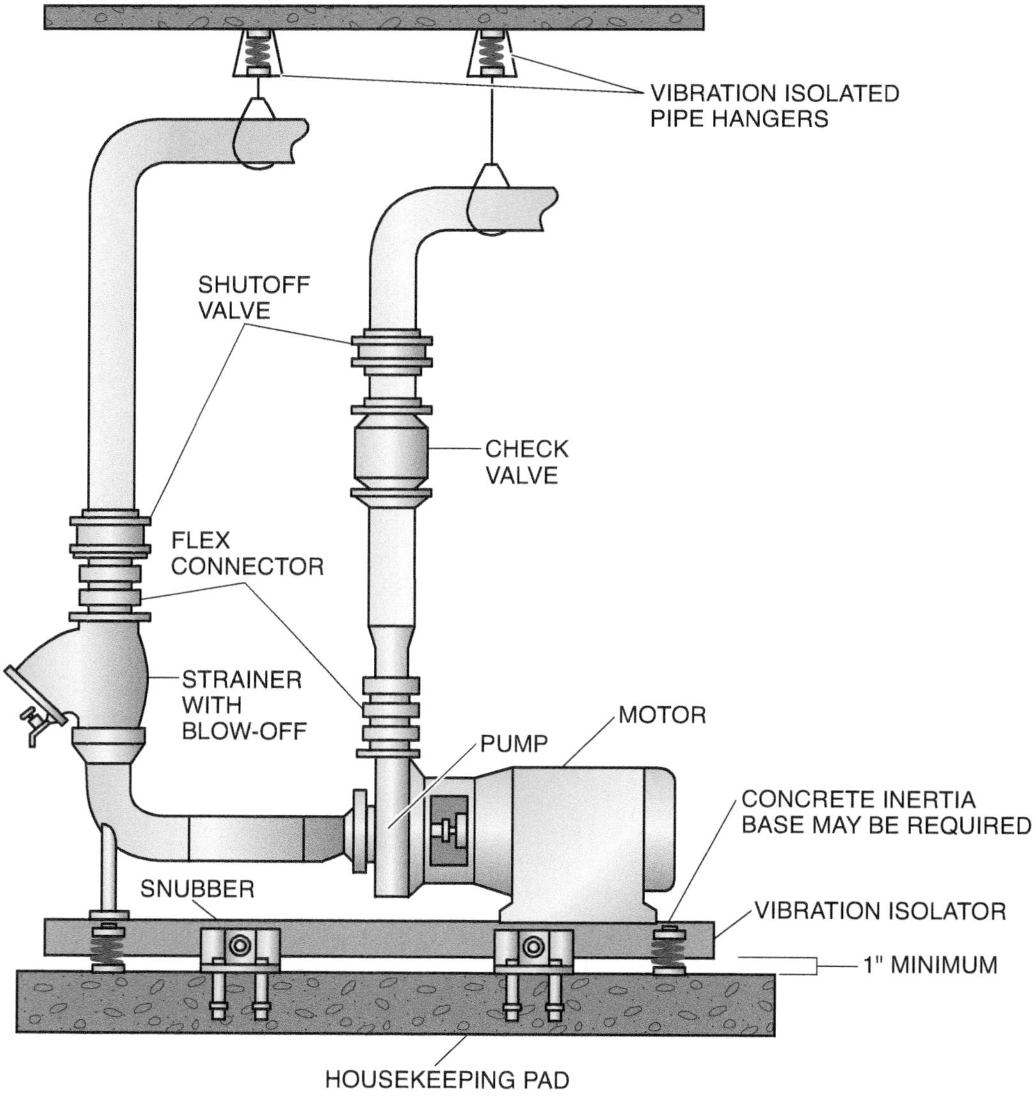

Figure 47 Water pump seismic restraints.

Figure 48 Combination seismic restraint.

Other vibration isolators or seismic restraints must be field-fabricated using common hardware items such as the following:

- Angle iron
- Metal straps
- Steel cable
- Threaded rod
- Nuts, bolts, washers, etc.

The size and quality of these materials is often spelled out in local codes or job specifications. For example, the SAE grade of bolts required for a restraint or the diameter of threaded rod required to suspend a fan coil may be specified.

Vibration isolators and seismic restraints use similar devices and methods to achieve similar results. An excellent resource for learning what to use for a particular application can be found in *FEMA 412, Installing Seismic Restraints for Mechanical Equipment* published by the Federal Emergency Management Agency. Tables based on equipment type offer suggestions on how to properly install a piece of equipment.

Did You Know?
Earthquake-Resistant Buildings

Many new skyscrapers being built in earthquake zones around the world are being built on giant suspension springs. This allows the building to sway instead of collapse when an earthquake occurs.

(A)

TYPE 1" DEFLECTION C, CI, CIP & CIW RATINGS

Size	Rated Capacity (lb)	Rated Defl (in)	Mount Constant (lb/in)	Spring Color/Stripe
A-45	45	1.60	28	Blue
A-75	75	1.50	50	Orange
A-125	125	1.33	94	Brown
A-200	200	1.15	174	Black
A-310	310	1.00	310	Yellow
A-400	400	1.00	400	Green
A-510	510	1.00	510	Red
A-625	625	1.00	625	White
B-65	65	2.10	31	Brown
B-85	85	2.10	40	White/Black
B-115	115	2.00	57	Silver
B-150	150	2.00	75	Orange
B-280	280	1.60	174	Green
B-450	450	1.31	344	Red
B-750	750	1.12	670	White
B-1000	1000	1.00	1000	Blue
B-1250	1250	1.00	1250	Gray
B-1650	1650	1.00	1650	Black
C-1000	1000	1.00	1000	Black
C-1350	1350	1.00	1350	Yellow
C-1750	1750	1.00	1750	Black
C-2100	2100	1.00	2100	Yellow*
C-2385	2385	1.00	2385	Yellow**
C-2650	2650	1.00	2650	Red*
C-2935	2935	1.00	2935	Red**
D-3500	3500	1.00	3500	Black*
D-4200	4200	1.00	4200	Yellow*
D-4770	4770	1.00	4770	Yellow**
D-5300	5300	1.00	5300	Red*
D-5870	5870	1.00	5870	Red**
E-7000	7000	1.00	7000	Black*
E-8400	8400	1.00	8400	Yellow*
E-9540	9540	1.00	9540	Yellow**
E-10600	10600	1.00	10600	Red*
E-11740	11740	1.00	11740	Red**
F-12600	12600	1.00	12600	Yellow*
F-14310	14310	1.00	14310	Yellow**
F-15900	15900	1.00	15900	Red*
F-17610	17610	1.00	17610	Red**
G-18900	18900	1.00	18900	Yellow*
G-21465	21465	1.00	21465	Yellow**
G-23850	23850	1.00	23850	Red*
G-26415	26415	1.00	26415	Red**

*with RED core spring   **with GREEN core spring

(B)

Figure 49 Vibration isolator selection information.

For example, assume an air handler is to be mounted on a concrete floor. For simple vibration isolation, open- or housed-spring devices can be used. If the installation must be resistant to seismic activity, housed-spring isolators cannot be used because they cannot resist uplift. If open-spring devices are used and the installation must be seismic-compliant, snubbers or bumpers must be used to restrict back-and-forth and side-to-side motion. Equipment manufacturers may provide specific instructions and/or kits for installing their equipment to be seismic compliant.

Seismic Building Codes

In many parts of the United States, concerns about earthquakes are not reflected in local building codes. However, in states such as California, all aspects of building construction must consider the effect an earthquake can have on a structure and what must be done during the construction of the building to reduce or eliminate those effects.

Additional Resources

FEMA 412, Installing Seismic Restraints for Mechanical Equipment, Latest Edition. Washington, DC, Federal Emergency Management Agency.

2.0.0 Section Review

1. Flexible vibration isolators installed in ducts are mounted in _____.
   a. vertical sections of duct
   b. horizontal sections of duct
   c. supply air ducts only
   d. vertical or horizontal ducts

2. Seismic restraints are placed on HVAC equipment to prevent _____.
   a. movement
   b. vibrations
   c. fires
   d. noises

3. The product catalog for an open-spring vibration isolator is likely to contain information about how much it can safely move during normal operation.
   a. True
   b. False

SECTION THREE

3.0.0 WIRING TERMINATIONS

Objectives

Identify common low- and line-voltage electrical termination hardware and explain how to terminate wiring connections.
 a. Identify common low- and line-voltage electrical terminating hardware.
 b. Explain how to properly terminate low- and line-voltage wiring connections.

Performance Task

3. Terminate low- and line-voltage wiring on a compressor contactor.

Trade Terms

AL-CU: An abbreviation for aluminum and copper, commonly marked on electrical connectors to indicate the device is suitable for use with conductors made with either metal.

Connector: A device used to physically and electrically connect two or more conductors.

Lug: A wiring terminal used on wire sizes No. 8 and larger. Also a term used to describe a screw-type terminal that simply applies pressure to a stripped wire, without the need for a crimp-on terminal.

Terminal: A device used for connecting wires to an object.

Termination: The connection of a wire to its destination.

A properly made termination should last as long as the insulation on the wire itself. A poorly made termination will always be a source of trouble because it will overheat under load and eventually fail.

The basic requirements for a good electrical termination include the following:

- It must be mechanically and electrically secure.
- It must be insulated as well as or better than the existing insulation on the conductors.
- These characteristics should last as long as the conductor is in service.

There are many different types of wiring terminations. The selection of the proper type for a given application often depends on how and where the termination is used. Terminations are normally made with a solder-less pressure connector to save time.

3.1.0 Connectors and Terminations

A variety of wire terminations for stranded wire are shown in *Figure 50*. These connectors are available in various sizes to accommodate wire sizes No. 22 AWG and larger. They can be installed with manual crimping tool. The range is normally stamped on the tongue of each terminal.

Crimp-type terminations for wires smaller than No. 8 AWG are normally made to accept at least two wire sizes and are often color-coded. For example, one manufacturer's color code is red for No. 18 or No. 20 wire, blue for No. 16 or No. 14 wire, and yellow for No. 12 or No. 10 wire. Crimp-type terminations for wire sizes No. 8 and larger, commonly called lugs, are made to accept one specific conductor size. Mechanical compression-type connectors and lugs are made to accommodate a range of different wire sizes.

Compression-type connectors for connecting conductors to screw terminals for low- and line-voltage circuits include those in which hand tools indent or crimp tube-like sleeves on the end of a conductor. Proper crimping action changes the size and shape of the connector and deforms the conductor strands enough to provide good electrical conductivity and mechanical strength.

Figure 51 shows the basic structure of a crimp connector. The crimp barrel receives the wire and is crimped to secure it in place. The Vs or dimples inside the barrel improve the wire-to-terminal conductivity and also increase the termination tensile strength. Most crimp connectors have nylon or vinyl insulation covering the barrel to reduce the possibility of shorting to adjacent terminals. The insulation is color coded according to the connector's wire range to reduce the problem of wire-to-connector mismatch. An inspection hole is provided at the end of the barrel to allow visual inspection of the wire position. For the smaller wire sizes, a sleeve is crimped over the conductor insulation in the process of crimping the barrel, providing strain relief for the conductor.

Conductor Terminations and Splices

Poor electrical connections are responsible for a large percentage of equipment burnouts and fires. Many of these failures are a direct result of improper terminations, poor workmanship, and the use of improper devices.

03313 Fasteners, Hardware, and Wiring Terminations

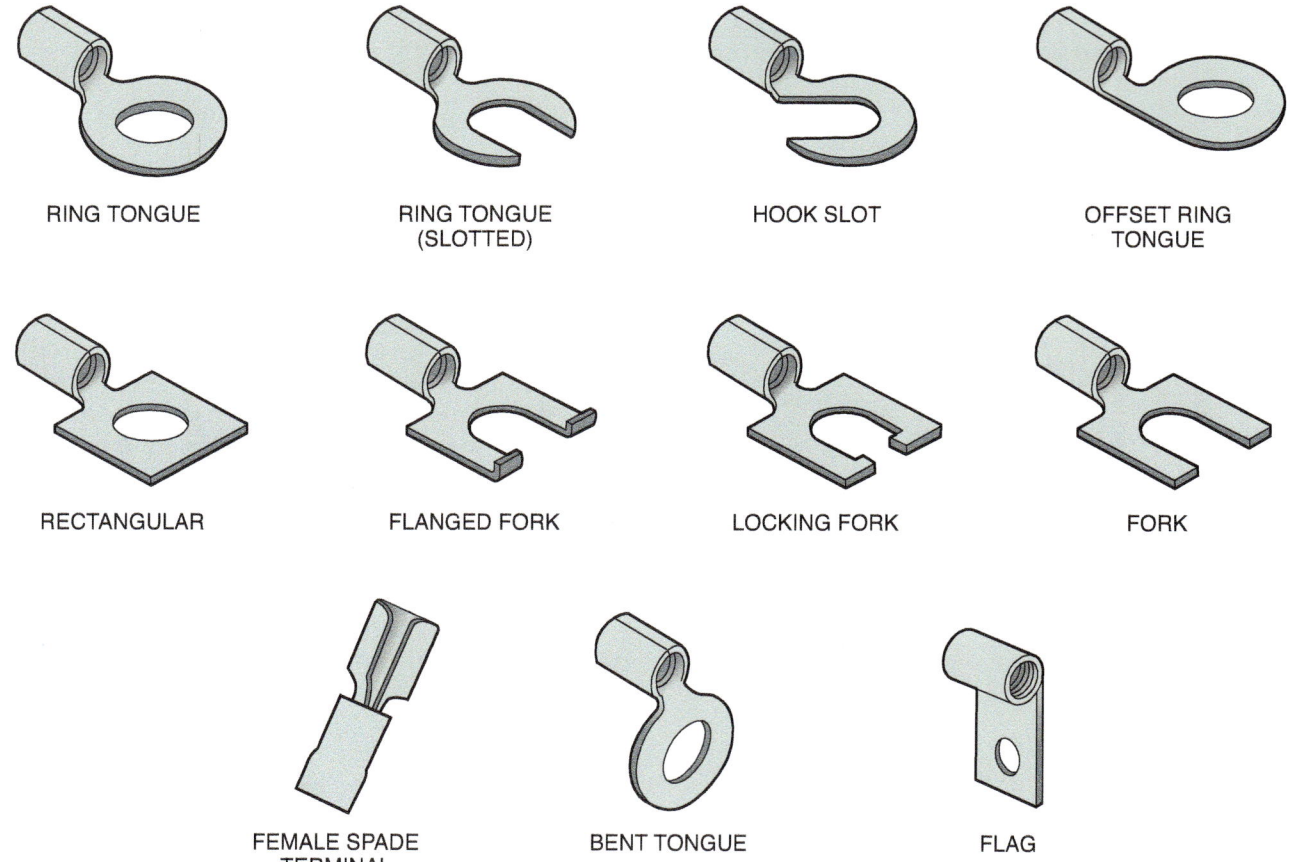

Figure 50 Crimp-on terminations.

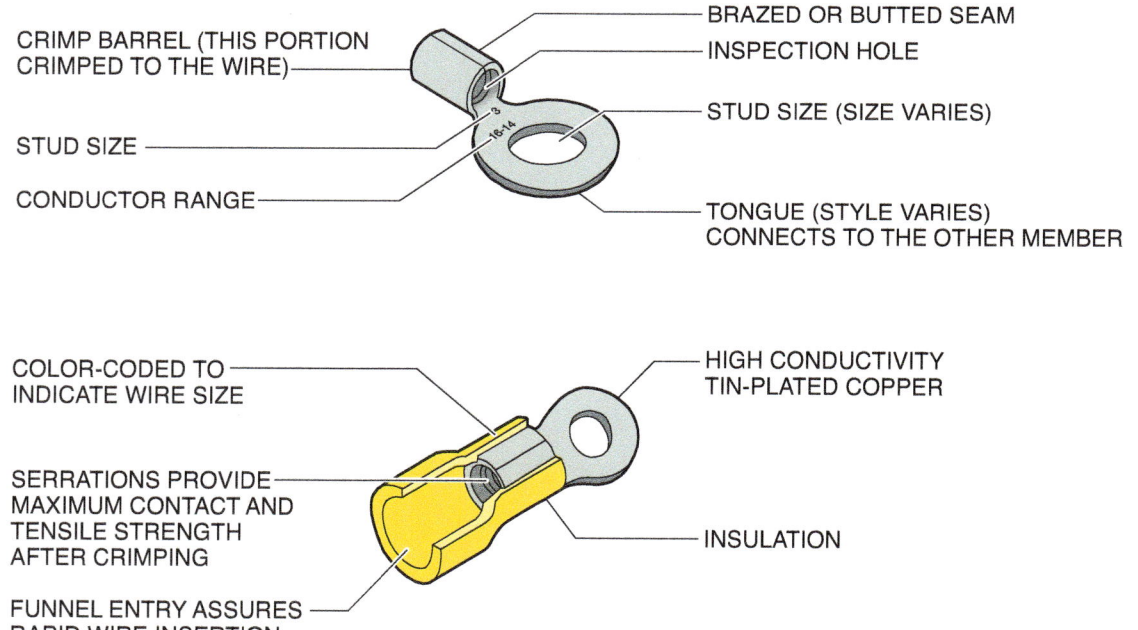

Figure 51 Basic crimp connector structure.

Aluminum has certain properties that are different from copper and special precautions must be taken to ensure reliable terminations and connections. Standard copper connectors cannot be safely used on aluminum wire. Aluminum connectors are designed to bite through the aluminum oxide film that coats aluminum conductors as the connector is applied to a conductor. The conductor should also be wire brushed and coated with an oxide-inhibiting compound to ensure a reliable joint. Never use connectors designed strictly for use on copper conductors on aluminum conductors. Connectors listed for use on both metals will be marked AL-CU.

3.2.0 Installing Terminations

Conductors must be stripped to the proper length and the correct termination selected before the termination is applied to the conductor.

3.2.1 Preparing Conductors for Termination

Before any connection can be made, the ends of the conductors must be properly cleaned and stripped. This ensures a low-resistance connection and avoids contaminating the termination. Stripping is the removal of insulation from the end of the conductor. Conductors should only be stripped using the appropriate stripping tool. This will help to prevent cuts and nicks in the wire, which can reduce the conductor area and weaken the conductor.

Poorly stripped conductors can result in damage that can lead to a stress concentration in the damaged area. Heat, rapid temperature changes, mechanical vibration, and motion can aggravate the damage, causing faults in the circuitry or even total failure. Lost strands are a problem in crimp-type terminals, while exposed strands present a safety hazard.

A variety of factors determine how precisely a conductor can be stripped, including the wire size and type of insulation. Wires should only be stripped using the appropriate stripping tool. The specific tool used depends on the size and type of wire being stripped. *Figure 52* shows a common type of wire stripper for small conductors. It can be used to cut, strip, and crimp wires from No. 22 through No. 10 AWG.

To use this tool, insert the conductor into the proper size knife groove, then squeeze the tool handles. The tool cuts the insulation, allowing the conductor to be easily removed without crushing its stripped end. If the knife groove used is for a smaller wire size than the wire being stripped,

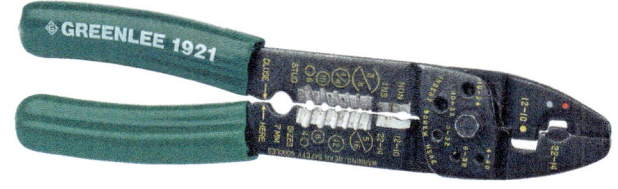

Figure 52 Wire stripper/crimper.

the conductors beneath the insulation will be damaged. With conductors with fine strands, several of the strands will likely be cut along with the insulation. The length of the strip is regulated by the amount of wire extending beyond the blades when it is inserted in the knife groove.

Figure 53 shows two types of production-grade stripping tools. The upper tool (A) can be used to strip conductors from No. 20 to No. 10 AWG. The lower tool (B) strips wires from No. 18 to No. 6 AWG. Note that the lower tool has front entry jaws for use in tight spaces.

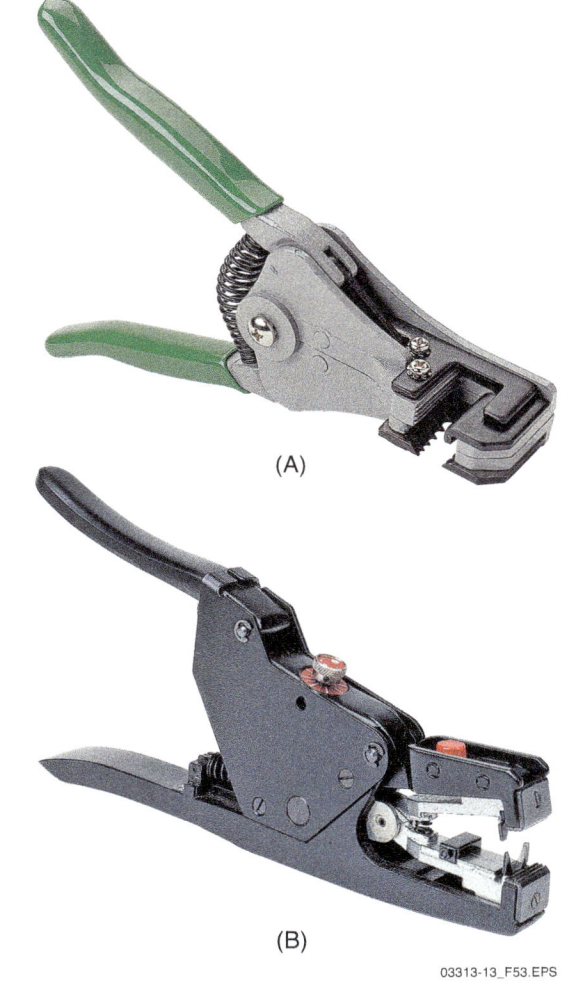

Figure 53 Production-grade wire strippers.

> **CAUTION**
> Never use a knife blade to strip conductors. The knife blade can nick or damage the conductor.

Figure 54 shows the positioning of the wire in the crimp barrel when stripped to the proper length. The conductor insulation must be in the belled mouth of the terminal. This relieves stress on the strands or wire and increases the strength of the connection. Allowing conductor strands to protrude out of the inspection hole more than $1/32$" will interfere with the terminal screw. Cutting the strands too short will reduce the contact surface area.

Figure 55 Manual crimping tool.

3.2.2 Installing a Crimp Connector

Most connectors used in low- and line-voltage circuits in HVAC equipment can be applied with a simple plier-type crimping tool (*Figure 55*). These hand tools are similar in construction to ordinary mechanics' pliers except the jaws are shaped for crimping.

To install a crimp connector, first insert the stripped end of the wire completely into the connector. Position the crimping tool in place over the connector, and then squeeze the tool to fully crimp the connector. Make sure that the crimping tool jaws are fully closed, indicating that a full compression crimp has been made.

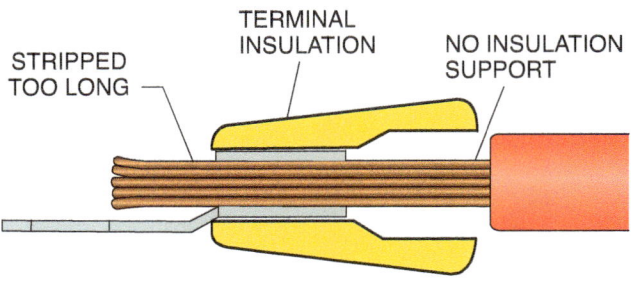

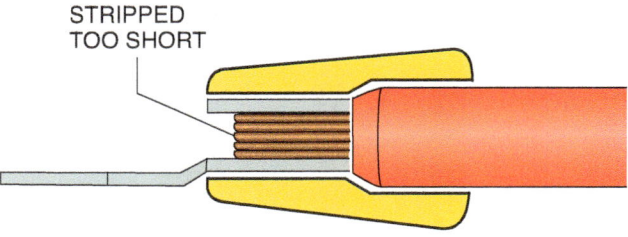

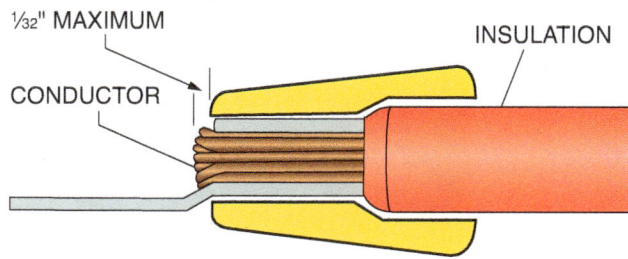

Figure 54 Proper stripping length.

Additional Resources

Air Conditioning Systems, Principles, Equipment, and Service, Latest Edition. Upper Saddle River, NJ: Prentice Hall.

3.0.0 Section Review

1. Crimp-type terminations for wire sizes smaller than No. 8 AWG are normally made to accept at least _____.

    a. a single wire size
    b. stranded wire only
    c. two wire sizes
    d. copper wires only

2. When is it acceptable to use a very sharp knife blade to strip insulation from a conductor?

    a. If the stripper blade is dull
    b. If the conductor is solid
    c. It is never acceptable
    d. If the conductor is copper

SUMMARY

HVAC technicians are constantly assembling mechanical equipment and taking it apart. Knowing what fastener to use and how to correctly install the fastener helps ensure that the equipment will operate without breakdowns.

Vibrations caused by equipment operation must be prevented from becoming objectionable or from causing damage. Vibration isolators and vibration dampers are used to minimize this problem. In areas where seismic activity occurs, mechanical equipment must be prevented from uncontrolled movement that could damage the equipment and/or cause personal injury when a seismic event occurs.

Being able to correctly apply terminations on conductors ensures that electrical equipment will experience trouble-free operation.

Review Questions

1. In a situation where stress is extremely high, which grade of bolt should be used?
   a. Grade 1
   b. Grade 2
   c. Grade 5
   d. Grade 8

2. Bolts and screws used on parts that must have a finished appearance are called _____.
   a. machine bolts
   b. machine screws
   c. cap screws
   d. stud bolts

3. A jam nut is used _____.
   a. when frequent removal is required
   b. to lock a standard nut
   c. when appearance is important
   d. to lock a split washer

4. Which type of lock washer is used with bolts and cap screws?
   a. Split ring
   b. External
   c. Countersunk
   d. Internal

5. When threading the bottom of a blind hole, use a _____.
   a. pipe tap
   b. plug tap
   c. bottoming tap
   d. taper tap

6. When installing a ½" Grade 8 steel bolt, it should be tightened to _____.
   a. 37 foot-pounds
   b. 60 foot-pounds
   c. 85 foot-pounds
   d. 92 foot-pounds

7. The force required to loosen a fastener is called _____.
   a. run-down resistance
   b. break-free resistance
   c. break-away torque
   d. loosen-up torque

8. How are concrete inserts installed?
   a. Set in place with epoxy
   b. Set in place with grout
   c. Screwed directly into concrete
   d. Imbedded in wet concrete

9. Blind rivets are used when a joint can only be accessed from one side.
   a. True
   b. False

10. Welded hermetic compressors used in residential systems minimize compressor vibration by being mounted on _____.
    a. rubber grommets
    b. open springs
    c. rubber pads
    d. vibration isolators

11. Bumpers are used as seismic restraints on equipment to prevent _____.
    a. tip-over
    b. up and down movement
    c. side-to-side movement
    d. swaying

12. Local building codes may specify the SAE grade of bolts that must be used in a seismic restraint.
    a. True
    b. False

13. The wire size range for a crimp-on terminal can be found _____.
    a. on the terminal packaging
    b. stamped on the tongue
    c. printed on the insulation
    d. printed on the conductor

14. Lugs are used with _____.
    a. stranded wires
    b. No. 8 and larger wires
    c. solid wires
    d. No. 10 and smaller wires

15. When crimping a conductor on a terminal, conductor strands should protrude out the inspection hole no more than _____.
    a. ¼"
    b. ⅛"
    c. 3/32"
    d. 1/32"

Trade Terms Introduced in This Module

AL-CU: An abbreviation for aluminum and copper that is commonly marked on electrical connectors to indicate the device is suitable for use with conductors made with either metal.

Break-away torque: The torque required to loosen a fastener. This is usually lower than the torque to which the fastener has been tightened.

Connector: A device used to physically and electrically connect two or more conductors.

Lug: A wiring terminal used on wire sizes No. 8 and larger. Also a term used to describe a screw-type terminal that simply applies pressure to a stripped wire, without the need for a crimp-on terminal.

Nominal size: A means of expressing the size of a bolt or screw. It is the approximate diameter of a bolt or screw.

Run-down resistance: The torque required to overcome the resistance of associated hardware such as locknuts and lock washers, when tightening a fastener.

Seismic activity: Movement of the earth such as during an earthquake.

Seismic restraint: A device or method used to prevent or restrict the movement of equipment during a seismic event.

Set or seizure: In the last stages of rotation in reaching a final torque, the fastener may lock up; this is known as setting or seizing. This is usually accompanied by a noticeable popping effect.

Tensile strength: The maximum stress a material can endure before being pulled apart.

Terminal: A device used for connecting wires to an object.

Termination: The connection of a wire to its destination.

Thread classes: Threads are distinguished by three classifications according to the amount of tolerance the threads provide between the bolt and nut, and are designated by a number and letter combination that indicates the degree of fit for a thread. Higher numbers indicate a tighter fit.

Thread series: The combinations of diameter and pitch in threads that are applied to specific thread diameters to identify the coarseness or fitness of a thread.

Thread standards: An established set of standards for machining threads.

Torque: Applying a specific amount of twisting force to a threaded fastener.

Vibration isolator: A device that prevents the transmission of vibrations from operating machinery to surrounding objects.

Additional Resources

This module presents thorough resources for task training. The following resource material is suggested for further study.

Air Conditioning Systems, Principles, Equipment, and Service, Latest Edition. Upper Saddle River, NJ: Prentice Hall.

FEMA 412, Installing Seismic Restraints for Mechanical Equipment, Latest Edition. Washington, DC, Federal Emergency Management Agency.

Figure Credits

Courtesy of Irwin Tools, Figures 12A, 13, 29B (photo)
webBikeWorld.com, Figure 16A
Courtesy of The Eastwood Company, Figure 16C
Anvil International, LLC, Exeter, NH, Figure 22A
Courtesy of DEWALT Industrial Tool Co., SA01
Courtesy of Stanley Tools, Figure 33

Courtesy of DiversiTech, Figure 37
Courtesy of CalDyn, Figures 39, 49A
Courtesy of QE Quality®, Figure 42
Courtesy of Kinetics Noise Control, Figure 48
Courtesy of Mason Industries, Figure 49B
Greenlee/A Textron Company, Figures 52, 53, 55

Section Review Answer Key

Answer	Section Reference	Objective
Section One		
1. a	1.1.1	1a
2. c	1.2.2	1b
3. c	1.3.1	1c
4. d	1.4.2	1d
5. b	1.5.2	1e
Section Two		
1. d	2.1.3	2a
2. a	2.2.0	2b
3. a	2.3.0	2c
Section Three		
1. c	3.1.0	3a
2. c	3.2.1	3b

NCCER CURRICULA — USER UPDATE

NCCER makes every effort to keep its textbooks up-to-date and free of technical errors. We appreciate your help in this process. If you find an error, a typographical mistake, or an inaccuracy in NCCER's curricula, please fill out this form (or a photocopy), or complete the online form at **www.nccer.org/olf**. Be sure to include the exact module ID number, page number, a detailed description, and your recommended correction. Your input will be brought to the attention of the Authoring Team. Thank you for your assistance.

Instructors – If you have an idea for improving this textbook, or have found that additional materials were necessary to teach this module effectively, please let us know so that we may present your suggestions to the Authoring Team.

NCCER Product Development and Revision
13614 Progress Blvd., Alachua, FL 32615

Email: curriculum@nccer.org
Online: www.nccer.org/olf

❏ Trainee Guide ❏ Lesson Plans ❏ Exam ❏ PowerPoints Other _____

Craft / Level: _____ Copyright Date: _____

Module ID Number / Title: _____

Section Number(s): _____

Description: _____

_____

_____

_____

_____

Recommended Correction: _____

_____

_____

_____

_____

Your Name: _____

Address: _____

_____

Email: _____ Phone: _____

Control Circuit and Motor Troubleshooting

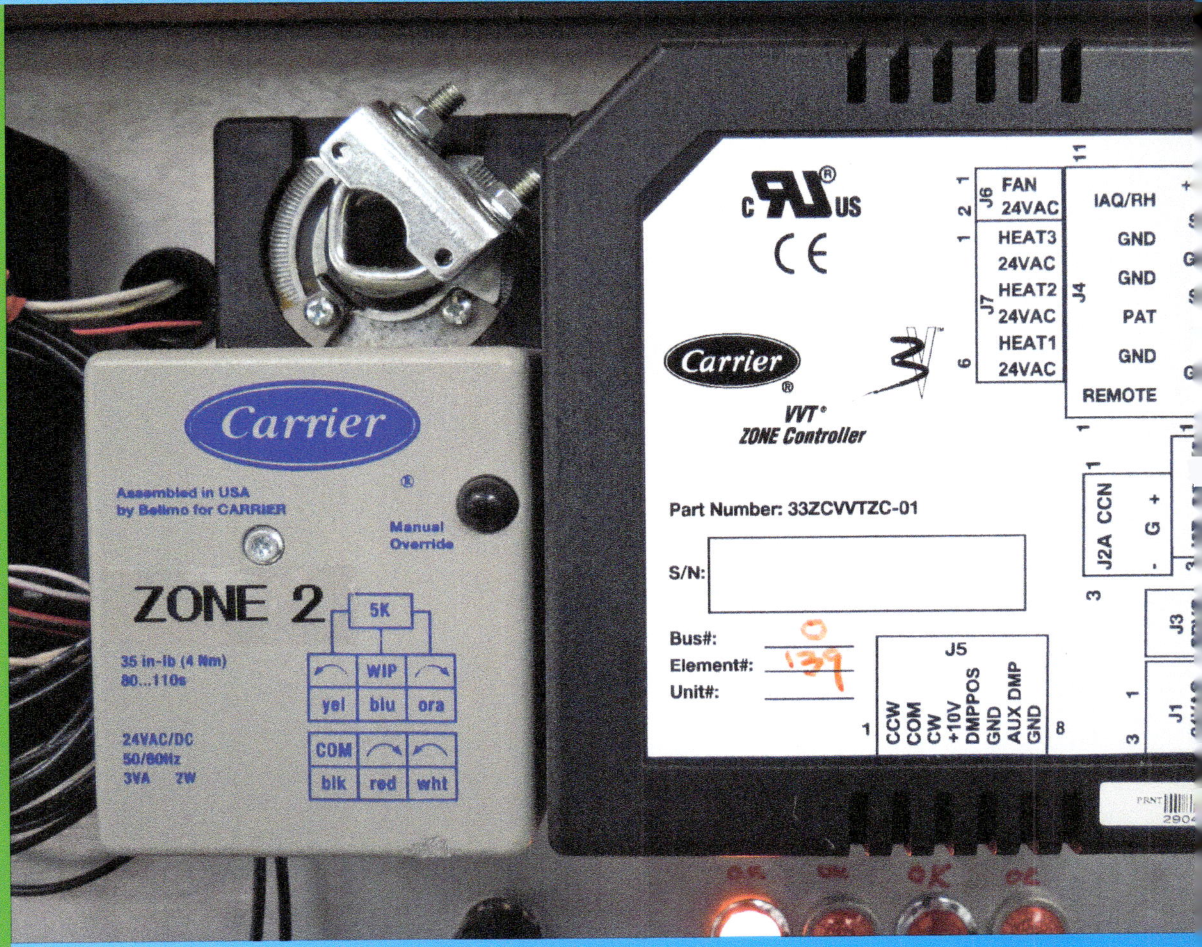

Overview

Most malfunctions that occur in HVACR equipment are caused by failure in the power distribution or control circuits. Technicians must have an understanding of how the various components function and be able to interpret the schematics, wiring diagrams, and other service literature. In addition, new technologies such as variable frequency motor drives and electronically commutated motors are adding more complexity. The successful troubleshooter can read schematics and wiring diagrams to understand the underlying operating sequences and apply that understanding to troubleshooting.

Module 03314

Trainees with successful module completions may be eligible for credentialing through the NCCER Registry. To learn more, go to **www.nccer.org** or contact us at 1.888.622.3720. Our website has information on the latest product releases and training, as well as online versions of our *Cornerstone* magazine and Pearson's product catalog.

Your feedback is welcome. You may email your comments to **curriculum@nccer.org**, send general comments and inquiries to **info@nccer.org**, or fill in the User Update form at the back of this module.

This information is general in nature and intended for training purposes only. Actual performance of activities described in this manual requires compliance with all applicable operating, service, maintenance, and safety procedures under the direction of qualified personnel. References in this manual to patented or proprietary devices do not constitute a recommendation of their use.

Copyright © 2018 by NCCER, Alachua, FL 32615, and published by Pearson, New York, NY 10013. All rights reserved. Printed in the United States of America. This publication is protected by Copyright, and permission should be obtained from NCCER prior to any prohibited reproduction, storage in a retrieval system, or transmission in any form or by any means, electronic, mechanical, photocopying, recording, or likewise. To obtain permission(s) to use material from this work, please submit a written request to NCCER Product Development, 13614 Progress Blvd., Alachua, FL 32615.

03314 V5

From *HVAC Level Three, Trainee Guide*. NCCER.
Copyright © 2018 by NCCER. Published by Pearson. All rights reserved.

03314
CONTROL CIRCUIT AND MOTOR TROUBLESHOOTING

Objectives

When you have completed this module, you will be able to do the following:

1. Identify and describe the operation of common HVACR control circuit devices.
   a. Identify and describe the operation of relays, contactors, and motor starters.
   b. Identify and describe the operation of common safety and control circuit devices.
2. Describe the operation, installation, and testing of various thermostats and temperature controls.
   a. Describe the operation of various thermostats and temperature controls.
   b. Identify and describe how to troubleshoot thermistors.
   c. Explain how to install and wire thermostats.
   d. Explain how to troubleshoot the functions of a thermostat.
3. Explain how to troubleshoot common control circuits and load components.
   a. Identify basic safety practices related to troubleshooting HVACR power and control circuits.
   b. Describe the operating sequence of simple heating and cooling systems.
   c. Explain how to approach HVACR problems and prepare for troubleshooting.
   d. Explain how to test line-voltage power sources.
   e. Explain how to troubleshoot control circuits and low-voltage power sources.
   f. Explain how to troubleshoot both resistive and inductive loads.
   g. Explain how to troubleshoot various hydronic control-system components.
4. Describe the operation of variable frequency drives (VFD) and their selection considerations.
   a. Describe the operation of a VFD.
   b. Identify VFD parameters that can be programmed.
   c. Describe the important considerations for the selection of a VFD.
   d. Explain dynamic motor braking processes.
5. Identify and describe how to service electronically commutated motors (ECMs).
   a. Identify and describe the operation of ECMs.
   b. Describe how to install and set-up an ECM.
   c. Describe how to troubleshoot an ECM.

Performance Tasks

Under the supervision of your instructor, you should be able to do the following:
1. Wire, check the operation of, and adjust the cycle rate of a thermostat.
2. Interpret control circuit diagrams.
3. Perform electrical tests and/or troubleshooting procedures on the following:
   - Single- and three-phase power sources
   - Fuses and circuit breakers
   - Resistive loads
   - Relays and/or contactors
   - Motor windings
   - Start and run capacitors
   - Start relays and thermistors

Trade Terms

Base speed
Cooling compensator
Deadband
Differential
Droop
Fault isolation diagrams

Insulated gate bipolar transistor (IGBT)
Invar®
Label diagrams
Ladder diagram
Wiring diagram

Industry-Recognized Credentials

If you are training through an NCCER-accredited sponsor, you may be eligible for credentials from NCCER's Registry. The ID number for this module is 03314. Note that this module may have been used in other NCCER curricula and may apply to other level completions. Contact NCCER's Registry at 888.622.3720 or go to **www.nccer.org** for more information.

Contents

- **1.0.0** HVACR Control Device Operation 1
  - **1.1.0** Relays, Contactors, and Motor Starters 1
    - 1.1.1 Basic Relays 2
    - 1.1.2 Time-Delay Relays 4
    - 1.1.3 Compressor Short-Cycle Timers 4
    - 1.1.4 Lockout Control Relays and Circuits 5
    - 1.1.5 Solid-State Relays 5
    - 1.1.6 Contactors and Motor Starters 6
    - 1.1.7 Switch and Contactor/Relay Contact Checks 7
  - **1.2.0** Control Circuit Switches and Safety Devices 8
    - 1.2.1 Pressure Switches 8
    - 1.2.2 Freezestats 8
    - 1.2.3 Outdoor Thermostat 9
    - 1.2.4 Furnace Fan Control 10
    - 1.2.5 Furnace Limit Controls 11
    - 1.2.6 Combination Fan/Limit Switch 12
- **2.0.0** Thermostats and Temperature Controls 14
  - **2.1.0** Thermostat and Temperature Control Operation 14
    - 2.1.1 Heating-Only Thermostats 15
    - 2.1.2 Cooling-Only Thermostats 16
    - 2.1.3 Heating-Cooling Thermostats 16
    - 2.1.4 Automatic-Changeover Thermostats 16
    - 2.1.5 Two-Stage Thermostats 17
    - 2.1.6 Programmable Thermostats 17
    - 2.1.7 Line-Voltage Thermostats 18
  - **2.2.0** Thermistors 18
  - **2.3.0** Thermostat Installation 21
    - 2.3.1 Installation Guidelines 21
    - 2.3.2 Thermostat Wiring 22
    - 2.3.3 Adjusting Cycle Rates 24
    - 2.3.4 Final Check 24
    - 2.3.5 Mercury-Thermostat Disposal 25
  - **2.4.0** Thermostat Troubleshooting 25
    - 2.4.1 Fan-Switch Operation Checks 26
    - 2.4.2 Cooling Operation Checks 26
    - 2.4.3 Heating Operation Checks 27
    - 2.4.4 Electronic Thermostats 29
- **3.0.0** Troubleshooting Circuits and Load Components 32
  - **3.1.0** Safety Practices 32
    - 3.1.1 Electrical Safety Precautions 33
    - 3.1.2 Precautions When Working with Solid-State Controls 33
    - 3.1.3 OSHA Lockout/Tagout Rule 34
    - 3.1.4 Lockout/Tagout Procedure 34
    - 3.1.5 Restoring Normal Operation 35

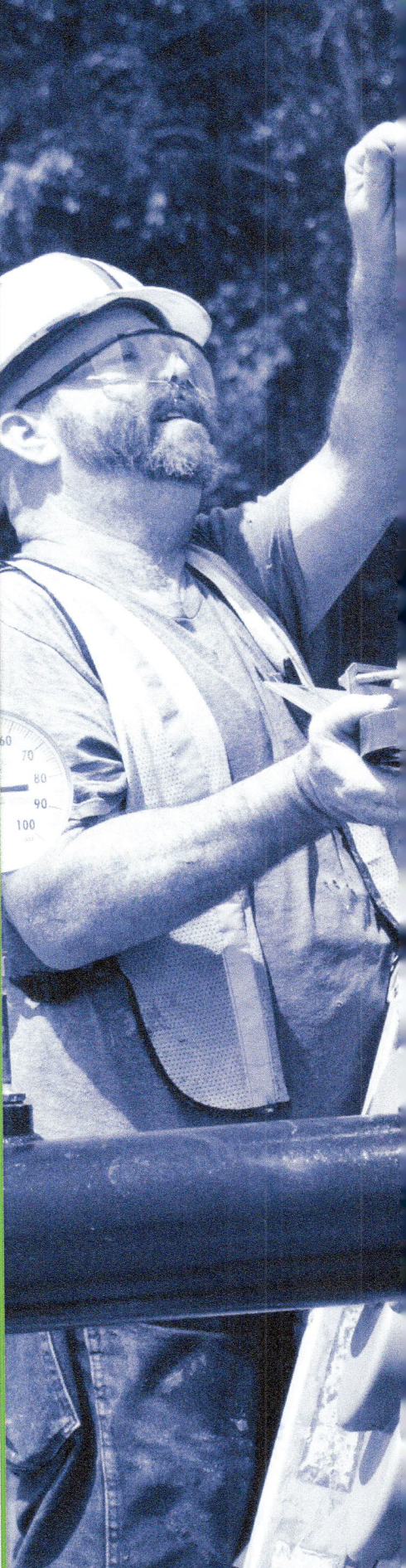

Contents (continued)

- 3.2.0 Basic Operating Sequences 35
  - 3.2.1 Heating and Cooling System Sequence 36
- 3.3.0 An Organized Approach to Electrical Troubleshooting 39
  - 3.3.1 Customer Interviews 39
  - 3.3.2 Physical Examination of the System 40
  - 3.3.3 Basic System Analysis 40
  - 3.3.4 Use of Manufacturers' Troubleshooting Aids 41
  - 3.3.5 Fault Isolation in the Equipment Problem Area 43
- 3.4.0 Testing High-Voltage Power Sources 43
  - 3.4.1 Effects of High and Low Voltage 43
  - 3.4.2 Voltage and Current Imbalance 45
  - 3.4.3 Fuse/Circuit Breaker Checks 45
- 3.5.0 Control Circuits and Low-Voltage Power Sources 48
  - 3.5.1 Isolating to a Faulty Circuit via the Process of Elimination 48
  - 3.5.2 Isolating to a Faulty Circuit Component 48
  - 3.5.3 Low-Voltage Power Source Testing 50
- 3.6.0 Resistive and Inductive Loads 51
  - 3.6.1 Resistive Load Testing 52
  - 3.6.2 Basic Motor Testing 52
  - 3.6.3 Open, Shorted, or Grounded Winding Checks 54
  - 3.6.4 Identifying Unmarked Terminals of a PSC/CSR Motor 55
  - 3.6.5 Start and Run Capacitor Checks 56
  - 3.6.6 Start Relay Checks 56
  - 3.6.7 Start Thermistor Tests 57
- 3.7.0 Hydronic Controls 57
  - 3.7.1 Aquastat 58
  - 3.7.2 Reset Controller 58
  - 3.7.3 Low-Water Cutoff 58
  - 3.7.4 Circulator Pump 59
  - 3.7.5 Zone Valves 59
  - 3.7.6 Zoned Hydronic System Operating Sequence 60
- 4.0.0 Variable-Frequency Drives 63
- 4.1.0 Basic VFD Operation 63
- 4.2.0 Programmable VFD Parameters 66
- 4.3.0 VFD Selection Considerations 66
  - 4.3.1 VFD-Operated Motors 67
  - 4.3.2 Selecting the Drive 67
  - 4.3.3 Environment 69
  - 4.3.4 Torque Requirements 69
  - 4.3.5 Duty Cycle 69
- 4.4.0 VFD Dynamic Braking 70
  - 4.4.1 Dynamic DC Braking of an AC Motor 70
  - 4.4.2 Dynamic Braking with an AC Drive 71
- 5.0.0 Electronically Commutated Motors 73
- 5.1.0 ECM Characteristics and Operation 73
  - 5.1.1 ECM Speed Control 74

5.2.0   ECM Installation and Setup 75
5.3.0   ECM Troubleshooting 76
    5.3.1   Motor Troubleshooting 76

Figures and Tables

Figure 1   Examples of plug-in relays 2
Figure 2   Normally open and normally closed relay contacts 2
Figure 3   Single-pole, single-throw (SPST) relay 3
Figure 4   Single-pole, double-throw relay 3
Figure 5   Double-pole, double-throw (DPDT) relay 4
Figure 6   Four-compressor control circuit with time-delay relays 5
Figure 7   Lockout relay used in an HVACR control circuit 5
Figure 8   Solid-state relay 6
Figure 9   Typical definite-purpose contactor 6
Figure 10 Motor starter with attached overload relay 7
Figure 11 Continuity and voltage tests 7
Figure 12 Pressure switches 9
Figure 13 Inducer pressure switch 10
Figure 14 Chilled-water freezestat 10
Figure 15 24-VAC furnace control circuit 11
Figure 16 Limit switches 12
Figure 17 Combination fan/limit switch wiring options 12
Figure 18 Bimetal sensing elements 14
Figure 19 Heating-only thermostat 15
Figure 20 Cooling-only thermostat 16
Figure 21 Cooling compensator 16
Figure 22 Standard heating-cooling thermostat 16
Figure 23 Heat pump thermostat 17
Figure 24 Wi-Fi-enabled, programmable thermostat 17
Figure 25 Remote-bulb thermostat 18
Figure 26 Thermistors 18
Figure 27 Example thermistor resistance chart 19
Figure 28 PTC thermistor start-assist device 20
Figure 29 TE expansion valve testing 20
Figure 30 Compressor-protection thermistor test 20
Figure 31 Thermostat mounting 22
Figure 32 Leveling a thermostat 22
Figure 33 Thermostat wiring 23
Figure 34 Mercury in a thermostat bulb 25
Figure 35 Examples of products containing mercury 25
Figure 36 Checking the function of a thermostat fan-switch
           (control circuit energized) 27
Figure 37 Thermostat cooling-function check 28

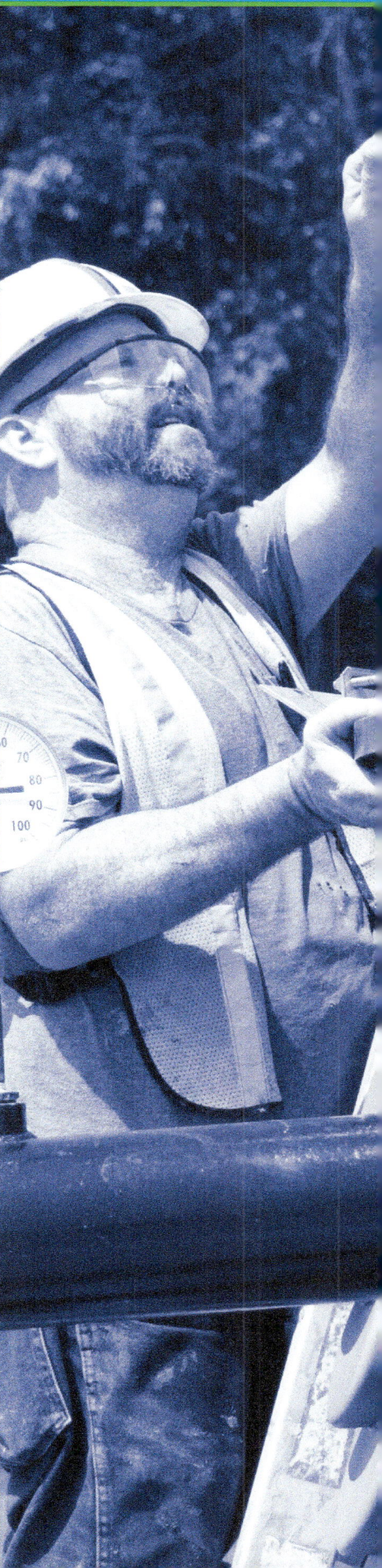

Figures and Tables (continued)

Figure 38	Thermostat heating-function check	28
Figure 39	Example of an electronic thermostat troubleshooting guide	30
Figure 40	ESD protection bag and tote box	34
Figure 41	Condensing unit with wall-mounted disconnect switch	35
Figure 42	Basic cooling-system control circuit	35
Figure 43	Typical cooling-system control circuit	36
Figure 44	Circuit diagram of a cooling/gas-heating system	37
Figure 45	Typical label diagram	41
Figure 46	Ladder diagram	41
Figure 47	Typical troubleshooting table	42
Figure 48	Typical fault-isolation diagram	43
Figure 49	Single-phase voltage checks	45
Figure 50	Three-phase voltage and current checks	46
Figure 51	Fuse tests	47
Figure 52	Circuit-breaker tests	47
Figure 53	Heat pump ladder diagram	49
Figure 54	Isolating to a faulty circuit component	50
Figure 55	Control-transformer checks	50
Figure 56	High-voltage and low-voltage circuit-separation on a ladder diagram	51
Figure 57	Resistive and inductive load resistance checks	52
Figure 58	PSC and CSR motors	53
Figure 59	PSC multi-speed motor	53
Figure 60	Three-lead, wye-connected, single-voltage, three-phase motor	54
Figure 61	Motor open- or shorted-winding test	54
Figure 62	Grounded winding test	55
Figure 63	Identifying unmarked terminals of a PSC/CSR motor	55
Figure 64	Capacitor tests	56
Figure 65	Start relay test	57
Figure 66	Start thermistor test	57
Figure 67	Simple hydronic system	58
Figure 68	Digital aquastat	58
Figure 69	Aquastat used for low-limit and circulator control in an oil-fired hydronic system	59
Figure 70	Typical reset controller outside-temperature sensor mounting	59
Figure 71	Low-water cutoff control	60
Figure 72	Circulator pump	60
Figure 73	Zone valve	60
Figure 74	Pulse-width modulated, variable-frequency drive system	64
Figure 75	VFD operating frequency range	65
Figure 76A	VFD application checklist (Part 1 of 2)	68
Figure 76B	VFD application checklist (Part 2 of 2)	69
Figure 77	Circuit for DC braking of an AC motor	71

Figures and Tables (continued)

Figure 78 Simplified schematic of an AC drive using dynamic braking 72
Figure 79 ECM major components 73
Figure 80 ECM motor rotation 74
Figure 81 ECM identification 74
Figure 82 ECM speed control 75
Figure 83 ECM speed-control interface board 76
Figure 84 Forming a drip loop 77
Figure 85 Input voltage check 78
Figure 86 ECM tester 79
Figure 87 ECM winding-resistance test 79

Table 1 Thermistor Values 21

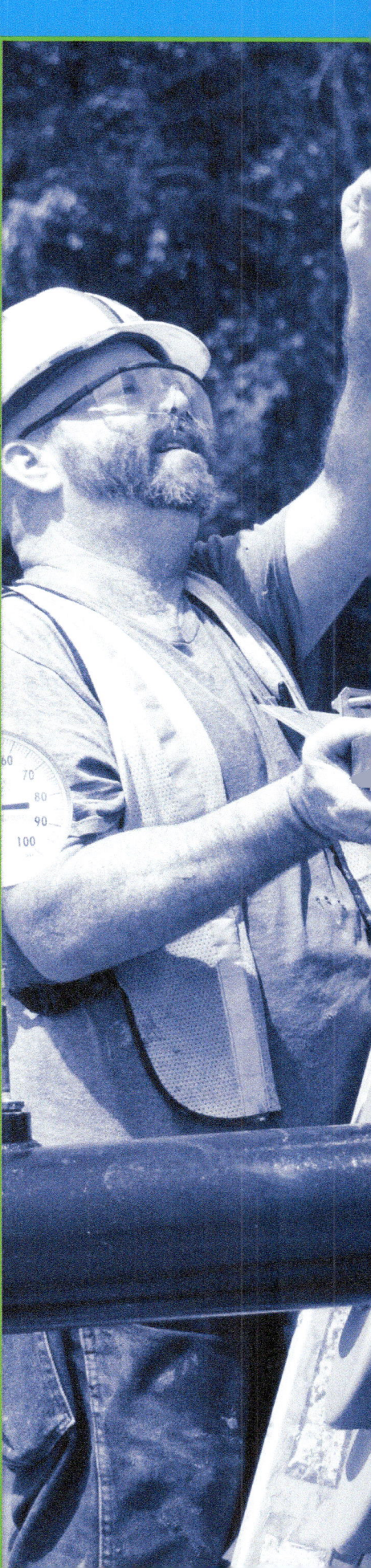

SECTION ONE

1.0.0 HVACR CONTROL DEVICE OPERATION

Objective

Identify and describe the operation of common HVACR control circuit devices.

a. Identify and describe the operation of relays, contactors, and motor starters.
b. Identify and describe the operation of common safety and control circuit devices.

Performance Tasks

3. Perform electrical tests and/or troubleshooting procedures on the following:
   - Relays and/or contactors

Trade Terms

Ladder diagram: An electrical diagram that provides a schematic diagram of circuits, with the load lines arranged like the rungs of a ladder between vertical lines representing the power source.

Wiring diagram: A troubleshooting aid, sometimes called a schematic, which provides a picture of what the unit does electrically and shows the actual external and internal wiring of the unit.

Most HVACR control systems are designed to automatically maintain the desired heating, cooling, and ventilation conditions. The controls for a small system such as a window air conditioner are very simple—a couple of control switches, a thermostat, and a couple of relays. As the system gets larger and provides more features, the controls become more complicated. For example, add gas heating to a packaged cooling unit, and the size and complexity of the control circuits more than double.

The use of electronic controls is now widespread across all HVACR product lines. It is rare to find a modern unit that does not contain a solid-state control. Manufacturers put these controls in their equipment to increase reliability and energy efficiency. Electronic controls also allow engineers to design more versatility into the products than was possible with electromechanical controls.

Large commercial systems may use pneumatic and electronic controls in conjunction with conventional electrical controls. These systems may have thirty or forty control devices, whereas a window air conditioner has just a handful.

The good news is that there are only a few kinds of control devices. Once different controls are recognized and the role each one plays is understood, it will not matter how many are used to control a system.

All automatic control systems have the following basic characteristics in common:

- One or more sensing elements (thermistor, thermostat, pressure switches, or humidistat) measure changes in temperature, pressure, and humidity.
- One or more control mechanisms translate these changes into energy that can be used by devices such as motors and valves.
- The connecting wiring, pneumatic piping, and mechanical linkages transmit the energy to the motor, valve, or other controlled devices that need to respond.
- The controlled devices then use a significant amount of energy to do work. For example, motors operate compressors, fans, or dampers. Valves control the flow of gas to burners or cooling coils and permit the flow of air in pneumatic systems. Valves also control the flow of liquids in chilled-water systems.
- The sensing elements in the control detect the change in conditions and again signal the control system.
- The controls then stop the motor, close the valves, or terminate the action of the component being used. As a result, the call for change is ended.

Although this may sound like a complex arrangement, it is actually quite simple. A need to change the environmental conditions is transmitted through the control system to devices that can make those changes happen.

1.1.0 Relays, Contactors, and Motor Starters

HVACR control system circuits contain relays, contactors, and motor starters. These devices are physically different in size and configuration, but their principles of operation are basically the same. Because of their extensive use in control and power distribution circuits, it is extremely important to understand how these devices operate. Without this understanding, it will be difficult to read schematics and troubleshoot circuits.

Refer to the *Appendix* at the back of this module for common wiring diagram symbols.

1.1.1 Basic Relays

A relay operates to stop or permit the flow of electricity to a load. Sometimes, a relay is used to reroute the flow of electricity in a different direction. Since thermostats and other control devices are unable to carry a significant electrical load, the load of some devices must be transferred to the somewhat-heavier electrical contacts of a relay. Relays are used to make and break other electrical circuits when they are themselves energized. Relays can be hard-wired into a circuit, or plug-in relays (*Figure 1*) can be used. Plug-in relays have a mating socket or base into which they fit, making troubleshooting and replacement easier. All wiring terminations are then made at the socket or base. Instead of having to connect and disconnect wiring, a new relay can simply be plugged in.

In today's equipment, relays that were once stand-alone components are now often soldered to a PC board. If a relay soldered on a board is suspected of being defective, the entire electronic control will have to be replaced.

In its basic form, the relay consists of two parts: an electromagnetic coil and a set of contacts. When the coil is energized with the proper voltage, it causes the position of the relay's contacts to change. Contacts are identified as being either normally closed or normally open. This refers to their position when the relay coil is de-energized. All relay contacts shown on schematic diagrams are shown with the relay in the de-energized position.

Figure 2 shows the open and closed contacts of a typical relay. It also shows the schematic symbol for normally closed and normally open relay contacts. Remember this is the position of the contacts when the relay coil is de-energized. A normally closed set of contacts will allow electric current to flow through the contacts when the relay is de-energized. A normally open set of contacts close only when the relay is energized.

Relays

The invention of the relay is credited to Samuel Morse, one of the inventors of the telegraph. Morse developed the relay to boost signal strength. An incoming signal activated an electromagnet, which closed a battery circuit, thereby transmitting the signal to the next relay.

Figure 3 shows the schematic symbol for a simple relay consisting of the relay coil and a set of contacts. As shown, this relay is a normally closed relay. It is also classified as a single-pole, single-throw (SPST) relay because it only has one set of contacts and one current path. When the relay coil is de-energized as shown, current present at terminal 3 can travel through the closed contacts to terminal 4 for subsequent application to the remainder of control circuit. Terminal 3 is the pole described by the words *single-pole*. Since there is only one possible current path, it is further described as single-throw. An electromechanical representation is also shown in *Figure 3*. Although it is never drawn this way on a diagram, it shows what is really happening inside the relay.

When voltage is applied across the coil of the relay via terminals 1 and 2, the coil energizes and the relay contacts open, preventing any current applied at terminal 3 from flowing through the contacts to terminal 4. Remember, when the coil is energized, the normally closed contacts open.

Figure 1 Examples of plug-in relays.

Figure 2 Normally open and normally closed relay contacts.

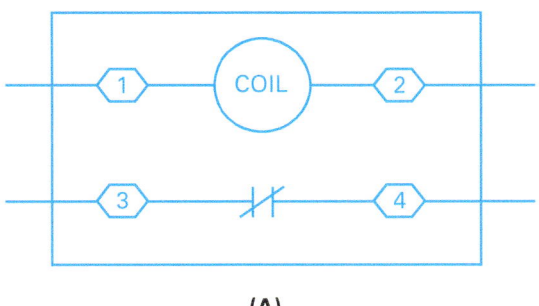

**(A)
SCHEMATIC SYMBOLS**

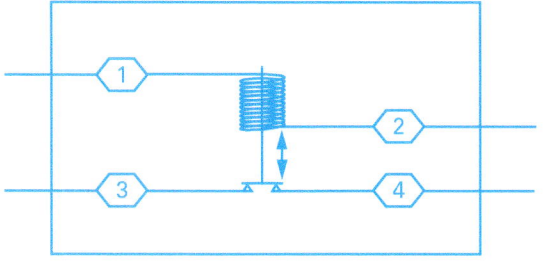

**(B)
ELECTROMECHANICAL REPRESENTATION**

Figure 3 Single-pole, single-throw (SPST) relay.

The last thing that needs to be determined about a relay is its coil voltage. Most relay coils used in HVACR control circuits operate on 24VAC, but that should never be assumed.

It is important to understand that no electrical connection exists between the coil of a relay and its contacts within the relay housing. Only in rare cases is there an internal connection. Opening and closing of the relay contacts happens because of electromechanical linkage between the coil and contacts. *Figure 3(B)* shows a mechanical representation of the relay just described. As shown, terminals 1 and 2 are still the coil connections, and terminals 3 and 4 are still the contact connections. When power is applied to the relay coil, an electromagnetic field is created around it. This electromagnetic field pulls the plunger up, causing the normally closed contacts to open. There are other ways to accomplish this in relays, but this diagram shows what is going on inside the relay mechanically.

Typically, the power applied to the coil might be 24VAC, while 240VAC may be applied through the contacts. This is one common situation but there are a number of other coil/contact voltage combinations. There are some applications in which the same voltage is applied to both the coil and the contacts of a relay. In this case, the connection between the two is always made using terminals outside of the relay. Again, it is important to remember that there is no electrical connection between the coil and contacts of a relay within the housing of most relays.

There are other types of relays. The counterpart of the first relay is the SPST normally open relay. This relay is similar to the SPST normally closed relay, except that when the coil is de-energized, the contacts are normally open. When the relay coil is energized, the contacts close.

The next relay is the single-pole, double-throw (SPDT) relay (*Figure 4*). In the SPST relay described earlier, the power applied at terminal 3 was applied to terminal 4, or not applied to

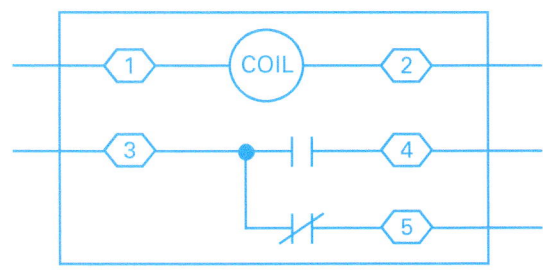

SINGLE-POLE, DOUBLE-THROW (SPDT) RELAY

Figure 4 Single-pole, double-throw relay.

terminal 4, depending on whether the contacts where closed or open, respectively.

With a double-throw relay, power can be directed to one of two different terminals. As shown, with the coil de-energized, power coming in on the common terminal 3 is applied through the normally closed set of contacts to terminal 5. If the coil is energized, the positions of the relay contact sets change, causing the normally open contacts to close and the normally closed contacts to open. This causes the power applied from terminal 3 to be redirected from terminal 5 to terminal 4 for subsequent application to a different branch of the unit's control circuits. Terminal 3 is still the single pole found in this relay.

The next relay is the double-pole, double-throw (DPDT) relay. As shown in *Figure 5*, another set of double-throw contacts is added to a basic SPDT relay to form this type of relay. When the coil is energized, it causes both sets of contacts to change position simultaneously. The two poles are terminals 3 and 6. It is important to remember that there is no electrical connection between the first set of contacts and the second set of contacts.

In *Figure 5*, the coil is designated as R1 and would be found on a wiring diagram labeled as such. The four sets of contacts would also be identified on a wiring diagram with the labels shown here.

With an understanding of the concept of adding contact sets, it is easy to see how poles can be added to a relay. Applications may be found that require the use of a 3PDT (three-pole, double-throw) relay or even a 4PDT (four-pole, double-throw) relay. Regardless of the relay coil/contact configuration, each pole is electrically isolated from the coil and from all other poles. When the coil is energized, the position of all the poles changes simultaneously.

Since each pole in a relay is electrically isolated from each other and from the coil, the poles from the same relay can be wired into different branches of the equipment's control circuits. Even though the coil and poles are physically housed in the same relay assembly, they are commonly shown in different areas of a schematic or ladder diagram used for troubleshooting. These diagrams are covered later in this module.

Because the relay coil and contacts are often shown in different areas of a schematic or ladder diagram, we need a method of designating which relay is associated with which contact set. An example is shown in *Figure 5*, where the coil is designated R1 (Relay 1) and all related contact sets are also identified with numbers that begin with R1. As shown, the first pole has contacts R1-1 and R1-2 (relay 1/contact set 1, and relay 1/contact set 2). The second pole has contacts R1-3 and R1-4. This method of relay designation is also used for three- and four-pole relays. For example, if a schematic or ladder diagram contains a normally open contact set designated R1-3, that means that the coil of relay R1 controls the contacts. If the R1 coil is energized, the R1-3 contacts will close. Likewise, all the other sets of R1 contacts are also going to change position.

There are many different designations for coils. For example, CR (control relay), CC (compressor contactor), or FR (fan relay) may be seen on diagrams. There are dozens of possible designations and they vary widely among manufacturers. Always consult the legend provided with the unit diagram, and be careful of assumptions about the meaning of an abbreviation.

1.1.2 Time-Delay Relays

The purpose of a time-delay relay is to delay some normal operation for a predetermined length of time. The length of the delay depends on the timing of the relay mechanism, and may vary from a fraction of a second to much longer. Many time-delay relays have an adjustable delay period. A common use for a time-delay relay is to delay the startup and shutdown of a heating blower to avoid blowing cold air on occupants, and to extract all the remaining heat from the heat exchanger at the end of the cycle.

Electrical systems containing several motors may also use time delays to start the motors one at a time to limit the inrush current. For example, the schematic drawing in *Figure 6* shows four compressor control circuits with time-delay relays for sequencing the starting of each motor. In this type of motor design, if all three stages of the thermostat are closed and electrical power is supplied to the units, compressor contactor coil 1 and time-delay relay 1 become energized. Compressor 1 will start immediately. After the time delay, the contacts of time-delay relay 1 close to energize compressor contactor coil 2, which starts compressor 2. Time-delay relay 2 also becomes energized, which eventually starts compressor 3. All four compressors are eventually started in a staged manner.

1.1.3 Compressor Short-Cycle Timers

Attempting to start a compressor against high head pressure can damage the motor. When a refrigerant system shuts down, it should not be restarted until the pressures in the system have had time to equalize. Short-cycling can be caused by a momentary power interruption or by an occupant changing the thermostat setting.

A compressor short-cycle protection circuit contains a timing function that prevents the compressor contactor from re-energizing for a specified period after the compressor shuts off. This type of relay is referred to as a delay-on-break model, since the delay period begins when the contactor control circuit opens. A delay-on-make model begins its timing period when the circuit is first energized. If a delay-on-make model is used, the timing period will not start until the system calls for the compressor to start. This results in an undesirable delay in most cases; the timing period should begin when the compressor shuts down.

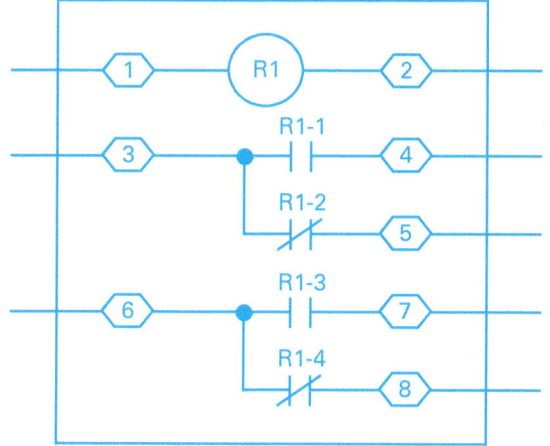

Figure 5 Double-pole, double-throw (DPDT) relay.

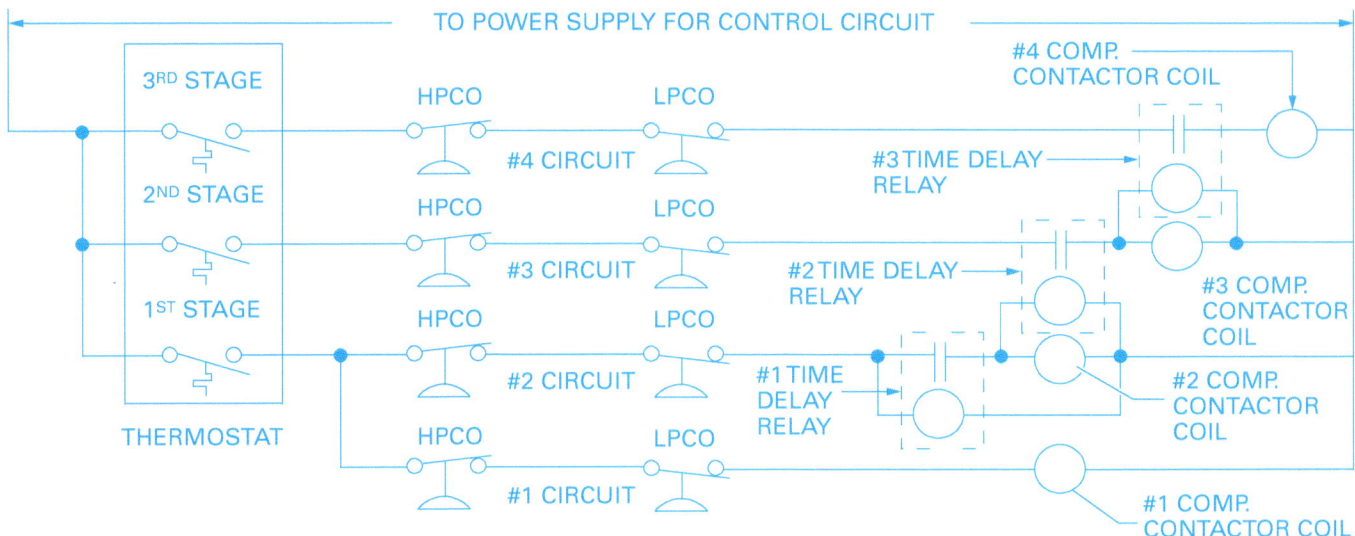

Figure 6 Four-compressor control circuit with time-delay relays.

Anti-short-cycle periods are typically 2 to 5 minutes. Short-cycle timers are available as self-contained modules that can be direct-wired into a unit. They are often sold as optional and after-market accessories. They are also a common built-in feature of electronic programmable room thermostats. The feature can usually be turned on and off through the programming or set-up menu.

1.1.4 Lockout Control Relays and Circuits

The purpose of the lockout relay in a control circuit is to prevent the automatic restart of the HVACR equipment. If the lockout relay has been activated, the system may be reset only by interrupting the power supply to the control circuit; for example, turning the room thermostat function switch OFF and then ON, or turning the main power switch off and then on again.

In the circuit in *Figure 7*, the lockout relay coil, due to its high resistance, is not energized during normal operation. This high resistance, or impedance, is by design. This type of relay is also called an impedance relay. However, when any one of the safety controls opens the circuit to the compressor contactor coil, current flows through the lockout relay coil, causing it to become energized and to open its contacts. These contacts remain open, keeping the compressor contactor circuit open until the power is interrupted after the safety control has reset. Performance depends on the resistance of the lockout relay coil being much greater than the resistance of the compressor contactor coil.

If the lockout relay becomes defective, it should be replaced with an exact duplicate to maintain the proper coil circuit resistance.

> **CAUTION**: Never put additional lockout relays, lights, or other load devices in parallel with the lockout relay coil. Doing so might defeat the lockout and cause equipment damage.

1.1.5 Solid-State Relays

Unlike the electromechanical relays just discussed, solid-state relays (*Figure 8*) have no moving parts. They use electronic devices such as triacs (a special type of current gateway) to control switching action and light-emitting diodes (LEDs) to report their status. Solid-state relays are used in general control applications, as well as in motor starting circuits.

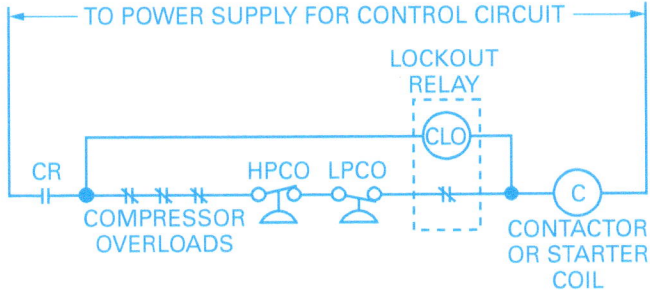

Figure 7 Lockout relay used in an HVACR control circuit.

Electronic Solid-State Relays

Some advantages of solid-state relays are longer life, higher reliability, high-speed switching, and high resistance to shock and vibration. The absence of mechanical contacts eliminates contact bounce, arcing when contacts open, and hazards from explosives and flammable gases.

1.1.6 Contactors and Motor Starters

Contactors (*Figure 9*) are a type of heavy-duty relay. They are used to start and stop high-current loads. They are used to start motors and compressors where overload protection is provided, is not required, or is provided separately. A contactor alone, like a relay, does not provide any level of overload protection.

A contactor is a normally open, single-throw device with one or more poles. It operates in the same way as a relay and is often represented on a schematic in a similar way. When the coil is energized, the movable contacts are closed against the stationary contacts, thus completing the circuit. When power is removed from the coil, the contacts open, stopping the flow of electric current to the related load.

Most HVACR contactors are definite-purpose contactors. Contactors that conform to higher NEMA standards were too expensive for most HVACR equipment manufacturers. The NEMA contactors were outlasting the rest of the equipment, but at too great a cost. Thus, the industry created a new class of contactors to satisfy the unique needs of HVACR applications. The definite-purpose contactor is not as durable and long-lasting as NEMA-rated contactors, yet it serves the industry very well.

Many definite-purpose contactors can be fitted with accessory contact sets. This allows the contactor to also perform like a relay. For example, when the contactor is energized and the main contacts close, an auxiliary set of contacts could open to remove power from a crankcase heater circuit.

A motor starter (*Figure 10*) is used to stop and start motors and provide overload protection. They are essentially contactors with an overload relay attached. Thermal overloads or current sensors in the overload relay sense excessive current flowing to a motor and protect it from overloading. If more current is flowing than the motor is designed to handle, the overload device breaks the contactor control circuit and de-energizes the motor. Once the overload relay has been tripped, a manual reset is required to restore normal operation.

Figure 8 Solid-state relay.

Figure 9 Typical definite-purpose contactor.

The starter shown in *Figure 10* requires the installation of a thermal element for each phase of power. These elements must be very carefully sized to accommodate the motor being protected. Manufacturers provide charts for this purpose. The selection is primarily based on two things: the full load current rating of the motor, and the temperature of the area where the starter is located. Since the overload device is thermal, the temperature of the installation area must be considered. If the area is considerably warmer than the area where the motor is, the starter may trip before the motor has reached its maximum load rating. Electronic overloads offer a range dial to set the desired overload current value.

Like contactors, motor starters often have auxiliary contact sets. Some may have one or more sets of auxiliary contacts built in to the main body. Combination starters combine motor overload protection with circuit protection and a means of disconnect. These typically include a standard starter and a fused or non-fused switch or circuit breaker in the same enclosure.

Auxiliary Contact Sets

Auxiliary contacts generally mount on one or both sides of a contactor. Rather than being electrically energized, like a relay, a tab on the movable part of the contactor physically engages the contact linkage and pushes the contacts closed. Many different configurations are available, and more than one set can often be added. The set shown here is double-pole, single-throw.

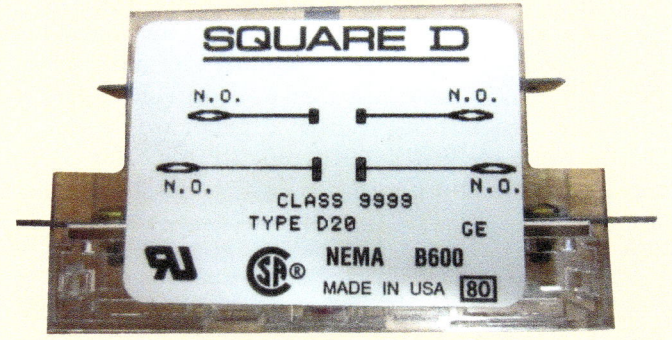

Figure Credit: Courtesy of Mara Industrial Supply

1.1.7 Switch and Contactor/Relay Contact Checks

Once a switch or contact has been identified as the possible cause of an electrical problem, the contacts should be tested to confirm their position. Switches and contacts can easily be tested by making a continuity measurement with power off (*Figure 11*) to determine whether the contacts are open or closed. If the switch contacts are open, the ohmmeter indicates an infinite resistance reading. If the switch contacts are closed and well-mated, the ohmmeter indicates continuity (near-zero or zero ohms).

Remember that the circuit must be de-energized to make a continuity check. As such, the check can only confirm the status of normally open and normally closed switch contacts. It does not reflect the status of the contacts when the system is powered up.

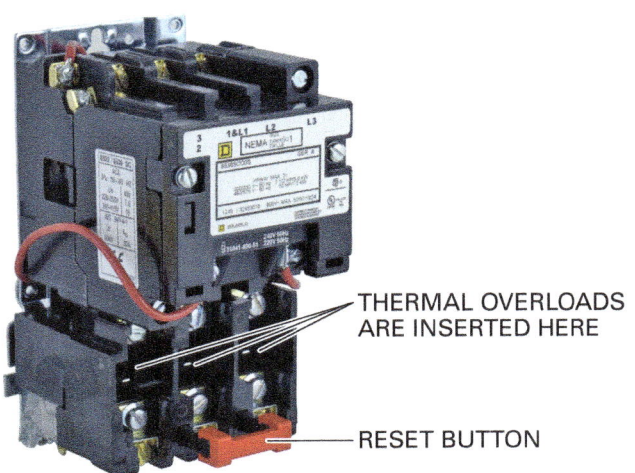

Figure 10 Motor starter with attached overload relay.

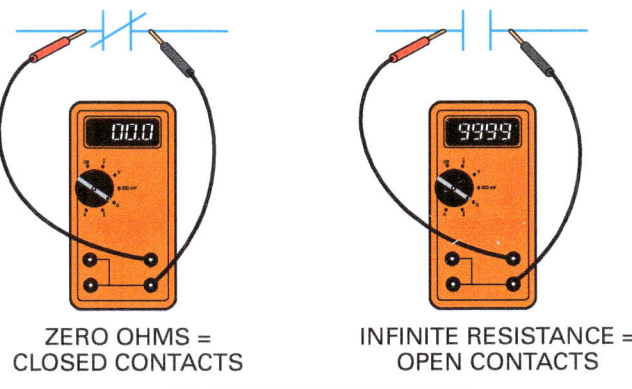

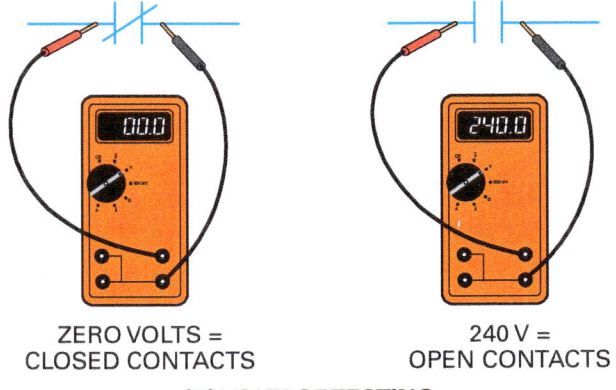

Figure 11 Continuity and voltage tests.

Relays, Contactors, and Motor Starters

Like any switching device, relays, contactors, and motor starters have a limited life. Normal failure modes include contact burning and sticking. Be careful when selecting replacement contactors because they have different ratings for different uses. Some are rated for inductive loads, others are rated for resistive loads, and still others are rated for both types of loads.

03314 Control Circuit and Motor Troubleshooting   Module Two   7

Switches and contacts can also be tested to determine contact position by placing voltmeter leads across energized contacts. This is also shown in *Figure 11*. If the switch contacts are open, the voltmeter indicates line voltage (240V in this example). If the switch contacts are closed, the voltmeter indicates zero volts. In this case, the circuit must be energized.

> **WARNING!**
> An energized circuit can deliver a severe electric shock. Always use caution when working with around energized circuits.

1.2.0 Control Circuit Switches and Safety Devices

Compressor control circuits normally include several different types of safety switches. These include pressure and temperature switches.

Many systems use one or more pressure switches in the control circuit. Some are safety devices designed to protect the compressor. Others are used to activate condenser fans, prove the operation of an induced draft fan, and similar functions. Smaller commercial refrigeration systems may even use a pressure switch to control the compressor instead of a thermostat.

In all HVACR systems, temperatures must be kept within a specified range. Electromechanical devices such as thermostats have been used for years to sense temperature. In modern equipment, thermistors have taken over the role of sensing temperature because of their versatility. Electromechanical thermostats only open or close a set of contacts at a preset temperature, limiting their versatility. Thermistors, on the other hand, provide actual and current temperature values to electronic devices, allowing the system to respond in a variety of ways. For example, an outdoor temperature thermistor can simultaneously provide outdoor temperature information to both a room thermostat and a defrost control board. The two devices use the temperature information in different ways. The room thermostat uses the information to display the outdoor temperature to the user, while the heat pump control board uses the information to decide if supplemental heat is required.

1.2.1 Pressure Switches

Most pressure switches use a bellows mechanism that presses against switch contacts to open or close them. While many pressure switches have a fixed pressure setting, others are adjustable (*Figure 12*). Differential pressure switches, like those used as oil-pressure safety switches, require two pressure inputs. They are designed to monitor the difference between two pressures and react accordingly. In an oil pressure safety switch, the difference between the pressure in the compressor crankcase and the oil pump outlet are monitored.

High-pressure safety switches are designed to open if the compressor discharge pressure is too high. Low-pressure safety switches are designed to open if the suction pressure is too low. Common causes of high head pressure are dirty condenser coils and a failed condenser fan motor. Common causes of low suction pressure include a loss of charge and low evaporator airflow. A pressure safety switch is normally closed, and is wired in series with the compressor contactor control circuit. While low-pressure safety switches often reset automatically, high-pressure safety switches typically require a manual reset.

Some manufacturers also use a type of low-pressure switch called a loss-of-charge switch that removes power to the equipment if the refrigerant charge is low. Loss-of-charge switches are often used in heat pumps. The setting of these switches is usually far below any normal operating pressures, and below the setting of most low-pressure safety switches.

Newer gas furnaces that use an induced draft fan often have a pressure switch in the heating control circuit to monitor and prove the operation of the inducer motor. Some furnaces do use a centrifugal switch in the back of the motor to prove operation. However, technically speaking, such a switch only proves the motor is turning; it does not prove that air is moving through the heat exchanger. A differential pressure switch can prove a sufficient amount of air is moving by comparing the pressure in the furnace enclosure to the pressure in the latter part of the heat exchanger.

A pressure switch is the more common type of furnace draft-inducer proving switch (*Figure 13*). It has tubing connected to the housing, and in some cases, to the burner enclosure. When the inducer motor is operating, a pressure is created that causes the pressure switch to operate, allowing the ignition sequence to continue. To check the operation of an inducer pressure switch, disconnect the pressure tube from the switch and set the furnace thermostat to call for heat. The furnace burner should not ignite.

1.2.2 Freezestats

A freezestat is an electromechanical safety switch designed to provide protection against freezing. For example, in a chiller, water temperature must not be allowed to fall to the point that it could

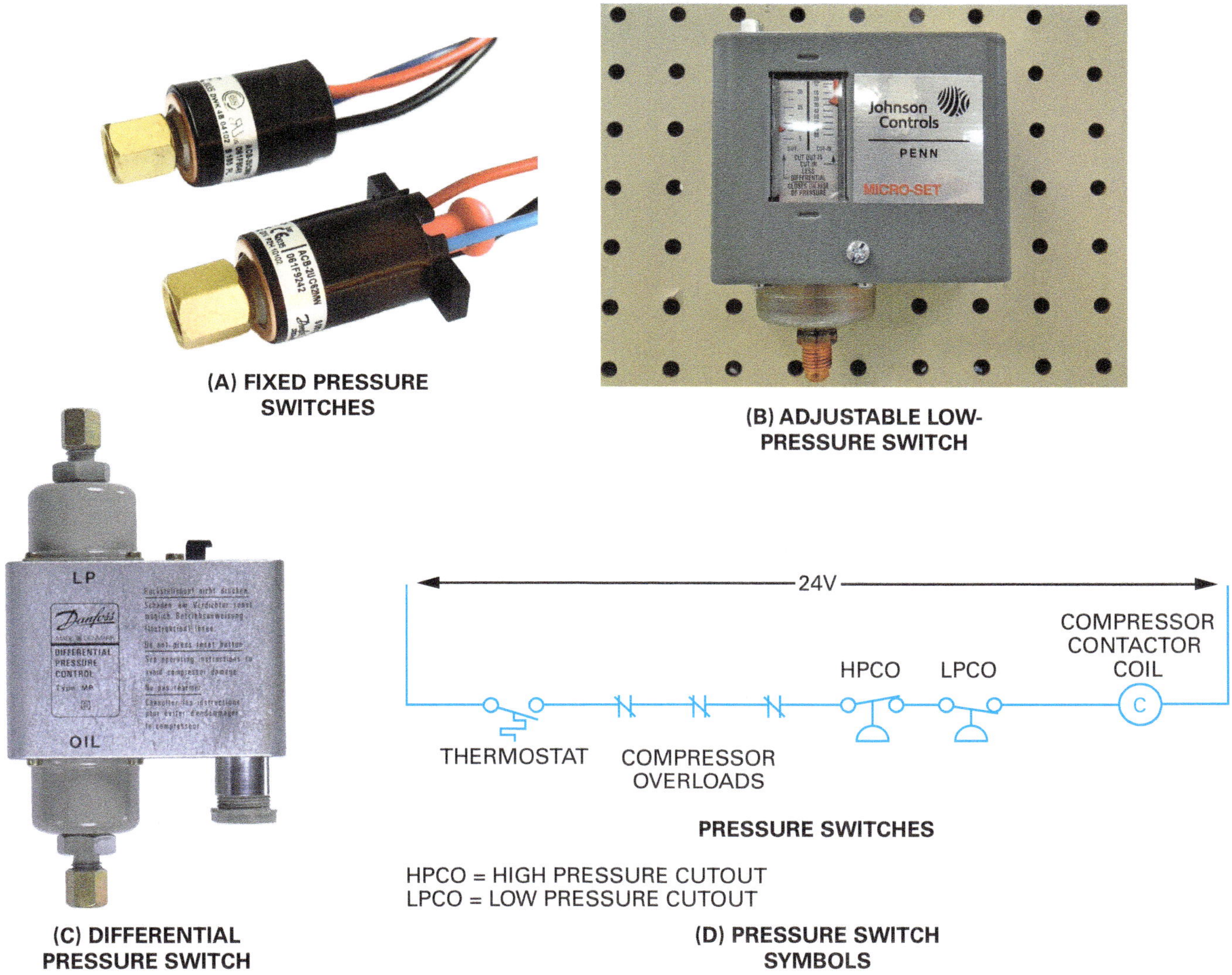

(A) FIXED PRESSURE SWITCHES

(B) ADJUSTABLE LOW-PRESSURE SWITCH

(C) DIFFERENTIAL PRESSURE SWITCH

(D) PRESSURE SWITCH SYMBOLS

HPCO = HIGH PRESSURE CUTOUT
LPCO = LOW PRESSURE CUTOUT

Figure 12 Pressure switches.

freeze and severely damage the chiller. The freezestat is usually a normally closed, bimetal switch. On a chiller, it is typically installed on the chilled water line leaving the chiller (*Figure 14*). They are typically set to open at a temperature no lower than 35°F (2°C).

Freezestats are also commonly found on water coils that are subjected to outside air. For example, a hot water coil may be found in an outside air duct. The coil is used to preheat the outside air before it mixes with return air and passes over the primary heating and cooling coils. A freezestat would be used to close a set of outside air dampers if the boiler fails and the coil is in danger of freezing.

Freezestats are simple devices with a sensing element inserted into a piping well. For coils, a freezestat typically has a very long capillary tube that is laced across the face of the coil surface. This type of capillary is called an averaging element, as it does not read the temperature at a single spot. The sensing element is active along its full length.

1.2.3 Outdoor Thermostat

Outdoor thermostats are widely used with HVACR systems. Like freezestats, they open/close at a predetermined temperature. Some common uses of outdoor thermostats include the following:

- Allowing the operation of heat pump supplemental electric heaters below a pre-determined outdoor temperature. Multiple outdoor thermostats can stage additional electric heaters on as the temperature continues to drop.
- Allow the operation of a compressor crankcase heater only when the outdoor temperature drops below a pre-determined temperature, typically about 65°F (18°C).

03314 Control Circuit and Motor Troubleshooting   Module Two 9

Figure 13 Inducer pressure switch.

- On a dual-fuel heat pump, it switches heating between heat pump and furnace operation as the outdoor temperature drops below, or climbs above, the heat pump's balance point.
- Provide both capacity and energy control for an air conditioner by staging additional compressor capacity on or off as outdoor temperature rises and falls.
- Provide capacity control for a furnace or boiler by modulating burner operation or staging additional burners on or off as outdoor temperature rises and falls.
- Prevent equipment operation when outdoor temperatures become excessively hot or cold. For example, air-to-air heat pumps should not be operated in the heating mode when outdoor temperature exceeds 65°F (18°C). Some air-to-air heat pumps use an outdoor thermostat to stop operation at below-zero temperatures when the heat pump is relatively ineffective.

1.2.4 Furnace Fan Control

A furnace fan control (*Figure 15*) is a temperature-actuated switch that, when heated, will close a set of contacts to start the furnace blower motor. The sensing element of the fan switch is positioned inside one of the heat exchangers where the temperature is the highest.

The fan switch is normally actuated by a bimetal element that opens or closes the contacts in response to a temperature change. The fan control may be set to bring the fan on at about 120°F and to stop the fan at about 100°F.

In operation, the burner provides heat to the heat exchangers for as long as a minute or two to warm the air before the fan is started. This avoids blowing cold air into the conditioned space. When the thermostat is satisfied, the main burner stops providing heat, but the fan continues to operate until the air temperature has been reduced. This removes excess heat from the heat exchangers and improves efficiency by using residual heat. In modern equipment, most manufacturers now use a timed on and off delay of the furnace blower motor instead of a temperature-actuated control.

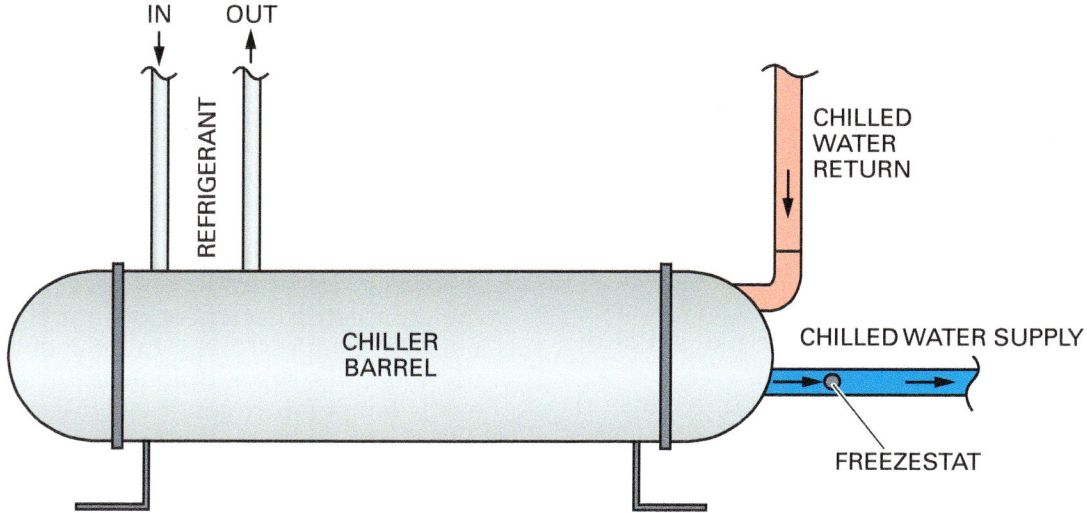

Figure 14 Chilled-water freezestat.

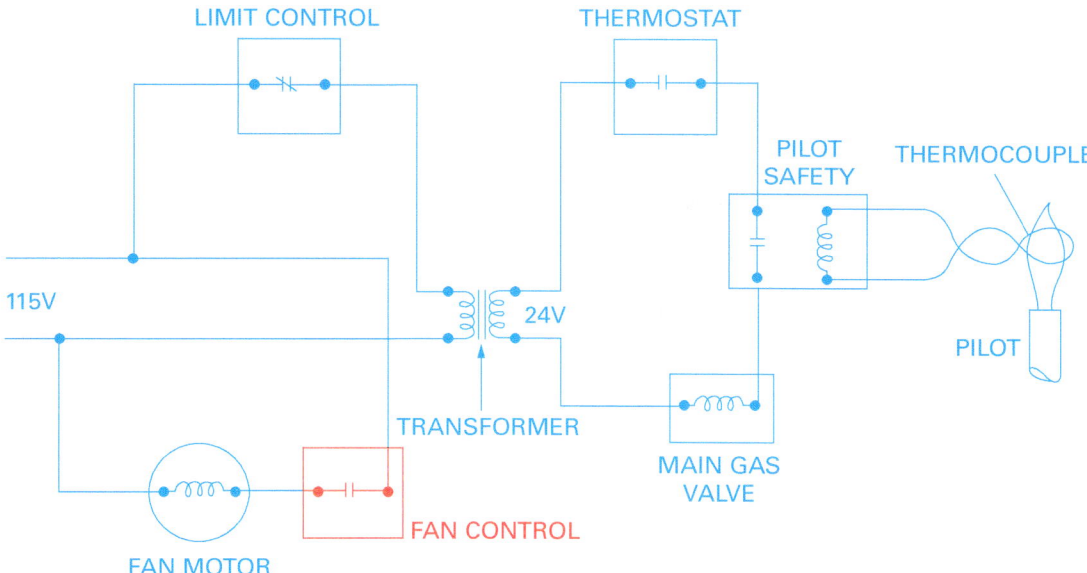

Figure 15 24-VAC furnace control circuit.

Air handlers that use electric resistance heaters do not use temperature-activated switches to activate the fan. A fan relay or the first sequencer is used to power the fan because air must immediately flow across the energized electric heaters to prevent them from overheating.

1.2.5 Furnace Limit Controls

A furnace limit control is also a temperature-actuated switch with a bimetal sensing element positioned near furnace heat exchangers or heating elements. This safety control is normally wired in series with the low-voltage control circuit of the furnace. If the temperature inside the heating chamber or plenum reaches approximately 200°F (93°C), power in the control circuit will be interrupted to prevent overheating.

On oil burners, the limit switch is commonly wired in series with the power supply serving the furnace primary control. In an oil burner, the power will be shut off to the primary control, which stops burner operation. The blower motor will continue to function, since it does not rely on power from the primary control to operate.

Electric-heat limit switches (*Figure 16*) consist of a bimetal sensing element that is enclosed in a disc-shaped housing that resembles a large coin. Radiant heat from the electric heating elements will cause the sensing element to open if there is inadequate airflow across the heating elements. In contrast to a gas or oil furnace that contains a single limit switch, a fan coil with multiple banks of electric heaters will contain several limit switches, all wired in series.

Limit switches protect from transient events. For example, a dirty air filter may cause a furnace to cycle on and off on the limit switch. When the dirty filter is replaced, the problem goes away because most furnace limit switches are automatic reset. There are other events that can occur to a furnace that require it to be shut down until the problem is corrected. For example, a soot-blocked heat exchanger may cause flames to roll out the front of the burner. This condition cannot be allowed to repeat for obvious reasons. To prevent such events from reoccurring, manual-reset limit switches are used.

A flame-rollout switch is an example of a manual-reset limit switch. The switch must be reset manually before operation will resume. A service technician will usually have to reset the switch to troubleshoot the furnace, observe the problem, and then correct it. Manual reset switches of any type are easily identified by the presence of a reset button, usually colored red. Fusible links, also

(A) AUTO-RESET LIMIT SWITCH (B) MANUAL-RESET LIMIT SWITCH

Figure 16 Limit switches.

called thermal fuses, contain a link that melts at a specific temperature. When the link melts, it opens the control circuit and shuts down the unit. Fusible links are rarely used on fossil fuel equipment. They are most often used as additional protection with limit switches on electric heat assemblies. Fusible links must be replaced, with one of the correct melting temperature, to restore operation. On air handlers with electric heating elements, the same types of devices provide similar protection.

1.2.6 Combination Fan/Limit Switch

A combination fan/limit switch uses a common bimetal element that is spiral-wound to activate a fan switch and a limit switch. Both switches are contained in the same housing. The element is inserted so that it is between sections of a furnace heat exchanger or installed in the warm air plenum directly above the furnace. Heat acts on the spiral-wound bimetal element, causing a rotating motion that activates the fan switch when the selected fan-on temperature is reached. If the fan fails to come on or airflow is restricted, the temperature will continue to rise until the bimetal element opens the limit switch, typically set around 200°F (93°C).

This type of combination switch often comes factory-equipped with a jumper that enables the device to operate on line voltage only (*Figure 17*). With the jumper removed, the fan switch operates on line voltage, but the limit switch is in the low-voltage circuit. Since the two sets of contacts can be isolated from each other, they can function in different circuits at different voltages.

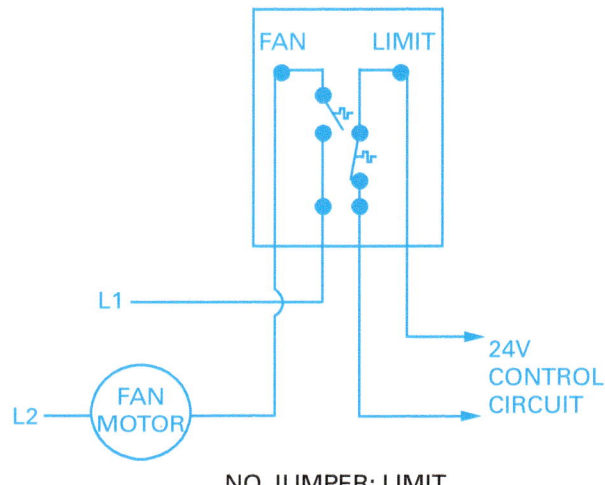

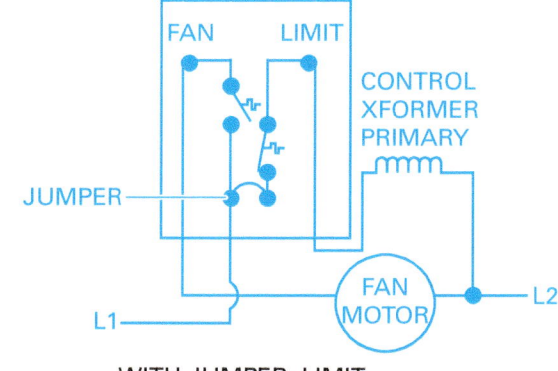

Figure 17 Combination fan/limit switch wiring options.

Additional Resources

Refrigeration and Air Conditioning: An Introduction to HVAC/R, Larry Jeffus. Current edition. New York, NY: Pearson.

1.0.0 Section Review

1. A technician measures 230VAC across a set of normally open relay contacts, and the relay coil is energized. What does this indicate?
    a. Normal operation.
    b. The relay contacts are grounded.
    c. There is continuity through the contacts.
    d. The contacts are still open; the relay is defective.

2. A furnace limit control is wired into a gas furnace in _____.
    a. parallel with the fan switch
    b. series with the fan switch
    c. series in the control circuit
    d. parallel with the transformer primary

SECTION TWO

2.0.0 THERMOSTATS AND TEMPERATURE CONTROLS

Objective

Describe the operation, installation, and testing of various thermostats and temperature controls.
a. Describe the operation of various thermostats and temperature controls.
b. Identify and describe how to troubleshoot thermistors.
c. Explain how to install and wire thermostats.
d. Explain how to troubleshoot the functions of a thermostat.

Performance Tasks

1. Wire, check the operation of, and adjust the cycle rate of a thermostat.

Trade Terms

Cooling compensator: A fixed resistor installed in a thermostat to act as a cooling anticipator.

Deadband: A temperature band, usually 3°F (1.6°C), that separates heating and cooling in an automatic changeover thermostat.

Differential: The difference between the cut-in and cut-out points of a control.

Droop: A mechanical condition caused by heat that affects the accuracy of a bimetal thermostat.

Invar®: An alloy of steel containing 36% nickel. It is one of the two metals in a bimetal device used in temperature controls.

The room thermostat is the temperature-control device commonly used in residential and light commercial HVACR systems to maintain indoor comfort. It does this by cycling the cooling equipment on if the temperature rises in the conditioned space, or by cycling the heating equipment on if there is a temperature drop. The type of room thermostat chosen, the location where it is installed, and how it is adjusted are factors that can affect its ability to maintain space temperature at or near its setpoint.

2.1.0 Thermostat and Temperature Control Operation

Programmable electronic thermostats using electronic sensing elements now dominate the market. However, thermostats with bimetal sensing elements are still being manufactured and used in some applications. Although not quite as precise as the electronic thermostat, bimetal devices are considerably less expensive and have proven to be effective and reliable for simple applications. Most bimetal thermostats will maintain the space temperature within ±2° of the setpoint. A good electronic thermostat will maintain the temperature within ±1°.

A bimetal element (*Figure 18*) is composed of two different metals bonded together; one is usually copper or brass. The other, a special metal called Invar®, contains 36% nickel. When heated, the copper or brass has a more rapid expansion rate than the Invar®, resulting in the shape of the element changing. The change in shape moves delicate linkage to open or close switch contacts, or the contacts are mounted on the element and open or close as the bimetal element curves.

While bimetal elements are constructed in various shapes, the spiral-wound element is the most compact in construction and the most widely used. In *Figure 18*, note that the thermometer on the face of an electromechanical thermostat operates on the same principle.

Thermostat switching actions should take place rapidly to prevent arcing, which could cause damage to the switch contacts. A magnet is often used in some bimetal thermostats to help close the contacts quickly and eliminate the arcing potential. These are referred to as snap-action thermostats.

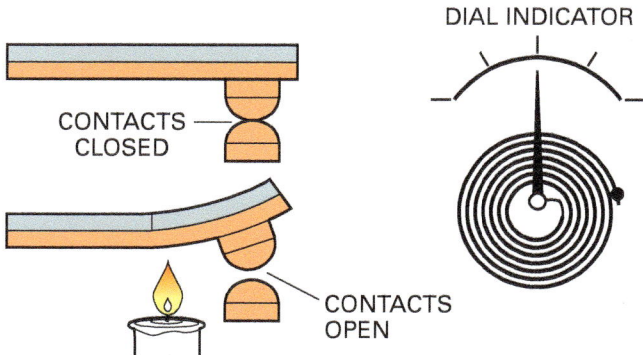

Figure 18 Bimetal sensing elements.

The majority of wall thermostats in the past used mercury encased in glass to provide switching with a contained arc. As the bimetal element changes shape, the mercury rolls from one end of its glass bulb to the other, opening or closing a circuit. Due to the hazardous nature of mercury, they have virtually been eliminated from the market. Mercury-bulb thermostats encountered should be replaced with an electronic model whenever possible.

> **WARNING!**
> Mercury is toxic. Even short-term exposure may result in damage to the lungs and central nervous system. Mercury is also an environmental hazard. Do not dispose of thermostats containing mercury bulbs along with common trash or garbage. Contact your local waste management authority, or the Thermostat Recycling Corporation (TRC) at **https://www.thermostat-recycle.org/**, for disposal/recycling instructions.

Most residential and commercial thermostats are of the low-voltage type. With low-voltage control circuits, there is less risk of electrical shock and less chance of fire from short circuits. Low-voltage components are also less expensive and less likely to produce arcing, coil burnout, and contact failure. However, line voltage thermostats are also used, usually to control simple electric heating equipment.

A very few self-generating systems still exist that generate about 750mV from the pilot flame. This is just enough power to operate the gas valve; these systems have no fan or other significant electrical loads. The thermostats in millivolt systems are similar in construction and design to 24V thermostats; however, they are not interchangeable.

Did You Know?
Mercury
The Thermostat Recycling Corporation (TRC) determined that roughly 15 tons of mercury was installed in thermostats in 2001. Since a thermostat holds roughly 4 grams, that's a lot of thermostats. Today, the amount of mercury used in thermostats annually is trivial. The TRC, a joint effort by industry titans, collects mercury-bulb thermostats. The mercury, a heavy metal, is recovered before disposing of the remaining parts. Many HVACR vendors and parts distributors participate in the TRC program.

Did You Know?
Millivolt Heating Systems
Millivolt heating systems were developed decades ago when natural gas was first brought to major cities. At that time, many homes were heated with coal using gravity-type duct systems. Those gravity furnaces did not require electrical power to operate since they did not use a blower to circulate the heated air. The millivolt system allowed utilities to retrofit existing coal furnace to burn natural gas. This kept costs to the customer low because the existing furnace could be used and electrical power did not have to be brought to the furnace. The millivolt system generated just enough power to open and close the gas valve. As late as the 1970s, these systems could still be encountered. No doubt a few of these relics still exist, but they are rare.

2.1.1 Heating-Only Thermostats

A wall-mounted heating-only thermostat typically contains a temperature-sensitive switch as shown in *Figure 19*. The thermostat shown here is the bimetal, snap-action type.

When the temperature in the conditioned space reaches the thermostat setpoint, the thermostatic switch will open. Because of residual heat in the heat exchangers, the continued operation of the fan, and the time it takes for the bimetal to respond to room conditions, the temperature can overshoot the setpoint. To avoid the discomfort that this condition might cause, the thermostat contains an adjustable heat anticipator that causes the thermostat to open before the temperature in the space reaches the setpoint. The heat anticipator is a small resistor in series with the switch contacts. The resistor gets warm when the heating circuit is energized. The anticipator warms the bimetal strip, causing the contacts to open earlier than they would otherwise. Heat anticipators are typically adjustable, and are set based on the current draw of the heating control circuit flowing through the thermostat.

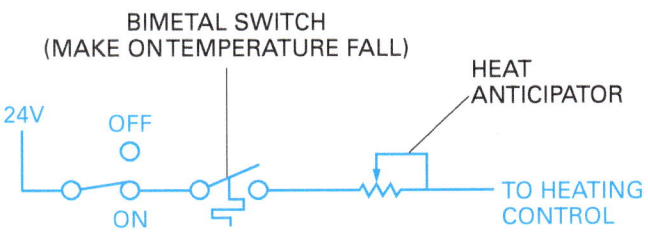

Figure 19 Heating-only thermostat.

2.1.2 Cooling-Only Thermostats

The cooling thermostat (*Figure 20*) is the opposite of a heating thermostat. When the bimetal coil heats up and unwinds, the switch closes its contacts and starts the compressor. The cooling-only thermostat contacts close on a rise of temperature, while the heating-only thermostat contacts open on a rise.

Cooling thermostats contain a device called a **cooling compensator** to help improve indoor comfort. The cooling compensator (*Figure 21*) is a fixed resistance in parallel with the thermostatic switch. No current flows through the compensator when cooling is on because it has a much higher resistance than the switch contacts. In this case, the contacts are essentially a short circuit. When the thermostat circuit is open, however, a small current can flow through the compensator and the contactor coil.

Because of the size of the compensator, the current is not enough to energize the contactor. The heat created by the current flowing through the compensator makes the thermostatic switch contacts close sooner than they would without the compensator. In this way, the cooling compensator accounts for the lag between the call for cooling and the time when the system actually begins to cool the space.

2.1.3 Heating-Cooling Thermostats

When heating and cooling are combined for year-round comfort, it is impractical to use a separate thermostat for each mode. Therefore, the two are combined into one heating-cooling thermostat (*Figure 22*).

A switch provides a means to change the operate mode. The same switch (the Mode button in *Figure 22*) can also be set to Off. The button in the center controls the function of the fan. In the Auto position, the indoor fan cycles on and off with the heating or cooling equipment. In the On position, the indoor fan runs continuously, regardless of whether the unit is in heating or cooling.

This thermostat is also equipped with a Filter Reset button. A timer in the thermostat can be set for 30 to 180 days. At the end of the selected period, the thermostat will remind the user to change the air filter. After it is changed, the user resets the reminder and the time interval starts over.

2.1.4 Automatic-Changeover Thermostats

The disadvantage of a manual-changeover heating/cooling thermostat is that the building occupant must determine whether heating or cooling is needed at a given time and set the thermostat

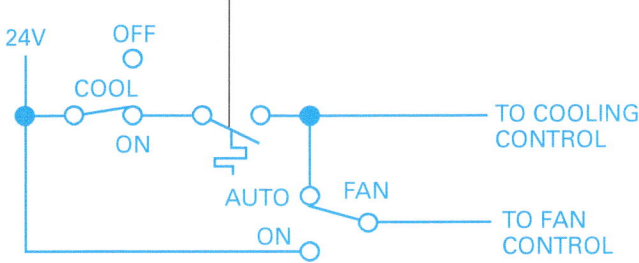

Figure 20 Cooling-only thermostat.

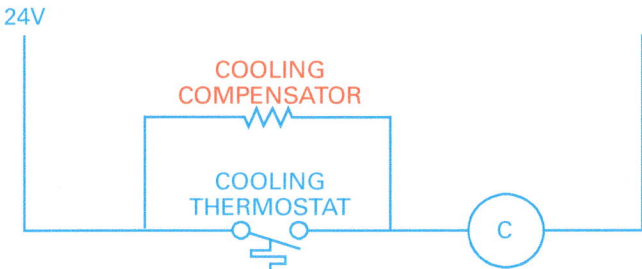

Figure 21 Cooling compensator.

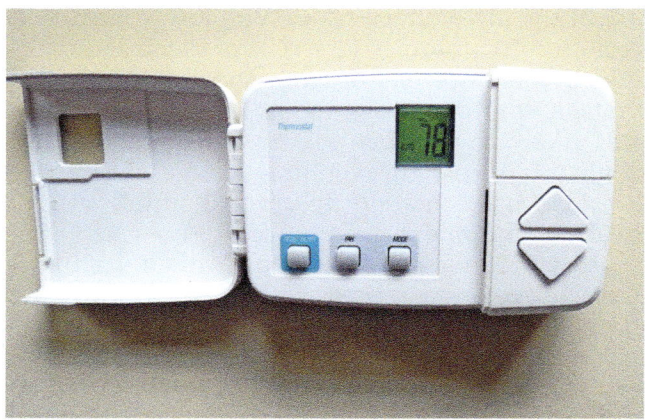

Figure 22 Standard heating-cooling thermostat.

switch accordingly. In some climates, that is very impractical; the need can change several times a day.

The automatic-changeover thermostat automatically selects the mode, depending on the heating and cooling setpoints. Automatic-changeover thermostats look the same as other thermostats. The one exception is the addition of another position on the function switch. In addition to Heat, Cool, and Off, an Auto position is available. This should not be confused with the Auto position on the Fan switch. The occupant can still select either heating or cooling. However, when the thermostat function switch is in the Auto position, the thermostat selects the operating mode. In electromechanical thermostats, all that is necessary is for one of the thermostatic switches to close, indicating that the conditioned space is too warm or too cold. With electronic thermostats, the temperatures at which heating or cooling is selected is determined by the user and entered into the thermostat. The logic in the microprocessor then determines which operating mode is needed to maintain those setpoints.

Older electromechanical thermostats contain a built-in mechanical *differential*, which is the difference between the cut-in and cut-out points of a thermostat. The differential is normally 2° to 3°F (1.1° to 1.6°C). For example, if the heating setpoint is 70°F (21.1°C), the furnace may turn on at 69°F (20.5°C) and run until the temperature is 71° to 72°F (21.6° to 22.2°C). Automatic changeover thermostats also have a minimum interlock setting, commonly known as the *deadband*. The deadband is a built-in feature that prevents the heating and cooling setpoints from being any closer together than 3°F (1.6°C) and fighting each other to control the space temperature. The deadband can be adjusted on many electronic models to suit unique applications. However, the 3°F deadband is a common default value and serves most applications well. Defeating a minimum deadband altogether through settings is not usually allowed by the software.

2.1.5 Two-Stage Thermostats

A two-stage indoor thermostat (*Figure 23*) is used to control equipment that has more than one stage of heating or cooling capacity. Examples of such equipment include heat pumps, furnaces with staged burners, and cooling equipment using multiple compressors or compressors with capacity control. For example, when a heat pump calls for heat, the first stage of the room thermostat controls the compressor, providing compression heat. If compression heat cannot satisfy the room thermostat, the temperature continues to drop and the second stage of the room thermostat, usually preset 1°–2°F (0.6°–1.1°C) from the first stage, energizes supplementary electric heat to satisfy the room thermostat.

2.1.6 Programmable Thermostats

Programmable thermostats are self-contained controls with all features and functions located in the unit mounted on the wall. Early programmable thermostats looked more like conventional thermostats and contained a motor-driven time clock that raised and lowered temperatures at desired intervals. This technology is now obsolete.

Modern electronic programmable thermostats (*Figure 24*) use microprocessors and integrated circuits to provide a wide variety of control and energy-saving features. Their control panels use touch-screen technology and soft buttons, and their displays are digital. Different thermostats offer various features; the more sophisticated and expensive the thermostat, the more features it generally offers. The following are some of the features available on electronic thermostats:

- *Better temperature control*—Temperatures are sensed electronically, allowing temperatures to be held within a much tighter range than possible with bimetal thermostats. Some versions have built-in humidity control.
- *Installation versatility*—Wireless versions of electronic thermostats are available for applications where it is difficult or impossible to pull thermostat cable. A built-in transmitter sends information to a remote receiver connected to the equipment being controlled. The model shown in *Figure 24* is capable of wireless communication.

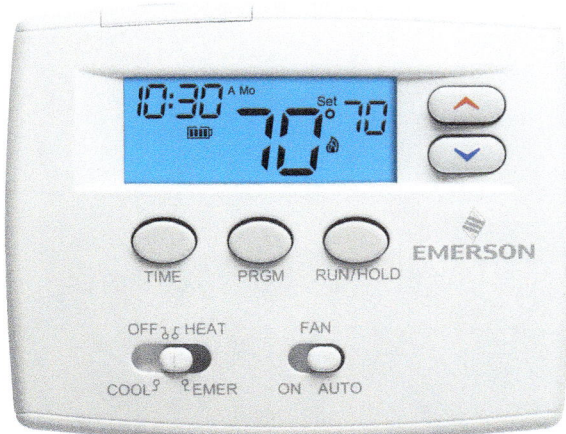

Figure 23 Heat pump thermostat.

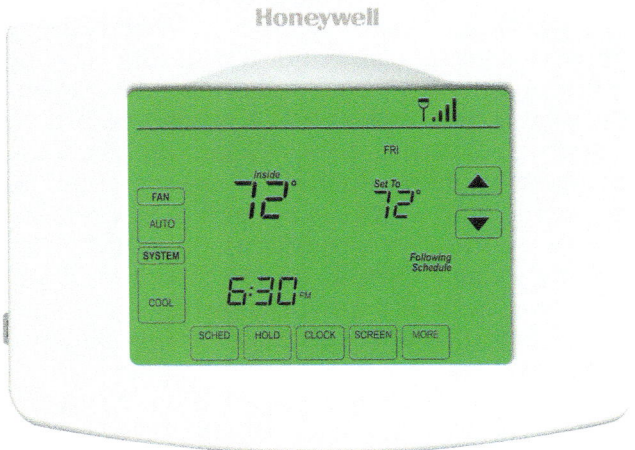

Figure 24 Wi-Fi-enabled, programmable thermostat.

- *Override control*—This feature allows the occupant to override the program when desired. For example, the night setback on Monday night could be overridden so the thermostat is not lowered before the football game is over.
- *Multiple programs*—This feature allows the occupant to design and select different schedules for different conditions. For example, a special schedule for a vacation away from home could be programmed.
- *Battery backup*—This feature prevents program loss in the event of a power failure.
- *Staggered startup for multi-unit systems* – This feature avoids excessive current drain. This is an important feature in office buildings, shopping malls, and hotels.
- *Maintenance tracking*—This feature indicates when maintenance is to be performed (for example, when to replace filters).
- *Wi-Fi connection*—Many of today's thermostats connect to the local network and can be accessed through both the network and its internet connection. This allows the thermostat's software to be updated easily, and provides thermostat information and control to a user on a worldwide scale. The thermostat in *Figure 24* is Wi-Fi enabled, thus the network signal strength shown in the upper right-hand corner of the display.

The potential savings available from programmable thermostats are significant. For example, it is estimated that a setback of 10°F (5.5°C) for both daytime and nighttime can result in a 20% energy savings. A 5°F (2.8°C) setback should yield a 10% energy savings.

2.1.7 Line-Voltage Thermostats

Most modern thermostats operate at low voltages. There are also thermostats that operate at line voltages (115 volts or higher). They are commonly used for controlling electric baseboard heaters.

Line-voltage thermostats may be controlled by a bimetal sensing element or a hydraulic sensing bulb. The latter controls the thermostat by means of pressure. *Figure 25* shows a remote-bulb thermostat that is actuated by pressure on a bellows. The sensing bulb contains refrigerant that increases in pressure as the temperature increases. The increasing pressure acts to expand the bellows, thus causing the switch to change position. Note that bellows-type thermostats were also used in low-voltage applications as well. Bellows-type thermostats are now obsolete.

Line-voltage thermostats often suffer from a mechanical condition known as thermal offset, commonly called droop. Droop causes the thermostat

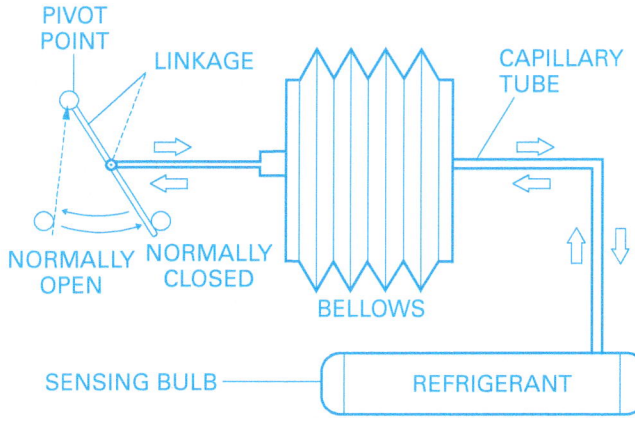

Figure 25 Remote-bulb thermostat.

control point to drift away from the selected setpoint. Because it is caused by changes in temperature within the thermostat itself, droop most often occurs in line-voltage thermostats. They are subject to excessive internal heating from elevated current flow through them. It may also occur in 24VAC thermostats that use anticipators.

2.2.0 Thermistors

Thermistors are temperature-sensitive semiconductor devices. Their resistance varies in a predictable way with variations in temperature. This allows them to be used in a variety of HVACR control applications. Thermistors have either a positive or a negative coefficient of resistance. If the resistance increases as temperature rises, it has a positive coefficient of resistance. If resistance increases as the temperature drops, it has a negative coefficient of resistance. Thermistors (*Figure 26*) come in various sizes and shapes to suit a variety of applications.

Thermistors are used to sense temperature changes. They are also used as motor protective devices. Some typical applications of thermistors

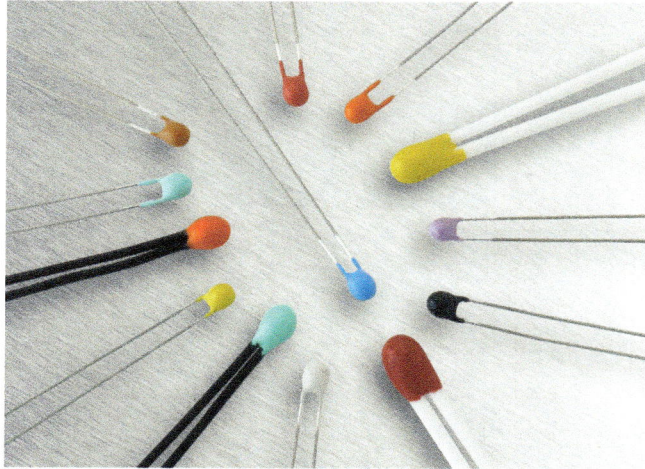

Figure 26 Thermistors.

are: electronic thermometers, room thermostats, duct sensors, electronic expansion valve sensors, heat pump temperature sensing, and selected control circuits.

Thermistors are used in temperature differential controls, such as the defrost control in some heat pumps. Two thermistors are used; one senses coil temperature and the other senses air temperature. Manufacturers often provide resistance charts based on temperature for checking thermistors used in their products (*Figure 27*).

Another use of thermistors is as a start-assist device for single-phase compressors (*Figure 28*). A ceramic thermistor with a steep-slope positive temperature coefficient (PTC) is wired in parallel with the compressor's run capacitor. This can increase the starting torque by 200 to 300%. Procedures for checking a start-assist thermistor are covered later in this module. Of course, this type of thermistor is much larger than a thermistor used to sense temperature. Also note that most temperature-sensing thermistors are negative temperature coefficient (NTC) instead of PTC.

When a thermistor is used as a sensing device for the operation of a thermal-electric (TE) expansion valve, both this expansion valve and the thermistor can be checked as follows:

- With the leads of the voltmeter placed across the terminals of the electric valve (*Figure 29*) and the system operating at or near peak load, the voltmeter reading should be within the range of 15V to 20V.

- With the system operating at lower loads, the reading should remain between 8V and 14V.

Note that the specifications can differ from valve to valve. Changes occur quickly in the industry. Therefore, the manufacturer's recommendations should always be consulted. The readings should be observed for several minutes to ensure consistency.

Like all thermistors, those used in the internal windings for temperature protection of three-phase motors can be checked by taking a resistance reading through them. An ohmmeter should be used and connected as illustrated in *Figure 30*.

A short or ground in the thermistor will be indicated by a reading of zero or a value approaching zero. An open would be indicated by a reading of infinity. The procedure is as follows:

Step 1 Shut off the disconnect switch and allow the equipment to cool.

Step 2 Connect the ohmmeter leads as indicated in *Figure 30* and record the resistance in ohms.

The values will change with a change in temperature, but at room temperature (75°F, or 24°C), the reading should be about 75 ohms. Again, the resistance of the winding-temperature thermistors will vary by manufacturer and application, so unit specifications should be consulted before condemning or accepting the devices.

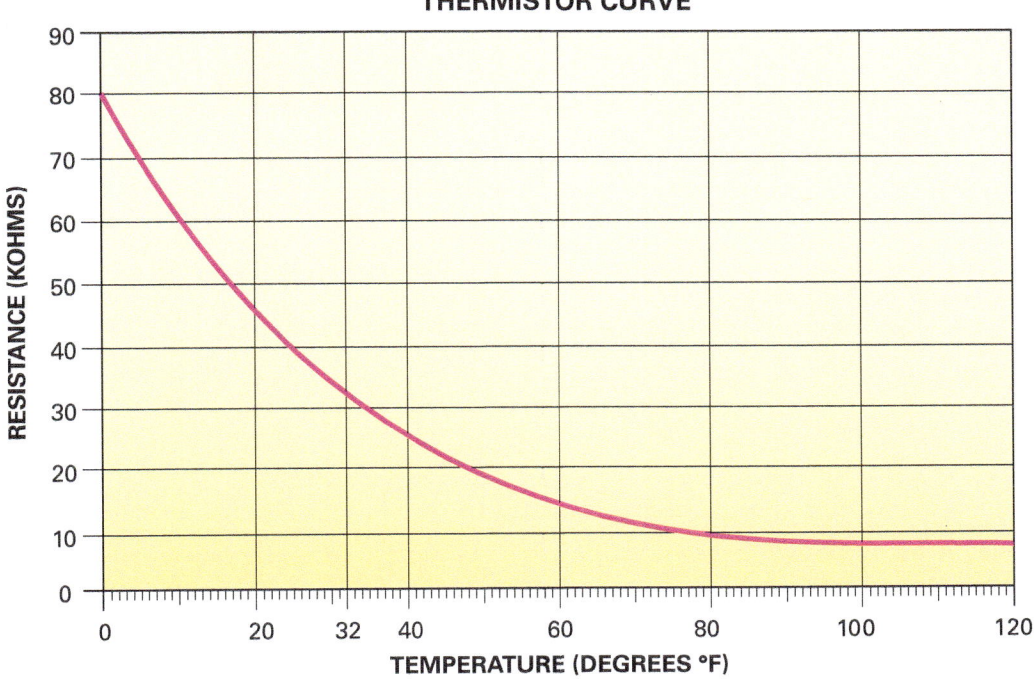

Figure 27 Example thermistor resistance chart.

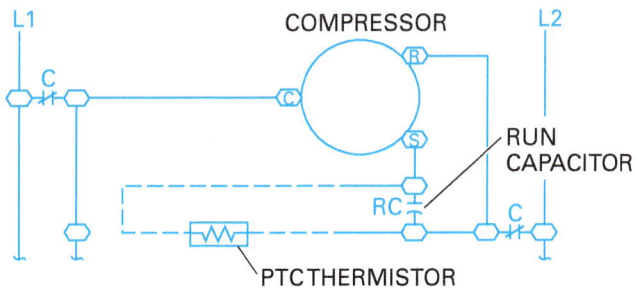

(A) START THERMISTOR WIRING

Thermistors will sometimes fail or provide incorrect resistance values. The built-in diagnostic feature of the system in which thermistors are connected will often evaluate information coming from the thermistors and report a fault when it senses an invalid input. A thermistor is not likely to fail on its own. If a failed thermistor is found, it could mean that it has been damaged, that the leads are damaged or loose, or that it is not making good thermal contact.

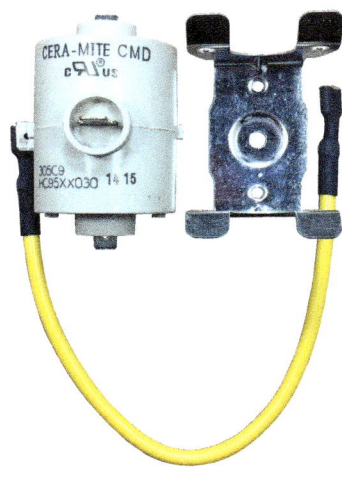

(B) START THERMISTOR KIT

Figure 28 PTC thermistor start-assist device.

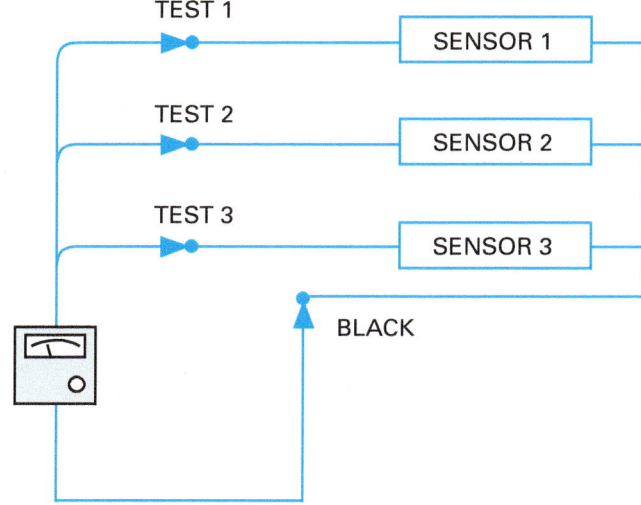

Figure 30 Compressor-protection thermistor test.

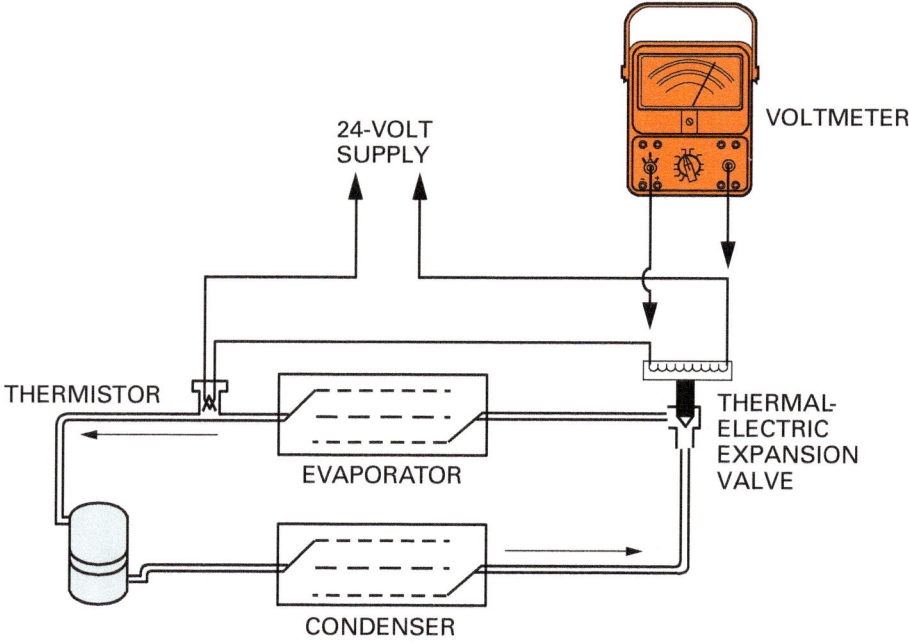

Figure 29 TE expansion valve testing.

The resistance of a thermistor itself can be tested with an ohmmeter. The manufacturer's literature will usually contain a graph or table that shows expected resistance readings for a range of temperatures. An example of a graph has been presented. *Table 1* is an example of the resistance information tabulated. All thermistors do not have the same characteristics, so it is important to acquire the right information for the system being serviced.

2.3.0 Thermostat Installation

Even the best thermostat cannot perform correctly if it is poorly installed. Selecting the proper location for the thermostat is the first step in any installation procedure.

2.3.1 Installation Guidelines

The thermostat must be installed in the conditioned space. For proper operation, electromechanical thermostats must be mounted level on a wall that is reasonably plumb. Electronic thermometers can function properly if they are not level or plumb, but they should be mounted in that manner for appearance sake.

The following practices should be observed when installing a thermostat:

- The installer must be a trained technician.
- The manufacturer's instructions must be followed. Failure to comply could lead to damage or a hazardous condition.
- Make sure the thermostat is suitable for the application.
- Check for proper operation in all modes when the installation is complete.

> **Think About It**
>
> ## Thermistors
>
> Thermistors are of two types: positive temperature coefficient (PTC) thermistors and negative temperature coefficient (NTC) thermistors. The resistance of a PTC thermistor increases with an increase in temperature; while the resistance of a NTC thermistor decreases with an increase in temperature. Based on the definitions given for PTC and NTC thermistors, what type of thermistor is being represented by the data shown in *Table 1*?

Table 1 Thermistor Values

Temperature (°F)	Resistance (Ohms)	Temperature (°F)	Resistance (Ohms)	Temperature (°F)	Resistance (Ohms)
−60	362,640	45	11,396	150	1,020
−55	297,140	50	9,950	155	929
−50	245,245	55	8,709	160	844
−45	202,841	60	7,642	165	768
−40	168,250	65	6,749	170	699
−35	139,960	70	5,944	175	640
−30	116,820	75	5,249	180	585
−25	98,420	80	4,644	185	535
−20	82,665	85	4,134	190	490
−15	69,685	90	3,671	195	449
−10	58,915	95	3,265	200	414
−5	50,284	100	2,913	205	380
0	42,765	105	2,600	210	350
5	36,475	110	2,336	215	323
10	31,216	115	2,092	220	299
15	26,786	120	1,879	225	276
20	23,164	125	1,689	230	255
25	19,978	130	1,527	235	236
30	17,276	135	1,377	240	219
35	14,980	140	1,244		
40	13,085	145	1,126		

CAUTION

Thermostats containing solid-state devices are sensitive to static electricity. Before handling the instrument, discharge any static electricity by touching a metal doorknob or any object that is grounded. Touch only the housing and similar nonmetallic areas when holding the device.

Use care when unpacking the new thermostat since rough handling can damage the thermostat. Save all literature and hardware for later reference and use.

Locate the thermostat five feet above the floor (the common mounting height) in an area with good air circulation at room temperature. Avoid the following locations and conditions:

- Drafty or stagnant spots
- Hot or cold air from ducts or diffusers
- Radiant heat from direct sunlight or hidden heat from appliances
- Concealed supply ducts and chimneys (to avoid radiant heat)
- Unheated areas behind the thermostat, such as an outside wall or garage
- Partition walls subject to vibration from operating equipment or slamming doors
- Outside walls

NOTE

The *National Electric Code* (*NEC*®) does not offer a mandatory mounting height for thermostats. However, for installations subject to the provisions of the Americans with Disabilities Act (ADA), thermostats cannot be mounted more than 48" above the finished floor.

To replace an existing thermostat, first turn off the power source and lock/tag it to also de-energize the control circuit. Before removing the existing thermostat, record the thermostat wire color and corresponding thermostat terminal designation. This helps prevent miswiring of the new thermostat.

Install the new wall plate or subbase (*Figure 31*). The subbase provides terminals for the connection of the wiring. If the wall plate or subbase must be mounted on an outlet box, use the proper adapter ring or cover plate. To prevent drafts in the wall from affecting the thermostat, pack sealing putty or insulation material around the thermostat wire where it penetrates the wall.

Before installing the thermostat on its subbase, ensure that the thermostat is level (*Figure 32*). It is best to hold the subbase on the wall and level it while marking the screw locations with a pencil or scribe. Once the screws are installed as shown

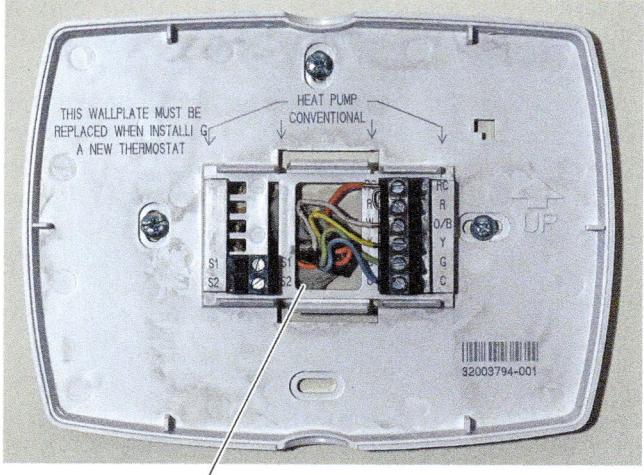

INSERT PUTTY SEALANT OR INSULATION MATERIAL AROUND THE WALL OPENING

Figure 31 Thermostat mounting.

here, level the thermostat again before tightening them. Thermostats have a specific place designed for placement of the level, as shown in the figure. As shown, there are two points across which the level is placed. Some have a flat area on the top to accommodate the level. You can tell that the top edge of this subbase is slightly rounded, so it is not suitable for leveling. As mentioned earlier, electronic thermostats are not affected by being out of level. Visually, however, a thermostat that is not level looks tacky and unprofessional.

2.3.2 Thermostat Wiring

The thermostat is connected through multi-conductor thermostat wire to a terminal strip or junction box in the air conditioning unit. A standard

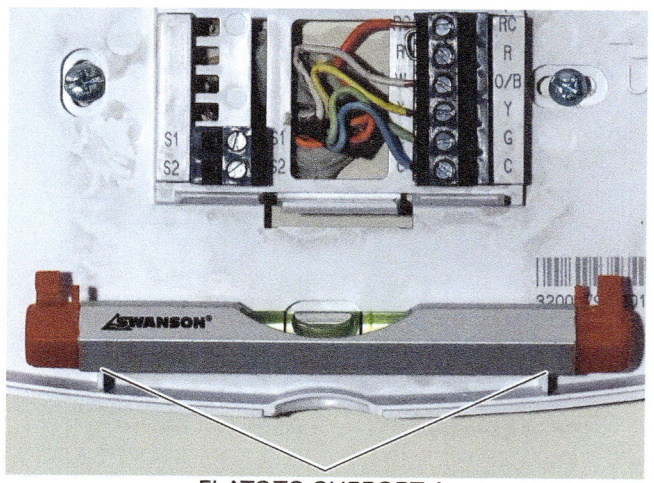

FLATS TO SUPPORT A POCKET LEVEL

Figure 32 Leveling a thermostat.

coding method is used in most of the HVACR industry to designate popular wiring terminals and wire colors. Note that more systems, including ductless units, communicate and control themselves differently. In this section, traditional control wiring practices are emphasized.

Figure 33 shows the color code and corresponding terminals of a heat/cool thermostat. These color code/terminal designations are standard among thermostat and equipment manufacturers. However, it is not safe to assume that any existing system is wired according to that color code. There are additional wire colors and terminal designations used with more complex thermostats. For example, when there are multiple stages of heating or cooling, those terminals are designated with the appropriate letter plus a number, such as W_1 and W_2. Wiring terminals are often provided for the connection of external temperature sensors.

In addition to the traditional terminals found on most room thermostats, many modern electronic thermostats also contain a C (Common) terminal. These modern thermostats require power for the display screen and other functions. By connecting the thermostat to R and C, 24VAC power is always available as long as the control transformer is energized.

Memorizing the common colors for HVACR thermostat wiring is not nearly as important as understanding what each wire does. It is important to try and maintain the color code when possible, and you will find that easy to do for the most common terminals. But it is far more important to understand the function of the terminals than to rely on color as a clue about the wires use. Color provides a way of distinguishing one wire from another—no more than that.

Thermostat wiring should be done in accordance with national and local electrical codes. Standard thermostat wire does not have the electrical-insulation characteristics as conductors used for higher voltages. Therefore, the *NEC®* does not allow Class 2 wiring to be placed inside conduit, junction boxes, or raceways with line-voltage wiring also present. Thermostat wiring also cannot be attached to the outside of conduit or raceways containing line-voltage conductors.

All wiring connections should be tight and sound. To avoid damaging the conductor, always use a stripping tool designed to strip small-gauge wire. Color-coded wiring should always be used for easy reference should system troubleshooting ever be required. Thermostat wires with multiple conductors are readily available from any HVACR parts distributor. 18-gauge thermostat wire is adequate for most installations. However,

THERMOSTAT WIRING CODES

TERMINAL DESIGNATION	WIRE COLOR	FUNCTION
C	BLACK OR BLUE	THE COMMON LEAD OF THE CONTROL TRANSFORMER
R	RED	POWER
G	GREEN	FAN CONTROL
Y	YELLOW	COOLING CONTROL Y1 = STAGE 1 Y2 = STAGE 2
W	WHITE	HEATING CONTROL W1 = STAGE 1 W2 = STAGE 2
O (B)	ORANGE (BROWN)	HEAT PUMP REVERSING VALVE CONTROL
E	N/A	RELATED TO THE EMERGENCY-HEAT FUNCTION OF A HEAT PUMP THERMOSTAT
L	N/A	USUALLY RELATED TO THE ILLUMINATION OF A LIGHT
S	N/A	TYPICALLY RELATED TO THE CONNECTION OF THERMISTORS FOR REMOTE TEMPERATURE SENSING

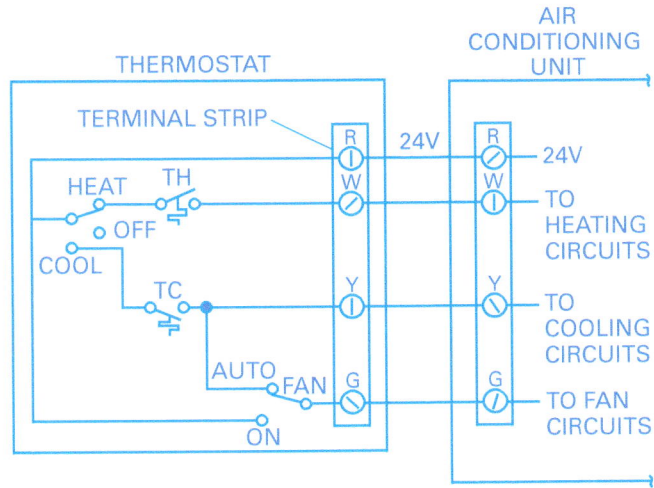

Figure 33 Thermostat wiring.

long runs of thermostat wire can produce a voltage drop that might interfere with equipment operation. If a run of thermostat wire is over 100' (30 m), use 16-gauge or heavier wire to reduce voltage drop. Splicing should be avoided whenever possible. When splicing is necessary, soldered splices are recommended. When wires are stapled to the structure, make sure that the staple does not go through the wire insulation. Also remember to seal the wire opening in the wall space behind the thermostat so it is not affected by drafts from the wall cavity. This problem is more significant if the thermostat must be mounted on an outside wall for some reason.

Some of today's thermostats with sophisticated capabilities are designed to be installed and wired in two pieces. The portion of the thermostat containing the display and interface is wall-mounted and only two wires are connected. These wires provide both DC power and a data link between the thermostat and a utility input/output (I/O) board. The utility I/O board can be mounted at the location of the indoor unit, and all the conventional thermostat-wiring terminations are made there. Alternatively, the thermostat and utility I/O board can be stacked together for a more conventional installation.

The utility I/O board interfaces with compatible smart electric meters and gateways, providing a 2-way communication path for the utility. This allows utilities that offer the service to directly control demand and alert the user to real-time power-pricing programs to save energy.

2.3.3 Adjusting Cycle Rates

Electronic programmable thermostats do not use traditional heat anticipation to control the length of time the burner operates. The traditional method applies a tiny amount of heat to the thermostat bimetal element to make it respond faster. Instead, they control cycle rate, which is defined as the number of times per hour that a burner operates. Different electronic thermostat manufacturers have different methods of setting the cycle rate. The following is an example of how the cycle rate is set for one particular thermostat:

- Consult the furnace manufacturer's literature for guidance and suggested cycle rates.
- Press the Up and Down buttons simultaneously for three seconds to open the menu, until a function number and option number are displayed.
- Press the Up and Down buttons simultaneously for one second until the desired function number is displayed. For this thermostat, function number 3 is the heating cycle-rate function.
- Once function 3 is displayed, press the Up or Down button to change the option number. Each option number represents a different cycle rate. This thermostat offers cycle rates from two to six per hour. Select the option recommended by the furnace manufacturer.
- Press the Up and Down buttons for three seconds to save the changes and exit the menu.

Like the traditional heat anticipator, adjustments can be made if the cycle rate does not satisfy the user. However, starting with the manufacturer's suggested setting, or even with the default setting of the thermostat manufacturer, often results in satisfactory operation.

2.3.4 Final Check

The last step is to confirm that the thermostat functions properly in all operating modes. With the furnace power on, place the thermostat in the Heat mode. Place the fan switch in the Auto position. Adjust the setpoint temperature to at least 5°F above the room temperature. The burner should come on followed by the blower after a short delay. Then adjust the setpoint temperature to 5°F below the room temperature. The main burner should shut off, but the blower will likely continue to run for one to two minutes. Next, set the thermostat to the cooling mode.

> **CAUTION**
> To avoid compressor damage, do not check cooling operation unless the outside temperature and crankcase heater operation is sufficient to ensure that liquid refrigerant has not condensed in the suction line near the compressor. This is to prevent the possibility of damage from a flooded start.

Set the thermostat to the Cool position, and the fan switch to Auto. Adjust the temperature setpoint to at least 5°F (2.8°C) below the room temperature. The cooling equipment should come on either immediately, or after any built-in start delay. The indoor fan should come on immediately. After the cooling system has started, adjust the setpoint to at least 5°F above the room temperature. The cooling equipment should turn off immediately or very soon. Depending on the equipment in use, the indoor fan may cycle off with the compressor, or it may run for an additional period for increased energy efficiency.

With the system switch now in the Off position, adjust the thermostat setpoint up and then down. The system should not respond for heating or cooling. Leave the mode selection switch set to Off. Then set the fan function switch to On. The indoor fan should come on and continue to run until the switch is placed back in the Auto position.

2.3.5 Mercury-Thermostat Disposal

Mercury is a toxic metal that was used, and still found, in functioning thermostats (*Figure 34*). It is estimated that of all the known mercury contamination in the United States, about 8% of it can be traced to mercury thermostats that have been disposed of improperly. Today, manufacturers are free to continue to make mercury thermostats as long as they have a program in place to collect and recycle old thermostats. Roughly half of the United States have laws in place to restrict the sale and distribution of thermostats containing mercury. These state laws have several things in common including the following:

- Mandatory recycling of all mercury thermostats
- Specific procedures to follow during disposal/recycling
- Designated entities to handle collection
- Specific methods for handling /shipping to prevent accidental spillage required

Since state laws vary, always consult a local hazardous-waste disposal authority for guidance. Across the nation, many HVACR wholesalers have voluntarily agreed to act as collection points for mercury thermostats. Contractors and the public can drop off mercury thermostats at these locations, free of charge. Some states offer a cash bounty for each mercury thermostat turned in.

Due to the hazard mercury represents, it is important to note that mercury is contained in many items (*Figure 35*), with fluorescent lamps a significant source of the material. All items containing mercury must be disposed of in accordance with local, state, and federal requirements.

2.4.0 Thermostat Troubleshooting

Troubleshooting thermostats involves visual checks, electrical checks, and electronic troubleshooting for electronic thermostats. When troubleshooting and servicing electronic thermostats, be sure to avoid damaging their microprocessors and integrated circuits. As a first step, visually check the thermostat for problems that could affect its operation and ensure the following:

- Location is well-chosen and the control is free from vibration.
- Check that function switches operate freely.
- Mercury-bulb thermostats must be level.
- Make sure the mercury bulb or bimetal element moves freely.

If the control circuit voltage (usually 24VAC) is inadequate, it can prevent normal system operation

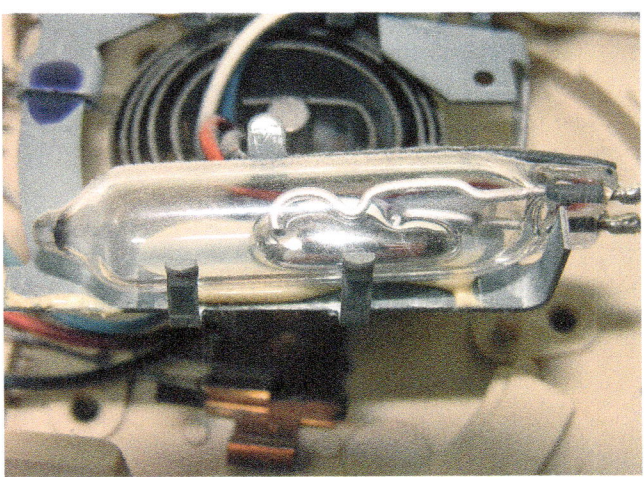

Figure 34 Mercury in a thermostat bulb.

Figure 35 Examples of products containing mercury.

and lead to the erroneous conclusion that the thermostat is defective. Measure the control circuit voltage between the R and C terminals on the furnace or air handler to check that adequate control power is available. If no voltage is measured, troubleshoot the control transformer circuit. If the voltage is lower than about 21V, it will affect the operation of control relays and other control components. Check the control transformer circuit for the cause of low or absent voltage.

Various functions of the thermostat can be checked by removing the thermostat from the wall and using jumper wires to simulate the different thermostat functions. Use an insulated jumper wire(s) with alligator clips connected across the thermostat's R (24V) terminal and fan (G), heat (W), or cool (Y) terminal, as applicable. The jumpers simulate the switches found within a thermostat. If a thermostat is suspected to be the cause of a system not operating properly, the thermostat should be removed from its subbase and the jumpers placed across the subbase terminals. The jumper wire testing can also be conducted at the indoor unit low-voltage terminal board as well. However, the thermostat should first be removed from the subbase to avoid any possibility of damaging it. If the use of jumpers confirms that the thermostat is defective, replace the thermostat. Do not attempt to field-repair the device.

> **NOTE**
> Use jumper wires in this manner only on low-voltage (<50VAC) control circuits.

> **CAUTION**
> While jumper wires can also be used in testing other controls, a jumper wire should be used for troubleshooting purposes only on safety controls. Do not bypass safety controls for any other purpose.

2.4.1 Fan-Switch Operation Checks

If the thermostat fan switch does not initiate indoor fan operation, move to the terminal board where the thermostat wiring is terminated. With the control circuit energized and the fan switch in the On position, use a multimeter to test for voltage between Common (C) and G at the indoor unit or furnace terminal board. If the circuit is intact, 24VAC will be read (*Figure 36*). Alternatively, or as confirmation, you can test for voltage between R and G—if the switch is closed and the circuit intact, there will be no voltage reading here. If the circuit is not intact, the problem is most likely the thermostat, but the integrity of the interconnecting wiring must also be considered.

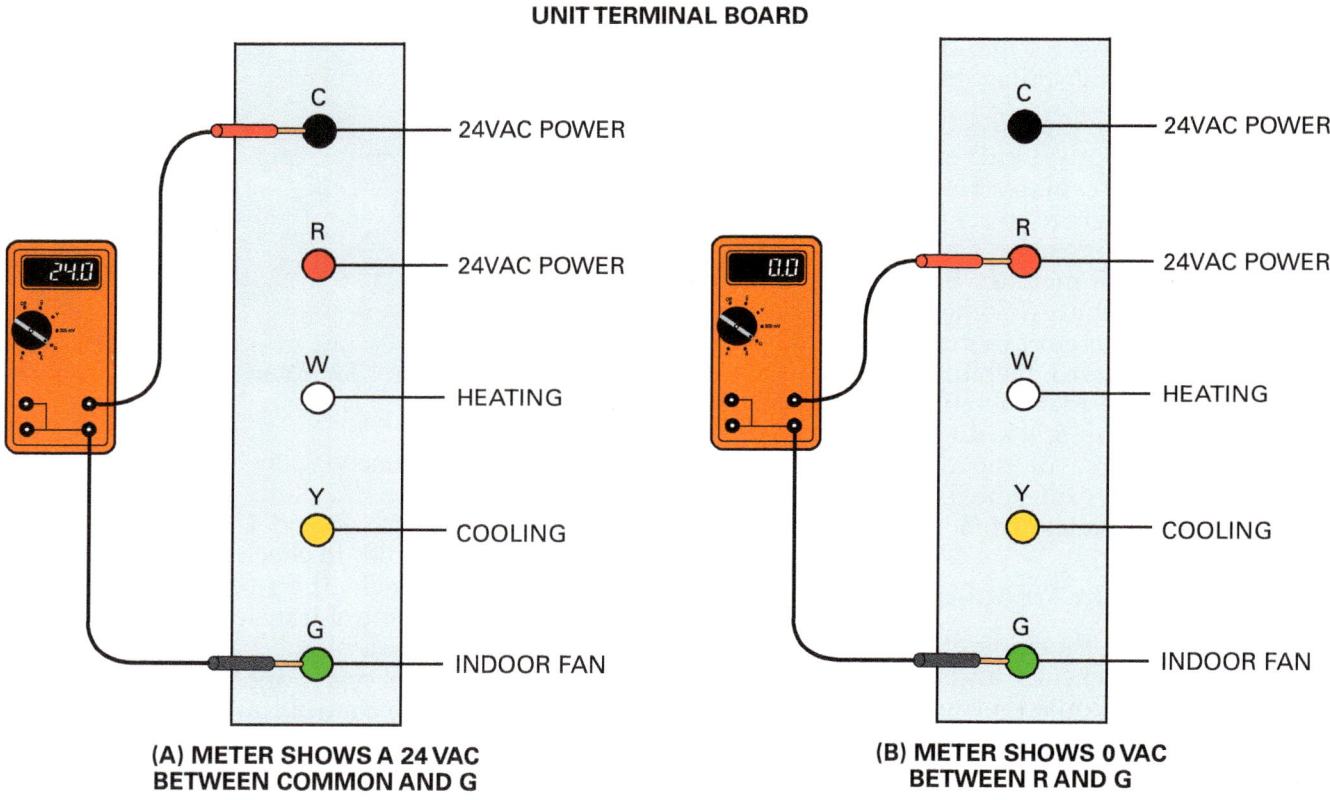

Figure 36 Checking the function of a thermostat fan-switch (control circuit energized).

2.4.2 Cooling Operation Checks

If the thermostat does not initiate cooling operation, move to the unit terminal board for testing. With the circuit energized, thermostat set to cooling mode, and the setpoint well below the room temperature, use a multimeter to test for the presence of voltage between C and Y. A completed circuit through the thermostat will result in a reading of 24VAC. You can also test for voltage between R and Y for confirmation—there should be no significant voltage here if the circuit is intact. This testing is shown in *Figure 37*.

Remember that the thermostat could potentially be timing out if the compressor short-cycle protection feature is enabled. Before accomplishing this test, make sure any programmed time delay has elapsed, or temporarily disable the feature from the menu.

2.4.3 Heating Operation Checks

If the thermostat does not initiate heating operation, begin testing once again at the unit terminal board. With the circuit energized, thermostat set to heating mode, and the setpoint well above the room temperature, use a multimeter to test for the presence of voltage between C and W. A completed circuit through the thermostat will result in a reading of 24VAC. You can also test for voltage between R and W for confirmation—there should be no significant voltage here if the circuit is intact. This testing is shown in *Figure 38*.

> ## Using Jumper Wires
>
> Jumper wires can be used to test circuits as well. For example, if the thermostat does not initiate indoor fan operation with the switch in the On position, remove the thermostat and connect a jumper from R to G. If the fan starts, the problem must be in the thermostat. If it does not start, you can virtually eliminate the thermostat from the list of possibilities; the problem is either in the wiring or in the unit itself.

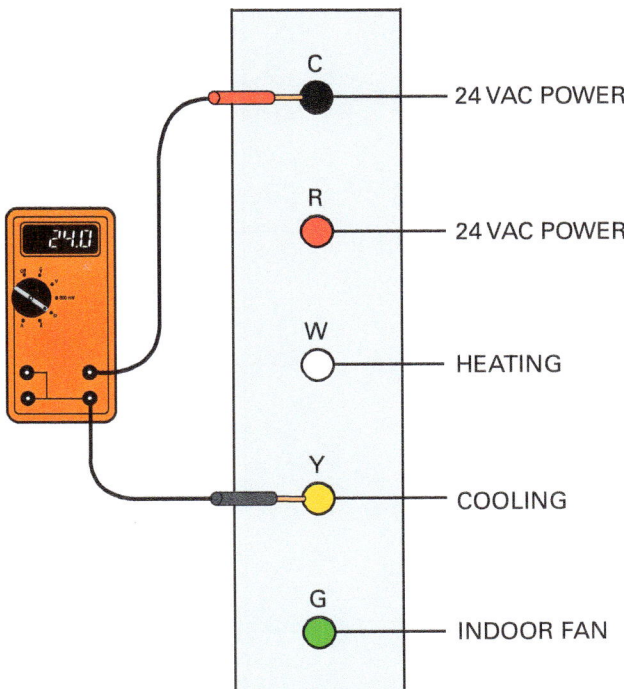

(A) METER SHOWS 24 VAC BETWEEN COMMON AND Y

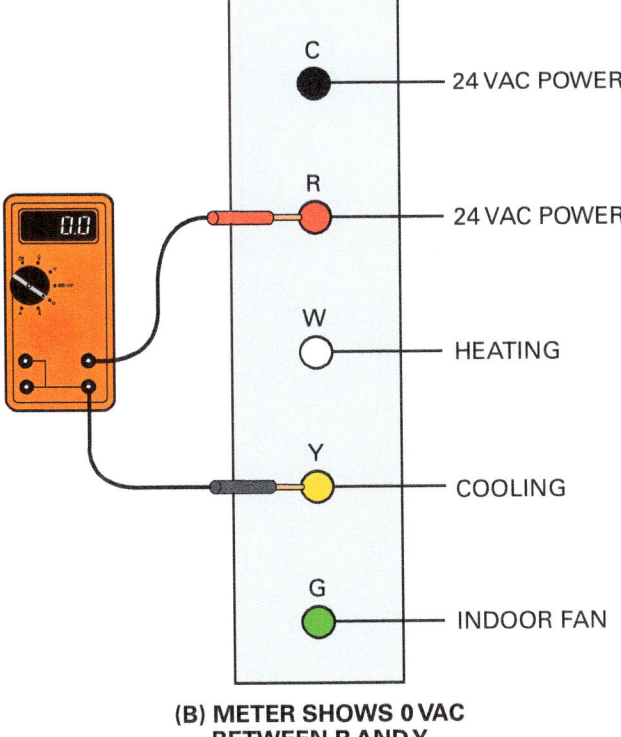

(B) METER SHOWS 0 VAC BETWEEN R AND Y

Figure 37 Thermostat cooling-function check.

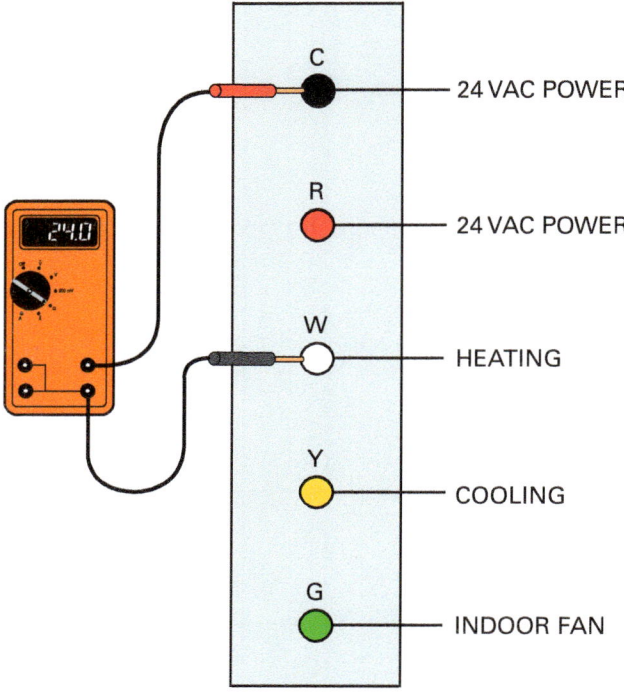

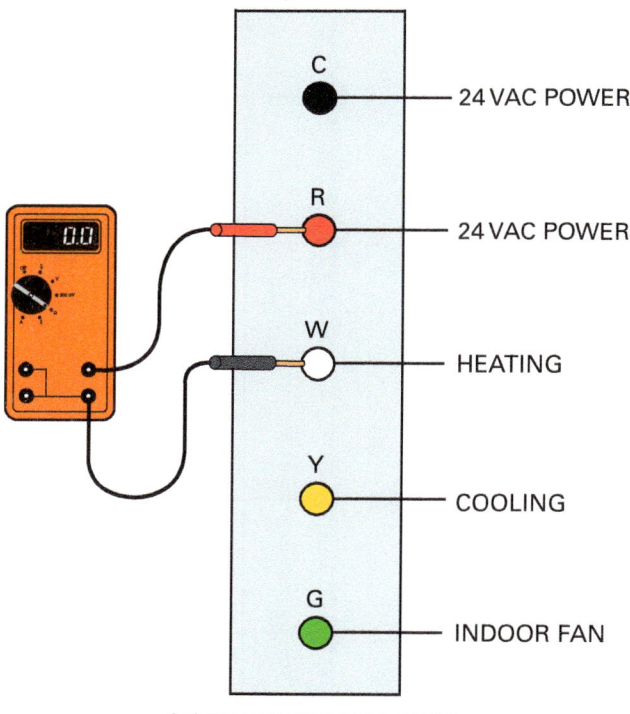

Figure 38 Thermostat heating-function check.

2.4.4 Electronic Thermostats

Due to their low cost, easy availability, and high reliability, electronic programmable thermostats are now the thermostat of choice in both residential and light commercial applications. These thermostats use microprocessors and integrated circuits to provide a wide variety of control and energy-saving features. Different thermostats offer different features; the more sophisticated the thermostat, the more features it generally offers.

Troubleshooting electronic thermostats is similar to troubleshooting electromechanical thermostats. If the thermostat is suspected as causing problems, a meter can be used for testing as described in the previous section. Since electronic thermostats require power to function, a loss of power can affect their operation. Power checks should include the following:

- High-voltage power is available at the heating/cooling equipment.
- Low-voltage (24V) power is available.
- Batteries (if used) are full strength.

Electronic thermostats have a variety of programming options. This makes them highly flexible and allows them to be applied to many system types. When troubleshooting suspected thermostat problems, it is important to first check its programming. A minor error in the setup can result in erratic operation and mislead technicians to suspect problems elsewhere in the system. Acquire the manufacturer's literature for the thermostat and review the setup before troubleshooting begins.

Electronic thermostats can be self-diagnosing to some degree. Error codes or other indications of a problem might be displayed on the screen. For instance, the image of a blinking battery on the screen often indicates a weak or dead battery. Other error codes identify faulty components such as an open or shorted temperature sensor. Because of their diversity and complexity, manufacturers of electronic programmable thermostats often provide aids in their service literature to use when troubleshooting their products. *Figure 39* is an example of a manufacturer's troubleshooting guide for an electronic thermostat.

TROUBLESHOOTING

Reset Operation

If a voltage spike or static discharge blanks out the display or causes erratic thermostat operation, reset the thermostat by pressing the ⌂, ⌄ and PRGM buttons simultaneously. If the thermostat has power, has been reset and still does not function correctly contact your heating/cooling service person or place of purchase.

Symptom	Possible Cause	Corrective Action
No Heat/No Cool/No Fan (common problems)	1. Blown fuse or tripped circuit breaker. 2. Furnace power switch to OFF. 3. Furnace blower compartment door or panel loose or not properly installed.	Replace fuse or reset breaker. Turn switch to ON. Replace door panel in proper position to engage safety interlock or door switch.
No Heat	1. System Switch not set to Heat. 2. Loose connection to thermostat or system 3. Heating System requires service or thermostat requires replacement.	Set System Switch to Heat and raise setpoint above room temperature. Verify thermostat and system wires are securely attached. Diagnostic: Set System Switch to Heat and raise the setpoint above room temperature. Within five minutes the thermostat should make a soft click sound. This sound usually indicates the thermostat is operating properly. If the thermostat does not click, try the reset operation listed above. If the thermostat does not click after being reset, contact your heating and cooling service person or place of purchase for a replacement. If the thermostat clicks, contact the furnace manufacturer or a service person to verify the heating system is operating correctly.
No Cool	1. System Switch not set to Cool. 2. Loose connection to thermostat or system. 3. Cooling System requires service or thermostat requires replacement	Set System Switch to Cool and lower setpoint below room temperature. Verify thermostat and system wires are securely attached. Same procedures as diagnostic for No Heat condition except set the thermostat to Cool and lower the setpoint below the room temperature. There may be up to a five minute delay before the thermostat clicks in Cooling if the compressor lock-out option is selected in the configuration menu (Item 6).
Heat, Cool or Fan Runs Constantly	1. Possible short in wiring. 2. Possible short in thermostat. 3. Possible short in Heat/Cool/Fan system. 4. Fan Switch set to Fan On.	Check each wire connection to verify they are not shorted or touching together. No bare wire should stick out from under terminal screws. Try resetting the thermostat as described below. If the condition persists, the manufacturer of your system or service person can instruct you on how to test the Heat/Cool/ system for correct operation. If the system operates correctly, replace the thermostat.
Furnace Cycles Too Fast or Too Slow (narrow or wide temperature swing)	1. The location of the thermostat and/or the size of the Heating System may be influencing the cycle rate.	Item 4 in the Configuration Menu is the adjustment that controls the cycle rate. If an acceptable cycle rate is not achieved using the FA (Fast) or SL (Slow) adjustment, contact a local service person for additional suggestions.
Cooling Cycles Too Fast or Too Slow (narrow or wide temperature swing)	1. The location of the thermostat and/or size of the Cooling System may be influencing the cycle rate.	The cycle rate for cooling is fixed and cannot be adjusted. Contact a local service person for suggestions.
Thermostat Setting and Thermometer Disagree	1. Thermostat thermometer setting requires adjustment.	The thermometer can be adjusted +/-4 degrees as listed in item 8 of the Configuration Menu. No other adjustment is possible.
Clock Loses or Gains Time	1. Loss of power to thermostat.	The thermostat will maintain its program in memory even with no power but the clock time will be incorrect when power is restored. See No Heat/No Cool/No Fan (common problems) above for items to check in the system.
Heat or Cool Starts Early	1. EMR activated.	See Configuration Menu (Item 3).
Thermostat Does Not Follow Program	1. AM or PM set incorrectly in program. 2. AM or PM set incorrectly on the clock. 3. Voltage spike or static discharge. 4. See "Heat or Cool Starts Early" above.	Check current clock and program settings including the AM or PM designations for each time period. If a voltage spike or static discharge occurs use the Reset Operation listed above.
Blank Display and/or Keypad Not Responding	1. Voltage Spike or Static Discharge.	If a voltage spike or static discharge occurs use the Reset Operation listed above.
Thermostat has HP/SS switch and Configuration Menu has selection for HP or SS	1. Earlier version of thermostat model.	If switch is present, it must be in HP position for proper fan operation. If selection appears in Configuration Menu, it must be set for HP.

Figure 39 Example of an electronic thermostat troubleshooting guide.

Sophisticated Front Ends

With the rise of variable refrigerant volume (VRV) systems in the industry, system and temperature controls are becoming increasingly sophisticated, but also more flexible and offering comprehensive system management. The small electronic thermostat shown here can control as many as 16 VRV units alone and offers full programmability, levels of security, and malfunction monitoring and alerts.

The tablet-style panel shown can be wall-mounted and controls hundreds of VRV units across a network, plus 10 outdoor units, through the graphical interface. It is classified as an integrated building management system (Intelligent Touch Manager™) that can control not only the VRV and ventilation equipment, but other equipment, sensors, alarms, and pumps. The device is also web-enabled and handles system-related e-mail.

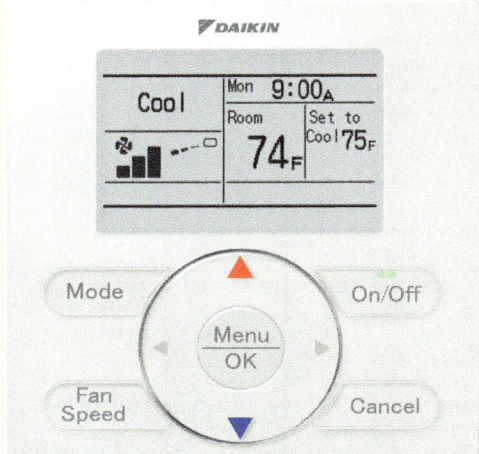

Figure Credit: Daikin Industries, Ltd.

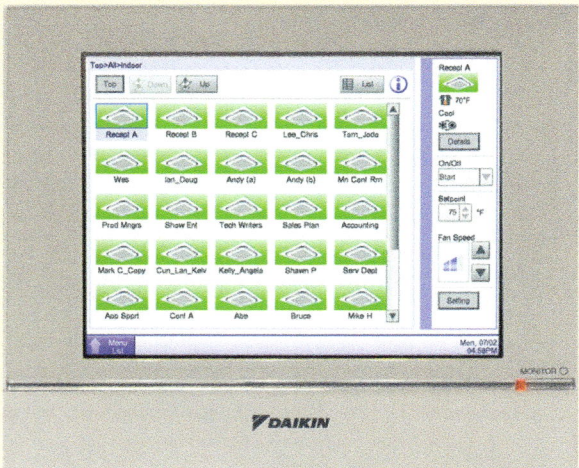

Figure Credit: Daikin Industries, Ltd.

Additional Resources

Refrigeration and Air Conditioning: An Introduction to HVAC/R, Larry Jeffus. Current edition. New York, NY: Pearson.

2.0.0 Section Review

1. The common deadband that prevents heating and cooling setpoints from being too close together is _____.

    a. 1°F (0.6°C)
    b. 2°F (1.1°C)
    c. 3°F (1.6°C)
    d. 5°F (2.8°C)

2. An open thermistor would be indicated by an ohmmeter reading of _____.

    a. infinity
    b. an abnormally high resistance
    c. an abnormally low resistance
    d. 10 ohms, consistently

3. Before handling an electronic programmable thermostat, always _____.

    a. read the instructions
    b. ground the thermostat
    c. install the batteries
    d. ground yourself

4. To ensure correct system and thermostat operation, the control circuit voltage should not be lower than _____.

    a. 21V
    b. 24V
    c. 200V
    d. 230V

SECTION THREE

3.0.0 TROUBLESHOOTING CIRCUITS AND LOAD COMPONENTS

Objective

Explain how to troubleshoot common control circuits and load components.
a. Identify basic safety practices related to troubleshooting HVACR power and control circuits.
b. Describe the operating sequence of simple heating and cooling systems.
c. Explain how to approach HVACR problems and prepare for troubleshooting.
d. Explain how to test line-voltage power sources.
e. Explain how to troubleshoot control circuits and low-voltage power sources.
f. Explain how to troubleshoot both resistive and inductive loads.
g. Explain how to troubleshoot various hydronic control-system components.

Performance Tasks

2. Interpret control circuit diagrams.
3. Perform electrical tests and/or troubleshooting procedures on the following:
   - Single- and three-phase power sources
   - Fuses and circuit breakers
   - Resistive loads
   - Relays and/or contactors
   - Motor windings
   - Start and run capacitors
   - Start relays and thermistors

Trade Terms

Fault isolation diagrams: Troubleshooting aids usually contained in the manufacturer's installation, startup, and service instructions for a particular product. Fault isolation diagrams are also called troubleshooting trees. Normally, fault isolation diagrams begin with a failure symptom and then guide the technician through a logical decision-action process to isolate the cause of failure.

Label diagrams: Troubleshooting aids usually placed in a convenient location inside the equipment. They normally depict a wiring diagram, a component arrangement diagram, a legend, and notes pertaining to the equipment.

Control circuits can be relatively simple, yet they are sometimes difficult for the new technician to understand. The function of most control circuits is simply to control the start and stop of various loads. To understand a basic control circuit, knowing when the load should be energized and de-energized is important. This is part of understanding a system's sequence of operation.

Troubleshooting HVACR systems covers a wide range of electrical and mechanical problems. Obviously, not all problems fit easily into these two categories. For example, a loose or corroded wire may cause a compressor in a cooling system to cycle on and off intermittently. Although the problem is electrical, it creates a mechanical refrigeration problem. Another example is a compressor that fails from shorted windings because of the acids formed during a previous burnout. This is an example of a mechanical refrigeration problem that leads to an electrical failure. Determining why a failure or breakdown occurred is often as important to success as determining what failed.

The methods used to troubleshoot electrical problems and components are very similar regardless of the type of HVACR equipment being serviced. This section describes those procedures that are common to troubleshooting electrical problems in most systems.

The sequence of operation determines the types of devices used in the control circuits of each HVACR system. When sitting down to design a control circuit, the first thing an engineer must know is the required sequence of operation—for example, what load is supposed to be energized and when it should occur. Then a control circuit sequence can be designed. Before reviewing operational sequences and troubleshooting, it is essential to be reminded of your responsibility to electrical safety.

3.1.0 Safety Practices

Practically all electric shocks are due to human error, rather than equipment failure. Nearly all deaths due to electric shock are due to the worker's failure to observe safety precautions, failure to repair equipment with electrical defects, or failure to remedy all defects found by tests and inspections. These same issues can also lead to damage to both electrical and electronic systems and components.

3.1.1 Electrical Safety Precautions

The following is a list of recommended safety practices:

- Always turn off power at the main disconnect before working on an electrical circuit or device. Lock and tag the power switch.
- Use a voltmeter to verify that the power to the unit is disabled. Remember that even though the power may be switched off, there is still potential at the input side of the shutoff switch.
- Never switch equipment on or off while standing in or touching a wet surface or area.
- Notify the local utility whenever a power line is touching the ground.
- If it is necessary to work on live electrical wiring, try to use one hand only. If a shock is received while using only one hand, current will probably flow through the hand and arm, then down the side and through the feet to the ground. If a shock is conducted through both hands and arms, the electrical path would be through the heart and lungs and could be fatal.
- Use protective equipment such as rubber gloves and insulated boots.
- Use tools with dielectric insulation.
- Remove metal jewelry such as rings and watches.

The *National Electrical Code® (NEC®)*, when used together with the electrical code for your local area, provides the minimum requirements for the installation of electrical systems. Always use the latest edition of the code as your on-the-job reference. It specifies the minimum provisions necessary for protecting people and property from electrical hazards. In some areas, different editions of the code may be in use, so be sure to use the edition specified by your employer.

3.1.2 Precautions When Working with Solid-State Controls

Solid-state equipment and/or components can be damaged if improperly handled or installed.

> **WARNING!** Solid-state equipment may contain components that produce potentially hazardous leakage current in the Off state. Power should be turned off and the device disconnected from the power source before working on the circuit or load.

> **CAUTION:** Some solid-state devices can be damaged by electrostatic charges. These devices should be handled in the manner specified by the manufacturer. Always ground yourself to discharge any stored static charge before handling these devices.

The following guidelines should be followed when handling, installing, or otherwise working with solid-state equipment or components:

- Microprocessors and other integrated circuit chips are very sensitive to, and can be damaged by, static electricity from sources such as lightning or people. When handling circuit boards, avoid touching the components, printed circuit, and connector pins. Always ground yourself before touching a board. Disposable grounding wrist straps are sometimes supplied with boards containing electrostatic-sensitive components. If a wrist strap is not supplied with the equipment, a wrist strap grounding system should be used. It allows handling of static-sensitive components without fear of electrostatic discharge. Store unused boards inside their special metallized shielded storage bags or a conductive tote box (*Figure 40*).
- Solid-state devices can be damaged by the application of reverse polarity or incorrect phase sequence. Input power and control signals must be applied with the polarity and phase sequence specified by the manufacturer.
- The main source of heat in many solid-state systems is the energy dissipated in the output power devices. Ensure that the device is operated within the maximum and minimum ambient operating temperatures specified by the manufacturer. Follow the manufacturer's recommendations pertaining to the selection of enclosures and ventilation requirements. For existing equipment, make sure that ventilation passages are kept clean and open.
- Moisture, corrosive gases, dust, and other contaminants can all have adverse effects on a system that is not adequately protected against atmospheric contaminants. If these contaminants collect on printed circuit boards, bridging between the conductors may result in a malfunction of the circuit. This could lead to noisy, erratic control operation or a permanent malfunction. A thick coating of dust could also prevent adequate cooling of the board or heat sink, causing a malfunction.

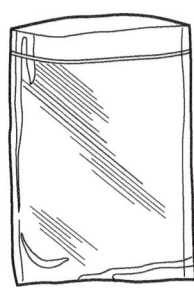 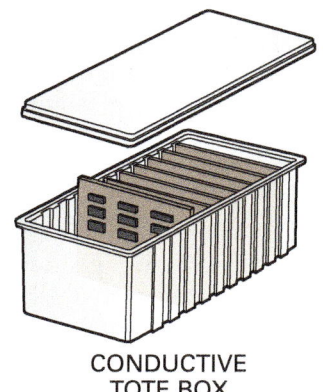

Figure 40 ESD protection bag and tote box.

- Excessive shock or vibration may cause damage to solid-state equipment. Follow all manufacturers' instructions for mounting.
- Electrical noise can affect the operation of solid-state controls. Sources of electrical noise include machines with large, fast-changing voltages or currents when they are energized or de-energized, such as motor starters, welding equipment, SCR-type adjustable speed devices, and other inductive devices. The following are basic steps to help minimize electrical noise:
  – Maintain sufficient physical separation between electrical noise sources and sensitive equipment to ensure that the noise will not cause malfunctions or unintended actuation of the control.
  – Maintain a physical separation between sensitive signal wires and electrical power and control conductors. This separation can be accomplished by using conduit, wiring trays, or methods recommended by the manufacturer.
  – Use twisted-pair wiring in critical signal circuits and noise producing circuits to minimize magnetic interference.
  – Use shielded wire to reduce the magnitude of the noise coupled into the low-level signal circuit by electrostatic or magnetic coupling.
  – Follow provisions of the *NEC®* with respect to grounding. Additional grounding precautions may be required to minimize electrical noise. These precautions generally deal with ground loop currents arising from multiple ground paths. The manufacturer's grounding recommendations must also be followed.

Ambient Temperature Surrounding Solid-State Devices

Do not overlook other sources of heat in or around an enclosure containing solid-state devices that might raise the ambient temperature to undesirable levels. Look for and avoid heat sources such as power supplies, transformers, radiated heat, adjacent furnaces, and sunlight.

3.1.3 OSHA Lockout/Tagout Rule

In addition to many other safety rules, 29*CFR* 1926.417, *Lockout and Tagging of Circuits*, was initially released years ago. This rule covers the specific procedure to be followed for the "servicing and maintenance of machines and equipment in which the unexpected energization or startup of the machines or equipment, or releases of stored energy, could cause injury to employees. This standard establishes minimum performance requirements for the control of such hazardous energy."

The purpose of the OSHA procedure is to make sure that machinery is isolated from all potentially hazardous energy, and tagged and locked out before employees perform any servicing or maintenance activities where the unexpected energization, startup, or release of stored energy could cause injury.

WARNING! The OSHA procedure provides only the minimum requirements for the lockout/tagout procedure. Employees should consult the lockout/tagout procedures in their local area as well as their company's own policies and procedures. Remember, lives depend on this procedure. It is critical that the correct procedures for a job site be used.

3.1.4 Lockout/Tagout Procedure

To prepare for a lockout/tagout, make a survey to locate and identify all isolating devices (*Figure 41*). Be certain which switch(es), valve(s), or other energy-isolating devices apply to the equipment to be locked and tagged. Remember that more than one energy source may be involved.

Figure 41 Condensing unit with wall-mounted disconnect switch.

The following procedure outlines the general steps for lockout/tagout (note that this procedure is provided only as an example). Each employer will designate who is qualified to conduct the procedure. Not every employee has the proper training to do so.

Step 1 Notify all affected employees of the lockout/tagout and why it is necessary. The authorized employee will know the type of energy that the machine or equipment uses and the associated hazards.

Step 2 If the machine or equipment is running, shut it down using the normal procedure. For example, press the stop button or open the toggle switch.

Step 3 Operate the switch, valve, or other energy-isolating device(s) so that the equipment is isolated from its energy source(s). Stored energy must be dissipated or restrained by repositioning, blocking, or bleeding down. Examples of stored energy are springs, elevated machine members, rotating flywheels, hydraulic systems, and air, gas, steam, or water pressure.

Step 4 Lock and tag the energy isolating devices with assigned individual lock(s) and tag(s).

Step 5 After making sure that no personnel are exposed, operate the start button or other normal operating controls to make certain the equipment will not operate. This will confirm that the energy sources have been disconnected. The equipment is now locked and tagged.

All equipment must be locked and tagged to protect against accidental operation when such operation could cause injury to personnel. Never try to operate any switch, valve, or energy-isolating device when it is locked and tagged.

> **CAUTION**
>
> Return operating controls to their neutral or Off position after testing.

3.1.5 Restoring Normal Operation

After service and/or maintenance is complete, the equipment is ready for normal operation. Use the following procedure to restore the machines or equipment to their normal operating condition.

Step 1 Check the area around the machines or equipment to make sure that all personnel are at a safe distance.

Step 2 Remove all tools from the machine or equipment and reinstall any guards.

Step 3 Again making sure that all personnel are in the clear, remove all lockout/tagout devices.

Step 4 Operate the energy isolating devices to restore energy to the machine or equipment.

3.2.0 Basic Operating Sequences

An automatic cooling circuit can be simple in nature, and it is a good place to begin an analysis of control circuits. *Figure 42* shows a very basic cooling system control circuit.

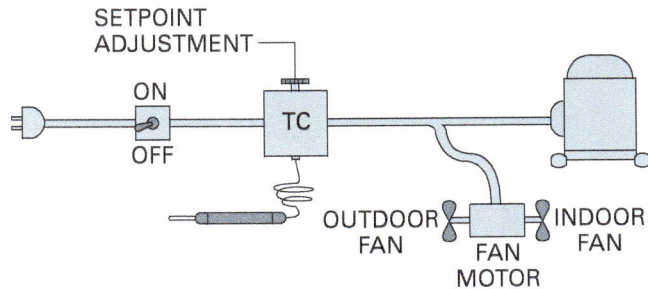

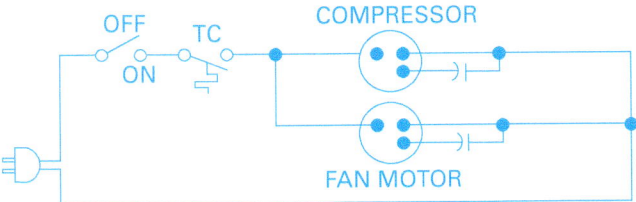

Figure 42 Basic cooling-system control circuit.

Power from the power plug is applied to the On-Off switch. When the switch is in the On position, power is passed to the cooling thermostat (TC). When the thermostat calls for cooling and closes, power is passed to both the compressor and fan motor, causing both motors to run. When the thermostat is satisfied and opens, the motors lose power and shut off. This describes the sequence of operation for this circuit.

Figure 43 shows a control circuit that is more typical of a basic comfort-cooling system. Some additional features have been added to this circuit. A transformer has been added because 24VAC is required for the control circuit. This makes the control circuit components and wiring less expensive, and easier to install and service. An indoor fan motor (IFM) has also been added to the circuit. Because the control circuit is now operating at 24VAC, the control relays (C and IFR) are added to provide control for the motors.

The following is the sequence of operation of the circuit shown in *Figure 43*.

- Line voltage is applied at terminals L1 and L2, making it present at the contactors (C), the indoor fan relay (IFR), and the primary winding of the system transformer at all times. With the transformer energized, the 24VAC output from its secondary is applied through one control circuit path to the On terminal of the Fan switch and by a second control circuit path to the thermostat On-Off switch.

- When the Fan switch is in the On position, power is applied through the first control circuit path to the coil of the indoor fan relay (IFR), energizing the relay. This causes the related IFR contacts to close, applying line voltage to the indoor fan motor (IFM) and causing it to run continuously. This mode of operation allows the occupant to use the indoor fan for ventilation without operating the compressor for cooling. Some occupants will allow the fan to operate continuously even as the compressor cycles on and off to cool the structure.

- With the On-Off switch closed, the thermostat calling for cooling, and the Fan switch placed in the Auto or On position, both the contactor coil (C) and the indoor fan motor coil (IFR) are energized. This causes the compressor, outdoor fan, and indoor fan motors to be turned on simultaneously as required for the cooling mode of operation. Most users with a cooling-only system choose to allow the indoor fan to cycle with the compressor.

> **NOTE**
>
> The negative effect of operating the indoor fan constantly while the compressor cycles to maintain temperature is related to humidity. As a cooling cycle is completed and the compressor shuts down, the evaporator and drain pan beneath are wet from condensation. A continuously running blower allows this moisture to be quickly re-evaporated by the air stream and returned to the conditioned space.

Regardless of the complexity, the basic control arrangement shown in *Figure 41* is at the heart of the control circuits used with cooling equipment. Any additional circuits will represent special features used to improve equipment safety or operating efficiency.

3.2.1 Heating and Cooling System Sequence

While cooling-only and heating-only systems exist, most HVACR systems provide both heating and cooling. *Figure 44* is the diagram of a combined heating/cooling unit. The circuit is more complex, but at its heart is the simple control circuit previously described. The additional features are not that complicated. Heating controls have been added near the bottom of the diagram, along with a heating-cooling thermostat. The cooling control has more extras, such as a compressor short-cycle protection circuit, a crankcase heater, and a current-sensitive overload device.

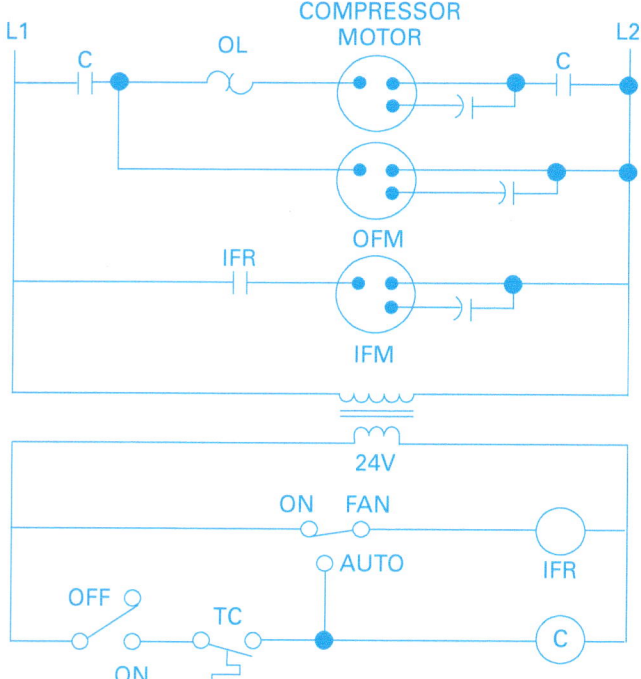

Figure 43 Typical cooling-system control circuit.

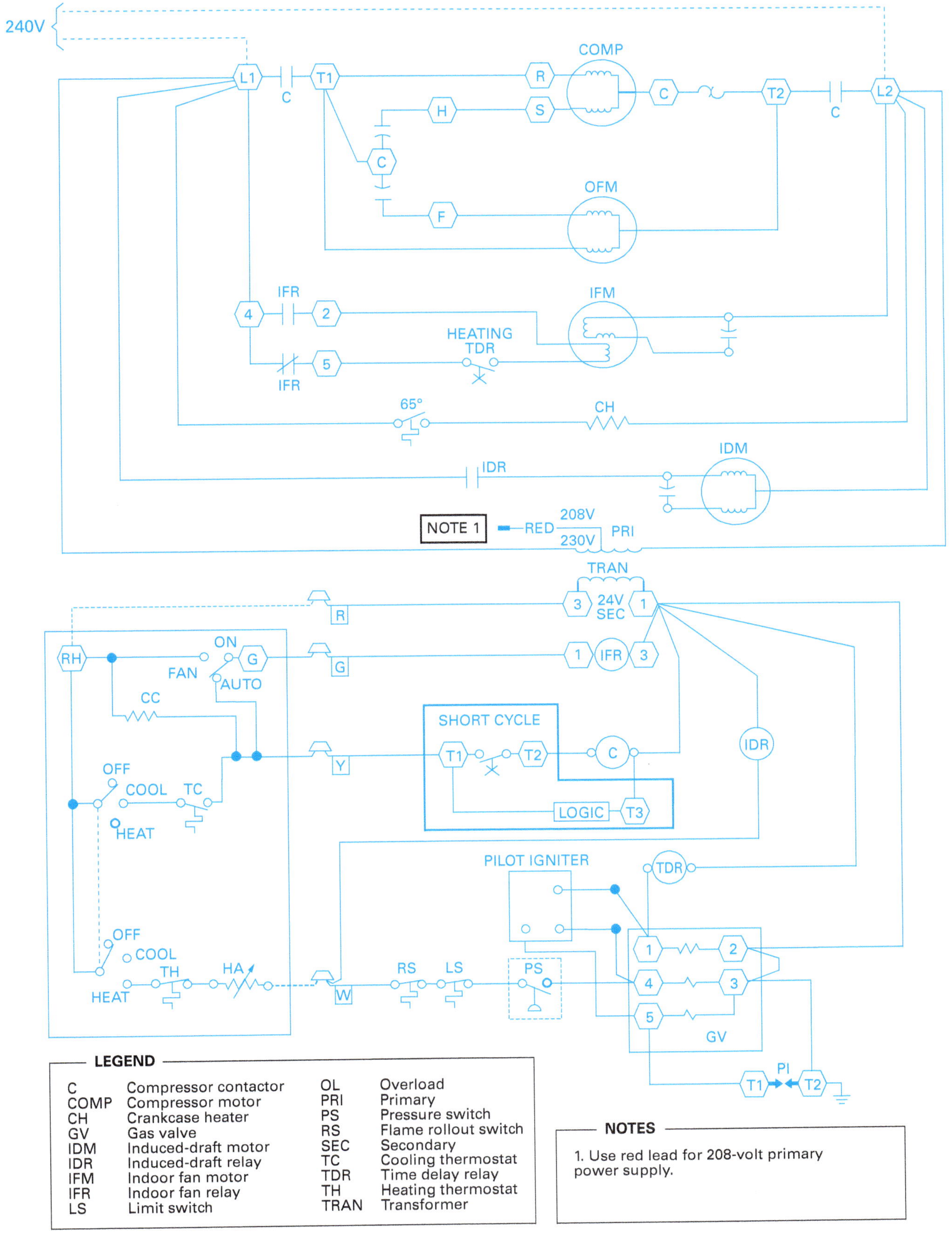

Figure 44 Circuit diagram of a cooling/gas-heating system.

 03314 Control Circuit and Motor Troubleshooting

The heating/cooling unit shown in *Figure 44* is typical of a product found in both residential and light commercial applications. Some of the features of the heating/cooling unit include:

- A common room thermostat provides temperature control in both operating modes.
- A two-speed indoor fan that runs on high speed for cooling and low speed for heating. In heating, the time-delay relay controls the indoor fan.
- Operation of the inducer fan must be proven before the gas valve is turned on. The inducer pressure switch (PS) closes when the induced-draft motor is running at the required speed. If the induced-draft fan stops, the pressure switch will open, disabling the gas supply. The induced-draft motor (IDM) is energized by the induced-draft relay (IDR) as soon as the thermostat closes.
- The heating section has two additional safety devices in series with the gas valve. A flame rollout switch (RS) will open if burner flames escape from the burner. This represents an unsafe condition. The limit switch (LS) is a heat-sensitive switch that disables burner operation if excessive heat builds up around the heat exchanger.

Refer to the circuit diagram in *Figure 44* for the following operating sequences.

Cooling Sequence—Assume that line-voltage and control circuit voltage is adequate and available to the unit. To start cooling operation in a heat/cool unit, set the room thermostat mode selector switch to Cool. Placing the mode selector switch in this position locks out all heating system functions. Set the fan function switch to Auto so that the indoor fan cycles on and off with the compressor.

As room temperature rises, thermostat contacts TC close, supplying 24VAC power to the coil of the indoor fan relay (IFR) and to the logic circuit of the short-cycle timer. The indoor fan relay energizes instantly, closing IFR contacts across terminals 2 and 4. This provides line-voltage power to the indoor fan motor (IFM), and it starts and runs on high speed. At the same time, IFR contacts across terminals 4 and 5 open. This isolates the low-speed windings of the indoor fan motor that are used for the heating mode, preventing any possibility of the motor having both windings energized.

Once the logic in the short-cycle timer times out (2 to 5 minutes), the contacts across terminals T1 and T2 close to energize the contactor (C). In the line-voltage side of the circuit, contactor contacts L1-T1 and L2-T2 close to supply power to the compressor (COMP) and outdoor fan motor (OFM). The outdoor fan motor is usually tied directly to compressor operation, except when one or more motors are staged on by pressure in large units.

The unit will continue to run until the room thermostat is satisfied. When TC contacts open, the contactor and indoor fan relay both de-energize, causing all motors to shut down.

Heating Sequence—Assume that line-voltage and control circuit voltage is adequate and available to the unit. To start heating operation, set the room thermostat mode selector switch to Heat. Placing the mode selector switch in this position locks out all cooling system functions. Set the fan function switch to Auto so that the indoor fan cycles on and off with the burners. Assume that no unsafe conditions exist and the flame rollout switch (RS) and limit switch (LS) are both closed.

As room temperature falls, thermostat contacts TH close, supplying 24VAC power to the coil of the induced-draft relay (IDR). In the line-voltage side of the circuit, the normally-open contacts of the induced-draft relay close, applying power to the induced-draft motor (IDM).

When the induced-draft motor comes up to speed, the pressure switch (PS) closes to provide power to the gas valve and pilot igniter. The gas valve feeds gas to the pilot assembly where a high-voltage arc ignites it. Once the pilot igniter module verifies that the pilot is sufficient to light the main burners, the gas valve opens and the main burners light.

When the gas valve is energized, the time-delay relay (TDR) is energized to begin timing. In the line-voltage side, TDR contacts are in series with normally closed contacts of the indoor fan relay (IFR) and in series with the indoor fan motor (IFM). The purpose of the time delay is to allow the air surrounding the heat exchanger time to warm up so that cold air is not blown into the structure. Once the delay times out, the indoor fan motor starts and runs on low speed. The normally-open contacts of the indoor fan relay (IFR) isolate the high-speed windings of the indoor fan motor.

The unit will continue to run until the room thermostat is satisfied. When TH contacts open, the gas valve and induced-draft relay are de-energized, shutting off the burners. The time-delay relay is also de-energized. A time-delay occurs after the relay is de-energized. This allows the indoor fan motor to continue to run after the burner shuts off to recover residual heat left in the heat exchanger.

3.3.0 An Organized Approach to Electrical Troubleshooting

Troubleshooting can be defined as a procedure through which a technician locates the source of a problem. Once the problem is identified, the technician then makes the repairs and/or adjustments to correct the cause of a problem so that it will not recur.

Troubleshooting can be divided into the five basic elements, listed here. The majority of problems can usually be identified quickly if a systematic approach is used. This includes the following:

- Customer interviews
- Physical examination of the system
- Basic system analysis
- Use of manufacturer's troubleshooting aids
- Fault isolation in equipment problem area

3.3.1 Customer Interviews

The troubleshooting procedure should begin with the technician learning all that can be learned about the customer's complaint by talking to the customer and the service dispatcher. Talking with the customer prior to working on the equipment is always recommended because it can provide valuable information on equipment operation that can aid in the troubleshooting process. It can also identify the source of a problem that is not related to the HVACR equipment, thereby eliminating unnecessary equipment maintenance.

> **Think About It**
>
> **Analyzing a Circuit Diagram**
>
> Use the diagram in *Figure 44* to determine the answers to these questions.
>
> 1. Does the induced-draft motor turn off if the rollout switch opens?
> 2. If you connect a voltmeter between L2 and terminal 4 of the IFR while the compressor contactor is de-energized, what voltage, if any, would the meter read?
> 3. Is the fan relay (IFR) energized or de-energized in the heating mode?

The first evidence of trouble with the HVACR system is often a complaint from an individual who is too cold, too hot, or may be bothered by drafts. In many cases, the problem behind such a complaint may not be an equipment or control system malfunction, but a personal comfort problem. If the system appears to be operating correctly but individuals are complaining about comfort problems, the technician should check for one or more of the following conditions before assuming the HVACR equipment is malfunctioning:

- *Air distribution and circulation problems*—Persons outside of the immediate area that is controlled by a thermostat may feel too hot or cold. The thermostat senses only the temperature at its location. Temperature levels in all the other areas controlled by the same thermostat are subject to variation that can be caused by poor air distribution and/or room air circulation.
- *False heat loads* – Direct sunlight on the thermostat or heat from lamps, appliances, and pipes can cause overcooling of a zone. Direct sunlight or artificial sources of radiant heat can also cause heating discomfort.
- *Covered grilles and diffusers* – Occupants frequently cover part or all of a discharge grille face, causing improper heating or cooling.
- *Occupant locations* – Occupants located adjacent to outside walls or windows may be subject to air infiltration or discomfort from a warm or cool wall.
- *Insufficient conditioned air supply* – May be caused by poor air distribution design or fan speed, dirty filters, or lack of the proper amount or size of return air outlets.
- *Overcrowding* – Overheating will result if more people or mechanical equipment occupy a conditioned area than the space was designed to hold.
- *System size* – Extreme weather conditions may exceed the capacity of the heating or cooling equipment.
- *Drafts* – Forced-air systems rely on the movement of air to deliver the desired conditioning. To many people, even a slight air motion may be uncomfortable. This problem may be alleviated by moving the occupant's work station or redirecting the airflow.
- *Stale air* – A stuffy or smoky atmosphere usually results from an insufficient fresh air supply, air that is too humid, or air that has inadequate exhaust. Stale air can also be caused by overcrowding or low air circulation.

03314 Control Circuit and Motor Troubleshooting

3.3.2 Physical Examination of the System

Many problems can be identified by simply using senses such as sight, touch, and smell to check the system. This can be done with power off or power on.

- Look for evidence of leaks and physical damage. Oil stains on coils or tubing may indicate a refrigerant leak.
- Look for dirt accumulation on filters and coils that can affect system operation.
- Listen for unusual sounds and feel for vibrations that could indicate a problem.
- Check for odors, especially for the odor of the heating fuel or burned components.

3.3.3 Basic System Analysis

The proper diagnosis of a problem requires knowledge of proper system operation. If a unit is unfamiliar, study the manufacturer's service literature to become familiar with the equipment's modes and sequence of operation.

The second part of the diagnosis is to find out what the unit is actually doing, and what symptoms are exhibited. Finding out what the unit is doing is accomplished both by carefully listening to the customer's complaints and by analyzing the operation of the unit. As applicable, this means making electrical, temperature, pressure, and/or airflow measurements at key points in the system. The set of measured values can then be compared with a set of typical readings for a

Pneumatic Controls

Pneumatic control systems use compressed air to supply energy for the operation of valves, motors, relays, and other pneumatic control components. They are used primarily in large commercial and industrial systems. Pneumatic control circuits consist of air piping, valves, orifices, and similar mechanical devices. Thermostats control the pressure in the air lines to downstream controls. Using pneumatic-electric switches, air can be used to control electrical devices.

Pneumatic control systems were once very popular in commercial office buildings, but are expensive and more challenging to service than electronics, which require very little periodic maintenance. However, they are still used today in older systems and in industrial control where it has some advantages. Shown here are pneumatic damper and valve actuators.

properly operating system as previously recorded on system operating logs or in the manufacturer's service literature. This process can often quickly pinpoint the system problem.

3.3.4 Use of Manufacturers' Troubleshooting Aids

To aid in the isolation of faults, many manufacturers provide troubleshooting information marked on the equipment or contained in the service instructions for a particular product. This information typically includes the following:

- Label diagrams
- Troubleshooting tables
- Fault isolation diagrams
- Diagnostic equipment and tests

Label diagrams (*Figure 45*) are usually placed in a convenient location inside the equipment, typically on the inside of a control circuit access panel. They normally show a component arrangement diagram, wiring diagram, legend, and notes pertaining to the equipment.

The component arrangement diagram shows where the components are physically located in the unit. It is useful because it helps locate and identify the components shown on the wiring diagram. The legend identifies the meanings of the symbols and abbreviations used on the label diagram.

The wiring diagram, sometimes called a schematic, provides a picture of what the unit does electrically and shows the actual external and internal wiring of the unit. Many label diagrams also contain a ladder diagram (*Figure 46*). Ladder diagrams do not usually show wire color and physical connection information. This makes it

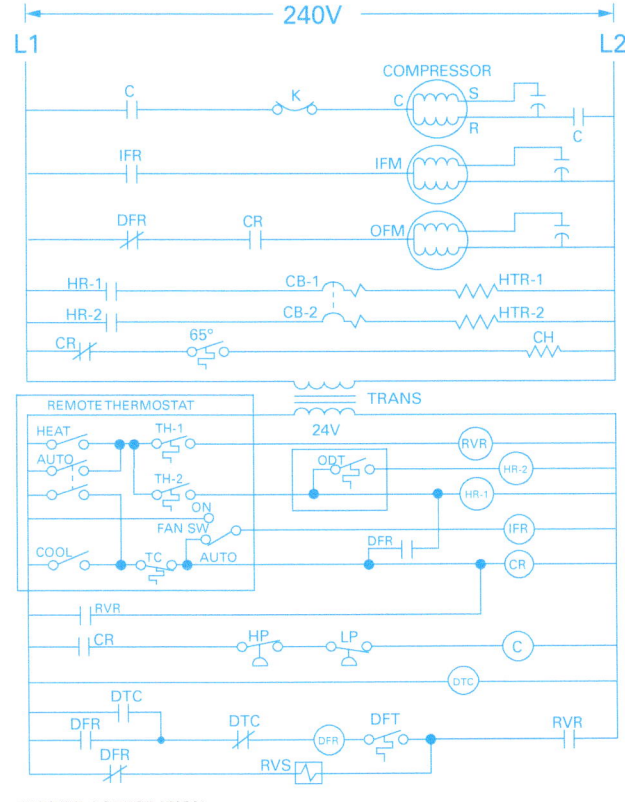

Figure 46 Ladder diagram.

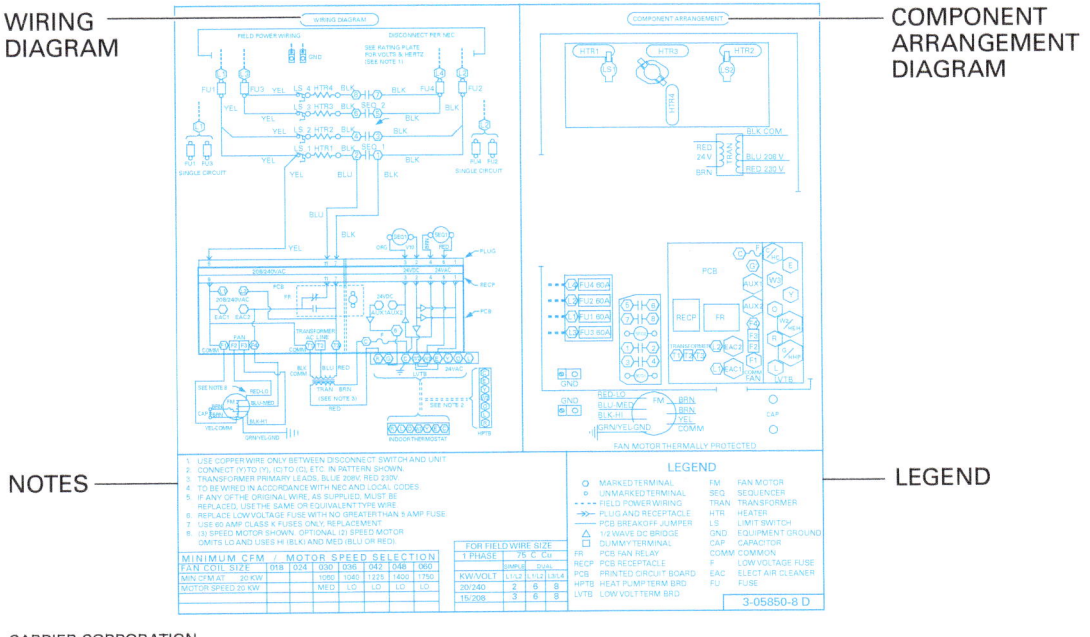

Figure 45 Typical label diagram.

03314 ♦ Control Circuit and Motor Troubleshooting · · · · · · · · Module Two 41 more useful by focusing on the functional, not the physical, aspects of the equipment. Wiring and ladder diagrams are the primary troubleshooting aids for isolating electrical problems.

Troubleshooting tables and/or fault isolation diagrams are usually contained in the manufacturer's installation, startup, and service instructions for a particular product. As shown in *Figure 47*, troubleshooting tables are intended to guide technicians to a corrective action based on observations of system operation. Fault-isolation diagrams (*Figure 48*), also called troubleshooting trees, normally start with a failure-symptom observation and take technicians through a logical decision-action process to isolate the failure.

Many manufacturers incorporate electronically controlled or semi-automatic testing features in their equipment to help isolate malfunctions. Depending on the equipment, these built-in diagnostic devices can be simple or complex. Some units contain microprocessor controllers that can run a complete check of all system functions, and then report back the results by means of a system numeric display or flashing LED display. Normally, these built-in test functions isolate a fault to a functional problem area. For example, if a test indicates a compressor failure, this means that the failure has been isolated to the compressor and its related control circuits and wiring. The technician must perform additional troubleshooting within the problem area to find out exactly where the fault is located.

In addition to built-in diagnostic equipment, many manufacturers have developed a series of stand-alone electronic module testers that are available to troubleshoot the different control modules commonly used in the manufacturer's product line. The module tester is usually plugged into the control module in the equipment under test.

Troubleshooting involves the testing of each module control circuit using a sequential

Malfunction	Probable Cause	Corrective Action
Compressor motor and condenser motor will not start, but fan/coil unit (blower motor) operates normally	Check the thermostat system switch to ascertain that it is set to COOL.	Make necessary adjustments to settings.
	Check the thermostat to make sure that it is set below room temperature.	Make necessary adjustments.
	Check the thermostat to see if it is level. Most thermostats must be mounted level; any deviation will ruin their calibration.	Remove cover plate, place a spirit level on top of the thermostat base, loosen the mounting screws, and adjust the base until it is level; then tighten the mounting screws.
	Check all low-voltage connections for tightness.	Tighten.
	Make a low-voltage check with a voltmeter on the condensate float switch; the condensate may not be draining.	The float switch is normally found in the fan/coil unit. Repair or replace.
	Low air flow could be causing the trouble, so check the air filters.	Clean or replace.
	Make a low-voltage check of the antifrost control.	Replace if defective.
	Check all duct connections to the fan-coil unit.	Repair if necessary.
Compressor, condenser, and fan/coil unit motors will not start		Adjust as necessary.

Figure 47 Typical troubleshooting table.

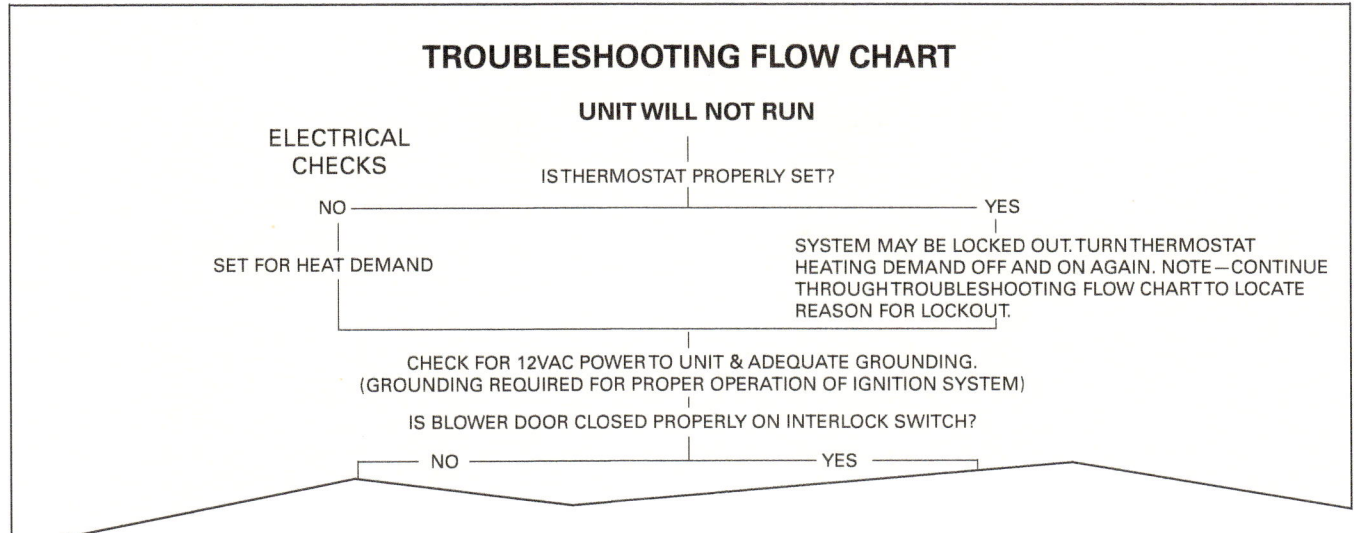

Figure 48 Typical fault-isolation diagram.

troubleshooting process that is performed per the manufacturer's instructions provided with the module tester.

3.3.5 Fault Isolation in the Equipment Problem Area

Once troubleshooting aids have isolated a problem to a functional equipment area, it may be necessary to make additional measurements and use a step-by-step process of elimination to isolate the specific cause of the problem. This is often the case for control and/or electrical issues.

3.4.0 Testing High-Voltage Power Sources

Input power distribution circuits serve as the power source for the entire unit. They operate at either single-phase or three-phase line voltages, and act to distribute the input power to the various loads in the unit. Power circuits usually consist of the field-installed power wiring from the main electrical service to a disconnect switch located near the unit, and from the disconnect switch to the unit. The input power and distribution circuits include protective devices such as fuses and/or circuit breakers.

All HVACR equipment is designed to operate within a specific range of system voltages including a safety factor, typically ±10%. This safety factor is added to compensate for temporary supply voltage fluctuations that might occur. Continuous operation of HVACR equipment outside the intended voltage range can damage motors and other loads. When the contactor closes in a cooling unit and applies power to inductive loads such as the compressor and outdoor fan motor, the voltage level may drop about 3% from the measured open-circuit voltage (contactor open). This is due to the increased current flow at startup. It is acceptable as long as the voltage does not drop below the manufacturer's stated minimum voltage.

3.4.1 Effects of High and Low Voltage

Too high or too low an operating voltage can cause overheating and possible failure of motors and other devices. The power supply voltage applied to the equipment should be measured and checked against the supply voltage indicated on the unit nameplate (*Figure 49*).

If low voltage exists at the equipment, the voltage should be measured at the electrical service entrance to make sure that a voltage drop does not exist in the branch circuit or feeder that supplies the HVACR equipment. If a voltage drop exists, it may be necessary to install larger conductors.

Operating voltages applied to motors in the equipment must be maintained within limits from the voltage value given on the motor nameplate. The voltage tolerances used for most HVACR motors are as follows:

- *Single-voltage rated motors* – The input supply voltage should be within ±10% of the nameplate voltage. For example, a motor with a nameplate single voltage rating of 230V should have an input voltage that ranges between 207V (–10% of 230V) and 253V (+10% of 230V).

Think About It

Wiring Diagram Exercise

Using the schematic diagram of the packaged heat pump included here, answer the following questions:

1. Where would voltmeter leads be placed to check for the correct incoming voltage to the unit?
2. Placing voltmeter leads across terminals R and X of the low-voltage terminal strip yields a reading of 24V. What does this indicate?
3. During defrost, the defrost relay (DR) is energized. During a normal defrost, voltmeter leads are placed across terminal 2 of the defrost relay and terminal X on the low-voltage terminal strip. What should the voltmeter read?
4. Assume a room thermostat is connected to the low-voltage terminal strip. Across which terminals on the low-voltage terminal strip would voltmeter leads have to be placed to confirm that a call for second-stage heat was coming from the room thermostat? What voltage should be measured?
5. During defrost, the defrost relay (DR) is energized. Knowing this, what happens to the outdoor fan motor during defrost?
6. It must be known if the demand defrost control is receiving the correct input power. Across which terminals of the demand defrost control should power be checked? What is the expected voltage reading?
7. The compressor in this unit occasionally won't start and trips the power supply circuit breaker. Compressor current draw during startup must be checked. What test instrument is required and where in the circuit would the instrument be attached or applied?
8. HC #1 and HC #2 relays control the operation of electric resistance heating elements. How do those relay coils receive power during a defrost cycle? Are both energized?
9. If the compressor fails in this heat pump, it is still possible to heat the home. How does the occupant make this happen? At what terminals on the low-voltage terminal strip would you expect to find 24VAC that proves the thermostat is calling for the electric heat to operate?
10. Since this unit energizes the defrost relay during heat pump defrost operation, what does this fact tell about the state of the reversing valve solenoid (RVS) during cooling operation? Is the RVS energized or de-energized during the cooling mode?

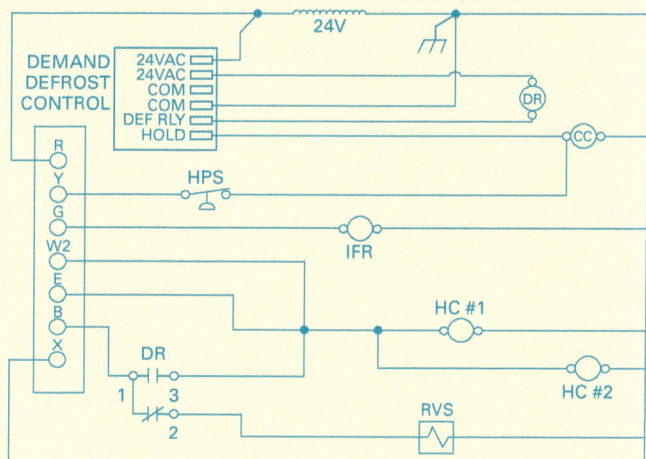

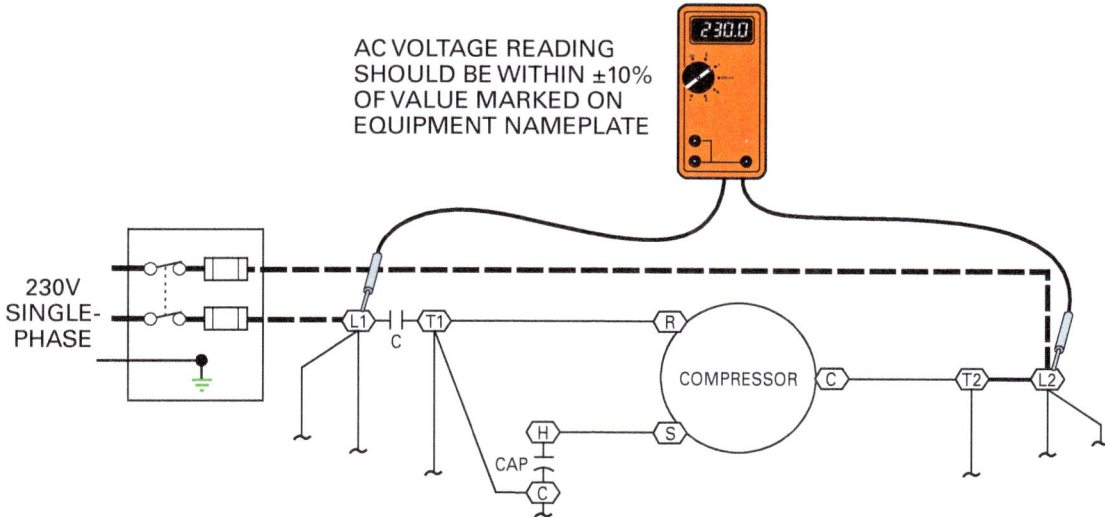

Figure 49 Single-phase voltage checks.

- *Dual-voltage rated motors* – The input supply voltage should be within ±10% of the nameplate voltage. For example, a motor with a nameplate dual voltage rating of 208/230V should have an input voltage that ranges between 187V (–10% of 208V) and 253V (+10% of 230V).

3.4.2 Voltage and Current Imbalance

Voltage imbalance becomes very important when working with three-phase equipment. A small imbalance in phase-to-phase voltage can result in a much greater current imbalance. This current imbalance increases the heat generated in the motor windings. Both current and heat can cause nuisance overload trips and may cause motor failure. For this reason, the voltage imbalance between any two legs of the supply voltage applied to a three-phase system or motor should not exceed 2%. If a voltage imbalance of more than 2% exists at the input to the HVACR equipment, correct the problem in the building or utility power distribution system before operating the equipment. *Figure 50* shows how the amount of voltage imbalance is determined in a three-phase system using the following formula:

$$\text{Percent imbalance} = \frac{\text{maximum deviation from average}}{\text{average voltage}} \times 100$$

The current imbalance in any one leg of a three-phase system should not exceed 10%. A current imbalance may occur without a voltage imbalance originating from the power source. This can occur when an electrical terminal, contact, etc. becomes loose or corroded, causing a high resistance in the leg. Since current follows the path of least resistance, the current in the other two legs will increase, causing more heat to be generated in the devices supplied by those legs. The current imbalance in a three-phase system is determined in the same way as voltage imbalance, but average current is substituted for average voltage.

> **NOTE**
> The system must be operating when voltage and current imbalance tests are performed. In the case of voltage, voltage drops occur only when the equipment is operating.

3.4.3 Fuse/Circuit Breaker Checks

Fuses and/or circuit breakers are normally the first components checked when a unit is totally inoperative. The best way to test a fuse is by measuring continuity (*Figure 51*). To check fuses, always open the unit disconnect switch, then remove the fuses using an insulated fuse puller. Test the fuses for continuity using an analog or digital multimeter (VOM/DMM). If a short (zero ohms) exists across the fuse, it is usually good. If an open (infinite resistance) exists across the fuse, it is defective. An abnormal overload condition usually causes a blown fuse, such as a short circuit within the equipment or an overloaded motor. Replacing a blown fuse without locating and correcting the cause can result in damage to the equipment.

Fuses can also be tested with the circuit energized. This is accomplished by shutting off power to the unit, then disconnecting the wires from the load side of each fuse. This eliminates the possibility of current being fed back to the meter through a short circuit within the unit. Set the multimeter to measure AC voltage on a range that is higher than the highest voltage expected. Turn on the

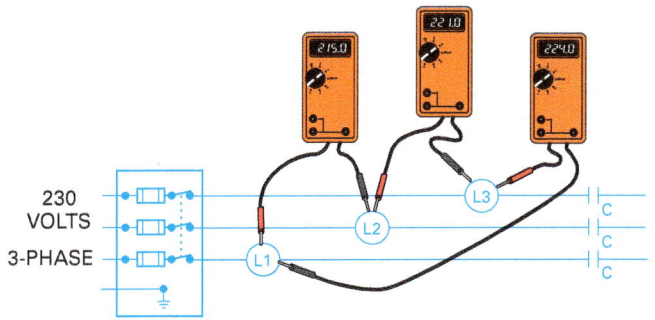
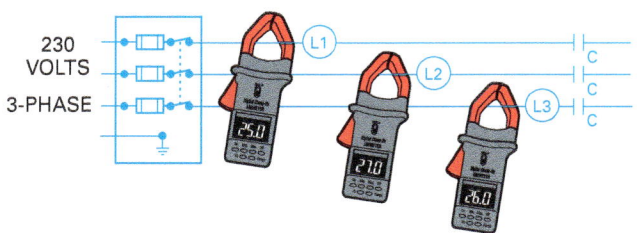

1) | PHASE | MEASURED READING |
   |-------|------------------|
   | L1 | 215V |
   | L2 | 221V |
   | L3 | 224V |

2) AVERAGE VOLTAGE = $\dfrac{215 + 221 + 224}{3}$ = 220V

3) INDIVIDUAL PHASE IMBALANCE FROM AVERAGE
   L1 TO L2 = 220 − 215 = 5V
   L2 TO L3 = 221 − 220 = 1V
   L3 TO L1 = 224 − 220 = 4V

4) 5V = MAXIMUM IMBALANCE

5) % IMBALANCE = $\dfrac{\text{MAXIMUM IMBALANCE}}{\text{AVERAGE VOLTAGE}} \times 100$

   % IMBALANCE = $\dfrac{5V}{220V} \times 100$ = 2.27% (OUT OF BALANCE)

MAXIMUM VOLTAGE IMBALANCE BETWEEN ANY TWO LEGS MUST NOT EXCEED 2%.

(A) CALCULATING VOLTAGE IMBALANCE

1) | PHASE | MEASURED READING |
   |-------|------------------|
   | L1 | 25A |
   | L2 | 27A |
   | L3 | 26A |

2) AVERAGE CURRENT = $\dfrac{25 + 27 + 26}{3}$ = 26A

3) INDIVIDUAL PHASE IMBALANCE FROM AVERAGE
   L1 TO L2 = 25 − 27 = 2A
   L2 TO L3 = 27 − 26 = 1A
   L3 TO L1 = 26 − 25 = 1A

4) 2A = MAXIMUM IMBALANCE

5) % IMBALANCE = $\dfrac{\text{MAXIMUM IMBALANCE}}{\text{AVERAGE CURRENT}} \times 100$

   % IMBALANCE = $\dfrac{2A}{26A} \times 100$ = 7.7% (IN BALANCE)

MAXIMUM CURRENT IMBALANCE BETWEEN ANY TWO LEGS MUST NOT EXCEED 10%.

(B) CALCULATING CURRENT IMBALANCE

Figure 50 Three-phase voltage and current checks.

power and place one of the multimeter test leads on the line side of a fuse. Touch the other test lead to the load side of another fuse. If voltage is measured on the load side of a fuse, the fuse is good; if not, the fuse is defective. Repeat this procedure so that all fuses are measured with one test lead on the load side and the other test lead on the input side of a different fuse. This method tests one fuse at a time. If the measurement is performed with both test leads on the load side of the fuses, and the multimeter shows no reading, it indicates that a fuse is blown, but not which one.

Before testing a circuit breaker (*Figure 52*), set the breaker to Off at the distribution panel. If required, remove the panel that covers the circuit breaker to expose the body and the wires connected to its terminals. Set the multimeter to measure AC voltage on a range that is higher than the highest voltage expected. Measure the voltage applied to the breaker input terminals:

- A to neutral or ground (single-pole breaker)
- A to B (two-pole breaker)
- A to B, B to C, and C to A (three-pole breaker)

Make sure that the breaker is closed by first setting it to the off position, then setting it to the on position. Measure the voltage at the circuit breaker output terminals:

- A1 to neutral or ground (single-pole breaker)
- A1 to B1 (two-pole breaker)
- A1 to B1, B1 to C1, and C1 to A1 (three-pole breaker)

The measured input and output voltages should be the same under a load. With no load applied, there will rarely be any difference. Defects in the breaker contacts, in the form of a voltage drop, will be revealed when a significant load is applied. If the output voltage is significantly lower than that measured at the input to the circuit breaker, visually inspect the circuit breaker for loose wires and terminals, or signs of overheating. If none are found, the circuit breaker should be replaced.

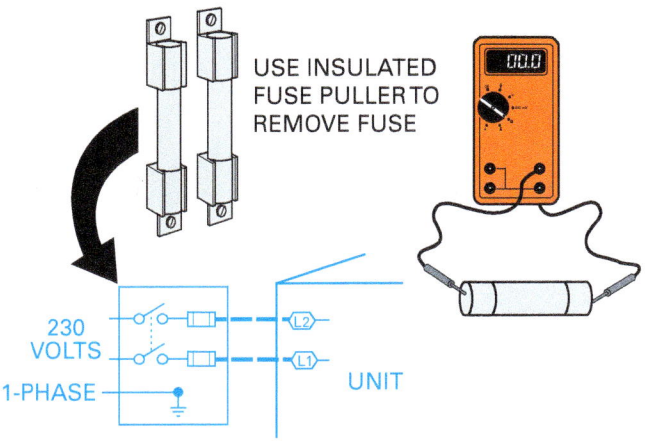

(A) CHECKING BY CONTINUITY (DE-ENERGIZED)

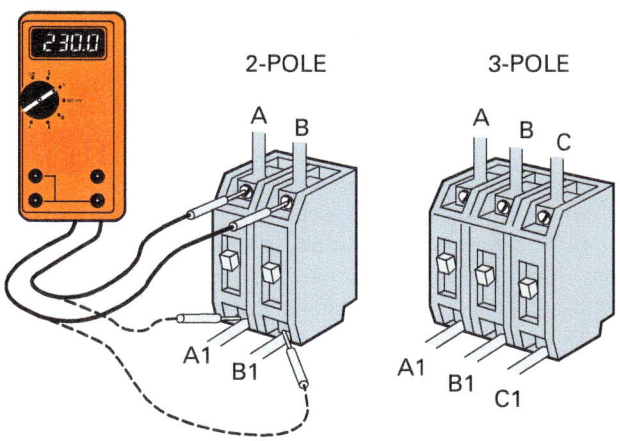

MEASURED INPUT AND OUTPUT VOLTAGES SHOULD BE THE SAME

(A) VOLTAGE TESTING

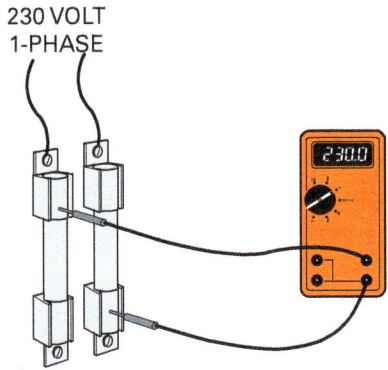

METER READS 230 V IF THE FUSE IS INTACT. IF ONE OF THE FUSES IS OPEN, IT WILL READ 0 V.

(B) CHECKING BY VOLTAGE (ENERGIZED)

Figure 51 Fuse tests.

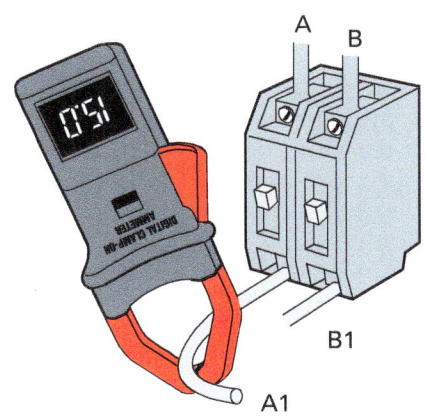

(B) CURRENT TESTING

Figure 52 Circuit-breaker tests.

If the circuit breaker shows signs of overheating, or trips when voltage is applied to the equipment, reset it, then check the current flow through the breaker using an AC clamp-on ammeter.

Set up the AC clamp-on ammeter to measure AC current on a range that is higher than the highest current expected. Check the ampere rating marked on the breaker. It is usually stamped on the breaker lever or body. One wire at a time, measure the current flow in the wires connected to the circuit breaker output terminals:

- A1 (single-pole breaker)
- A1, B1 (two-pole breaker)
- A1, B1, and C1 (three-pole breaker)

If the circuit breaker trips at a current below its rating or is not tripping at a higher current, the circuit breaker should be replaced. Be sure that the breaker is not being tripped because of high ambient temperature.

> **NOTE**
> Some service entrance panels will only accept replacement circuit breakers of the same brand as the panel manufacturer. Other service entrance panels will accept generic circuit breakers made by other manufacturers.

Circuit Breakers

Circuit breakers are available in a wide variety of sizes and types to suit various applications. Some types of circuit breakers are listed as heating, air conditioning, and refrigeration (HACR) circuit breakers. HACR breakers have a built-in time delay that allows a higher-than-rated current to momentarily flow in the circuit. This compensates for the large starting current drawn by such loads. The equipment being protected must also be marked by the manufacturer as suitable for protection by this type of breaker.

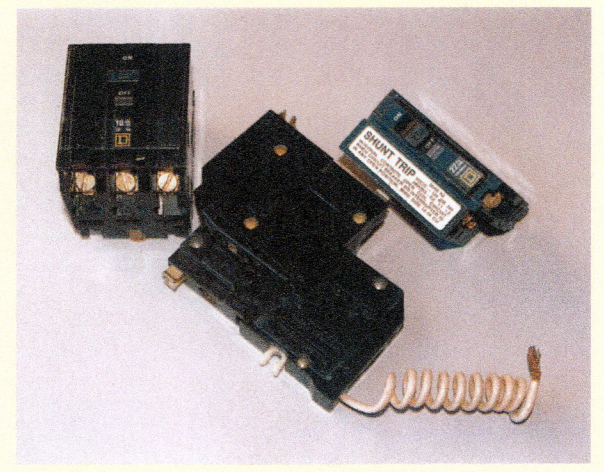

3.5.0 Control Circuits and Low-Voltage Power Sources

Control circuits provide a link between loads and the input power. Control circuits start, stop, or otherwise control the operation of a load. They usually contain one or more control devices such as relays, switches, and thermostats that work to apply or remove power from the loads. The more complex the system, the more control devices it will have. When a load such as a compressor motor is not working, it must be determined whether the problem is in the load itself, or in the circuits controlling the load.

Control circuits generally operate at 24VAC. This low voltage is obtained by using a control transformer to step down the line voltage. Because most control circuits operate at 24VAC, they are often called the low-voltage circuits. In some larger systems using three-phase line voltages of 240VAC and higher, a control voltage of 120VAC or higher may be used for some of the control devices. For this reason, always measure the control circuit voltage. Never assume the control circuit voltage is a low voltage.

3.5.1 Isolating to a Faulty Circuit via the Process of Elimination

Isolating a faulty functional circuit (input power distribution circuit, load circuit, or control circuit) can be relatively easy based on an analysis of the equipment operation and a process of elimination. Also, talking to the customer prior to working on the equipment is always recommended because it can provide valuable information that can aid in the troubleshooting process.

For example, on a service call on the heat pump system shown in *Figure 53*, the customer complains that the unit is switched to the cooling mode and running, but is blowing warm air instead of cool air. A preliminary check of system operation reveals that the compressor is not running when the thermostat is calling for cooling.

Begin fault isolation through the process of elimination. Because the indoor fan motor runs, it is possible to immediately eliminate the input power distribution circuits as the source of the problem. Next, find out if the compressor will run in the reverse cycle heating mode by setting the thermostat to call for heating. If it is too warm, thermostat TH-1 can be jumped to simulate a call for heating. For this example, assume the compressor runs. The compressor load circuit and everything in the compressor contactor energizing path, including the control relay (CR) can now be eliminated.

This isolates the problem to the only devices left in the control circuit that are unique to the cooling mode. By studying the diagram in *Figure 53*, those components are identified as the cooling thermostat (TC) and the related Cool control switch. The thermostat can be eliminated as the cause of the problem if the unit works when the control switch is set to the Auto mode.

3.5.2 Isolating to a Faulty Circuit Component

Once the source of an electrical problem has been isolated to the malfunctioning load or control circuit, the next step is to make a series of voltage measurements across the components in the malfunctioning circuit to find the faulty component. In *Figure 54*, the measurements can start from the line- or control-voltage side of the circuit and move toward the load device, such as a motor or a relay coil. Measurements are made until either no voltage is observed, or until the voltage has been measured across all the components in the

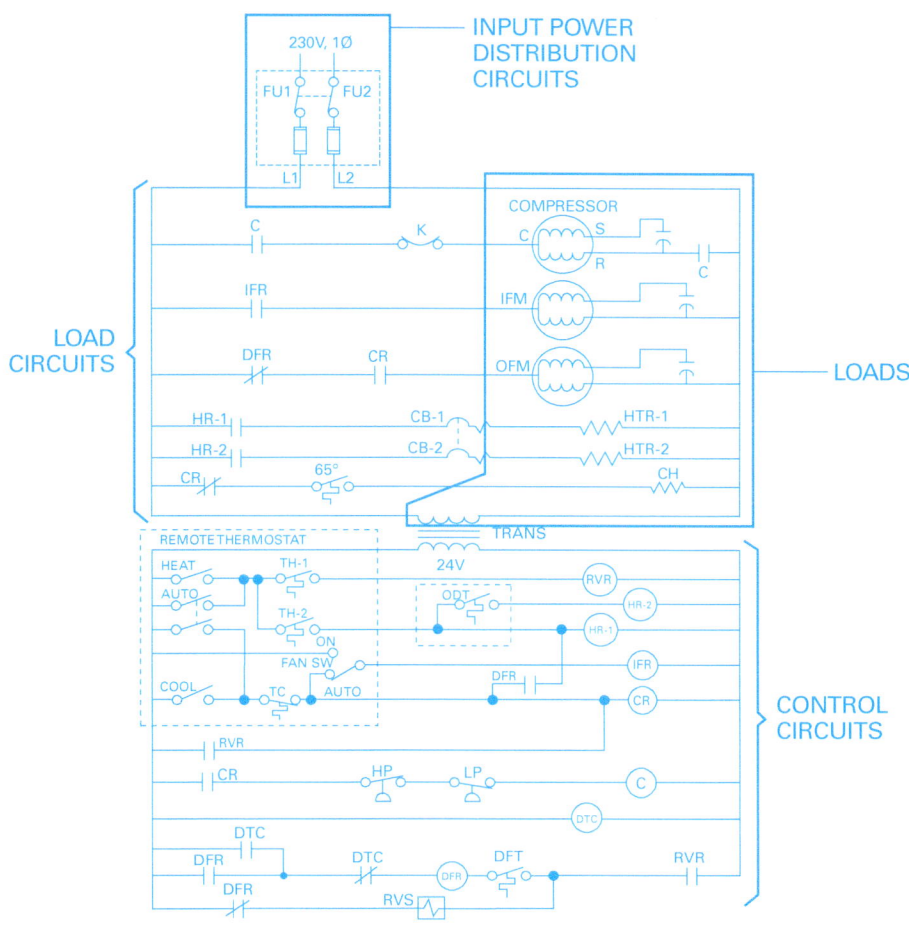

Figure 53 Heat pump ladder diagram.

circuit. Note that when there are many devices in the circuit under test, the measurements can be made by starting at the midpoint in the circuit (divide by two), then working towards either the source of voltage or the load device, depending on whether voltage was or was not measured at the mid-point. Based on these voltage measurements, one of the following situations should exist:

- Open low-pressure switch contacts.
- Open relay contacts – check related coil control circuit.
- Load device (contactor coil) is defective.

At some point within the circuit under test, no voltage will be indicated on the meter. This pinpoints an open component or set of switch contacts between the last measurement point and the previous measurement point. The upper portion of *Figure 54* is an example of this situation. In this case, the contacts of the low-pressure switch are open, preventing the compressor contactor coil (C) from energizing.

If the open is caused by a set of contactor or relay contacts, determine if the related contactor or relay coil is not being energized or is defective. The mid-portion of *Figure 54* is an example of this situation. In this case, the contacts (CR) in the control circuit are open, preventing the compressor contactor (C) from energizing. These contacts close when the control relay coil (CR) is energized. Before assuming that the problem is caused by the open contacts (CR), troubleshoot the control circuit containing the related relay coil (CR) to find out if the coil is energized or de-energized.

If the coil is de-energized, further troubleshoot its control circuit to find out why. For example, if the thermostat cooling switch contacts are open, the relay coil (CR) will not be energized.

If voltage is measured at the contactor coil, motor, or other load device, and the device is not working, the load device is most likely at fault. Turn off the power to the unit, then disconnect the device from the circuit and test it to confirm that it is defective. The lower portion of *Figure 54* shows an example of this situation, in which 24V power is applied to the compressor contactor (C), but it is not energized. In this case, the contactor is probably defective.

3.5.3 Low-Voltage Power Source Testing

Control transformers are usually checked by measuring the voltages across the secondary and primary windings (*Figure 55*). Typically, the secondary winding is measured first. The multimeter should be set to measure AC voltage on a range that is higher than the control voltage expected. If the voltage measured across the secondary winding is within ±10% of the required voltage, (21.6 to 26.4V) the transformer is good. If voltage is lower than 21.6V or no voltage is measured at the secondary winding, the voltage across the primary winding must be measured.

Some low-voltage transformers are equipped with more than one primary connection to accommodate different primary voltages. For example, a transformer may have a connection for a 208V power supply and a 240V power supply. The factory connection will be for one of the two voltages. It is the start-up technician's responsibility to check that this connection is correct. For example, if the transformer is factory-connected for 240V and the unit's supply voltage is 208V, the secondary voltage will be incorrect.

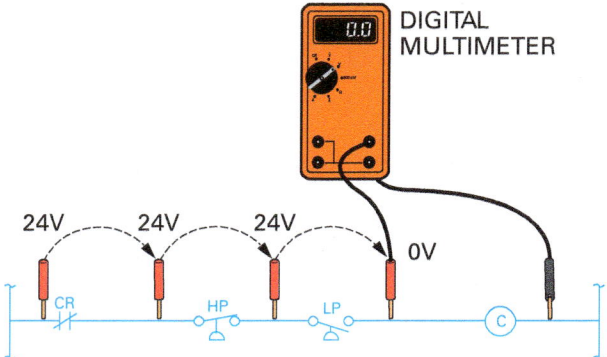

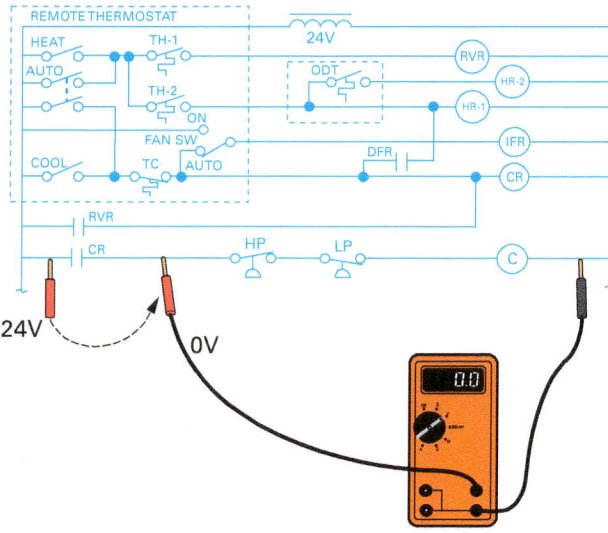

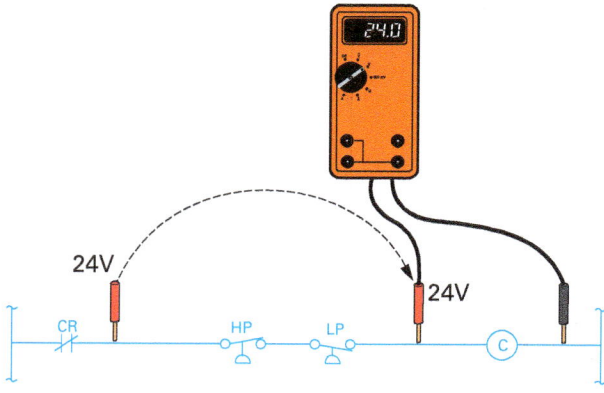

Figure 54 Isolating to a faulty circuit component.

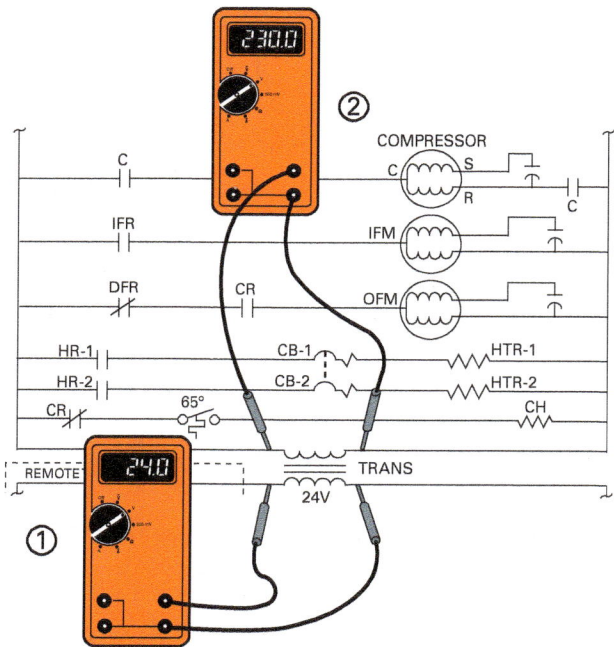

Figure 55 Control-transformer checks.

If the voltage measured at the primary winding is within ±10% of the required voltage, and the secondary voltage is low or zero, check the following:

- Check the transformer for overheating or physical damage.
- Check that all transformer electrical connections are tight.
- If the primary has more than one voltage selection lead, check that the lead matching the unit's supply voltage is connected.
- Check the fuse/circuit breaker in the transformer's secondary. If the fuse is external, replace it. Reset the circuit breaker if tripped. If the fuse is internal to the transformer, replace the transformer.

Continuity checks can confirm if either set of windings is open. To perform a continuity check, shut off power and disconnect one lead on each winding of the transformer (primary and secondary) to isolate the transformer from the rest of the circuit. Use an ohmmeter set on a low scale to measure continuity. Replace any transformer that has open windings. If the transformer appears to be fused internally, do not attempt to remove and replace the fuse. It is generally soldered into the circuit. It is best to simply replace the transformer, saving field labor that may not be productive in the end.

A transformer-protection circuit breaker may be built-in to the transformer. It has a small reset button that pops up to indicate the breaker has tripped. However, this type of protection is not common on smaller transformers. Smaller transformers—those below 75 VA—are more likely to have a fuse buried in the protective wrap around the windings.

3.6.0 Resistive and Inductive Loads

Loads are devices that consume power to do work. Compressor motors, fan motors, heater elements, and the primary winding of transformers are all loads normally found in cooling and heating units. Because the input power distribution circuits and the load circuits are both energized and operate at the input voltage level, they are often called the high-voltage circuits (*Figure 56*).

Electric crankcase heaters and electric heating elements are just two examples of resistive loads found in HVACR equipment. Inductive loads include contactor, relay, and motor starter coils. They also include control transformers, solenoid valves, and some gas valves. Compressor motors, fan motors, and other motors are also inductive loads. As a general rule, HVACR systems contain more inductive loads than resistive.

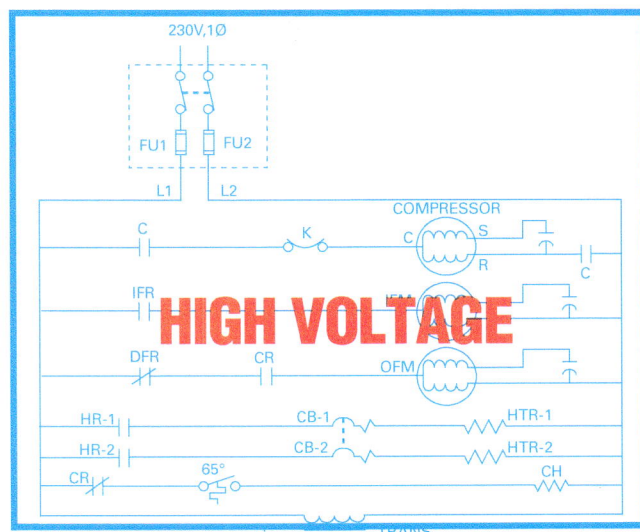
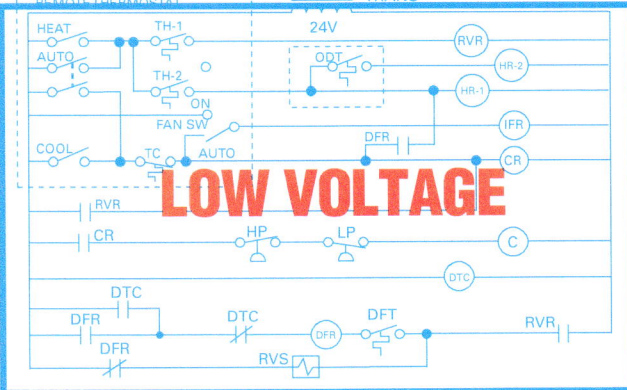

CARRIER CORPORATION

Figure 56 High-voltage and low-voltage circuit-separation on a ladder diagram.

Once a resistive or inductive load has been identified as the probable cause of an electrical problem, it should be tested to confirm its electrical condition. The best way to test a resistive or inductive load is by measuring the resistance across the terminals of the device. Before measuring resistance, make sure to electrically isolate the component being measured by disconnecting the wiring leads. This is important to achieve an accurate resistance reading, and to ensure power is not accidently applied to the ohmmeter. In *Figure 57*, a reading of zero ohms indicates a shorted load, while a reading of infinite resistance indicates an open load. In either case, the device should be replaced. When the multimeter indicates a measurable resistance (continuity), it usually indicates that the device is good. If a low resistance is measured on a contactor, relay, or starter coil, place one meter probe to ground or to the unit frame. Touch the other probe to each coil terminal. If a resistance is measured from either terminal to ground, it is grounded and must be replaced.

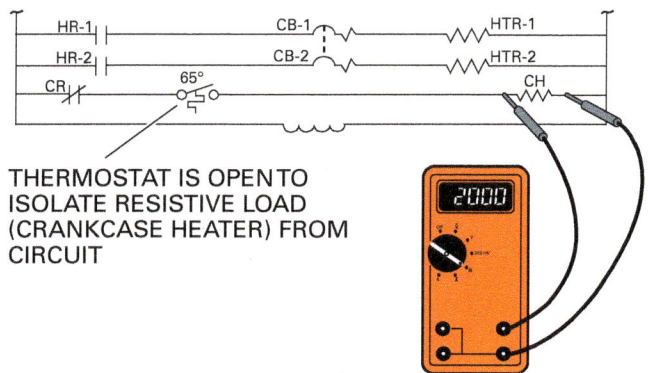

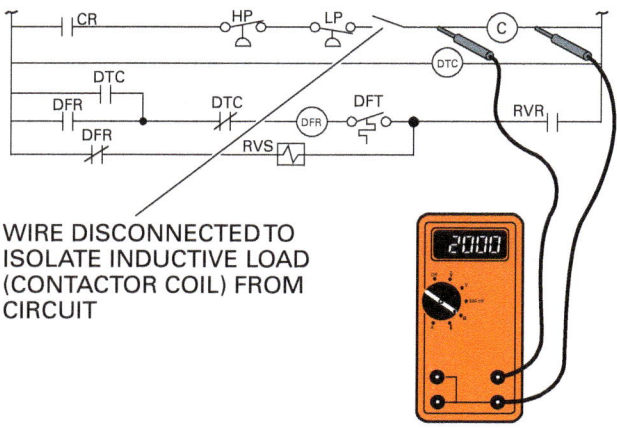

METER READS
MEASURABLE RESISTANCE = GOOD LOAD
ZERO RESISTANCE = SHORTED LOAD
INFINITE RESISTANCE = OPEN LOAD

Figure 57 Resistive and inductive load resistance checks.

3.6.1 Resistive Load Testing

Resistive loads used in HVACR tend to be heating elements. Crankcase heaters, electric heating elements, and even thermostat heat anticipators are all examples of common resistive heating loads. The thermistor is an example of a resistive device that is not a load. With the exception of thermistors, the resistance of heating devices is fixed. For example, a crankcase heater has a certain resistance value in ohms that, when subjected to a given voltage, will produce a certain amount of heat.

Resistive loads in HVACR equipment rarely fail by changing their resistance. They tend to fail by becoming open circuits—the electrical path opens. For example, when a section of a heating element melts, the resistive circuit is no longer intact. Once open, an electric heating element does not produce heat. The lack of heat is often the first symptom that a heating element has failed. Assuming power is still available to it, a clamp-on ammeter placed around one of the power leads supplying the heater can confirm if it is drawing current. If a resistive load has voltage available, but is not drawing current, it is probably open. That can be confirmed by shutting off power, disconnecting at least one side of the heater from the circuit, and checking continuity with an ohmmeter.

Manufacturers rarely provide the exact resistance value of resistive loads. The voltage and power consumption (watts) of a crankcase heater is often provided, but not the resistance value. On the rare occasion that the resistance of a crankcase heater is needed, measure the resistance of a similar heater that is known to be good. Usually though, you are looking for continuity or a lack of it to determine its condition.

3.6.2 Basic Motor Testing

The operation and uses of motors in HVACR equipment has been studied previously, but is reviewed briefly here. Six types of single-phase motors are commonly used in HVACR equipment: shaded-pole; split-phase; permanent split capacitor; capacitor start; capacitor start, capacitor run; and electronically commutated motors (ECM). Permanent split capacitor (PSC) and capacitor start, capacitor run (CSR) motors are most often used in single-phase hermetic compressors because of their good running characteristics and high efficiency. Indoor and outdoor fan motors and blower motors are usually single- or multi-speed PSC motors. Shaded-pole motors are typically used in low-torque applications such as small direct-drive fan and blower motors. The use of electronically commutated motors (ECMs) is now very common in fan and blower motor applications. Complete coverage of ECMs is provided later in this module.

> ### Measuring Resistance
> The actual resistance value measured for functional resistive and inductive loads can vary widely depending on the type of device. Ideally, the exact resistance value for the device can be found in the manufacturer's service literature. Another way to judge the resistance reading is by comparing the resistance of the device being tested with that of a similar device that is known to be good.

> **WARNING!**
> If damaged, the terminals of hermetic and semi-hermetic compressor motors have been known to blow out when disturbed in a pressurized system. To avoid injury, do not disconnect or connect wiring at the compressor terminals. When testing compressors, do not place test probes on the compressor terminals. Instead, use terminal points downstream from the compressor. To be safe, measurements and connecting/disconnecting wiring should only be done at the compressor terminals with the power off. The capacitors used in motor circuits can hold a high-voltage charge after the system power is turned off. Always discharge capacitors before handling them.

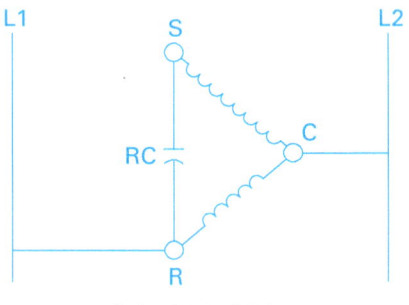

(A) PSC MOTOR

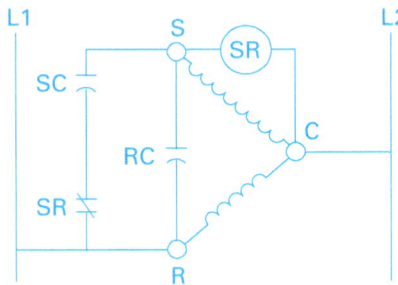

(B) CSR WITH A START RELAY

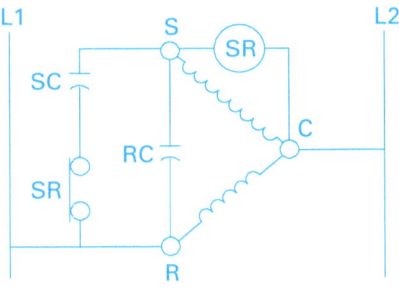

(C) CSR WITH A CENTRIFUGAL SWITCH

Figure 58 PSC and CSR motors.

Both PSC and CSR motors (*Figure 58*) have at least three external terminals leading to two internal windings. The main or run winding (R) contains relatively few turns of heavy wire. The start winding (S) contains a greater number of turns of lighter wire. The point where the two windings meet internally is called the common (C). The arrangement of the motor windings used in both the PSC and CSR motors are the same. The configuration of the motor as a PSC or CSR motor is determined by the run and/or start circuit components used with the motor. The PSC motor has a run capacitor permanently connected across the run and start windings. The CSR uses an extra capacitor called a start capacitor to aid in starting. As shown, the start capacitor and the contacts of a start relay (SR) are connected in parallel with the run capacitor. When the motor is turned on and reaches about 75% of full speed, these contacts open and remove the start capacitor from the circuit. The start relay method of removing the start capacitor from the circuit is commonly used with hermetic and semi-hermetic compressor motors. A start relay can be used with all CSR motors; however, in non-compressor applications, a centrifugal switch is frequently used to disconnect the start capacitor from the circuit when the motor comes up to speed. Start capacitor failures are often the cause of compressor and other motor problems.

Multi-speed PSC motors (*Figure 59*) used to drive fans and/or blowers in HVACR equipment are capable of operating at two or more speeds. The motor's speed can be changed by switching the motor leads, terminal taps, or through the use of speed control switches or relays. In many systems, the motor speed is selected by the control circuits, determined by the mode of operation.

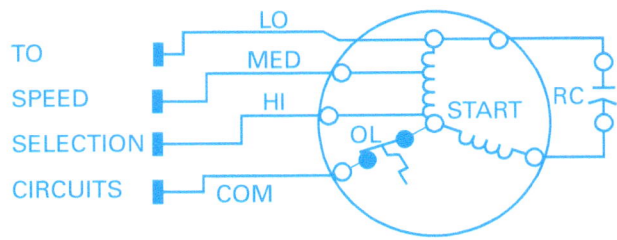

Figure 59 PSC multi-speed motor.

Normally, slower fan speeds are used with heating modes of operation, and higher speeds are used with cooling modes of operation. Fan speeds may be reduced in the dehumidification mode as well, if a system is equipped for that mode.

There are many types of multi-speed motors. As shown, the speed is changed by connecting the line voltage either to the low-speed tap (LO), medium-speed tap (MED), or high-speed tap (HI) of the motor. The specific taps used are selected when installing the unit. A control relay with contacts located in the motor load circuit is normally used to prevent more than one motor winding speed tap from being energized at the same time, a condition that would destroy the motor.

Because multi-speed motors use tapped windings, series-connected winding sections, and/or other wiring configurations that enable operation at different speeds, they may fail in such a way that the motor will not run at one or more speeds, but will run at other speeds. When troubleshooting multi-speed motors, it is important to eliminate the speed selection circuits external to the motor as the cause of the problem before condemning the motor itself.

Three-phase motors (*Figure 60*) are generally used when high starting torque is needed or when the motor requirements are greater than 1 horsepower. All have at least three internal windings, with each winding having an equal resistance and the same number of wire turns. Six- and nine-lead, three-phase motors are also found in large applications where part winding start is necessary to reduce the initial inrush current at motor startup. Three-phase motors have good starting and running characteristics and high efficiency. Three-phase motors require no external starting relays or capacitors.

3.6.3 Open, Shorted, or Grounded Winding Checks

Motors are tested for open, shorted, and/or grounded windings with the equipment power off, the capacitors discharged, and the motor leads disconnected from all the related components, including the run and start capacitors and the start relay.

Testing for shorted or open windings is done by measuring the resistance of the windings with the ohmmeter portion of a multimeter. Be sure to use a meter that has an accurate low range (R × 1Ω) resistance scale because the resistance of some undamaged windings can be as low as one-half ohm. Perform the test with the multimeter set to measure resistance on the R × 1Ω scale. Make sure that the meter is zeroed. One lead of the multimeter is connected to one of the motor leads, as shown in *Figure 61*. Touch the other lead to the remaining motor leads, one lead at a time, and observe the meter indication. If the multimeter reads a measurable resistance, the windings are probably good. If the multimeter reads zero resistance at one or more leads, the motor has a shorted winding; if it reads an infinite resistance, it has an open winding. Some rules of thumb commonly used to judge the condition of windings are as follows:

- If testing a single-phase PSC/CSR compressor motor, the resistance of the start winding is typically three to five times that of the run winding. For non-compressor motors, the resistances can vary widely depending on the design of the motor.
- When testing a single-phase multi-speed motor run winding, the highest resistance is normally measured between the common lead and the low (LO) speed lead, and the lowest resistance between the common lead and the high (HI) speed lead. The resistance measured between the common lead and the medium speed lead (MED) should be somewhere between that measured for the LO and HI leads.

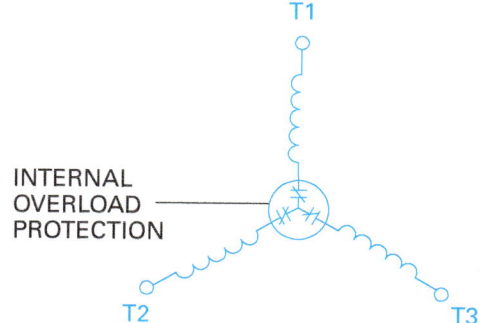

Figure 60 Three-lead, wye-connected, single-voltage, three-phase motor.

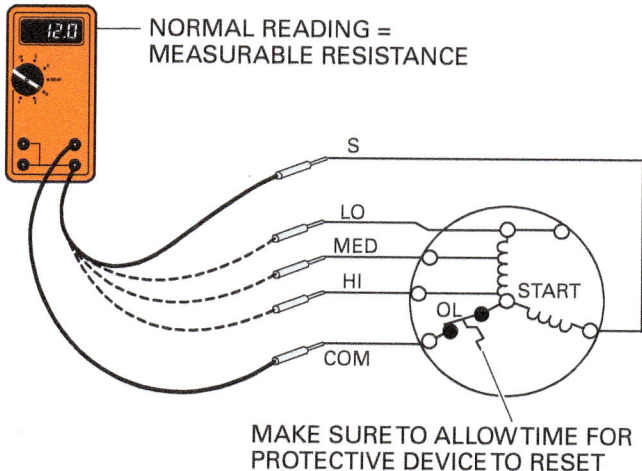

- INFINITE RESISTANCE = OPEN WINDING(S)
- ZERO OHMS = SHORTED WINDING(S)

Figure 61 Motor open- or shorted-winding test.

- If testing a three-phase motor, the motor windings are usually judged to be good if the resistance measured across each winding is nearly identical to the other two windings.

When checking a motor with an internal motor protection device, always make sure that the motor has had adequate time to cool off so that the protective device has time to reset. Note that this may take an hour or more.

Testing a motor for grounded windings is done by measuring the resistance of the windings with a multimeter. Perform the test with the multimeter set up to measure resistance on the R × 10,000Ω scale. Connect one lead of the multimeter to a good ground connection, such as the motor or compressor frame or compressor discharge/suction line. Poor electrical contact because of a coat of paint, layer of dirt, or corrosion can cause an inaccurate measurement and hide a grounded winding. The other meter lead is then touched to each of the motor leads, one lead at a time, while watching the meter indication (*Figure 62*). An infinite or high resistance should be measured from each lead to ground. If a high resistance reading is indicated, it should not be less than 1,000Ω per volt. For example, on a 230V motor, the resistance should not be less than 230,000Ω (230V × 1,000Ω/V = 230,000Ω). This indicates that the motor winding is not grounded.

If a low or zero resistance, or a measurable resistance that is less than 1,000Ω per volt is measured, this usually indicates that the motor winding is grounded. However, if testing a hermetic or semi-hermetic compressor motor, note that erroneous resistance readings to ground can be measured if liquid refrigerant is present in the compressor shell. In this instance, it is recommended that the refrigerant be recovered from the system. Then retest the compressor before condemning it.

3.6.4 Identifying Unmarked Terminals of a PSC/CSR Motor

Sometimes the terminals on a single-phase motor are not marked, or are hard to identify. The terminals can be identified by using a multimeter to measure the resistance of the motor windings. First, the multimeter is used to find the two terminals across which the highest resistance is measured (*Figure 63*). These are the run (R) and start (S) terminals. The remaining terminal is the common (C) terminal. Next, put one lead of the multimeter on the common (C) terminal, and find which of the remaining

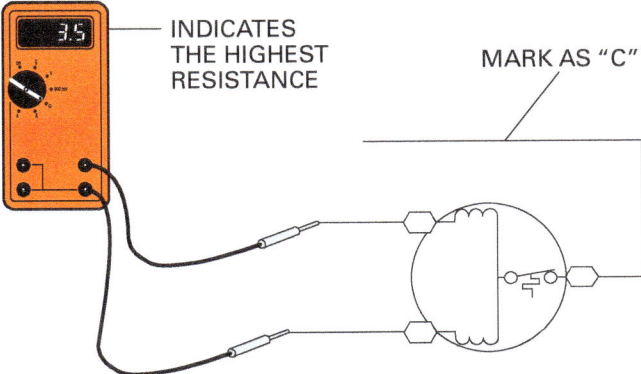

IDENTIFY THE COMMON (C) TERMINAL

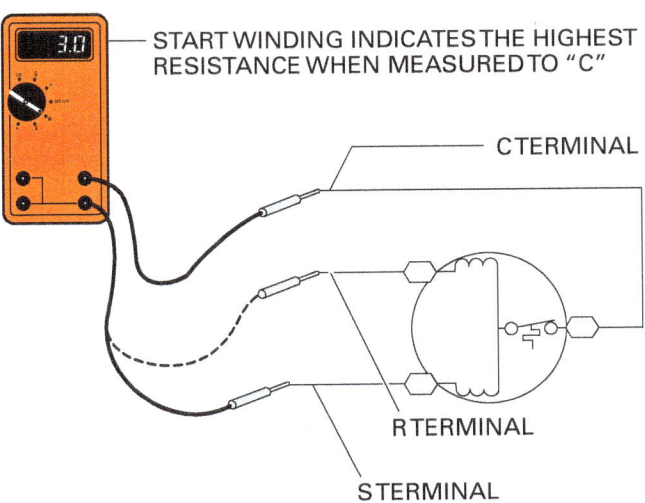

IDENTIFY THE START (S) AND RUN (R) TERMINALS

Figure 63 Identifying unmarked terminals of a PSC/CSR motor.

Figure 62 Grounded winding test.

terminals gives the highest resistance reading. This is the start winding (S) terminal. The remaining terminal is the run winding (R) terminal. The C to R resistance reading will be the lowest.

3.6.5 Start and Run Capacitor Checks

The start and/or run circuits on single-phase motors use capacitors. Capacitors affect the wattage, amperage draw, torque, speed, and efficiency of a motor. Run capacitors are connected in the motor circuit as it runs; therefore, they are referred to as continuous-duty capacitors. They are usually larger in physical size, but have lower capacitance ratings than start capacitors. Because run capacitors remain in the circuit while the motor runs, they are typically filled with a dielectric fluid that acts to dissipate heat. A shorted capacitor may provide a visual indication of its failure. The pop-out hole at the top of a start capacitor may appear bulged or blown. A run capacitor may be bulged and/or leaking. If a capacitor is found to be defective, always replace it with one specified by the manufacturer.

If the exact capacitance value of a capacitor is required, use a capacitor tester (*Figure 64*). Follow the tester manufacturer's instructions to perform the test. Typically, the measured microfarad value for a start capacitor should be ±20% of the value shown on the capacitor label. If the measured value is outside the range of ±20%, replace the capacitor. For a run capacitor, the measured value should be ±10% of the value shown on the capacitor. If the measured value is outside the range of ±10%, replace the capacitor.

3.6.6 Start Relay Checks

The start relay is used to remove the start capacitor from the motor starting circuit when the motor reaches about 75 to 80% of its operating speed. Start relays are made that can be actuated by either current or voltage. The start relays used with HVACR equipment motors are normally voltage-actuated relays (potential relays); therefore, the remainder of this discussion will cover the testing of a voltage-actuated start relay.

Start relays tend to fail with their contacts closed. This results in the start capacitor remaining in the start circuit, causing the motor's start winding to overheat and fail. It may also result in failure of the start capacitor. When it is necessary to replace a start relay, an identical replacement must be used. Substitution of a relay with a different pickup voltage can cause damage to the start capacitor or motor start winding. Also, the replacement relay must be positioned and wired exactly as the original.

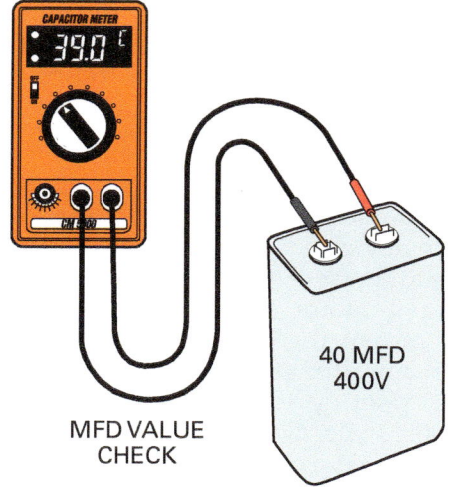

Figure 64 Capacitor tests.

Start relays can be tested by measuring the motor start winding current with a clamp-on ammeter. The use of an analog clamp-on ammeter is recommended because it is easier to observe the current reading. If using a digital clamp-on ammeter, one with a MIN/MAX current capability must be used. Testing begins by first finding the full load amps (FLA) rating for the motor as marked on the compressor or motor nameplate. The clamp-on ammeter is then set up to measure AC current on a range scale that is higher than the motor FLA.

With the power turned off, the clamp-on ammeter jaws are placed around the wire that connects the motor start capacitor to the start relay contacts, as shown in *Figure 65*. Power is turned on to the unit while watching the clamp-on ammeter indication to observe the current flow in the start capacitor circuit as the motor starts.

When the start relay is operating properly, the clamp-on ammeter current indication should momentarily indicate current flow, and then fall back to zero as the motor comes up to speed. This shows that the start relay is good because its contacts have opened.

If current continues to be read on the clamp-on ammeter after the motor is up to speed, the relay contacts are stuck closed. This means the relay is defective and should be replaced. The start capacitor and/or motor start winding can be damaged when the start relay contacts are stuck closed.

If no current is shown on the clamp-on ammeter, the relay contacts may be stuck open, the related start capacitor may have failed open, or the related wiring may be open. In this instance, the relay contacts can be checked for continuity with the unit power turned off. Contacts that are stuck open will measure infinite resistance.

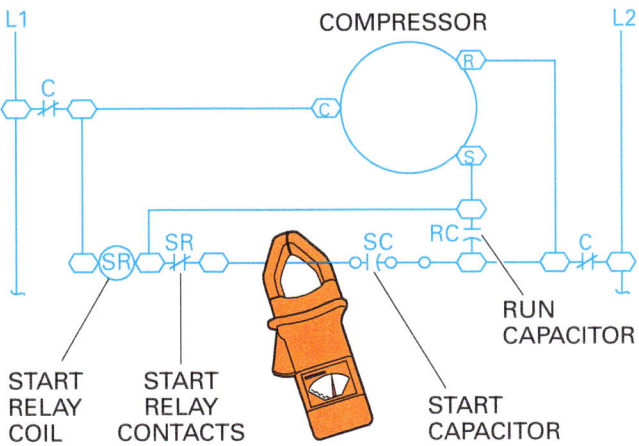

Figure 65 Start relay test.

3.6.7 Start Thermistor Tests

Start thermistors can be used to provide additional starting torque for PSC compressors. The start thermistor is a temperature-sensitive device that changes its electrical resistance as temperature changes. Positive temperature coefficient (PTC) thermistors increase their resistance with an increase in temperature. PTC thermistors are commonly used in the start circuits of PSC motors. *Figure 66* shows a PSC compressor motor with a PTC start thermistor. As shown, the PTC thermistor is placed across (parallel to) the run capacitor.

At room temperature, the PTC thermistor resistance is very low, about 25Ω to 50Ω. When the compressor is turned on, the application of voltage provides an initial surge of high current through the start winding, because the low resistance of the PTC thermistor is effectively bypassing (shorting) the run capacitor. This surge results in increased motor starting torque. The temperature increase created by the high current causes the PTC thermistor resistance to increase very rapidly to several thousand ohms, blocking current flow and effectively removing the thermistor from across the run capacitor. The motor then runs as a normal PSC motor. A small leakage current through the thermistor keeps the thermistor heated and its resistance high while the motor is running. Circuit operation remains this way until the motor is turned off. After a cool-down period, the thermistor's resistance will once again be the low value needed to start the motor.

The thermistor is tested with the equipment power off, the capacitors discharged, and the thermistor under test isolated from the remainder of the circuit. Testing the thermistor is done by making a resistance measurement. Before attempting to measure the resistance, wait at least 10 minutes to allow the thermistor to cool to ambient temperature. The cold resistance of any PTC thermistor should be about 100 to 180% of the thermistor ohm rating. For example, a thermistor rated at 25Ω should have a cold resistance of 25Ω to 45Ω. If the PTC thermistor resistance is much lower or more than 200% higher than its rating, the thermistor should be replaced.

Note that, if the resistance of a start thermistor is acceptable and yet the compressor still does not start consistently, a start capacitor and start relay may have to be installed to ensure consistent compressor starting energy is provided. PTC start thermistors are not as powerful a starting aid as a capacitor and start relay.

3.7.0 Hydronic Controls

In many parts of the United States, hot water or hydronic heat is very popular. With this type of heat, hot water (or in some cases, steam) is circulated through pipes to terminals in different parts of the building. Boilers are pressurized vessels and have unique safety and installation requirements. The nature of hydronic heat dictates specialized components and controls not found in other areas of HVAC.

A simple hydronic system, such as the one shown in *Figure 67*, contains a hot-water boiler where the water is heated; a circulating pump to move the heated water throughout the system; a tank to absorb water as it expands when heated; a device called an aquastat that controls the water temperature; terminals to exchange heat between the water and the air; and a relief valve to bleed excess pressure so that boiler pressures do not reach

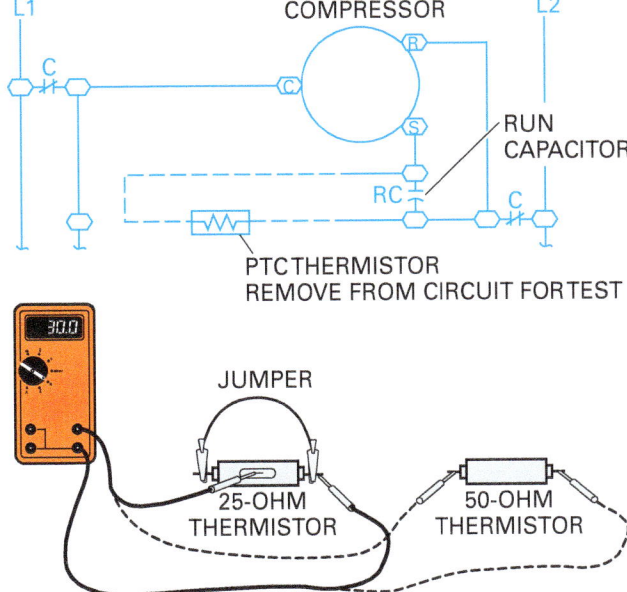

Figure 66 Start thermistor test.

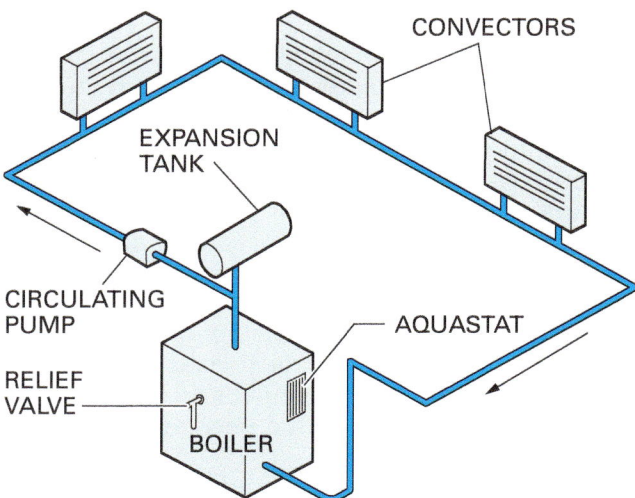

Figure 67 Simple hydronic system.

explosive levels. The water may be heated by an oil or gas burner. These burners and many of the controls associated with them are similar to the burner controls on gas or oil-fired warm air furnaces.

3.7.1 Aquastat

The aquastat performs several important functions. An aquastat, such as the one shown in *Figure 68*, is mounted on or near the boiler and has a temperature-sensing element that is inserted in a well in the boiler jacket. In some boilers, the water temperature is not maintained unless there is a call for heat. That activates the controls and the aquastat is then in control of boiler water temperature. The burner will operate until the room or zone thermostat is satisfied, or the water in the boiler reaches the high-temperature setting of the aquastat. If the call for heat continues and boiler water temperature falls, the aquastat will restart the burner when the temperature reaches its low-temperature setting.

Figure 68 Digital aquastat.

More complex aquastats provide additional functions. In addition to an operating temperature range, they may contain a low-limit setting that prevents the water in the boiler from getting too cold. This feature is necessary if domestic hot water is provided by a heat exchanger, called a tankless coil, that is inserted in the boiler jacket. This type of aquastat can also contain a pump/circulator control that prevents its operation until the water is warm enough to provide heat from the terminals.

Other specialized aquastats, especially those used on oil-fired boilers, have the oil-burner primary control built into the aquastat. *Figure 69* shows how an aquastat with a circulator and high and low limits would operate on an oil-fired boiler. If the water temperature is too low, contacts R-B remain closed, allowing the water to heat up but with no circulation. As the water heats up, R-W closes and R-B opens. If there is a call for heat, power from the switching relay feeds through the closed high limit switch to keep the burner running and through R-W to power the circulator pump.

3.7.2 Reset Controller

To achieve maximum comfort from a hydronic heating system, some systems incorporate an aquastat device called a reset controller. This device monitors the outdoor temperature and adjusts the boiler water temperature for maximum energy savings while still providing comfort. It allows the boiler water temperature to rise on cold days to supply more heat, but limits the boiler water temperature to lower levels on milder days when less heat is required. In addition to a temperature sensor in the boiler jacket, the reset controller contains an outside temperature sensor (*Figure 70*) that is typically mounted on the north side of the structure away from direct sunlight.

3.7.3 Low-Water Cutoff

Loss of water in a boiler can have catastrophic results. The boiler may produce large amounts of steam, overheat, and build up explosive pressures. To prevent this, a low-water cutoff device may be installed in the boiler. The simplest devices are nothing more than a float that activates a switch if the water level drops. The switch shuts off burner operation to prevent overheating and/or activates an alarm or warning light. Electronic versions (*Figure 71*) have a probe that is inserted in the boiler to monitor the water level. Poor water quality caused by contamination will affect the operation of electronic low-water cutoffs. These safety devices also shut off burner operation and/or sound an alarm when the water level drops too far.

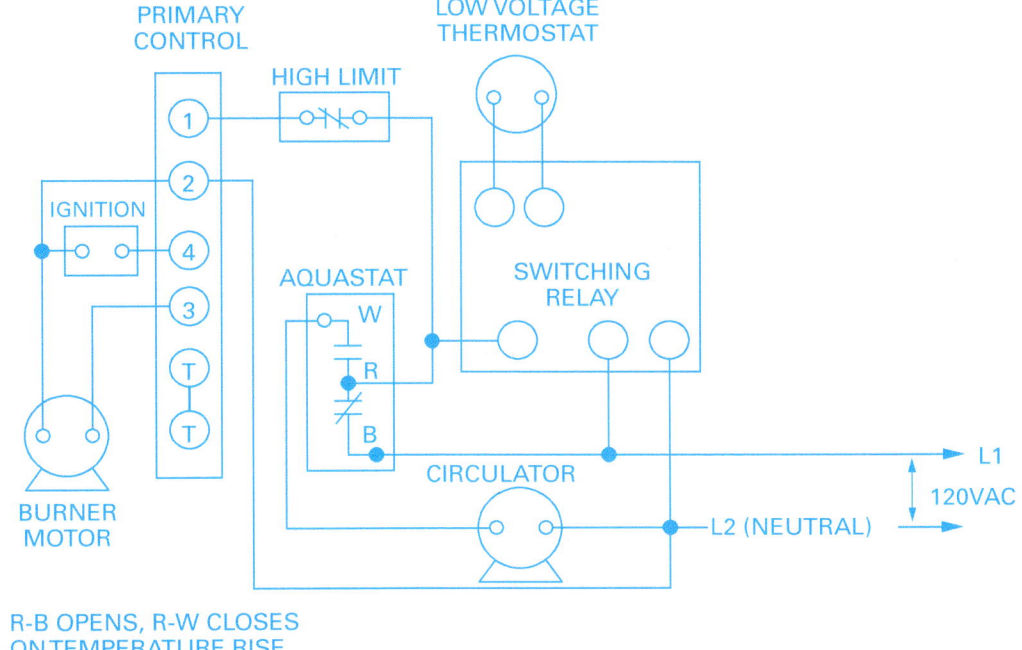

Figure 69 Aquastat used for low-limit and circulator control in an oil-fired hydronic system.

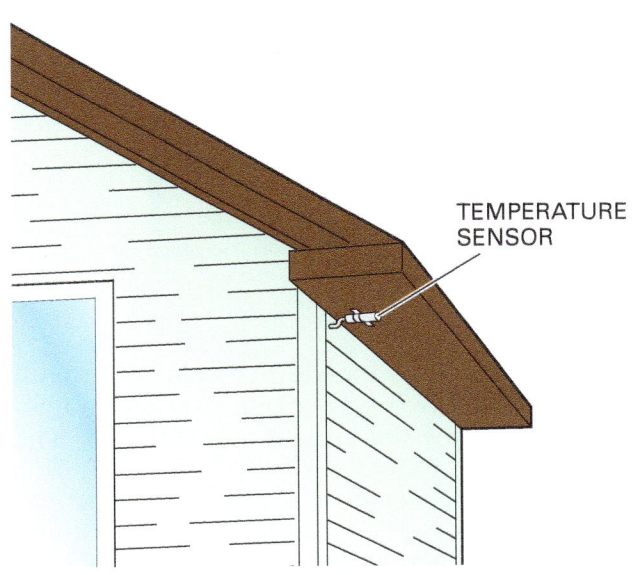

Figure 70 Typical reset controller outside-temperature sensor mounting.

3.7.4 Circulator Pump

The circulator pump in a hydronic system performs a job similar to the blower in a forced air furnace. The major difference is that the pump moves water instead of air. Circulating pumps can be mounted on the floor or in the supply or return water line. However, the preferred location for the pump is in the supply line leaving the boiler. Many pump motors are direct coupled to the pump's centrifugal impeller through a shaft seal in the impeller housing. However, many smaller pumps are magnetic and have no shaft seal that could leak. Magnets indirectly couple the impeller and motor, with a non-magnetic wall between them. The impeller is completely sealed inside a non-magnetic pump casing as a result. Circulator operation can be controlled by the aquastat or by the zone valves (if so equipped). Some residential hydronic systems have one circulator pump. However, in most zoned installations, each zone has its own, smaller pump (*Figure 72*). Some older circulator pumps require periodic lubrication of the motor and pump, while newer models are sealed units with little or no maintenance required.

3.7.5 Zone Valves

Zoned control is easily accomplished with hydronic heating systems due to the simplicity of the zone valve. This device is simply an electrically operated valve. It is placed in the supply line to the zone and opens when the thermostat in the zone calls for heat. *Figure 73* shows a typical zone valve.

In residential applications, zone valves are typically powered by the 24VAC control circuit. In commercial applications, line voltage may be used. Most valves are motorized so that on a call for heating from the zone, the valve opens. Some valves are equipped with an end-switch that closes as the valve opens. This switch closure starts the circulator pump so that water can flow to the zone that is calling for heat. There are a number of ways to control burner and circulator operation on a gas or oil-fired boiler. Much of it depends on the

Figure 71 Low-water cutoff control.

Figure 72 Circulator pump.

Figure 73 Zone valve.

aquastat used, but the installer does have some control over the sequence of operation.

3.7.6 Zoned Hydronic System Operating Sequence

In many ways, the operating sequence of a hydronic system is similar to the operating sequence of a forced-air heating system. The gas and oil burners used in both systems operate similarly and often use identical components. For example, the hot surface ignition system used in a furnace is identical to the one used in a residential boiler. As with any HVACR system, knowing its operating sequence is important when seeking to find the source of a problem. The following sequence of operation assumes a simple residential packaged boiler (gas-fired) equipped with an aquastat with low- and high-temperature limits. The structure is divided into three heating zones, each controlled by its own 24V room thermostat.

When power is first applied to a cold boiler with no call-for-heat from any of the zones, the low-limit contacts of the aquastat are closed. The closed contacts allow the burner to operate. This warms the boiler water to a preset temperature. When a call-for-heat signal does come, the boiler water is warm enough to provide immediate heat. The burner will cycle on and off to maintain this minimum water temperature as long as power is available to the boiler.

Assume there is a call-for-heat from the smallest of the three zones (a master-suite bathroom). The room thermostat in the zone signals the zone valve for the bathroom to open. As the motorized zone valve opens, it closes a set of end-switch contacts on the zone valve that provide a 24V signal to energize the circulating pump. The pump moves heated water to the terminal device in the bathroom. At this point, the boiler can react in the following ways:

- Depending on the thermostat setpoint in the bathroom and boiler water temperature, there may be enough heat available in the boiler water to satisfy the room thermostat in that zone without burner operation.
- If the thermostat setpoint in the bathroom is set high and/or additional zone thermostats also call for heat, boiler water temperature will drop and the aquastat low-limit will start burner operation to maintain boiler water temperature.
- If the thermostat setpoint in the bathroom is set high and that is the only zone calling for heat after the burner has come on, that small zone may not be able to rapidly dissipate all the heat the burner is providing. This will cause the boiler water temperature to rise to the point that the aquastat high-limit will shut the burner off.

In any system with multiple zones, the burner will cycle on and off based on water temperature within the boiler and the fluctuating zone heating capacity demands. Troubleshooting the components in a hydronic system is similar to troubleshooting components in a forced-air system. Energized temperature-actuated switches can be checked for correct operation using a voltmeter. Check for switch continuity with power off using an ohmmeter. The circulating pump motor can be checked like any other HVACR motor.

Additional Resources

Electrical Motor Controls for Integrated Systems. Gary Rockis and Glen A. Mazur. Current edition. Homewood, IL: American Technical Publishers, Inc.

Electric Motors and Drives: Fundamentals, Types, and Applications. Austin Hughes and Bill Drury. Current edition. Atlanta, Ga: Newnes.

3.0.0 Section Review

1. To maximize electrical safety when working on live electrical circuits, _____.
   a. try to work with one hand only
   b. keep one foot on an insulated surface
   c. lay insulation material under your knees
   d. wear metal jewelry to dissipate the current

2. Refer to *Figure 44*. What effect does a limit switch (LS) have on the operation of the induced-draft motor?
   a. It prevents its operation.
   b. It allows its operation after the switch resets.
   c. It has no effect.
   d. It allows its operation after the pressure switch resets.

3. The purpose of the legend on a label diagram is to _____.
   a. isolate electrical problems
   b. identify the meaning of symbols
   c. locate components in the equipment
   d. isolate section of a ladder diagram

4. With equipment that operates on three-phase power, the current imbalance in any one leg should not exceed _____.
   a. 2%
   b. 5%
   c. 8%
   d. 10%

5. A furnace provides indoor airflow in both heating and cooling modes. The furnace operates normally in the heat mode, but the blower does not run in cooling. Which of the components below is most likely at fault?
   a. Blower motor capacitor
   b. Furnace control board
   c. Cooling fan relay
   d. Low-pressure switch

6. When measuring the resistance of motor windings, always _____.
   a. use the R × 10Ω scale
   b. use the R × 10,000Ω scale
   c. zero the meter
   d. ground the meter

7. What is the primary purpose of an aquastat?
   a. Maintain boiler water temperature
   b. Maintain a safe water level
   c. Prevent boiler freeze-up
   d. Monitor outdoor temperature

Section Four

4.0.0 Variable-Frequency Drives

Objective

Describe the operation of variable-frequency drives (VFD) and their selection considerations.
a. Describe the operation of a VFD.
b. Identify VFD parameters that can be programmed.
c. Describe the important considerations for the selection of a VFD.
d. Explain dynamic motor-braking processes.

Trade Terms

Base speed: The rating given on a motor's nameplate at which the motor will develop rated horsepower at its rated load and voltage.

Insulated gate bipolar transistor (IGBT): A type of transistor that has low losses and low gate-drive requirements. This allows it to be operated at higher switching frequencies.

A variable-frequency drive (VFD), sometimes referred to as an adjustable-frequency drive (AFD), converts three-phase 60Hz input power to an adjustable frequency and voltage output to control the speed of an AC motor. The use of VFD-controlled systems provides energy savings by eliminating most of the losses associated with mechanical or electromechanical methods of adjusting speed used in the past. Many other benefits are derived from using a VFD. These include the following:

- *Controlled starting* – Provides for limited starting current, reduction of power line disturbance on starting, and low power demand on starting.
- *Controlled acceleration* – Provides for soft-start, adjustable acceleration based on time or load, and reduced motor size for pure inertial load acceleration.
- *Adjustable operating speed* – Allows process to be optimized or changed, provides energy savings, allows process start at reduced speed by programmable controller.
- *Adjustable torque-limiting* – Current limit is available for quick and accurate torque control, protecting the machinery, product, or process from damage.
- *Controlled stopping* – Soft stop, timed stopping, and fast reversal with much less stress on an AC motor than plug reverse.

4.1.0 Basic VFD Operation

VFDs change the speed of AC squirrel cage induction motors to meet system requirements by converting three-phase 60 Hz input power to an adjustable frequency and voltage. The following formula determines the speed of the motor:

Motor speed = 120 × frequency ÷ motor poles

For a specific motor, the number of poles is a constant since they are built into the motor. Examination of the formula shows that the speed of the motor is proportional to the applied frequency. As the frequency increases, the motor speed increases; as the frequency decreases, the motor speed decreases. To maintain a constant motor torque, VFDs are designed to automatically maintain a constant relationship between the voltage and frequency of the excitation output applied from the VFD to the motor terminals. This relationship is called the volts per hertz ratio (V/Hz). If this process is done properly, the speed of the motor can be controlled over a wide variation in shaft speed (0 rpm through twice nameplate) with the proper torque characteristics for the application.

VFDs are made in a wide variety of sizes and designs. Motor drive technology has changed greatly over the last decade. Today, digitally controlled, programmable, microprocessor- based, pulse-width-modulated (PWM) variable frequency drives are commonly used. Their use provides a significant improvement in the elimination of dirty power problems such as harmonics and electro-magnetic interference (EMI). These problems were generated by the analog SCR rectifier VFD units that were the state-of-the-art only a few years ago. For this reason, the remainder of this section will focus on the newer PWM type of VFD device.

Figure 74 shows a block diagram of a typical PWM VFD system. As shown, it consists of three basic parts: the operator control panel, the VFD controller, and an AC motor. The control panel provides an interface between the operator and the drive system. At the control panel, the operator can input commands for starting, stopping, and changing the speed and/or direction of the motor. He or she can also monitor various VFD and motor status signals. Typically, the control panel is removable and can be mounted externally and connected via a cable to the drive. Most VFDs have an input/output (I/O) interface board that allows communication with personal computers (PCs) or programmable logic controllers.

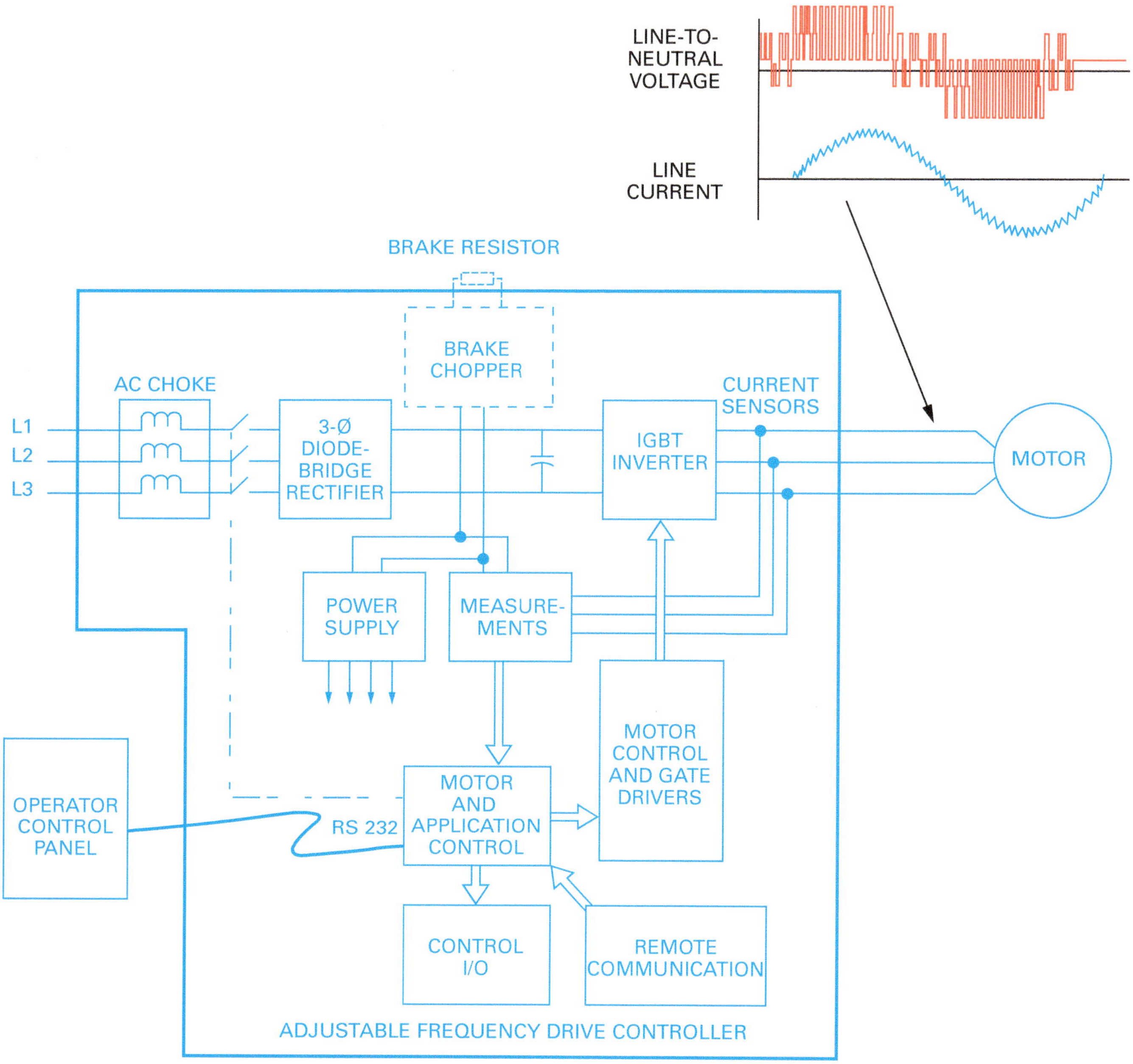

Figure 74 Pulse-width modulated, variable-frequency drive system.

The motor used with an VFD is typically a standard NEMA Design A or Design B, squirrel-cage induction motor rated for 240V or 480V, three-phase, 60 Hz operation. In response to the AC power waveform applied from the VFD, the motor drives the load.

The VFD controller converts three-phase, 60-Hz input power to an adjustable frequency and voltage output for controlling the speed of the AC motor. It typically receives 240V or 480V input power. The actual range of frequencies used is programmable and depends on the motor application. It also regulates the output voltage in proportion to the output frequency to provide a constant voltage-to-frequency ratio as required by the characteristics of the motor.

The three-phase AC-choke and the DC-link capacitor form an LC filter that, together with the three-phase, diode bridge-type full-wave rectifier, produces a regulated DC voltage for application to the inverter circuit. Note that the circuitry between the rectifier output and the inverter circuit input is sometimes referred to as the DC intermediate bus. The AC choke levels out any high-frequency (HF) disturbances from the utility to the drive and HF disturbances caused by the drive to the utility. It also improves the waveform of the input current to the drive.

Since the inverter section is powered by a fixed DC voltage, the maximum amplitude of its output waveform is fixed. The insulated gate bipolar transistor (IGBT) switches in the inverter circuit convert the DC voltage input into a three-phase, PWM, symmetrical AC voltage output for application to the AC motor. The frequency and amplitude of the waveform that is produced are determined by the duration and precise timing intervals of the gates applied to the IGBT switches from the gate driver circuit. The effective value of the output voltage applied to the motor is determined by the width of the zero voltage intervals in the output waveform. Use of IGBTs provides the high-speed switching necessary for PWM inverter operation. They are capable of switching on and off several thousand times a second.

VFDs typically have one range of output frequencies over which the ratio of volts to hertz remains constant and another range of output frequencies over which the voltage remains constant while the frequency varies (*Figure 75*). The frequency at which the transition from one range to the other occurs is called the base frequency. Between the base frequency and the maximum frequency points, the rated output voltage remains constant. Below the base frequency point, the output voltage is less than the rated voltage and is based on the volts per hertz ratio (V/Hz). Depending on motor characteristics, these two ranges can correspond to constant torque (constant V/Hz) and constant power (constant voltage with an adjustable frequency).

The motor and application control circuit consists of a programmable microprocessor and related circuits. The microprocessor controls the motor according to measured feedback signals and in accordance with parameter settings and commands. These commands may be applied from the control panel or from a remote device via the control input/output (I/O) circuits. The motor and application control circuit sends signals to and receives feedback signals from the motor control/gate driver circuits. The motor control/gate driver circuits calculate, then output switching position signals to the IGBT switches in the inverter.

When required, dynamic braking can be used with a VFD. This can be done by using a brake chopper circuit and external dynamic brake resistor that each operates to absorb motor regenerative energy for stopping the load and to limit the energy flowing back to the drive.

Dynamic braking is accomplished by continuing to excite the motor from the drive after the motor turns off. This causes a regenerative current at the drive's DC intermediate bus circuit. The dynamic brake resistor is then placed across

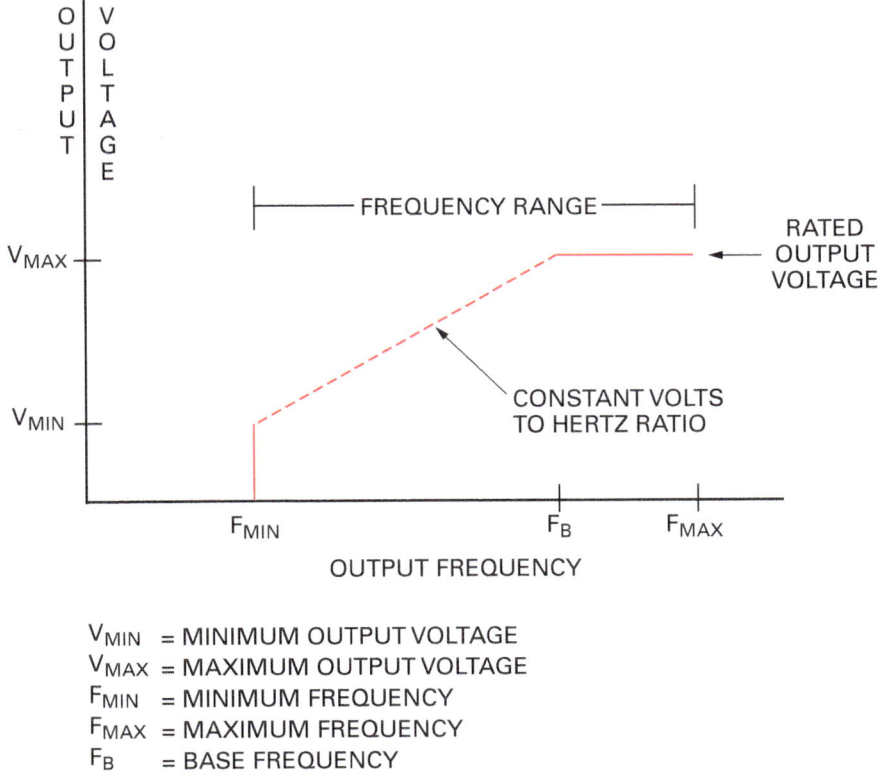

Figure 75 VFD operating frequency range.

> ### Adjustable Frequency Drives
>
> *GOING GREEN*
>
> Adjustable frequency drives are available for use with a wide range of motors and applications. These drives save energy by eliminating the losses associated with traditional speed control.
>
>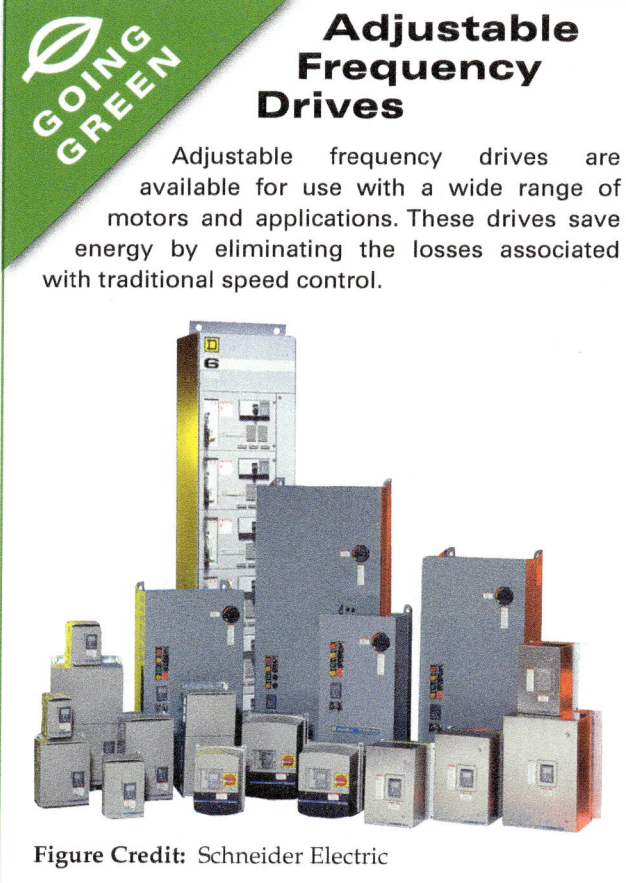
>
> **Figure Credit:** Schneider Electric

the DC bus to dissipate the returned power. The brake resistors are switched into the circuit under the control of the brake chopper. Dynamic braking cannot occur in an induction AC drive unless the power circuit and control circuit are operational. For this reason, a mechanical brake, actuated in case of a power failure, is often required to stop the motor. The resistor(s) used for dynamic braking must be capable of absorbing at least six times the stored rotational energy of the motor at base speed.

4.2.0 Programmable VFD Parameters

The parameters that can be programmed and/or monitored when using a VFD can vary widely based on the specific design and complexity of the VFD, the VFD manufacturer, and the application for which the VFD is to be used. Some of the parameters that can be programmed and/or monitored are briefly described here. Follow the instructions in the manufacturer's user manual for the specific VFD unit being used.

- *Minimum and maximum frequency* – This defines the frequency limits for the VFD operation.
- *Acceleration and deceleration time* – Acceleration time defines the time required for the output frequency to accelerate from the set minimum frequency to the set maximum frequency. Deceleration time defines the time required to decelerate from the set maximum frequency to the set minimum frequency.
- *Current limit* – This determines the maximum motor current the VFD will provide on a short-term basis.
- *V/Hz ratio* – Typically there are two settings: linear and squared. The linear setting is used with applications requiring constant torque. In this mode, the voltage supplied to the motor changes linearly with the frequency from zero hertz to the nominal frequency of the motor. With the squared setting, the voltage to the motor changes following a squared curve from zero hertz to the nominal frequency of the motor. In this mode, the motor runs under-magnetized (below the normal frequency) and produces less torque and electromechanical noise. This mode is used in applications where the torque demand from the load is proportional to the square of the speed, such as with centrifugal fans and pumps.
- *V/Hz optimization* – This mode provides an automatic torque boost. The voltage to the motor changes automatically, which allows the motor to produce sufficient torque to start and run at low frequencies. The voltage increase depends on the motor type and horsepower. Automatic torque boost can be used in applications where starting torque due to starting friction is high, such as with a conveyor.
- *Nominal voltage of the motor* – This is the rated voltage from the nameplate of the motor.
- *Nominal frequency of the motor* – This is the frequency from the nameplate of the motor.
- *Nominal speed of the motor* – This is the speed from the nameplate of the motor.
- *Nominal current of the motor* – This is the current value from the nameplate of the motor.
- *Supply voltage* – This is set to a range compatible with the nominal voltage of the supply.
- *Basic frequency reference* – This selects the basis for the frequency reference such as a voltage reference supplied by a potentiometer or a current reference supplied by a transducer or other source.

4.3.0 VFD Selection Considerations

Selecting a VFD for a particular application requires an understanding of the capabilities of various drive units and the characteristics and/or problems encountered with different types of loads. For this

PWM VFD Outputs to a Motor over Long Cable Runs

When a PWM VFD drive output is applied to a motor over long cable runs, the difference in the motor and cable impedance can cause voltage overshoots to be present at the motor terminals. These overshoots can greatly exceed 1,000V and have the potential to damage the motor's insulation. For small motors, this typically occurs with cable lengths that exceed 33', and for larger motors, with cable lengths that exceed 100'. For situations involving the use of long cable runs, install a reflective wave trap or similar filter at the motor terminals to prevent overshoot.

reason, evaluating the needs of a drive system and selecting the components for use in the system is a task normally done by an electrical engineer, a drive manufacturer application engineer, or other trained and qualified persons. Most VFD manufacturers offer training that is specific to their product line to help both technicians and engineers understand how to make proper selections and service their product when necessary. *Figure 76A* and *Figure 76B* show a typical checklist used by one manufacturer to gather the data needed to evaluate the requirements for a VFD system. The remainder of this section briefly describes some of the other factors that can affect the VFD selection process.

4.3.1 VFD-Operated Motors

In most applications, the motors used with VFDs are standard NEMA Design A or Design B motors. The use of a motor other than a Design A or Design B usually requires the addition of special drive circuitry. The design letter assigned to a motor by NEMA identifies certain characteristics such as its slip factor and locked rotor torque. Operating self-ventilated motors at reduced speeds may cause abnormal temperature rises. As the speed of a motor is reduced below its 60-Hz base speed, motor cooling becomes less effective because of the reduced speed of the self-cooling fan. This limitation determines the minimum allowable motor speed for continuous torque operation. Derating or forced ventilation may be necessary to achieve the rated motor torque output at reduced speeds.

VFD Environment

Heat rejection is a major concern with VFDs. The heat produced in the VFD cabinet can be substantial and can cause failure of the VFDs, SCRs, and IGBTs if their operating temperature limits are exceeded. For this reason, operating VFDs at higher altitudes often requires that the unit capacity be derated.

Variable Frequency Drives

For retrofit VFD applications, many VFD manufacturers require that existing motors be replaced to ensure that the VFD unit is properly sized and not affected by previous motor abuse or rewinding.

Using VFDs with older existing motors can cause an extra heating effect on the motor and subject the motor winding insulation to high-frequency-induced voltage stress. The load capacity of older motors should be derated to at least the next lower horsepower rating, particularly if the motor has a Class F temperature rise or a 1.0 service factor.

4.3.2 Selecting the Drive

The basic requirement of VFD sizing is to match the torque-versus-speed capability of the VFD to the torque-versus-speed requirement of the driven load.

If the load torque exceeds 150% for constant torque drives or 110% for variable torque drives during starting or intermittently while running the drive, oversizing of the drive may be required. The minimum and maximum motor speeds for the application will determine the drive's base speed.

The full-load current ratings of typical VFDs are matched to typical full-load, three-phase motor current ratings in current *NEC®* tables. Generally, a VFD of a given horsepower rating will be adequate for a motor of the same rating, but the actual motor current required under operating conditions is the final determining factor for VFD sizing. If the motor will be run at full load, the VFD output current rating must be equal to or greater than the motor nameplate current. If the motor is oversized to provide a wide speed range, the VFD should be sized to provide the current required by the motor at the maximum

03314 Control Circuit and Motor Troubleshooting

Module Two 67

VFD APPLICATION CHECKLIST

Motor

New _____ Existing _____ Horsepower: _____ Base Speed: _____ Voltage: _____

FLA: _____ LRA: _____ NEMA Design: _____ Gearbox/Pulley Ratio: _____

Service Factor: _____

Load

Application: _____

Load Type: Constant Torque _____ Variable Torque _____ Constant Horsepower _____

Load inertia reflected to motor: _____

Required breakaway torque from motor: _____

Running load on motor: _____

Peak torques (above 100% running): _____

Shortest/longest required accel. time: _____ / _____ secs up to _____ Hz from zero speed

Shortest/longest required decel. time: _____ / _____ secs down to _____ Hz from max. speed

Operating speed range: _____ Hz to _____ Hz

Time for motor/load to coast to stop: _____ secs

VFD

Source of start/stop commands: _____

Source of speed adjustment: _____

Other operating requirements: _____

Will the motor ever be spinning when the VFD is started? _____

Is the load considered to be high inertia? _____

Is the load considered to be hard to start? _____

Distance from VFD to the motor: _____ feet

Type of VFD (V/Hz, Flux Vector, Closed Loop Vector): _____

Options desired: _____

_____

_____

Other special requirements/conditions: _____

Figure 76A VFD application checklist (Part 1 of 2).

VFD APPLICATION CHECKLIST (Continued)

Power Supply

Supply Transformer: _____ kVA and _____ % Z or short circuit current at drive input: _____ amps

(If the drive does not include a built-in line reactor and the available feeder short circuit current is more than 100 times the drive FLA rating, a 1% line reactor or a drive isolation transformer is required.)

Total horsepower of all drives connected to supply transformer or feeder: _____ hp

Is a drive transformer or line reactor desired? _____

Any harmonic requirements? _____ % Voltage THD: _____ % Current THD: _____ IEEE 519: _____

Total non-drive load connected to the same feeder as drive(s): _____ amps

Service

Start-up Assistance: _____   Customer Training: _____

Preventive Maintenance: _____   Spare Parts: _____

Additional Issues

Will the VFD operate more than one motor? _____

Will the power supply source ever be switched with the VFD running? _____

Is starting or stopping time critical? _____

Are there any peak torques or impact loads? _____

Will user-supplied contactors be used on the input or output of the VFD? _____

Does the user or utility system have PF capacitors that are being switched? _____

Will the VFD be in a harsh environment or high altitude? _____

Does the utility system experience surges, spikes, or other fluctuations? _____

Figure 76B VFD application checklist (Part 2 of 2).

operating torque. Motor oversizing should generally be limited to an increase of one horsepower. Gross oversizing is not recommended and will result in wasted energy.

4.3.3 Environment

The environment in which the motor operates must be examined to properly select the equipment and enclosure. The ambient temperature, cooling air supply, and the presence of gas, moisture, and dust should all be considered when choosing a drive, its enclosures, and any protective features.

4.3.4 Torque Requirements

The starting, peak, and running torques should be considered when selecting a drive. Starting torque requirements can vary from a small percentage of the full load to a value several times that of full-load torque. The peak torque varies because of a change in load conditions or the mechanical nature of the machine. The motor torque available to the driven machine must be more than that required by the machine from start to full speed. The greater the excess torque, the more rapid the acceleration potential.

4.3.5 Duty Cycle

Selecting the proper drive depends on whether the load is steady, varies, follows a repetitive cycle of variation, or has pulsating torques. Certain applications may require continuous reversals, long acceleration times at high torque due to inertia loads, frequent high rate acceleration, or cyclic overloads, which may result in severe motor heat-

03314 **Control Circuit and Motor Troubleshooting**   Module Two 69

ing if not considered in the selection of the drive. The duty cycle, which is defined as a fixed repetitive load pattern over a given period of time, is expressed as the ratio of on-time to the cycle period. When the operating cycle is such that the drive operates at idle, or at reduced load for more than 25% of the time, the duty cycle becomes an even more important factor in selecting the proper drive.

4.4.0 VFD Dynamic Braking

Braking provides a means of quickly stopping an AC motor when it is necessary to stop it faster than is normally possible by removing the power and letting the motor wind down. Braking can be accomplished in several ways. The method used depends on the application, available power, and circuit requirements. Some common methods of motor braking are:

- Dynamic braking (DC electric braking)
- Dynamic braking (AC drives)
- Electromechanical braking

Electromechanical braking is accomplished using solenoid-operated friction brakes. Friction brakes are used for applications such as printing presses, small cranes, overhead doors, hoisting equipment, and machine tool control. Since electromechanical braking is not used in HVACR applications, only dynamic braking is discussed in this section. Note that braking is not required in every VFD application.

4.4.1 Dynamic DC Braking of an AC Motor

One common method of braking an alternating current induction motor is by using a DC electric braking circuit. This is done by momentarily connecting a DC voltage to the stator winding (*Figure 77*). When DC is applied to the stator winding of an AC motor, the stator poles become electromagnets. Current is induced into the windings of the rotor as the rotor continues to spin through the magnetic field. This induced current produces a magnetic field around the rotor. The magnetic field of the rotor is attracted to the magnetic field produced in the stator. The attraction of these two magnetic fields produces a braking action in the motor. An advantage of using this method of braking is that motors can be stopped rapidly without having to use brake linings or drums. DC braking cannot be used to hold a suspended load, however. Mechanical brakes must be employed when a load must be suspended, as with a crane or hoist.

The DC brake circuit in *Figure 77* operates as follows:

- Magnetic starter (M) is controlled by a standard Stop/Start pushbutton station with memory. Timing relay (TR) is an off-delay timer. Its normally open, timed-open (NOTO) contact is used to apply power to the braking relay (BR) after the Stop pushbutton is pressed.
- The off-delay of timing relay TR must be adjusted so that its contacts will remain closed until after the motor comes to a stop. At this time, the contacts open, disconnecting the DC power to the motor winding.
- When the Stop pushbutton is pressed, start contactor M de-energizes, thus removing three-phase input power from the motor. Also, braking relay BR is energized through the NC contact M and NOTO contact TR. This causes its normally-open contacts to close, energizing a DC supply consisting of a step-down transformer and full-wave bridge rectifier circuit. It also connects two of the motor leads (T1 and T3) to the output of the DC supply, initiating the motor braking action.
- A transformer with tapped windings is often used to adjust the amount of braking torque applied to the motor. Current-limiting resistors could be used for the same purpose. This allows for a low- or high- braking action depending on the application. The higher the applied DC voltage, the greater the braking force.
- An electrical interlock is provided by the NC contacts of starter M and the NC contacts of braking relay BR. This interlock prevents the motor starter and braking relay from being energized at the same time. This is required because the AC and DC power supplies must never be connected to the motor simultaneously. Total interlocking should always be used on electrical braking circuits. Total interlocking is the use of mechanical, electrical, and pushbutton interlocking.

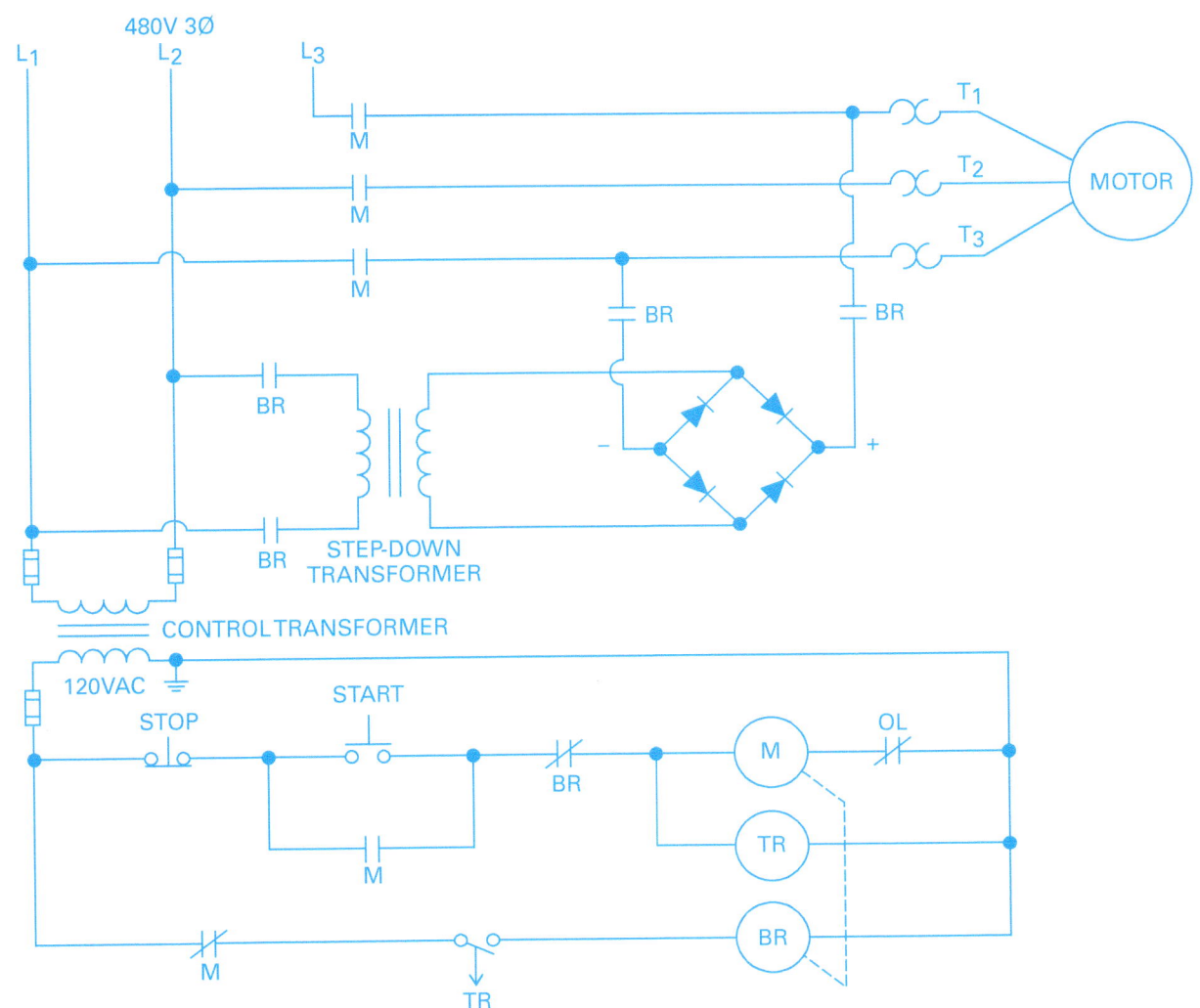

Figure 77 Circuit for DC braking of an AC motor.

4.4.2 Dynamic Braking with an AC Drive

Dynamic braking can be used with AC motors controlled by VFDs. Torque will always act to cause the motor's rotor to run towards synchronous speed. If the synchronous speed is suddenly reduced, negative torque is developed in the motor. When this occurs, the motor acts like a generator by converting mechanical power from the shaft into electrical power that is returned to the AC drive unit. This is like driving a car downhill in a lower gear so that the car's engine acts as a brake. For an AC drive to operate without damage during this condition, a means must exist to deal with the electrical energy returned to the drive by the motor.

Figure 78 shows a simplified diagram of an AC drive unit that uses dynamic braking. Electrical energy returned by the motor can cause voltage in the drive DC bus to become excessively high when added to the existing supply voltage. This excess voltage can damage the drive components. To alleviate this problem, a braking resistor is connected across the DC bus through an IGBT switching transistor. The braking resistor is added and removed from the circuit by the switching action of the IGBT switch. Energy returned by the motor is impressed on the DC bus. When this energy reaches a predetermined level, the IGBT switch is turned on by the control logic. This places the braking resistor across the DC bus, causing the excess energy to be dissipated by the resistor, thus reducing the DC bus voltage. When the DC bus voltage is reduced to a safe level, the IGBT is turned off, removing the resistor from the DC bus. This process allows the motor to act as a brake, slowing the connected load.

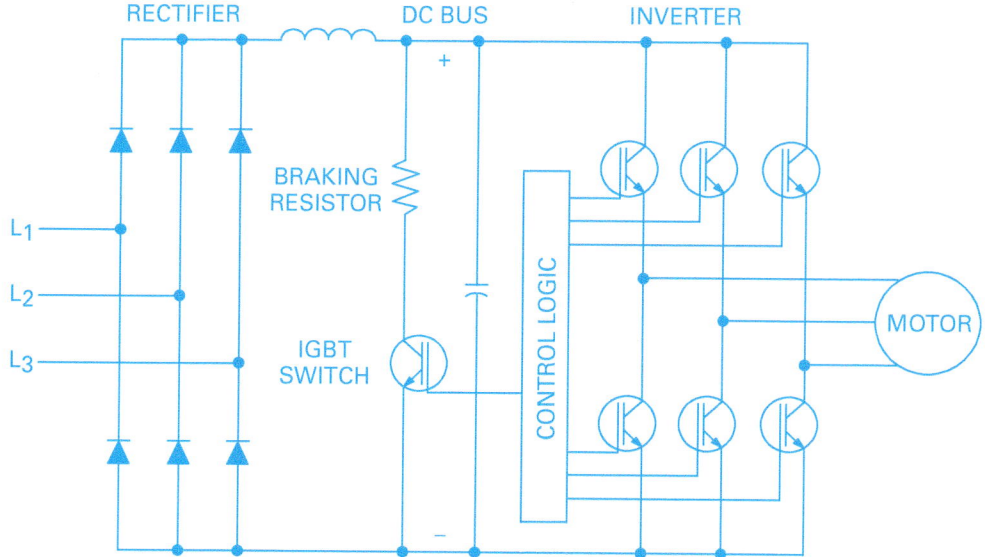

Figure 78 Simplified schematic of an AC drive using dynamic braking.

Additional Resources

Adjustable Frequency Drives, Application Guide, Latest edition. Milwaukee, WI: Cutler-Hammer.

Electric Motors and Drives: Fundamentals, Types, and Applications. Austin Hughes and Bill Drury. Current edition. Atlanta, Ga: Newnes.

4.0.0 Section Review

1. If done properly, the volts per hertz (V/Hz) ratio in a VFD can allow a motor to run at twice its nameplate _____.

    a. current
    b. speed (rpm)
    c. voltage
    d. service factor

2. When a VFD is in the volts per hertz (V/Hz) optimization mode, the mode provides _____.

    a. an automatic torque boost
    b. a 10% reduction in torque
    c. linear voltage increases
    d. time for motor acceleration

3. It is appropriate to oversize the VFD drive used with variable-torque drives if the load torque exceeds _____.

    a. 100%
    b. 110%
    c. 125%
    d. 150%

4. How is dynamic DC braking applied to slow a motor?

    a. DC voltage is applied to electric brake calipers.
    b. DC voltage is applied to the stator of an AC motor.
    c. DC voltage is applied to the rotor of an AC motor.
    d. AC motor power is replaced with DC power.

Section Five

5.0.0 Electronically Commutated Motors

Objective

Identify and describe how to service electronically commutated motors (ECMs).
a. Identify and describe the operation of ECMs.
b. Describe how to install and set-up an ECM.
c. Describe how to troubleshoot an ECM.

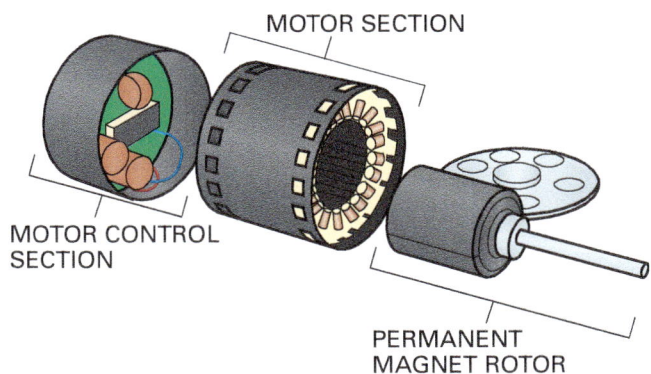

Figure 79 ECM major components.

Electronically commuted motors (ECMs) are now widely used in residential and commercial HVACR equipment. These motors offer several benefits to the end user including the following:

- *Increased energy savings* – These motors consume a fraction of the power of a conventional PSC blower motor.
- *Improved comfort* – When coupled with modulating or multi-stage heating and cooling products, they adjust fan speed to match heating or cooling capacity. Fan speeds can be slowed down to allow for better dehumidification without overcooling the conditioned space.
- *Quieter operation* – ECMs start at low speed and slowly ramp up to the required fan speed. When the thermostat is satisfied, they slowly ramp down the speed before shutting off. Continuous fan operation can be at a very low speed, allowing for good air circulation without excess air noise.

5.1.0 ECM Characteristics and Operation

ECMs are three-phase, direct-current (DC) motors that are brushless. They also have a permanent magnet rotor. Most ECM motors consist of two sections; the motor section and the motor control (*Figure 79*). The motor control section has two functions. It first takes single-phase 120VAC or 240VAC power and converts it to DC. The motor can operate at either of these AC voltages simply by adding or removing a jumper on the five-pin connector in the motor control section. The five-pin connector provides the motor with its primary power supply. The control section then converts the DC power to three-phase power to actually turn the motor. The motor control section houses the electronics that control the speed of the motor.

The permanent magnet rotor is turned by sequentially energizing the three sets of windings in the stator to move the rotor in a clockwise direction. The signal applied to each stator is a square-wave voltage produced from rectified and smoothed DC. To advance the rotor, the square-wave voltage is first applied to the A-B windings (*Figure 80*) that moves the rotor clockwise. Power is then applied to the B-C windings, and then to the A-C windings to complete a full revolution of the rotor.

ECM motors are easy to identify. They look like conventional PSC motors with a section added on to the motor on the end opposite the shaft (*Figure 81*). The added-on section is the motor control section. The size of the motor control section has shrunk over time as more advance designs were introduced. With one popular brand of ECMs, all versions contain a five-pin connector where power is applied. The addition or deletion of a jumper on the five-pin connector determines the motor's AC operating voltage. With that same manufacturer, the connector used to carry the various speed control signals between the motor and the HVACR unit can be either a 16-pin connector (earlier versions) or a four-pin connector (later versions). Unlike conventional PSC motors, power is always available to the motor through the five-pin connector. It is only after additional speed control inputs are received through the 16-pin or four-pin connectors that the motor will start and run. Although power is applied to the motor system, it cannot function without instructions in the form of electrical signals.

ECMs have another characteristic that can help identify them. When turning the motor shaft by hand, points of high and low resistance can be felt. That is caused by the magnetism in the rotor.

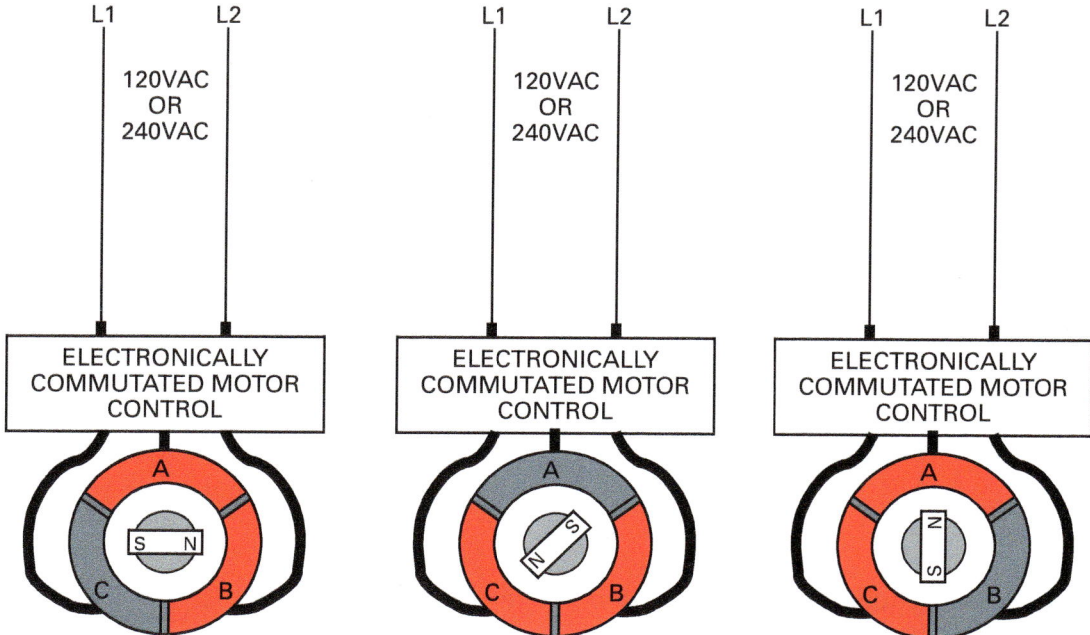

Figure 80 ECM motor rotation.

Figure 81 ECM identification.

ICM vs. ECM

Early HVACR systems literature and manufacturers referred to these motors as ICM (integrated control module) motors. The words themselves indicate that a module was incorporated into a motor for its control. Today they are referred to as ECMs (electronically commutated motors). However, the ICM acronym might still be encountered on some products and wiring diagrams.

Rock and Roll!

When an ECM motor is first started, it may rock back and forth a few times. This is considered normal operation, as it determines which direction the motor should be turning.

5.1.1 ECM Speed Control

The speed of an ECM is controlled by the electronics in the motor control section and the input signals it receives from the electronic control within the furnace or air conditioner in which the ECM is installed. For example, a furnace control board provides different speed-control signals to the motor speed control based on the furnace's operating mode. If the furnace has staged heating capacity (high- and low-fire modes), for example, it must provide a signal for a lower fan speed during low-fire operation. During installation, the furnace control board must be configured so that the correct fan speeds are available for the furnace's various operating modes. Different manufacturers accomplish this differently.

Figure 82 shows how an ECM installed in a two-stage furnace responds to a call-for-heat. As room temperature drops, the first stage of the room thermostat sends a call-for-heat signal to the furnace control board. Since the call is for first-stage heat, the furnace control board responds by sending a signal to the ECM requesting operation at low speed. These signals are typically in the form of a series of pulses. The frequency of the pulses tells the motor at what speed to operate. If the room thermostat calls for second-stage heat, the furnace control board would send a new signal to the motor to increase its speed.

5.2.0 ECM Installation and Setup

Other than replacement scenarios, ECMs are factory-installed by the manufacturer. There is no setup required for the motor itself. Any setup is done on the equipment that contains the ECM. The setup is very important, as they are installed in a variety of systems with varying airflow needs.

For example, a 100,000 Btuh gas furnace with an ECM may be installed with a 2, $2\frac{1}{2}$, or 3-ton cooling unit. Each cooling unit capacity has its own airflow requirements based on cooling capacity (350 to 450 cfm/ton). If that same gas furnace has staged heating, each stage of heating has its own airflow requirements as well. If the system has a dehumidification mode, airflow requirements for that mode are much lower than needed for heating or cooling. Continuous fan operation also can have a dedicated and low fan speed. The control that sends signals to the ECM must be setup to ensure that all the necessary functions are accounted for.

The furnace control board on a deluxe furnace or air handling unit needs the capability to accommodate all those diverse airflow requirements. How that is accomplished varies by manufacturer. Airflow setup can take the form of rotary selection switches, dip switches, or jumper wires. Some high-end systems may even allow setup to be done with a user interface such as a laptop computer, smart phone or tablet-style device. *Figure 83* shows an interface board used on a packaged heat/cool unit equipped with an ECM. On the board shown, equipment capacity and the size of auxiliary electric heaters are factors that help determine blower speed. The amplitude of the voltage signal and the terminal on the motor to which they are applied are based on the signals received from the room thermostat and/or other input devices.

> **CAUTION**
> Always remove power from the equipment before connecting or disconnecting ECM electrical connectors. Do not mate motor connectors with the circuit energized.

If an ECM must be replaced, use an exact replacement. The only motor setup required then is to check the five-pin connector jumper to ensure the voltage is correct for the application. The jumper must be installed for 120V operation and removed for 240V operation. Note, however, that the two connectors—the power connector and the communications connector—must be compatible with the connectors provided on the equipment. The connectors are keyed such that they can only be mated one way and with the opposite gender. This helps eliminate any error in connections. Forcing the connectors together incorrectly will likely result in motor and/or controller damage once they are energized. If an alternate motor must be used, the manufacturer's instructions must be followed carefully to ensure that the motor and the unit controller are compatible and communicating with each other. Note that there are no universal replacements in the ECM world. Even when an exact replacement is provided, it is important to read and apply the instructions provided with the motor. Changes in design or wiring may have taken place since the original, now-failed motor was placed in service.

When a new motor is installed, be sure to position the motor to allow the wiring connections to hang down between the 4 o'clock and 8 o'clock positions (*Figure 84*). This helps prevent any moisture from entering the connectors. As the wires exit the motor, form a drip loop that encourages water to roll down the wiring rather than into the motor.

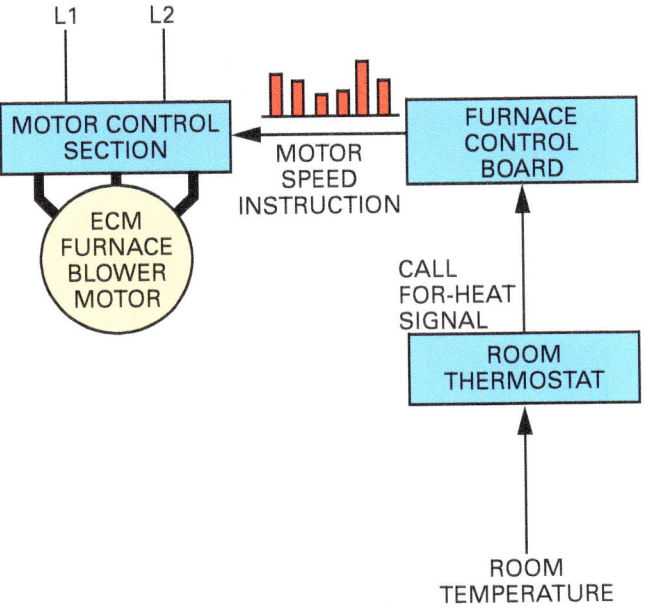

Figure 82 ECM speed control.

Once the new motor is installed, check the operation thoroughly in all operating modes to ensure the motors response to the conditions is correct. With furnaces, ensure that the temperature rise in the heating mode remains within the allowable range.

5.3.0 ECM Troubleshooting

Troubleshooting an ECM is likely to happen as part of a routine service call, especially one where comfort is an issue. Where comfort and airflow are involved (as opposed to complete failure), always check for obvious problems first, such as the following:

- System airflow problems such as dirty air filters, closed duct dampers or blocked supply registers or return grilles. It may be necessary to measure system airflow to confirm that the proper volume of air is being delivered.
- Low refrigerant charge.
- Room thermostat setpoint.
- Connections on the ECM are tight and secure. Both connectors are keyed to ensure a correct connection.
- If the equipment control board has a self-diagnostic feature or displays fault codes, use those features to help identify the problem.

If no obvious problems are found, check the heat/cool equipment to confirm the control board has been properly configured for all airflow contingencies. It is common for some manufacturers to have default settings on their controls as they come from the factory. Those default settings may not be appropriate for a specific equipment application. An incorrect setup in an ECM-equipped system can result in poor system performance.

5.3.1 Motor Troubleshooting

The following troubleshooting procedures were developed by one major manufacturer of ECMs.

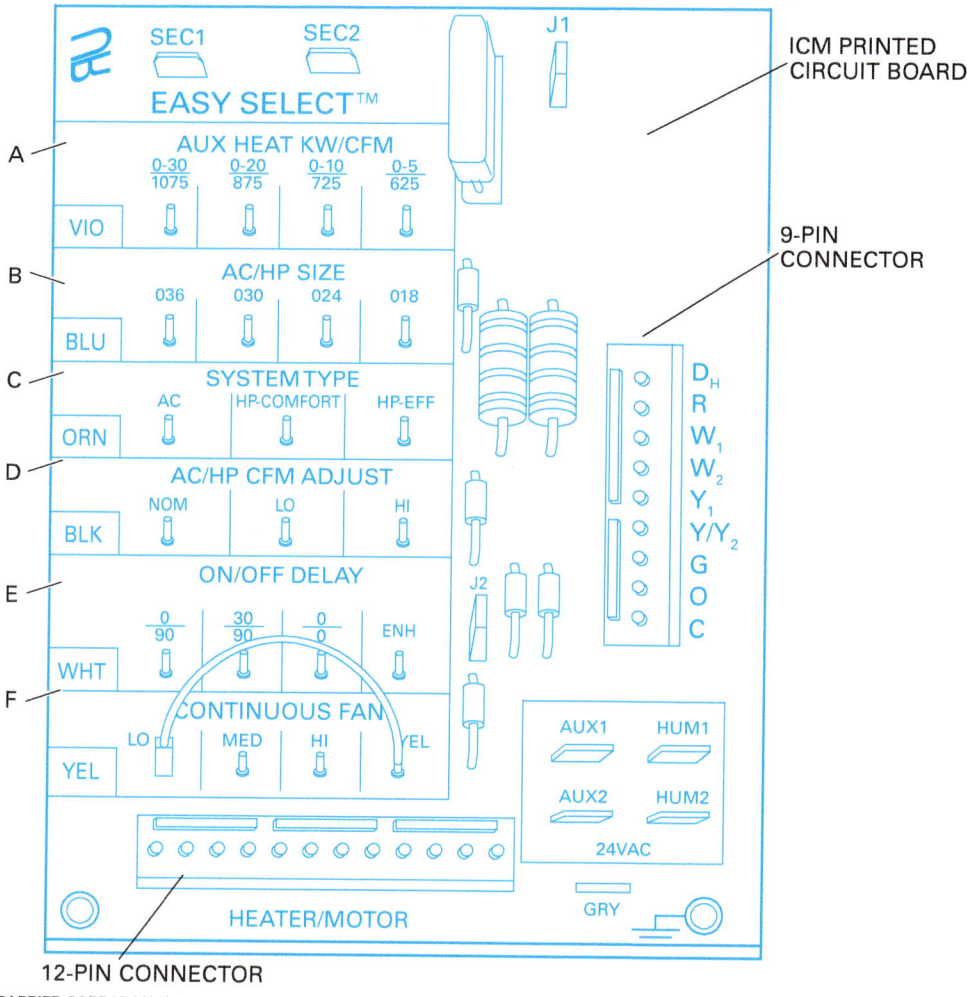

Figure 83 ECM speed-control interface board.

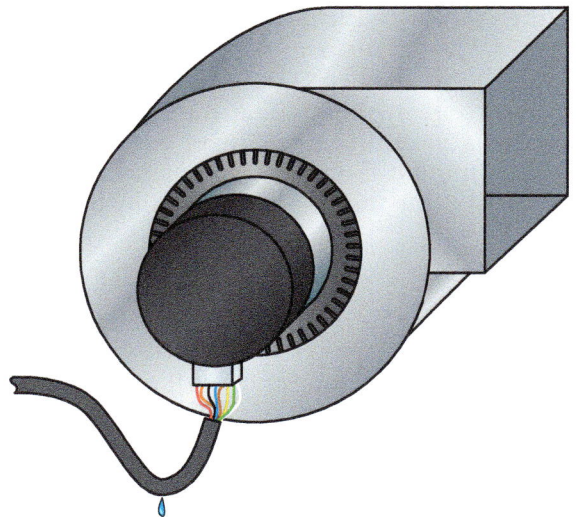

Figure 84 Forming a drip loop.

Other motor manufacturers have similar procedures. These procedures are general in nature and are not meant to replace specific troubleshooting procedures provided by the equipment manufacturer.

> **CAUTION**
> Always remove power from the equipment before connecting or disconnecting ECM electrical connectors. Do not mate motor connectors with the power applied.

If the motor is not running, disconnect power from the air handler. Then remove the five-pin connector and restore power. Check for the correct voltage at the connector pins (*Figure 85*). When using a 120VAC power supply, the polarity of the voltage must also be correct. The applied voltage should be ±15% of the desired voltage (120V or 240V). Note that the motor must always be well grounded for proper operation.

Equipment manufacturers often supply charts listing the voltages that should be read across the 16-pin or four-pin connectors of the motor. Voltages can be AC or DC. They also may supply the conditions under which the various voltages should be read. These signals and voltages can be tested when specific information is available. This process determines whether the equipment control module is functional and sending the expected signals.

Devices that simulate communication input signals are available to test the motor (*Figure 86*). The tester shown can be used with motors with 16-pin or four-pin connectors. The tester simulates signals from the control board in the air handler.

Here is how the tester is used to test an ECM.

- Turn off power to the air handler. Disconnect the 16-pin or four-pin connector from the motor but leave the five-pin power supply connector attached to it. Turn off the tester.
- Connect the alligator leads on the tester to the R and C terminals on the control transformer or terminal board to obtain 24VAC. This powers the tester.
- Connect the 16-pin or the four-pin connector on the tester to the matching connector on the motor control section. Both connectors may be provided on the tester to allow it to work with different motors.
- Turn the tester on.

If the motor runs when the tester is turned on, the problem lies elsewhere. Potential causes include a failed air handler control board, or a loose connection. If the motor does not run, the problem may be within the motor control section or the motor itself. Remove the tester and restore all normal connections when done.

If the motor did not run, testing must proceed to the motor and its integrated control. Since the motor control section is bolted to the motor, those bolts can be removed and both sections can be checked separately. The motor section can usually remain attached to its mount in the blower assembly during this procedure. The motor section is tested as follows:

- Turn off power to the air handler. Disconnect the 16-pin or four-pin communications connector from the motor as well as the five-pin power supply connector. Then wait at least 5 minutes before proceeding to the next step.

> **WARNING!**
> After disconnecting power and/or communications connectors from an ECM motor, wait at least five minutes before accessing the control section. This allows time for the capacitors to discharge. If the capacitors are not allowed sufficient time to discharge first, an electrical shock may result.

- Unbolt the two sections and disconnect the three-pin connector between the two sections. Place the motor control section aside.
- The three pins on the connector attach to each of the three-phase motor leads. Use an ohmmeter set on its lowest scale to check the resistance of each motor winding (*Figure 87*). Each winding should measure less than 20 ohms. There should be no more than a ±10% difference between the three resistance readings.

- Set the ohmmeter on its highest scale to test for grounded motor windings. Place one lead on the frame of the motor (ground) and the other lead on each of the pins of the three-pin connector. Ideally, all readings to ground will show no continuity (open circuit). Replace any motor that reads less than 100,000 ohms to ground.

If the motor windings are intact and not grounded, the motor control section is likely defective. The motor can be reused and a new control section installed. The motor can remain bolted to its mount in the blower section during the replacement. Of course, the entire ECM must be removed if the motor section is defective.

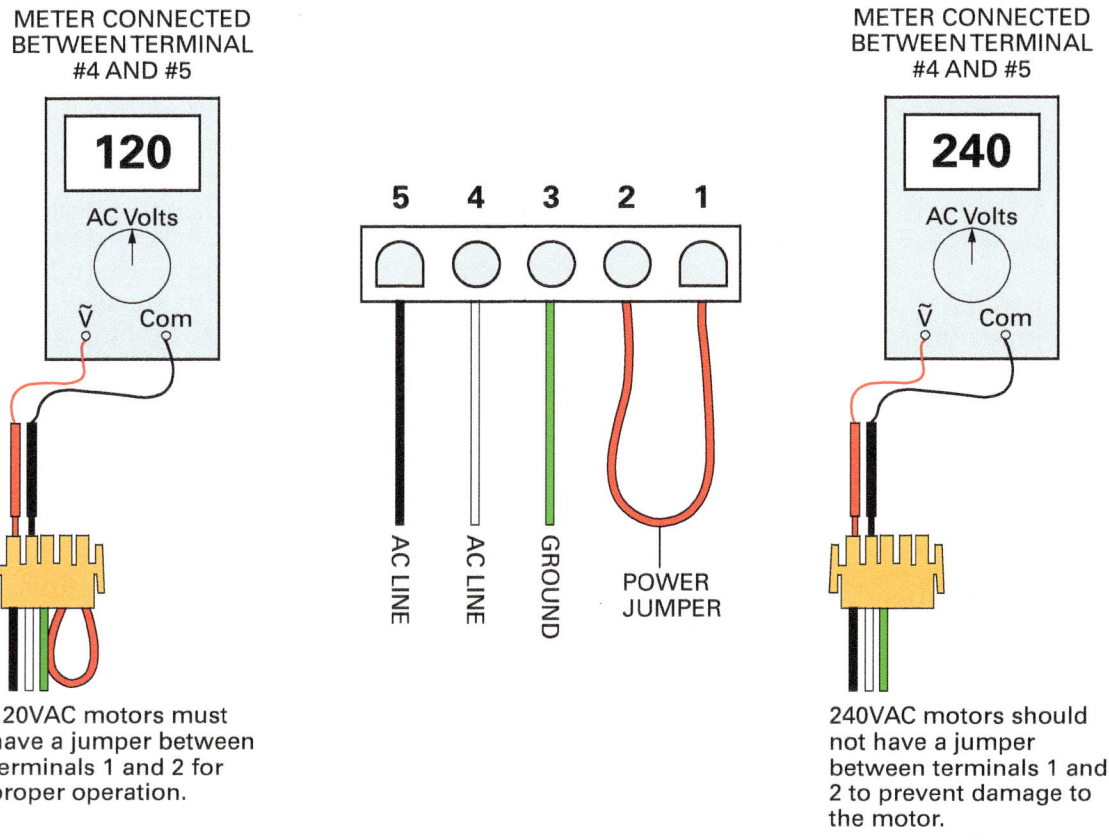

Figure 85 Input voltage check.

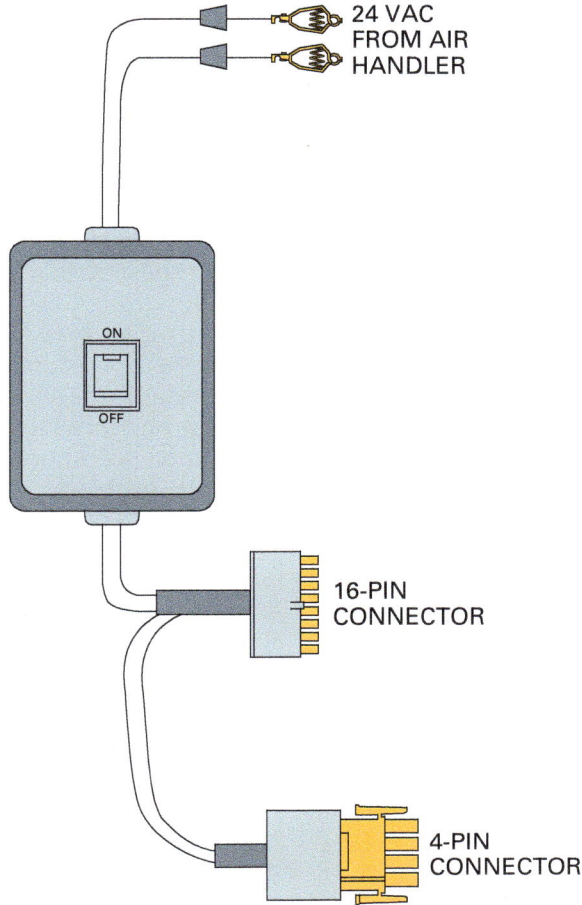

Figure 86 ECM tester.

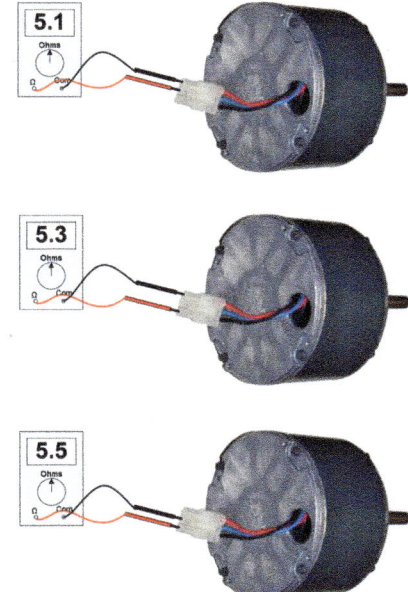

Figure 87 ECM winding-resistance test.

Additional Resources

The ECM Textbook. 2007. Fort Wayne, IN: GE-ECM by Regal-Beloit.

Understanding Electrically Commutated Motors. Christopher Mohalley. 2014. Rolling Meadows, IL: Refrigeration Service Engineers Society.

5.0.0 Section Review

1. The signal applied to the stator in an ECM is a _____.
   a. pure sine wave
   b. modulated AC signal
   c. current-limited AC signal
   d. square-wave voltage

2. Installing an ECM requires the technician to position the motor wiring connections _____.
   a. at the bottom of the motor
   b. at the top of the motor
   c. only to the left of the motor
   d. parallel with the motor shaft

3. When measuring the motor winding resistance of an ECM, the readings should not vary more than _____.
   a. ±1%
   b. ±10%
   c. ±15%
   d. ±20%

Summary

HVACR systems contain various types of electric motors. Many devices and control schemes control these motors. The ability to analyze HVACR control systems is a critical skill for the service technician, as a large number of problems that occur are control-circuit faults.

In the past, control circuit devices were electromechanical. Modern equipment relies more and more on electronic devices. Variable frequency drives are now used to efficiently change the speed of large AC motors. Air handlers and furnaces used in residential and light commercial applications use electronically commutated motors (ECMs) for more efficient operation.

No matter how complex the control system, it consists of individual components that are combined to form control functions such as heating, cooling, fan control, and defrost. If you know how the components work, you can figure out how they fit together and how the system functions as a whole. Once you have these skills, you can troubleshoot any system.

Effective troubleshooting is a process by which the HVACR technician listens to a customer's complaint, performs an independent analysis of a problem, and then initiates and performs a systematic step-by-step approach to troubleshooting that results in the correction of the problem.

Review Questions

1. The purpose of a compressor short-cycle timer is to _____.
   a. prevent compressor restart before system pressures equalize
   b. ensure the compressor runs for at least five minutes
   c. keep track of how often the compressor turns on
   d. run the compressor every five minutes to prevent freeze-up

2. The purpose of a lockout relay is to _____.
   a. prevent unauthorized equipment use
   b. prevent compressor operation until the fan runs
   c. protect the compressor motor from current surges
   d. prevent equipment restart when a safety control resets

3. The purpose of a motor starter is to _____.
   a. jump-start the compressor if power fails
   b. vary the speed of a motor as conditions change
   c. check motor current when the motor is running
   d. power the motor and provide overload protection

4. Which of the following is not an application for an outdoor temperature thermostat?
   a. Control compressor capacity
   b. Dual-fuel heat pump switchover
   c. Chiller protection
   d. Control crankcase heater operation

5. A limit control is provided on a furnace to _____.
   a. ensure the furnace provides adequate heat
   b. shut off the furnace fan motor
   c. prevent the furnace from overheating
   d. sense the presence of carbon monoxide

6. Low-voltage (24V) control circuits are used in HVACR systems because _____.
   a. step-down transformers produce only 24V output
   b. 24V power is safer and cheaper
   c. 24V power is readily available from the electric company
   d. it is the same voltage used to operate the compressor and fan motors

7. Which of the following is *not* true regarding a cooling compensator?
   a. It compensates for the lag between the call for cooling and when cooling starts.
   b. It is a variable resistor just like the heat anticipator.
   c. It causes the cooling thermostat to close sooner than it otherwise would.
   d. It is usually placed in parallel with the cooling thermostat switch.

8. The differential in a thermostat is _____.
   a. the difference between a thermostat's cut-in and cut-out points
   b. the difference between the cooling and heating setpoints
   c. normally set apart at least 12°F
   d. the difference between heat anticipator and heating setpoints

9. Which of the following is *not* a common feature on an electronic thermostat?
   a. Battery backup to prevent loss of programs
   b. Motor-driven heat anticipator
   c. Program override
   d. Digital readouts

 03314 Control Circuit and Motor Troubleshooting

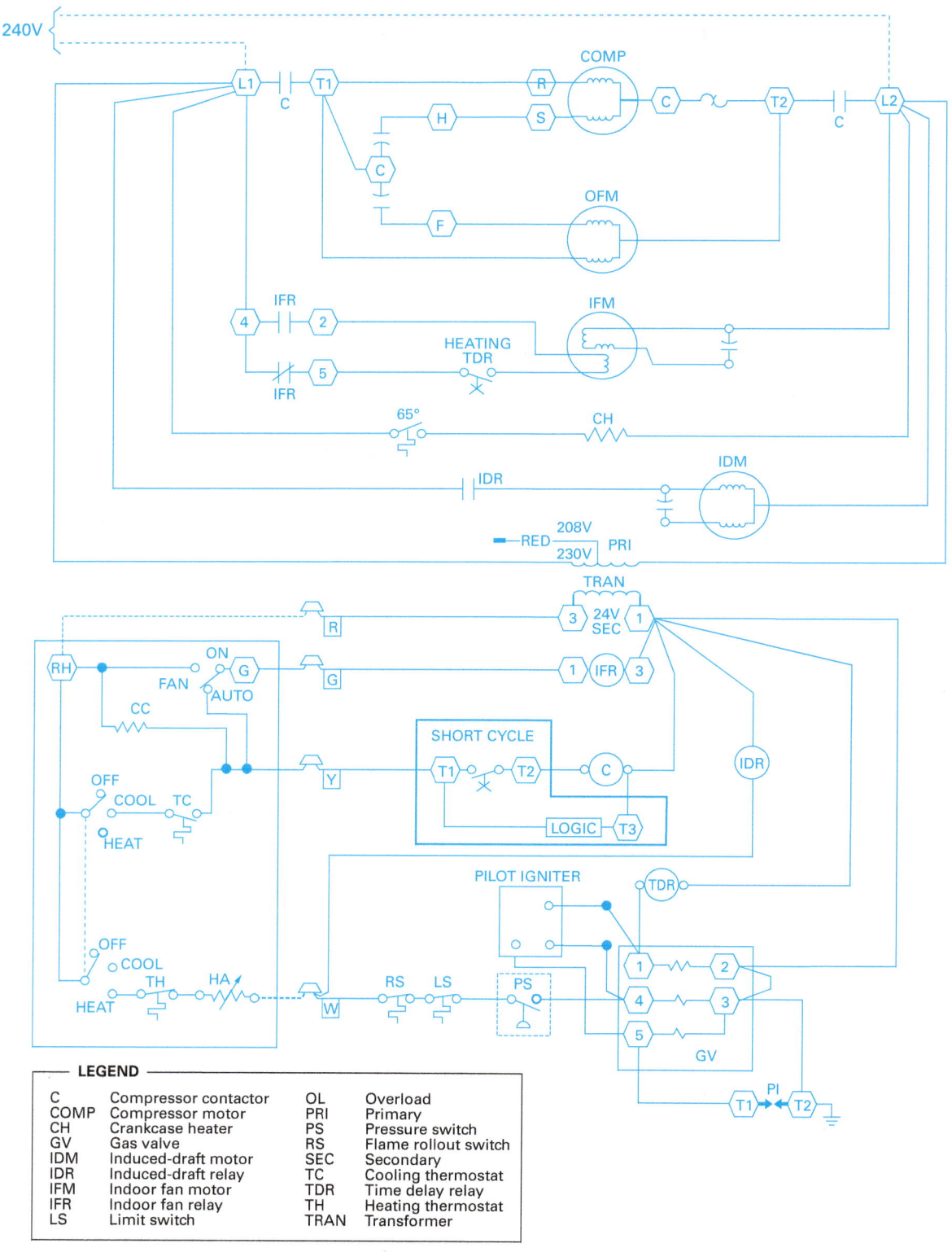

Figure RQ01

10. A good location for a thermostat is _____.
    a. anywhere that it will receive direct sunlight
    b. about five feet above the floor on an outside wall
    c. about five feet above the floor on an inside wall
    d. in a corner where there is minimum air circulation

11. In a standard thermostat-wiring scheme, the R terminal is connected to _____.
    a. the fan control
    b. 24VAC power source
    c. the reversing valve
    d. the cooling circuits

12. In the electrical control circuit of a combined heating/cooling system, the outdoor fan motor _____.
    a. usually has a continuous-on mode for ventilation
    b. runs at low speed in the heating mode
    c. is on whenever the compressor is on
    d. is off whenever the compressor is on

13. Refer to *Figure RQ01*. Which component prevents the compressor from starting immediately on a call-for-cooling?
    a. Time-delay relay
    b. Indoor fan relay
    c. Contactor
    d. Short-cycle timer

14. Refer to *Figure RQ01*. Which component allows the indoor fan to cycle on and off with the compressor?
    a. Time-delay relay
    b. Fan function switch
    c. Contactor
    d. Short-cycle timer

15. Refer to *Figure RQ01*. Which component prevents the simultaneous energizing of the low- and high-speed windings of the indoor fan motor?
    a. TDR
    b. IFR
    c. PS
    d. IDR

16. Refer to *Figure RQ01*. Which component prevents the IFR from being energized during a call-for-heat?
    a. Time-delay relay
    b. Flame rollout switch
    c. Pressure switch
    d. Room thermostat

17. Refer to *Figure RQ01*. When is the time-delay relay energized?
    a. On a call-for-cooling
    b. When the gas valve energizes
    c. On a call-for-heating
    d. When the compressor starts

18. A troubleshooting aid normally provided with a label diagram is the _____.
    a. troubleshooting tree
    b. troubleshooting table
    c. wiring diagram
    d. fault isolation diagram

19. When testing a relay coil, the multimeter indicates a measurable resistance. This normally means that the coil is _____.
    a. shorted
    b. good
    c. open
    d. partially shorted

20. When checking a 440V motor for grounded windings, the minimum resistance that should be measured between any lead and ground is _____.
    a. 1,000 ohms
    b. 230,000 ohms
    c. 440,000 ohms
    d. infinity

21. When troubleshooting the operation of a compressor start relay circuit by measuring the start winding current draw with a clamp-on ammeter, no current is measured on the clamp-on ammeter when the compressor is first energized. This likely means that the start _____.
    a. relay contacts are open
    b. relay contacts are closed
    c. relay coil is open
    d. capacitor is shorted

22. With a variable-frequency drive (VFD) motor, its speed is proportional to the _____.
    a. applied frequency
    b. voltage amplitude
    c. power factor
    d. motor winding resistance

23. Why are insulated gate bipolar transistors (IGBT) used in VFD controllers?
    a. Low cost
    b. Rapid heat dissipation
    c. High switching speed
    d. Low power consumption

24. An ECM can be described as a _____.
    a. single-phase motor
    b. direct-current motor
    c. three-phase AC motor
    d. split-phase AC motor

25. Voltages measured at the four-pin or 16-pin connector on an ECM are _____.
    a. AC only
    b. DC only
    c. filtered AC only
    d. AC or DC

Trade Terms Introduced in This Module

Base speed: The rating given on a motor's nameplate at which the motor will develop rated horsepower at its rated load and voltage.

Cooling compensator: A fixed resistor installed in a thermostat to act as a cooling anticipator.

Deadband: A temperature band, usually 3°F (1.6°C), that separates heating and cooling in an automatic changeover thermostat.

Differential: The difference between the cut-in and cut-out points of a control.

Droop: A mechanical condition caused by heat that affects the accuracy of a bimetal thermostat.

Fault isolation diagrams: Troubleshooting aids usually contained in the manufacturer's installation, startup, and service instructions for a particular product. Fault isolation diagrams are also called troubleshooting trees. Normally, fault isolation diagrams begin with a failure symptom and then guide the technician through a logical decision-action process to isolate the cause of failure.

Insulated gate bipolar transistor (IGBT): A type of transistor that has low losses and low gate-drive requirements. This allows it to be operated at higher switching frequencies.

Invar®: An alloy of steel containing 36% nickel. It is one of the two metals in a bimetal device used in temperature controls.

Label diagrams: Troubleshooting aids usually placed in a convenient location inside the equipment. They normally depict a wiring diagram, a component arrangement diagram, a legend, and notes pertaining to the equipment.

Ladder diagram: An electrical diagram that provides a schematic diagram of circuits, with the load lines arranged like the rungs of a ladder between vertical lines representing the power source.

Wiring diagram: A troubleshooting aid, sometimes called a schematic, which provides a picture of what the unit does electrically and shows the actual external and internal wiring of the unit.

Appendix

Common Schematic Symbols

SWITCHES						
DISCONNECT	MAGNETIC CIRCUIT BREAKER	THERMAL CIRCUIT BREAKER	LIMIT SPRING RETURN			MAINTAINED POSITION
			NORMALLY OPEN	NORMALLY CLOSED	NEUTRAL	
			DOUBLE THROW	DOUBLE POLE	TRIPLE POLE	

Liquid Level		Vacuum & Pressure		Temp. Actuated		Air or Water Flow	
NORMALLY OPEN (1)	NORMALLY CLOSED (2)	NORMALLY OPEN (1)	NORMALLY CLOSED (2)	NORMALLY OPEN (1)	NORMALLY CLOSED (2)	NORMALLY OPEN (1)	NORMALLY CLOSED (2)

CONDUCTORS		FUSES	COILS			
NOT CONNECTED	CONNECTED	OR	RELAYS, TIMERS, ETC.	OVERLOAD THERMAL	SOLENOID	TRANSFORMER

PUSHBUTTONS			
SINGLE CIRCUIT	DOUBLE CIRCUIT	MUSHROOM CIRCUIT	MAINTAINED CONTACT
NORMALLY OPEN / NORMALLY CLOSED			

TIMER CONTACTS CONTACT ACTION IS RETARDED WHEN COIL IS:				GENERAL CONTACTS STARTERS, RELAYS, ETC.		
ENERGIZED		DE-ENERGIZED				
NORMALLY OPEN	NORMALLY CLOSED	NORMALLY OPEN	NORMALLY CLOSED	OVERLOAD THERMAL	NORMALLY OPEN	NORMALLY CLOSED

(1) **Make on rise**
(2) **Make on fall**

NCCER – *HVAC Level Three* 03314

COILS			
AUTOMATIC TRANSFORMER	REACTORS		ADJUSTABLE
	IRON CORE	AIR CORE	SHOWN WITH IRON CORE

DIODE RECTIFIERS		MOTORS & COMPRESSORS		
HALF-WAVE	FULL-WAVE	THREE-PHASE	SINGLE-PHASE	
			TWO LEADS	CAPACITOR START/RUN

RESISTORS	
FIXED	VARIABLE (OR)

THERMOCOUPLE	LAMPS	BATTERY	GROUND		CAPACITOR	
			ELECTRICAL	MECHANICAL	FIXED	ADJUSTABLE
					X-SIDE NEAR GROUND	

Additional Resources

This module presents thorough resources for task training. The following reference material is recommended for further study.

Refrigeration and Air Conditioning: An Introduction to HVAC/R, Larry Jeffus. Current edition. New York, NY: Pearson.

Electrical Motor Controls for Integrated Systems, Gary Rockis and Glen A. Mazur. Current edition. Homewood, IL: American Technical Publishers, Inc.

Electric Motors and Drives: Fundamentals, Types, and Applications, Austin Hughes and Bill Drury. Current edition. Atlanta, Ga: Newnes.

Adjustable Frequency Drives, Application Guide, Latest edition. Milwaukee, WI: Cutler-Hammer.

The ECM Textbook, 2007. Fort Wayne, IN: GE-ECM by Regal-Beloit.

Understanding Electrically Commutated Motors, Christopher Mohalley. 2014. Rolling Meadows, IL: Refrigeration Service Engineers Society.

Figure Credits

Courtesy of Crydom, Inc., Figure 8

Schneider Electric, Figures 9, 10, SA08

Courtesy of Mara Industrial Supply, SA01

Courtesy of Danfoss, Figure 12A and 12C

Courtesy of White-Rodgers/Emerson Climate Technologies, Figures 16B, 23, 39

Courtesy of Honeywell International, Figure 24

U.S. Sensor Corp., Figure 26

Shortys HVAC Supplies, LLC, Figure 28B

© iStock.com/PhotoEuphoria, Figure 34

© itakdalee/ Shutterstock.com, Figure 35

Daikin Industries, Ltd., SA02, SA03

Courtesy of Pinnacle Heating and Cooling, Figure 41

Carrier Corporation, Figures 45, 46, 53, 56, 83

Hydrolevel Company, Figure 68

Xylem Applied Water Systems, Figure 71

Courtesy of Taco, Inc., Figures 72, 73

Courtesy of www.thedealertoolbox.com, Figures 79, 81, 85, 87

Section Review Answer Key

Answer	Section Reference	Objective
Section One		
1. d	1.1.7	1a
2. c	1.2.6	1b
Section Two		
1. c	2.1.4	2a
2. a	2.2.0	2b
3. d	2.3.1	2c
4. a	2.4.0	2d
Section Three		
1. a	3.1.1	3a
2. c	3.2.1	3b
3. b	3.3.4	3c
4. d	3.4.2	3d
5. c	3.5.1	3e
6. c	3.6.3	3f
7. a	3.7.1	3g
Section Four		
1. b	4.1.0	4a
2. a	4.2.0	4b
3. b	4.3.2	4c
4. b	4.4.1	4d
Section Five		
1. d	5.1.0	5a
2. a	5.2.0	5b
3. b	5.3.1	5c

NCCER CURRICULA — USER UPDATE

NCCER makes every effort to keep its textbooks up-to-date and free of technical errors. We appreciate your help in this process. If you find an error, a typographical mistake, or an inaccuracy in NCCER's curricula, please fill out this form (or a photocopy), or complete the online form at **www.nccer.org/olf**. Be sure to include the exact module ID number, page number, a detailed description, and your recommended correction. Your input will be brought to the attention of the Authoring Team. Thank you for your assistance.

Instructors – If you have an idea for improving this textbook, or have found that additional materials were necessary to teach this module effectively, please let us know so that we may present your suggestions to the Authoring Team.

NCCER Product Development and Revision
13614 Progress Blvd., Alachua, FL 32615

Email: curriculum@nccer.org
Online: www.nccer.org/olf

❏ Trainee Guide ❏ Lesson Plans ❏ Exam ❏ PowerPoints Other _____

Craft / Level: _____ Copyright Date: _____

Module ID Number / Title: _____

Section Number(s): _____

Description: _____

_____

_____

_____

Recommended Correction: _____

_____

_____

_____

Your Name: _____

Address: _____

_____

Email: _____ Phone: _____

Troubleshooting Cooling

OVERVIEW

The refrigeration circuit is a closed-loop system, so it is not always easy to determine the location of a problem in the system. Since the operation of the refrigerant circuit can be affected by external conditions, it can also be challenging to determine if a problem is within the system or the result of an external influence. A thorough understanding of the refrigeration circuit provides the basis for matching specific symptoms to probable causes. An experienced service technician is able to recognize refrigeration cycle failure patterns quickly and efficiently and knows how to apply effective solutions.

Module 03210

Trainees with successful module completions may be eligible for credentialing through the NCCER Registry. To learn more, go to **www.nccer.org** or contact us at 1.888.622.3720. Our website has information on the latest product releases and training, as well as online versions of our *Cornerstone* magazine and Pearson's product catalog.

Your feedback is welcome. You may email your comments to **curriculum@nccer.org**, send general comments and inquiries to **info@nccer.org**, or fill in the User Update form at the back of this module.

This information is general in nature and intended for training purposes only. Actual performance of activities described in this manual requires compliance with all applicable operating, service, maintenance, and safety procedures under the direction of qualified personnel. References in this manual to patented or proprietary devices do not constitute a recommendation of their use.

Copyright © 2018 by NCCER, Alachua, FL 32615, and published by Pearson, New York, NY 10013. All rights reserved. Printed in the United States of America. This publication is protected by copyright, and permission should be obtained from NCCER prior to any prohibited reproduction, storage in a retrieval system, or transmission in any form or by any means, electronic, mechanical, photocopying, recording, or likewise. To obtain permission(s) to use material from this work, please submit a written request to NCCER Product Development, 13614 Progress Blvd., Alachua, FL 32615.

03210 V5

From *HVAC Level Three*, Fourth Edition. NCCER.
Copyright © 2013 by NCCER. Published by Pearson Education. All rights reserved.

03210
Troubleshooting Cooling

Objectives

When you have completed this module, you will be able to do the following:

1. Describe the operation of the refrigeration cycle and identify problems that can occur.
   a. Identeify the primary components of the refrigeration circuit and explain their individual function.
   b. Describe the typical refrigeration cycle.
   c. Explain how to analyze refrigeration-circuit operating conditions.
   d. Identify possible causes for specific abnormal pressures and temperatures.
   e. Explain how condenser and evaporator airflow affects the refrigeration cycle.
   f. Identify and describe problems related to fixed metering devices.
   g. Identify and describe problems related to TXVs and distributors.
   h. Identify and describe other problems related to the refrigerant circuit.
2. Explain how to troubleshoot and replace a compressor.
   a. Identify common problems that can lead to compressor failure.
   b. Explain how to troubleshoot compressor mechanical problems.
   c. Explain how to replace a hermetic compressor.
   d. Describe the additional steps that may be required to replace a compressor following an electrical failure.

Performance Task

Under the supervision of your instructor, you should be able to do the following:

1. Demonstrate the ability to isolate and determine the solution for at least four of the following types of malfunctions:
   - Compressor failures
   - System-related compressor problems
   - Refrigerant undercharge or overcharge
   - Evaporator and condenser problems
   - Metering device problems
   - Refrigerant lines and accessories
   - Non-condensables and refrigerant circuit contamination

Trade Terms

Burnout
Capacitance boost
Floodback

Flooded starts
Slugging

Industry-Recognized Credentials

If you are training through an NCCER-accredited sponsor, you may be eligible for credentials from NCCER's Registry. The ID number for this module is 03210. Note that this module may have been used in other NCCER curricula and may apply to other level completions. Contact NCCER's Registry at 888.622.3720 or go to **www.nccer.org** for more information.

Contents

- **1.0.0** Refrigeration Cycle Operation and Troubleshooting 1
  - **1.1.0** Refrigeration Cycle Components and Functions 2
  - **1.2.0** Refrigeration Cycle for a Typical Air Conditioning System 3
  - **1.3.0** Analyzing Operating Conditions 5
    - 1.3.1 Preliminary Inspection 5
    - 1.3.2 Analyzing System Operating Conditions 6
  - **1.4.0** Abnormal Pressures and Temperatures 7
    - 1.4.1 Low or Excessive Refrigerant Charge 7
  - **1.5.0** Evaporator and Condenser Problems 10
    - 1.5.1 Evaporator Airflow Problems 10
    - 1.5.2 Condenser Airflow Problems 11
  - **1.6.0** Fixed Metering Device Problems 12
  - **1.7.0** Thermostatic Expansion Valve Problems 13
  - **1.8.0** Other Refrigerant Circuit Problems 14
    - 1.8.1 Refrigerant Line and Accessory Problems 14
    - 1.8.2 Air and Moisture Contamination 16
    - 1.8.3 Acid Contamination 17
- **2.0.0** Compressor Troubleshooting and Replacement 21
  - **2.1.0** Problems for Compressors 21
    - 2.1.1 Slugging 21
    - 2.1.2 Floodback 23
    - 2.1.3 Flooded Starts 23
    - 2.1.4 Overheating 24
  - **2.2.0** Compressor Mechanical Problems 25
    - 2.2.1 Valves and Rings 25
    - 2.2.2 Seized Compressors 26
    - 2.2.3 No Pressure Change 27
    - 2.2.4 Troubleshooting Scroll Compressors 28
    - 2.2.5 Troubleshooting Rotary Compressors 28
  - **2.3.0** Compressor Replacement 28
  - **2.4.0** Compressor Replacement Following an Electrical Failure 30
    - 2.4.1 Mild Burnout Procedures 31
    - 2.4.2 Severe Burnout Procedures 31
    - 2.4.3 Changing Oil 33

Figures and Tables

Figure 1 Basic refrigeration cycle 2
Figure 2 Typical air conditioning cycle for both
HFC-410A and HCFC-22 refrigerants 4
Figure 3 Causes of abnormal refrigerant pressure 9
Figure 4 Ice on an evaporator coil 11
Figure 5 Cleaning a layered condenser coil 12
Figure 6 Suction temperature variations that indicate hunting 13
Figure 7 Multi-circuit evaporator fed by a distributor 14
Figure 8 Refrigerant lines and accessories 15
Figure 9 Polyol ester (POE) lubricating oil 17
Figure 10 Refrigerant vapor test kit 18
Figure 11 Oil test kit 19
Figure 12 Compressor troubleshooting 22
Figure 13 Liquid line solenoid... 23
Figure 14 Checking current with an ammeter.. 26
Figure 15 Starting a seized PSC compressor with a capacitance boost 26
Figure 16 Replaceable-core filter-drier shell and core 32
Figure 17 Suction line filter-drier 32
Figure 18 Semi-hermetic compressors 34
Figure 19 Oil pump 35

Table 1 Typical Temperature-Pressure Data for the
Example Shown in Figure 2 8

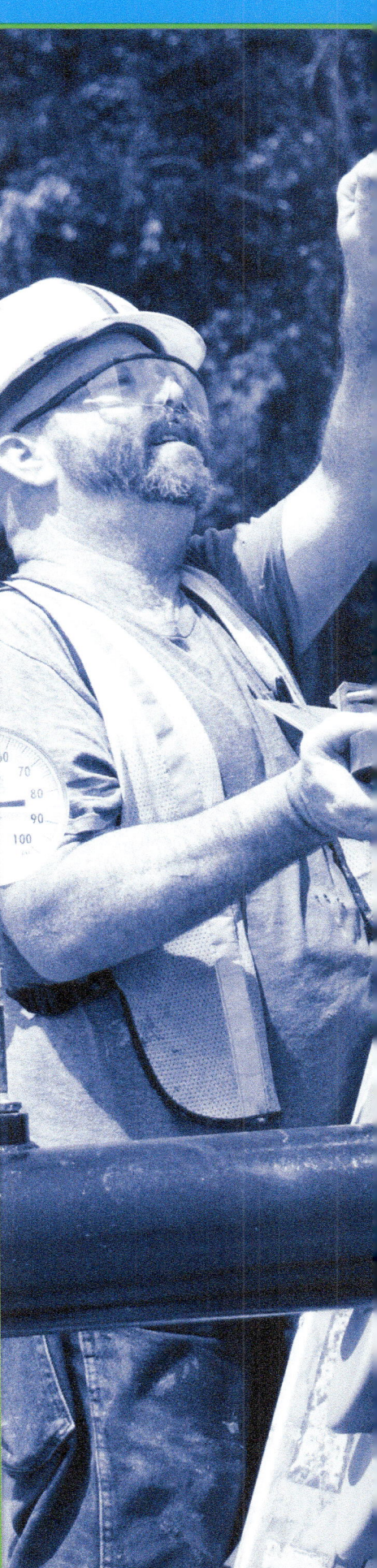

SECTION ONE

1.0.0 REFRIGERATION CYCLE OPERATION AND TROUBLESHOOTING

Objective

Describe the operation of the refrigeration cycle and identify problems that can occur.

a. Identify the primary components of the refrigeration circuit and explain their individual function.
b. Describe the typical refrigeration cycle.
c. Explain how to analyze refrigeration-circuit operating conditions.
d. Identify possible causes for specific abnormal pressures and temperatures.
e. Explain how condenser and evaporator airflow affects the refrigeration cycle.
f. Identify and describe problems related to fixed metering devices.
g. Identify and describe problems related to TXVs and distributors.
h. Identify and describe other problems related to the refrigerant circuit.

Performance Tasks

1. Demonstrate the ability to isolate and determine the solution for at least four of the following types of malfunctions:
   - Compressor failures
   - System-related compressor problems
   - Refrigerant undercharge or overcharge
   - Evaporator and condenser problems
   - Metering device problems
   - Refrigerant lines and accessories
   - Non-condensables and refrigerant circuit contamination

Trade Terms

Burnout: The condition in which the breakdown of the motor winding insulation causes the motor to short out or ground electrically. The breakdown of the insulating coating typically results in the formation of strong acids.

Floodback: A condition in which there is liquid refrigerant in the suction vapor being returned to an operating compressor.

An understanding of the basic refrigeration cycle is critical in order for technicians to effectively troubleshoot cooling systems. There are many types of systems used to provide cooling for personal comfort, food preservation, and industrial processes. Each of these systems uses a mechanical refrigeration system to produce the cooling.

The operation of the mechanical refrigeration system is basically the same for most vapor compression systems. Items that vary from system to system are the type of refrigerant used, the size and style of the components, pressures and temperatures, and the installed locations of the components and refrigerant lines.

The basic refrigeration cycle and its components are covered in other modules of this curriculum. A review is presented here as a foundation for troubleshooting cooling systems. Note that the information in this module is based primarily on an air-cooled system.

1.1.0 Refrigeration Cycle Components and Functions

A basic refrigeration system (*Figure 1*) includes the following components:

- *Evaporator* – A heat exchanger in which the heat from the area or item being cooled is absorbed and transferred to the refrigerant.
- *Compressor* – Creates the pressure differential needed to make the refrigeration cycle work. The compressor is referred to as the heart of the system.
- *Condenser* – A heat exchanger in which the heat absorbed by the refrigerant is transferred to the cooler outdoor air, or another cooler substance.
- *Metering device* – Controls the flow of refrigerant to the evaporator. It provides a pressure drop that lowers the boiling point of the refrigerant just before it enters the evaporator. In *Figure 1*, the metering device is a thermostatic expansion valve, commonly called a TXV or a TEV.

Also shown in *Figure 1* is the refrigerant piping used to connect the basic components to provide the path for refrigerant flow. Together, the components and lines form a closed refrigeration system. The arrows show the direction of flow for the refrigerant through the closed system. The lines are identified as follows:

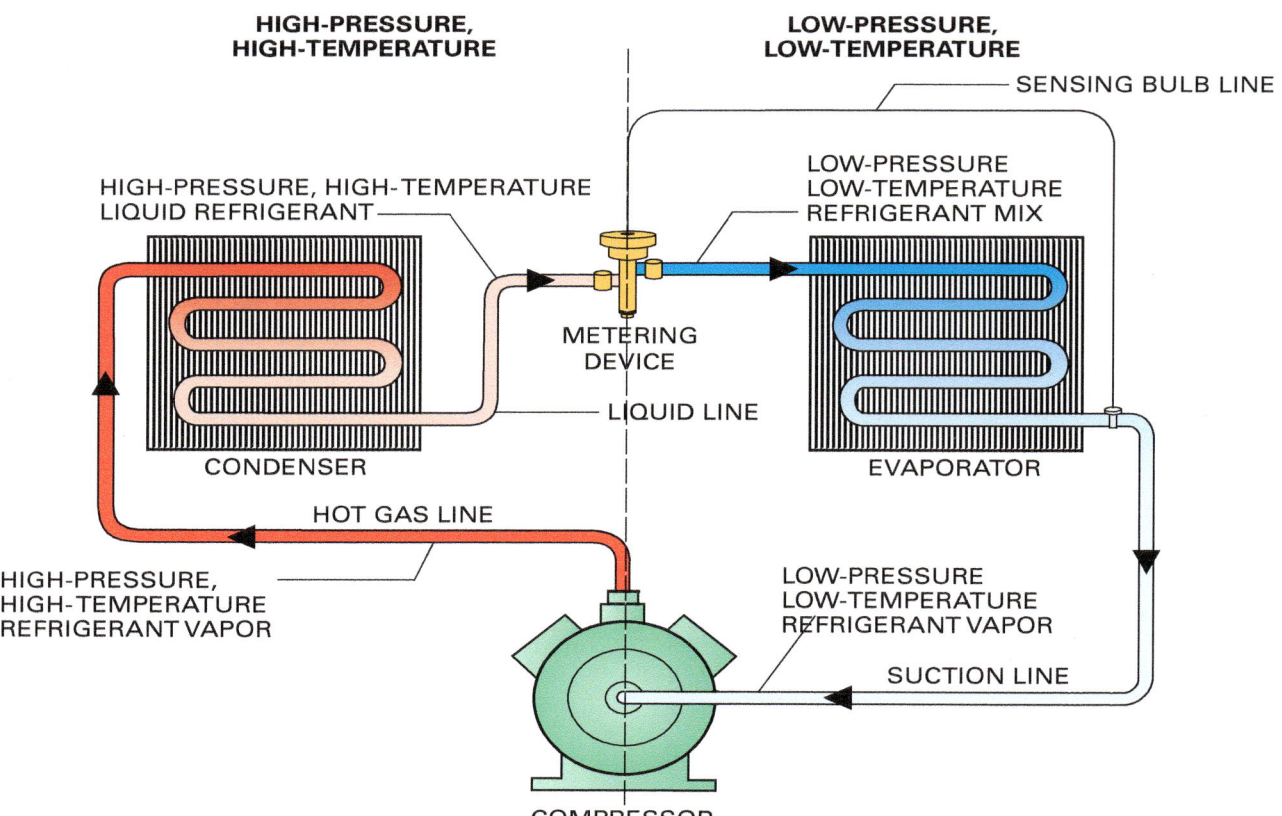

Figure 1 Basic refrigeration cycle.

- *Suction line* – The tubing that carries the low-temperature refrigerant vapor from the evaporator to the compressor.
- *Hot gas line* – The tubing that carries hot refrigerant vapor from the compressor to the condenser; also called the discharge line.
- *Liquid line* – The tubing that carries liquid refrigerant, formed in the condenser, to the metering device.

A properly operating refrigerant circuit for common applications is described as follows:

Upon a call for cooling, the control circuit starts the compressor. The pressure difference between its inlet and outlet begins to build immediately, causing refrigerant to flow. The incoming vapor is compressed, creating a high-pressure, high-temperature vapor stream. This vapor travels to the condenser via the hot gas line.

The condenser is a series of tubing coils through which the refrigerant flows. As cooler air moves across the condenser tubing, the hot refrigerant vapor cools. As it gives up heat to the outside air, it cools to the condensation point, where it begins to change from a vapor into a liquid. Eventually all the refrigerant vapor condenses to a liquid. Any decrease in temperature below the saturation point is referred to as *subcooling*. This high-temperature, high-pressure liquid travels through the liquid line to the input of the metering device.

The metering device regulates the flow of refrigerant into the evaporator. It also decreases the refrigerant pressure significantly. This results in a dramatic drop in refrigerant temperature. By restricting refrigerant flow, the metering device causes the high-pressure liquid refrigerant from the condenser to become the low-temperature, low-pressure refrigerant needed to absorb heat in the evaporator. As the refrigerant exits the metering device, it remains mostly liquid, but some vapor is present.

The evaporator receives the low-temperature, low-pressure liquid/vapor refrigerant mix from the metering device. The evaporator is a series of tubing coils that exposes the cooler refrigerant to the warmer air passing over them. Heat from the air is transferred through the evaporator tubing to the cooler refrigerant, causing it to boil and vaporize. Remember that even though it is boiling, it is not hot because refrigerants boil at such low temperatures. A low-temperature, low-pressure refrigerant vapor finally exits the evaporator. Any increase in vapor temperature that occurs after all the liquid has boiled off is called *superheat*.

The vapor then returns to the inlet of the compressor, where the cycle repeats.

1.2.0 Refrigeration Cycle for a Typical Air Conditioning System

A basic air conditioning system is shown in *Figure 2*. The components are divided into two sections. The high-pressure side includes all the components in which the pressure of the refrigerant is at or above the condensing pressure. This is often referred to as *head pressure, discharge pressure*, or *high-side pressure*. The low-pressure side includes all the components in which the pressure is at or below the evaporating pressure. This is often called the *suction pressure* or *low-side pressure*. The vertical dashed line represents the division between the two sides of the system. The major pressure changes occur at the compressor and the metering device.

This section describes a typical air conditioner operating on either HCFC-22 or HFC-410A. These are the two most commonly used refrigerants in today's comfort-cooling systems. This example demonstrates the temperature/pressure relationships typical of those at key points in the circuit. For this example, assume an indoor air temperature of 75°F and an outdoor air temperature of 95°F. The temperatures and pressures used in the example are representative values only. They will vary due to equipment and load conditions.

Load conditions are based on climatic conditions and the heat load generated within the space. The humidity level of the indoor air also plays a key role in the load. Since a great deal of energy is required to condense airborne moisture, humid air indoors increases the refrigeration load. Condensing moisture on the evaporator coil surface represents a latent load, which is a substantial part of the total refrigeration load.

The numbered descriptions below correspond to the numbers in the black circles shown in *Figure 2*. Follow along on the figure as the system is described. Notice that the temperature at each point remains the same for both refrigerants. The first pressure (highlighted in green) represents the pressure of HCFC-22 at these conditions; the second pressure (highlighted in red) represents that of HFC-410A.

1. After compression, the highly superheated gas from the compressor flows through the hot gas line to the condenser. The hot gas may be around 170°F at 243/390 psig, and the refrigerant superheat would be around 55°F as a result (170°F – 115°F saturation temperature for these pressures = 55°F). Typically, compressor-discharge superheat will range between 40°F and 150°F. All the superheat must be removed before the refrigerant vapor can be condensed into a liquid.

2. Because the refrigerant in the condenser is still hotter than the warmer outside air passing over the condenser, it quickly gives up superheat as it flows through. The temperature falls to 115°F in the early condenser passes. In this example, 115°F represents the saturation temperature of the refrigerant. At this temperature, and only at this temperature for the pressures shown in *Figure 2*, liquid and vapor can coexist. As heat continues to be transferred from the vapor to the cooler outside air, the vapor begins to condense into a liquid. After the refrigerant has traveled about three quarters of the way through the condenser, all the refrigerant has condensed into a liquid.

3. During the remaining travel through the condenser, the liquid refrigerant continues to cool (subcooling). This lowers its temperature to about 105°F. In other words, the liquid refrigerant is subcooled 15°F (115°F saturation temperature – 105°F = 15°F).

4. Subcooled liquid refrigerant from the condenser flows through the liquid line to the metering device. Note that, since the liquid line is uninsulated, the subcooling may increase or decrease due to line length, environment, and elevation change. For this example, no change in temperature has occurred.

5. The metering device controls the flow of liquid refrigerant to the evaporator. Subcooled liquid from the condenser enters at a temperature of 105°F and a pressure of 243/390 psig. A very significant pressure drop occurs at this point. This pressure drop reduces the pressure of the liquid refrigerant from 243/390 psig to 69/118 psig at the outlet. At this new pressure condition, the refrigerant temperature should be 40°F and in its saturated state. For this to happen, some of the liquid refrigerant must flash to its vapor state instantly, absorbing a great deal of heat in the process to cool the remaining refrigerant liquid. This results in a refrigerant mixture of about 80-percent liquid and 20-percent vapor, and the refrigerant is now at saturation (no superheat and no subcooling exists). The low-temperature, low-pressure mixture from the metering device then enters the evaporator coil.

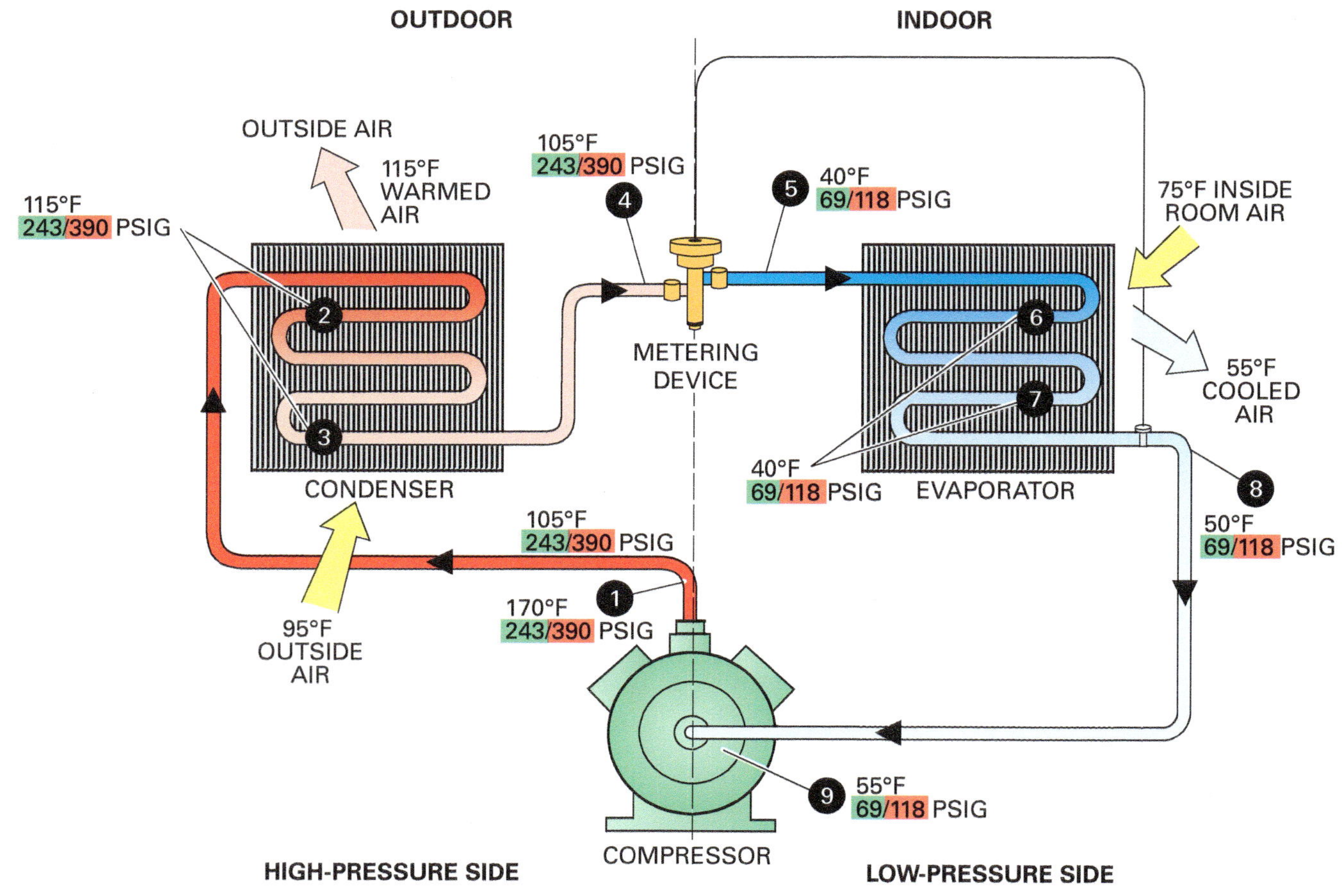

Typical Air Conditioning Cycle

OUTDOOR | **INDOOR**

- 115°F 243/390 PSIG (point 2, condenser)
- OUTSIDE AIR 115°F WARMED AIR
- 105°F 243/390 PSIG (point 4)
- 40°F 69/118 PSIG (point 5)
- 75°F INSIDE ROOM AIR
- 55°F COOLED AIR
- 40°F 69/118 PSIG (points 6, 7, evaporator)
- 50°F 69/118 PSIG (point 8)
- 105°F 243/390 PSIG (compressor outlet)
- 170°F 243/390 PSIG (point 1)
- 95°F OUTSIDE AIR
- 55°F 69/118 PSIG (point 9)

HIGH-PRESSURE SIDE — CONDENSER, COMPRESSOR
LOW-PRESSURE SIDE — EVAPORATOR

Note: Lower pressure is HCFC-22/higher pressure is HFC-410A.

TEMPERATURE/PRESSURE CHART

TEMPERATURE °F	PRESSURE HCFC-22	PRESSURE HFC-410A	TEMPERATURE °F	PRESSURE HCFC-22	PRESSURE HFC-410A
27	51.2	91.0	50	84.0	142.2
28	52.4	92.9	55	92.6	155.5
29	53.6	94.9	60	101.6	169.6
30	54.9	96.8	65	111.2	184.5
31	56.2	98.8	70	121.4	200.4
32	57.5	100.9	75	132.2	217.1
33	58.8	102.9	80	143.6	234.9
34	60.1	105.0	85	155.7	253.7
35	61.5	107.1	90	168.4	273.5
36	62.8	109.2	95	181.8	294.4
37	64.2	111.4	100	195.9	316.4
38	65.6	113.6	105	210.8	339.6
39	67.1	115.8	110	226.4	364.1
40	68.5	118.1	115	242.7	389.9
41	70.0	120.3	120	259.9	416.9
42	71.5	122.7	125	277.9	445.4
43	73.0	125.0	130	296.8	475.4
44	74.5	127.4	135	316.6	506.9
45	76.0	129.8	140	337.3	540.1
46	77.6	132.2	145	358.9	575.1
47	79.2	134.7	150	381.8	611.9
48	80.8	137.2			
49	82.4	139.7			

Figure 2 Typical air conditioning cycle for both HFC-410A and HCFC-22 refrigerants.

6. The refrigerant liquid/vapor mixture enters the evaporator. The pressures of 69/118 psig correspond to the 40°F saturation point for each refrigerant (see the PT chart below the diagram in *Figure 2*). The 40°F saturation point used here is typical of the temperatures normally used for evaporators in comfort cooling. Since there is a mixture of refrigerant and liquid present leaving the metering device, the refrigerant is in its saturated state on the way in.

7. Because the refrigerant flowing through the evaporator is cooler than the room air passing over the evaporator, it absorbs heat, causing the liquid refrigerant to boil and become a vapor. After traveling roughly 90 percent of the way through the evaporator tubing, all the refrigerant should have boiled away, existing only as a vapor. It will begin to increase in temperature (superheat) only after this change of state is complete.

8. During the remaining 10 percent of travel through the evaporator, the vapor continues to absorb heat from the warmer air, raising its temperature to 50°F. In other words, the vapor is now superheated 10°F (50°F – 40°F = 10°F). This superheated vapor flows through the suction line and is directed back toward the inlet of the compressor. The cooled room air is re-circulated by the evaporator fan back into the room at a temperature of roughly 55°F.

9. The superheated vapor entering the compressor may experience an additional 3°– 5°F (or more, in some cases) of superheat along the way, because the vapor in the suction line absorbs heat from the warmer surrounding air as it travels from the evaporator to the compressor. This can be a significant variable from one system to another. Like the liquid line, it depends largely on the line length and its environment. Once it enters the compressor, the refrigerant has completed the refrigeration cycle and is ready to repeat the process.

The pressures and temperatures in this circuit walk-through illustrate just one example of system operating characteristics. There are many possible combinations of temperatures and pressures that can represent proper operation for each refrigerant and application.

1.3.0 Analyzing Operating Conditions

When problems arise in the refrigeration cycle, it can be difficult to determine the cause. Often what seems to be the problem is only a symptom. Also, many problems produce the same or similar symptoms. Isolating the problem requires a logical, systematic approach, which generally begins with a preliminary inspection. An analysis of the operating characteristics follows the inspection.

1.3.1 Preliminary Inspection

A preliminary inspection of the system is performed using the senses of sight, hearing, touch, and smell to identify potential problems. A preliminary inspection is performed as follows:

> **WARNING!** Be sure all electrical power to the equipment is turned off. Open, lock, and tag disconnects. Use extreme caution around pressurized or hot components. Follow all safety instructions labeled on the equipment and provided in the manufacturer's service manual.

Step 1 Make a quick visual inspection of the system, looking for any obvious signs of leakage or malfunction. Look for any fault codes that may be displayed on indoor and outdoor unit control boards.

Step 2 Turn off the equipment. Lockout and tag equipment so it cannot be energized.

Step 3 Look for the following piping problems:
- Oversized or undersized refrigerant lines or excessive elbows or fittings
- Long or uninsulated suction line that might develop excessive superheat
- Liquid line running through an unconditioned space (hot or cold) that might affect subcooling
- Buried lines that might cause refrigerant to condense
- Extremely long liquid line that might hold an excessive amount of refrigerant

Step 4 At the evaporator and condenser, check for the following:
- Evaporator or condenser coils mounted above the compressor that might allow liquid refrigerant to descend to the compressor during the off cycle.
- Fin collars corroded
- Fins or coils dirty or damaged
- Supply plenum dirty
- Filters dirty or missing

- Fan belts at improper tension
- Blowers and fans dirty
- Incorrect blower speed
- Evaporator coil shows signs of frost formation
- For future reference, determine if a variable-speed motor is being used on the indoor blower

Step 5 As applicable, at the compressor:
- Check that the service valves are fully open.
- Check that the mounting bolts are positioned per manufacturer's instructions (usually loosened).
- Check that the crankcase heater is working.
- On open or semi-hermetic compressors, inspect cylinder heads to see if they are scorched or blistered from excessive heat.
- Inspect for rust streaks indicating condensation from cold return gas.
- Ensure the oil level is at the proper height in the sight glass (semi-hermetic compressors).

The preliminary inspection may immediately reveal at least one or more problems with the system. A large puddle of oil under a fitting, for example, shows that at least one significant leak is present. Some items noted may need to be resolved before more in-depth testing begins. Never assume that a system has only one problem. The preliminary inspection may not reveal the most significant problem, but it may reveal one or more issues of concern. In addition, it provides the technician with information to be factored in as more detailed testing begins.

1.3.2 Analyzing System Operating Conditions

To determine if the system or compressor has a functional problem, the actual conditions that exist in the system must be known. If the compressor is operable, system operation should be monitored, and the critical parameters measured. This is necessary so that the actual conditions can be compared against a set of normal system operating parameters.

Check the system operating conditions and record the values for system parameters. The following steps describe this procedure:

Step 1 Connect a gauge manifold to the system gauge ports. Be prepared to go to several different system locations to take air and refrigerant line temperature measurements.

Step 2 Check and record the voltage applied to the compressor contactor.

Step 3 Provide power to the system and set the thermostat to call for cooling.

Step 4 Start the system and listen for abnormal sounds.
- Check for excessive vibration of the compressor, piping, motors, and fans.
- Note any compressor knocks or rattles, which may indicate liquid refrigerant is being drawn into the cylinder(s). Such noise may also indicate that mechanical damage to the compressor has already occurred.

Step 5 Visually inspect the operating compressor for abnormalities. Check the compressor oil sight glass (if equipped). Heavy foaming at the sight glass should clear within 5 to 10 minutes after startup. If it does not, it may indicate refrigerant floodback is occurring, or that an abnormal amount of refrigerant is condensing in the oil during the off cycle.

Step 6 Measure and record the input voltage and current at the compressor contactor. The voltage should be compared to the voltage recorded before startup to see if there is any significant voltage drop under load.

> **WARNING!** Use extreme caution around rotating, pressurized, or hot components. Follow all safety instructions labeled on the equipment and provided in the manufacturer's service manual.

> **WARNING!** Danger exists if the compressor terminals are damaged and the system is pressurized. Disturbing the terminals to take measurements could cause them to blow out, causing injury. When making voltage, current, or continuity checks on a hermetic or semi-hermetic compressor in a pressurized system, always take measurements at contactors and other test points inside the unit.

- The measured voltage should be within ±10 percent of the motor nameplate value.
- In a three-phase motor, the voltage imbalance between any two phases should not exceed 2 percent. Any current imbalance between phases must not exceed 10 percent.

Step 7 Check that all fans, motors, and pumps are operational and moving the proper amounts of air or water.

Step 8 As applicable, measure and record or calculate the following operating parameters:
- Indoor dry-bulb and wet-bulb temperatures
- Outdoor dry-bulb temperature
- Suction line pressure and actual temperature
- Saturated suction temperature
- Superheat at both the outlet of the evaporator and the inlet of the compressor
- Discharge pressure
- Saturated discharge temperature
- Discharge line temperature
- Liquid line pressure and actual temperature
- Subcooling at both the condenser outlet and the metering device inlet
- Liquid line temperature entering the metering device
- Oil pressure (semi-hermetic compressors)
- Compressor temperature at the bottom of the cylinder heads (semi-hermetic compressors)
- Compressor temperature at the top and bottom of the motor barrel (semi-hermetic compressors)
- Crankcase or motor shell temperature
- Compression ratio

Once the actual conditions are known, they can be compared against a set of normal operating parameters to determine if there is a problem. *Table 1* shows an example of the types of system parameters that should be recorded. This data reflects the normal operation parameters for the system described in *Figure 2*. Remember that readings vary widely among systems because of equipment application, ambient conditions, and type of refrigerant used.

Refer to any existing operating logs or the manufacturer's service manual to find typical readings for the specific equipment you are servicing. Better yet, if the same kinds of systems are serviced for various customers, or the same system is periodically serviced for one customer, capture the necessary data when the system is operating properly. It will be a valuable reference for later use if it becomes necessary to troubleshoot the system.

1.4.0 Abnormal Pressures and Temperatures

Now that you have the information determined by the system operation analysis, begin by consulting the manufacturer's troubleshooting aids to identify the likely causes of the problem. Most refrigeration-cycle problems result in abnormal system pressures and/or temperatures. *Figure 3* summarizes the common system problems that relate to high and low pressures. Note that a low head pressure in combination with a high suction pressure and low compressor current normally indicates that the compressor has leaky valves or worn piston rings.

The information provided in the manufacturers' troubleshooting aids helps to isolate the problem to the specific failed component, especially when the problem is related to the evaporator, condenser, and the air or water flow circuits. For more complex problems, or where multiple problems exist, the troubleshooting aids may only provide some clues for further investigation. For example, the cause may be listed as a loss of charge, restriction in the refrigeration system, or non-condensables in the system. For this type of problem, further system analysis using a process of elimination is usually needed to isolate the problem.

1.4.1 Low or Excessive Refrigerant Charge

An incorrect refrigerant charge is a common problem. It is important for technicians to understand how an incorrect charge affects system operation so that the problem is more easily recognized.

Loss of refrigerant is a frequent problem. The customer may complain of deteriorating cooling performance over a period of days or weeks. Low refrigerant usually results from leaks that occur due to poor installation, physical damage, or a factory defect. Some leaks are small enough that the loss of charge only becomes significant over the course of months.

A low charge can also be the result of incorrect charging techniques at startup or during a previous service call. For example, if a system with a TXV metering device is charged to only a 5°F subcooling value on a very cool day with a low

Table 1 Typical Temperature-Pressure Data for the Example Shown in *Figure 2*

System Parameter	Typical Operational HFC-410A Air Conditioning System	Typical Operational HCFC-22 Air Conditioning System	Air Conditioning System Under Test
Suction Pressure (psig)	118 psig	69 psig	
Suction Line Temperature (°F)	55°F	55°F	
Saturated Suction Temperature (°F)	40°F	40°F	
Superheat (°F) – TXV	10°F	10°F	
Discharge Pressure (psig)	390 psig	243 psig	
Discharge Line Temperature (°F)	170°F	170°F	
Saturated Discharge Temperature (°F)	115°F	115°F	
Liquid Temperature Leaving Condenser (°F)	105°F	105°F	
Liquid Line Temperature Entering Expansion Device (°F)	105°F	105°F	
Subcooling (°F)	10°F	10°F	
Air Temperature Entering Evaporator (DB°F)	75°F	75°F	
Air Temperature Leaving Evaporator (DB°F)	55°F	55°F	
Air Temperature Drop Across Evaporator (DB°F)	20°F	20°F	
Outdoor Temperature (DB°F)	95°F	95°F	
Outdoor Temperature Rise Across Condenser (DB°F)	20°F	20°F	
Compressor Compression Ratio (Absolute Discharge Pressure/Absolute Suction Pressure)	3.05 to 1	3.08 to 1	
Compressor Oil Pressure (psig)	135 psig	86 psig	

evaporator load, it will likely not have enough refrigerant to operate properly in higher temperatures and evaporator loads. As the refrigerant flow rate increases, subcooling diminishes until there is none.

With a low charge, suction and discharge pressures will be lower than normal with a high superheat, especially with a fixed metering device. A fixed metering device is more sensitive to the refrigerant charge. If the system has a sight glass, it will likely show bubbles. However, some refrigerants, like HFC-410A, may show bubbles in a sight glass at any time. Systems with an automatic expansion valve (such as some window and through-wall units) may not have low suction pressure because the valve may be operating wide open, trying to raise the evaporator pressure. In some cases, condenser subcooling will be very low because there is little refrigerant to subcool.

Once you determine that the cause of a system problem is a low charge, find and repair any leaks. The methods used for leak detection and the subsequent evacuation/dehydration and charging of the system are covered in other modules of this curriculum. Never simply add refrigerant to a system without looking for leaks and

HIGH HEAD PRESSURE

1. Air or noncondensibles in the system
2. Obstructions in the condenser coil fins, such as dirt, etc.
3. Overcharge of refrigerant
4. Recirculation of condenser air
5. Higher-than-ambient temperature air entering condenser
6. Wrong rotation of condenser fan blade

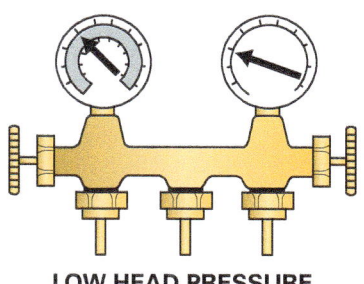

LOW HEAD PRESSURE

1. Low refrigerant charge
2. Defective compressor valves
3. Low ambient temperature

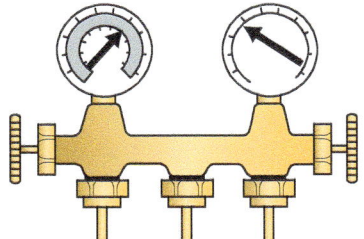

HIGH SUCTION PRESSURE

1. Overcharge of refrigerant
2. Damaged or incorrect metering device
3. Damaged or leaking compressor valves
4. Worn compressor piston rings
5. Leaking oil separator

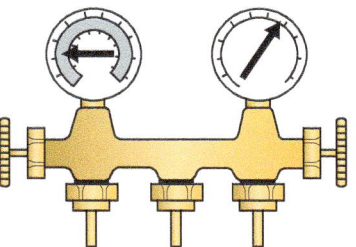

LOW SUCTION PRESSURE

1. Loose or broken blower belt
2. Obstructed or dirty evaporators
3. Dirty air filters
4. Low refrigerant charge
5. Dirty or faulty expansion valve (leaking around push rod)
6. Recirculation of evaporator air (compare return air temperature with conditioned space temperature)
7. Restriction in refrigerant system
8. Restricted or undersized duct work
9. Wrong rotation of evaporator blower

Figure 3 Causes of abnormal refrigerant pressure.

repairing them, unless there is reason to believe that the undercharge condition is the result of a previous error. It is important to be certain—repeated service calls to recharge a leaking system and failure to repair the problem will result in customer frustration and dissatisfaction.

The symptoms of a refrigerant overcharge are similar from system to system. Both the suction and discharge pressures are typically higher. The superheat may be low or nonexistent on a system with a fixed metering device, possibly causing floodback. A system with an automatic expansion valve will not have a high suction pressure because it is designed to maintain a constant suction pressure. A high discharge temperature from the compressor may also be a problem, as well as a high compression ratio.

If it is determined that excess refrigerant charge is the cause of a system problem, recover the excess refrigerant as needed to achieve a correct system charge. When checking the refrigerant charge, focus your attention on subcooling for TXV-based systems, and on superheat for systems with a fixed metering device.

Refrigerant Charging

When charging a system with HFC-410A refrigerant, always charge with liquid refrigerant. When adding a charge to an operating system, charge with liquid into the suction side of the system through a flow restrictor installed in the hose of the gauge manifold. Alternatively, use the gauge manifold valve as a metering device.

Remember that many blends tend to fractionate. As a result, even when subcooling is present, bubbles may appear in a liquid line sight glass. This is especially true when the system is lightly loaded, and the refrigerant is moving through the system rather slowly. Sight glasses are of little value in charging blends, but most provide a helpful moisture indicator.

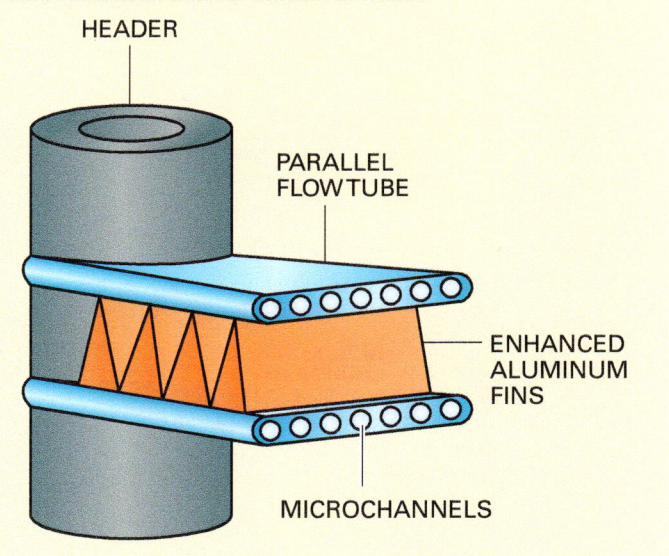

1.5.0 Evaporator and Condenser Problems

Both the evaporator and the condenser often experience airflow problems. These must be corrected for the system to operate properly.

1.5.1 Evaporator Airflow Problems

To perform properly, a blower in a cooling system must move about 350 to 450 cubic feet of air per minute (cfm) per ton of cooling across the evaporator coil. Multi-speed and variable-speed motors should be checked and set up properly during the initial installation and checked again during servicing. Too much or too little air can cause comfort problems. For example, too much air across the evaporator coil (moving too quickly) typically results in poor humidity control. Decreasing the blower speed usually corrects this problem.

Incorrect airflow can also cause operational problems. For example, a dirty evaporator coil will cause low suction pressure and low superheat, due to reduced airflow and poor heat transfer. The problem of too little airflow is far more common than too much airflow. Low airflow is generally indicated in refrigerant circuit performance by low suction pressures and, especially with fixed metering devices, a low superheat condition. The airflow can be increased by increasing the blower speed, providing that none of the following conditions are the real cause of the reduced airflow:

- The system air filter, blower wheel, or evaporator coil is dirty.
- The blower is rotating in the wrong direction.
- The belts on belt-driven blowers are loose or worn or the belt tension is incorrect.
- There is loose insulation inside the ductwork.
- The air distribution system is poorly designed and undersized, causing excessive resistance to air movement.

A problem with reduced airflow can often appear as the result of a low refrigerant charge. Many technicians who fail to fully investigate a problem are fooled by some of the most common symptoms they encounter when they connect the gauges. The first symptoms of a low refrigerant charge are a low suction pressure coupled with a low head pressure. In a hurried attempt to complete the work, they may immediately add refrigerant. But there are other signs that will indicate a very different problem.

In the situation described above, the technician can simply touch the base of a hermetic compressor. If it is very cool or cold, it is unlikely that an undercharge is the problem. This cool crankcase temperature suggests floodback is occurring. A check of the superheat will show that it is very low or even non-existent, confirming floodback. At this point, the technician should examine the evaporator and airflow. Very often, poor evaporator flow from a dirty filter will be found. In other cases, the coil may be frozen over (*Figure 4*).

As the low superheat and low suction pressure condition persists, the evaporator temperature drops below freezing, and ice begins to form. Once it begins, it progresses more quickly as the evaporator surface is increasingly blocked. Since the system is not able to satisfy the demand for cooling, the system continues to run, further

Figure 4 Ice on an evaporator coil.

increasing the thickness of the ice. Although the evaporator coil cannot be easily seen in many installations, the ice often begins to show itself on the outside of the coil cabinet or on the refrigerant lines. Eventually, it can form around the inlet of the compressor, and can even engulf a compressor over time.

When looking for low airflow and/or low evaporator-load problems, remember that electronically controlled variable-speed motors are frequently used today with indoor blowers. When checking airflow in a system using a variable-speed motor, ensure that the blower is operating at the expected speed before checking the airflow. During some normal operating modes, such as dehumidification, variable-speed motors do not operate at their maximum speed. The presence of a variable-speed motor should be determined during the preliminary inspection of the system.

Going Green: Microchannel Coils

Microchannel coils are a new way to get more from less. Coil manufacturers have developed coils that eliminate the standard copper or aluminum round tubing altogether. In their place are flat tubes that have multiple small passages inside. The idea is to increase the refrigerant surface area in contact with the heat transfer surface, while also minimizing the refrigerant volume. Microchannel coils have increased efficiency while maintaining or reducing the size of the coils overall. Condenser coils were first fitted with microchannel coils, but they are also now being used for evaporator coils.

1.5.2 Condenser Airflow Problems

For the system to operate properly, the condenser coil must be able to reject the heat absorbed by the refrigerant in the evaporator. In addition, the condenser must reject the heat of compression. Therefore, the condenser must be capable of rejecting more heat than the evaporator is able to absorb. Roughly 15 percent of the condenser's heat rejection capacity is required to reject the heat of compression.

When sufficient heat rejection does not occur, the system pressures and temperatures typically rise. If the problem is severe enough, protective devices will shut the compressor off. Some possible causes for poor condenser performance in air-cooled applications are as follows:

- Failure of the condenser fan(s) to operate due to motor or control circuit failure.
- The condenser fan is rotating in the wrong direction.
- The fan blade is not positioned properly in the fan orifice.
- The wrong fan blade has been used, or the pitch of a replacement blade is incorrect.
- Hot condenser-air discharge is being recirculated through the coil instead of being carried away. The usual cause is a unit installed too close to buildings, or under decks or overhangs.
- Higher-than-ambient air temperature is entering the condenser coil from other warm air sources, such as exhaust vents, dryer vents, and adjacent condensing units.
- Grass clippings or other airborne debris are present on the surface of the coil or embedded deeply within, reducing airflow.
- Airflow is blocked by shrubs and plants.
- A replacement motor has the wrong rpm.

Many commercial condenser coil assemblies are constructed of two separate coils in layers (*Figure 5*). The coil fins of any condenser coil are set close together, but dust and dirt can easily make its way in between the coil fins, where they either pass through or become trapped. With layered coil construction, much of the debris that would normally find its way through the coil fins and out the opposite side becomes trapped against the face of the rear coil. This happens easily, as the fins of the two coils are not aligned. Cleaning only the face of the coil often results in little or no improvement. To properly clean these layered coils, the two layers must be very carefully separated enough to wash between them.

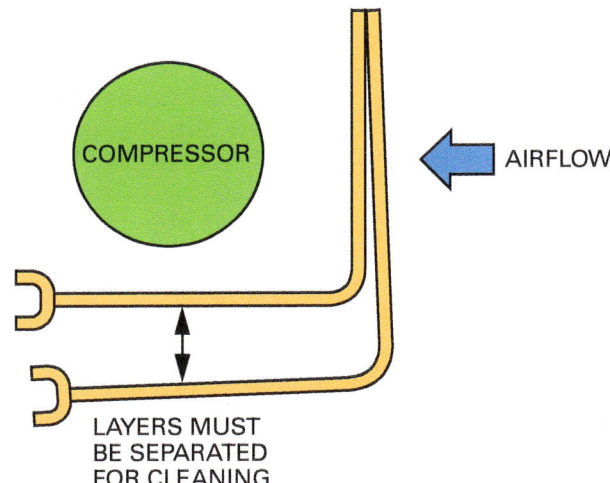

Figure 5 Cleaning a layered condenser coil.

1.6.0 Fixed Metering Device Problems

Two primary problems occur with fixed metering devices: they are either the wrong size for the application, or they become partially or completely restricted. On systems with a single fixed-orifice device, clogging is rare due to the size of the opening. However, some commercial systems that use several fixed-orifice metering devices on the same circuit are usually smaller in size.

Normally, the sizes of both fixed-orifice and capillary tube devices are factory-selected for use with a system. If a capillary tube or fixed orifice is too small in diameter, the evaporator will be starved for refrigerant. Excess liquid will then build up in the condenser. The effect initially will be inadequate cooling, with the potential for a high head pressure initially. If the unit runs for a while, the liquid trapped in the condenser will cool and the pressure will drop to normal. If a capillary tube is too short or the diameter too great, excess liquid will be fed to the evaporator. This condition causes low superheat and could lead to compressor flooding.

For any given pressure difference across a fixed metering device, it can pass a larger volume of liquid refrigerant than vapor because the

Metering Devices

Because of the need to meet higher energy efficiency standards, some manufacturers are abandoning the use of fixed metering devices on the indoor coils of some products, switching entirely to TXVs. TXVs also provide much better performance in a wider variety of operating conditions.

Figure Credit: Courtesy of Parker Hannifin, Sporlan Division

Ice on Suction Lines and Compressors

For technicians working with comfort-cooling systems, the appearance of ice on suction lines and compressor inlets indicates trouble. Most comfort systems are designed to operate at evaporator temperatures well above freezing. Ice on suction lines and compressor inlets is not unusual in refrigeration circuits, however.

For a freezer to maintain a box temperature of 10°F (–12°C), a common evaporator temperature would be around 0°F (–18°C). Even if the system is operating with a superheat value of 10°F (5.5°C), the suction line temperature will still be 22°F (–5.5°C)—well below the freezing mark. Therefore, ice is going to form somewhere. In fact, if a freezer system operates for a significant amount of time each day, the absence of ice on the suction lines suggests a problem.

This is a perfect example of why technicians must understand their trade, and the value of taking superheat readings and making pertinent observations. To evaluate a subsystem, one must first see the big picture.

When working on a freezer, scrape off any ice from the pipe surface in the immediate area before measuring the surface temperature. Otherwise, the reading will be inaccurate.

liquid has a much greater density. This makes a fixed metering device very sensitive to excess oil and poor subcooling. Excess oil in circulation displaces some of the liquid refrigerant and reduces the actual weight of refrigerant flowing through the tube. This can starve the evaporator. Excess flashing due to a shortage of refrigerant or a liquid line restriction will cause both liquid and vapor to enter the metering device, also resulting in a starved evaporator coil.

1.7.0 Thermostatic Expansion Valve Problems

TXVs are often blamed for system problems caused by other faults. This is because TXVs are often misunderstood. Troubleshooting TXV problems should always begin with a measurement of superheat. If too little refrigerant is fed to the evaporator, the superheat will be high. If too much refrigerant is fed to the evaporator, the superheat is low. Although you may think these symptoms are caused by improper TXV control, often the problem is elsewhere. Problems often blamed on TXVs include the following:

- Overfeeding
- Underfeeding
- Erratic operation (called *hunting*)

Floodback can be caused by the valve sticking in an open position. It may also be caused by the following valve or system problems:

- Oversized valve
- Improper superheat setting on the valve
- Improper location of the sensing bulb on the suction line
- Poor thermal contact between the sensing bulb and the suction line
- A very light system load (well below design)
- Wrong type of valve power element for the system refrigerant

Starving can be caused by the TXV sticking closed or by a loss of charge in the power element. The valve may also be restricted, either at the inlet strainer (if equipped) or internally. An improper superheat setting, or an undersized valve can also cause the evaporator to be starved for refrigerant. However, neither of these conditions represents a valve failure, but rather indicate errors made by the installer or a service technician. The evaporator coil can be starved for the following additional reasons:

- Shortage of refrigerant
- Plugged filter-drier
- Kinked or obstructed liquid line

- Exceptionally tall liquid-line riser
- Undersized or exceptionally long liquid line
- Plugged refrigerant distributor
- Improper valve location
- Plugged equalizer line

Hunting describes consistent changes in refrigerant flow rates as the TXV attempts to find the proper valve position to maintain the superheat setting. All TXVs may hunt to some degree, but not in an easily noticeable way. Hunting is characterized by swings in superheat of several degrees up and down, rather rapidly.

Hunting can only be verified by making several suction line temperature or pressure measurements over a period of time. If there is a repetitive pattern, such as that shown in *Figure 6*, the valve is hunting. Excessive hunting can be caused by sticking or binding of the TXV internal parts. It can also occur if the system load on the TXV drops below 30 percent of its capacity. Other than faults in the valve itself, the following conditions can cause excessive hunting:

- Oversized expansion valve
- Very light load
- Long liquid refrigerant line
- Rapid changes in condensing pressure or temperature
- Rapid load changes
- Intermittent flashing in the liquid line

It is important to remember that TXVs, by design, will try to maintain the superheat setting under any operating condition. When the system charge is low, the superheat will likely be low as well. However, the TXV may be able to mask this symptom somewhat by simply opening wider and trying to feed enough refrigerant to reduce the superheat. A small loss of refrigerant charge, therefore, may not be obvious in the superheat reading. As long as some subcooling exists, and possibly even longer, the TXV can likely compensate for the shortage.

If the system is overcharged, the TXV is even better at masking this problem. It will only open as far as necessary to maintain the superheat; any

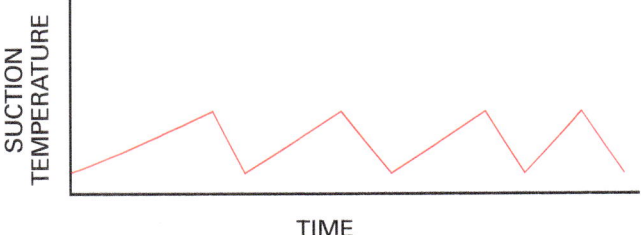

Figure 6 Suction temperature variations that indicate hunting.

excess refrigerant simply backs up in the condenser or the receiver (if equipped). It may take a considerable overcharge to cause the high-side pressure to rise to the point of being obvious. In this case, an excessive amount of subcooling will indicate that there is more refrigerant in the system than necessary. No other obvious symptoms may be noted unless the overcharge is massive.

Some system problems caused by TXVs are the result of poor installation. Make sure the sensing bulb is securely fastened on the top of a clean, straight section of the suction line, close to the evaporator outlet. On all but the smallest of suction lines, the bulb should be placed at the 10 or 2 o'clock position on a horizontal run of pipe. Also, check the insulation on the bulb to make sure outside air temperatures are not affecting it. Over time, some light corrosion can take place between the bulb and the line, slowing heat transfer. The contact surfaces should be clean.

A distributor is installed downstream of the metering device (usually a TXV) and is designed to ensure that refrigerant leaving the TXV gets evenly distributed to the various circuits in the evaporator coil. Distributors (*Figure 7*) do not function properly at loads well below their design flow capacity. If the distributor flow rate falls below 50 percent of its capacity, the liquid and vapor leaving the TXV will not be properly mixed. As it then passes through the distributor, the liquid will go to the bottom circuits and the vapor will go to the top circuits. A way to check for this problem is to look for high temperatures at the top of the evaporator and low temperatures at the bottom.

1.8.0 Other Refrigerant Circuit Problems

There are several other problems that are common to the refrigerant circuit. These problems are related to the following:

- Interconnecting refrigerant lines
- Refrigerant circuit accessories
- Contaminants and non-condensables

1.8.1 Refrigerant Line and Accessory Problems

A restriction in a refrigerant liquid line or related accessory (*Figure 8*) reduces or prevents the system refrigerant from getting to (or through) the evaporator. This reduces system capacity. Problems with lines or accessories are rarely, if ever, identified by too much refrigerant flow.

The cause of problems in a system with a restriction is sometimes diagnosed as the system being low on refrigerant charge. This is because the symptoms of low suction and discharge pressures and high superheat are similar. The subcooling temperature can be used to help determine if the problem is a restriction rather than a low charge. Systems with a major restriction anywhere in the liquid line usually show normal or increased subcooling at the condenser, with little or no subcooling at the metering device inlet. This is because the mass flow of refrigerant in the system slows down. Because of the restriction, most of the system liquid refrigerant remains in the condenser and/or receiver and subcools further. The liquid refrigerant stays in the condenser

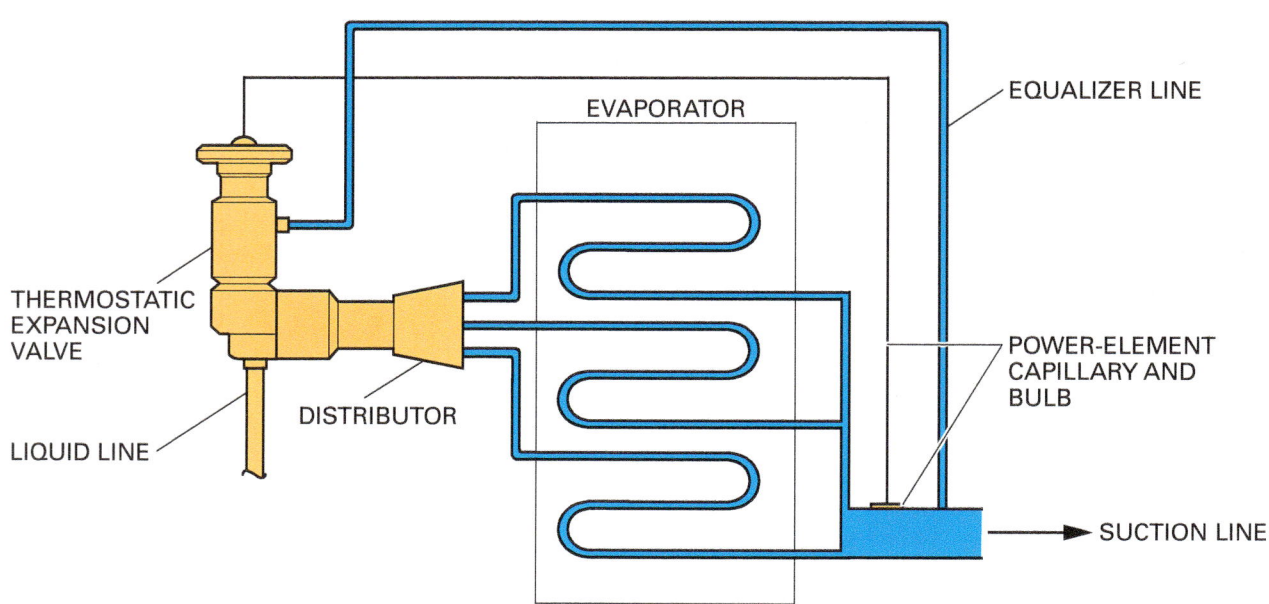

Figure 7 Multi-circuit evaporator fed by a distributor.

longer, resulting in greater subcooling because it has more time to be cooled.

Refrigerant tubing can become restricted from physical damage that causes the tubing to be pinched or bent. This often occurs on systems that have been running normally for a long time. Physical damage to the tubing often goes undetected because it can be hidden under insulation or behind a fixture.

When a partial restriction occurs in a liquid line, it acts like a metering device and drops the pressure and temperature of the refrigerant at the point of the restriction. It is a good practice to consider any accessories that are installed in a system as a potential point of restriction. A restriction could be detected by measuring the pressure drop across the accessory or tubing suspected to have the restriction; however, this is not always practical because gauge ports are not usually available.

An easier way is to measure the temperature drop on each side of the accessory or tubing suspected to contain the restriction. Split HVAC systems are more susceptible to refrigerant

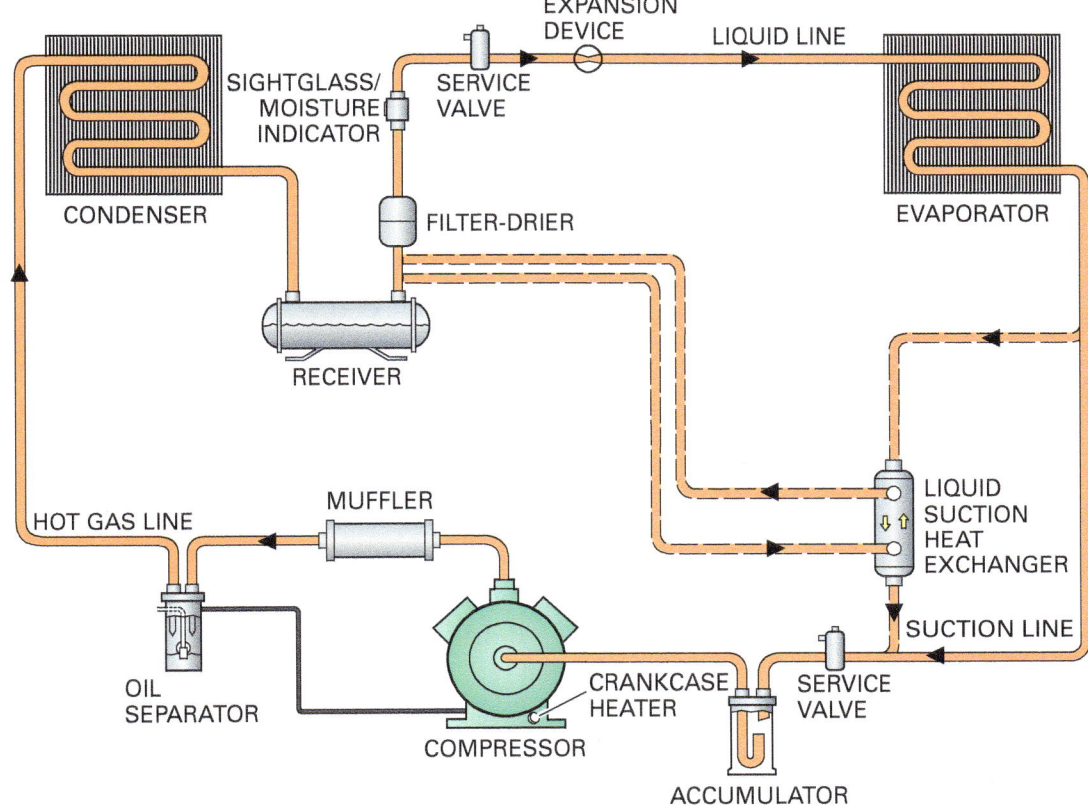

Figure 8 Refrigerant lines and accessories.

Capillary Tubes

Capillary tubes, like the one shown here, were once widely used in systems up to five tons. With the introduction of the fixed-orifice metering device, the capillary tube is now mainly used in room air conditioners and packaged terminal air conditioners (PTACs). They also continue to be used in domestic refrigerators and freezers.

Figure Credit: Courtesy of Henry Technologies, Inc.

line restrictions and damage because the field-installed interconnecting tubing may be routed through areas where they are subject to accidents, abuse, or vandalism. The refrigerant lines of packaged HVACR products are all contained within the security of the package, so they are less susceptible to damage. Restrictions in packaged systems are far more likely to be found at an accessory.

Depending on the severity of the restriction, a technician can sometimes feel a temperature change in the line, or even observe frost or sweat just downstream of the problem area. If the restriction is very slight, temperature measurement with a thermometer is usually the only way to find the location of the restriction.

A plugged or undersized filter-drier is a common cause of a restriction. On a newly installed system, the restriction may occur shortly after startup when it may become plugged by solid contaminants that entered the system during installation. If a system is equipped with a sight glass installed downstream of the filter-drier, a restricted filter-drier will cause flash gas (bubbles) to appear in the sight glass. To confirm that a filter-drier is restricted, always measure the temperature drop across it, since a restriction here can go undetected using the touch method. The amount of pressure drop across a filter-drier can also be an indication of a restriction. The pressure drop should not exceed 5 psi.

Restrictions can be caused by expansion valves that do not open all the way. Normally, TXVs either work or do not work. However, a partial restriction can occur in the valve if it loses part of the charge in the thermal bulb, or if the inlet strainer becomes dirty. Also, if moisture is present and circulating in a system that operates at a saturation temperature below freezing, it can freeze at the metering device and restrict refrigerant flow. This is a problem that progresses as the system operates. Once the system shuts down for a short period, the frozen moisture melts, remaining in the circuit to progressively freeze again during the next cycle.

Test for loss of charge in the TXV bulb by removing it from the suction line and holding it tightly in the hand. If the valve fails to feed the evaporator more refrigerant, reducing the measured superheat, the bulb has probably lost some charge.

Another potential restriction point in split-system products is the service valves. During normal operation, ensure that both the liquid line and suction line service valves are 100 percent open. During installation, technicians sometimes fail to open these valves fully, causing a restriction. During certain service procedures, these valves must be closed, but make sure to open one or both valves 100 percent after the service procedure is complete.

Beyond restrictions, refrigerant lines can be affected by their environment. For example, a liquid line significantly exposed to a very hot environment can lose a lot of subcooling. Conversely, if it is routed through a cold environment, it may gain some beneficial subcooling. Liquid lines experience a 1 psi pressure loss for every 2.3 feet that they rise in height. A riser that is 23 feet in height loses 10 psi. This significant change can erase any subcooling that was gained in the condenser.

Suction lines can also experience these positive or negative changes. However, they are less likely to change due to the insulation that is normally applied. When troubleshooting the refrigerant circuits of split systems, it is extremely important to make superheat and subcooling measurements at each end of the line. Superheat should be checked at the evaporator outlet and at the compressor, and the two values compared. Subcooling should be checked at the condenser outlet and at the metering device inlet for comparison. When pressure readings cannot be taken at both locations, be sure to compensate by considering any normal pressure drop in the lines in the calculations.

1.8.2 Air and Moisture Contamination

Refrigeration systems should contain only refrigerant and oil. Anything else must be considered a contaminant. Air and moisture are the most common contaminants in a refrigeration or air conditioning system. Air is a non-condensable gas, which means that it does not change into a liquid at common operating temperatures and pressures. Air entering the system can also contain moisture in the form of water vapor. When air is in a system, it accumulates in the top of the condenser, taking up valuable condensing surface area. This results in high head pressure and an increase in the condensing temperature.

Under the heat of compression, moisture will react with the refrigerant to form acids. These acids cause corrosion of metals and breakdown of the insulation on the motor windings and wiring. Moisture in the refrigerant can also cause oil sludge, which reduces the lubricating properties of the oil and plugs oil passages and screens in the compressor. Moisture can also freeze at the metering device, as mentioned previously.

Air and moisture can enter a system in the following ways:

- Improper or no evacuation/dehydration process at installation or after system servicing
- Refrigerant circuit leaks while under a vacuum
- Failure to install a filter-drier at installation or after servicing
- Failure to purge hoses prior to charging refrigerant

Symptoms of system contamination include the following:

- Sealed-tube refrigerant test shows acid and/or moisture
- Refrigeration oil acid test shows acid
- Discoloration of the moisture indicator in a liquid line sight glass
- System valves stick
- Clogged metering device
- Poor compressor operation
- Compressor failures
- Overheating of compressor
- High compressor current draw
- High head pressure (non-condensable contaminants)

If a system using extremely hygroscopic lubricants, such as POE (*Figure 9*) and PAG types, is left open and exposed to the air, the oil can absorb large amounts of moisture. This moisture cannot be removed with a vacuum pump. Since these oils attract moisture, it is essential that refrigeration circuits not be left open to the atmosphere for any significant length of time.

In a system that uses a TXV, the problem of non-condensables in the system is often diagnosed as an overcharge of refrigerant because the symptoms are similar.

1.8.3 Acid Contamination

Acid is not introduced into a system. Rather, it is formed inside an operating system by the reaction of air or moisture with the refrigerant. It can also be formed during a compressor failure, as the insulation on motor windings breaks down in the extreme heat. Acid creates sludge and varnish, which can plug oil passages and metering devices, and it can restrict the strainers in the compressor and filter-driers, as well as other refrigeration system components.

Acid most likely exists in a system after the compressor has failed as the result of shorted or grounded motor windings, which are called burnouts. Both conditions are fatal. Burnouts

Figure 9 Polyol ester (POE) lubricating oil.

Preventing Non-Condensables in a System

Do you want to prevent non-condensables in an HVACR system? The answer is simple. If the sealed refrigeration system must be opened for any reason, close it as quickly as possible. Evacuation removes non-condensables most of the time. Always use a well-maintained vacuum pump and evacuate the system to 500 microns before recharging. Allow the pump to run for a significant length of time at this vacuum level; boiling liquid water to vapor doesn't happen instantly. Also, when installing refrigerant tubing, make sure to purge the tubing with nitrogen before use to remove any moisture caused by condensation trapped within the tubing.

One piece of advice from an industry veteran: Treat cans of oil, as well as the refrigerant circuit, as if it is full of angry bees that you do not want to release! If it must be opened, do so quickly, and close it just as quickly.

are classified as either mild or severe based on the amount of acid present in the system. A mild burnout tends to yield little acid because it occurs suddenly, causing the motor to be de-energized and stop before the contaminants created by the burnout leave the compressor.

A severe burnout usually occurs over a longer period, allowing considerable acid and other contaminants to be produced and pumped through the system while the compressor is still running. These situations may also be referred to as a *starting burn* or a *running burn*. If the compressor is trying to start but not pumping as the motor fails, little acid will be circulated. However, if it is actively pumping during the burning process, a great deal of acid can be released into the circuit.

> **WARNING!**
> When working on any system containing refrigerant, eye and hand protection must be worn at all times to prevent personal injury. This is especially true when working with acid-contaminated systems. Acid in contact with eyes or skin can cause severe burns.

Before replacing a compressor that has failed because of a burnout, the system must be tested for acid contamination. A sealed-tube acid/moisture test kit (*Figure 10*) is connected to a system service port to obtain a sample of the refrigerant. Oil test kits (*Figure 11*) require that the test be performed on a sample of the compressor oil. In either case, follow the test kit manufacturer's instructions for using the kit and for determining the level of contamination.

BEFORE

AFTER
(POSITIVE RESULT)

SEALED-TUBE TEST KIT

Figure 10 Refrigerant vapor test kit.

Refrigerant Smell Can Indicate a Compressor Burnout

While an acid test kit can be used to tell if acid is present, your nose is also able to detect it. Refrigerant from a system that has had a moderate to severe compressor motor burnout will have a very strong, pungent odor. Once you've smelled it, you will always be able to recognize it. The burning of some refrigerants can create a material known as phosgene gas, once used in warfare to sicken enemy troops. It can be formed inside the circuit during a severe compressor burnout. It can also be formed when using a torch in the presence of escaping refrigerant gas, which can easily happen while replacing a compressor.

Figure 11 Oil test kit.

Additional Resources

Refrigeration and Air Conditioning, An Introduction to HVAC/R, Fourth Edition. Larry Jeffus. Air Conditioning and Refrigeration Institute. Upper Saddle River, NJ: Prentice Hall.

System Diagnostics and Troubleshooting Procedures. 2002. John Tomczyk. Mt. Prospect, IL: Esco Press.

1.0.0 Section Review

1. Which of the following components takes high-pressure, high-temperature refrigerant vapor and cools it to allow the formation of liquid refrigerant?
   a. Compressor
   b. Evaporator
   c. Metering device
   d. Condenser

2. Subcooled liquid refrigerant from the condenser flows through the liquid line to the _____.
   a. evaporator
   b. metering device
   c. compressor
   d. muffler

3. During a preliminary inspection of the evaporator and condenser coils, which of the following would logically be done?
   a. Check that the crankcase heater is working.
   b. Look for corroded fin collars.
   c. Inspect for scorching or blistering indicating excessive heat.
   d. Inspect for an excessive number of elbows and other fittings.

4. One of the possible causes of a low suction pressure would be _____.
   a. a loose or broken blower belt
   b. defective compressor valves
   c. an excessive refrigerant charge
   d. the recirculation of condenser air

5. On a unit with a fixed metering device, low airflow is generally indicated by a _____.
   a. high superheat condition
   b. low superheat condition
   c. high subcooling condition
   d. low subcooling condition

6. A capillary tube that is too short in length will result in _____.
   a. high pressures
   b. low subcooling
   c. high superheat
   d. low superheat

7. On a unit equipped with a TXV, overfeeding refrigerant can be caused by _____.
   a. a plugged filter-drier
   b. a shortage of refrigerant
   c. poor thermal contact between the bulb and suction line
   d. an excessively long liquid line riser

8. One advantage packaged cooling units have over split systems is that _____.
   a. they can operate at a lower superheat
   b. they never need a filter-drier
   c. the refrigerant lines are protected by the packaging
   d. they can operate properly with little or no subcooling

Section Two

2.0.0 Compressor Troubleshooting and Replacement

Objective

2. Explain how to troubleshoot and replace a compressor.
   a. Identify common problems that can lead to compressor failure.
   b. Explain how to troubleshoot compressor mechanical problems.
   c. Explain how to replace a hermetic compressor.
   d. Describe the additional steps that may be required to replace a compressor following an electrical failure.

Performance Tasks

1. Demonstrate the ability to isolate and determine the solution for at least four of the following types of malfunctions:
   - Compressor failures
   - System-related compressor problems
   - Refrigerant undercharge or overcharge
   - Evaporator and condenser problems
   - Metering device problems
   - Refrigerant lines and accessories
   - Non-condensables and refrigerant circuit contamination

Trade Terms

Capacitance boost: A procedure used to start a stuck PSC compressor. It involves momentarily connecting a start (boost) capacitor across the run capacitor of the stuck compressor to increase the starting torque in an attempt to start the stuck compressor.

Flooded starts: A condition in which slugging, foaming, and inadequate lubrication occur at compressor startup due to oil in the compressor having absorbed refrigerant during the off cycle. When the compressor starts and the pressure in the crankcase drops quickly, the refrigerant boils rapidly, foaming the oil.

Slugging: The entrance of liquid refrigerant and/or a significant amount of oil into a compressor cylinder. Slugging can occur at startup due to condensed refrigerant and/or collected oil near the compressor inlet.

Many cooling system problems will result in compressor failure if they are not corrected. If the compressor fails and the cause has not been found and eliminated, the replacement compressor will probably fail. *Figure 12* summarizes some of the symptoms exhibited by the compressor for common electrical and refrigeration-cycle problems.

Electrically troubleshooting the compressor motor is not significantly different than troubleshooting other motors. The measurements and checks used to electrically test compressor motors are provided in another module of this curriculum. This section presents symptoms and solutions for other types of compressor problems that are frequently encountered.

2.1.0 Problems for Compressors

Compressor problems include slugging, floodback, flooded starts, and overheating. Although overheating can be a symptom of an internal problem, it can also result from an external problem. Slugging, floodback, and flooded starts are problems that are not the fault of the compressor.

2.1.1 Slugging

Slugging occurs when a compressor tries to compress liquid refrigerant, oil, or both, instead of superheated gas. Slugging is generally considered to involve a large single dose of liquid returning to the compressor at once. If slugging occurs, it will usually occur at startup or during a rapid change in system operating conditions.

Slugging can usually be detected by a loud knocking noise at the compressor. In severe cases, it can result in the cylinder cracking, bolts shearing, or parts actually separating from the compressor. Due to the extreme stress it causes, it often results in blown gaskets and broken valves, piston rods, or crankshafts. This type of damage can happen as a result of a single slugging event.

Common causes of slugging include the following:

- An overcharge of refrigerant.
- An oversized, damaged, or failed metering device.
- An overcharge of oil.
- A failed crankcase heater.
- Condensed refrigerant is present in any cold part of the system, such as the evaporator, during the off cycle. Buried refrigerant lines or those passing through cold spots can allow the refrigerant to condense back into a liquid at shutdown.

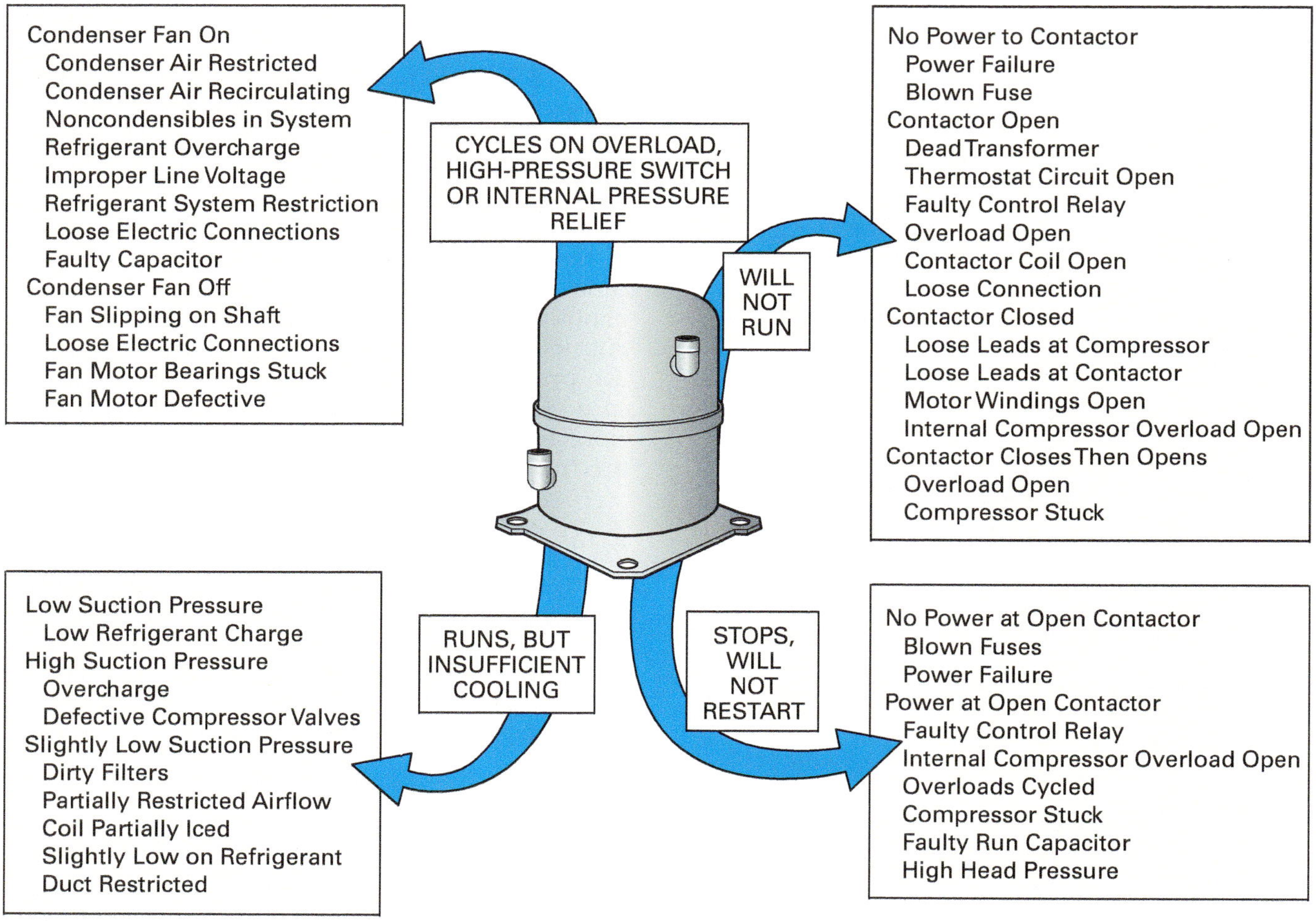

CYCLES ON OVERLOAD, HIGH-PRESSURE SWITCH OR INTERNAL PRESSURE RELIEF
- Condenser Fan On
  - Condenser Air Restricted
  - Condenser Air Recirculating
  - Noncondensibles in System
  - Refrigerant Overcharge
  - Improper Line Voltage
  - Refrigerant System Restriction
  - Loose Electric Connections
  - Faulty Capacitor
- Condenser Fan Off
  - Fan Slipping on Shaft
  - Loose Electric Connections
  - Fan Motor Bearings Stuck
  - Fan Motor Defective

WILL NOT RUN
- No Power to Contactor
  - Power Failure
  - Blown Fuse
- Contactor Open
  - Dead Transformer
  - Thermostat Circuit Open
  - Faulty Control Relay
  - Overload Open
  - Contactor Coil Open
  - Loose Connection
- Contactor Closed
  - Loose Leads at Compressor
  - Loose Leads at Contactor
  - Motor Windings Open
  - Internal Compressor Overload Open
- Contactor Closes Then Opens
  - Overload Open
  - Compressor Stuck

RUNS, BUT INSUFFICIENT COOLING
- Low Suction Pressure
  - Low Refrigerant Charge
- High Suction Pressure
  - Overcharge
  - Defective Compressor Valves
- Slightly Low Suction Pressure
  - Dirty Filters
  - Partially Restricted Airflow
  - Coil Partially Iced
  - Slightly Low on Refrigerant
  - Duct Restricted

STOPS, WILL NOT RESTART
- No Power at Open Contactor
  - Blown Fuses
  - Power Failure
- Power at Open Contactor
  - Faulty Control Relay
  - Internal Compressor Overload Open
  - Overloads Cycled
  - Compressor Stuck
  - Faulty Run Capacitor
  - High Head Pressure

Figure 12 Compressor troubleshooting.

Slugging

Burying refrigerant lines may sound like a good idea, but it is not a preferred practice. Since the temperature is usually cooler underground, refrigerant in buried lines will condense during shutdown and cause slugging when the compressor starts. For this reason, most HVACR equipment manufacturers prohibit or severely restrict the use of buried refrigerant lines with their equipment. Other steps may be required, such as the installation of a suction accumulator.

- Slugs of oil are trapped in the suction line because the suction gas does not have enough velocity to return the oil to the compressor. Normally, the oil and refrigerant mix together effectively. The oil is circulated through the system in very small drops as it is being swept along by the velocity of the refrigerant vapor. If it gets trapped in the system piping and then returns all at once, it causes slugging. This condition also occurs in systems that use compressors equipped with capacity-control devices, especially when the compressor runs unloaded for long periods.
- Poor piping design allows liquid refrigerant to return to the compressor during the off cycle.

2.1.2 Floodback

Floodback is the continuous return of liquid refrigerant or significant liquid droplets in the suction vapor to the compressor during operation. Compared to slugging, it is a more progressive situation that can eventually develop into slugging. Floodback dilutes the oil, resulting in crankcase foaming and overheating of bearing surfaces. If bad enough, it can damage the pistons, rings, and valves because liquid refrigerant washes the oil off the bearing surfaces. Flooding is most often caused by the uncontrolled flow of refrigerant that results from one or more of the following conditions:

- An oversized thermostatic expansion valve.
- A broken, improperly located, improperly mounted, or poorly insulated TXV power element and/or sensing bulb.
- A low superheat setting.
- Low evaporator load.
- An overcharge of refrigerant, especially on systems using a fixed metering device.

2.1.3 Flooded Starts

Flooded starts are caused by the oil in the compressor crankcase absorbing refrigerant. This is more relevant during long compressor shutdowns at low temperatures. Refrigerant will migrate to and condense in a cool or cold compressor shell, where it mixes with the oil. The amount absorbed into the oil depends on the temperature of the oil and the pressure in the crankcase.

On startup, the refrigerant-rich oil is pumped through the oil pump of the compressor, resulting in marginal lubrication of the bearings. Also, as the crankcase pressure drops suddenly at startup, the refrigerant will flash from a liquid to a gas, causing the oil to foam. This foaming can restrict the oil passages and cause the oil pressure to build. A great deal of this foam can also be carried over into piston cylinders.

The problem of flooded starts can be minimized by first making sure that the system has a proper refrigerant charge and the correct amount of oil in the crankcase. Make sure the crankcase heater is working or add one, if necessary. Crankcase heaters are used to warm the oil during the off cycle and prevent the condensation of refrigerant in the crankcase. Piping should also be designed to prevent liquid refrigerant from accumulating in the suction line and then flowing by gravity to the compressor inlet while it remains off.

Another way to prevent the migration of refrigerant through the system during the off cycle is to use a liquid line solenoid valve (*Figure 13*). The valve can be wired to operate on a call for cooling along with the compressor. Once the need for cooling is satisfied, the valve closes as the compressor shuts down.

Figure 13 Liquid line solenoid.

A Real Situation

In a real-life example of compressor failures, a number of new 20-ton packaged rooftop units were experiencing rapid-fire compressor failures. It did not take long to determine why.

In this industrial plant, a great deal of heat was being generated in a building with a 28-foot (8.5 m) roof height. Included in the operation was an industrial oven that added significantly to the heat. The heat collected in the top of the building, as there were no fans or other devices to mix the building air and keep it from stratifying. Temperatures at the roof level exceeded 125°F (52°C), as measured in the return-air duct by the technician.

The return-air grilles for each unit were positioned at the roof level, to save on the air-distribution system cost. The technician, upon seeing the return-air temperature, checked the superheat at each compressor in multiple units. The lowest superheat reading exceeded 50°F (28°C). Several exceeded 60°F (33°C), due to the extreme heat load.

In industrial buildings like this one, there is little or no reason to attempt to cool the entire volume of the building. The people and primary equipment are on the ground. Although it was costly with production in progress, the technician convinced the client to lower both the return and supply ducts down to the 16-foot (5 m) level. Heat that rose above this level was intentionally allowed to stratify and remain there. Not only did the compressors begin to survive and produce more effective cooling, the energy cost fell dramatically.

On some systems, the solenoid may close before the compressor is shut down. This allows the compressor to pump some or all the refrigerant remaining in the evaporator and suction line out and into the condenser and/or receiver before it shuts down. This prevents any significant amount of refrigerant from condensing in these areas. The compressor eventually shuts down when it reaches the setting of a pressure switch installed for this purpose. When a solenoid valve is used in this way, it is often referred to as a pump-down solenoid.

2.1.4 Overheating

Compressors normally generate heat and are designed to handle this normal heat. When the compressor is overheating, the cause must be determined. When discharge temperatures from the compressor exceed 250°F (121°C), the oil begins to break down and the situation becomes progressively worse. Overheating can damage internal components and cause moving parts to seize due to thermal expansion. The lubricating oil can also burn and form a residue known as coke. A sample of oil can provide some visual clues. Dark, black oil is a sign that the compressor has been running too hot. The oil will also show this dark, black color following a severe burnout, due to contaminants from the winding insulation and as well as coke residue.

It is important to understand that most compressor types depend on the cool refrigerant gas returning to absorb a great deal of the heat produced. The heat is then rejected from the system by the condenser. As a result, a system that is operating with a high superheat, for any possible reason, can cause the compressor to overheat.

How Much Oil?

One concern for technicians during a compressor change is related to the amount of oil in the system. A system may have been operating under unusual conditions, such as undercharged or at reduced capacity, for a long time before the compressor failure. As a result, much of the oil is likely trapped in the circuit somewhere rather than returning to the compressor crankcase. Some of the oil remains out in the piping, coils, and accessories at all times.

With a semi-hermetic compressor, a sight glass helps to determine how much oil may be missing. However, with hermetic compressors it is much more difficult to tell. They can be drained after removal to compare their holding capacity to the contents, but all the oil may not come out. On systems that experience multiple compressor failures, the volume of oil in the system must become a matter of concern. Unfortunately, there is no foolproof way of determining exactly how much oil is in the circuit except by tracking what went in.

Figure Credit: Courtesy of Emerson Climate Technologies (Photo)

Some compressors are protected from overheating by an internal or external thermal limit switch. Internal limit switches also provide some back-up protection for three-phase compressors. When a three-phase compressor tries to start with only two phases of power provided, it will become hot very quickly. The thermal limit switch, also referred to as a *thermal overload*, will then break the electrical circuit to protect it. Without other forms of protection against single-phasing, this cycle can repeat itself many times. The limit switch closes; the compressor tries to start; it does not rotate, but heat builds quickly; the limit switch opens. After several cycles like this, the entire mass of the compressor becomes very warm. It can take some time for it to cool after a technician arrives to stop the cycle.

If a compressor overheats, check for a high superheat. This may be coupled with a low suction pressure. A high compression ratio can also cause overheating. For example, a system with a TXV may have a condenser coil that is badly clogged with grass clippings or other debris. Although the TXV will try to compensate and the superheat may not be excessively high, the compression ratio may be extremely high due to the increase in high-side pressure. The compression ratio increases as the differential between the high and low pressures grows.

2.2.0 Compressor Mechanical Problems

Compressors experience mechanical problems of their own. Some failures are initiated or caused by external problems. Others may be the result of mechanical flaws or normal wear. When a compressor fails due to external problems such as a failed metering device, the technician will not be able to discover this until an operable compressor is installed. To ensure the proper operation of a new compressor, it is important for the system to be thoroughly tested and any problems solved before the replacement is made.

A visual examination of an oil sample may provide clues regarding the condition of the compressor. As noted previously, dark, black oil indicates significant overheating, assuming the compressor has not already experienced a burnout. However, oil with gray or silver coloration, or oil that appears to have tiny pieces of glitter in it, indicates mechanical wear. Minute metal particles flaking off of piston rings, cylinder walls, or bearings will cause this change in color.

2.2.1 Valves and Rings

In a reciprocating compressor, valves and rings provide a seal between the high-pressure and low-pressure sides. If they are damaged, the compressor must be replaced, or the valves repaired. Valves can often be field-repaired or replaced in semi-hermetic and open-drive models. Piston rings cannot be field-repaired or replaced. Fortunately, they are far less likely to be damaged than the valves.

Suction and discharge pressure checks can be used to test for this condition. Bad valves or rings may exist if the suction pressure will not pull down or discharge pressure will not build up when the system is properly charged and under normal load. Suction valves are damaged more often than discharge valves.

Another check can be made by measuring the compressor's running current with an ammeter (*Figure 14*). If, under loaded conditions, the running current is considerably lower than normal, and the suction/discharge pressures are abnormal (high suction pressures and/or low discharge

Crankcase Heaters

Crankcase heaters come in a variety of sizes (wattage) and shapes. A belly-band heater wraps tightly around the bottom of a welded hermetic compressor shell. Other types of heaters are inserted in a well in the base of the compressor shell that allows the heater to be immersed in the oil. Some equipment is designed so that power is applied to the crankcase heater at all times, while others have various thermostatically controlled schemes that apply power to the heater only when temperatures drop below a certain point.

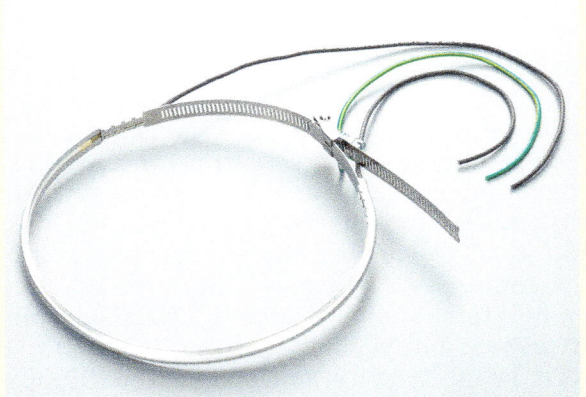

Figure Credit: Courtesy of Backer Springfield

Figure 14 Checking current with an ammeter.

pressures), faulty valves or rings should be suspected. During this type of test, ensure that there are no compressor capacity control devices in operation that could cause misleading results.

Many equipment manufacturers specify that power must be applied to their equipment for up to 24 hours before the initial startup or after a prolonged winter shutdown. This is done to allow the crankcase heater enough time to drive any accumulated refrigerant out of the oil in the compressor shell to prevent a flooded start. If this is not done, the compressor may be damaged or destroyed the first time it is started.

A scroll compressor may exhibit the same combination of symptoms that indicate valve problems in a reciprocating compressor. However, a scroll compressor has no valves at all. When the same symptoms are encountered in a scroll compressor, it could indicate a crack or similar damage in the orbiting scroll plate.

2.2.2 Seized Compressors

A seized, or stuck, compressor hums but will not start, and typically draws locked rotor amperage (LRA) for at least a few seconds. Normally, the LRA, which can be four to six times the normal running amperage, lasts for a fraction of a second when the motor first starts. Once the motor begins to rotate, it's starting current drops quickly. When a compressor is mechanically seized, the current will remain at or near its LRA value until a safety device intervenes.

A seized compressor is a mechanical condition. However, there are also electrical problems that can prevent the compressor motor from rotating.

Before condemning a seized compressor, check for the following conditions:

- Three-phase compressor contactor not making good contact on all poles (single-phasing)
- Defective or open start relay, preventing the start capacitor from assisting
- Start and/or run capacitor open or weak (single-phase compressors only)
- Unequal system pressures (especially in units using PSC compressor motors)
- Low supply voltage

If the supply voltage is within ±10 percent of the motor's nameplate rating, but a single-phase PSC motor fails to start, a start kit may have to be installed to increase the motor starting torque to correct the problem. This is more often required on systems using a TXV metering device than a fixed restrictor.

If electrical tests prove a PSC compressor has no open, shorted, or grounded windings, and if all external factors check out to be good, a technician can attempt to start a seized PSC compressor with a capacitance boost (*Figure 15*). Capacitance boosting involves momentarily connecting a start capacitor (200 to 300 MFD range) across the compressor's existing run capacitor when power to the compressor is turned on. Since the start capacitor is placed in parallel with the run capacitor, the values of the capacitors are combined,

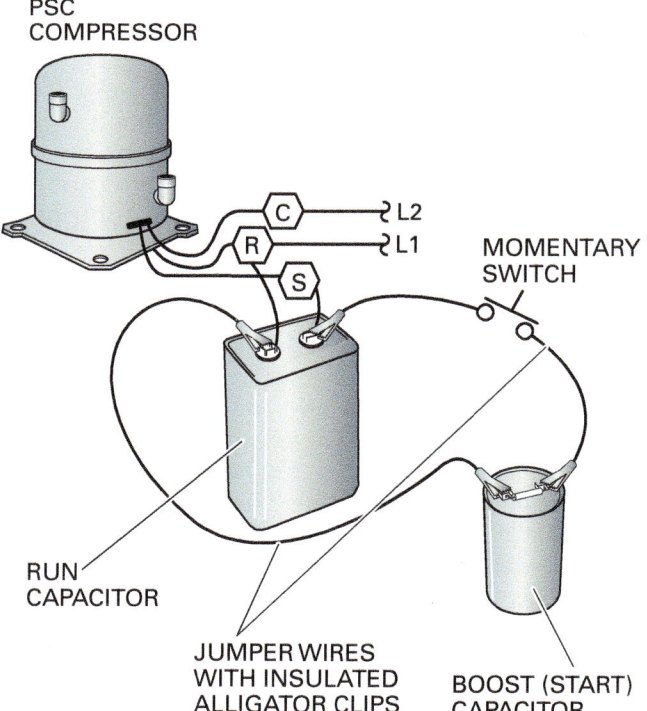

Figure 15 Starting a seized PSC compressor with a capacitance boost.

providing increased capacitance and torque to start the PSC motor. Once started, the compressor should continue to run at full speed with the start capacitor removed from the circuit. It is important that the start capacitor be disconnected from the circuit; leaving it in the circuit for more than a few seconds will cause it to overheat and fail altogether.

> **WARNING!**
> To avoid the possibility of electric shock, do not handle live wires. Switch the power off and lock it out before connecting the wires. Also, bridge the start capacitor to ensure that it is discharged before connecting wires to it.

After the operating pressures equalize, try restarting the compressor again without the capacitance boost. If the compressor starts the first time and starts on subsequent attempts, a permanently installed start capacitor and start relay are probably not needed. If the compressor will not start without a capacitance boost, permanently install a correctly sized start capacitor and start relay. Start kits designed for use with most compressors are available from HVAC distributors.

> **CAUTION**
> Always use the correct start kit specified for the compressor. Failure to use the appropriate start kit may cause damage to the compressor motor. Also, never use a megohmmeter to check motor windings, or attempt to start a compressor, while the refrigerant circuit is in a vacuum.

Checking for a Grounded Condition

A motor can be grounded by a winding contacting some portion of the case. In this case, the resistance reading from a winding to ground will be very low. However, resistance readings can be several thousand ohms and still be considered shorted. A good rule of thumb is that the resistance to ground should be greater than 1,000Ω per operating volt. For example, a motor that operates on 240V should have a minimum resistance of 240,000Ω to ground. For better accuracy, consult the manufacturer to get the expected resistance values. Both refrigerant and oil contamination can also create low resistance-to-ground readings. To improve these readings, change the refrigerant oil and driers.

Capacitors

Use a capacitance checker to determine the capacitance of run and start capacitors. The needle deflection of an analog or digital ohmmeter will not indicate the correct capacitance. A capacitor should operate within 10 percent of its microfarad rating.

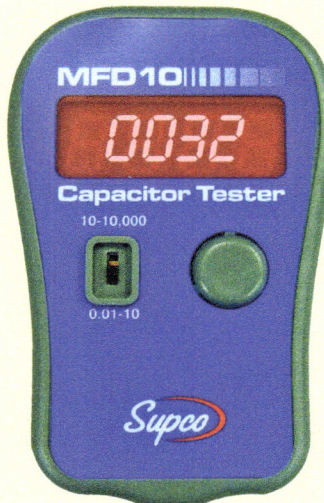

Figure Credit: Courtesy of Sealed Unit Parts Co., Inc.

If a compressor has been changed, a different run or start capacitor rating may be required. This is sometimes overlooked when compressors are replaced. The new compressor may be equipped with a different motor than the previous model, and capacitors of a different microfarad value may be needed. Every time a compressor or motor is changed, even when there is no change in capacitance required, the capacitor is best changed with it. The terms of many warranties require installing a new capacitor when a compressor or motor is replaced.

Three-phase motors can sometimes be unstuck by temporarily reversing or interchanging any two of the leads. After the motor has started, always turn off the power, then reconnect the compressor leads as shown on the unit wiring diagram.

2.2.3 No Pressure Change

Occasionally, a compressor will appear to do absolutely nothing when it is energized. There may be very little audible sound coming from it. A check of the refrigeration gauges will show no activity at all. Yet voltage is being applied to the compressor terminals, and no safety device is, or has been, activated.

These are the symptoms of a compressor with a broken drive or crankshaft. An ammeter placed on the compressor leads will show that some, but very little, power is being consumed. The current reading will be very low, but present. This is merely the power required to spin the motor without any applied load. When the shaft is broken, the motor is free to spin, and there is no movement in any other part of the compressor.

2.2.4 Troubleshooting Scroll Compressors

Basic troubleshooting for scroll compressors is similar to that of other compressors. Expect a pressure differential between the suction and discharge sides of the compressor.

Like reciprocating compressors, scroll compressors sometimes require start-assist components, especially in the low-voltage starting conditions that often occur in the heat of summer. Because scroll compressors run backwards momentarily at shutdown, it is important to use short-cycle timers, either those provided by most digital thermostats or separate delay-on-break time-delay relays. Short-cycle timers provide the scroll compressor with the time required for pressures to equalize and rotation to stop completely, allowing the compressor to start without a load and rotate in the proper direction.

If the compressor is restarted too quickly, rotation may continue in the opposite direction. Unlike reciprocating compressors, scroll compressors cannot pump properly when their rotation is reversed. If rotating in the wrong direction, they will pump poorly and overheat. It is important to note that three-phase models must be properly phased to ensure proper rotation.

2.2.5 Troubleshooting Rotary Compressors

Rotary compressors sometimes become oil-locked. This is a condition in which the rotor will not turn due to the tight clearances in the compressor. Oil lock can occur in new compressor motors. To fix an oil-locked compressor, try to start the motor several times. If this does not work, install a soft- or hard-start kit. If the kit does not fix the problem, the compressor must be replaced.

Another problem experienced by rotary compressors is that of stuck vanes. When a vane sticks, there is no pumping action and the motor draws low amperage. The rotor is turning, but the vane(s) is not moving in and out to create compression pockets. Before replacing the compressor, try removing the suction line from the compressor shell.

Use a heat gun to warm the tubing and suction area on the compressor. The heat may cause the stuck vane to release. Before reconnecting the suction line, momentarily start the compressor with the palm of your gloved hand over the suction fitting. If the vane is no longer stuck, the compressor will pull on the palm of your hand. This indicates that the vane is now working and creating suction pressure.

A worn, leaking vane will cause pressure readings that are lower than normal on the high side and higher than normal on the suction side.

2.3.0 Compressor Replacement

The procedures used to change out a compressor differ somewhat depending on whether the cause of the compressor failure is electrical or mechanical. There are many types and variations of compressors. The procedures in this section describe the methods used to replace a welded hermetic-type compressor. Regardless of the type of compressor being replaced, the guidelines given in these procedures normally apply. Always consult the system and replacement compressor manufacturer's service literature for the specific system being serviced and the compressor being used.

When replacing a compressor because of a mechanical problem, use the following basic guidelines:

> **WARNING!** Be sure all electrical power to the equipment is turned off. Open, lock, and tag disconnects. Use extreme caution around pressurized or hot components. Follow all safety instructions labeled on the equipment and provided in the manufacturer's service manual. Failure to do so may result in personal injury.

Step 1 Test the new compressor motor for open windings, grounds, or shorts before removing the failed compressor. Ensure that it is the correct voltage for the application.

Step 2 Turn off the equipment. Lockout and tag equipment so that no one can start it.

Step 3 Recover the refrigerant, regardless of its condition. Recycle or reclaim the refrigerant as its condition dictates.

Step 4 Disconnect and tag all wiring from the compressor.

> **WARNING!**
> Danger exists if the compressor terminals are damaged and the system is pressurized. Disturbing the terminals to make measurements could cause them to blow out, causing injury. When making voltage, current, or continuity checks on a hermetic or semi-hermetic compressor in a pressurized system, always make measurements at terminal boards and points of test away from the compressor. Once refrigerant has been recovered and the system is no longer under pressure, measurements can be made at the compressor terminals.

Step 5 Use a tubing cutter to cut the system refrigerant tubing connected to the compressor. Do not use a hacksaw; it can introduce chips into the system that can cause damage. Tape the open lines in the equipment to prevent dirt or moisture from entering.

> **WARNING!**
> Never use a torch to cut the compressor lines. Oil vapor in the lines can flare up and cause severe burns.

Step 6 Remove the mounting bolts, and then remove the old compressor. Be sure to get help to avoid injury caused by heavy lifting.

Step 7 Unpack the new compressor and read any manufacturer's literature that accompanies it. Replacement compressors often have new features or devices that should be used. Compare nameplates between the old and new compressors to be sure the new compressor is the correct type and voltage.

Step 8 Remove the vibration-isolating mounts from the old compressor and install them on the feet of the new one.

Step 9 Mount the new compressor in the equipment and bolt it down. Be sure to get adequate help to avoid injury caused by heavy lifting. Remember that the mounting bolts of most compressors are not tightened all the way, except when the equipment is being prepared for shipment. The bolts should be tightened until they nearly contact the rubber vibration isolators. This allows the isolators to function properly. Semi-hermetic compressors usually have spring isolators under the feet, and the mounting bolts must remain loose to allow the springs to function.

Step 10 Measure, cut, and fit new piping components for the compressor suction and discharge lines.

Step 11 Braze the system refrigerant lines to the compressor. Clean any flux from the joints and paint them for protection. Ensure that any flux used is not allowed to enter the refrigerant lines. A very low-pressure nitrogen purge should be maintained to prevent copper oxides and the possibility of oil burning inside the refrigerant lines.

Step 12 Remove the existing liquid line filter-drier and replace it with one that is one size larger. If the system is not equipped with a filter-drier, install one. Make sure that only one liquid line drier is installed in the system. A suction line drier may need to be added temporarily if contaminants are expected. The use of suction line driers is discussed more fully in a later section.

Step 13 Leak test the system, then use the proper evacuation techniques.

Step 14 While the evacuation process is ongoing, use the time to consult the equipment wiring diagram and connect all wiring to the new compressor. Replace the contactor if it has been damaged and replace the capacitor(s) if present.

Step 15 Recharge the system with the correct type and weight of refrigerant. If the charge is not weighed in, ensure a sufficient amount of refrigerant is added to the system for operation while the charge is adjusted at startup.

Step 16 Start the system and allow it to run to stabilize the system pressures. Refer to the equipment manufacturer's service instructions and follow them to make any adjustments in the refrigerant charge as deemed necessary. Check and record all the important operating information.

Liquid Refrigerant Charging

When adding a full refrigerant charge following a compressor replacement, charging with liquid from the cylinder will generally be necessary. If vapor is drawn, it is highly unlikely that a sufficient amount of refrigerant can be installed before the refrigerant cylinder pressure equalizes with the system. With blend refrigerants, liquid charging is always a necessity. Using a device in the charging hose (like the liquid charger shown here from YELLOW JACKET®) between the gauge manifold and the system connection ensures that liquid is metered and vaporized at a rate the compressor can handle.

Figure Credit: Courtesy of Ritchie Engineering Company, Inc., YELLOW JACKET Products Division

2.4.0 Compressor Replacement Following an Electrical Failure

Compressor electrical failures are quite common for a number of reasons. Problems with the incoming power, such as surges from lightning or low voltage conditions, place a great deal of stress on the motor windings. As a rule, electrical failures related to open windings do not create any more problems than a mechanical failure. This is also true when an internal wiring device such as an overload fails to close. However, when an internal path to ground is formed, arcing occurs, and an astonishing amount of contaminants can be produced very quickly inside the compressor. Electrical failures involving windings that are shorted to ground are often called burnouts. A burnout refers to the breakdown of the motor winding insulation due to the arcing and high temperatures that are generated at the moment of failure. Burnouts are classified either as mild or severe.

A mild burnout usually occurs very suddenly, causing the motor to stop before the contaminants created by the burnout leave the compressor. Few contaminants, if any, are produced and little or no chemical reaction of the refrigerant and oil occurs. Mild burnouts often occur as the motor tries to start. If the compressor does not start and is not pumping when the event occurs, little contamination escapes the compressor shell. Another factor in a mild burnout is the overcurrent protection device(s) in the circuit. If a fuse or circuit breaker opens quickly, the amount of time power is applied and damage is occurring is significantly reduced.

Severe burnouts usually happen over a longer period of time. The worst tend to occur while the compressor is operating and moving refrigerant. If the process begins slowly and the overcurrent protection is slow to respond, considerable amounts of contaminants are produced and pumped throughout the refrigerant circuit. This is sometimes referred to as a running burn.

> **WARNING!**
> When working on a system suspected of having a compressor burnout, wear appropriate personal protective equipment, including rubber gloves and eye protection. Contaminated refrigerant oil may contain heavy concentrations of acid. Do not allow acid to come in contact with the skin or eyes as severe burns may result.

Before replacing a compressor that has failed because of a burnout, it must first be determined if the burnout was mild or severe. To do this, check the system for acid, using one of several acid/moisture refrigerant test kits or oil test kits that are available. Acid/moisture refrigerant test kits can typically be connected to a system service port to obtain a sample of the refrigerant. Oil test kits require taking a sample of the system oil.

For systems with hermetic compressors, an oil sample may be very difficult to obtain. If an oil sample trap has been installed in the suction line for this purpose, then a sample can be easily obtained. However, as practical as it may seem, this is very rare. In either case, follow the test kit manufacturer's instructions for using the kit and for determining the amount of contamination in the system.

Oil Sample Traps

Although they can be very handy, oil sample traps are very rarely encountered. To some technicians, this is one of those items that is too much trouble to install when an installation is being done. However, when an oil sample is needed it is too late to install a sample trap. It can be a valuable tool when troubles begin.

A simple oil sample trap can be constructed in the suction line near the compressor. Locate the trap in a horizontal section of line, preferably at least several inches away from a vertical section. Braze in a tee fitting, with the branch looking straight down. Using the necessary bushings and adapters, adapt the branch of the tee to a $3/8"$ female flare fitting. Then, install a short, flared ball valve, flared sight glass, and a second short, flared ball valve. Cap the final ball valve.

The sight glass provides a handy reservoir to capture oil as it flows by in the suction line, and also allows the technician to see the contents. The two valves can be used to isolate the reservoir from the system and collect the sample. After a sample is collected, the valves can also be used to evacuate the reservoir before reopening it to the system to collect another sample.

> **WARNING!** Do not cut into or use a torch on a line with any system pressure in it. There is a potential for explosion as well as the production of deadly phosgene gas. The system pressure must be 0 psig before repairs are attempted.

Another way to determine the type of burnout in a hermetic compressor is to cut the suction and discharge lines and check for carbon by running a clean, lint-free swab inside the lines. In a semi-hermetic compressor, the easiest way to tell the type of burnout is to remove the cylinder head. If you find carbon using either method, it is a good indication that a severe burnout has occurred. With a severe burnout, the refrigerant will also have a very strong rotten-egg smell.

2.4.1 Mild Burnout Procedures

Since most contaminants produced by a mild burnout remain within the compressor itself, the procedure for changing the compressor is the same as that used when there is a mechanical failure. An oversized liquid line filter-drier may be chosen to provide extra surface area to capture any particles and contaminants. During the compressor replacement process, also install an oil sample trap for future oil testing. The liquid line filter-drier should be replaced again after several hours of operating time.

After the compressor installation is completed, the system should be triple-evacuated before charging it with refrigerant. The triple-evacuation method is described in another module in this curriculum.

2.4.2 Severe Burnout Procedures

Servicing a system after a severe burnout requires not only compressor replacement, but also a thorough cleanup of the entire system to remove all harmful contaminants. In many cases, the compressor replacement is only the beginning of the work to be done. A proper system cleanup can be a time-consuming and very costly process.

Successive failures on the same system can often be traced to improper cleanup. Unless the cleanup is performed correctly, a repeat failure usually occurs soon. The following is an overview of the procedure for replacing a compressor after a severe burnout. It is essential to follow the system and/or compressor manufacturer's guidelines for replacing a compressor under these conditions. Note that the warranty usually depends on following specific steps that may differ from this example.

Step 1 Recover the system refrigerant and remove the compressor. This refrigerant should be sent away for reclamation, as it is very unlikely that any field recycling process can successfully remove the contaminants.

Step 2 Remove the liquid line filter-drier.

Step 3 Purge the system piping with dry nitrogen in the direction opposite to normal refrigerant flow.

Step 4 Remove, clean, and/or replace the metering devices, receiver, and accumulator, if equipped.

Step 5 Install the new compressor.

03210 Troubleshooting Cooling

Step 6 Add or replace the liquid line filter-drier. Note that a reversible filter-drier is required in the liquid line of a heat pump. For sealed-core driers, use one size larger than recommended. For larger systems, it may be best to install a replaceable-core filter-drier shell (*Figure 16*) so that the core can be replaced one or more times once the system is returned to service. Install isolation ball valves on each side of the filter-drier with an access valve to recover refrigerant from this section of the line. The access valve will also be used to evacuate this section and the new filter-drier when it is replaced. This step eliminates the need to recover the entire system charge to change the filter-drier later. Most replaceable-core filter-drier shells have a fitting in the cap for the installation of an access valve.

Step 7 Add a suction line filter-drier with input and output pressure taps (*Figure 17*). The taps allow for the pressure drop to be tested while the system is in operation. A replaceable-core type filter-drier may be required to reach the necessary capacity for the system. Per *ANSI/AHRI Standard 730*, suction line filter-driers should be selected for a maximum pressure drop of 3 psi in a permanent installation, and 8 psi for temporary use.

(A) FILTER-DRIER SHELL

(B) REPLACEABLE CORE

Figure 16 Replaceable-core filter-drier shell and core.

> **NOTE**
>
> Even in the case of a running burn, a limited amount of contamination typically reaches the suction line. This is because the refrigerant is being pumped out of the compressor through the discharge line. Suction line filter-driers can often be removed completely after several hours of system operating time. If a replaceable-core shell is used, the cores can eventually be removed and replaced with simple filters, or they can be left empty. Suction line filter-driers must be larger than liquid line models due to the extreme pressure drop they can create in the vapor stream. They should not be left in the system indefinitely. If they are, they should be checked periodically to ensure the pressure drop never exceeds 3 psi.

Step 8 Triple-evacuate the system.

Step 9 Recharge the system with new virgin or reclaimed refrigerant.

Figure 17 Suction line filter-drier.

> **Drier Cores**
>
> A wide variety of drier cores are offered by various manufacturers for specific system uses. Some are designed to capture and hold more water than a normal core; others are designed specifically for POE oils; still others are designed to do a better job of capturing acids. Always consult the manufacturer's literature for the filter-drier to determine the proper size and best core type for your specific needs.
>
> Drier cores are fragile and should be treated gently. If subjected to vibration and impacts, the cores can chip or crack, even inside their packaging.

Step 10 Run the system for one hour. Immediately check the pressure drop through the suction line filter-drier on startup to provide a baseline. Stop the system and change the suction line filter-drier any time the pressure drop shows a significant increase above the baseline pressure drop.

Step 11 After one hour has elapsed, stop the system and change the liquid and suction line filter-driers or drier cores. If changing a semi-hermetic compressor, replace the oil.

Step 12 Run the system for two more hours.

Step 13 fter two hours have elapsed, stop the system and test for acid or moisture contamination to make sure the system is clean. If the test is negative, change the liquid line filter-drier. Remove the suction line filter-drier or drier cores from the system. Never leave a suction line filter-drier in a system for an extended period. For an extra measure of caution, a suction line filter-drier or core can be left in the circuit, but only if the pressure loss through the clean filter is below 3 psi. Even then, note that this additional pressure drop affects system energy use. Liquid line filter-driers may be permanently left in a clean system.

Step 14 If acid or moisture is still present in the system at Step 13, change the oil (semi-hermetic compressors only) and repeat Steps 12 and 13 as necessary to achieve a clean system.

2.4.3 Changing Oil

After a semi-hermetic compressor has been changed due to a severe burnout, the oil will likely need to be changed one or more times. Although it would be good to change the oil in a hermetic compressor as well, it is impractical to do. The refrigeration oil, since it is always in a liquid state, tends to capture particles, acids, and other contaminants that enter the system.

Before removing oil from a compressor, test a sample of the oil for the presence of contamination. Use an acid test kit and follow the test kit manufacturer's instructions to determine the amount of acid contamination in the system, if any. The test provides a baseline for future test comparison as the cleanup process continues.

Depending on the compressor model, the oil can be removed one of two ways: through an oil drain plug, or through an oil fill port (*Figure 18*). Some compressors do not have an accessible drain plug. For these compressors, the oil fill port must be used.

Remove the oil from a semi-hermetic compressor with a drain plug using the following general guidelines:

Step 1 If possible, run the compressor fully loaded. Then, close the suction service valve and reduce the crankcase pressure to 0 psig as the compressor runs. Note that the low-pressure switch may have to be bypassed temporarily, depending on where it is connected. Otherwise, a refrigerant recovery process may be required to reduce the crankcase pressure to 0 psig.

Step 2 Stop the compressor and completely isolate it from the system by closing the discharge service valve.

Step 3 Loosen the drain plug until the oil begins to seep out around the plug threads. Removing the plug too quickly may allow the oil to rapidly drain uncontrollably. Drain the oil into a suitable container so that it is properly captured and can be properly disposed of. After the oil has been completely drained, reinstall the plug.

When there is no accessible drain plug provided, the oil must be removed through an oil fill plug hole. Remove the oil from a compressor with only an oil fill plug hole using the following general guidelines:

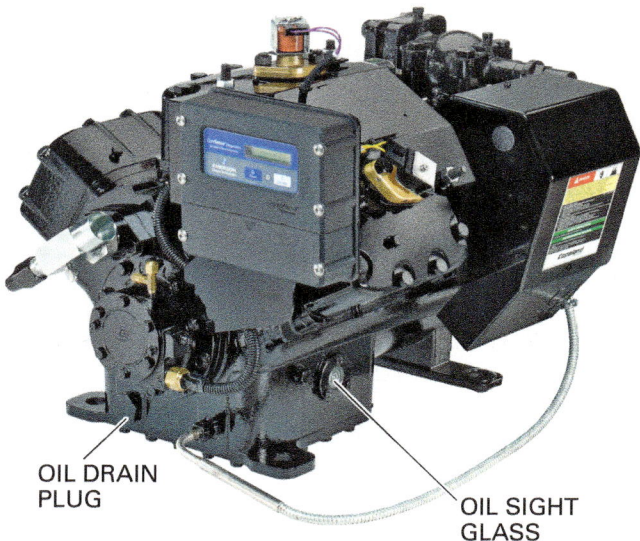

OIL DRAIN PLUG

OIL SIGHT GLASS

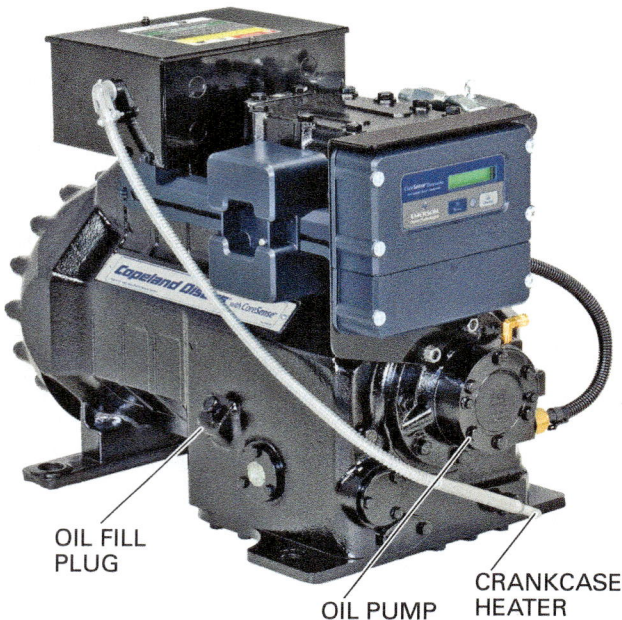

OIL FILL PLUG

OIL PUMP

CRANKCASE HEATER

Figure 18 Semi-hermetic compressors.

WARNING!

When removing oil, be sure to wear rubber gloves and eye protection to prevent possible injury. Contaminated refrigerant oil may contain heavy concentrations of acid. Do not allow contact with your skin or eyes as severe burns may result. Also ensure that the crankcase pressure has been relieved.

Step 1 If possible, run the compressor fully loaded. Then, close the suction service valve and reduce the crankcase pressure to 0 psig. Note that the low-pressure switch may have to be bypassed.

Step 2 Stop the compressor and isolate it from the system by closing the discharge service valve.

Step 3 Remove the oil fill plug. Use a refrigeration oil pump (*Figure 19*) equipped with a flexible siphon kit to remove the oil from the compressor. Insert the siphon hose into the oil fill opening and pump the oil into a suitable container. Continue to remove the oil in this manner until all the oil is removed.y

Oil is charged into a compressor after it has been drained or when oil must be added to make up for some loss. Use a refrigeration oil pump such as the one shown in *Figure 19* to charge oil. The pump shown can also be used to add oil into an operating compressor, since it is able to develop enough pressure to overcome common crankcase pressures.

Before charging oil into the compressor, always refer to the manufacturer's instructions for the compressor being serviced. Make sure to use the correct amount and type of oil specified by the manufacturer. This is very important; do not use substitutes unless authorized to do so. This is especially true with the synthetic oils that are used with the newer refrigerants. Their characteristics can vary widely from one manufacturer to another.

Figure 19 Oil pump.

The procedure for charging oil into an empty compressor is described in the steps that follow:

Step 1 If the compressor has an oil drain plug, it should now be reinstalled and tight. The oil fill plug should be removed. Some compressors may have a special access valve that can connect directly to the pump.

> **CAUTION**
>
> POE oil in a closed container is free of moisture but it will readily absorb moisture if exposed to air. POE oil should not be used from a container that has been open for any significant length of time or from one that contains used oil. POE oil should never be stored in an open container.

Step 2 Mount and secure the refrigeration oil pump on or next to (depending on the pump style) the container of new oil. Lay the end of the charging hose in a suitable container to capture purged oil. Purge the oil pump line by operating the pump until oil appears without air bubbles. Then quickly connect the end of the hose to the compressor fill opening.

Step 3 Open the service valve (if applicable). While watching the oil level in the compressor sight glass, pump oil into the compressor crankcase from the container of oil as needed.

Step 4 Close the service valve and disconnect the pump.

Step 5 To remove any moisture that may have entered the compressor, evacuate the compressor to 500 microns.

Step 6 Open the suction and discharge service valves, and then restart the compressor.

Step 7 Run the system for about 20 minutes. Full-load operation is necessary to ensure that refrigerant velocities in the piping system are sufficient to entrain and move oil back to the compressor. Then recheck the oil level at the sight glass. Add additional oil if necessary, using the pump instructions for an operating compressor, if possible.

> **NOTE**
>
> The correct sight glass oil level varies from one compressor manufacturer to another. Refer to the manufacturer's instructions.

Additional Resources

Refrigeration and Air Conditioning, An Introduction to HVAC/R, Fourth Edition. Larry Jeffus. Air Conditioning and Refrigeration Institute. Upper Saddle River, NJ: Prentice Hall.

System Diagnostics and Troubleshooting Procedures. 2002. John Tomczyk. Mt. Prospect, IL: Esco Press.

2.0.0 Section Review

1. If a compressor runs but does not provide sufficient compression, the likely fault is _____.
    a. an open thermostat control circuit
    b. a faulty run capacitor
    c. defective compressor valves
    d. open motor windings

2. When replacing a compressor, even if it is the same basic model, _____.
    a. always install a start capacitor and relay
    b. a change in capacitor rating may be needed
    c. install a new crankcase heater
    d. drain the oil from the new compressor

3. Before putting a new compressor into position, _____.
    a. compare the nameplate information between the old and new compressors
    b. cut and fit new piping components for the compressor
    c. leak test the system
    d. connect the wiring to the new compressor

4. The worst burnouts, producing the most contaminants, typically occur _____.
    a. when the compressor has not yet started
    b. while the compressor is shut down
    c. when a fuse or circuit opens very quickly
    d. when the compressor is running and pumping

SUMMARY

Effective troubleshooting of a refrigerant circuit is a process that requires patience, experience, and careful thought. The refrigerant circuit can, at times, be confusing. A systematic, step-by-step approach that begins with listening carefully to the complaint and the events that led up to the failure always yields the best results. It important to remember that more than one issue can affect the operation of the system at any one time.

For example, a poorly maintained system may suffer from both a clogged condenser and evaporator coil but remain reasonably functional until the TXV power element fails. This set of conditions is not unusual. A technician must be able to tell whether a given device is functioning properly and recognize the symptoms arising from the improper operation of any part of the equipment.

Troubleshooting the compressor motor is no different than troubleshooting any other motor, except that it is hidden from view. Like any other motor, it has a difficult time starting under a load. Although mechanical problems with compressors are not uncommon, external issues often cause compressors to fail. When this happens, it is crucial that the cause be found and resolved prior to installing the new compressor.

Review Questions

1. The condition of the refrigerant entering an evaporator can best be described as _____.
   a. low-temperature, low pressure
   b. low-temperature, high-pressure
   c. high-temperature, low-pressure
   d. high-temperature, high-pressure

2. The refrigerant entering the evaporator, after passing through a metering device, is usually about _____.
   a. 20-percent liquid
   b. 50-percent liquid
   c. 20-percent vapor
   d. 50-percent vapor

3. Air contamination in a refrigeration circuit will likely show a symptom of a _____.
   a. high suction pressure
   b. low suction pressure
   c. low head pressure
   d. high head pressure

4. When all else seems to function properly, a system with a high suction pressure and a low head pressure suggests _____.
   a. the compressor has leaky valves and/or worn piston rings
   b. a clogged filter drier
   c. a metering device that is too small
   d. a clogged condenser coil

5. When both the suction pressure and head pressure are low, how might a technician determine that the problem is low evaporator airflow and not a low refrigerant charge?
   a. The compressor will likely be very hot.
   b. The subcooling will be normal.
   c. The condenser coil will be frozen.
   d. The compressor crankcase will likely feel cold.

6. Hunting can be caused by _____.
   a. an undersized TXV
   b. a very light load
   c. subcooling below 15°F
   d. a clogged filter-drier

7. If a refrigerant liquid line is routed through an unusually hot area on its way to the metering device, how might this affect it?
   a. It can increase its subcooling.
   b. It can decrease its subcooling.
   c. It can increase its superheat.
   d. It can decrease its superheat.

8. A 1 psi pressure loss occurs in a liquid line riser for every _____.
   a. inch of rise
   b. foot of rise
   c. 2.3 feet of rise
   d. 3.5 feet of rise

9. One problem that may not be evident when a system first starts, but develops progressively as the system operates, is _____.
   a. moisture freezing at the metering device
   b. a plugged filter-drier
   c. a low refrigerant charge
   d. a failed condenser fan motor

10. A running burn is indicated by _____.
    a. a failed start capacitor and/or relay
    b. a failed metering device
    c. a significant amount of acid in the system
    d. only a limited amount of acid in the system

11. Slugging is defined as _____.
    a. an overcharge of refrigerant
    b. an overcharge of oil
    c. the consistent return of liquid refrigerant in small amounts
    d. liquid refrigerant and/or oil entering the compressor cylinder

12. What causes oil to foam in a compressor crankcase?
    a. The flashing of liquid refrigerant to vapor as the pressure suddenly decreases
    b. The flashing of liquid refrigerant to vapor as the pressure suddenly increases
    c. The condensing of refrigerant as the pressure suddenly increases
    d. The condensing of refrigerant as the pressure suddenly decreases

13. The migration of refrigerant to the compressor crankcase during the off cycle can be prevented with a(n) _____.
    a. evaporator pressure regulator
    b. liquid line solenoid
    c. TXV
    d. fixed-restrictor metering device

14. Capacitance boosting involves _____.
    a. placing an additional run capacitor is series with the first
    b. placing a run capacitor in series with the start winding
    c. temporarily replacing the run capacitor with a start capacitor
    d. placing a start capacitor in parallel with the run capacitor

15. When sizing a suction line filter-drier, a permanent installation should be sized for no more than a _____.
    a. 1 psi pressure drop
    b. 2 psi pressure drop
    c. 3 psi pressure drop
    d. 4 psi pressure drop

Trade Terms Introduced in This Module

Burnout: The condition in which the breakdown of the motor winding insulation causes the motor to short out or ground electrically. The breakdown of the insulating coating typically results in the formation of strong acids.

Capacitance boost: A procedure used to start a stuck PSC compressor. It involves momentarily connecting a start (boost) capacitor across the run capacitor of the stuck compressor to increase the starting torque in an attempt to start the stuck compressor.

Floodback: A condition in which there is liquid refrigerant in the suction vapor being returned to an operating compressor.

Flooded starts: A condition in which slugging, foaming, and inadequate lubrication occur at compressor startup due to oil in the compressor having absorbed refrigerant during the off cycle. When the compressor starts and the pressure in the crankcase drops quickly, the refrigerant boils rapidly, foaming the oil.

Slugging: The entrance of liquid refrigerant and/or a significant amount of oil into a compressor cylinder. Slugging can occur at startup due to condensed refrigerant and/or collected oil near the compressor inlet.

Additional Resources

This module presents thorough resources for task training. The following resource material is suggested for further study.

Refrigeration and Air Conditioning, An Introduction to HVAC/R, Fourth Edition. Larry Jeffus. Air Conditioning and Refrigeration Institute. Upper Saddle River, NJ: Prentice Hall.

System Diagnostics and Troubleshooting Procedures. 2002. John Tomczyk. Mt. Prospect, IL: Esco Press.

Figure Credits

Courtesy of Ritchie Engineering Company, Inc., YELLOW JACKET Products Division, Module opener, SA07, Figure 19

Courtesy of Joshua West, Figure 4

Courtesy of Parker Hannifin, Sporlan Division, SA02

Courtesy of Henry Technologies, Inc., SA03

Courtesy of QuikProducts by Mainstream Engineering Corp., Figure 10

Courtesy of Highside Chemicals, Inc., Figure 11

Courtesy of Emerson Climate Technologies, Figures 13, 17, 18, SA04 (photo)

Backer Springfield, SA05

Courtesy of Sealed Unit Parts Co., Inc., SA06

Courtesy of Danfoss, Figure 16A and B

Section Review Answer Key

Answer	Section Reference	Objective
Section One		
1. d	1.1.0	1a
2. b	1.2.0	1b
3. b	1.3.1	1c
4. a	1.4.0; Figure 3	1d
5. b	1.5.1	1e
6. d	1.6.0	1f
7. c	1.7.0	1g
8. c	1.8.1	1h
Section Two		
1. c	2.0.0; Figure 12	2a
2. b	2.2.2	2b
3. a	2.3.0	2c
4. d	2.4.0	2d

NCCER CURRICULA — USER UPDATE

NCCER makes every effort to keep its textbooks up-to-date and free of technical errors. We appreciate your help in this process. If you find an error, a typographical mistake, or an inaccuracy in NCCER's curricula, please fill out this form (or a photocopy), or complete the online form at **www.nccer.org/olf**. Be sure to include the exact module ID number, page number, a detailed description, and your recommended correction. Your input will be brought to the attention of the Authoring Team. Thank you for your assistance.

Instructors – If you have an idea for improving this textbook, or have found that additional materials were necessary to teach this module effectively, please let us know so that we may present your suggestions to the Authoring Team.

NCCER Product Development and Revision
13614 Progress Blvd., Alachua, FL 32615

Email: curriculum@nccer.org
Online: www.nccer.org/olf

❏ Trainee Guide ❏ Lesson Plans ❏ Exam ❏ PowerPoints Other _____

Craft / Level: _____ Copyright Date: _____

Module ID Number / Title: _____

Section Number(s): _____

Description: _____

_____

_____

_____

_____

Recommended Correction: _____

_____

_____

_____

_____

Your Name: _____

Address: _____

_____

Email: _____ Phone: _____

Troubleshooting Heat Pumps

Overview

This module reviews the operation of heat pumps and presents methods and procedures for troubleshooting them. Heat pumps contain more control devices and have more complex circuits than a cooling-only system. Depending on the malfunction, the cooling and heating functions are usually examined separately, although some problems affect both modes of operation. Because heat pumps are unique, some of the applicable troubleshooting tasks are unique as well.

Module 03311

Trainees with successful module completions may be eligible for credentialing through the NCCER Registry. To learn more, go to **www.nccer.org** or contact us at 1.888.622.3720. Our website has information on the latest product releases and training, as well as online versions of our *Cornerstone* magazine and Pearson's product catalog.

Your feedback is welcome. You may email your comments to **curriculum@nccer.org**, send general comments and inquiries to **info@nccer.org**, or fill in the User Update form at the back of this module.

This information is general in nature and intended for training purposes only. Actual performance of activities described in this manual requires compliance with all applicable operating, service, maintenance, and safety procedures under the direction of qualified personnel. References in this manual to patented or proprietary devices do not constitute a recommendation of their use.

Copyright © 2018 by NCCER, Alachua, FL 32615, and published by Pearson, New York, NY 10013. All rights reserved. Printed in the United States of America. This publication is protected by copyright, and permission should be obtained from NCCER prior to any prohibited reproduction, storage in a retrieval system, or transmission in any form or by any means, electronic, mechanical, photocopying, recording, or likewise. To obtain permission(s) to use material from this work, please submit a written request to NCCER Product Development, 13614 Progress Blvd., Alachua, FL 32615.

03311 V5

From *HVAC Level Three, Trainee Guide*. NCCER.
Copyright © 2018 by NCCER. Published by Pearson. All rights reserved.

03311
TROUBLESHOOTING HEAT PUMPS

Objectives

When you have completed this module, you will be able to do the following:

1. Compare heat pumps to standard cooling systems and describe their operating cycles.
    a. Compare heat pump systems to standard cooling systems and identify the different types.
    b. Describe the three operating cycles of common heat pumps.
2. Describe the sequence of operation for the common operating modes.
    a. Describe the sequence of operation for the cooling mode.
    b. Describe the sequence of operation for the three heating modes.
    c. Describe the sequence of operation for the defrost mode.
    d. Describe the sequence of operation of dual-fuel systems.
    e. Describe the use of microprocessor controls in heat pump systems.
3. Explain how to check and/or troubleshoot various functions and components of heat pump systems.
    a. Explain how to check field and factory wiring.
    b. Explain how to check and troubleshoot heat pump thermostats.
    c. Explain how to test thermistors.
    d. Explain how to check the various valves found in heat pumps.
    e. Explain how to check defrost control circuits.

Performance Tasks

Under the supervision of your instructor, you should be able to do the following:

1. Demonstrate the ability to isolate and determine the solution for various electrical and mechanical malfunctions in heat pumps.
2. Initiate the defrost cycle of a heat pump.

Trade Terms

Demand defrost
Direct-acting valve
Electronically commutated motor (ECM)
Field wiring
Pilot-operated valve

Industry-Recognized Credentials

If you are training through an NCCER-accredited sponsor, you may be eligible for credentials from NCCER's Registry. The ID number for this module is 03311-18. Note that this module may have been used in other NCCER curricula and may apply to other level completions. Contact NCCER's Registry at 888.622.3720 or go to **www.nccer.org** for more information.

Contents

1.0.0 Pump Operating Cycles 1
  1.1.0 Heat Pump Systems 1
    1.1.1 Heat Pump Operation 1
    1.1.2 Heat Pump Classifications... 2
  1.2.0 Heat Pump Operating Modes 5
    1.2.1 Cooling Mode 5
    1.2.2 Heating Mode 5
    1.2.3 Defrost Mode 5
2.0.0 Heat Pump Operating Sequences 7
  2.1.0 Cooling Sequence of Operation 7
  2.2.0 Heating Sequence of Operation 7
  2.3.0 Defrost Sequence of Operation 10
  2.4.0 Dual-Fuel Heat Pump Operation 13
  2.5.0 Microprocessor Controls in Heat Pumps 14
    2.5.1 Variable-Speed Heat Pumps 16
3.0.0 Testing and Troubleshooting Heat Pumps 19
  3.1.0 Checking Wiring 21
    3.1.1 Field Wiring 21
    3.1.2 Factory Wiring 21
  3.2.0 Checking Heat Pump Thermostats 23
    3.2.1 Communicating Thermostats 25
    3.2.2 Room Thermostat Troubleshooting Sequence............. 25
    3.2.3 Heat Pump Outdoor Thermostats 26
  3.3.0 Testing Thermistors... 26
  3.4.0 Troubleshooting Heat Pump Valves................................... 26
    3.4.1 Solenoid Valves 26
    3.4.2 Reversing Valves 27
    3.4.3 Check Valves.. 29
  3.5.0 Troubleshooting Defrost Controls 31
    3.5.1 Checking Inputs and Outputs... 33
    3.5.2 Defrost Troubleshooting Tips 34

Figures and Tables

Figure 1   Heat pump operation in the heating mode. 1
Figure 2   Refrigerant flow in an air-to-air heat pump. 3
Figure 3   Geothermal heat pump system. 4
Figure 4   Frost on a heat pump. 5
Figure 5   Cooling sequence of operation. 8
Figure 6   Compression heat with supplemental heat, sequence of operation. 9
Figure 7   Emergency heat sequence of operation. 11
Figure 8   Defrost sequence of operation. 12
Figure 9   Dual-fuel heat pump operation. 14
Figure 10 Packaged heat pump schematic. 15
Figure 11 Input and output signals. 17
Figure 12 Variable-speed, inverter-driven scroll compressor. 17
Figure 13 Cooling sequence of operation. 20
Figure 14 Wi-Fi enabled programmable heat pump thermostat. 22
Figure 15 Example of a configuration menu. 24
Figure 16 Outdoor thermostat check. 26
Figure 17 Thermistor on a refrigerant line. 26
Figure 18 Thermistor resistance-versus-temperature chart. 27
Figure 19 Reversing valves. 28
Figure 20 Refrigerant flow – cooling cycle. 28
Figure 21 Refrigerant flow – heating cycle. 28
Figure 22 Reversing valve temperature testing. 29
Figure 23 Check valve construction. 30
Figure 24 Magnet test. 31
Figure 25 Defrost speed-up terminals. 32
Figure 26 Compressor run time selection for defrost control. 32

Table 1   Example Thermostat Fault Codes 24
Table 2   Check Valve Troubleshooting 31

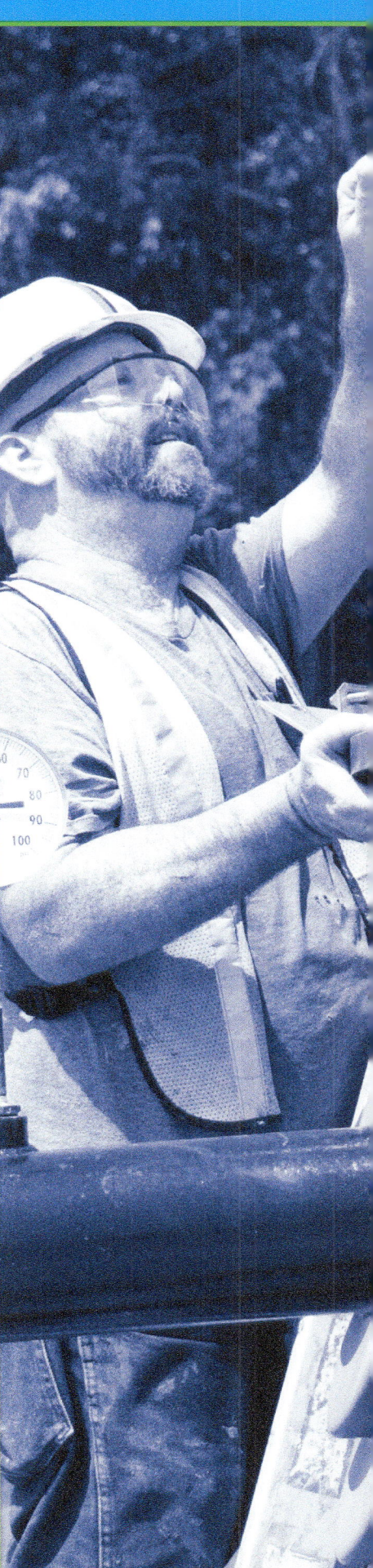

SECTION ONE

1.0.0 HEAT PUMP OPERATING CYCLES

Objective

Compare heat pumps to standard cooling systems and describe their operating cycles.
   a. Compare heat pump systems to standard cooling systems and identify the different types.
   b. Describe the three operating cycles of common heat pumps.

Servicing a heat pump can be more complicated than a service call for a cooling-only system or a furnace. Heat pump control circuits are more complex because they not only control heating and cooling cycles, but a defrost cycle as well.

There are several types of heat pumps, most of which are air-to-air split systems or packaged units. Some types of heat pumps use water as a heat transfer medium. They can also be used to heat domestic hot water.

As with cooling-only systems, the foundational operating principle of a heat pump is a compressor-based refrigeration cycle. While some of the troubleshooting procedures covered in other modules of this curriculum also apply to heat pumps, many troubleshooting tasks are unique to these systems.

1.1.0 Heat Pump Systems

A heat pump is a combination heating and cooling unit. It produces cooling in the same manner as a conventional air conditioner, and then reverses the cycle to produce heat (*Figure 1*). Both systems have the following components in common:

- Compressor
- Condenser and evaporator coils
- Indoor blower fan
- Condenser fan
- Service valves or service ports
- Pressure and temperature controls
- Refrigerant and refrigerant lines

In addition to typical air conditioning components, the heat pump requires a reversing valve and additional metering devices and control circuits. Because heat pumps perform both heating and cooling functions, location identifies the coils. In other words, the condensing and evaporator coils on an air conditioner are identified as the outdoor and indoor coils on a heat pump.

The main advantage of a heat pump is that it is less expensive to operate than electric resistance heat. Depending on the operating conditions, a heat pump may be able to provide three times as much heat for the same electrical cost as electric heat. Depending on the climate and related fuel costs, a heat pump may or may not be less costly to operate than gas heat. For combined heating and cooling, it is likely to be more cost effective than other methods.

1.1.1 Heat Pump Operation

Even at very cold temperatures, there is some heat in the air. The air-to-air heat pump operates on the principle of extracting heat from the outdoor air and transferring it to the indoor air. Heat can be extracted from water, as well (water-to-air heat pump).

The reversing valve sets heat pumps apart from cooling-only units. This device is critical to heat pump operation. It is used to reverse the flow of refrigerant through the system when the unit is switched to the heating mode. In this mode, the outdoor coil becomes the evaporator and the indoor coil becomes the condenser.

The control circuits are almost always arranged so that the reversing valve is de-energized in the heating mode. This way, if there is a reversing valve failure, the heating function will continue to operate. In the summer, a reversing valve failure

---

GOING GREEN

Heat Pump Swimming Pool Heaters

In mild climates where swimming pools can be used year-round, heat pumps are used as pool heaters. Heat is extracted from the air and transferred to the swimming pool by a refrigerant-to-water heat exchanger. This method of heating swimming pools is an energy-efficient alternative to conventional fossil-fuel pool heaters.

03311 Troubleshooting Heat Pumps   Module Four 1

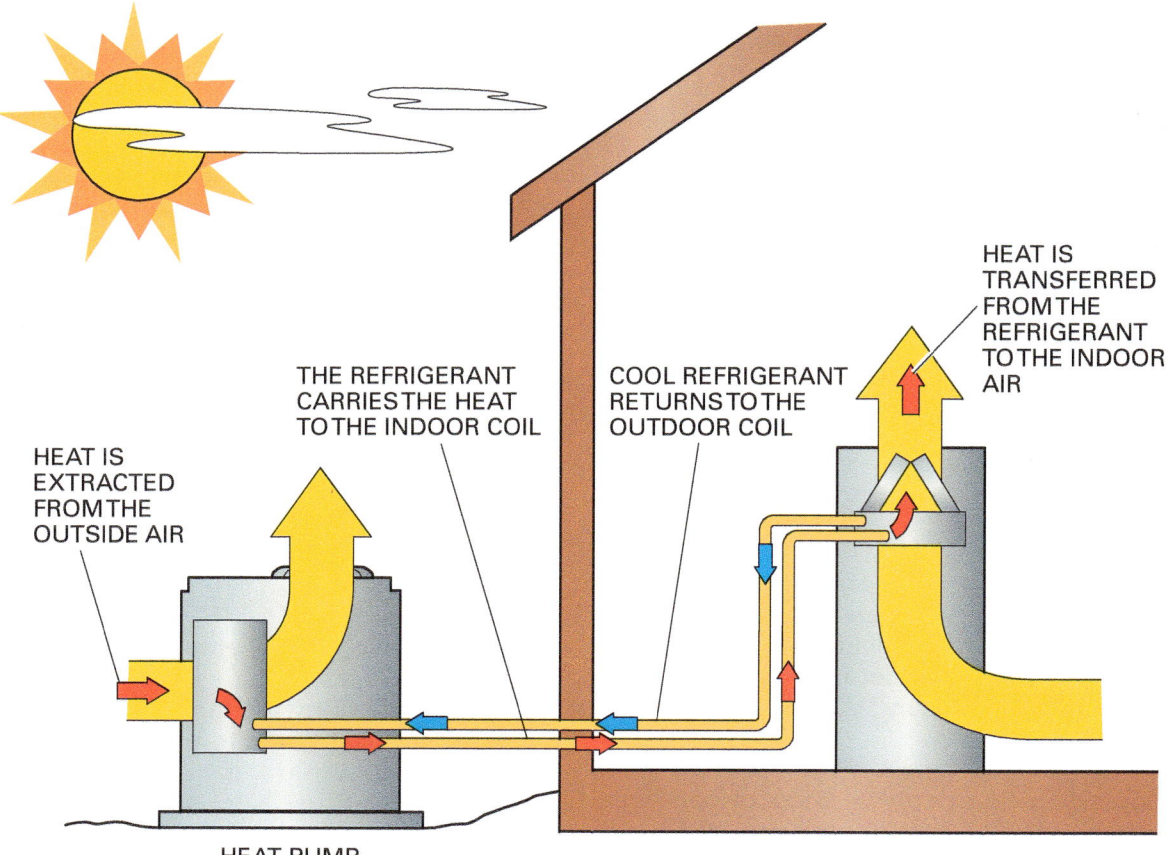

Figure 1 Heat pump operation in the heating mode.

will leave the structure without cooling. And since the defrost cycle is based on the cooling mode, a reversing valve failure will result in a frozen outdoor coil.

Another key difference between cooling-only systems and heat pumps is the metering device arrangement. The heat pump in *Figure 2* has two metering devices. In the arrangement shown, the devices are fixed-restrictor metering devices. Built-in or separate check valves are used to block or allow refrigerant flow, depending on the mode of operation.

In this unique situation, the restrictor is truly fixed; it cannot move inside the housing. In other applications, fixed-orifice, piston-type metering devices are used. In this case, check valves are not required because the piston acts as its own check valve. Check valves may be used with TXVs, although some TXVs also provide this function internally.

1.1.2 Heat Pump Classifications

Heat pumps are classified according to their heat source and the medium to which the heat is transferred (heat sink). The most common types of heat pumps are air-to-air and air-to-water.

In a common air-to-air system, air is the medium used to transfer heat both indoors and outdoors. In the cooling mode, heat is transferred to the outdoor air. In the heating mode, heat is extracted from the outdoor air and transferred indoors.

An air-to-water system also uses air to heat and cool indoors, but water is used as the heat source or heat sink. The refrigerant is circulated through one side of a heat exchanger, and water is circulated through the other side. The source of the water can vary. Some systems rely on a connected boiler and cooling tower to warm or cool the water. Others may rely on water drawn from a lake or pond, but this is rare today for a variety of reasons.

From the standpoint of temperature, ground water is an excellent heat source, but its use is not always practical. Two wells are typically required—water to circulate through the heat exchanger is drawn from one, and the water is then rejected into the second well. Installing two wells with some distance between them may not be possible for the location, and maintaining the wells can often be a problem. Because of these limitations, air is the predominant heat source in residential and small commercial heat pump installations.

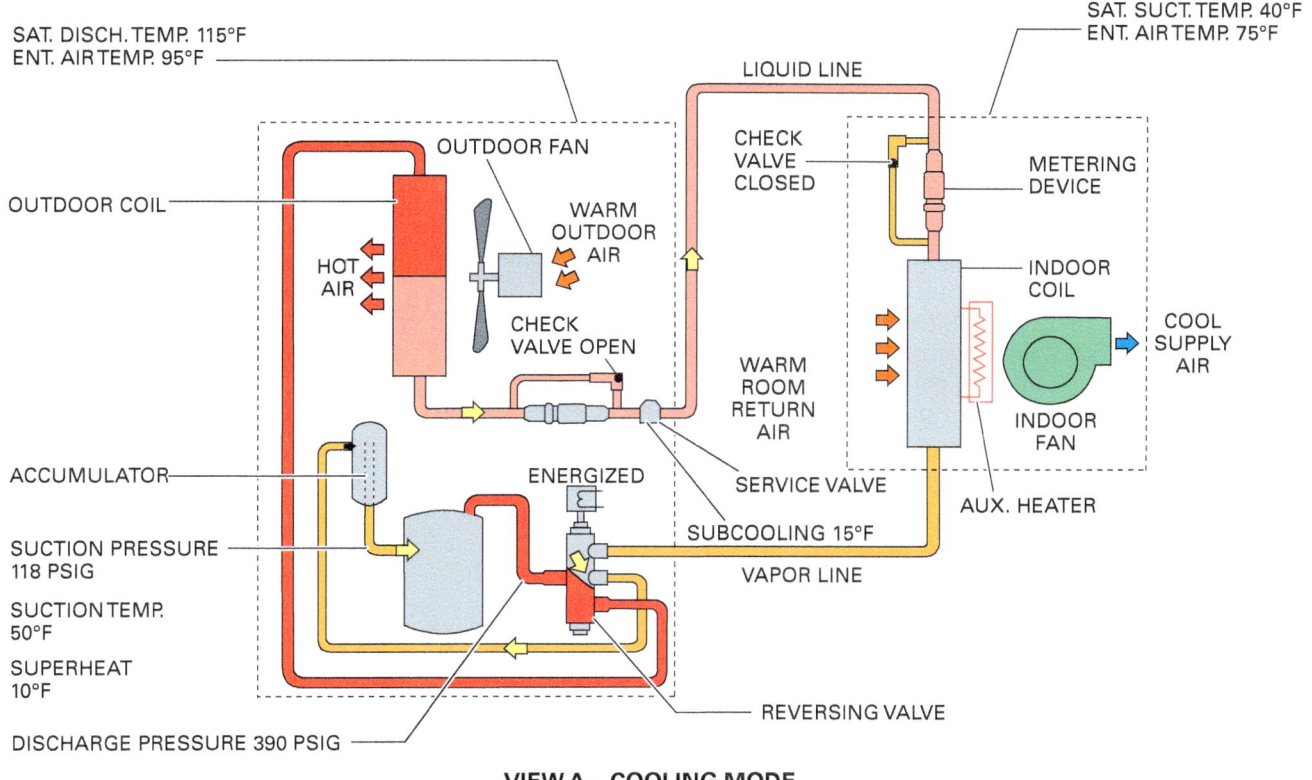

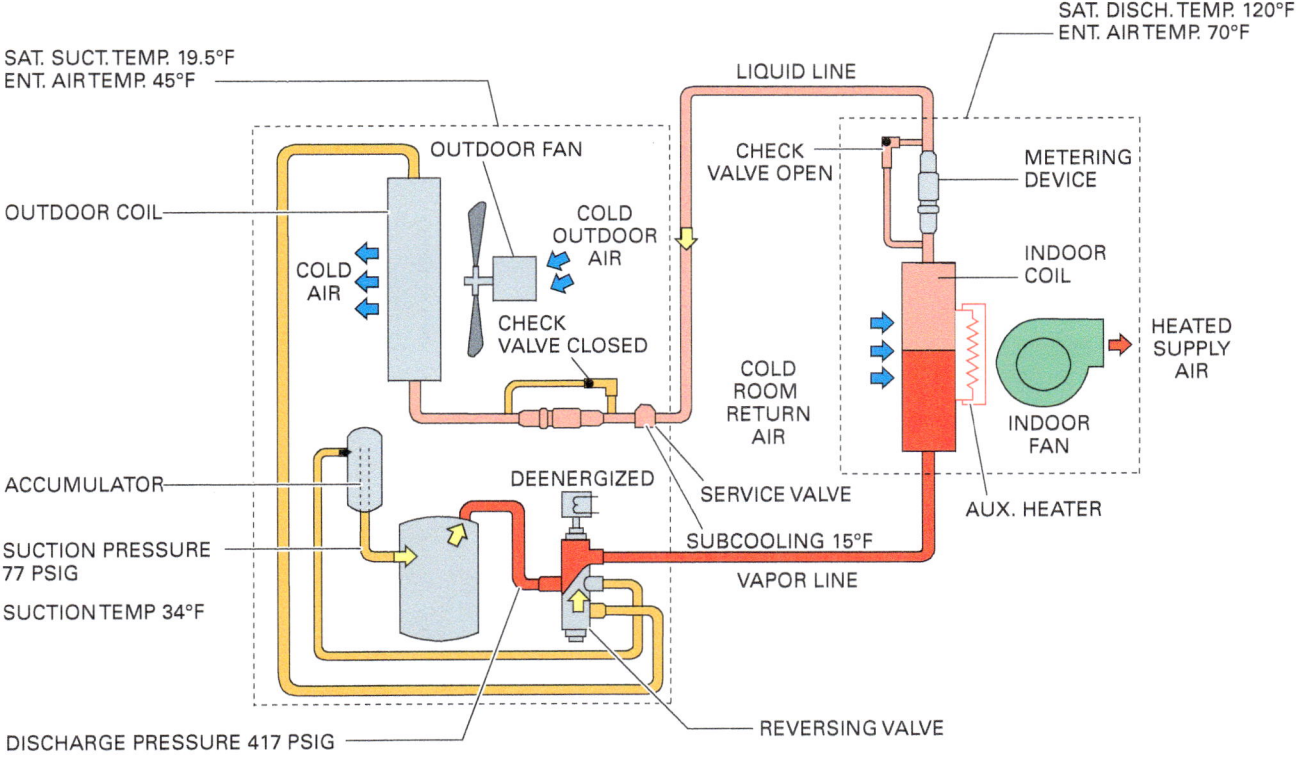

Figure 2 Refrigerant flow in an air-to-air heat pump.

Because the ground temperature is quite stable not far below the surface, air-to-water heat pump systems may use a buried coil to transfer heat to or from the ground. Water is then circulated through the buried coils (*Figure 3*). These systems are referred to as geothermal heat pumps. The ground is used as both the heat source and the heat sink. Since moist soil transfers heat far better than dry soil, the moisture level in the ground affects heat transfer efficiency. The water (often a glycol-based anti-freeze solution) is circulated through a heat exchanger located in the heat pump.

A typical geothermal heat pump is a packaged product that contains the compressor, refrigerant-to-fluid heat exchanger, indoor coil, and blower motor. Because pumping requirements vary from one installation to another, the pump that circulates the fluid through the ground coil is usually an external component.

In addition to air and water, other heat sources may be used for heat pumps. Examples are waste heat from selected industrial processes, exhaust air from ventilation, solar energy, and heat extracted from refrigerated spaces. These are commonly used in addition to the primary heat source, rather than as the sole source.

Figure 3 Geothermal heat pump system.

Geothermal Heat Pumps

Geothermal systems can be installed several different ways, depending on the property and its geology. The illustration on the left shows a vertical closed loop system. Closed loop systems can be installed horizontally or vertically in the ground, or even in a body of water.

Open loop systems use two separate water wells, as shown in the diagram on the right. However, open loop systems are not generally considered to be geothermal.

CLOSED VERTICAL LOOP SYSTEM

OPEN LOOP SYSTEM
(TWO SEPARATE WELLS)

1.2.0 Heat Pump Operating Modes

A heat pump has three primary operating cycles or modes: cooling, heating (reverse-cycle or compression heat), and defrost.

In addition to these three primary modes, compression heat may operate with supplemental electric heaters and an emergency heat mode. When selected, the emergency heat mode locks out compressor operation. It will then enable supplemental electric heaters to heat the structure. The emergency heat mode would be used if a component failure that hinders compression heating is confirmed or suspected.

1.2.1 Cooling Mode

In most heat pumps, the reversing valve is energized when the thermostat calls for cooling and de-energized on a call for heating. This guarantees that heat is available even if there is a failure in the reversing valve control circuits. When the unit is in the cooling mode, the refrigerant flow is the same as that of any cooling unit.

Refer again to *Figure 2*, View A. Note the path through the reversing valve. The check valve used with the indoor coil metering device is closed, forcing the refrigerant to be metered. The cold, low-pressure refrigerant flowing through the indoor coil (evaporator) absorbs heat from the conditioned space and is boiled into a superheated vapor.

Hot, high-pressure refrigerant gas leaving the compressor is pumped through the outdoor coil (condenser), where the heat is rejected. The check valve used with the outdoor coil metering device is open, allowing refrigerant to bypass the metering device.

The refrigerant pressures and temperatures shown in the figure are based on HFC-410A and are typical of those in a normally operating air conditioning system.

1.2.2 Heating Mode

In the heating mode (*Figure 2*, View B), the reversing valve changes position. The refrigerant leaving the compressor is routed in the opposite direction from that of the cooling cycle. Instead of flowing through the outdoor coil, the hot, high-pressure refrigerant vapor leaving the compressor flows to the indoor coil. When this happens, the indoor coil acts as a condenser. The check valve used with the indoor coil metering device is pushed open, allowing refrigerant to bypass the metering device.

Heat is extracted from the refrigerant because the air in the conditioned space is cooler than the refrigerant in the coil. The cooled refrigerant then flows to the outdoor coil. At this point, the check valve used with the outdoor coil metering device is pushed closed, forcing the refrigerant to flow through the metering device. Acting as an evaporator, the outdoor coil absorbs heat from the relatively warmer outdoor air.

Heat provided in this way is known as reverse-cycle heat or compression heat. Notice the significant differences in the pressures and temperatures at key points in the system compared to the cooling mode.

An accumulator and crankcase heater, which are optional accessories on cooling units, are essential on most heat pumps. The crankcase heater keeps liquid refrigerant from condensing in the crankcase of the compressor, where it could cause slugging at startup. The accumulator captures liquid refrigerant that gets into the suction line during operation in the heating mode.

The two coils—indoor and outdoor—have different internal volumes. As a result, less refrigerant is needed in the circuit during operation in the heating mode, when the indoor coil becomes the condenser. This excess refrigerant charge makes its way to the suction line and into the accumulator. The liquid refrigerant slowly vaporizes in the accumulator and returns to the compressor.

1.2.3 Defrost Mode

At low outdoor temperatures, some amount of frost will likely form on the outdoor coil of the heat pump while it is running in the reverse-cycle heating mode. The amount of frost that collects is directly related to the moisture content of the air. As the frost builds, it will restrict airflow and significantly reduce heat transfer. The heat pump shown in *Figure 4* has gone beyond an acceptable level of frost accumulation.

Figure 4 Frost on a heat pump.

The frost is melted by reversing the flow of refrigerant so that the hot, high-pressure refrigerant from the compressor discharge flows through the outdoor coil. The defrost cycle is automatic. The start and termination of the cycle may be based on elapsed time or the coil temperature, or it may be a combination of the two.

The defrost mode is basically the same as the cooling mode. The unit is operating as a cooling unit, sending the hot refrigerant to the outdoor coil. The primary difference is that the condenser fan remains off during a defrost cycle. Since the object is to transfer heat to the coil and melt frost and ice, the condenser fan would work against the cycle. During a defrost cycle, it is common for steam to be seen rising from the melting frost. A tremendous amount of heat is transferred from the refrigerant to the ice/frost due to the latent process involved (ice changing to water).

Since the unit is technically operating in the cooling mode, the discharge air temperature can be quite cold. During the defrost mode, a supplementary electric heater or the burner of a fossil-fuel furnace is usually turned on to prevent the occupied areas from becoming cold.

Additional Resources

NCCER Module 03211, *Heat Pumps*.

1.0.0 Section Review

1. Which of the following valves is required to control the flow of refrigerant when using a fixed-restrictor metering device?

   a. Reversing valve
   b. Check valve
   c. Gate valve
   d. Pressure-reducing valve

2. When frost builds up on a heat pump's outdoor coil, _____.

   a. the outdoor fan speed increases
   b. the outdoor coil becomes more efficient
   c. the heat pump shuts off
   d. heat transfer is significantly reduced

Section Two

2.0.0 Heat Pump Operating Sequences

Objective

Describe the sequence of operation for the common operating modes.
a. Describe the sequence of operation for the cooling mode.
b. Describe the sequence of operation for the three heating modes.
c. Describe the sequence of operation for the defrost mode.
d. Describe the sequence of operation of dual-fuel systems.
e. Describe the use of microprocessor controls in heat pump systems.

Trade Terms

Electronically commutated motor (ECM): A variable-speed electric motor based primarily on a DC-driven permanent magnet design that responds to microprocessor inputs to achieve the desired speed.

A technician troubleshooting a heat pump must understand the operating sequence of the pump in every mode. The control circuits for a heat pump are generally more complicated than those of a cooling-only or heating-only system because there are more controls and more operating modes.

For most air-to-air heat pumps, there are five different operating modes:

- Cooling mode
- Heating mode with compression heat only
- Heating mode with compression heat and supplemental heat
- Defrost mode
- Emergency heating (compression heat disabled)

2.1.0 Cooling Sequence of Operation

In the cooling mode, a heat pump acts like a conventional air conditioner. Use the wiring diagram in *Figure 5* to follow the heat pump sequence of operation in cooling mode, as follows:

- When the room thermostat function switch is placed in the cooling mode, the reversing valve (1) is energized. In this position, refrigerant will be directed to the outdoor coil which acts as a condenser, and to the indoor coil which acts as an evaporator. Generally, the reversing valve will remain energized and in this position, even when the system is not actively cooling.
- As the room temperature rises, room thermostat contacts (2) close to generate a call-for-cooling signal. Assuming the discharge temperature switch and high-pressure switch (3) are closed, 24-volt control circuit power is supplied from the Y terminal to energize the compressor contactor (4).
- With the contactor contacts closed, power is supplied to start the compressor and outdoor fan motor (5).
- At the same time the compressor is energized, a 24-volt signal from the G terminal on the room thermostat energizes the indoor blower relay (6), starting the indoor blower motor (7).
- The unit will continue to operate until the room thermostat is satisfied.

2.2.0 Heating Sequence of Operation

Use the wiring diagram in *Figure 6* to follow the sequence of heat pump operation in heating mode, as follows:

- The reversing valve is de-energized in the heating mode. In the de-energized position, refrigerant is directed to the outdoor coil which acts as an evaporator, and to the indoor coil which acts as a condenser.
- As the room temperature drops, the thermostat (1) closes to generate a call-for-heat signal. Assuming the discharge temperature switch and high-pressure switch (2) are closed, 24-volt control circuit power is supplied from the Y terminal to energize the compressor contactor (3).
- With the contactor contacts closed, power is supplied to energize the compressor and outdoor fan motor (4). At the same time the compressor is energized, a 24-volt signal from the G terminal on the room thermostat energizes the indoor blower relay (5), starting the indoor blower motor (6).
- At this point, the unit is supplying compression heat only. If the heating load is low and the outdoor temperature is above the balance point, the unit should be able to satisfy the room thermostat on compression heat alone.

03311 Troubleshooting Heat Pumps   Module Four   7

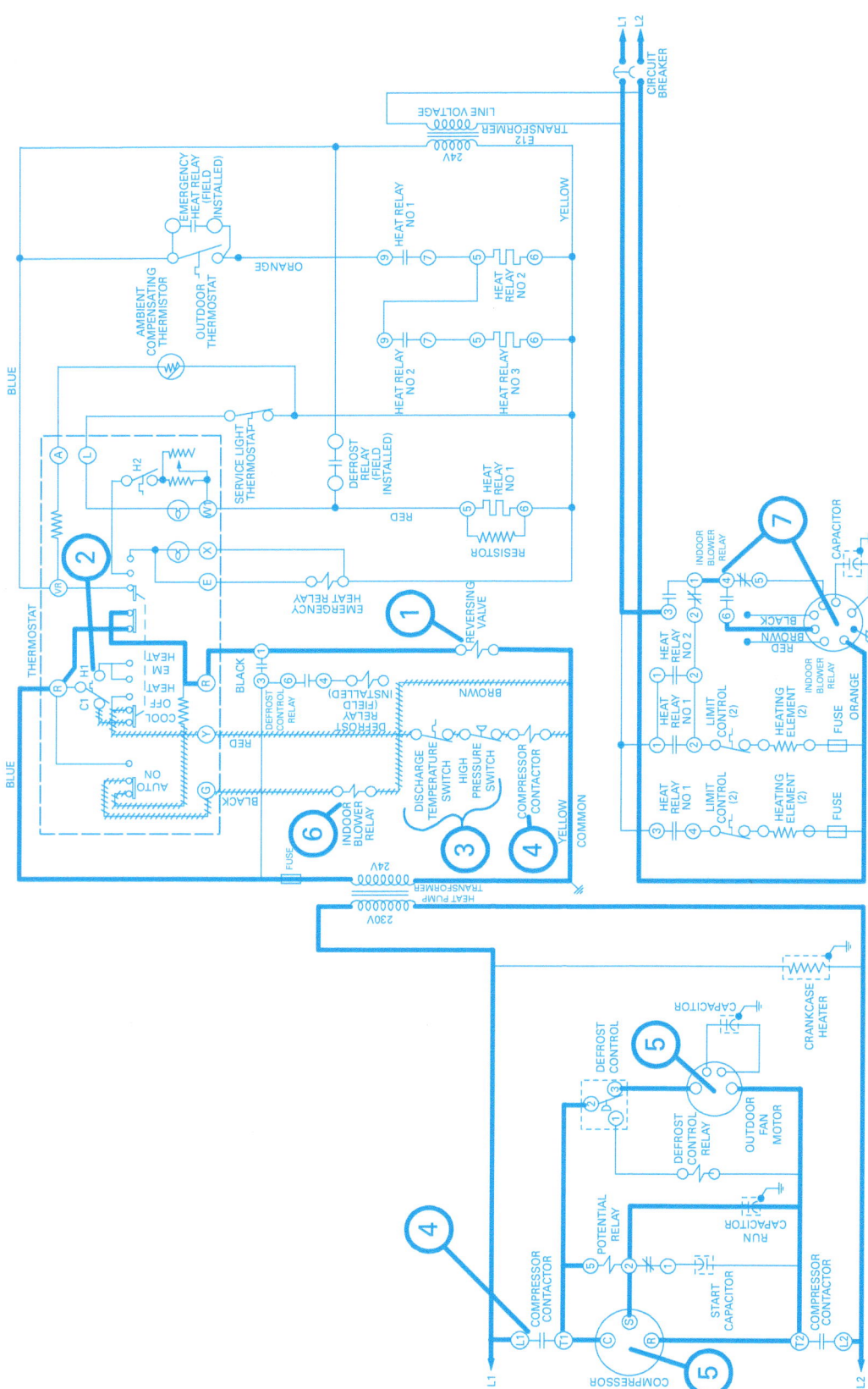

Figure 5 Cooling sequence of operation.

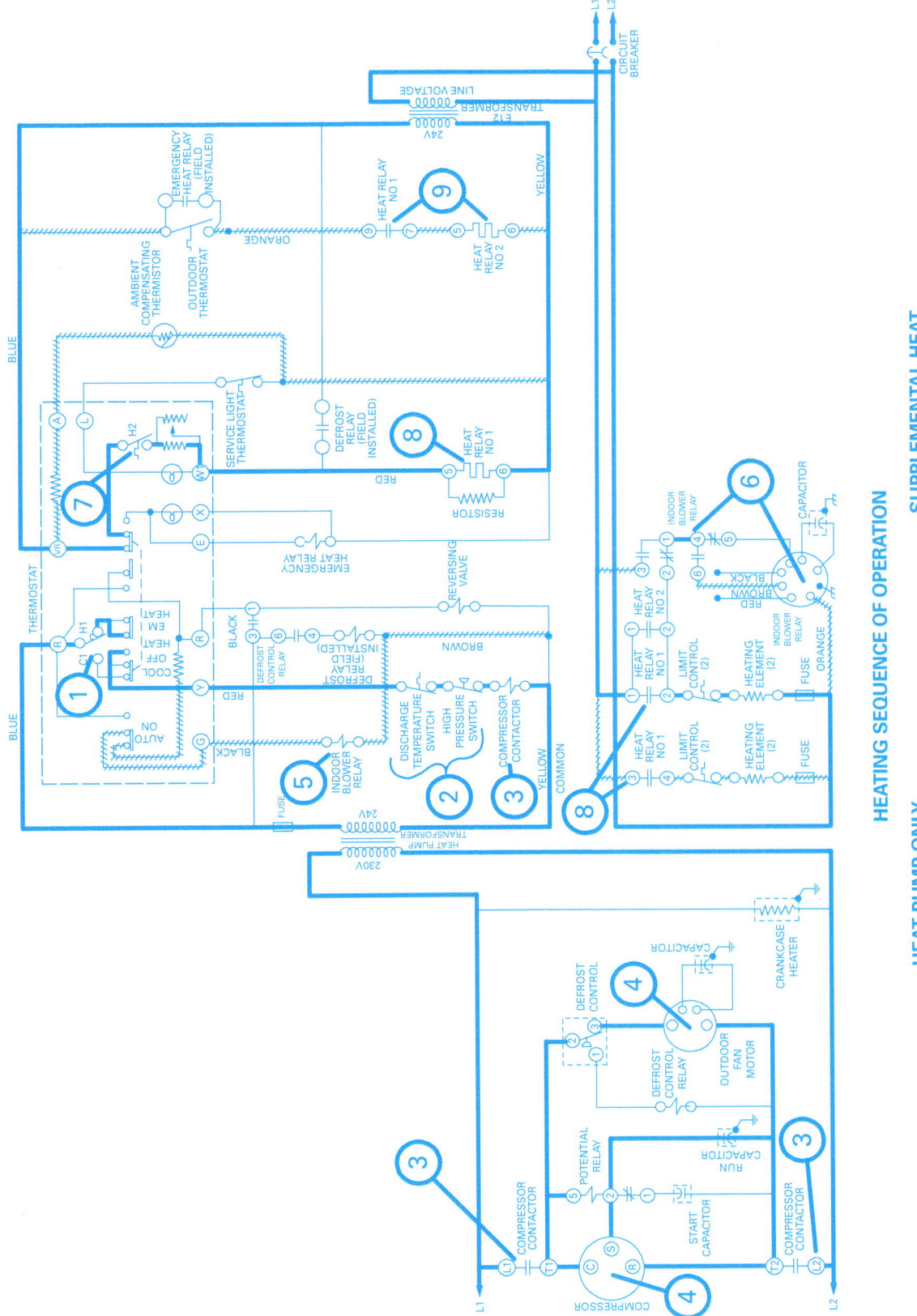

Figure 6 Compression heat with supplemental heat, sequence of operation.

- If compression heat alone is unable to satisfy the room thermostat, the second stage of the room thermostat (7) closes as the room temperature continues to drop. This energizes the W1 terminal and supplemental electric heat by first energizing heat relay No. 1 (8).
- Assuming the outdoor thermostat is closed, heat relay No. 2 (9) is energized through the auxiliary contacts of heat relay No. 1. The outdoor thermostat prevents the second stage of electric heat from operating above a preset outdoor temperature. The compressor and supplemental electric heater(s) will continue to operate until the room thermostat is satisfied.

If the compressor fails to operate, no compression heat will be available. On some systems, a service light on the room thermostat (1) will alert the occupants that there is a problem. This light is usually responding to a lock-out relay or similar device.

Use the wiring diagram in *Figure 7* to follow the sequence of heat pump operation in emergency heat mode, as follows:

- To provide heat until the compressor problem is resolved, the room thermostat function switch can be placed in the Emergency Heat mode (2) to enable emergency heat. Once in this mode, an emergency heat signal light (3) is energized on the room thermostat and the emergency heat relay (4) is also energized from the E terminal of the thermostat. The light reminds the owner that the system is operating in emergency heat mode, which provides less heat and is costlier to operate. Once the thermostat is switched to Emergency Heat, only the supplemental or auxiliary heat source and the blower operate. The compressor does not function at all.
- One set of contacts (4) of the emergency heat relay bypasses the outdoor thermostat for the electric heat, enabling all available stages of supplemental heat to be energized. Without compression heat, both stages of electric heat are available to maintain comfort. Once in the emergency heat mode, heat relay No. 1 (5) can energize all available stages of electric heat with the help of heat relay No. 2.
- The unit will operate until the room thermostat is satisfied. Some thermostats may make the electric heat respond to the first stage of heating, in place of compression heat. On thermostats that do not provide this function, the electric heat continues to respond to the second stage. The result is that the room temperature lags behind the thermostat setting by 2° to 3°F (1.1° to 1.6°C).

Once in emergency heat, it is important to find and correct the problem that is preventing compressor operation.

2.3.0 Defrost Sequence of Operation

During normal heating operation, an air-to-air heat pump will accumulate frost on the outdoor coil. If allowed to build up, the frost restricts airflow across the coil, reducing heat transfer. To remove the accumulated frost, the heat pump must periodically defrost the outdoor coil.

Use the wiring diagram in *Figure 8* to follow the sequence of heat pump operation in defrost mode, as follows:

- Before a defrost cycle can be initiated, the unit must be operating in the heating mode. An idle heat pump does not start up with the intention of defrosting itself. Defrost is initiated and controlled by a defrost control (1). The unit must be running with the compressor contactor closed in order for this circuit to be energized. This means that both the compressor and indoor fan are running as the defrost cycle begins. Both the indoor fan and the compressor must continue running throughout the defrost cycle.
- There are various strategies used to initiate defrost. Some systems check the outdoor coil temperature on a timed basis to see if a defrost cycle is required. Very old systems may enter defrost on a timed basis only, without monitoring temperatures. Once the defrost mode is initiated by the control, a set of contacts changes position. Contacts between terminals 2 and 3 open, de-energizing the outdoor fan motor. Contacts between terminals 2 and 1 close, energizing the defrost control relay.
- Once energized, the defrost control relay (3) energizes the reversing valve and a second defrost relay. This second relay is normally located at the indoor unit to enable the electric heat. To prevent cold air from being blown into the conditioned space, this defrost relay energizes heat relay No. 1 (4).
- With the heat relay energized, line-voltage power is available to the supplemental electric heat (5), which warms and tempers the air being discharged into the conditioned space. Without the supplemental heat, the system will deliver cold air to the structure as long as the defrost mode is active. Note that the outdoor temperature control may still be open, leaving only one stage of electric heat energized during the defrost cycle. Some systems may bypass the outdoor temperature control during defrost.

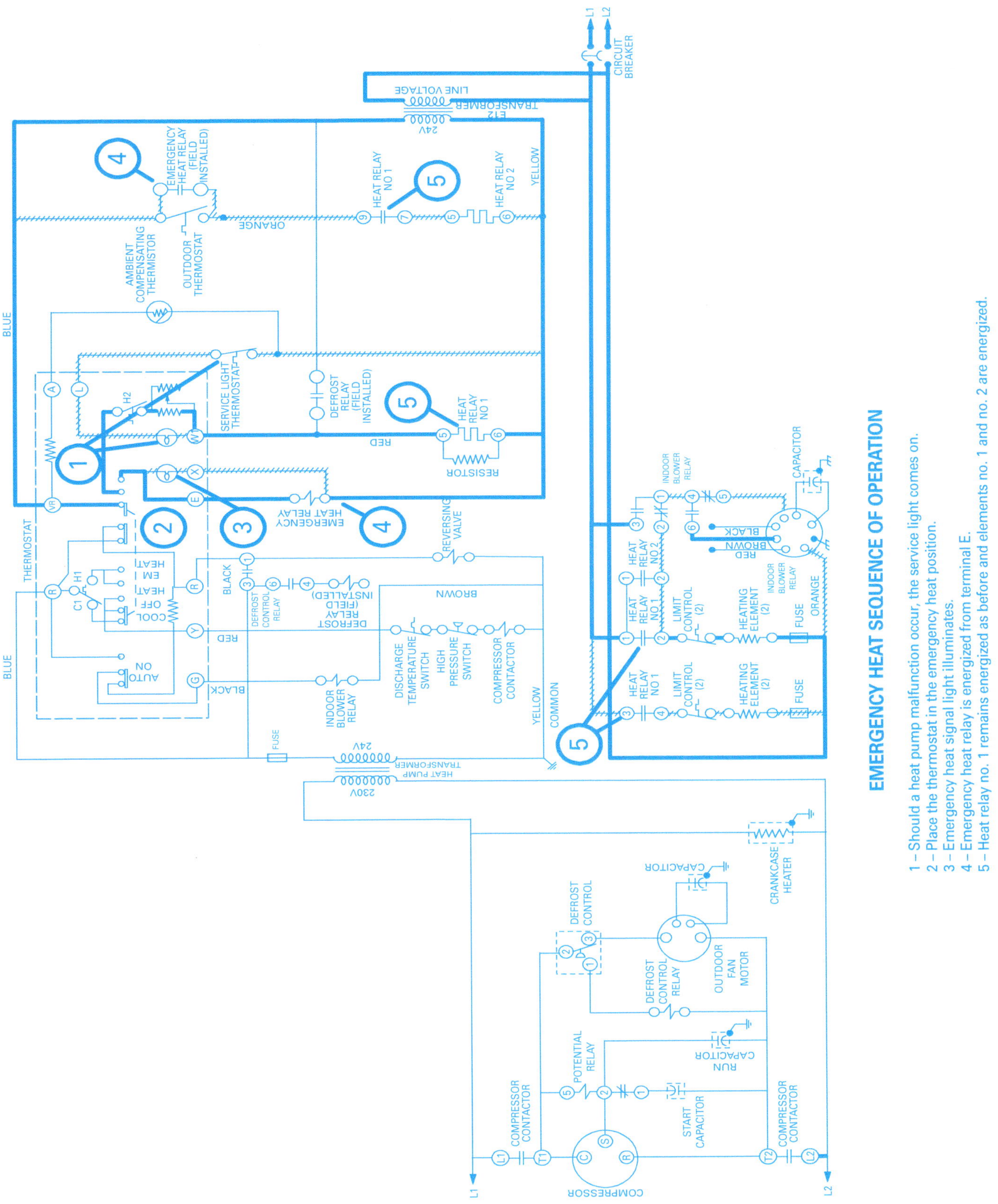

Figure 7 Emergency heat sequence of operation.

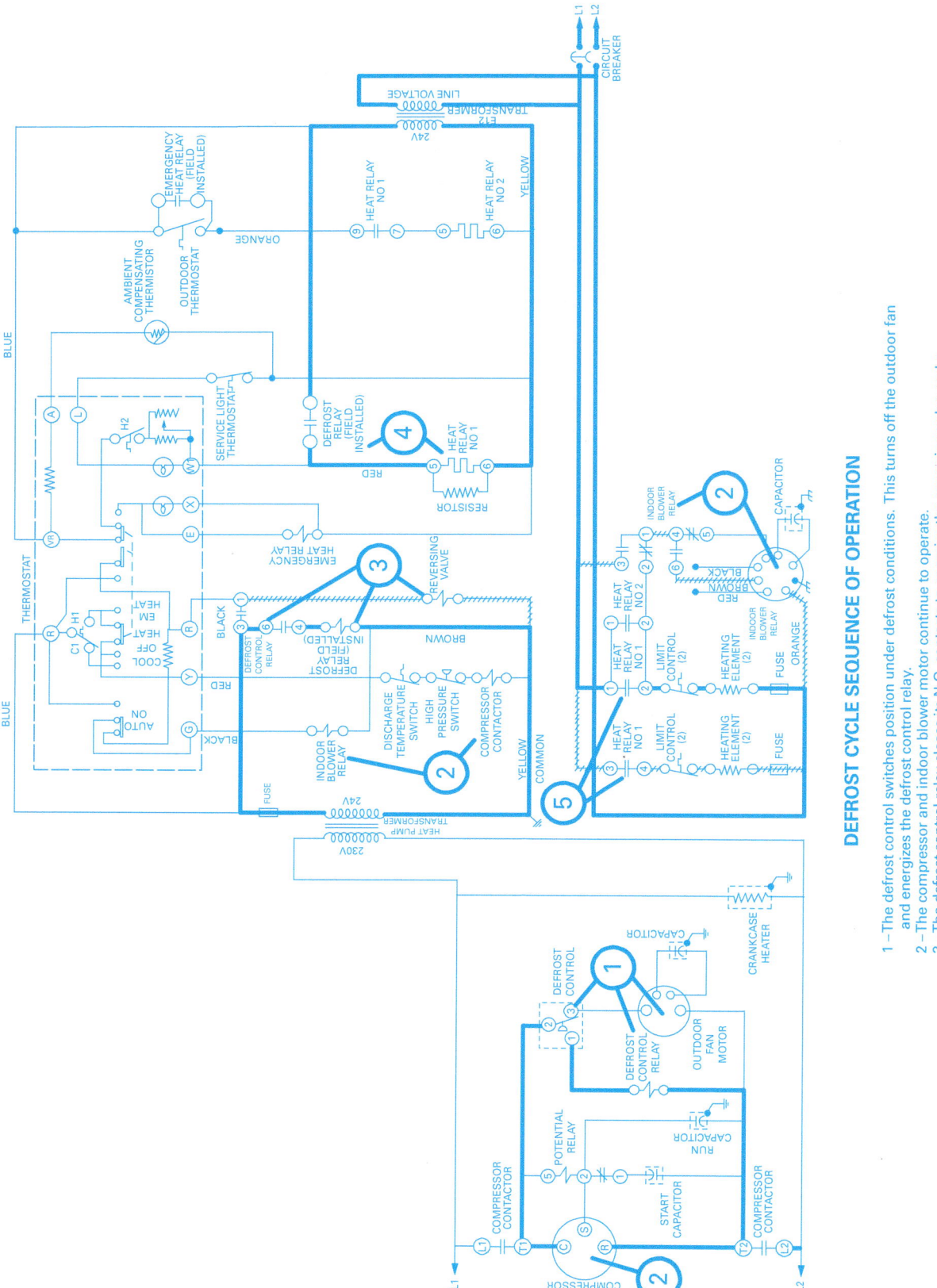

Figure 8 Defrost sequence of operation.

- With the reversing valve now energized, the unit has changed to its cooling mode, sending hot refrigerant vapor to the outdoor coil to melt accumulated frost. The compressor and indoor blower motor (2) continue to operate non-stop throughout the defrost cycle.
- Once it has been determined that the coil is free of frost, defrost is terminated by the defrost control. Termination is usually based on both time and temperature. The defrost cycle will terminate when a specific temperature has been reached at the outdoor coil (monitored by a thermostat or thermistor), or when a maximum amount of time has elapsed. Strategies for defrost termination vary among manufacturers.

Once the defrost cycle terminates, the heat pump returns to operation in the heating mode.

2.4.0 Dual-Fuel Heat Pump Operation

A popular, energy-efficient use of air-to-air split-system heat pumps is as an add-on to a fossil-fuel furnace. In cooling, the heat pump operates like any other split-system air conditioner, with the furnace acting as the air handler. It pushes air through the indoor coil, which is mounted downstream of the furnace.

In the heating mode, the heat pump operates under certain conditions, while the fossil-fuel furnace maintains comfort when other conditions are present. The primary factor in changeover is the balance point. Above the balance point, a properly sized heat pump can handle the heating needs of the structure without supplemental heat. The balance point often falls just above the freezing point outdoors. As a result, only a limited amount of frost tends to build on the outdoor coil, especially since the heat pump is cycling off occasionally above the balance point. By operating the fossil-fuel heating equipment below the balance point, the heat pump only operates when it is most efficient and with a very limited need for energy-wasting defrost cycles.

In *Figure 9*, an outdoor thermostat or thermistor is used to switch the system between compression heat and fossil-fuel heat. At low outdoor temperatures (generally below the balance point) where the heat pump is less efficient and lacks heating capacity, the gas or oil furnace heats the structure. Above the balance point, the more efficient heat pump heats the structure.

The temperature at which the changeover takes place between the two methods of heating can vary. The balance point is only one possible factor, although it is the primary one. Another factor that may affect the chosen point for changeover is the cost of electricity compared to the cost of the fossil fuel used. In most cases, a temperature slightly above the actual balance point of the structure is chosen as the point where compression heat is locked out and fossil fuel is used to heat the structure. This ensures that indoor comfort is maintained at lower outdoor temperatures.

Unlike a conventional all-electric heat pump where compression and supplemental electric heat are often supplied at the same time, only one form of heat is available with dual-fuel systems; compression heat or fossil-fuel heat, but not both together. Since the heat pump indoor coil is downstream of the furnace, the heat pump compressor would quickly overload with the furnace blowing hot air across its already-hot coil. Reversing the order of the coil and furnace would not help. The furnace would exceed its high-temperature limits if hot air from the heat pump coil were provided to its inlet.

The one exception to this situation applies during the defrost cycle. Instead of using supplemental electric heaters to temper the indoor air, the system's gas or oil burner can be used to supply heat and place a load on the heat pump system. However, the indoor coil is usually mounted downstream of the burner. This results in very hot air passing over the indoor coil while it is serving as the evaporator. In summer conditions, this would quickly overload the compressor. However, with frost/ice on the outdoor coil, there is an enormous heat sink available, preventing the discharge pressures from rising too high. The furnace heat is transferred to the refrigerant where it moves to the outdoor coil. The large amount of heat carried by the refrigerant results in a very fast and thorough defrost cycle.

> **CAUTION**
> Using the fossil-fuel unit of a dual-fuel system during the defrost cycle is not possible or allowed in every system. To prevent system damage, consult the manufacturer before proceeding. The better solution is to size the heat pump such that the balance point is above freezing, with the fossil-fuel unit taking control of the heating needs at this same point. Defrost cycles under these conditions are rarely, if ever, needed.

For many years, the dual-fuel arrangement was only available in a split-system configuration. Today, packaged heat pumps are available that also contain gas burners, allowing users of packaged products to benefit from this unique energy-saving combination. Aside from the differences just described, an air-to-air heat pump installed with a fossil-fuel furnace is no different in its operation than any other air-to-air heat pump.

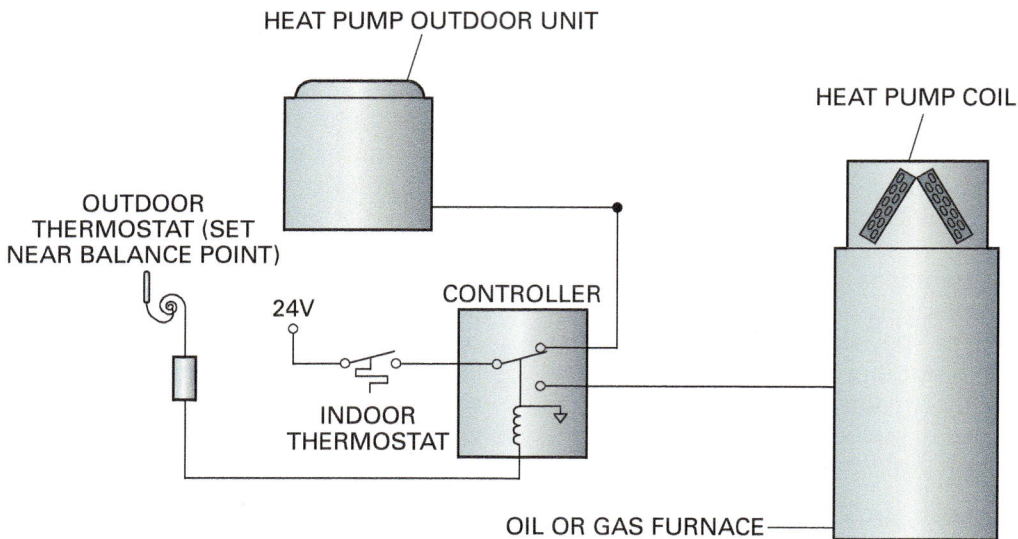

Figure 9 Dual-fuel heat pump operation.

> ### Economic Balance Point
>
> **GOING GREEN**
>
> An air-to-air heat pump's balance point is the outdoor temperature below which the heat pump cannot supply enough compression heat to maintain indoor comfort. The conventional wisdom when installing a dual-fuel heat pump is to set the changeover point from compression heat to fossil-fuel heat at a temperature a few degrees above the heat pump's balance point. While this is acceptable, it may not make sense economically.
>
> For the best energy savings, the changeover should be done at the economic balance point, which can be a different temperature from the thermodynamic balance point. The price of electricity varies across the United States, as does the price of natural gas, propane, and fuel oil. When you know the costs of the different forms of energy available, it is possible to determine the most economical temperature at which to make the changeover.
>
> Heat pump manufacturers have work sheets available that enable the economic balance point to be calculated. As long as the economic balance point is not below the actual balance point of the structure, comfort will be maintained for the lowest cost.

2.5.0 Microprocessor Controls in Heat Pumps

The operating sequences described earlier are typical for heat pumps using electromechanical controls. They are ideal for explaining basic electrical operating sequences. While millions of electromechanical heat pumps are still in use, they are rapidly being replaced with heat pumps that use electronic controls and variable-speed motor technology. Electronic controls can do everything that can be done with electromechanical controls better and more reliably. Electronic controls have few, if any, moving parts which can be prone to fail after repeated use. Consider the example of a motor-driven defrost timer turning cams that activate switches compared to the efficiency of an electronic timer with no moving parts.

Figure 10 is the schematic diagram of a modern packaged heat pump that contains electronic controls and a variable-speed blower motor. Even though what is going on inside the control boards cannot be seen, it is possible to determine a lot about the sequence and timing of operations from this schematic.

The unit contains two control boards, a defrost board (DB) and an integrated control motor (ICM) board that controls the operation of the variable-speed indoor blower motor. Looking at the defrost board helps to reveal how it functions. A timing (logic) circuit is in series with the compressor contactor coil and the low- and high-pressure switches in the Y circuit. This indicates that the defrost board contains a compressor start-delay timer.

Another timing circuit is in series with the defrost thermostat. This indicates that defrost is based on elapsed time and outdoor coil temperature. Elapsed time is field-selectable and is done based on the position of dip switches, as shown on the schematic.

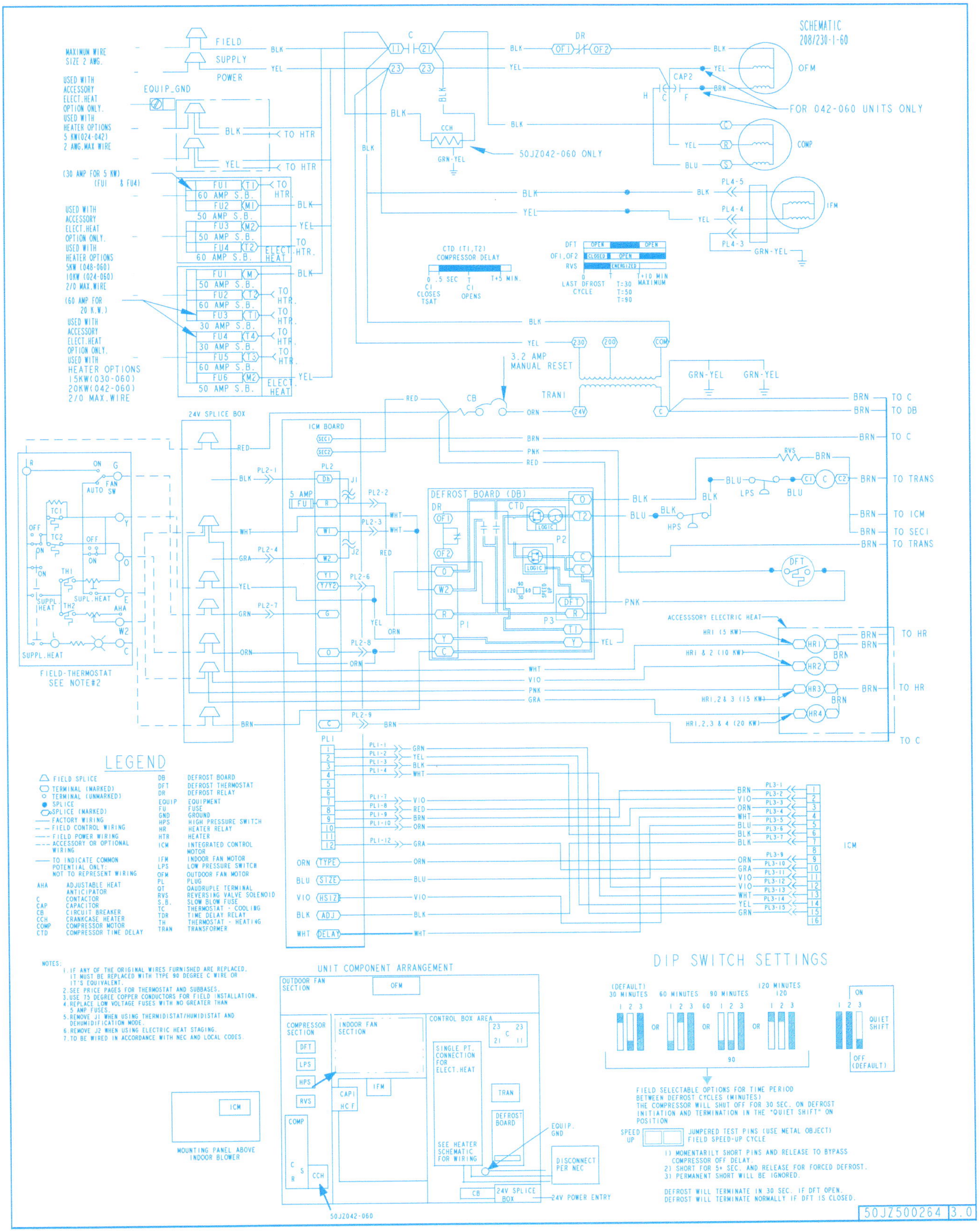

Figure 10 Packaged heat pump schematic.

 03311 Troubleshooting Heat Pumps

A direct connection between the room thermostat cooling contacts and O on the defrost board indicates that the reversing valve is energized in cooling. Open relay contacts within the defrost board imply that they close during defrost to energize the reversing valve and apply power to the electric heat relays (HR) by providing a current path from R to W2. Closed relay contacts across OF1 and OF2 open during defrost to stop outdoor fan operation. The same contacts can be seen in series with the outdoor fan motor (OFM) in the high-voltage side (upper part) of the schematic.

The ICM board is used to control the variable-speed indoor fan motor (IFM). Looking at the schematic, one could conclude that there are four electric motors in the unit: three in the high-voltage side (outdoor fan motor, compressor motor, and indoor fan motor), and one in the low-voltage side, labeled as the ICM. The indoor fan motor and the ICM are actually one and the same.

The indoor fan motor has line-voltage power applied at all times, even when the contactor is de-energized. This is characteristic of an *electronically commutated motor (ECM)*. To run, it must also receive various low-voltage control signals from the ICM board through the 16-pin connector on the motor. This connector is labeled as the ICM in the lower part of the diagram. This results in the indoor fan motor having two physical connections: one for line-voltage (upper part of schematic), and another that carries the low-voltage input signals (lower part of schematic). The ICM board generates the low-voltage signals based on the inputs it receives from the room thermostat.

During cooling operation, signals from the thermostat generate output signals from the ICM board that tell the fan to run at its highest speed. If continuous fan operation is required, the lack of a Y signal (cooling) and the presence of a G signal (fan operation only) cause the ICM board to signal the fan to run at a much lower speed for ventilation purposes only. The ICM board generates the appropriate speed control signals based on the position of jumper wires across pins on the board. These selections are made during the initial system startup.

The startup selections made are partially based on system capacity. The cooling fan speed of a 3-ton system, for example, would be higher than the selected cooling fan speed of a 2-ton system. The electric heater size in kilowatts (kW) also determines fan speed in the heating mode; larger heaters require more airflow. When properly configured, the ICM board will allow the fan to operate at different speeds based on the equipment's operating mode. Configuration procedures must be obtained by reading the manufacturer's installation instructions.

Always read the manufacturer's literature for the special features offered by an electronic control. The training programs available from manufacturers are also very helpful, providing the specifics for a particular product or family of products. Technicians are often intimidated when they see the number of different components found on electronic control boards. In fact, HVACR technicians do not need to know how each component on a control functions, nor do they have to troubleshoot electronic controls to the component level. What they do need to know is whether a control is functioning the way it is supposed to. If it is not, the control must be replaced.

Most electronic controls respond to input signals of some sort. If the control has the correct input signals, it responds by generating appropriate output signals. If a technician can confirm that the input signals are correct but the control is not responding with the correct output signals, there is a good chance it is defective.

On the other hand, if inputs and outputs from a control are as they should be, and the device being controlled is not responding, the problem is likely with the device and not the electronic control. For example, if the control board is sending power to the reversing valve to change its position but the position does not change, the reversing valve coil should be tested; the control board is not at fault.

Always check the device in question before replacing it. Something as simple as a loose connection may prevent normal operation. Manufacturers often provide the input and output signals expected from an electronic control. One example of this type of information is shown in *Figure 11*.

2.5.1 Variable-Speed Heat Pumps

Traditional heat pumps with single-speed compressors are still readily available and represent the majority of heat pumps installed today. In the ongoing search for greater efficiency, however, technology has enabled the development of extremely efficient systems that can provide a seasonal energy-efficiency ratio (SEER) of 20+.

Variable-speed compressors (*Figure 12*) modulate motor speed to more closely track capacity to demand. These compressors use inverter technology for speed control, instead of older multi-speed motor technology. In addition, the most efficient systems not only use ECMs to power the indoor blower, but also to power the outdoor condenser fan.

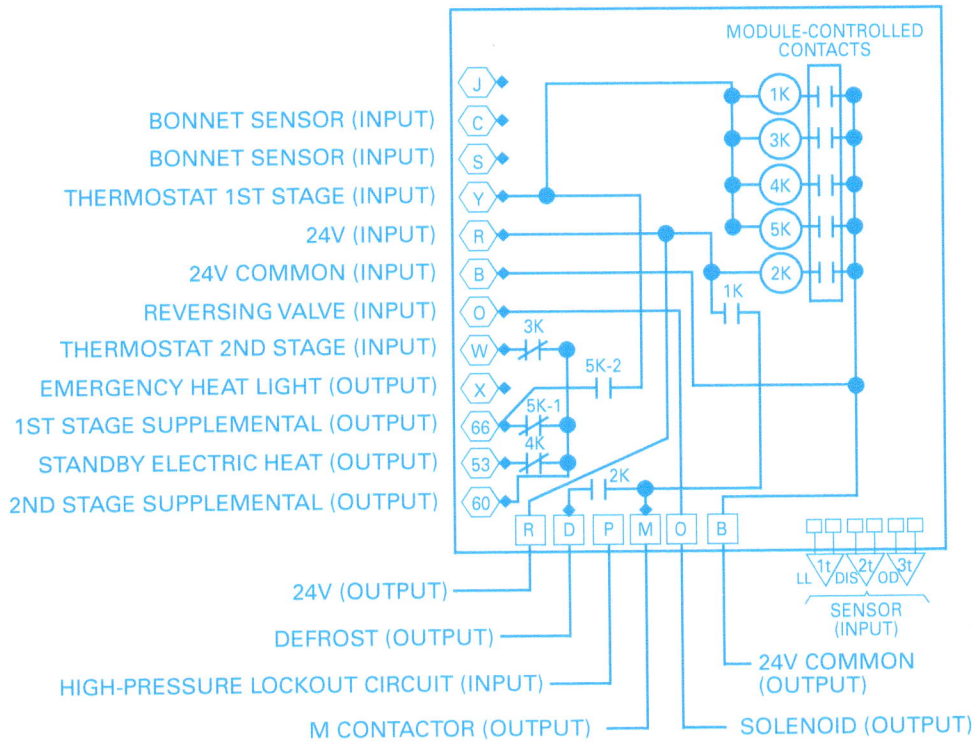

Figure 11 Input and output signals.

Figure 12 Variable-speed, inverter-driven scroll compressor.

Troubleshooting these advanced systems is beyond the scope of this module. As you might expect, they are more complex than traditional systems and require careful study of the manufacturer's sequence of operation, wiring diagrams, and troubleshooting guides for successful installation and repair.

ECM vs. ICM Motors

ECMs and ICM motors are really the same thing. Earlier versions of HVACR equipment that used these motors referred to them in literature and wiring diagrams as Integrated Control Module (ICM) motors. ICM actually referred to the electronic module that drove the motor, which is now typically integrated into the motor itself, as shown here. In later years, the term ECM was commonly adopted to describe these motors.

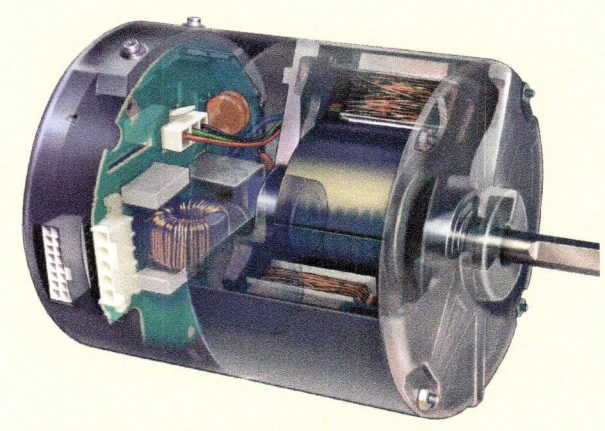

Figure Credit: www.thedealertoolbox.com

Additional Resources

NCCER Module 03211, *Heat Pumps*.

2.0.0 Section Review

1. The indoor blower motor is usually energized during a call for cooling _____.
   a. before the compressor is energized
   b. after the compressor is energized
   c. at the same time the compressor is energized
   d. after the outdoor fan is energized

2. A heat pump system should be able to manage the heating load without supplemental heat if _____.
   a. the indoor temperature setting is above the balance point
   b. the outdoor temperature is above the balance point
   c. ice/frost is never allowed to form on the outdoor coil
   d. variable-speed blowers with ECMs are used

3. Without the assistance of supplemental heat while in the defrost mode, a heat pump _____.
   a. still adds heat to a structure
   b. will deliver cold air to the structure
   c. must disable compressor operation
   d. would de-energize the reversing valve

4. In a dual-fuel heat pump, the indoor coil is usually mounted _____.
   a. in a basement
   b. near the air handler
   c. in the return duct
   d. downstream of the furnace

5. One reason electronic controls are more reliable than electromechanical controls is because they _____.
   a. are hermetically sealed
   b. have few, if any, moving parts
   c. operate on DC voltage
   d. draw very low current

Section Three

3.0.0 Testing and Troubleshooting Heat Pumps

Objective

Explain how to check and/or troubleshoot various functions and components of heat pump systems.
a. Explain how to check field and factory wiring.
b. Explain how to check and troubleshoot heat pump thermostats.
c. Explain how to test thermistors.
d. Explain how to check the various valves found in heat pumps.
e. Explain how to check defrost control circuits.

Performance Tasks

1. Demonstrate the ability to isolate and determine the solution for various electrical and mechanical malfunctions in heat pumps.
2. Initiate the defrost cycle of a heat pump.

Trade Terms

Demand defrost: A defrost control strategy that typically does not depend on time to determine when a defrost cycle is needed. Instead, the system responds directly to current conditions, such as temperature or pressure drop through the outdoor coil, to determine when frost is present.

Direct-acting valve: A valve in which the solenoid coil operates the valve stem directly to open or close.

Field wiring: Wiring that is installed during installation of the system, such as a thermostat hookup to a unit.

Pilot-operated valve: Valves in which a solenoid operates a smaller valve, which then ports more pressure and/or volume to a larger valve to open or close it. Large valves use this strategy due to the limitations in the magnetic field of solenoid coils.

Although heat pumps are more complex than cooling units, they have operating modes that share common functions. If a problem such as a disabled compressor is observed in the cooling mode, check to see if the same problem occurs in the heating or defrost mode. If it does, the problem is most likely in the compressor or one of its direct control or protective devices. If it does not, then the problem is most likely in some device that is unique to the cooling mode.

Once the operating sequence of a given heat pump is known, a process of elimination can often isolate a failed component merely by observing how the unit operates in each mode. These observations can often be done without using test equipment. Some examples of this process are described in this section.

Refer to *Figure 13*. If the compressor does not start in the cooling mode but works in the reverse-cycle heat and defrost modes, the malfunction would likely be found in the room thermostat. All safety control contacts must be closed, or the compressor would not run at all. The circuitry of the thermostat that initiates a call for cooling may be faulty. This would be a good place to begin troubleshooting in this scenario.

What is the likely cause if the unit runs in heating but does not defrost and refuses to switch to the cooling mode of operation? A dysfunctional reversing valve is a strong possibility, because heating takes place with the valve de-energized. To operate in cooling, the reversing valve must be energized. If the unit does not go into defrost while in the heating mode, it is a further indication that the reversing valve has a problem, because it must be energized during defrost.

Understand, however, that the valve itself may be fine. The solenoid coil is more likely to fail than the valve itself, or a wiring problem may exist. Testing and/or changing the solenoid coil is very simple. Replacing an entire reversing valve is a tedious, invasive, and expensive task.

Many manufacturers provide troubleshooting guidance in their product literature. The *Appendix* of this module provides examples of troubleshooting charts and guides, showing how such resources differ among manufacturers. While troubleshooting aids like these can be helpful, some can be either too general or too complicated. In some cases, vital information is lost in translation. When troubleshooting, there is no substitute for a thorough understanding of the operating sequence of the product.

 03311 Troubleshooting Heat Pumps

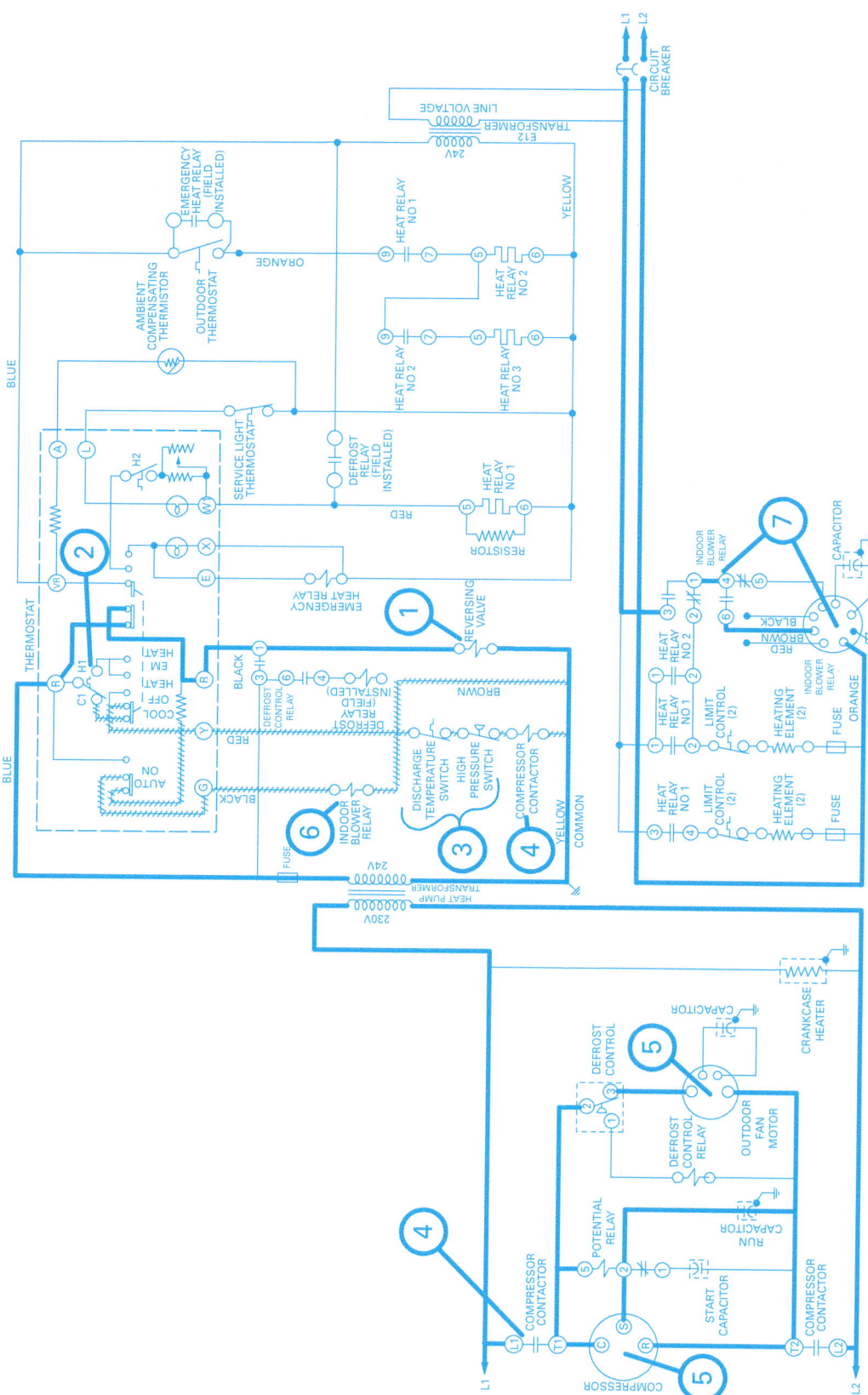

Figure 13 Cooling sequence of operation.

> **Restricted Airflow**
>
> The supply air temperature of a heat pump in the heating mode is typically lower than that of electric, gas, and oil heating systems. If the air being discharged into the conditioned space from a heat pump is warmer than 120°F (49°C), it could indicate restricted airflow. Otherwise, the supply air temperature will rarely approach this temperature.
>
> A heat pump is typically designed to provide airflow of 450 to 500 cfm per ton. Undersized ductwork, a dirty air filter, or a dirty indoor coil can cause reduced airflow, which in turn will cause an increase in compressor discharge temperature. High temperatures can cause the compressor to overheat and trip on the overload protection.

> **Field Wiring Problems**
>
> Field wiring problems are most often found at connection points. This is because HVACR installers may fail to use the correct devices or procedures to join the wires. For example, it is common to encounter low-voltage thermostat wires that have been simply twisted together and taped. A correctly sized wire nut should be used to make this type of connection.

3.1.0 Checking Wiring

The wiring for heat pumps and similar equipment can generally be classified as *field wiring* or factory wiring. Factory wiring is the line-voltage and low-voltage wiring that is installed and terminated at the point of manufacture. Field wiring is the wiring installed and terminated by the installer.

3.1.1 Field Wiring

Control circuit field wiring must be connected by at least 18 AWG copper wire, or the relays and contactors may chatter or fail to operate due to voltage drop. As a general rule, if the wiring circuits are longer than 100' (30 m), wire heavier than 18 AWG copper is needed to prevent excessive voltage drop. 16 AWG wire is the next available size. It is very rare that control circuit loads require a larger wire size than this. Remember that a low secondary voltage caused by insufficient wire size or a similar issue will not be evident until the circuit is under load.

Low primary voltage at the transformer can cause relays and contactors to chatter or fail to operate. Therefore, the primary voltage applied to the transformer must be checked. The secondary voltage, typically 24VAC, is dependent upon the primary voltage applied. If the primary voltage is low, the secondary voltage will be low as well. If the primary voltage is more than 10 percent below its rating, erratic control operation can result. This situation will usually cause the secondary control voltage to fall below 22VAC.

When looking for wiring problems, check for loose connections at the terminals and wire nuts. The additional resistance caused by a poor connection can cause erratic operation. Control circuit fuses or breakers (if any) should also be checked before checking the field wiring.

If the terminations appear sound, disconnect each end of a wire in a specific circuit and measure the resistance with an ohmmeter. Excessive resistance often indicates loose connections or corroded terminals. Open circuits will be indicated by a reading of infinity. To check thermostat wiring in the wall and other inaccessible locations, join two wires with a wire nut and check for continuity at the opposite end of the run.

3.1.2 Factory Wiring

Factory wiring in the unit can be checked in the same manner as field wiring. However, the field wiring should be checked first. When factory wiring errors occur, the malfunction should show up when the unit is checked out during installation. Factory wiring problems are less likely to occur after the system has been in operation over the short term. Over time, however, all terminations tend to deteriorate and need to be checked. Factory connections are generally of better quality since they are completed under ideal conditions using precise and consistent processes.

Check the equipment wiring using the wiring diagram attached to the unit. Trace the circuits by observing the actual location of the electrical devices, terminal designations, and wire color codes as designated on the wiring diagram.

3.2.0 Checking Heat Pump Thermostats

Heat pump thermostats should be checked to ensure that they are a correct match for the system. Thermostats should not be mounted in a location where extraneous cooling or heating sources will affect it. Although digital thermostats are not affected by an unlevel condition, they should be mounted level as an indication of the proficiency of the installer. The wall penetration behind it

Aluminum-to-Copper Wire Connections

Joining aluminum and copper wires requires special techniques and connectors. If correct procedures are not followed, the aluminum-copper connection can overheat or fail. For this reason, many HVACR equipment manufacturers prohibit the connection of aluminum wires to their products. Even some products designated for making aluminum-to-copper terminations, particularly wire nuts, aren't always successful.

Products designed for Al/Cu connections are purple or have purple markings to help identify them. The terminal block shown here is a good example. Once the connections are made, the plastic box is closed, minimizing contact with water and fresh air that hasten corrosion.

Figure Credit: King Innovation®

should be plugged to prevent drafts inside the wall from affecting the accuracy of the thermostat. Heat pump thermostats installed today are almost exclusively digital. There are several reasons for this. Most importantly, many older thermostats contained mercury, which is a highly toxic substance. When replacing an old mercury-bulb thermostat, it must be treated as hazardous waste and disposed of properly.

Digital room thermostats offer more precise temperature control. This allows the full range of features designed into modern heat pumps to be fully utilized. The thermostat shown in *Figure 14* is typical of the newest generation of controls. It offers the versatility of being used for heating, cooling, and heat pump operation. The installer configures the thermostat for the desired mode of operation. Within the heating mode, some thermostats offer the versatility of programming for heat pump operation with auxiliary electric heaters or with a dual-fuel system. Notice the signal-strength indicator in the upper right-hand corner of the display; this thermostat is Wi-Fi enabled, and can be accessed from the local network or via the internet.

The traditional mode-selection subbase switches are often replaced with a mode-selection soft button or touch-sensitive screen. Adjustable heat anticipators have been replaced by a method to select cycle rates directly. Within the thermostat itself, the traditional hard-wired electromechanical circuits are replaced with electronic circuits. Miniature relays provide the switching that was previously done by the subbase switches.

Because the thermostats are electronic devices, they must be powered to operate. This can be done with batteries and the 24VAC available from the control circuit. If control circuit power is used, a 24VAC common lead is typically provided and connected to the common (C) terminal on the thermostat. Most models have batteries that maintain programming for an extended period of time if the power is interrupted.

To check a heat pump room thermostat, it is best to begin by determining the function controlled by each terminal. These terminal designations are fairly standardized in industry, but occasional variations will be found. In the case

Figure 14 Wi-Fi enabled programmable heat pump thermostat.

of the thermostat shown in *Figure 14*, the subbase terminals are related to the following circuits or functions:

- *Terminal C* – Wire color is often blue or brown; it is the 24VAC common lead from the control-transformer secondary circuit.
- *Terminal R* – Wire color is red; it is the 24VAC power lead from the control transformer secondary circuit. There may be two R terminals on a thermostat, marked as R_C and R_H. Some systems, especially older systems, may have two control transformers in operation; one serves the heating equipment and one serves the cooling equipment. Separating the R terminals keeps the two transformer circuits apart. When a single transformer is used (very common), a jumper is placed across the two R terminals.
- *Terminal Y* – Wire color is yellow; it energizes the compressor control circuit. The Y terminal may be followed by a number (Y1, Y2, etc.) to indicate the cooling stage it controls.
- *Terminal G* – Wire color is green; it energizes the indoor fan relay.
- *Terminal O or B* – Wire color is usually orange or blue; these terminals are energized when modes are selected. When cooling is the selected mode of operation, O is energized and remains so. B is energized when the heating mode is selected and O becomes de-energized. The O terminal is commonly used to power the heat pump reversing valve. Some manufacturers, however, do use terminal B and energize the reversing valve in the heating mode.
- *Terminal W* – Wire color is white; it controls the heat source relay for supplementary heat. The W terminal is often followed by a number (W1, W2, etc.) to show the different stage of heat each terminal enables.
- *Terminal S* – Wire color is optional; it may be red and white when two-conductor thermostat wire is used. The S terminals are used to connect external temperature sensors.
- *Terminal E or X* – Any color wire; it connects to the emergency heat circuit.
- *Terminal L* – Any color wire; provides power for the Emergency Heat icon on the thermostat. The icon is powered constantly when the thermostat is switched to Emergency Heat. Otherwise, it can be powered by heat pump control circuits that indicate a fault has occurred.

This wiring description applies to the specific thermostat shown in *Figure 14*. It does not necessarily apply to all thermostats.

Another common characteristic of digital thermostats is a configuration menu (*Figure 15*). The configuration menu can have many possible entries and options, each of which can significantly affect the behavior and performance of the thermostat. It is always a good idea to check the configuration menu, especially when a system doesn't appear to be broken but simply doesn't perform as expected.

In most heat pumps, the reversing valve is energized in cooling and defrost, but de-energized in heating. This is so that heat can be available if the reversing valve solenoid or something in the circuit fails while in the heating mode. However, there are some heat pumps designed to operate with the reversing valve energized in heating. For digital thermostats, this is generally one of the settings found in the configuration menu.

To determine if the reversing valve is energized during cooling, use the following procedure:

- If the reversing valve is wired directly to the room thermostat through the O terminal, it is energized in cooling. The O terminal is always a terminal energized immediately when the cooling mode is selected.
- If the reversing valve is wired directly to the room thermostat through the B terminal, it is energized in heating. The B terminal is historically energized immediately when the heating mode is selected.
- Check for normally open contacts of a defrost relay in the reversing valve solenoid circuit. The relay contacts close during defrost in the heating mode to energize the reversing valve solenoid. Being in defrost is like being in the cooling mode as far as reversing valve operation is concerned.

3.2.1 Communicating Thermostats

There are a growing number of thermostats that operate using a communication bus rather than through common 24VAC wiring terminals. The Carrier Infinity™ system, for example, uses a two- or four-wire approach to control wiring. The terminals are designated as A, B, C, and D. In the four-wire product, the A and B wires are used for communication signals to travel back and forth, while the C and D terminals provide 24VAC power. The primary equipment must also be configured to accept this type of control arrangement.

Troubleshooting these units requires a different strategy than troubleshooting standard controls. When problems are encountered, be sure

CONFIGURATION MENU

Screen Reference Number	SS1 MS2	HP1 HP2	Press key	Displayed Factory (Default)	Press ▷ or ◁ to select from listed options	Comments
1	●	●	⚠	MS 2	HP 1, HP 2, SS 1	Selects Multi-Stage (MS2, No Heat Pump), Heat Pump 1 (HP1, 1 compressor), Heat Pump 2 (HP2, 2 compressor or 2 speed compressor), or Single Stage
2	●	●	⚠	(GAS)	ELE	GAS setting: furnace controls blower. ELE setting: thermostat controls blower
3		●	⚠	0b (0)	b	Selects Reversing Valve (This item is only to appear if HP1 or HP2 is selected above.)
4	●	●	⚠	Days, (7) P	5 or 0	Programs per week (7 days, 5-1-1 days or non-programmable)
5	●		⚠	Cool-Off-Heat-Auto	Cool-Off-Heat, Heat Off, Heat, Cool-Off, Auto Off	System switch configuration in non heat pump mode
		●	⚠	Cool-Off-Heat-Em-Auto	Cool-Off-Heat-Em, Off-Em-Auto	System switch configuration, heat pump mode
6	●	●	⚠	(On) E	OFF	Selects Energy Management Recovery. Not available if 4 is 0
7	●	●	⚠	(FA) Heat, Cr	SL	Selects Adjustable Anticipation, cycle rate, Heat
8	●	●	⚠	(FA) Cool, Cr	SL	Selects Adjustable Anticipation, cycle rate, Cool
9		●	⚠	(FA) Cr/AU, Em	SL	Selects Adjustable Anticipation, cycle rate auxiliary, (This item is only to appear if HP1 or HP2 is selected above)
10	●	●	⚠	(OFF) SC	On	Selects Stage Cycle Completion On (OFF)
11	●	●	⚠	(OFF) CL	On	Selects Compressor Lockout
12	●	●	⚠	(On) dL	OFF	Selects Continuous Display backlight.
13	●	●	⚠	0 (Temperature)	5, LO to 5, HI	Selects Adjustable Ambient Temperature Display [range -5 (LO) to +5 (HI)].
14	●	●	⚠	°F	°C	Selects °F/°C Display (temperature units in Fahrenheit or Celsius)
15	●	●	⚠	(On) b	OFF	Selects audible Beeper On/Off.
16	●	●	⚠	(On) dS	OFF	Selects Daylight Saving Time calculation
17	●	●	⚠	(On) Heat, AS	OFF	Selects Automatic Schedule for comfort temperature Programming, heat mode. Not available if 4 is 0
18	●	●	⚠	(On) Cool, AS	OFF	Selects Automatic Schedule for comfort temperature Programming, cool mode. Not available if 4 is 0
19	●	●	⚠	(OFF) CS	On	Selects Cool Savings Feature On of Off.
20	●	●	⚠	CS	1-2-3-4-5-6	Selects amo...

Figure 15 Example of a configuration menu.

to acquire the service information for the specific model in use.

Communicating models are designed to interact with the equipment at a deeper level than common digital thermostats. Digital thermostats, even those that are Wi-Fi enabled, do not provide detailed information about the system. Most of the features they provide come from information gathered by the thermostat alone, and do not come directly from the equipment.

The HVACR equipment must be designed and configured for communicating thermostats in order to reap the benefits of their features. For example, when connected to matching equipment, most will display fault codes that translate to system problems. *Table 1* shows a small sample of the fault codes they might display. This information is made available to the thermostat from the equipment solid-state control board(s). Each product is different however, and using the proper documentation is necessary when troubleshooting.

Table 1 Example Thermostat Fault Codes

Code	Fault
Code 14	Blower motor lockout
Code 15	Gas heat lockout
Code 16	Furnace ignition lockout
Code 18	Abnormal flame signal response
Code 21	Control fuse open
Code 25	No line voltage source
Code 30	Contactor coil shorted
Code 33	Failure of high-stage cooling (three times)
Code 34	Failure of low-stage cooling (three times)
Code 40	High-pressure switch open
Code 41	Low-pressure switch open

3.2.2 Room Thermostat Troubleshooting Sequence

Electromechanical room thermostats are simply a series of manual switches that are placed in various combinations of positions to get a desired result. Digital thermostats provide the same function, but in a different way. One way to determine if a room thermostat is faulty is to remove it from its subbase, or disconnect it from the interface board, and use jumper wires at the traditional terminals to simulate the circuit closures.

The R terminal is the hot terminal, so connecting it directly to each of the functional terminals will energize the appropriate component or function. With the 24VAC control power source energized, place jumpers across the following terminals or thermostat wires to check for operation:

- R to G should energize the indoor fan.
- R to Y should energize the compressor and outdoor fan motor.
- R to O or B should energize the reversing valve.
- R to W and G should energize the resistance auxiliary heat if the indoor fan relay contacts are closed.
- R to X2 and G should energize the emergency heat controls (heat pump) if the indoor fan relay contacts are closed.

If a component fails to function with the jumper in place, the problem is not in the thermostat. If the component does function with the jumper, but fails to function with the thermostat installed, the problem is somewhere in the thermostat or its configuration.

Thermostat troubleshooting can also take place at the furnace or fan coil unit. For example, if the thermostat is calling for cooling, there should be 24VAC at the furnace C and Y terminals. Before condemning any thermostat, verify that all switch positions are correct and that any installer programming has been properly completed. Incorrect configuration of a digital thermostat will make it appear to be malfunctioning when in fact it is operating as instructed.

3.2.3 Heat Pump Outdoor Thermostats

Electromechanical outdoor thermostats are used with air-to-air heat pumps in the following ways:

- Lock out supplementary electric heaters above an outdoor temperature
- Lock out compression heat at very low outdoor temperatures
- Switch between compression heat and a furnace in dual-fuel applications
- Stage banks of supplementary electric heaters as the outdoor temperature drops

Electromechanical outdoor thermostats are still in widespread use, but thermistors have replaced them in newer equipment.

Check the operation and calibration of an outdoor thermostat as follows:

Step 1 Connect an ohmmeter as shown in *Figure 16*. Set the sensing bulb in a container with a slushy ice-and-water mixture. The mixture, after a few minutes, should be at or very near 32°F (0°C). This can be verified with an electronic thermometer. Slowly rotate the outdoor thermostat dial until the ohmmeter indicates that the contacts are closed. The dial should indicate about 32°F.

Step 2 By rotating the dial one way and then the other several times while watching the ohmmeter (with many thermostats, the open and close action of the contacts are audible), a technician can determine if the contacts are operating in the 32°F (0°C) range with reasonable accuracy. Note that laboratory-level accuracy is not required of an outdoor thermostat. However, if more than a ±5°F (±2.8°C) variation is found to exist between the ice solution and the dial, the thermostat should be calibrated or replaced.

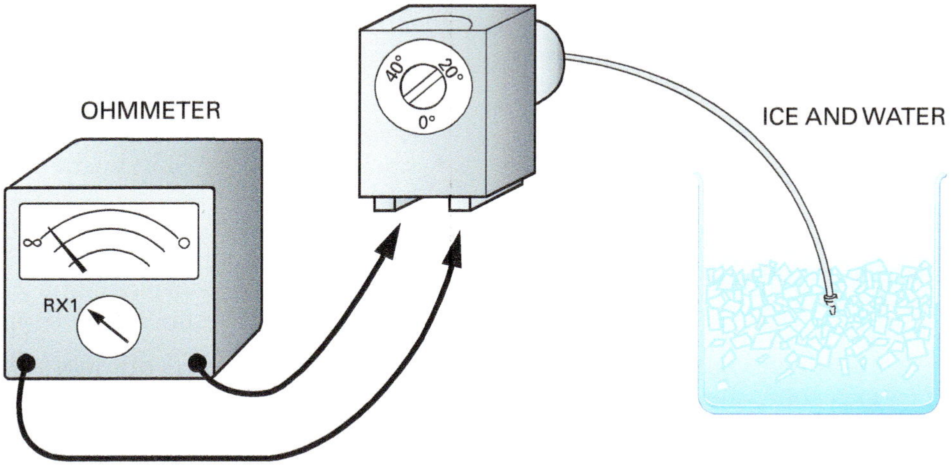

Figure 16 Outdoor thermostat check.

3.3.0 Testing Thermistors

Thermistors (*Figure 17*) are used in modern heat pumps to sense outdoor air temperature, coil and line temperature, and other temperature values that are helpful in controlling and monitoring an HVACR system. Thermistors are much more versatile than electromechanical thermostats.

Thermistor calibration can be checked in a manner similar to the procedure used to check remote-bulb thermostats. Perform the following steps:

Step 1 Turn off power to the unit, disconnect the thermistor, and immerse it in an ice water solution.

Step 2 Place an accurate electronic thermometer sensor in the ice water and stir the solution until 32°F (0°C) registers on the thermometer.

Step 3 Place ohmmeter leads across the thermistor leads and read the resistance. Compare the resistance at 32°F to the resistance that the thermistor should read at that temperature by consulting a resistance-versus-temperature chart (*Figure 18*). In the example shown, the resistance measured at this temperature should be about 33,000 ohms.

An ice bath is not always necessary. If the temperature that the thermistor is attempting to measure is accurately known (air temperature or the surface temperature of a refrigerant line, for example), enter that value into the chart to find the resistance for that temperature. Then check the manufacturer's literature to determine if the resistance measured is within acceptable limits.

Figure 17 Thermistor on a refrigerant line.

3.4.0 Troubleshooting Heat Pump Valves

Depending on its operating mode, refrigerant in a heat pump flows to different destinations from different directions. To direct the refrigerant to where it needs to be, various types of valves are needed.

3.4.1 Solenoid Valves

When the coil is energized on a typical solenoid valve, a magnetic field is produced that pulls the valve stem and opens or closes the valve. There are two types of solenoid valves:

- **Direct-acting valve** – The solenoid coil pulls the valve port open (or draws it closed) directly by lifting the pin. Power limitations of the solenoid coil restrict the port size of a given valve, however. A large valve, especially one that must operate against significant pressure differentials, may be impossible to open with electromagnetic force alone.

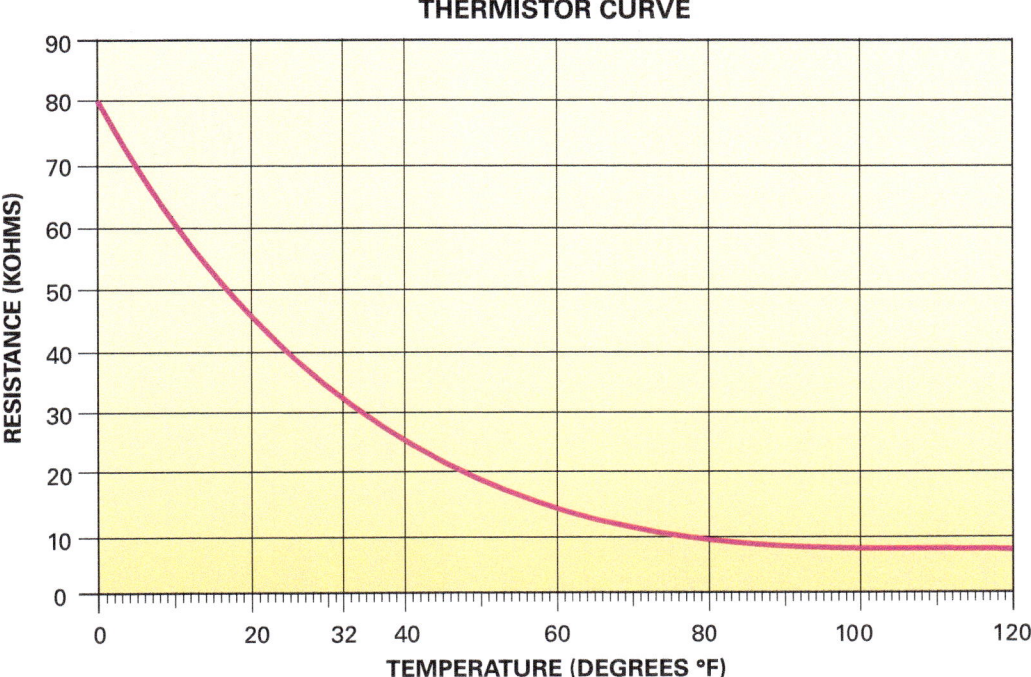

Figure 18 Thermistor resistance-versus-temperature chart.

- **Pilot-operated valve** – The solenoid coil operates a plunger covering a small pilot port hole. When the port hole is open, system pressure trapped between a larger piston and the port hole escapes. The resulting pressure differential on the two sides of the piston causes it to move toward (or away from) the pilot port hole; this opens (or closes) the main port hole on the other side of the piston. Heat pump reversing valves are usually pilot-operated valves.

When troubleshooting solenoid valves, look for the following common problems:

- Valve does not open
- Valve does not close
- Valve produces excess noise

When checking or testing solenoid valve coils, perform the following procedure:

Step 1 Inspect for loose or improperly installed wires.

Step 2 Test supply voltage to the coil with a voltmeter. The voltage should fall within +10 percent and –15 percent of the rated voltage.

Step 3 Place the tip of a steel screwdriver on the solenoid valve plunger when it is energized. The magnetic force of the solenoid should attract the screwdriver, indicating that the magnetic attraction of the coil is active.

Step 4 Inspect the coil for burning, which can result from too high of an ambient temperature, too much moisture, or too much humidity. Energizing the solenoid coil for more than a few seconds without a magnetic load imposed on it can damage the coil. This occurs whenever the solenoid coil assembly is removed from the valve. If the coil needs to be energized while removed from the valve stem, place a steel screwdriver or similar object through the hole to impose a magnetic load on it.

Step 5 Test the coil for continuity with it disconnected from the circuit.

Assuming the solenoid coil is functional but the valve refuses to move, internal sticking or seizing may be the problem. Before condemning or disassembling a valve for internal inspection, check the pressure differential across the valve. Solenoid valves have a maximum pressure differential specification. If the differential is too high, the valve cannot move. Pilot-operated valves usually have both a minimum and a maximum pressure differential specification, since they depend on system pressure differences to physically change position.

3.4.2 Reversing Valves

A reversing valve (*Figure 19*) has a slide mechanism that directs the refrigerant to different paths, depending on the operating mode. *Figure 20* and *Figure 21* show the path of the refrigerant in the cooling and heating modes, respectively. In the defrost mode, the valve is positioned the same as in the cooling mode, directing the hot discharge gas to the outdoor coil.

The following procedure is one way to test a reversing valve:

Step 1 Inspect the valve and solenoid coil for dents, deep scratches, or cracks. These may cause binding.

Step 2 Check the electrical system to be sure the solenoid has power and can be energized. On some units, this can be determined by the sound of the plunger moving in and out.

Step 3 If the coil cannot be energized, check other components in the electrical system for the source of the problem. Check the coil resistance with the wiring disconnected to be sure it is not open or shorted.

Step 4 Check the system pressure differential. Remember that a reversing valve is a pilot-operated valve. For a reversing valve to operate, a minimum operating pressure differential must exist. While some valves will shift when the differential is as low as 10 psig, others may require more than a 100-psig differential. Most reversing valve manufacturers do not guarantee the movement of the valve unless a 100-psig differential exists. Also, check the system for the proper refrigerant charge. If the charge is too low, the pressure differential may be insufficient.

Figure 19 Reversing valves.

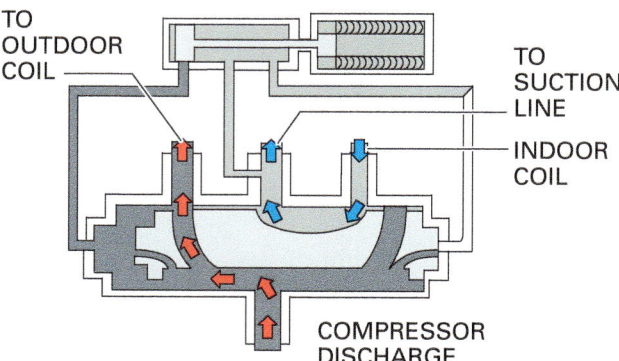

Figure 20 Refrigerant flow – cooling cycle.

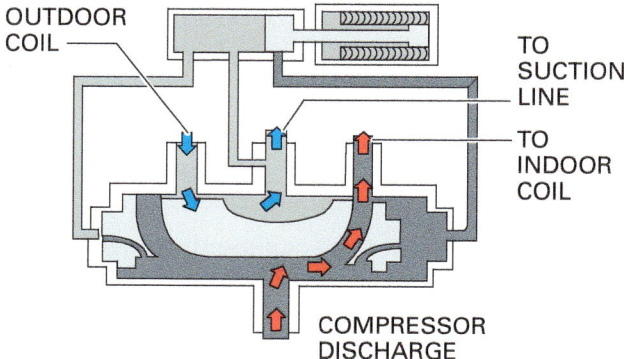

Figure 21 Refrigerant flow – heating cycle.

> **CAUTION**
> While reversing valves appear to be rugged, they can suffer from mishandling. Handle reversing valves with care. Even very small dents or dings can result in binding and/or internal leakage. When replacing a reversing valve, steps must be taken to avoid overheating the internal seals and gaskets during brazing.

The reversing valve may also be tested for internal leakage by measuring temperatures. *Figure 22* shows the temperature measurement points for testing a reversing valve. To avoid error, temperature readings should be made at least 5" (13 cm) from the valve body. The temperature probes should be insulated for accuracy. Note that infrared thermometers may not provide the needed

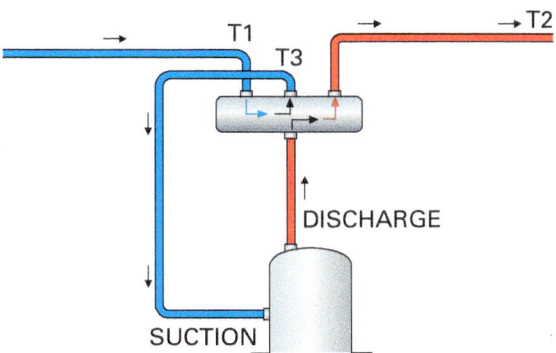

(A) CHECKING TEMPERATURE DIFFERENCE – COOLING MODE

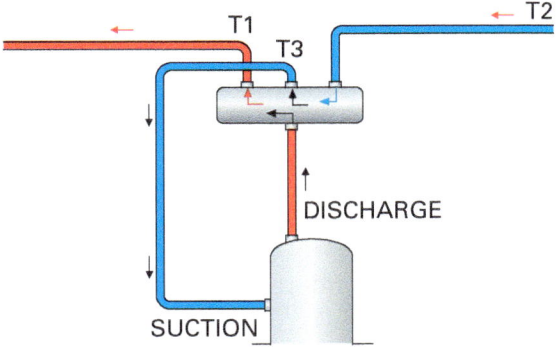

(B) CHECKING TEPERATURE DIFFERENCE – HEATING MODE

Figure 22 Reversing valve temperature testing.

accuracy to determine the valve condition; therefore, it is recommended that contact thermometers be used.

The following steps should be used to test for internal leaking of a reversing valve by measuring temperature:

Step 1 Set the thermostat to start the system in the cooling mode.

Step 2 Attach the thermometers at T1, T2, and T3 (refer to *Figure 22*).

Step 3 If the temperature difference between measurements at T2 and T3 is greater than 3°F (1.7°C), the valve is leaking internally and must be replaced.

Step 4 Set the thermostat to turn on heating.

Step 5 If the difference between T1 and T3 is greater than 3°F (1.7°C), the valve is leaking internally and must be replaced.

The current refrigerant used in residential heat pumps (HFC-410A) produces much higher pressures than the refrigerant it replaced (HCFC-22). Components used with HFC-410A heat pumps are rated for those higher pressures. If a heat pump that uses HFC-410A requires a replacement reversing valve, select the replacement according to the manufacturer's specifications. It is crucial to use a valve that is properly rated for the application and the refrigerant in use.

3.4.3 Check Valves

If a unit has two sets of capillary tubes for metering devices, there will be two check valves. Some units, especially older models, may also have thermostatic expansion valves that require external check valves. During the heating cycle, one check valve is used to bypass the indoor metering device. The check valve on the outdoor coil is forced closed, sending the refrigerant through the metering device. During the cooling cycle, the outdoor check valve allows refrigerant to bypass the outdoor metering device. The indoor check valve closes, forcing refrigerant to flow through the metering device. Modern heat pump expansion valves have the check valve and bypass built into the valve itself.

Check valves will open to bypass refrigerant and close to force refrigerant to be metered. Note that these check valves have no solenoid coil or other operating device. They are simple ball-check devices. The force of refrigerant trying to flow in one direction or the other determines the position of the ball. In one direction, the ball is forced against a seat that does not allow refrigerant to pass (closed). In the other, the ball is forced against a stop that allows refrigerant to pass around it on all sides (open). In its open position, the valve offers a low resistance (bypass) path for the refrigerant to travel.

When a check valve is suspected to be defective, test the heat pump equipment in both the heating and cooling cycles. If the equipment maintains normal head and suction pressures during either the heating or cooling cycle but pressures are abnormal in the other cycle, a sticking check valve is the likely cause.

When testing check valves during cold weather, cover the outdoor coil surface area with cardboard or plastic while conducting cooling-cycle tests. By blocking the outdoor coil until the head pressure is elevated to a normal level for a summer day, cooling cycle conditions can be reasonably simulated.

Figure 23 illustrates typical check valve construction. When a check valve becomes faulty, a number of predictable problems will usually be encountered. *Table 2* provides a general overview of some problems associated with specific check valve failures.

Keeping It Cool

The overheating of refrigerant piping components has always been an issue. Many components that must be soldered or brazed into the piping system, such as reversing valves and thermostatic expansion valves, are sensitive to the extreme heat applied during this process. Some components can be disassembled before the installation, leaving only the metal body exposed to the heat. However, this is not common for reversing valves.

Heat sinks are materials that have a high capacity to absorb heat. A variety of products are available to help absorb the heat that would otherwise be transferred to the component. Rather than bathe the component in the paste or gel, apply it in a ring around the tubing between the heat source and the component. The product should be applied with good contact and without voids which would allow the heat to simply pass it by. It can be very effective to wrap a wet cloth around the main body of a reversing valve and then apply a heat sink in the proper locations.

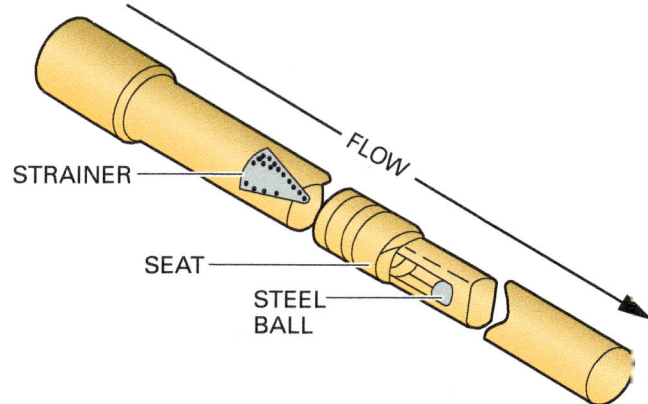

Figure 23 Check valve construction.

One very easy test can be conducted to see if a check valve ball is stuck in place. Since the ball is typically steel, the test involves the use of a magnet (*Figure 24*). Follow these steps to perform a magnet test:

Step 1 Shut down the equipment, allowing head and suction pressures to equalize. Once the pressures are equal, slide a strong magnet back and forth along the check valve body.

Step 2 If the ball valve is stuck, nothing will be heard. If the ball valve is free, a clear clicking sound will be heard as the magnet moves the ball back and forth in its housing.

If it is determined that the check valve must be replaced, follow these steps:

Step 1 Properly recover or isolate the refrigerant.

Step 2 Remove the faulty check valve.

Step 3 Avoid bending or deforming the valve body when installing the replacement. Use care when handling the valve.

Step 4 Avoid overheating the check valve or applying excessive brazing material when brazing. Wrap the valve body with a wet rag to keep the valve from overheating during brazing.

Metering Device Advances

Check valves in heat pumps have become far less common. Fixed-orifice metering pistons have been used for years, combining a metering device and check valve in a single component. Special heat pump thermal expansion valves (TXVs) were also developed that have the check valve and bypass features built into the valve body.

Note that these advances are more about reducing manufacturing costs and increasing the flexibility of components than about eliminating the check valve. Although check valves can be fouled after compressor burnouts and other events that add contaminants to the circuit, TXV passages can also be fouled the same way.

Table 2 Check Valve Troubleshooting

Check Valve Problems	Description
Outdoor check valve sticks closed during a cooling cycle	There will be low suction pressure, high discharge pressure, and high superheat at the indoor coil. The compressor is hot and may trip the internal overload protector. The indoor unit has a thermostatic expansion valve, a defective thermal element on this valve can lead to similar symptoms.
Outdoor check valve sticks closed during the cooling mode	The equipment operates at normal pressures in the heating cycle.
Indoor check valve sticks closed during a heating cycle	There is low suction pressure, high discharge pressure, and high superheat at the outdoor coil. The compressor is hot and may trip the internal overload protector.
Outdoor unit has a thermostatic expansion valve	A defective thermal element can lead to similar symptoms.
Indoor check valve is stuck closed during the heating cycle	The equipment operates at normal pressures in the cooling cycle.
Indoor check valve sticks open during the cooling cycle	There is high suction pressure and liquid flooding back to the compressor. During the heating cycle, the equipment operates at normal pressures.
Outdoor check valve sticks open during a heating cycle	There is high suction pressure, low superheat, and liquid flooding back to the compressor. The compressor is cool or cold. During the cooling cycle, equipment operates at normal pressures.
Indoor check valve leaks during the cooling cycle	There is a slightly high suction pressure and a slightly low head pressure. Depending upon the rate of the leak, there could be cool or normal temperatures in the compressor. In the heating cycle, the equipment operates at normal pressures.
Outdoor check valve leaks during the heating cycle	There is slightly high suction pressure and slightly low head pressure. The compressor will be near normal temperatures or slightly cool, and the equipment operates at normal pressures during the cooling cycle.

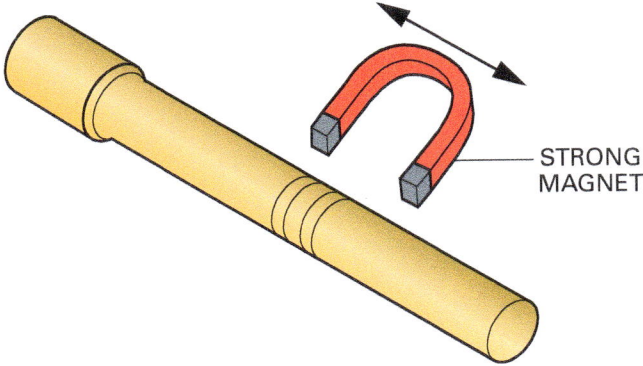

Figure 24 Magnet test.

3.5.0 Troubleshooting Defrost Controls

Most defrost controls encountered by technicians will be solid-state packaged controls that will be replaced if they are determined to be defective. These controls usually rely on a combination of compressor run time and outdoor coil temperature. However, some controls use only the outdoor coil temperature to initiate and terminate defrost. To avoid delays while troubleshooting, most defrost controls provide a speed-up feature that allows the technician to initiate and observe a defrost cycle. Otherwise, having to wait until the system initiates a defrost cycle on its own could take hours. Methods used to initiate the speed-up feature vary from manufacturer to manufacturer, but most are similar.

The most popular method is to use a coin or the tip of a screwdriver blade to jumper across two specific pins or terminals on the board. The two pins must be shorted together continuously for about five seconds, in most cases. In *Figure 25*, the speed-up terminals (as they are commonly called) are labeled TEST. Other boards may label them as SPEED-UP. Once the self-testing feature is initiated, the system enters the defrost mode, allowing the technician to observe the sequence.

If an electromechanical defrost control is used, technicians can also manipulate the unit to start a defrost cycle. The mechanical timer usually has a manual dial that allows a technician to advance the time. The action of the timer to initiate a defrost cycle is usually audible as the internal contacts switch and the cycle begins. The temperature switch is built into these older controls, using a refrigerant-filled remote bulb and a capillary tube connected to the timer. These controls have become rare, however.

It is important to ensure that the defrost controls are properly set up before troubleshooting a problem. Most time/temperature defrost controls allow the installer to select the time between defrosts based on the local climate. For example, in cold areas of consistent freezing fog or drizzle a shorter interval is selected, since frost forms quickly under these conditions. In warmer, drier climates a longer interval is more suitable, since frost accumulates more slowly.

Figure 26 shows a jumper wire permanently attached to the board, which is placed over one of three pins to complete a circuit. The pins on this board are labeled 30, 60, and 90 minutes. Most heat pump units and replacement boards ship with the setting at 30 minutes, as shown here. What is being set is the amount of compressor run time required before a defrost cycle initiates. On warmer days, a heat pump may go for hours between defrost attempts before it accumulates 30 minutes of run time.

The time/temperature defrost method is used in most heat pumps, but other methods may be used. With a time/temperature cycle, the selected period must pass before a defrost cycle can begin. At the end of that period, the system will enter defrost if the coil temperature is below a fixed value. The temperature setting of the coil thermostat or thermistor is usually around 30°F. If the coil is not cold enough to indicate the presence of frost/ice, the unit continues operating in the heating mode, and the period begins again. If the temperature is below the set value, a defrost cycle begins and continues until either the coil temperature rises above the set temperature or a certain amount of time has elapsed (usually 10 minutes).

Other methods for defrost include the use of a pressure switch that monitors the pressure drop across the outdoor coil. As frost builds up on the coil and less air flows through it, the pressure drop increases until it reaches a pre-set point. At that point, the defrost cycle is triggered. Defrost control that is based solely on need without time involved is called a **demand defrost** system.

Another demand defrost strategy is to use two separate thermistors. One monitors the coil temperature, while the other monitors the ambient temperature. When the outdoor temperature is above a preset level (40°F [4°C], for example) a defrost cycle is never initiated. Below that

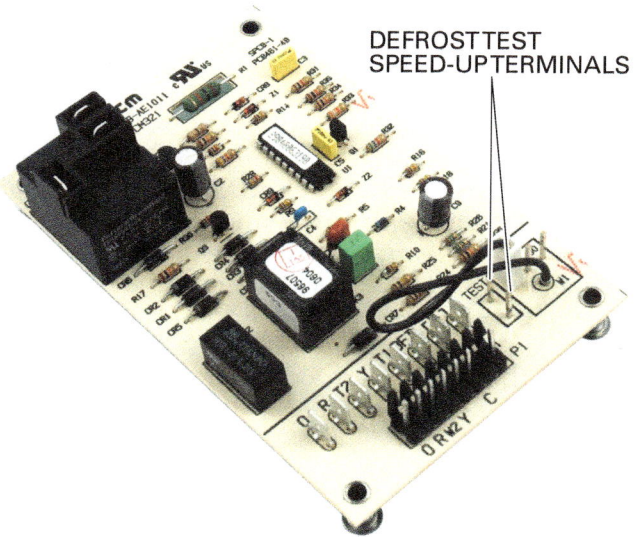

Figure 25 Defrost speed-up terminals.

Figure 26 Compressor run time selection for defrost control.

Wind Baffles

If a heat pump is installed in an area that is subject to high winds—such as a roof-mounted unit—it may be necessary to install baffles around the outdoor coil. Baffles, or wind screens, should be located on the side of the unit from which the prevailing or swirling winds approach, in order to shield the coil from the wind. Wind blowing through a defrosting coil has almost the same effect as leaving the outdoor fan running during defrost, slowing the process considerably.

---

temperature, the coil temperature is monitored and evaluated along with the outdoor temperature. The defrost control board decides when the conditions indicate the likely presence of frost, and then it initiates a defrost cycle.

3.5.1 Checking Inputs and Outputs

All electronic controls, including defrost controls, respond to input signals. If an electronic defrost control has the correct input signals, it responds by generating appropriate output signals or relay contact closures. If it is confirmed that the input signals are correct, but the control is not responding with the correct output signals, there is a good chance the defrost control is defective.

However, if the inputs and outputs from a defrost control are normal and the device being controlled (such as a reversing valve) is not responding, the problem is most likely with the reversing valve or its coil and not the defrost control.

The defrost control manages the various input/output signals using miniature relays that are part of the control. Some typical defrost control board inputs include the following:

- *24V power* – If the defrost control does not have power, it will not operate.
- *Timing input* – With defrost controls that are time/temperature based, the timing function is built into the board.
- *Temperature input (binary/switch closure)* – With time/temperature defrost systems, this input takes the form of an open or closed set of outdoor coil thermostat contacts. Closed contacts provide an input indicating that defrost is needed. Open contacts indicate that the outdoor coil temperature is not low enough to initiate defrost. When those same contacts open *during* defrost, it signifies the outdoor coil temperature is warm enough to terminate defrost. The presence of open/closed contacts can be verified with a voltmeter (power on) or an ohmmeter (power off). Note that the switch will not close at the same temperature that it opens. A differential is built into the temperature switch. For example, a defrost thermostat that closes at 30°F (–1°C), may re-open at 40°F (4.4°C). The same switch is typically used to terminate the defrost cycle as well as initiate it.
- *Temperature input (electronic)* – If a heat pump uses one or more thermistors to sense system temperatures, their input signal(s) would be in the form of a resistance that varies based on the temperature being monitored. Thermistor inputs can be checked with an ohmmeter and the manufacturer's resistance-versus-temperature chart (like the one in *Figure 18*). These temperature

---

Defrost Board Replacement

When replacing a defrost board, it is important to use a board that is designed for the unit being serviced. Boards differ in programming, so using an incorrect board can result in immediate defrost problems. A new board can be ordered from the equipment distributor by heat pump model and serial number to ensure the board features are correct.

Some component manufacturers specialize in building boards that are also specifically designed for certain heat pump models but can be purchased directly from HVACR parts houses. These after-market boards usually perform well and offer competitive pricing. Technicians must be sure that the correct replacement board is selected. The after-market defrost board shown here is designed to replace several specific Trane heat pump boards.

Figure Credit: Courtesy of ICM Controls

inputs are used to terminate the defrost cycle as well as initiate it.

- *Pressure input* – The pressure drop across an outdoor coil (indicating a frosted coil) may be used to initiate and terminate defrost. The pressure switch opens and closes a set of contacts, so the input to the board is binary. The pressure switch is used to terminate the defrost cycle as well as initiate it.

Some typical defrost control output signals include the following:

- *24V power* – The reversing valve solenoid is typically powered by 24VAC. Once defrost is initiated, this output voltage is supplied to the solenoid valve to energize it and removed to terminate defrost. 24V is also provided to energize supplemental electric heaters used to temper the indoor air during defrost. These voltages can typically be measured at output terminals on the defrost control board.
- *Line-voltage power* – The outdoor fan motor is cycled off during defrost. There are usually terminals on the defrost control where the outdoor fan motor is connected, allowing it to control power to the outdoor fan motor. Voltage can be measured at the terminals.

Keep in mind that different manufacturers use varying methods to initiate and terminate defrost. However, they all depend on the defrost control receiving the correct input signals, in whatever form they may be. With the correct input signals, a properly functioning electronic defrost control will deliver the correct output signals.

3.5.2 Defrost Troubleshooting Tips

There are several things that can interfere with the successful completion of a defrost cycle. Even when defrosting is effectively completed, it may create discomfort for the occupants if the supplemental heat does not work. The sequence of operation and construction of heat pumps has varied widely over the years. As always, it is important to review the manufacturer's literature for the specific heat pump model being repaired.

Some common defrost problems and possible solutions are described as follows:

- *The defrost cycle activates but does not readily melt the frost buildup* – The problem could be that the outdoor fan is not being turned off during defrost. If so, check the switch or relay contacts that control the outdoor fan. A strong wind blowing across the outdoor coil could also be a factor, but this would likely be an intermittent problem rather than a consistent one. An incorrect refrigerant charge can also affect defrost performance; check the refrigerant charge according to the manufacturer's instructions.
- *Cold air is being blown into the conditioned space during defrost* – This problem indicates that the electric heater or furnace burner is not turning on when it should. In a system with supplemental electric heat, if the heater works when staged on by the thermostat, it means the problem is likely related to the defrost relay or the defrost control board, depending on the design. If the electric heater does not work in either situation, check the heating element and its associated thermal protection.

  Since dual-fuel systems disable the heat pump at or near the balance point, significant frost buildup is rare. Therefore, defrost cycles are infrequent, and are usually conducted at temperatures near or above freezing. A few dual-fuel systems do allow the furnace burner to operate during defrost if a cycle is initiated. If cold air is blowing into the conditioned space with a dual-fuel system and the burner should be able to use the furnace by design, check to be sure the furnace operates properly with the outdoor temperature below the balance point, or by switching the system to Emergency Heat. If it does, the problem is likely related to the defrost relay, defrost control board, or the interconnecting wiring. If the furnace does not operate in its normal modes, then attention should be focused on the furnace controls.
- *The defrost cycle activates when there is no frost or ice on the coil, or the cycle will not terminate even when all the frost is clearly gone* – With a time/temperature defrost scheme, this proves that there is nothing wrong with the timing cycle or the cycle would never start. This problem is often caused by a failed defrost thermostat that has remained closed. The system will then initiate a defrost cycle at the end of every timing cycle, and the cycle will continue for the full amount of time allowed by the defrost control (usually 10 minutes). A defective thermistor, or a failed or out-of-calibration pressure switch can cause the same symptoms.
- *Ice tends to develop and remain on the lower part of the outdoor coil* – With this problem, the operation of all the defrost controls may appear to be normal. However, it can be caused by problems with the defrost thermostat or thermistors. If they are not in their proper location or not making good surface contact, the defrost control board may terminate the cycle before the task is done. One of the most likely causes, though,

is an incorrect refrigerant charge. If the refrigerant charge is low, the hot refrigerant entering the outdoor coil will condense very quickly, delivering all its latent heat to the top half of the coil. Since the vast majority of the heat that melts the ice is the latent heat from the change of state, there is not enough heat remaining in the refrigerant to thaw the remainder of the coil. When this problem is encountered, check the refrigerant charge first.

Depending on how the outdoor coil is circuited, the lower portion may always be the last part of the coil to thaw. In some cases, it may be that the defrost cycles are simply not occurring frequently enough. If all aspects of defrost cycle operation are normal and the refrigerant charge is correct, try increasing the frequency of the cycles by adjusting the compressor run time on the defrost control board.

> **NOTE**
> Refer to the heat pump troubleshooting guides in the *Appendix* of this module. Become familiar with different troubleshooting guide styles offered by various manufacturers.

Additional Resources

NCCER Module 03211, *Heat Pumps*.

3.0.0 Section Review

1. If a heat pump outdoor unit is located 150' (46 m) from the air handler containing the 24VAC transformer, the control circuit copper wire size should be _____.
   a. 24 AWG
   b. 22 AWG
   c. 18 AWG
   d. 16 AWG

2. Which feature is likely to be found on a digital heat pump thermostat?
   a. Mercury-bulb switch contacts
   b. Subbase function switches
   c. Heat anticipator
   d. Mode selection soft button

3. Thermistors are tested by comparing their _____.
   a. temperature to a resistance value
   b. resistance to a current value
   c. voltage to a resistance value
   d. current to a voltage value

4. Reversing valves can be checked for internal leakage using a(n) _____.
   a. ohmmeter
   b. thermometer
   c. stethoscope
   d. leak detector

5. A defrost control may be responsible for controlling line-voltage power to the _____.
   a. thermostat
   b. indoor blower
   c. outdoor fan motor
   d. compressor contactor

SUMMARY

The control circuits for a heat pump are more complex than those of a cooling-only system because the heat pump performs more functions. Many heat pumps also have supplementary electric heaters or dual-fuel capability, as well as emergency heat functions, making the control circuits seem even more complex. This is often an advantage to a troubleshooter, however; the ability to switch between functions makes it easy to eliminate significant portions of the control circuits when attempting to isolate a problem.

Technicians troubleshooting heat pumps must be familiar with the techniques and tools used to troubleshoot electrical devices and cooling equipment. A thorough knowledge of the electrical sequence of operation is also essential.

Methods used for mechanical troubleshooting are similar to those used for standard cooling systems. However, temperature and pressure readings will be different in the heating mode than in the cooling mode. Because a heat pump is less tolerant of variations in the refrigerant charge, a refrigerant leak or a slight charging error can have a significant impact on the system's performance.

Review Questions

1. If a heat pump works normally in the heating mode but will not provide cooling or a defrost mode, which component is most likely defective?
   a. Compressor
   b. Room thermostat
   c. Reversing valve
   d. Compressor contactor

2. Air-to-air heat pumps produce heat by _____.
   a. circulating indoor air through an electric heating element
   b. circulating low-pressure, low-temperature refrigerant through the indoor coil
   c. extracting heat from the outdoor air and transferring it to the indoor air
   d. circulating hot, high-pressure refrigerant through the outdoor coil

3. In addition to the reversing valve, the other major difference between a heat pump and a cooling-only system is that _____.
   a. the heat pump requires a head pressure control
   b. the cooling unit requires a filter-drier
   c. heat pumps use only fixed-orifice metering devices
   d. heat pumps require two metering devices

4. In the cooling mode, the reversing valve is normally _____.
   a. energized
   b. de-energized
   c. bypassed
   d. back-seated

5. When a heat pump is operating in the heating mode, the refrigerant bypasses the _____.
   a. indoor coil metering device
   b. outdoor coil metering device
   c. indoor coil
   d. outdoor coil

6. In the reverse-cycle heating mode, _____.
   a. the indoor coil acts as an evaporator
   b. the indoor coil acts as a condenser
   c. the outdoor coil acts as a condenser
   d. defrost is only needed in very cold climates

7. Refer to *Figure RQ01*. What happens in the cooling mode if the discharge temperature switch fails to close?
   a. The indoor fan does not run.
   b. The indoor blower relay de-energized.
   c. The outdoor fan does not run.
   d. The contactor will chatter.

8. Refer to *Figure RQ02*. If the outdoor thermostat is not closed and this system is in the heating mode, a likely symptom would be _____.
   a. no compressor operation
   b. less supplemental heat available
   c. no supplemental heat available
   d. no defrost operation

9. Refer to *Figure RQ03*. Defrost is initiated by the _____.
   a. defrost control
   b. outdoor thermostat
   c. defrost relay
   d. room thermostat

10. Which of the following conditions is not common and proper during the defrost mode?
    a. The outdoor fan shuts off.
    b. The indoor fan runs.
    c. An electric heater is energized.
    d. The compressor shuts off.

11. If a heat pump blows cool air into the conditioned space during the colder parts of the heating season, a possible cause is _____.
    a. a low refrigerant charge
    b. that the supplemental electric heaters are not on during defrost
    c. the outdoor fan is not turning off during defrost
    d. the indoor thermostat is defective

12. During dual-fuel heat pump operation, the compressor and furnace burner can operate together _____.
    a. any time there is a call for heat
    b. only if the outdoor temperature is above 40°F (4.4°C)
    c. only if the outdoor temperature is below 40°F (4.4°C)
    d. during periods of defrost

13. Refer to *Figure RQ04*. The variable-speed indoor fan motor receives low-voltage input signals by way of _____.
    a. a 16-pin connector
    b. IFM terminal PL-4
    c. ICM terminal PL-2
    d. the 24VAC splice box

14. If a heat pump will not initiate defrost and you suspect the reversing valve, the quickest way to verify your suspicion is to _____.
    a. see if the reversing valve switches when the operating mode is changed
    b. see if the reversing valve switches when the system is turned off
    c. bypass the reversing valve and see if the unit switches into defrost
    d. wait for the next defrost cycle

15. Low primary voltage at the control circuit transformer may cause _____.
    a. thermostat droop
    b. relay chatter
    c. fuse failure
    d. temperature swings

16. Why is a 24VAC common lead connected to a digital thermostat?
    a. To power the thermostat.
    b. To power the transformer.
    c. To charge the thermostat battery.
    d. To maintain battery polarity.

17. The heat pump component that is controlled through room thermostat terminal O is the _____.
    a. indoor fan relay
    b. compressor contactor
    c. electric heat relay
    d. reversing valve

18. Which of the following is true of a pilot-operated valve?
    a. It is non-magnetic.
    b. The solenoid coil opens the main valve port.
    c. The solenoid coil operates a plunger which opens a smaller valve port.
    d. It must be manually operated.

19. The function of a reversing valve will be affected by all the following, *except* _____.
    a. a defective solenoid
    b. low system refrigerant charge
    c. a crack or dent in the valve body
    d. high superheat settings

20. If the outdoor check valve sticks closed during the cooling cycle, there will be _____.
    a. a high suction pressure and high superheat at the outdoor coil
    b. a low suction pressure and low superheat at the outdoor coil
    c. a low suction pressure and high superheat at the indoor coil
    d. abnormal temperatures and pressures in the heating cycle

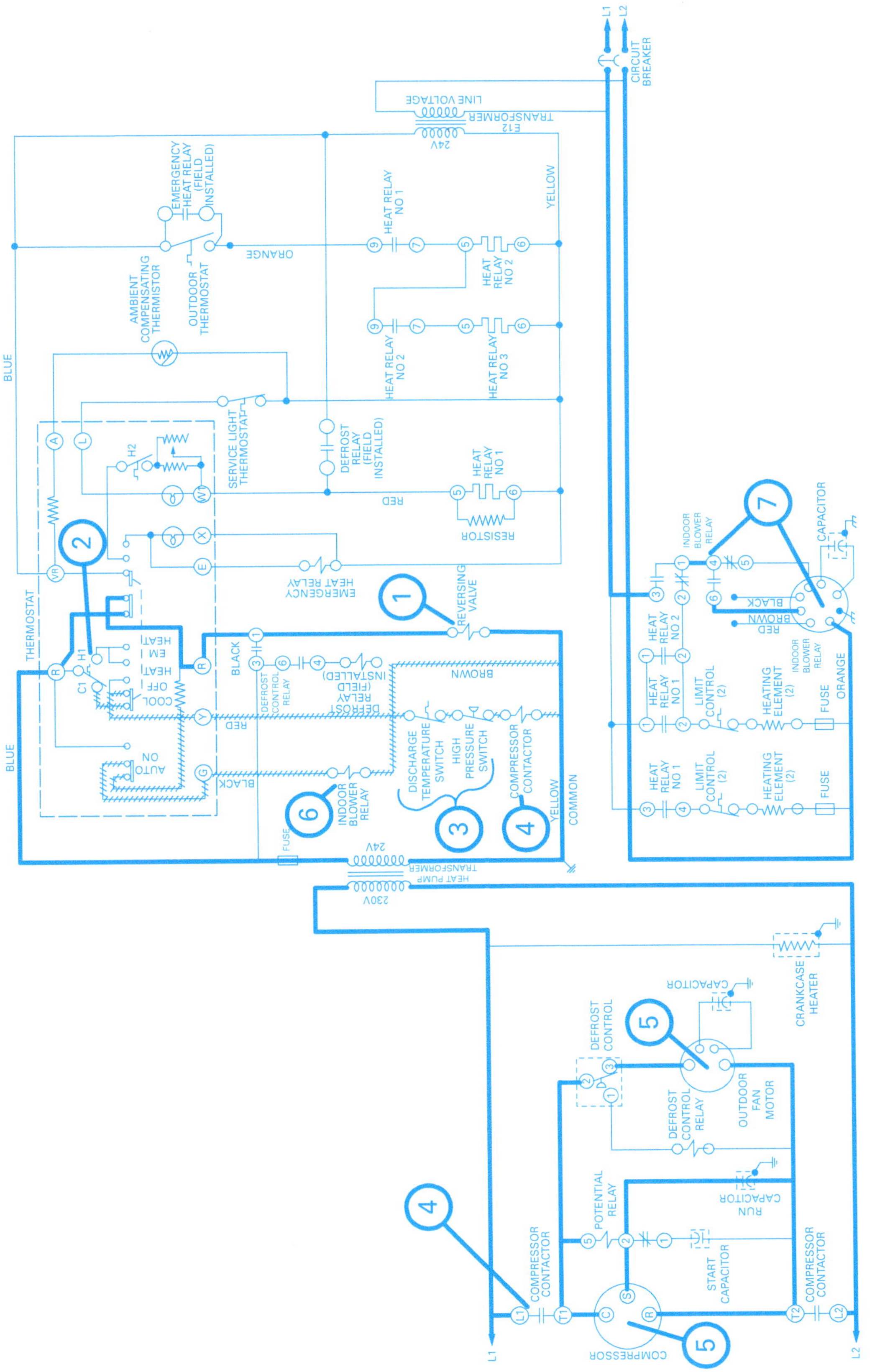
L1
L2
CIRCUIT BREAKER
LINE VOLTAGE
TRANSFORMER E12
24V
EMERGENCY HEAT RELAY (FIELD INSTALLED)
AMBIENT COMPENSATING THERMISTOR
OUTDOOR THERMOSTAT
ORANGE
HEAT RELAY NO 1
HEAT RELAY NO 2
HEAT RELAY NO 3
YELLOW
BLUE
SERVICE LIGHT THERMOSTAT
DEFROST RELAY (FIELD INSTALLED)
RED
RESISTOR
EMERGENCY HEAT RELAY
REVERSING VALVE
THERMOSTAT
BLACK
DEFROST CONTROL RELAY
DEFROST RELAY (FIELD INSTALLED)
BROWN
DISCHARGE TEMPERATURE SWITCH
HIGH PRESSURE SWITCH
COMPRESSOR CONTACTOR
INDOOR BLOWER RELAY
COMMON
FUSE
HEAT PUMP TRANSFORMER
230V
INDOOR BLOWER RELAY
CAPACITOR
LIMIT CONTROL (2)
HEATING ELEMENT (2)
CRANKCASE HEATER
DEFROST CONTROL
OUTDOOR FAN MOTOR
POTENTIAL RELAY
RUN CAPACITOR
START CAPACITOR
COMPRESSOR

Figure RQ01

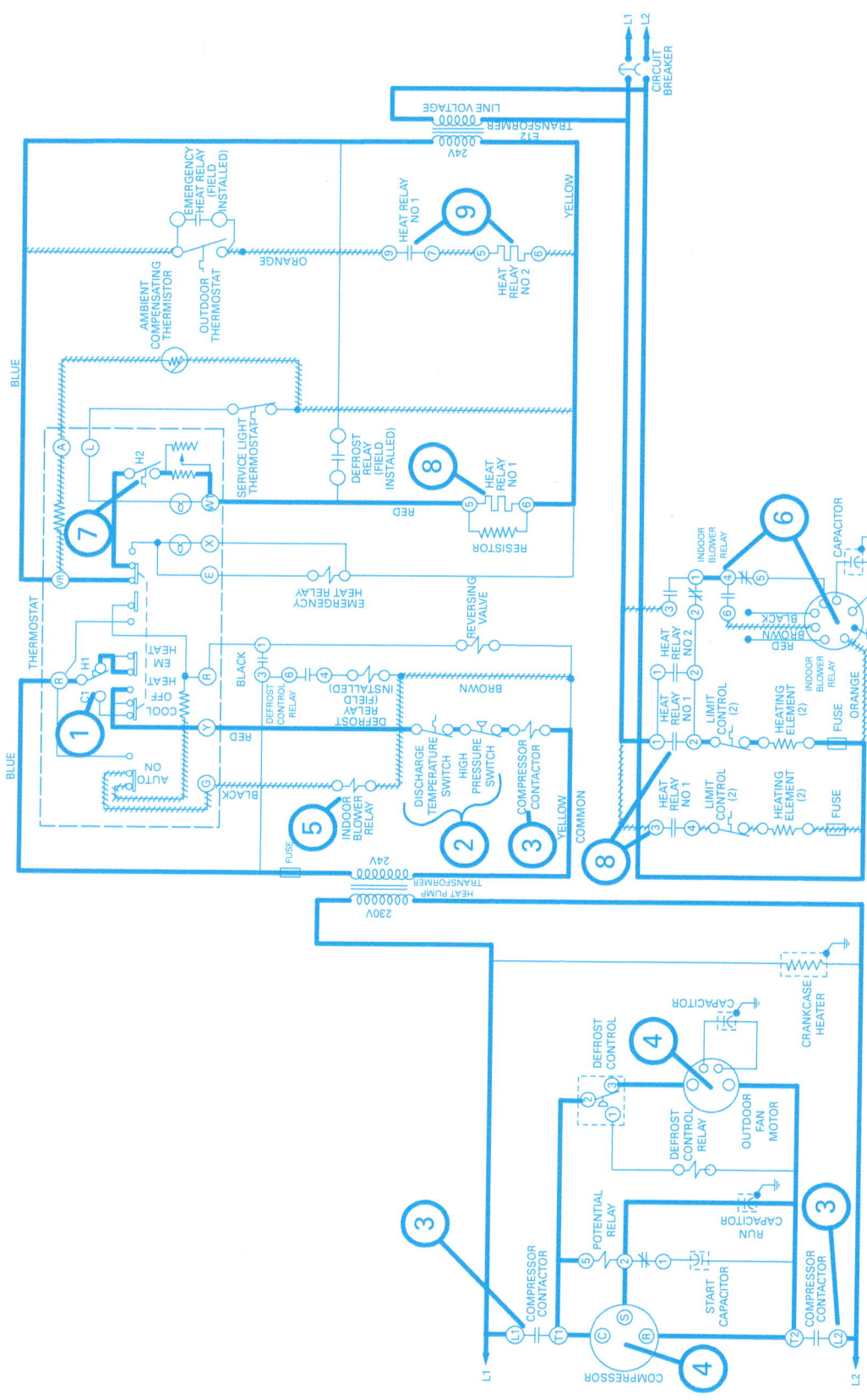

Figure RQ02

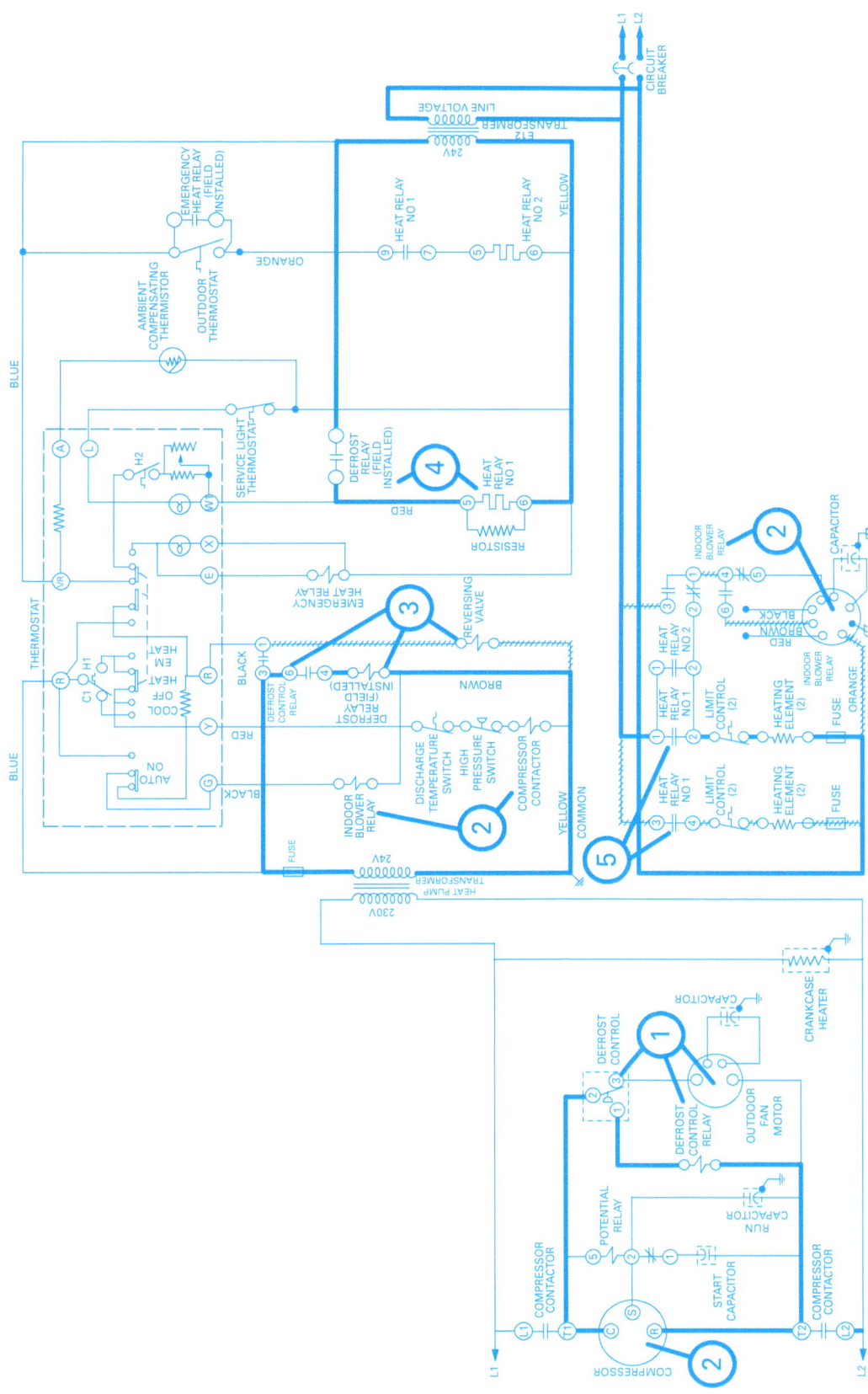

Figure RQ03

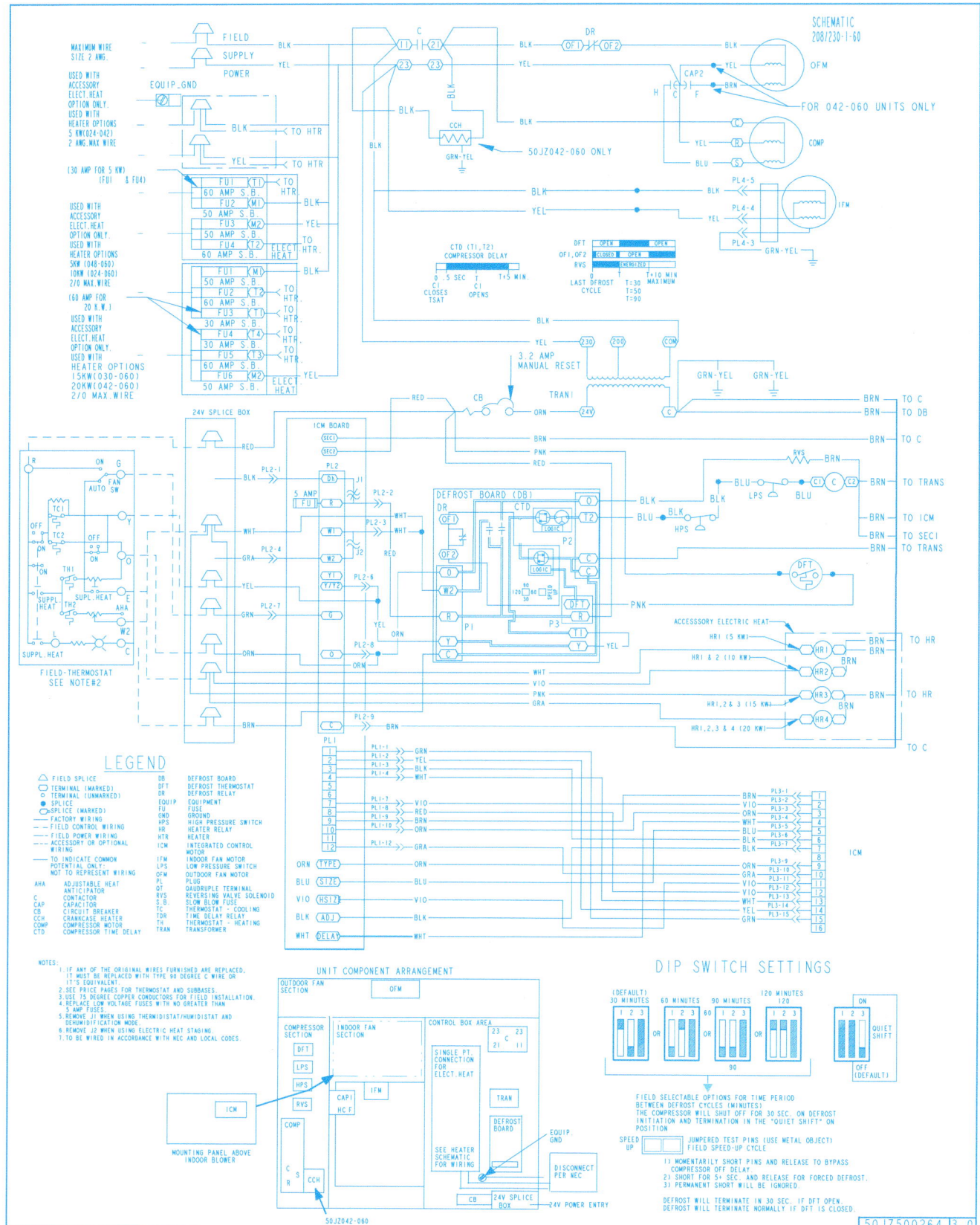

Figure RQ04

Trade Terms Introduced in This Module

Demand defrost: A defrost control strategy that typically does not depend on time to determine when a defrost cycle is needed. Instead, the system responds directly to current conditions, such as temperature or pressure drop through the outdoor coil, to determine when frost is present.

Direct-acting valve: A valve in which the solenoid coil operates the valve stem directly to open or close.

Electronically commutated motor (ECM): A variable-speed electric motor based primarily on a DC-driven permanent magnet design that responds to microprocessor inputs to achieve the desired speed.

Field wiring: Wiring that is installed during installation of the system, such as a thermostat hookup to a unit.

Pilot-operated valve: A valve in which a solenoid operates a smaller valve, which then ports more pressure and/or volume to a larger valve to open or close it. Large valves use this strategy due to the limitations in the magnetic field of solenoid coils.

Appendix A

SAMPLE TROUBLESHOOTING CHART

GENERALLY THE CAUSE Make these checks first.	**OCCASIONALLY THE CAUSE** Make these checks only if the first checks failed to locate the trouble.	**RARELY THE CAUSE** Make these checks only if other checks failed to locate the trouble.
PROBLEM: Compressor and outdoor fan motor do not start		
Power failure Blown fuse Faulty wiring – Motor to line side of contactor – Load side, contactor to motor terminal – Control circuit Loose terminals – Motor to line side of contactor – Load side, contactor to motor terminal – Control circuit Compressor overload Contactor coil Pressure control – Internal overload in compressor Refrigerant charge low High head pressure Low suction pressure Run capacitor Start capacitor – Compressor motor – Compressor stuck	Low voltage – Motor to line side of contactor – Load side, contactor to motor terminal – Control circuit – Defective contacts in contactor Control circuit thermostat	Unbalanced power supply Outdoor fan motor Indoor fan motor Refrigerant overcharge Noncondensibles Defective power element – Outdoor expansion valve – Indoor expansion valve Thermostat setting
PROBLEM: Outdoor fan motor will not start		
Faulty wiring – Load side, contactor to motor terminal Loose terminals – Load side, contactor to motor terminal Defrost relay Outdoor fan motor Outdoor fan motor capacitor	Defective control, timer, or relay	
PROBLEM: Compressor hums but will not start		
Faulty wiring – Motor to line side of contactor – Load side, contactor to motor terminal Loose terminals – Motor to line side of contactor – Load side, contactor to motor terminal Compressor overload Potential relay Start capacitor	Unbalanced power supply Low voltage – Load side, contactor to motor terminal Defective contacts in contactor Compressor motor Compressor bearing defective Compressor stuck	Refrigerant overcharge High head pressure Noncondensibles Restrictions

GENERALLY THE CAUSE Make these checks first.	**OCCASIONALLY THE CAUSE** Make these checks only if the first checks failed to locate the trouble.	**RARELY THE CAUSE** Make these checks only if other checks failed to locate the trouble.
PROBLEM: Compressor cycles on overload		
Faulty wiring – Motor to line side of contactor – Load side, contactor to motor terminal Loose terminals – Motor to line side of contactor – Load side, contactor to motor terminal Low voltage – Motor to line side of contactor – Load side, contactor to motor terminal Unbalanced power supply Potential relay Run capacitor Start capacitor	Defective contacts in contactor Compressor overload Compressor motor Refrigerant charge low Refrigerant overcharge High head pressure High suction pressure Noncondensibles Excessive load cooling Defective reversing valve	Compressor valve defective Compressor oil level Fins dirty or plugged – Indoor section of system – Outdoor section of system Fan belt slipping Coil air short circuiting Air volume low (for cooling) Blower belt slipping Air volume low Air filters dirty
PROBLEM: Compressor off or cycling on low-pressure control		
Refrigerant charge low Low suction pressure Defective power element Push rod packing leak – Outdoor expansion valve Indoor fan motor Outdoor fan motor Air volume low Air filters dirty Small or restricted ductwork Restrictions	Indoor fan relay Indoor fan motor Sensing element loose or poorly located Cycle too long (clock timer) Defective power element Fins dirty or plugged Fan belt slipping Low-temperature coil air (for cooling)	Pressure control Out of adjustment (de-ice control) Dirty internally Push rod packing leak – Indoor expansion valve Coil air short circuiting Air stratification in conditioned space
PROBLEM: Compressor off on high-pressure control		
Defrost relay Outdoor fan motor Indoor fan relay Indoor fan motor Refrigerant overcharge Cycle too long (clock timer) Coil Defective control timer or relay Fins dirty or plugged Fan belt slipping Coil air short circuiting Blower belt slipping Air volume low Air filters dirty	Out of adjustment (de-ice control) Defective reversing valve Air volume low (for cooling) Fins dirty or plugged Small or restricted ductwork	Pressure control Temperatures Noncondensibles Air stratification in conditioned space Auxiliary heat upstream from indoor fan

GENERALLY THE CAUSE Make these checks first.	OCCASIONALLY THE CAUSE Make these checks only if the first checks failed to locate the trouble.	RARELY THE CAUSE Make these checks only if other checks failed to locate the trouble.
PROBLEM: Compressor noisy		
Hold down bolts Compressor valves defective Compressor oil level Refrigerant overcharge	Run capacitor Leaking or defective check valve – Indoor section of system – Outdoor section of system Superheat setting TXV stuck open Loose thermal bulb – Indoor section of system	Compressor bearings defective
PROBLEM: Compressor loses oil		
Refrigerant leak Low suction pressure Refrigeration piping incorrect	TXV stuck open Restrictions	Superheat setting Refrigeration lines too long
PROBLEM: Unit runs in heat cycle—pumps down in cooling or defrost cycle		
Defective reversing valve Defective power element Restrictions	Check valve sticking closed Dirty internally	Push rod packing leak – Indoor section of system
PROBLEM: Unit runs normal in one cycle—high suction pressure in other cycle		
Leaking reversing valve Leaking or defective check valve – Outdoor section of system Loose thermal bulb – Outdoor section of system	Superheat setting Leaking or defective check valve – Indoor section of system Loose thermal bulb – Indoor section of system	TXV stuck open – Outdoor section of system – Indoor section of system
PROBLEM: Head pressure too high		
Refrigerant overcharge Temperatures Noncondensibles Fins dirty or plugged Fan belt slipping Air volume too low (for cooling) – Outdoor coil Blower belt slipping Air volume too low – Indoor coil Air filters dirty	Excessive load cooling Coil air short circuiting Fins dirty or plugged Refrigeration piping incorrect	Compressor oil level Small or restricted ductwork Thermostat setting
PROBLEM: Head pressure too low		
Refrigerant charge low Low suction pressure Leaking reversing valve Low-temperature coil air (for cooling)	Compressor valves defective	

GENERALLY THE CAUSE Make these checks first.	OCCASIONALLY THE CAUSE Make these checks only if the first checks failed to locate the trouble.	RARELY THE CAUSE Make these checks only if other checks failed to locate the trouble.
PROBLEM: Suction pressure too high		
Compressor valves defective High head pressure Excessive load cooling Reversing valve leaking	Refrigerant overcharge Temperatures Leaking or defective check valve – Indoor section of system – Outdoor section of system Loose thermal bulb – Outdoor section of system TXV stuck open – Indoor section of system	TXV stuck open – Outdoor section of system Superheat setting Loose thermal bulb – Indoor section of system
PROBLEM: Suction pressure too low		
Refrigerant charge low Fan belt slipping Blower belt slipping Air volume low Air filters dirty Indoor fan motor Indoor fan relay	Temperatures Leaking or defective check valve – Outdoor section of system – Indoor section of system Superheat setting Dirty internally Defective power element Fins dirty or plugged Coil air short circuiting Small or restricted ductwork Check valve sticking closed Restrictions	Low head pressure Push rod packing leak – Outdoor section of system – Indoor section of system Air stratification in conditioned space Refrigeration piping incorrect
PROBLEM: Blower will not start		
Power failure Blown fuse Run capacitor Faulty wiring – Motor to line side of contactor – Load side of contactor to motor – Control circuit Indoor fan motor	Indoor fan relay	Low voltage – Motor to line side of contactor – Load side, contactor to motor terminal – Control circuit Thermostat
PROBLEM: Indoor coil frosting or icing		
Low suction pressure Blower belt slipping Air volume low Air filters dirty – Indoor section of system Indoor fan motor Indoor relay Small or restricted ductwork	Refrigerant charge low Temperatures Reversing valve sticking closed Dirty internally	

 03311 Troubleshooting Heat Pumps   Module Four   47

GENERALLY THE CAUSE Make these checks first.	OCCASIONALLY THE CAUSE Make these checks only if the first checks failed to locate the trouble.	RARELY THE CAUSE Make these checks only if other checks failed to locate the trouble
PROBLEM: Compressor runs continuously — no cooling		
Compressor valve defective Refrigerant charge low Reversing valve leaking Reversing valve defective Fins dirty or plugged (outdoor) Blower belt slipping Air volume low Air filters dirty Small or restricted ductwork – Indoor section of system	High suction pressure Noncondensibles Superheat setting Dirty internally Push rod packing leak – Indoor section of system Restrictions	Fins dirty or plugged (indoor) Coil air short circuiting Air volume low (for cooling) Low-temperature coil air Defective power element Check valve sticking closed – Indoor section of system Check valve leaking or defective Air stratification in conditioned space
PROBLEM: Compressor runs continuously — cooling		
Faulty wiring (control circuit) Thermostat (control circuit)	Thermostat location	Thermostat setting Air duct not insulated
PROBLEM: Liquid refrigerant flooding back to compressor in cooling cycle — TXV system		
Superheat setting – Indoor section of system Loose thermal bulb – Indoor section of system	Refrigerant overcharge Temperatures TXV stuck open – Indoor section of system Check valve leaking or defective – Indoor section of system	Fins dirty or plugged Blower belt slipping Air volume low Air filters dirty Small or restricted ductwork Air stratification in conditioned space
PROBLEM: Liquid refrigerant flooding back to compressor in cooling cycle — cap tube system		
Refrigerant overcharge Indoor fan motor Indoor fan relay Blower belt slipping Air volume low Air filters dirty	High head pressure Fins dirty or plugged Small or restricted ductwork Check valve leaking or defective	Temperatures Air stratification in conditioned space
PROBLEM: Compressor runs continuously — no heating		
Refrigerant charge low Reversing valve leaking Contactor stuck	Defective control, timer, or relay	Superheat setting Dirty internally Push rod packing leak – Outdoor section of system Air stratification in conditioned space
PROBLEM: Compressor runs continuously — heating		
Faulty wiring (control circuit) Thermostat Thermostat location Contactor stuck	Thermostat setting	Air duct not insulated

GENERALLY THE CAUSE Make these checks first.	**OCCASIONALLY THE CAUSE** Make these checks only if the first checks failed to locate the trouble.	**RARELY THE CAUSE** Make these checks only if other checks failed to locate the trouble.
PROBLEM: Unit runs in cooling cycle—pumps down in heating cycle		
Defective reversing valve Defective power element	Dirty internally Fins dirty or plugged Fan belt slipping Coil air short circuiting Check valve sticking closed Restrictions	Push rod packing leak – Outdoor section of system
PROBLEM: Unit cycles on low-pressure control when switching cycle or at end of defrost cycle		
Defective low pressure control Refrigerant charge low Defective reversing valve Defective power element Push rod packing leak – Outdoor section of system – Indoor section of system		
PROBLEM: Defrost cycle initiates—no ice on coil		
Refrigerant charge low De-ice control out of adjustment Sensing element loose or poorly located Defective control, timer, or relay	Faulty wiring (control circuit) Fins dirty or plugged Fan belt slipping Coil air short circuiting Air volume low (for cooling)	Cycle too long (clock timer) Defective reversing valve Superheat setting
PROBLEM: Defrost cycle initiates but won't terminate defrost relay		
Defrost relay Refrigerant charge low De-ice control out of adjustment Defective control, timer, or relay	Faulty wiring (control circuit) Compressor valves defective Reversing valve defective	
PROBLEM: Reversing valve will not shift		
Faulty wiring (control circuit) Loose terminal (control circuit) Defective reversing valve Refrigerant charge	Low voltage (control circuit) Defrost relay Compressor stuck	
PROBLEM: Ice buildup on lower part of outdoor coil		
Defrost relay Refrigerant charge low De-ice control out of adjustment Sensing element loose or poorly located	Compressor valves defective Cycle too long (clock timer) Reversing valve leaking Reversing valve defective Superheat setting – Outdoor section of system	Loose thermal bulb – Outdoor section of system
PROBLEM: Liquid refrigerant flooding back to compressor in heating cycle—TXV system		
Superheat setting – Outdoor section of system Loose thermal bulb – Outdoor section of system	Leaking or defective check valve TXV stuck open – Outdoor section of system	Refrigerant overcharge Temperatures

03311 Troubleshooting Heat Pumps

Module Four 49

GENERALLY THE CAUSE Make these checks first.	OCCASIONALLY THE CAUSE Make these checks only if the first checks failed to locate the trouble.	RARELY THE CAUSE Make these checks only if other checks failed to locate the trouble.
PROBLEM: Liquid refrigerant flooding back to compressor in heating cycle — cap tube system		
Refrigerant overcharge High head pressure	Leaking or defective check valve	Temperatures
PROBLEM: Auxiliary heat on — indoor blower off		
Faulty wiring – Load side, contactor to motor terminal – Control circuit Loose terminals – Load side, contactor to motor terminal – Control circuit Fan relay Indoor fan motor	Thermostat	
PROBLEM: Excessive operating costs		
Compressor stuck Refrigerant charge low De-ice control out of adjustment Reversing valve leaking Reversing valve defective Blower belt slipping Air volume low Air filters dirty Air duct not insulated No outdoor thermostat Outdoor thermostat adjustment	Refrigerant overcharge Fins dirty or plugged Fan belt slipping Coil air short circuiting Thermostat location Air stratification in conditioned space Refrigeration piping incorrect Unit incorrect size for application	Faulty wiring – Motor to line side of contactor Loose terminals – Motor to line side of contactor Noncondensibles Superheat setting TXV stuck open Dirty internally Defective power element Thermostat setting
PROBLEM: Unit short cycles on defrost control		
Refrigerant charge low De-ice control out of adjustment Defective control, timer, or relay Fan belt slipping	Defective power element Fins dirty or plugged	
PROBLEM: Outdoor blower does not stop in defrost cycle		
Fan relay IOD or ID section Defrost relay		

Appendix B

SAMPLE TROUBLESHOOTING GUIDE

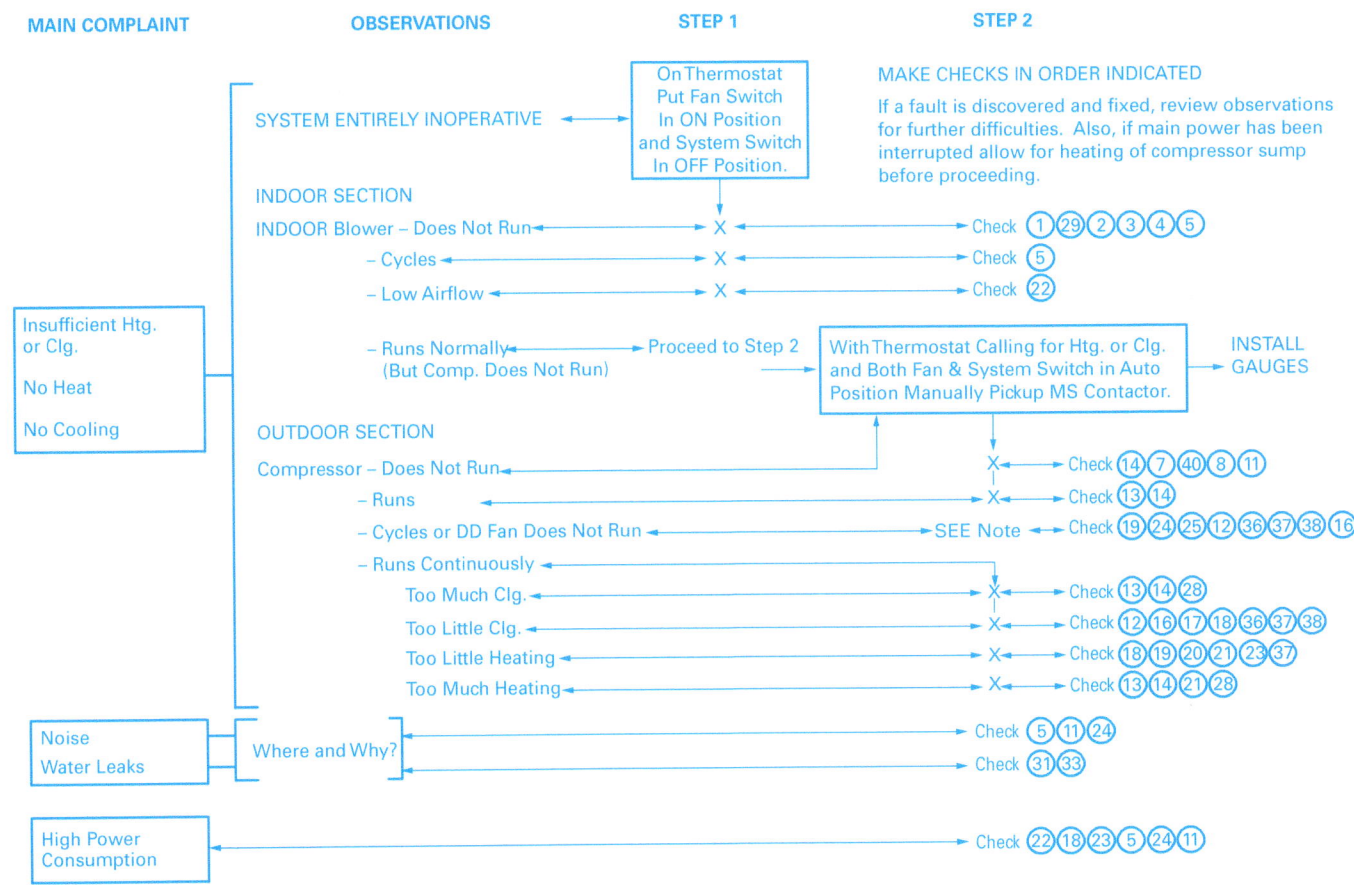

Appendix C

HVACR Common Wiring Diagram Abbreviations

AH	Supplementary Heater	HVTB	High-Voltage Terminal Board
AR	Auxiliary Relay	IOL	Internal Overload Protector
BH	Supplementary Heater Contactor	1K	Economizer Relay
BHA	Supplementary Heater Contactor – Auxiliary	2K	Compressor Lock-out Relay
BLR	Balancing Relay	LA	Lock-in Relay
CA	Cooling Anticipator	LAS	Low Airflow Sensing
CB	Circuit Breaker	LB	Lock-in Relay
CF	Fan Capacitor	LPCO	Low-Pressure Cutout
CFA	Outdoor Fan Capacitor	LT	Light
CH	Supplementary Htr. Contactor	LTS	Low-Temperature Switch
CL	Contactor Relay	LVTB	Low-Voltage Terminal Board
CN	Wire Connector	MAC	Mixed Air Control
COC	Changeover Control	MPM	Motor Protection Module
CR	Run Capacitor	MPP	Min. Position Potentiometer
CPR	Compressor	MPS	Motor Protection Sensor
CSR	Capacitor Switching Relay	MS	Compressor Motor Contactor
D	Defrost Relay	MSA	Compressor Motor Contactor – Auxiliary
DFC	Defrost Control	MTR	Motor
DFT	Defrost Timer	ODA	Outdoor Temperature Anticipator
DH	Supplementary Htr. Contactor	ODF	Outdoor Fan Relay
DR	Defrost Relay	ODS	Outdoor Temperature Sensor
DS	Diaphragm Switch	ODT	Outdoor Thermostat
DT	Defrost Termination Switch	OFT	Outdoor Fan Thermostat
EDC	Evaporator Defrost Control	RH	Emergency Heat Relay
EDR	Evaporator Defrost Relay	RHS	Resistance Heat Switch
ETB	Economizer Terminal Board	S-FU	1/16 Amp Fuse
ER	Electric Heater Relay	SEN	Sensor
F	Indoor Fan Relay	SC	Switchover Valve Solenoid
FL	Flexible Link	SM	System Switch (Room Thermostat)
FP	Feed-Back Potentiometer	T	Thermistor
FR	Contactor Relay	TCO	Temperature Limit Switch
FA	Contactor Relay	TDL	Discharge Line Thermostat
FM	Manual Fan Switch	TDR	Time Delay Relay
FS	Fan Signal Relay	TM	Compressor Motor Thermostat
FST	Fan Switch Terminal	TNS	Transformer
FTB	Fan Terminal Board	TS	Heating & Cooling Thermostat
FU	Fuse	TSC	Cooling Thermostat
FUA	Compressor Fuse	TSH	Heating Thermostat
FUB	Transformer Fuse		
H	Heating Relay		
HA	Heating Anticipator		
HPCO	High-Pressure Cutout		
HTR	Heater		

Additional Resources

This module presents thorough resources for task training. The following resource material is suggested for further study.

NCCER Module 03211, *Heat Pumps*.

Figure Credits

© Verena Matthew/Dreamstime, Module opener
Jason Pearce of www.jasonpearce.com, Figure 4
Carrier Corporation, Figure 10, Review Questions Figure 4
www.thedealertoolbox.com, SA02
Courtesy of Emerson, Figure 12
King Innovation®, SA03
Courtesy of Honeywell International, Figure 14
Courtesy of Emerson Climate Technologies, Figure 15
Sterling Sensors Ltd., Figure 17
Courtesy of Invensys Controls, Figure 19
Courtesy of ICM Controls, Figures 25, 26, SA05

Section Review Answer Key

Answer	Section Reference	Objective
Section One		
1. b	1.1.1	1a
2. d	1.2.3	1b
Section Two		
1. c	2.1.0	2a
2. b	2.2.0	2b
3. b	2.3.0	2c
4. d	2.4.0	2d
5. b	2.5.0	2e
Section Three		
1. d	3.1.1	3a
2. d	3.2.0	3b
3. a	3.3.0	3c
4. b	3.4.2	3d
5. c	3.5.1	3e

NCCER CURRICULA — USER UPDATE

NCCER makes every effort to keep its textbooks up-to-date and free of technical errors. We appreciate your help in this process. If you find an error, a typographical mistake, or an inaccuracy in NCCER's curricula, please fill out this form (or a photocopy), or complete the online form at **www.nccer.org/olf**. Be sure to include the exact module ID number, page number, a detailed description, and your recommended correction. Your input will be brought to the attention of the Authoring Team. Thank you for your assistance.

Instructors – If you have an idea for improving this textbook, or have found that additional materials were necessary to teach this module effectively, please let us know so that we may present your suggestions to the Authoring Team.

NCCER Product Development and Revision
13614 Progress Blvd., Alachua, FL 32615

Email: curriculum@nccer.org
Online: www.nccer.org/olf

❏ Trainee Guide ❏ Lesson Plans ❏ Exam ❏ PowerPoints   Other _____

Craft / Level: _____   Copyright Date: _____

Module ID Number / Title: _____

Section Number(s): _____

Description: _____

_____

_____

_____

_____

Recommended Correction: _____

_____

_____

_____

_____

Your Name: _____

Address: _____

_____

Email: _____   Phone: _____

Troubleshooting Gas Heating

OVERVIEW

Gas furnaces and other gas heating devices utilize a variety of controls. Most of these controls are safety devices that either shut the system down in the event of a problem or prevent it from operating if certain conditions are not met in a specified sequence. Service technicians must understand how these devices interact to control the system. They must also learn to recognize combustion-related problems, and be able to properly test the combustion system using specialized test instruments.

Module 03209

Trainees with successful module completions may be eligible for credentialing through the NCCER Registry. To learn more, go to **www.nccer.org** or contact us at 1.888.622.3720. Our website has information on the latest product releases and training, as well as online versions of our *Cornerstone* magazine and Pearson's product catalog.

Your feedback is welcome. You may email your comments to **curriculum@nccer.org**, send general comments and inquiries to **info@nccer.org**, or fill in the User Update form at the back of this module.

This information is general in nature and intended for training purposes only. Actual performance of activities described in this manual requires compliance with all applicable operating, service, maintenance, and safety procedures under the direction of qualified personnel. References in this manual to patented or proprietary devices do not constitute a recommendation of their use.

Copyright © 2018 by NCCER, Alachua, FL 32615, and published by Pearson, New York, NY 10013. All rights reserved. Printed in the United States of America. This publication is protected by copyright, and permission should be obtained from NCCER prior to any prohibited reproduction, storage in a retrieval system, or transmission in any form or by any means, electronic, mechanical, photocopying, recording, or likewise. To obtain permission(s) to use material from this work, please submit a written request to NCCER Product Development, 13614 Progress Blvd., Alachua, FL 32615.

03209 V5

From *HVAC Level Three, Trainee Guide*. NCCER.
Copyright © 2018 by NCCER. Published by Pearson. All rights reserved.

03209
TROUBLESHOOTING GAS HEATING

Objectives

When you have completed this module, you will be able to do the following:

1. Describe how to troubleshoot the components related to gas heating.
    a. Describe the control circuits and typical sequence of operation of various gas heating units.
    b. Describe the operation and troubleshooting process for ignition devices.
    c. Describe the operation and troubleshooting process for flame sensors.
    d. Identify common problems associated with system airflow.
2. Identify infrared gas heaters and describe how they operate.
    a. Identify various types of infrared gas heaters.
    b. Describe the operating characteristics of infrared gas heaters.
3. Explain how to conduct a combustion analysis on a gas furnace.
    a. Identify combustion analysis equipment and the combustion byproducts that are of importance to the analysis.
    b. Describe the combustion analysis process and how to interpret basic results.

Performance Tasks

Under the supervision of your instructor, you should be able to do the following:

1. Using the proper tools, instruments, and control circuit diagrams, isolate and correct malfunctions in a gas heating system.
2. Complete a combustion analysis on a gas furnace or boiler.

Trade Terms

Flame rectification
Modulating gas valve
Net stack temperature

Reverberator
Submerged flame

Industry Recognized Credentials

If you are training through an NCCER-accredited sponsor, you may be eligible for credentials from NCCER's Registry. The ID number for this module is 03209. Note that this module may have been used in other NCCER curricula and may apply to other level completions. Contact NCCER's Registry at 888.622.3720 or go to **www.nccer.org** for more information.

Contents

1.0.0 Troubleshooting Gas Heating Components 1
  1.1.0 Gas Heating Control Circuits 1
    1.1.1 Electronic Controls in Heating Systems 2
    1.1.2 Natural-Draft Furnaces 6
    1.1.3 Induced-Draft Furnaces 6
    1.1.4 Condensing Furnaces 10
    1.1.5 Gas-Fired Boiler Operation 11
  1.2.0 Troubleshooting Pilot Re-Ignition and Direct Ignition Devices 13
    1.2.1 Spark Igniters 13
    1.2.2 Hot-Surface Igniters 14
  1.3.0 Troubleshooting the Flame Sensor 14
    1.3.1 Testing a Combined HSI/Flame Sensor 15
  1.4.0 Airflow-Related Problems 16
    1.4.1 Air Distribution System and Blowers 16
    1.4.2 Induced-Draft System 17
2.0.0 Infrared Gas-Fired Heaters 19
  2.1.0 Infrared Heater Types 19
    2.1.1 Low-Intensity Infrared Gas-Fired Heaters 19
    2.1.2 High-Intensity Infrared Gas-Fired Heaters 19
  2.2.0 Infrared Heater Operation 20
3.0.0 Gas Combustion Analysis 24
  3.1.0 Gas Combustion Byproducts 24
    3.1.1 Gas Combustion Analysis Equipment 26
  3.2.0 Conducting a Combustion Analysis 26
    3.2.1 Measuring Flue Gas Temperature 27
    3.2.2 Sampling Flue Gas 28
    3.2.3 Analyzing the Combustion Efficiency Test 28
    3.2.4 Other Combustion Efficiency Information 28

Figures

Figure 1 Gas furnace ignition devices 2
Figure 2 Gas unit control board 3
Figure 3 Gas furnace control board and wiring schematic 4
Figure 4 Gas furnace controllers 5
Figure 5 Natural-draft gas heating system 7
Figure 6 Field wiring diagram 8
Figure 7 AFUE regional map for the continental United States 9
Figure 8 Typical 80-percent AFUE induced-draft gas furnace 10
Figure 9 Typical 90-percent AFUE condensing gas furnace 11
Figure 10 Gas-fired boiler control circuit 12
Figure 11 Checking HSI source voltage 15
Figure 12 Flame sensor 15
Figure 13 Flame sensor test 15
Figure 14 Testing a combined HSI/flame sensor 16
Figure 15 Low-intensity infrared tubular heaters 20
Figure 16 High-intensity infrared heater 20
Figure 17 Multi-zone control panel 20
Figure 18 Low-intensity infrared heater wiring diagram 22
Figure 19 Byproducts of combustion 25
Figure 20 Combustion analyzer 26
Figure 21 Sampling hole locations 27
Figure 22 Electronic combustion-analyzer display 28
Figure 23 Combustion-efficiency calculator chart 28

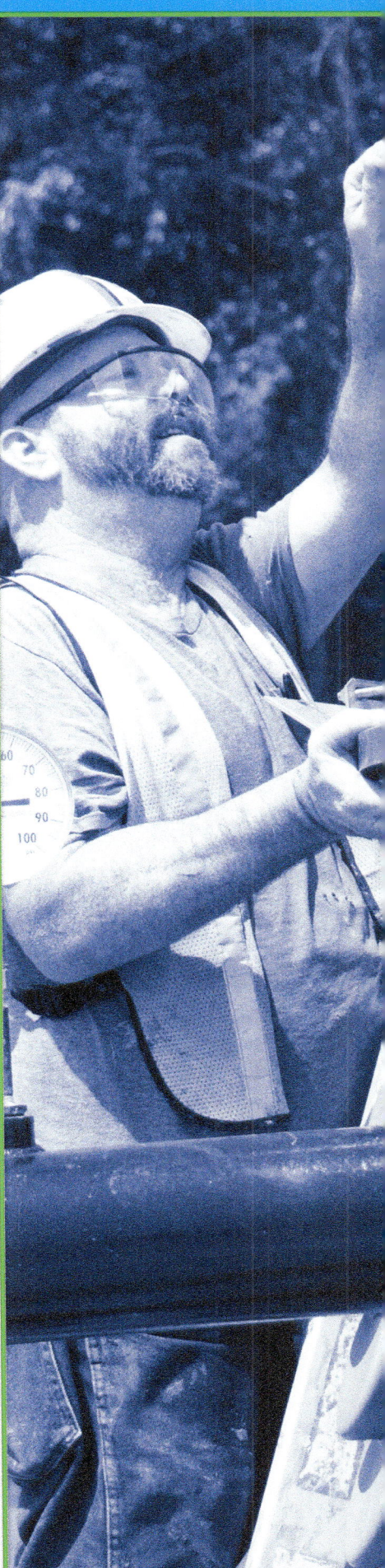

Section One

1.0.0 Troubleshooting Gas Heating Components

Objective

Describe how to troubleshoot the components related to gas heating.

a. Describe the control circuits and typical sequence of operation of various gas heating units.
b. Describe the operation and troubleshooting process for ignition devices.
c. Describe the operation and troubleshooting process for flame sensors.
d. Identify common problems associated with system airflow.

Performance Task

1. Using the proper tools, instruments, and control circuit diagrams, isolate and correct malfunctions in a gas heating system.

Trade Terms

Flame rectification: A process in which exposure of a sensing rod to a flame produces a tiny direct current which can be used to control the gas valve.

Modulating gas valve: A gas valve that precisely controls burner capacity by adjusting the flow of gas to burners in small increments.

To properly troubleshoot a gas heating system, the technician must consider three separate but related subsystems: controls, air or water flow, and gas combustion. The ability to read circuit diagrams, use electrical test equipment, recognize and correct combustion-related problems, and measure and adjust airflow are all necessary skills for troubleshooting gas-fired heating equipment effectively.

1.1.0 Gas Heating Control Circuits

Control circuits for gas-fired systems come in many varieties. It is rare to find two furnace models that are precisely the same electrically, even if they are made by the same company. Although packaged electronic controls have allowed manufacturers to standardize control circuits, variations from model to model will be found, if only in the programming of the microprocessor. Manufacturers often make small changes from one furnace series to another.

This module provides examples of circuit diagrams from different furnace manufacturers, along with a description of their operating sequences. As you study these circuits to reinforce and expand your knowledge of furnace controls, it will help you to analyze the control circuit of an unfamiliar product to locate the cause of a malfunction. Note that these example circuits represent only a few of the circuits found in furnaces today. It is important to develop the skill of evaluating all circuits, not just those represented by these examples.

In addition to gas furnaces and boilers, gas heating can also be provided by packaged heating and cooling units. One major difference is that packaged units are installed outdoors and need no vent piping. The control circuits of packaged systems and gas furnaces are roughly the same.

If a gas heating system is not working at all, the ignition system and/or gas valve may be malfunctioning. In older furnaces with standing pilots, thermocouples are used to indicate the presence of a pilot flame to the main gas valve. Thermocouples develop a voltage in the range of 30mV from the heat of the pilot flame. Modern furnaces use **flame rectification** with intermittent-pilot spark igniters or direct-ignition devices such as hot-surface igniters (HSIs) (*Figure 1*). To prove a flame is present, flame rectifiers typically develop a small current from the heat of the flame, rather than a voltage like a thermocouple.

Indoor Comfort

What is indoor comfort? To most consumers, comfort means hot air coming from the registers in winter and cold air coming from them in summer. True indoor comfort is more than just hot or cold. Today's equipment manufacturers are taking advantage of new technologies that allow them to provide improved temperature and humidity control, improved air quality, and energy efficiency levels that were only dreamed of a few years ago. These include the widespread use of electronic controls, capacity control using staged gas valves and new compressor technology, and the use of variable-speed blower motors. Technicians must have a working knowledge of these new technologies when servicing modern gas heating systems.

HSIs may wear out or crack after several years of use. Older HSIs are constructed of a material called silicon carbide. When replacing a hot-surface igniter, consider replacing it with a more durable and reliable type made of silicon nitride.

This section provides information on furnace controls. Troubleshooting the combustion system in gas heating equipment begins with observing the ignition sequence and the mechanical operation of the unit. These observations may help to identify the source of a problem immediately or with minimal testing. The troubleshooting guide provided in *Appendix A* outlines common problems that occur in gas heating systems, along with their probable causes.

1.1.1 Electronic Controls in Heating Systems

Microprocessor-based electronic controls are now common across all HVAC product lines. It is rare to find a modern heating unit that does not contain some kind of digital/electronic control. Manufacturers utilize these controls in their equipment to increase reliability and energy efficiency. Digital microprocessor controls also allow engineers to design more versatility into the products than was possible with electromechanical controls.

Microprocessor control boards are checked for correct operation by making sure that the board is receiving correct data, such as voltage signals from the room thermostat and open/closed circuit indications from devices such as limit switches. If all inputs are correct, the control should provide appropriate outputs.

For example, if the control is receiving a room thermostat input to W (the typical 24VAC control terminal used for heating) on the control, and all other inputs are within an acceptable range, the control will provide output signals to energize the draft inducer, igniter, gas valve, and blower motor. If the correct input and output signals are present and the furnace does not run, the problem is likely related to the control board. Various input and output signals are typically measured with a multimeter.

Figure 2 shows a printed circuit board containing the microprocessor control circuits for a packaged gas heating/electric cooling unit. *Figure 3* shows the wiring diagram for a gas forced-air furnace. Note that *Figure 3* is a combination diagram, with a connection diagram on the left and a schematic diagram on the right. Having both types of diagrams helps the troubleshooting process by allowing the technician to see the control structure from two points of view.

Highlights of the operating sequence for an induced-draft gas forced-air furnace are as follows:

- As in all furnaces, the sequence is initiated by a call for heat from the room thermostat in the form of a 24VAC signal. The control board often begins the combustion process by starting the draft inducer through a board-mounted relay.
- Before the microprocessor sends a signal to the igniter to initiate combustion, it will first verify that each safety control is in its correct position. For example, the microprocessor might look for answers to the following questions before sending a signal to energize the igniter and a signal to open the gas valve:
  - Is there continuity through the rollout and limit switches, indicating they are closed?
  - Is the pressure switch closed to indicate that the induced-draft motor is running?

  If these conditions are met, the microprocessor will energize the igniter and signal the gas valve to open.
- After the igniter and gas valve have been energized, the microprocessor looks for a current signal from the flame sensor to verify that there

(A) SPARK IGNITION

(B) HOT SURFACE IGNITER

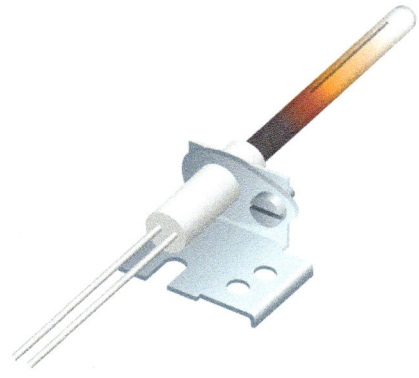
(C) NITRIDE IGNITER

Figure 1 Gas furnace ignition devices.

Figure 2 Gas unit control board.

is a steady, strong flame. If so, the ignition process ends. If not, the microprocessor may be programmed to retry for ignition a number of times over a period of several minutes. After each attempt, the microprocessor will impose a delay that gives the induced-draft fan time to purge the heat exchangers, preventing gas buildup. One electronic furnace control on the market will attempt ignition 33 times over a 15-minute period before it stops. At that point, the microprocessor locks out ignition and a fault indicator begins to flash. The system is then disabled until the power is manually reset. Note that systems using LP gas have different lockout schemes, since LP vapors are heavier than air and tend to fall rather than rise easily out of the heat exchanger area.

> **WARNING!**
> LP gas, also known as propane, is heavier than air. While natural gas tends to rise and dissipate easily, LP gas tends to collect in low places and remain there. Use extreme caution when dealing with LP gas leaks or when purging the gas lines of operating equipment.

- Once the burner flame is proven, the microprocessor fan logic will be notified and the blower time delay will begin. This delay gives the heat exchangers time to warm up before the blower starts circulating air.

The operating sequence just described is common to many gas furnaces that use electronic controls. Differences are generally related to special features that can be programmed into the microprocessor-controlled system. Some of these features include the following:

- Some systems have a feature to prevent nuisance trips of the high-limit switch. Furnace blowers run for a period after the wall thermostat is satisfied and combustion has stopped to extract more heat from the heat exchanger. If the blower stops too quickly, built-up heat in the heat exchanger could trip the high-limit switch. The microprocessor will automatically adjust the blower-off delay on the limit switch if the unit is tripping off after the wall thermostat has been satisfied. It will also flash a fault message to notify maintenance that this has happened.
- Variable-speed blowers are common features in high-efficiency furnaces. Some are programmed to automatically adapt to changing conditions such as the opening and closing of zone dampers in a zoned system.
- Most microprocessor-controlled systems can diagnose and isolate common faults. Many automatic diagnostics include a failure message to indicate that the control board/module itself has failed. This must be checked before the board is replaced because other problems, such as an intermittent power loss, might also cause this message.
- Some gas heating controls can automatically adjust the length of the high-fire and low-fire cycles of two-stage burners to maintain indoor comfort.

Flame Sensors

The flame sensor on a gas furnace is placed in the burner flame and senses a very small electrical current (in the microamp range) that flows through the burner flame. The furnace's electronic control senses this current and allows the burner to continue operating as long as the current is above a certain value. Over time, an oxide coating can build up on the flame sensor, reducing current flow and preventing furnace operation. Cleaning the sensing rod with fine steel wool can remove the oxide coating and restore normal operation.

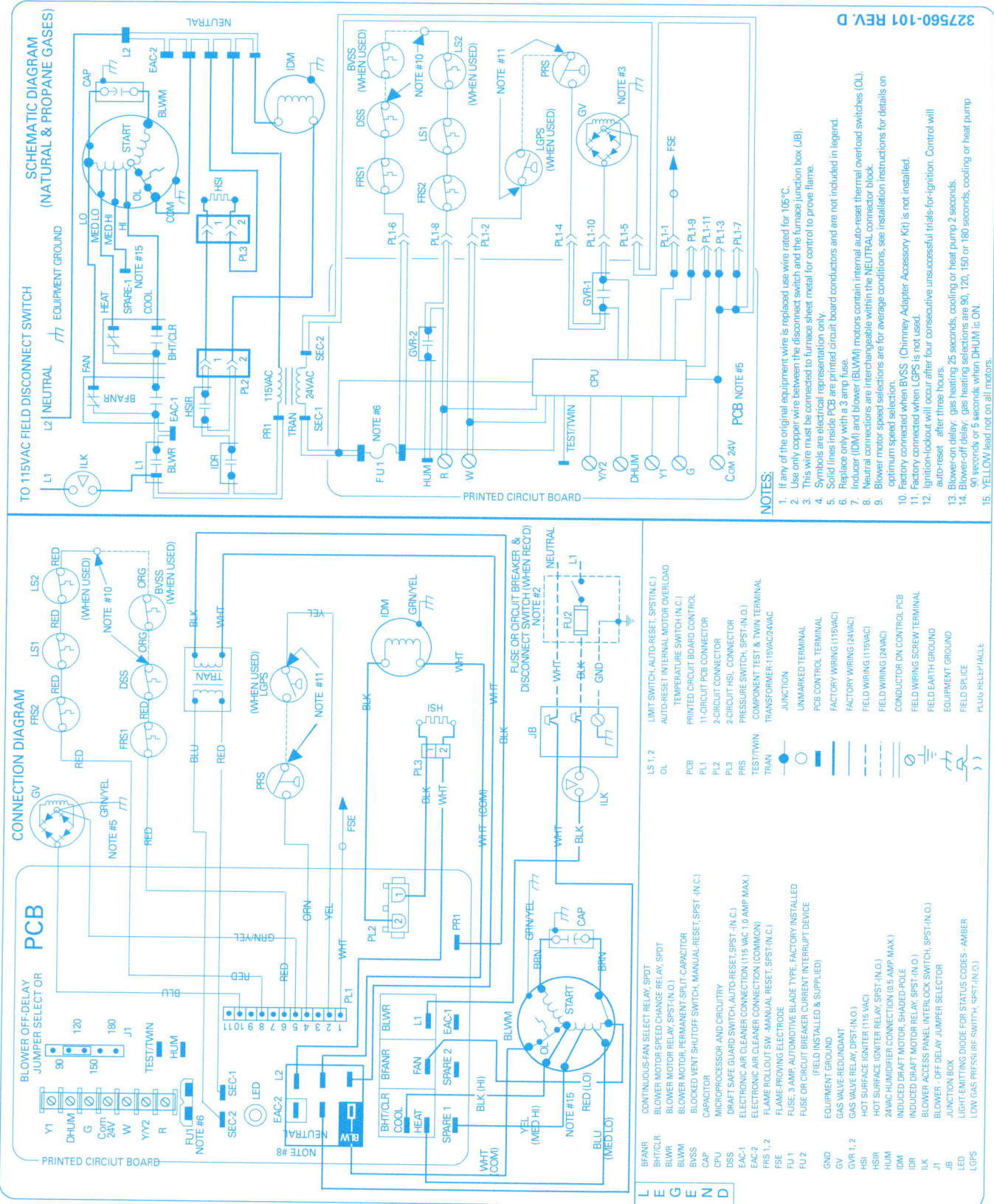

Figure 3 Gas furnace control board and wiring schematic.

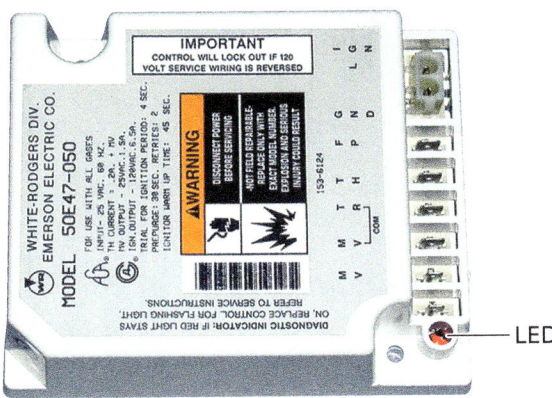

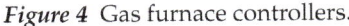
LED ON = FAILED CONTROLLER
LED FLASHING = EXTERNAL COMPONENT FAILURE
CONTROLLER WITH MINIMAL DIAGNOSTICS

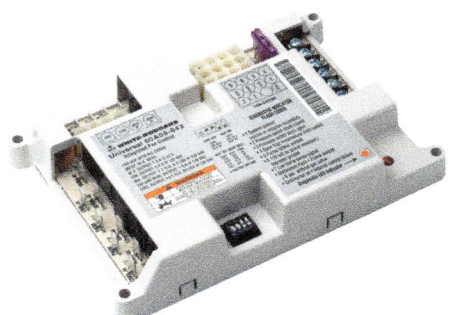

DIAGNOSTIC INDICATOR FLASH CODES

1. System lockouts (Retries of lockouts exceeded)
2. Pressure switch stuck closed
3. Pressure switch stuck open
4. Open high-temperature limit switch
5. Flame rollout sensed
6. 115 VAC power reversed/improper ground
7. Low flame sense signal

Continuous flash – flame sensed 0.5 sec without gas valve
Continuous on – internal control failure

Figure 4 Gas furnace controllers.

Electronic gas-furnace controls (*Figure 4*) are usually equipped with a diagnostic aid in the form of multiple or single light-emitting diodes (LEDs). Controllers with a single LED only light the LED when the ignition sequence is interrupted by a fault. The LED usually flashes in a code corresponding to the type of failure detected. For example, two flashes might mean the pressure switch is stuck closed, while three flashes could mean the pressure switch is stuck open. Controllers with multiple LEDs usually light the LEDs in a sequence corresponding to the ignition sequence of the controller during operation. If the ignition sequence fails, the last LED lighted is usually the last valid action completed. Controllers with LEDs have labels stating the meaning of the LEDs or listing the flash codes.

To isolate the failure, voltages at the designated (or next) component in the sequence can be monitored by the technician as the ignition sequence is restarted. Consult the manufacturer's literature to verify the actual ignition sequence and components involved.

Older solid-state gas furnace controllers may have limited diagnostics, or none at all. To determine an ignition sequence failure, the furnace must be observed during the ignition sequence to visually check where the sequence is halting. Then, voltages representing inputs from, or outputs to, components associated with the sequence failure can usually be monitored at test points on a connection block. This is done to isolate the malfunctioning component while the controller is recycled. Consult the manufacturer's literature for these types of controllers to obtain the ignition sequence and test point information.

> ## Think About It
> ### Circuit Diagram Interpretation Exercises
> Test your knowledge of microprocessor-controlled heating systems with these exercises. Use the diagram in *Figure 3* as a reference.
>
> 1. Which component on the control board controls blower speed in the heating and cooling modes?
> 2. If the draft inducer (IDM) runs and the igniter (HSI) energizes, but the gas valve (GV) does not energize, what should you do?
>   a. Replace the gas valve.
>   b. Replace the control board.
>   c. Check to see if 24VAC is available at the gas valve.
>   d. Check to see if there is a call-for-heat signal from the room thermostat.

Soot Buildup

If excessive soot is noticed on the secondary heat exchanger of a condensing furnace, there may be a problem with the gas pressure or combustion air source. Perform a gas pressure check.

1.1.2 Natural-Draft Furnaces

Natural-draft furnaces are rarely sold anymore. Government mandated energy efficiency requirements for gas furnaces cannot easily be met with furnaces using natural-draft venting technology. Induced-draft furnaces allow for the use of more efficient heat exchanger designs, thus placing the hot air in contact with a larger amount of surface area. Greater surface area results in a higher degree of heat exchange and a more efficient furnace. This has made natural-draft furnaces virtually obsolete. However, natural-draft systems are likely to be encountered, as they were still being installed in the 1990s.

Figure 5 represents the control circuits and sequence of operation for a packaged air conditioning unit with natural-draft gas heat. *Figure 6* is the wiring diagram for this unit. Because gas heating is the focus here, the cooling system is only partially shown.

This type of unit is sometimes called a year-round air conditioner (YAC). It may also be known as a packaged gas-electric unit (gas heat, electric cooling). A packaged electric-electric unit would be an air conditioning package with an electric resistance-heat package.

The basic operating sequence of any natural-draft unit is described as follows:

Step 1 The wall thermostat calls for heat.

Step 2 The system verifies (proves) that the pilot is lit.

Step 3 Assuming all safety controls are closed, the gas valve is turned on and the gas ignites.

Step 4 The blower comes on after the heat exchanger heats and closes the fan switch.

Step 5 The wall thermostat opens. The blower continues to operate until the heat exchanger cools, opening the fan switch.

The heating side of the unit shown in *Figure 5* has the following variations from the basic sequence:

- It uses an electronic pilot igniter rather than a standing pilot.
- There is a blower time delay at the beginning and end of the cycle.

1.1.3 Induced-Draft Furnaces

Annual Fuel Utilization Efficiency (AFUE) identifies the percentage of energy produced by the furnace/boiler that is available for heating as determined under tightly controlled test conditions established by the manufacturer. The remaining heat is lost via the flue system. *Figure 7* is a map of the United States showing the AFUE ratings for products destined for the indicated regions as of 2017.

The US DOE is in the process of updating the AFUE regulations pertaining to gas-fired furnaces. Changes can occur at any time, so it is important for contractors to pay attention to changes in the requirements for their region.

The 80-percent AFUE gas furnace shown in *Figure 8* is a common induced-draft heating unit. Ignition is accomplished with an HSI, the burner flame is monitored by a flame sensor, and the sequencing/monitoring of the furnace is accomplished by a sealed, solid-state controller containing micro-miniature relays. This furnace is similar to one represented by the wiring diagram in *Figure 3*. Refer to *Figure 3* as you follow the operating sequence of a typical 80-percent AFUE gas furnace provided here.

Step 1 The electronic control board (CPU) is powered through terminals Com and R. Power to the R terminal first flows through the 3-amp fuse (FU1). A 24VAC signal from the room thermostat is applied to the W terminal of the control board. This happens when the thermostat heating contacts close, applying power from the R terminal to the W terminal of the thermostat.

Electronic Ignition

The development of electronic ignition for gas furnaces was a major step for all furnaces, but it may have been even more significant for outdoor units. Because outdoor units are exposed to winds and have very short flue vents, standing pilots are blown out quite easily. To restart the unit, the pilot would have to be relit manually.

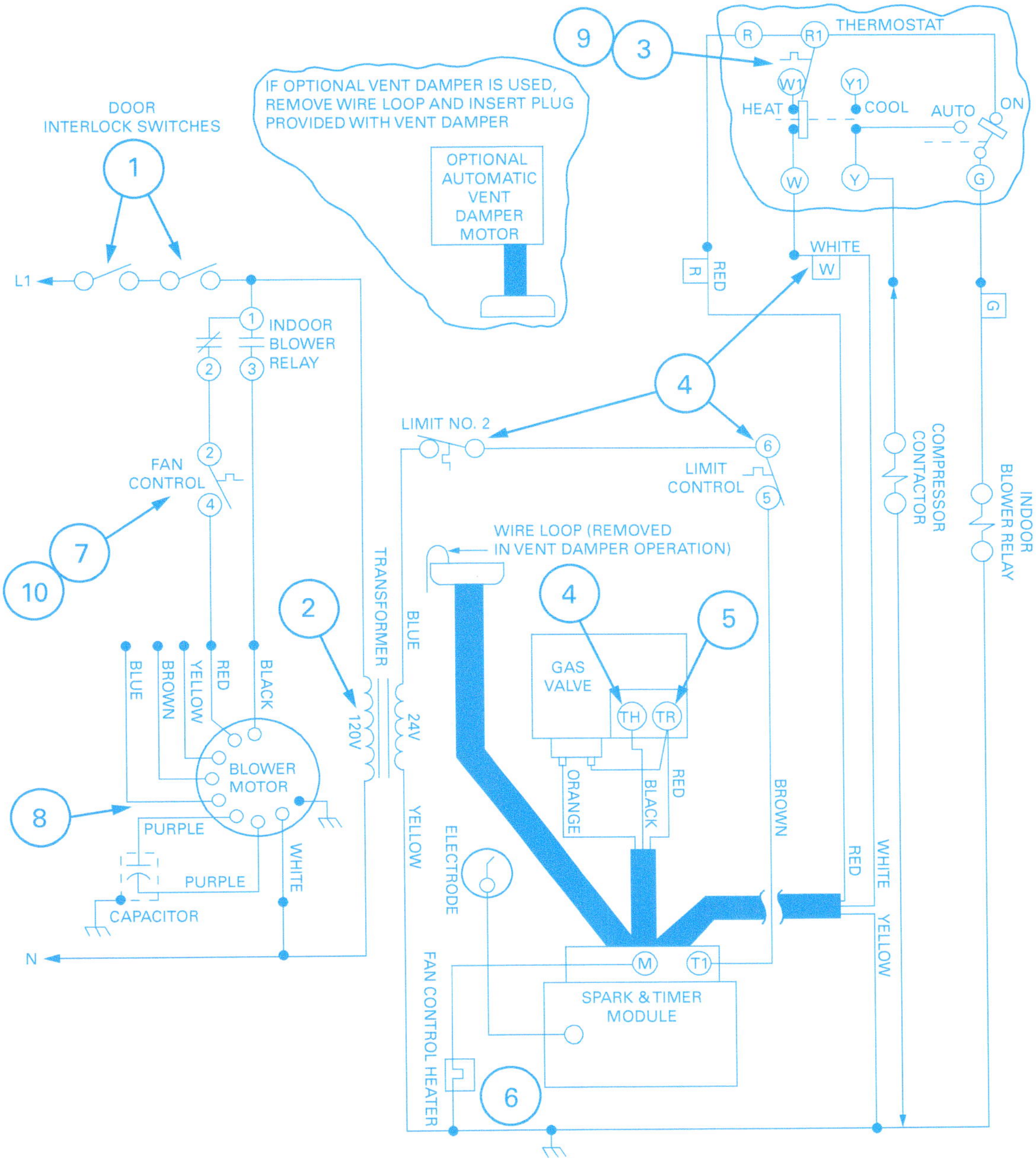

1. Line potential feeds through the door interlocks (if used). Access panels must be in place to energize machine.

2. Transformer provides 24V control circuit.

3. On a heating demand, the thermostat heating contacts close.

4. The 24V control circuit is complete to the pilot valve through limit controls, W leg of the thermostat, and internal ignitor circuits.

5. After the pilot flame has been proven, the main gas valve is energized. Main burners are ignited by the pilot flame.

6. As the main gas valve is energized, the fan control heater is also activated after a brief delay.

7. After a brief delay, the heater provides sufficient heat to close the fan control contacts.

8. This energizes the blower motor on the selected speed.

9. As heating demand is satisfied, the thermostat heating contacts open. This de-energizes the ignition control, gas valve, and fan control heater.

10. The blower motor continues running until the furnace temperature drops below fan control set point.

Figure 5 Natural-draft gas heating system.

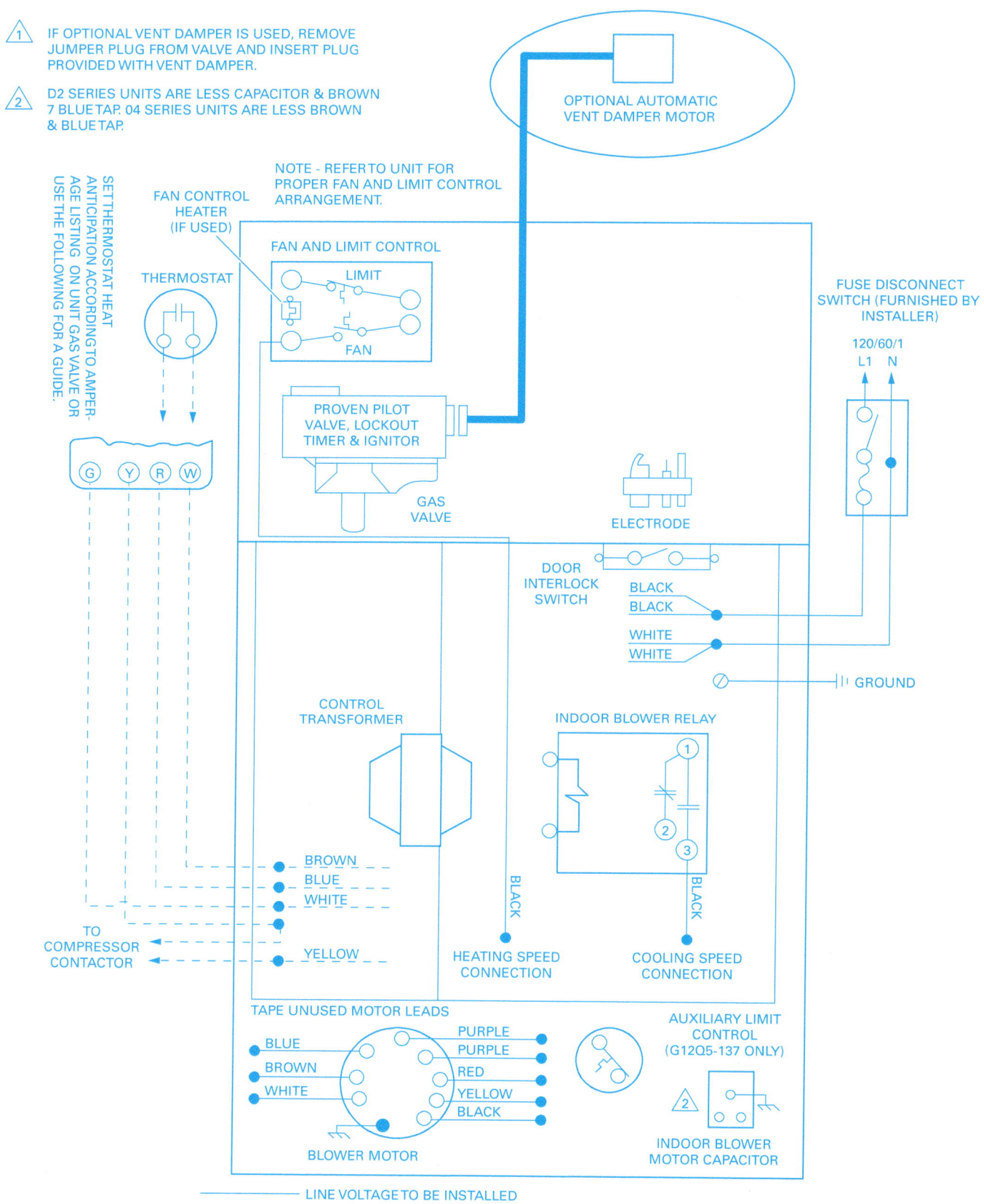

Figure 6 Field wiring diagram.

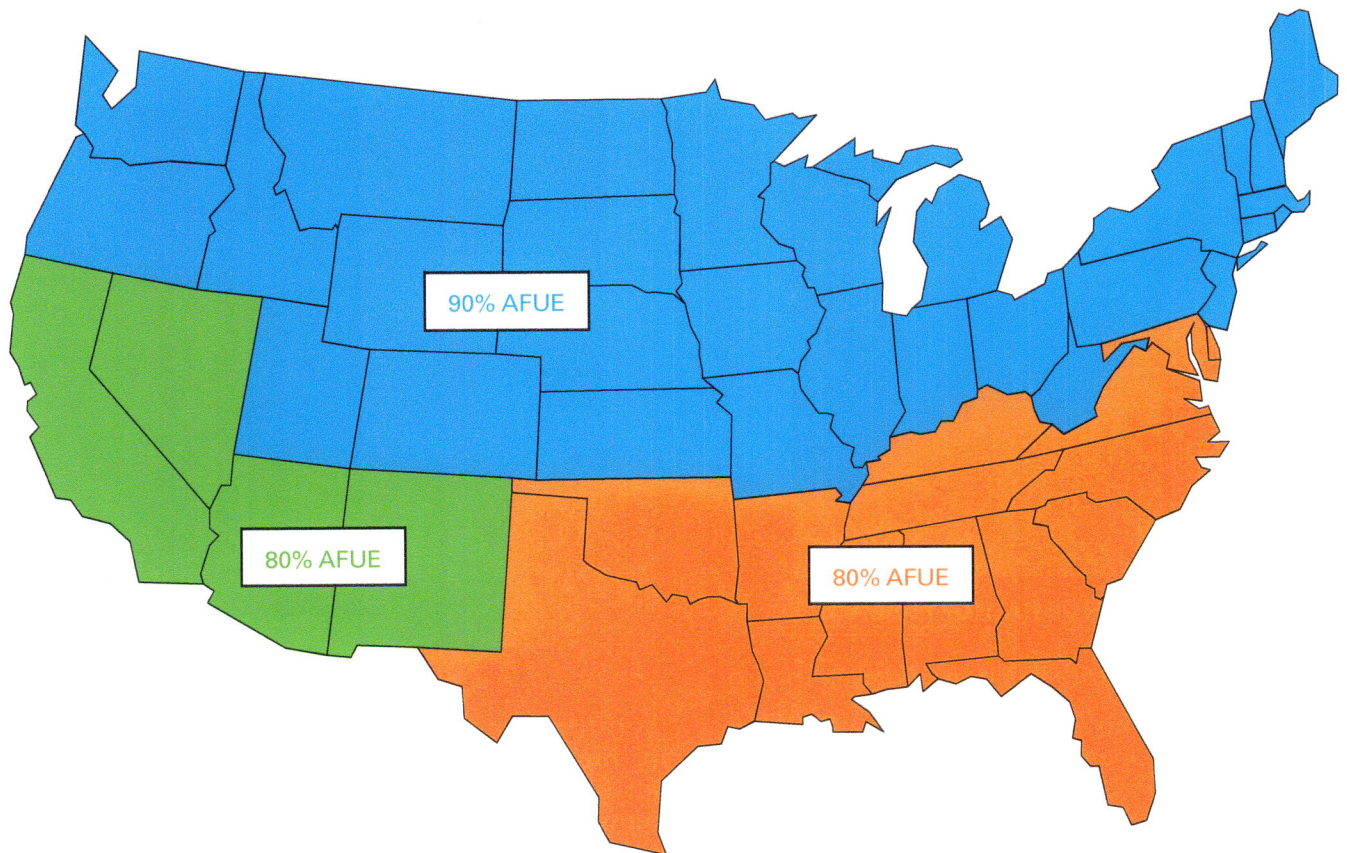

Figure 7 AFUE regional map for the continental United States.

Step 2 The electronic control board (CPU) performs a self-check and verifies that the pressure switch (PRS) contacts are open. If the pressure switch contacts are open, the draft inducer motor (IDM) can then start. It will run for 15 seconds to purge the heat exchanger. This causes the pressure switch contacts to close, proving that the inducer is running and that airflow through the heat exchanger is adequate for safe combustion.

Step 3 At the end of the purge period, the HSI is energized. After the igniter warms up, the gas valve relay (GVR) contacts close, applying power to the gas valve (GV), to energize it. The igniter will light the gas that is now flowing to the burners. The flame sensor (FSE) has two seconds to prove that an adequate burner flame is present. The igniter is de-energized as soon as an adequate flame is proven. Modulating gas valves can provide very precise levels of indoor comfort. They are often combined with a variable-speed blower motor that matches airflow to burner output. The gas valve can typically modulate between 40- to 100-percent burner capacity in 5-percent increments. A microprocessor-based furnace control tells the gas valve at what level to modulate the burner flame based on various values, such as indoor and outdoor temperature and the length of previous burner cycles. Note that the unit represented by *Figure 3* does not use a modulating gas valve.

Step 4 Failure to sense an adequate burner flame will cause the control board to de-energize the gas valve and attempt to re-establish a flame. If these attempts fail, the control board will lock out any more attempts to establish a flame. Interrupting the line voltage to the furnace or the control power to the module will reset the control.

Step 5 When the burner flame is proven, the blower motor (BLWM) is energized 25 seconds after the gas valve opens. The blower motor operates on HEAT speed. The motor speed for heating is field-selectable to meet the needs of the air distribution system.

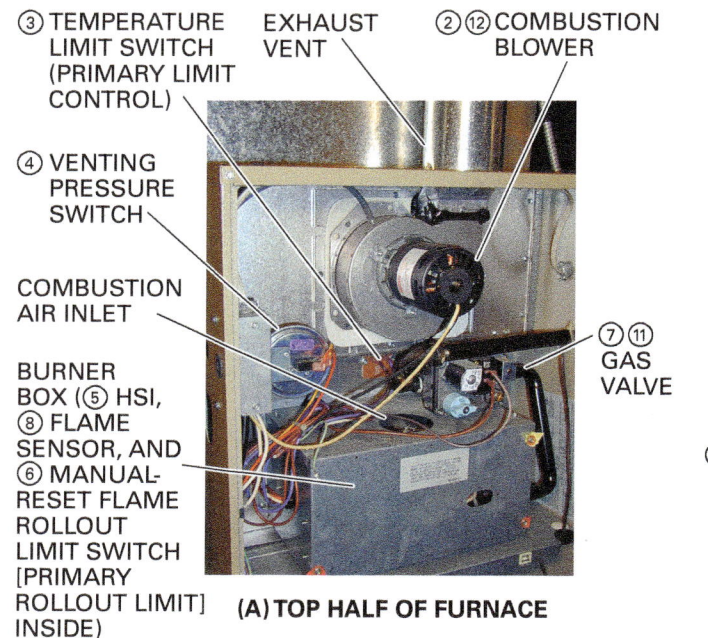
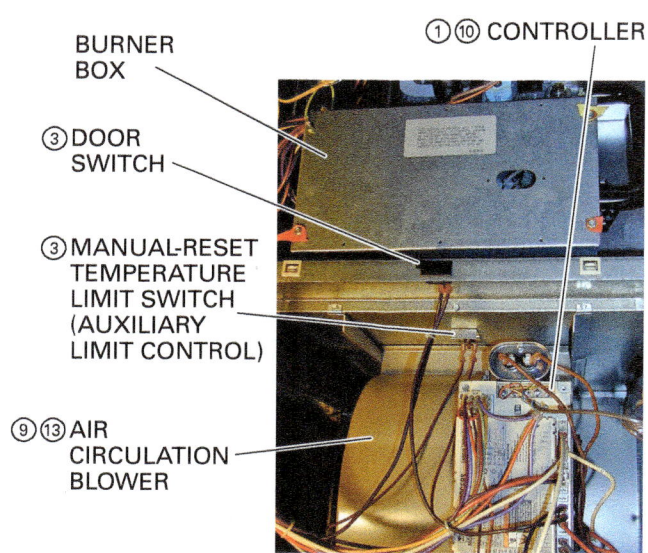

Figure 8 Typical 80-percent AFUE induced-draft gas furnace.

Step 6 If the room thermostat is satisfied, its contacts open and the 24VAC signal to W ceases. This de-energizes the gas valve, causing the burner flames to extinguish. The inducer motor will run for a short period to purge any combustion byproducts from the heat exchanger. The blower motor will continue to run for up to 180 seconds. The blower-off delay time is field-selectable.

> **NOTE**
> The high-limit switch prevents the furnace from overheating, which is a dangerous condition. Over the life of a properly operating and maintained furnace, the high-limit switch may never open. The most common cause of a furnace opening the high-limit switch is insufficient airflow, which can be caused by clogged air filters, insufficient duct sizing, or a duct layout that impedes airflow.

If a flame-rollout limit switch has been tripped, the cause of the flame rollout must be corrected and the switch must be manually reset before the gas valve will function. Flames may roll out of the combustion chamber and into the furnace vestibule if the heat exchanger is blocked or the draft inducer blower wheel is loose or deteriorated. If the draft inducer pressure switch has failed closed, the control module will not be aware of these conditions.

If the furnace shuts down repeatedly because of high-limit switch activation, and the supply air filters, ducts, and blower are clean, check the temperature rise of the furnace. If it is normal, the switch may be defective. If it is not normal, too many registers or air terminals may have been closed, or the blower speed may require adjustment. *Appendix B* provides an example of an induced-draft gas furnace troubleshooting flowchart.

1.1.4 Condensing Furnaces

A condensing furnace (*Figure 9*) has both a primary and secondary heat exchanger. The ability of these units to extract additional heat from flue gases produces AFUEs that typically exceed 90 percent. Because the secondary heat exchanger extracts most of the remaining heat from the vent gases, condensation of some of the combustion products occurs there.

This acidic condensate is drained from the secondary heat exchanger and routed through a condensate drain trap to an appropriate drain, either directly or by use of a condensate pump. Some local codes require that a replaceable acid neutralizer cartridge be installed in the condensate line before it is routed to a drain. The typical sequence of operation, along with the combustion devices, controller, and monitoring devices, is essentially the same as the 80-percent AFUE gas furnace. However, the condensate pump serving the furnace may be equipped with a float switch

Door Interlock Switch

Door interlock switches have traditionally been used to disconnect power when the door of a cabinet containing electrical circuits is opened. This helps to minimize the potential for electrical shock. On gas furnaces, the door interlock serves a second purpose: it prevents furnace operation if the blower door is removed. Keeping the door in place during furnace operation prevents harmful products of combustion from being drawn into the duct system by the blower. During troubleshooting procedures, it is acceptable to temporarily tape the door interlock switch closed. Any tape must be removed when the troubleshooting procedure is complete. A suitable jumper wire may also be used. Place it so that it hangs outside of the compartment, so that it cannot be forgotten when the repairs are complete.

that opens if the pump fails to operate and fills. The switch is wired into the furnace control circuit as an additional safety device and can shut down the furnace. This is an additional control that would not be found in the control circuit of induced-draft furnaces, and not likely shown on the furnace wiring diagram.

Most manufacturers do not recommend operating condensing furnaces with return air temperatures below 55°F (13°C) for prolonged periods. Condensation may occur in the primary heat exchanger, causing it to corrode and fail prematurely. The owners of such furnaces should be instructed to set the room temperature thermostat no lower than 60°F (16°C) if the building will be unoccupied for an extended period.

1.1.5 Gas-Fired Boiler Operation

The operating sequence for a gas-fired boiler differs from that of a forced-air furnace. A typical boiler control circuit is shown in *Figure 10*.

The operating sequence for a typical small packaged gas-fired boiler is described as follows:

Step 1 A call for heat from the room thermostat closes the circuit between the two T terminals on the aquastat, energizing the 1K relay. What is not shown on the right-hand connection diagram is that there is a small transformer on the aquastat board, and one leg of that power is applied to one of the two T terminals. A thermostat closure completes the circuit and applies that power to the second T terminal. This completes the circuit to the 1K relay coil. Again, having both a connection diagram and a schematic or ladder diagram is extremely helpful.

Step 2 1K relay contacts 1K1 and 1K2 close. Assuming the high-limit switch is closed, 120VAC power is applied to the circulator pump and the draft inducer.

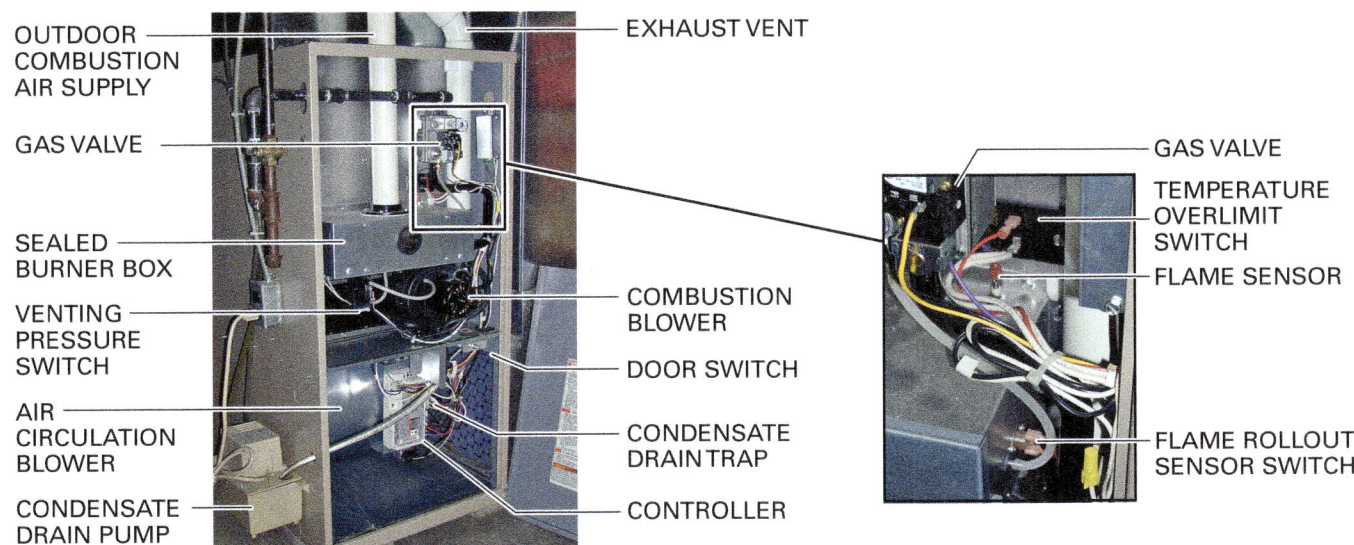

Figure 9 Typical 90-percent AFUE condensing gas furnace.

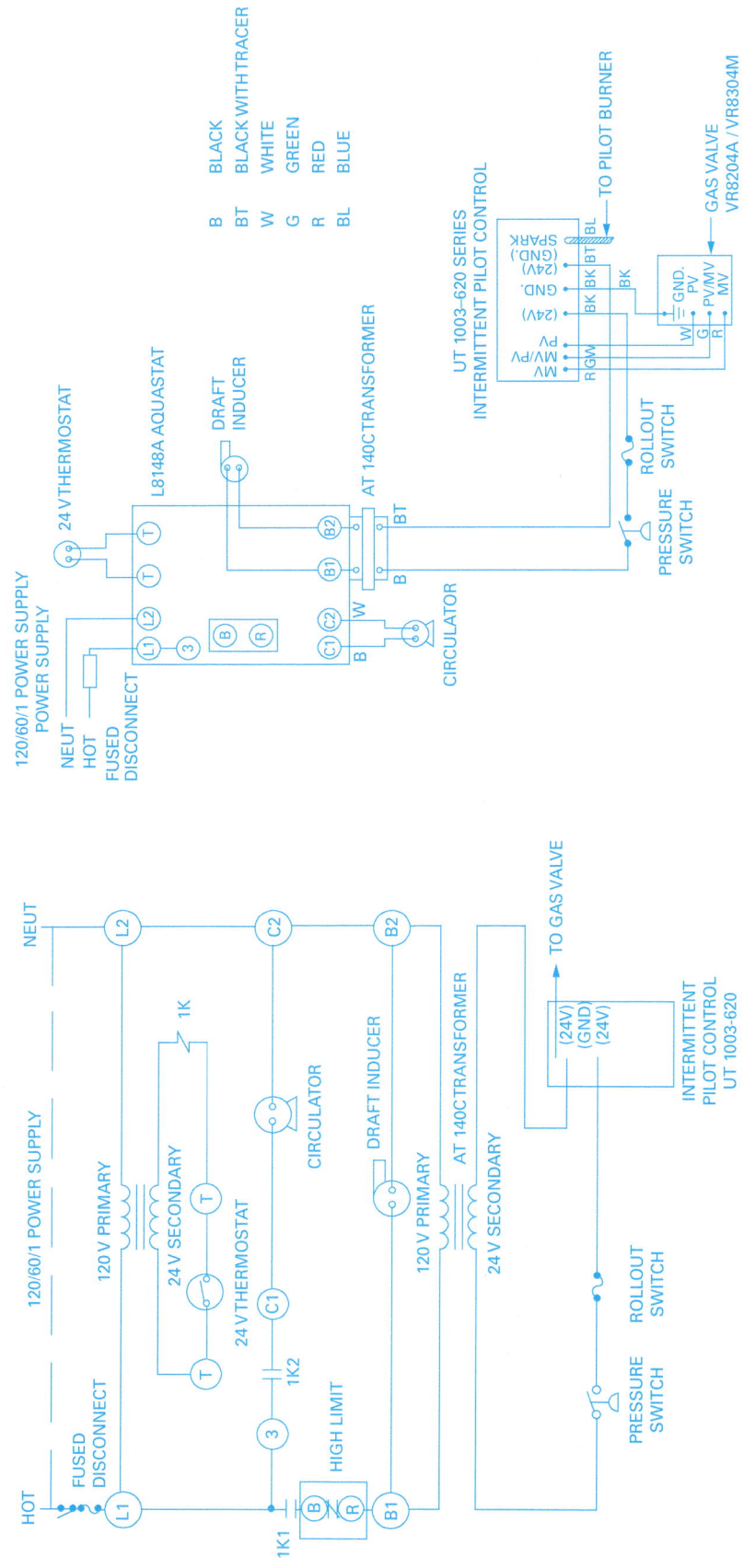

Figure 10 Gas-fired boiler control circuit.

Step 3 The draft inducer comes up to speed. If there is no vent blockage, the pressure switch closes, indicating proper airflow through the combustion chamber. This applies 24VAC power to the intermittent pilot control through the flame rollout switch.

Step 4 The intermittent pilot control opens the pilot gas valve, allowing the spark generated by the pilot control to light the pilot.

Step 5 When an adequate pilot flame is proven to the pilot control, the spark ceases.

Step 6 The main gas valve now opens and the pilot ignites the main burners.

Step 7 The main burner heats water within the boiler. If the water temperature exceeds the high-limit setpoint, burner operation will cease, but the circulator will continue to run until the room thermostat is satisfied. As a general rule, the circulator responds solely to the thermostat and is not otherwise involved in the operation of the boiler. When the water temperature drops below the setpoint, burner operation will resume, assuming the room thermostat is still calling for heat.

Step 8 If the vent system becomes blocked, insufficient draft will fail to close the pressure switch or keep it closed, stopping burner operation. The draft inducer stays powered as long as the room thermostat continues to call for heat and the limit switch contacts remain closed. If the vent blockage clears, burner operation resumes. If the draft inducer is running but no combustion is taking place, it is often related to a blocked vent or a failed pressure switch.

Step 9 When the room thermostat is satisfied, the 1K relay de-energizes, opening the 1K1 and 1K2 contacts to stop the burner, circulator pump, and draft inducer operation.

The Incredible Shrinking Furnace

As late as the 1980s, commonly available upflow gas furnaces were up to 60 inches (152 cm) in height. Due to improvements in heat exchanger technology, several manufacturers now offer upflow furnaces that are less than 36 inches (91 cm) in height.

1.2.0 Troubleshooting Pilot Re-Ignition and Direct Ignition Devices

Some systems use spark igniters to light a pilot flame when the thermostat calls for heat. Pilotless systems use either an HSI or a direct spark igniter. Both of these devices ignite the gas directly at the burner; as the name implies, there is no pilot.

The spark ignition system has a flame sensor that sends a signal to the gas valve proving combustion whenever the pilot or burners are lighted. The sensor operates through flame rectification, changing the AC input to DC. Some spark ignition assemblies use a separate flame-rectification rod. In others, the rod used for the ignition spark also acts as the flame rod.

When an HSI is used, the flame-sensing rod, or flame rod, may not be part of the igniter assembly. However, it will be located somewhere in the burner area and positioned in the main burner flame. Some HSI devices incorporate the flame rod into the HSI assembly. Regardless of the type of igniter used, if the flame sensor is not producing an output, the gas supply will shut off.

When troubleshooting spark igniters or HSIs in an electronically controlled system, keep in mind that the system will lock out the ignition circuit if it fails to establish and maintain a flame after a number of tries or a certain amount of elapsed time. Usually, either the control voltage or the primary power must be reset before the system will operate again once a lockout occurs. Remember that ignition devices require a good ground to operate. If the ignition device is poorly grounded, it will not create a spark.

LP/propane-fired equipment has different lockout schemes. As noted earlier, propane vapors are heavier than air, and appropriate safety precautions must be implemented.

1.2.1 Spark Igniters

You should be able to see and hear whether the spark igniter is working. If you can hear the spark, but the gas is not igniting, the ceramic insulator may be cracked. This allows the spark to be misdirected toward another surface outside of the pilot-gas envelope. If there is no spark at all, the spark gap may need to be readjusted according to the manufacturer's instructions. This is a relatively common problem.

If the gap is correct but there is still no spark, or the spark is weak or intermittent, the problem is most likely in the ignition module or the high-voltage wire to the igniter. If there is a spark but no pilot flame, the pilot orifice may be plugged.

During the summer, small insects may build webs or cocoons inside the orifice.

> **WARNING!**
> Be very careful when troubleshooting spark igniters. They are operated by a very high voltage pulse (10,000 to 20,000 volts) from an ignition transformer. Do not touch the spark igniter with your fingers or a screwdriver, or attempt to disconnect the igniter without first turning off the power.

1.2.2 Hot-Surface Igniters

Older HSIs are made of silicon carbide. Also called carborundum, this synthetic compound has a high electrical resistance. When electrical power is applied through the material, it heats up rapidly, becoming very hot and bright red. The carborundum element of the igniter is very brittle and can easily be cracked or broken if handled carelessly or bumped during equipment cleaning. If a bright white line is visible anywhere on an operating HSI element, the element is cracked, but hasn't completely failed. Contrary to popular belief, any oils transferred to the element during handling will not harm it, but it must be handled delicately. Silicon nitride materials are used to construct many HSIs today, and they are considered to be more durable and reliable than silicon carbide HSIs.

If an HSI fails to glow at the proper time during an ignition sequence and the required preceding actions (inducer motor running, pressure switch closed, and so forth) have occurred, then one or more of the following has occurred:

- The element is open because it is cracked, broken, burned out, or the HSI wiring is defective.
- No power is being supplied to the element.

> **WARNING!**
> Testing an HSI involves measuring 120VAC power. Use caution to avoid contacting the HSI element, any live wires, or bare electrical contacts when making these measurements. Make sure the 120VAC power to the equipment is turned off before disconnecting and connecting wiring or bare contacts.

Observe the HSI element without removing it from the burner section. If it is not visibly cracked, broken, or burned out, monitor the 120VAC power applied to the HSI during the ignition sequence at its inline connector inside the equipment (*Figure 11*).

Cycle the equipment through its ignition sequence while monitoring the HSI voltage. If the proper voltage is detected, the HSI has failed and must be replaced. If multiple failures of the same HSI have occurred, make sure that the voltage does not exceed 125VAC. High operating voltages can significantly reduce the life of an HSI.

In other cases, the HSI may glow but will be faint and fail to reach a temperature hot enough to ignite the gas. This indicates the HSI has lost some of its resistance after many cycles. The resistance can be tested, and should conform to the manufacturer's specifications. A common range of resistance is 40-90 ohms, but ensure that you have the correct range for the specific igniter in use. Controllers will typically not admit gas unless the resistance has a minimum value. If the resistance measured is outside of the required range, replace the HSI.

If no power is found at the inline connector, the power source (inline fuse, controller, or separate relay) for the HSI has failed and must be replaced. In this case, make sure the HSI leads are not damaged or shorted to any metal part of the furnace before reapplying power to the HSI.

1.3.0 Troubleshooting the Flame Sensor

In direct-ignition systems, the flame rod is not an integral part of the igniter. It is installed in the path of the burner flame (*Figure 12*). It uses flame rectification to produce a tiny direct current in the microamp (µA) range. This current must be sensed by the gas-fired equipment controller, or the main gas valve will be turned off.

Figure 13 shows the method for testing a flame sensor. A multimeter capable of reading DC in the low-microamp range (less than 10µA) is connected in series between the flame sensor electrode and the flame-proving circuit. Then the power is turned on and the thermostat is set to call for heat. After the burners ignite, the current reading should be within the range specified by the manufacturer. Typically, the normal range is 0.5–4.5µA DC. Follow the manufacturer's specifications for this value.

If the current reading is absent or too low, it could be caused by one of the following problems:

- The flame-sensing rod is incorrectly positioned.
- The ignition module is not properly grounded.
- The sensor electrode is not making good electrical contact.
- The sensing rod has become oxidized; clean the rod with fine steel wool.
- The controller or flame sensor insulation/wiring is defective.

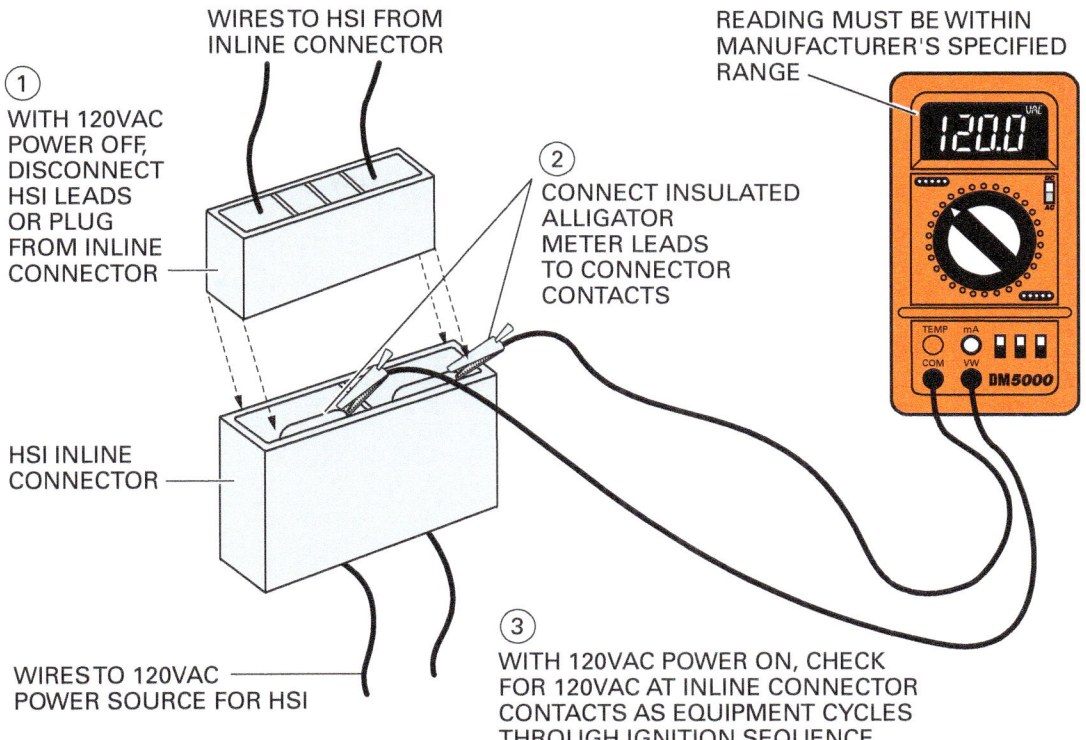

Figure 11 Checking HSI source voltage.

Special equipment is available for testing spark generators, flame sensors, gas valves, and other ignition devices. One manufacturer provides optional accessories for a digital multimeter that permit voltage and current measurements in the very low ranges necessary for testing ignition devices.

A furnace manufacturer might provide a furnace tester for use with specific furnace models. The tester can be plugged into the furnace. Switches on the tester can then be manipulated to simulate various inputs.

1.3.1 Testing a Combined HSI/Flame Sensor

Instead of utilizing a separate HSI and flame sensor, many gas-fired systems use a controller that switches the connections to the HSI so that it functions as the igniter for a prescribed time, then as a flame sensor using flame rectification. The controller accomplishes this internally by removing the AC current from the HSI after the ignition sequence, opening one of the HSI leads, then substituting a high-resistance circuit and an AC signal in the other lead to the HSI element. The element can then function as a separate flame sensor.

As the HSI ages, an insulating oxide coating slowly forms on the surface of the element. This coating is normal, but it may reduce the strength

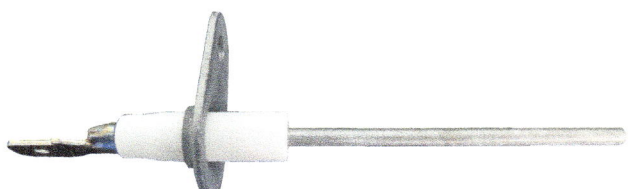

Figure 12 Flame sensor.

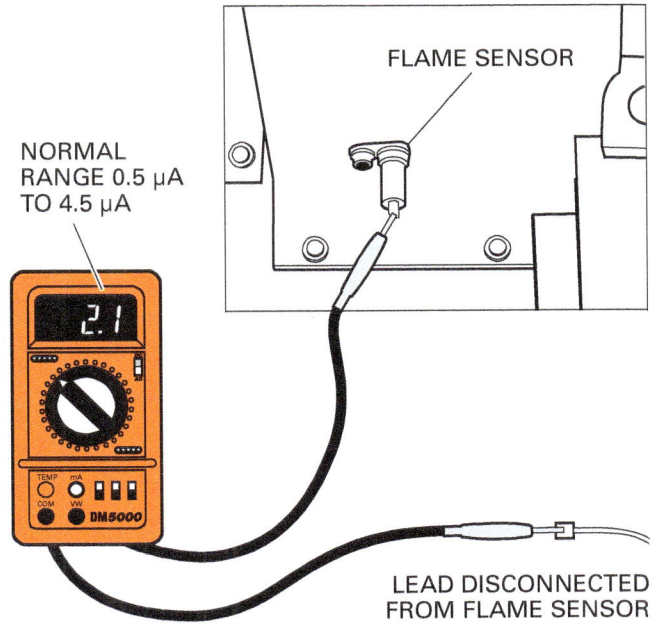

Figure 13 Flame sensor test.

of the flame signal to the controller. If the gas ignites, but immediately or occasionally goes out before the thermostat is satisfied (nuisance trips), a weakened flame-sensing signal may be the cause. Always check the continuity of the burners and their flame carryover tube to ground, as well as the controller ground connection. If the grounds are satisfactory, visually check the HSI element for contamination. If not visibly contaminated, check that the HSI element is positioned so that 3/4" to 1" (19–25 mm) of the element is continuously immersed in flame.

If desired, the flame signal can be checked using a test circuit, as shown in *Figure 14*. The meter used must have a DC microammeter range capable of displaying about 10µA.

After turning off all primary 120VAC power, connect the test circuit at a convenient point (inline connector) between the controller and the

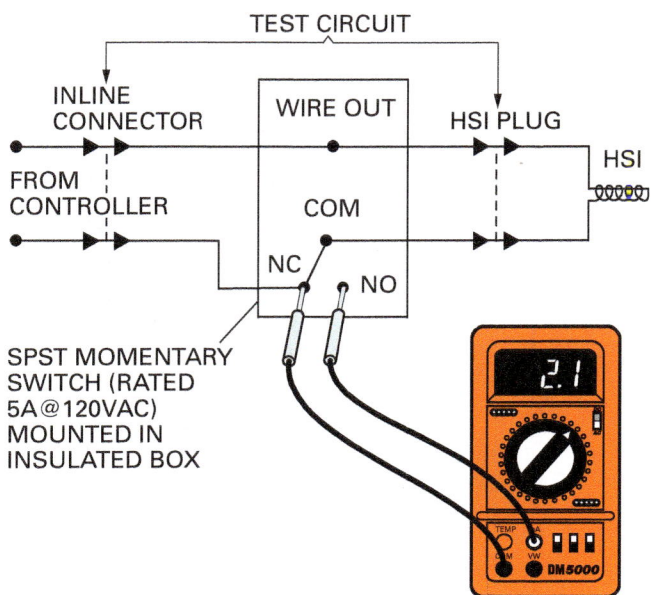

Figure 14 Testing a combined HSI/flame sensor.

1.4.0 Airflow-Related Problems

Forced-air gas furnaces can malfunction if there are problems in the air distribution system. There are several ways that airflow problems can create problems for a gas-fired furnace. Proper combustion airflow is essential for efficient furnace operation and reliability.

1.4.1 Air Distribution System and Blowers

In addition to component failure, there are three types of problems that can occur in the supply and return air system:

- *Obstructed airflow* – The most likely cause of an airflow problem is a dirty filter. Dirty heat exchangers or a dirty blower wheel will also reduce airflow and prevent the appliance from providing adequate heating. If internally insulated ductwork is used, it is possible for the insulation to come loose and block airflow. Insufficient airflow can cause nuisance trips of the temperature limit switch because heat is allowed to accumulate around the heat exchangers. Even when the airflow is not sufficiently obstructed to overheat the furnace, it will cause a loss of efficiency. Another possible source of obstructed airflow is a cooling coil installed in the ductwork. If air filters are removed and/or filter maintenance is neglected, dirt can build up on the fins of the coil, blocking airflow. Always access the interior of the ductwork to check the cleanliness of the finned coils when other possible causes are not present.
- *Ductwork* – Problems with the ductwork size are likely to occur only if the replacement

> **WARNING!**
> Testing a combined HSI/flame sensor involves measurement of circuits operating with 120VAC power. Avoid contacting the HSI element, any live wires, or bare electrical contacts when making the measurements. Make sure the 120VAC power to the equipment is turned off before disconnecting and connecting wiring or bare contacts.

> **CAUTION**
> Do not push the momentary contact switch when testing a combined HSI/flame sensor until after the igniter function de-energizes; otherwise, the meter could be damaged.

HSI. Depending on the gas equipment manufacturer, the test circuit may require various inline connector plugs to interface to the equipment.

Reapply power and cycle the system through an ignition sequence. When the gas ignites and after the igniter function de-energizes, push and hold the momentary contact switch to read the DC flame signal on the meter. If no reading occurs, the switch may be in the wrong lead from the controller. In that case, turn off the primary 120VAC power, reverse the test leads to the controller, and repeat the process.

Check that the flame signal is within the prescribed limits for the controller manufacturer. If the flame signal is out of limit, or if the flame goes out before a reliable reading can be obtained, replace the HSI and retest. If the problem reoccurs, replace the controller and retest. If the problem is solved with replacement of the controller, reinstall the original HSI and retest.

appliance has significantly greater or smaller capacity and the ductwork was not resized. Such systems will often be noisy, especially if the ductwork is too small.

- *Blower motor* – If the appliance has previously been operating properly and the blower wheel has been periodically cleaned, it is unlikely that the blower size or speed is the source of a problem. However, a blower motor may have difficulty starting or may be beginning to fail, providing intermittent or short-term service only.

Electronically controlled variable-speed motors (ECMs) are frequently used as furnace blower motors. These are different from multi-speed motors. While multi-speed motors may operate at several different fixed speeds, variable-speed motors can modulate the speed through a wide range. When checking airflow in a furnace that uses a variable-speed blower motor, always ensure that the blower is operating at the correct speed for the conditions before checking airflow. For example, if the space temperature is more than 3°F (1.7°C) lower than the thermostat setting, the motor may be running faster than when the two temperatures are very close together. Variable-speed motors come up to speed slowly and may reduce speed slowly before coming to a stop.

The example troubleshooting chart in *Appendix A* identifies symptoms and likely causes of air system problems.

Heating and cooling generally require different volumes of air. If a unit with a single-speed blower is used for both heating and cooling, it may be inadequate in one mode or the other. For example, a high-speed blower that is suitable for cooling may prevent the heat exchangers from getting warm enough to do their job. In such cases, it may be necessary to replace the motor and its controls with a multi-speed motor and controls that automatically choose the correct blower speed based on the mode of operation.

1.4.2 Induced-Draft System

Today's high-efficiency furnaces and boilers use draft inducers to help move products of combustion through the heat exchangers and the vent system. Problems related to air movement through the heat exchangers and vent piping are usually the result of inadequate combustion air, a blockage in the furnace or vent, or an incorrectly sized vent. This assumes that the draft inducer motor is operating properly, the combustion blower has not deteriorated, and devices to prove the airflow are set properly and functional.

Induced-draft furnaces have more restrictive heat exchangers that are designed to extract more heat from the flue gases. The inducer helps the flue gases overcome any restriction so that they can be properly vented. In 80-percent AFUE furnaces and boilers, the flue gas that exits the heat exchanger should be at a neutral pressure. Natural currents established in the vent then act to remove the flue gas from the structure. The draft inducer is designed primarily to overcome the restriction of the heat exchanger only, not the vent system. However, in 90-percent AFUE condensing furnaces, the inducer does impart a positive pressure to the flue gases, forcing them out the vent system which is sealed and pressurized.

If a blockage occurs in the furnace or vent, safety devices in the appliance will prevent operation until the problem is corrected. In most sequences of operation, the first event that occurs after a call for heat is the startup of the draft inducer motor. A pressure switch is typically used to prove that the inducer is running. The closure of the pressure switch tells the system that there is a sufficient flow of air through the heat exchanger for the rest of the ignition sequence to occur. If the inducer runs but no ignition occurs, look for blockage or restrictions in the air supply, heat exchanger, or vent.

A defective or improperly adjusted pressure switch can also prevent ignition from occurring. In addition to the pressure switch, there are other manual-reset safety switches that open if the vent becomes blocked or flames roll out the front of the burner.

Due to the high level of sensitivity required of the draft inducer pressure switch, it is best to avoid attempting any field adjustments. Although many switches are adjustable, they are typically factory set and calibrated. This is often indicated by a small amount of colored cement placed on the threads of the adjusting screw. Field adjustment of the switch may result in unsafe operation. If a pressure switch is not operating properly, it is best to replace it with a factory provided part. Otherwise, very sensitive and properly calibrated pressure measurement instruments that can be tied into the same pressure sensing line are required.

Condensing furnaces have some unique venting problems. Most condensing furnaces use a sealed combustion system in which outside air is brought in through a dedicated plastic pipe and flue gases are vented out a second plastic pipe. The size of both pipes is critical. Failure to size the pipes according to the manufacturer's instructions may result in a restriction that would prevent the inducer from closing the pressure switch. Improper routing of one or both pipes can also cause inadequate airflow. Dips, sags, or an improper pitch in the vent may cause condensate water to pool in the vent pipe, which causes a restriction. If the combustion air intake and vent are not properly terminated, ice may form at the vent outlet and snow may be drawn into the air intake. Both conditions create restrictions that prevent or interfere with furnace operation.

Additional Resources

Refrigeration and Air Conditioning, An Introduction to HVAC/R, Fourth Edition. Larry Jeffus. Air Conditioning and Refrigeration Institute. New York, NY: Pearson

1.0.0 Section Review

1. In an induced-draft furnace, on which control board terminal is the heating control voltage typically applied?
    a. C
    b. G
    c. W
    d. Y

2. Hot-surface igniters can be made from _____.
    a. silicon carbide
    b. Nichrome
    c. sodium nitride
    d. aluminum nitrate

3. The output signal of a flame sensor is typically a(n) _____.
    a. stabilized AC voltage
    b. rectified DC current
    c. variable resistance
    d. oscillating waveform

4. In most sequences of operation, the call-for-heat signal first energizes the _____.
    a. HSI
    b. gas valve
    c. pressure switch
    d. inducer motor

Section Two

2.0.0 Infrared Gas-Fired Heaters

Objective

Identify infrared gas heaters and describe how they operate.
a. Identify various types of infrared gas heaters.
b. Describe the operating characteristics of infrared gas heaters.

Trade Terms

Reverberator: A wire or corrugated-metal screen used to cover the ceramic grids in a gas-fired radiant heater, which boosts the radiant-heat output of the heater by converting convective heat produced by the grid into radiant heat.

Submerged flame: A method of gas-fired radiant heater operation that uses burning gas within two-layer, porous-ceramic grids.

Infrared heaters are used in buildings with a high volume of air exchange resulting from numerous external door operations, or in high-bay buildings where forced-air heating is impractical. The basic purpose of infrared heaters is to heat objects, not air. Since air is not being heated, opening a large door does not result in a significant loss of heat. This makes infrared heaters ideal for commercial applications in factories, foundries, large workshops, vehicle repair garages, warehouses, aircraft hangers, and sports arenas. Smaller versions are available for limited use in residential applications.

It is important to note that infrared heat functions on a line-of-sight principle. If the person or object is not within view of the heater, they cannot absorb the infrared rays.

2.1.0 Infrared Heater Types

Infrared heaters are available in both low- and high-intensity types. High-intensity gas-fired heaters have been in use since the 1950s. They are usually mounted high on a ceiling, use an open flame to heat a ceramic surface, and are typically unvented. Low-intensity heaters use a steel tube in which the flame is enclosed, often with a draft inducer to overcome the resistance of the tube. This heats the tube, which then emits radiant heat. Low-intensity heaters are typically vented.

2.1.1 Low-Intensity Infrared Gas-Fired Heaters

Figure 15 shows typical gas-fired, tubular, low-intensity infrared heaters (also known as *indirect radiant heaters*), available in straight-tube and U-tube configurations. They are mounted at a relatively high level in a building. A gas flame is directed from a burner assembly through the steel tube, heating the tube. The tube's hot outer surface produces infrared radiant energy focused by a stainless steel reflector onto the floor, occupants, and other surfaces or objects under the heater. The effect is similar to that caused by the sun's infrared rays. Surfaces, occupants, and objects under the heater are warmed, creating a comfortable environment without producing drafts.

Tubular heaters are available in either vented or non-vented versions. Most models require some form of mechanical assistance for venting due to the resistance of the long tubular heat exchanger. If more than one heater is connected to a single vent, a vacuum is created by a single induced-draft blower located at the vent's exit to the outdoors. Single-vented or non-vented units may be assisted (power-vented) by a blower located at the burner assembly itself. Venting can be made through the roof or horizontally through a wall.

In a non-vented unit, combustion products are discharged into the heated space for additional heat. These heaters are permitted in areas with a high outdoor air infiltration rate, or when exhaust fans change the building air a set number of times over a specific period. The burner assembly operates the same as the burner assembly in an induced-draft gas furnace. Only a minimal amount of heat is lost due to the constant air exchange.

Safety devices used in low-intensity infrared gas-fired heaters are similar to those used in induced-draft gas furnaces. The pilot ignition devices used are also similar.

2.1.2 High-Intensity Infrared Gas-Fired Heaters

These non-vented units (*Figure 16*), also called *direct-fired radiant burners*, are used at high levels off the floor as spot or space heaters. The units operate by burning gas within two-layer, porous-ceramic grids, referred to as a **submerged flame**. The grids attain a red-heat level of about 1,600°F (870°C) and produce intense infrared radiant energy. In some cases, the grids can also be closely

03209 Troubleshooting Gas Heating

Module Five 19

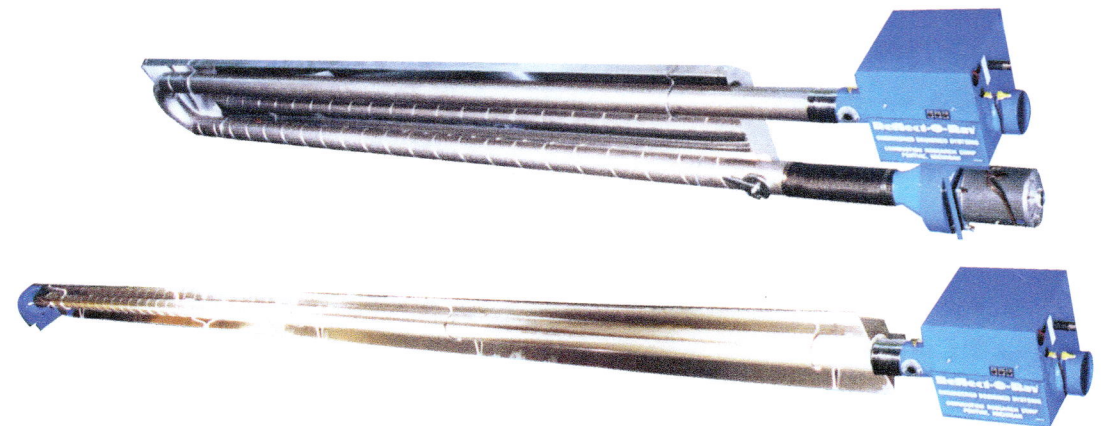

Figure 15 Low-intensity infrared tubular heaters.

Figure 16 High-intensity infrared heater.

covered by an optional wire screen, called a reverberator, that increases the infrared radiation when heated.

High-intensity infrared gas-fired units must be mounted at a specified angle to allow the combustion products to flow upward off each grid and into the room. Reflectors around the perimeter of the grids focus the radiant energy to the desired location. These non-vented units are permitted for use in areas with a high outdoor air infiltration rate or when exhaust fans change the building air a set number of times over a specific period. This ensures that the byproducts of combustion are diluted and eventually removed from the space. The gas burner, located behind the grids in a plenum, operates in the same way as the burner in a natural-draft gas furnace.

As with low-intensity units, the safety devices used in high-intensity infrared gas-fired heaters are similar to those used in induced-draft gas furnaces.

2.2.0 Infrared Heater Operation

Multiple commercial infrared heaters may be controlled from a central multi-zone control panel (*Figure 17*). Panels typically include digitally programmed time switches that can be set up according to a variety of heater control scenarios. If needed, a built-in temperature sensor can also be used in conjunction with the programmed switches for improved temperature control.

Conditioned-space temperature control with gas-fired infrared heaters is different from temperature control with more conventional heating systems. Infrared heaters are typically used for spot heating in an otherwise unheated area. A simple On/Off switch can often provide adequate temperature control. For example, a worker at a desk on an unheated loading dock can turn the heat on when the area is occupied, then turn off the heat when the area is vacated.

Infrared heaters may use low-voltage or line-voltage thermostats for temperature control. However, conventional thermostats may not offer the best temperature control for these heaters, due to the following conditions:

- The spot area being heated is often surrounded by cold air which can affect the thermostat.
- Line-of-sight radiant heat energy can heat the thermostat; a radiant shield may need to be installed to prevent this.
- Thermostats installed near open doors, such as found on a loading dock, can be exposed to cold drafts.

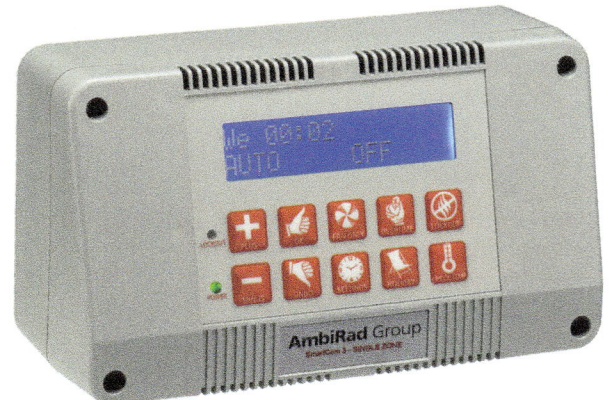

Figure 17 Multi-zone control panel.

> ### Clean Natural Gas
>
> The use of natural gas as fuel can help to reduce carbon emissions that contribute to global warming. Since natural gas has a higher energy density compared to other carbon-based fuels, it produces 30 percent less carbon emissions than oil and 45 percent less carbon emissions than coal for a given weight of fuel burned. In addition, natural gas does not produce the ash particles and other airborne contaminants that coal and oil produce. Recent advances in drilling technology now allow access to this relatively clean fuel.

- Workstation machinery may generate heat that can affect the thermostat.

Figure 18 shows a wiring diagram for a low-intensity tubular heater that is controlled by a thermostat. The components used in the heater and its operating sequence are very similar to that of a conventional gas-fired furnace.

The operating sequence of the unit shown in *Figure 18* is described as follows:

Step 1 On a call-for-heat, the inducer (blower) starts running. When it reaches operating speed, the air switch closes and provides an input to activate the ignition module.

Step 2 After a pre-purge period of about 30 seconds, the ignition module energizes the spark igniter. Simultaneously, the gas valve is energized for about 15 seconds to allow the burner to ignite.

Step 3 If burner ignition occurs, the flame-sensing rod sends a signal to the ignition module. This input to the ignition module allows the gas valve to remain energized.

If no burner flame is detected, the ignition module shuts off the gas valve and the inducer runs for 30 seconds to purge any unburned gas from the system. The ignition module will attempt to restart the burners several times on natural gas systems. If no restart occurs, the ignition module will lock out for 1 hour or until it is manually reset by removing power for at least 5 seconds. Units operating on LP gas typically do not retry ignition; lockout occurs on the first failure.

If a 24VAC thermostat is used for temperature control, the inducer will run for 30 seconds after the burners shut off to purge products of combustion from the system. This feature is de-activated if a line-voltage thermostat is used for temperature control.

> **WARNING!**
> Infrared heaters are often suspended from high ceilings. Maintenance or repair of these units may require the use of a ladder, scaffold, or self-propelled aerial lift. When working at heights above 6 feet (2 m), a technician must wear a personal fall arrest system (PFAS).

WIRING DIAGRAM ALL MODELS SERIES U

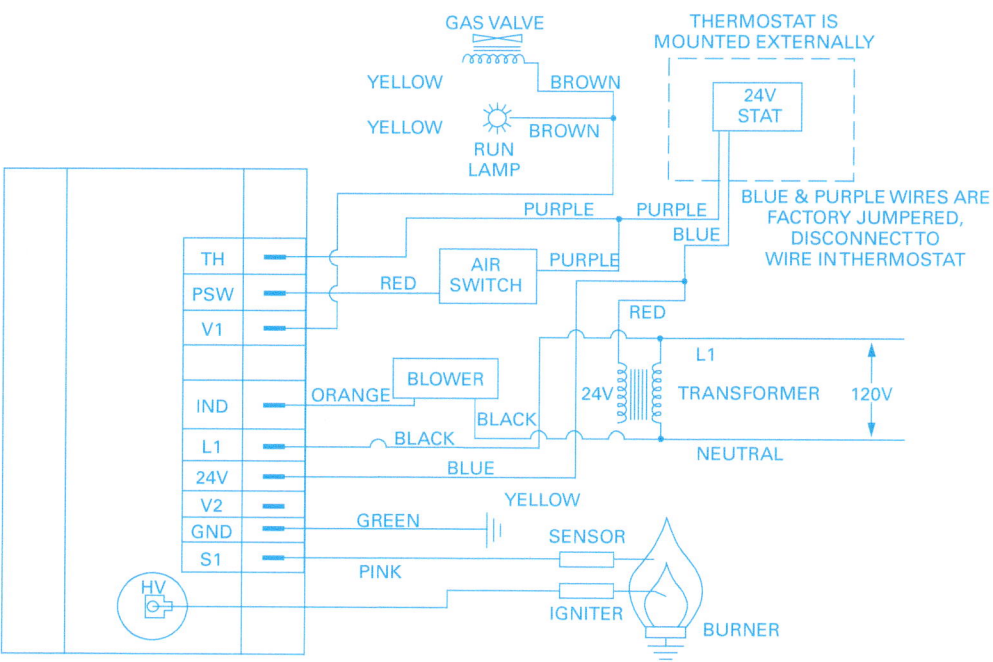

(A) 24 VAC THERMOSTAT

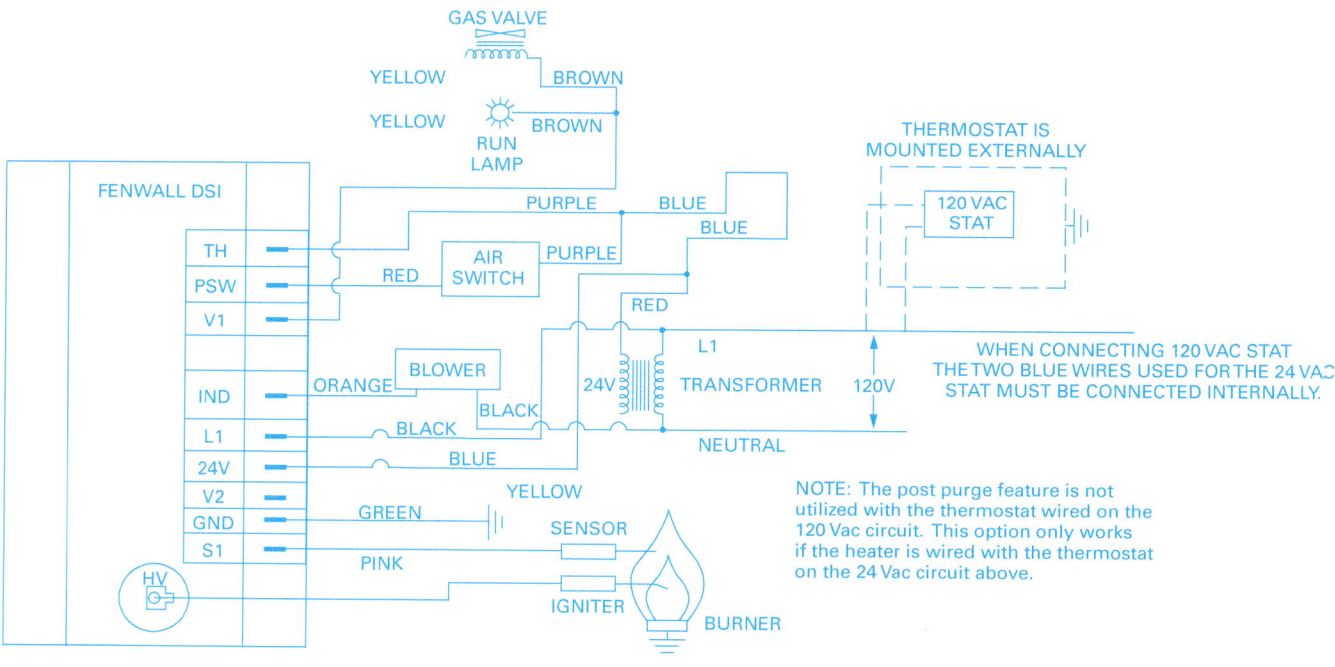

(B) 120 VAC THERMOSTAT

Figure 18 Low-intensity infrared heater wiring diagram.

Additional Resources

Refrigeration and Air Conditioning, An Introduction to HVAC/R, Fourth Edition. Larry Jeffus. Air Conditioning and Refrigeration Institute. New York, NY: Pearson.

2.0.0 Section Review

1. The ceramic grid in a high-intensity infrared heater operates at a temperature of about _____.

    a. 500°F (260°C)
    b. 1,200°F (650°C)
    c. 1,600°F (870°C)
    d. 2,000°F (1,090°C)

2. The simplest method of controlling the operation of an infrared space heater is to use a(n) _____.

    a. line-voltage thermostat
    b. low-voltage thermostat
    c. multi-zone panel
    d. On/Off toggle switch

SECTION THREE

3.0.0 GAS COMBUSTION ANALYSIS

Objective

Explain how to conduct a combustion analysis on a gas furnace.
a. Identify combustion analysis equipment and the combustion byproducts that are of importance to the analysis.
b. Describe the combustion analysis process and how to interpret basic results.

Performance Task

2. Complete a combustion analysis on a gas furnace or boiler.

Trade Terms

Net stack temperature: A value needed to perform a combustion efficiency analysis, calculated by subtracting room temperature from the temperature measured in a furnace's vent or vent connector.

A combustion analysis of an operating gas-fired furnace or boiler provides several benefits, including the following:

- Reduced fuel consumption. A burner operating at maximum efficiency burns less fuel.
- Undesirable exhaust emissions, such as carbon monoxide (CO) and other noxious gases, are reduced when a burner is operating properly.
- A furnace or boiler in which the burner is properly adjusted is much safer. For example, CO is less likely to form and soot is less likely to build up.

Combustion analysis entails measuring the flue gas temperature and sampling the flue gas using specialized test instruments. After the temperature is measured and levels of certain gases in the flue gas are known, combustion efficiency can be determined.

Use the listed AFUE of the furnace or boiler as a guide for expected efficiencies. It is common to expect combustion efficiency numbers on new and properly tuned equipment to be slightly lower than the AFUE number, which is determined under ideal laboratory conditions. Expect lower numbers on older equipment and equipment in need of service or repairs as well.

Furnace efficiency stated as AFUE and combustion efficiency are terms that are often used interchangeably. While similar, the two terms have slightly different meanings, and using them interchangeably can lead to confusion. When the AFUE is determined for a furnace or boiler, the product is tested in a laboratory where conditions are tightly controlled and the furnace is cycled on and off under part-load conditions. A field combustion efficiency analysis is done under real-world conditions with all its variables. The furnace or boiler must be operating continuously under steady-state conditions. A steady-state efficiency test may be a more accurate description of a combustion efficiency test.

Because of test condition differences, it is unrealistic for a combustion efficiency test to produce efficiency numbers that match AFUE numbers generated under precise laboratory conditions. Actual field tests have shown that combustion efficiency tests produce efficiency numbers several points lower than the posted AFUE of a product. These differences should not be of significant concern to service technicians. Be aware that matching the numbers of any stated AFUE value when performing a field combustion analysis is unlikely.

3.1.0 Gas Combustion Byproducts

Air used for combustion is composed of about 21 percent oxygen (O_2), about 78 percent nitrogen, and a small percentage of other gases. In an ideal situation, the O_2 in the air combines with hydrocarbons in the fuel to produce the combustion products carbon dioxide (CO_2) and water vapor. In fact, the byproducts of combustion (*Figure 19*) also contain some O_2 because more air is supplied to the combustion process than is required. This excess air is also provided to ensure that only insignificant quantities of toxic carbon monoxide (CO) can form.

Byproducts of combustion from a furnace operating normally contain CO_2, water vapor, and a small amount of O_2. Nitrogen is not required for combustion, so it passes through the burner. Even though it theoretically passes through the burner unchanged, the heat of the combustion process can form small amounts of nitrogen oxides that are considered pollutants. Nitrogen oxides in the atmospheric sciences are given the generic symbol NO_x. Flue gas temperature also affects combustion efficiency. If the flue gas is too hot, it indicates that heat is exiting through the vent instead of being used to heat the structure.

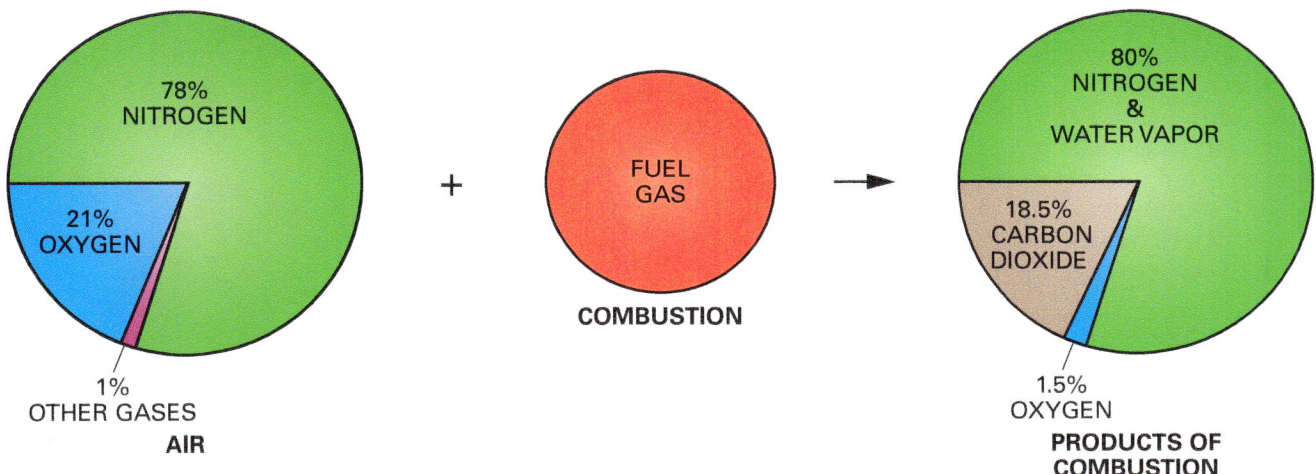

Figure 19 Byproducts of combustion.

Some states and regional pollution control authorities require testing of gas-fired heating equipment to determine levels of CO, CO_2, and NO_x. The NO_x compounds form a group of highly reactive gases. Many nitrogen oxides are colorless and odorless.

However, one common pollutant, nitrogen dioxide (NO_2), along with smoke and other particles in the air, can often be seen as a reddish-brown smog layer over many urban areas. NO_x pollutants form when fuel is burned at high temperatures. The primary manmade sources of NO_x are motor vehicles, electric utilities, and other industrial, commercial, and residential sources that burn fuels. NO_x can contribute to acid rain and is known to cause degradation of surface water quality. It also has a much higher global warming potential than CO_2.

> **NOTE**
> Another nitrogen-oxide compound is nitrous oxide (N_2O), which is not a product of combustion nor is it included in the NO_x category of air pollutants. It is a relatively inert compound commonly used for propellants in aerosols, oxidizers in rocket fuels, and as a mild anesthetic (laughing gas).

The flue gas of an efficient gas furnace or boiler normally operating will have the following characteristics:

- Flue gas temperature will be such that there is adequate heat available for proper operation of the vent, but not excessive to the point that too much heat is being lost through the vent.
- CO_2 and O_2 levels in the flue gas will be at acceptable levels indicating that enough excess air is being supplied to the burners.
- Only minute quantities (less than 25 parts per million) of CO should be present in the flue gas. The presence of any significant amount of CO is an indication of incomplete combustion. If significant CO is present, ensure that the furnace is receiving adequate combustion air and check for other problems that may be causing the level to be high.
- Traces of nitrogen and sulfur oxides may be present.

Natural-draft and induced-draft furnaces and boilers must be installed so that they receive adequate air for combustion. Furnaces or boilers installed in confined spaces must have air openings into the confined space large enough to supply adequate air. Opening sizes are based on the input rating of the furnace. If the building has tight construction, outside air may need to be ducted into the confined space.

> **GOING GREEN**
> **Low-NO_x Furnaces**
> NO_x byproducts are considered to be an air pollutant and their emissions are subject to regulation in many parts of the country. For example, residential gas furnaces and gas water heaters installed in California must be designed to reduce these emissions. Since NO_x byproducts are produced at higher burner temperatures, one solution to reduce these emissions is to cool the burner flame. This is often done by positioning metal rods, metal screens, or metal tabs into the burner flame. The metal acts as a heat sink, cooling the flames so that fewer pollutants are produced. In the United States, all major furnace manufacturers now offer low-NO_x furnaces.

Having an adequate supply of combustion air for condensing furnaces (90-percent AFUE and higher) is not the problem that it is with other furnaces. Condensing furnaces use a sealed combustion system that takes in outdoor air for combustion through a dedicated intake. Indoor air is not used for combustion. One of the reasons for their added efficiency is that heated room air is not being removed to satisfy the venting process.

3.1.1 Gas Combustion Analysis Equipment

Both mechanical and electronic gas burner combustion test equipment are available to conduct an analysis. Mechanical testing equipment is usually a tool kit that contains indicators for CO_2, O_2, and CO in flue gas samples. The kit also contains a thermometer for measuring flue gas or supply air temperatures; a draft gauge to measure the pressure in the vent; a gas pressure manometer to test the gas pressure to the burner manifold; and a slide-rule style combustion efficiency calculator. Blank combustion efficiency calculation forms are often provided with mechanical test kits.

The electronic combustion analyzer shown in *Figure 20* can be used for both gas- and oil-burning appliances. It is one of several makes of electronic combustion analyzers currently available. Digital electronic combustion analyzers are far more versatile than mechanical test kits, though both will produce similar test results. Digital electronic analyzers can provide a combustion analysis in a fraction of the time that it would take with a mechanical analyzer. The kits with these units typically contain multi-function or dedicated sampling probes that are plugged into the analyzer to obtain various readings. Results are available instantly on a digital display and can produce real-time graphs of steady-state operation. Data can be saved, printed out, or downloaded into a computer or mobile device. Many have wireless communication features.

The basic steps for a combustion analysis described in the next section apply to both mechanical and electronic combustion analyzers. Before attempting to use any combustion analyzer, read the instruction manual and become familiar with the various features of the device.

3.2.0 Conducting a Combustion Analysis

Before conducting a combustion analysis on a gas-fired furnace or boiler, inspect the equipment to see if there is any problem that might affect the analysis. Correct any problems before proceeding. Things to look for during a preliminary inspection include the following:

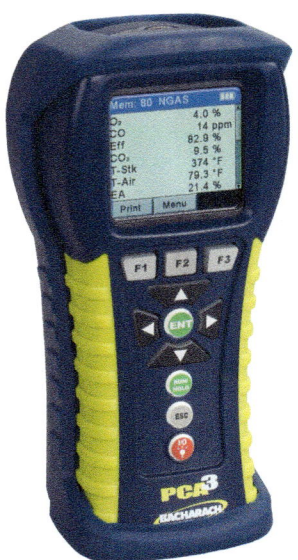

Figure 20 Combustion analyzer.

- Plugged or dirty supply air filters.
- Rust or holes in vent connectors; holes in furnace heat exchangers.
- Soot buildup that could indicate incomplete combustion.
- Restricted or blocked air intakes into an equipment room or a blocked or restricted outside air intake.
- Blocked or restricted combustion air intake and/or vent pipes on condensing furnaces.

WARNING! Use caution when conducting a combustion analysis. Vent connector pipes can be hot. To prevent burns, avoid touching them.

Prepare the equipment for the combustion analysis by drilling a sampling hole in the vent pipe to sample the flue gas and measure the flue gas temperature. *Figure 21* shows the location of the sampling hole in different types of furnaces and boilers.

Each type of appliance has special requirements for the sampling hole, described as follows:

- No sampling hole needs to be drilled when checking a natural-draft furnace. Instead, the flue gas and flue gas temperature is sampled at the outlet of each clam-shell heat exchanger, at the draft diverter. For best results, sample each heat exchanger outlet and then average the values of all samples.
- Natural-draft boilers equipped with an external draft diverter should have the sampling hole below the draft diverter and as close to the boiler vent connector collar as possible.

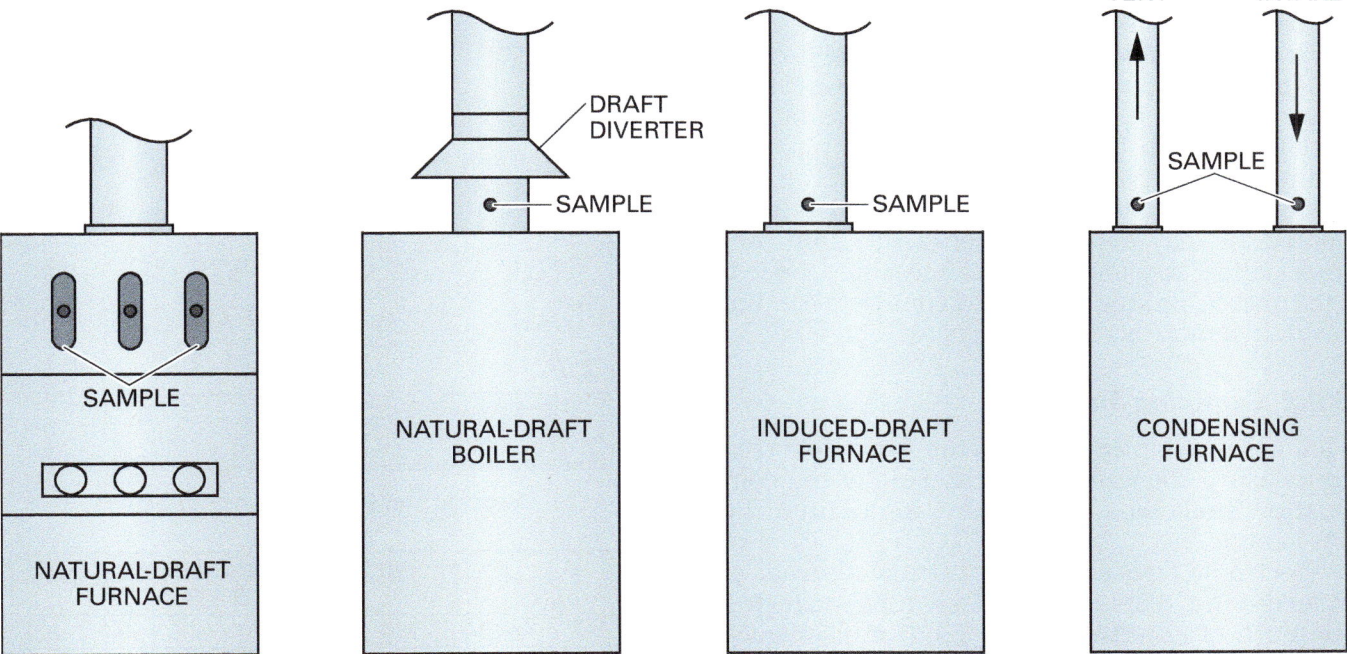

Figure 21 Sampling hole locations.

- 80-percent AFUE furnaces and boilers that use induced-draft technology should have the sample hole drilled directly above the vent connector collar on the furnace.
- Condensing furnaces and boilers should have two sample holes; one in the vent pipe as it leaves the furnace and one on the air intake pipe just before it enters the furnace. If codes or the manufacturer's literature prohibit drilling holes in the plastic pipes, samples can be taken at the outdoor terminations of these two pipes.

> **NOTE** After completing the analysis, seal all sampling holes made in vent connectors with a high-temperature sealant.

After the sample holes have been drilled, set the room thermostat high enough so that the furnace or boiler will run for the duration of the combustion analysis. Allow the burners to operate for at least five minutes to allow the furnace or boiler to reach stability.

3.2.1 Measuring Flue Gas Temperature

Flue gas temperature is a key component of a combustion analysis. The as-read flue-gas temperature cannot be used for the calculation, since the temperature of the flue gas is the sum of the temperature created by combustion and the temperature of the air as it enters the burner. Temperature rise created by combustion is the important value. For example, if the flue gas temperature is 400°F (204°C), that temperature is actually the sum of the room temperature (70°F or 21°C, for example) and the temperature rise due to combustion (330°F or 183°C).

The temperature needed to calculate burner efficiency is known as **net stack temperature**. Net stack temperature is calculated by subtracting room temperature (or temperature in the air intake pipe as it enters the furnace) from the measured flue gas temperature.

To find net stack temperature, insert the thermometer probe into the sample hole. Many probes have a self-sealing feature that allows for a tight seal. Slide the probe back and forth until the highest temperature is reached. On electronic analyzers, pressing a button will record the value. Otherwise, write down the highest stack temperature.

To obtain the net stack temperature, measure the temperature in the room where the furnace or boiler is located, and then subtract that temperature from the flue gas temperature. Electronic analyzers automatically perform this calculation.

For condensing furnaces that do not use indoor air for combustion, measure the temperature of the air entering the furnace through the combustion air intake pipe to calculate net stack temperature. If codes prohibit drilling sampling holes in the plastic air intake and vent pipes of condensing furnaces, measurements may be taken at the terminations of these two pipes. However, measurements taken at the terminations will not be as accurate as those taken close to the furnace. As the pipes move through the structure to or away from the furnace, there may be some heat exchanged with the air outside of the pipe. For example, if the piping is routed through the attic, combustion air may become warmer while the vented air may become cooler. These gains and losses affect any temperatures measured close to the furnace. Taking them into consideration will result in a more accurate analysis.

3.2.2 Sampling Flue Gas

Flue gas is sampled through the same hole used to measure stack temperature. Follow the combustion analyzer's instructions to obtain a flue gas sample.

Electronic devices quickly calculate burner efficiency and display it with all other values at the push of a button (*Figure 22*). If a mechanical analyzer is used, you will need to manually record the measured levels of O_2 and CO_2 and then make some additional calculations.

When the net stack temperature and levels of O_2 and CO_2 are known, these values can be entered into a chart or slide-rule style calculator to determine burner efficiency (*Figure 23*). For example, with a net stack temperature of 380°F, a measured CO_2 level of 8.4 percent, and a measured O_2 level of 6 percent, the calculated burner efficiency is 79.5 percent.

At the completion of the combustion analysis, make any adjustments to the equipment to improve efficiency (if required) and restore the equipment to its previous operating status. Make a final test to ensure there were positive results.

3.2.3 Analyzing the Combustion Efficiency Test

When the combustion analysis is complete, the results should be assessed and corrective action taken to solve any problems. Some problems that the analysis may reveal, along with suggested corrective actions, are as follows:

- Higher than normal flue gas temperatures may be the result of an over-fired burner or a high temperature rise across the heat exchanger. Check burner input and measure temperature rise to confirm and/or correct the problems.
- Lower than normal CO_2 and O_2 levels may indicate the burner is not receiving adequate combustion air. Check for and correct any combustion air problems.

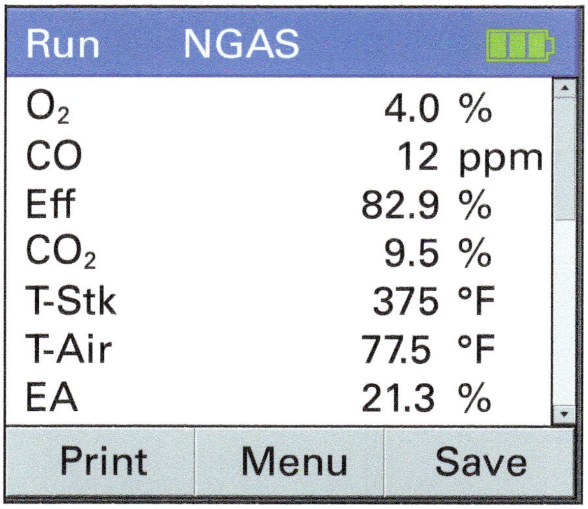

Figure 22 Electronic combustion-analyzer display.

% O_2	% CO_2	NET STACK TEMPERATURE DEGREES °F						
		340	360	380	400	420	440	460
5.0	9.0	81.1	80.5	80.0	79.5	78.9	78.4	77.8
5.5	8.7	80.8	80.3	79.7	79.2	78.6	78.1	77.5
6.0	8.4	80.6	80.0	79.5	78.9	78.3	77.8	77.2
6.5	8.2	80.3	79.7	79.2	78.6	78.0	77.4	76.8
7.0	7.9	80.0	79.4	78.8	78.2	77.6	77.0	76.4
7.5	7.6	79.7	79.1	78.5	77.9	77.3	76.7	76.0
8.0	7.3	79.4	78.8	78.1	77.5	76.9	76.2	75.6
8.5	7.0	79.1	78.4	77.8	77.1	76.4	75.8	75.1
9.0	6.7	78.7	78.0	77.3	76.6	76.0	75.3	74.6

Figure 23 Combustion-efficiency calculator chart.

- Higher than normal CO levels may indicate incomplete combustion caused by a lack of combustion air or a burner that is damaged in some way.

3.2.4 Other Combustion Efficiency Information

Both electronic and mechanical combustion analyzers can provide other valuable information, including the following:

- Vent draft can be measured through the sampling hole in the vent of natural-draft and 80-percent AFUE furnaces and boilers. Draft on gas furnaces is a function of vent height and diameter. If the draft is excessive, it may cause excess heat to be moved out the vent, decreasing combustion efficiency.
- A smoke spot tester in mechanical combustion analysis kits can detect smoke in the flue gases. This device is used when doing a combustion analysis on oil burners, but is of little use with gas-fired appliances.

- Electronic combustion analyzers can often detect other gases that may be present in flue gas, including NO_x and sulfur, or CO. Many states and localities have limits on NO_x emissions. The combustion analysis can determine whether these emissions are within acceptable limits.

Paperwork

When you complete a combustion analysis, always present a copy of your before-and-after analysis to the customer. Even if they do not completely understand the process, a printed copy of your analysis boosts the credibility of your work and is an indication of your professionalism.

Additional Resources

Refrigeration and Air Conditioning, An Introduction to HVAC/R, Larry Jeffus. Fourth Edition. Air Conditioning and Refrigeration Institute. New York, NY: Pearson.

3.0.0 Section Review

1. Which substance would you *not* expect to find in any significant quantity in the flue gas of a normally operating gas furnace?
   a. Water vapor
   b. O_2
   c. CO_2
   d. CO

2. The flue gas sampling hole in a natural-draft boiler should be drilled _____.
   a. in a clam-shell heat exchanger
   b. directly below the draft diverter
   c. directly above the inducer motor
   d. in the combustion air intake pipe

SUMMARY

Troubleshooting gas heating appliances requires skill in reading and interpreting electrical diagrams as well as the ability to evaluate and test the ignition, combustion, and air distribution systems. The HVAC technician must have good observation skills and a working knowledge of the many gas-heating systems and accessories available. Technicians must also possess a thorough understanding of the safety precautions required when working with gas heating equipment.

Although all gas heating controls have elements in common, the controls are implemented by various means. Packaged gas-heating units include natural-draft and induced-draft units. Gas furnace types include induced-draft furnaces and low- and high-intensity infrared gas-fired heaters. It is important for technicians to be able to analyze the various types of control circuits used in these devices.

The HVAC technician must understand the operation of various types of control devices, including igniters, gas valves, and flame sensors. Combustion analysis of an operating gas-fired furnace or boiler provides several benefits, including reduced fuel consumption; elimination of undesirable exhaust emissions such as CO and other noxious gases; and a furnace or boiler that is safer to operate.

Review Questions

1. The most durable hot-surface igniters are made with _____.
   a. silicon carbide
   b. silicon dioxide
   c. silicon chloride
   d. silicon nitride

2. The ignition device used in the system depicted in *Figure 3* is a(n) _____.
   a. hot-surface igniter
   b. glow plug
   c. spark igniter
   d. intermittent pilot

3. Natural-draft furnaces are no longer available for new installations because they are _____.
   a. expensive to consumers
   b. not energy efficient
   c. too noisy
   d. not profitable

4. The ignition device used in the system depicted in *Figure 5* is a _____.
   a. hot-surface igniter
   b. glow coil
   c. spark igniter
   d. standing pilot

5. Pilotless heating systems use either a direct spark igniter or a(n) _____.
   a. indirect spark igniter
   b. bimetal igniter
   c. thermocouple igniter
   d. hot-surface igniter

6. Which of the following is *not* a problem to expect with a spark igniter?
   a. Faulty cold junction
   b. Incorrect spark gap
   c. Ignition module failure
   d. Cracked ceramic insulator

7. When testing a flame-sensing rod, the reading should be _____.
   a. 12VAC
   b. 30Ω
   c. 0.5–4.5mA
   d. 0.5–4.5μA

8. Oxidation on a flame-sensing rod could _____.
   a. prevent further corrosion of the rod
   b. enhance the thermal conductivity of the rod
   c. cause a low current reading
   d. have a negligible effect on flame sensing

9. The basic purpose of infrared heaters is to heat _____.
   a. objects
   b. air
   c. water
   d. coolant

10. High-intensity infrared gas-fired heaters are also called _____.
    a. vented units
    b. indirect radiant heaters
    c. direct-fired radiant burners
    d. tubular heaters

11. Optional wire screens used to cover the grids of high-intensity infrared gas-fired heaters are called _____.
    a. U-tubes
    b. radiators
    c. vents
    d. reverberators

12. A lack of adequate combustion air is indicated if the flue gas contains _____.
    a. CO_2
    b. moisture
    c. CO
    d. O_2

13. Which mathematical function is used to calculate net stack temperature?
    a. Addition
    b. Subtraction
    c. Multiplication
    d. Division

14. What temperature should you seek when the temperature probe is inserted into the vent connector?
    a. Lowest
    b. Average
    c. Highest
    d. Most stable

15. Which of the listed quantities is *not* required to calculate combustion efficiency?
    a. Percentage flue gas O_2
    b. Net stack temperature
    c. Percentage flue gas CO_2
    d. Percentage atmospheric O_2

Trade Terms Introduced in This Module

Flame rectification: A process in which exposure of a sensing rod to a flame produces a tiny direct current which can be used to control the gas valve.

Modulating gas valve: A gas valve that precisely controls burner capacity by adjusting the flow of gas to burners in small increments.

Net stack temperature: A value needed to perform a combustion efficiency analysis, calculated by subtracting room temperature from the temperature measured in a furnace's vent or vent connector.

Reverberator: A wire or corrugated-metal screen used to cover the ceramic grids in a gas-fired radiant heater, which boosts the radiant-heat output of the heater by converting convective heat produced by the grid into radiant heat.

Submerged flame: A method of gas-fired radiant heater operation that uses burning gas within two-layer, porous-ceramic grids.

Appendix A

GAS FURNACE TROUBLESHOOTING GUIDE

GENERALLY THE CAUSE Make these checks first.	OCCASIONALLY THE CAUSE Make these checks only if the first checks failed to locate the trouble.	RARELY THE CAUSE Make these checks only if other checks failed to locate the trouble.
PROBLEM: No heat – Burner fails to start		
Power failure Blown fuses or tripped circuit breaker Open disconnect switch or blown fuse Thermostat set too low Thermostat switch not in proper position Pilot Manual fuel valve Pilot valve	Control transformer Limit controls (manual reset) Pilot safety control Gas valve Thermocouple Fuel lines Gas meter	Faulty wiring Loose terminals Low voltage Thermostat faulty Improper burner adjustment Plugged burner orifices Gas pressure Thermostat not level (mercury bulb thermostat only)
PROBLEM: No heat – Inducer starts but burner fails to ignite		

Combustion air switch
Inducer motor
Primary or safety controls
Gas valve
Flame detector or electrode
Improper burner adjustment
Manual fuel valve
Pilot valve
	Ignition devices
	Regulator
	Plugged vent or air supply
	   (induced draft only)
	Plugged pressure switch line
	   at inducer (induced draft only)
		Low voltage
		Fuel lines
		Gas pressure
		Plugged burner orifice
		Condensate trap broken or empty
		   (condensing furnace only)

PROBLEM: Insufficient or no heat – Burner starts and fires, then switches off when safety switch trips		
Safety controls Flame detector or electrode Improper burner adjustment Dirty air filter Dirty or plugged heat exchanger	Combustion air switch Inducer motor	Faulty wiring Loose terminals Fuel lines Gas pressure
PROBLEM: Insufficient or no heat – Burner cycle too short		
Limit controls Thermostat not level (mercury bulb thermostat only) Thermostat location Blower belt broken or slipping Dirty filters Thermostat heat anticipation incorrect	Fan control Blower bearings Blower motor speed Low air volume Dirty blower wheel	Faulty wiring Loose terminals Low voltage Blower wheel Ductwork small or restricted Restrictions
PROBLEM: Insufficient or no heat – Burner runs continuously		
Blower belt slipping Dirty blower wheel Dirty filters Gas pressure	Orifice size Low air volume Improper burner adjustment Plugged burner ports Input too low Undersized furnace	Ductwork small or restricted Displaced or damaged baffles Insufficient insulation Excessive infiltration

03209 Troubleshooting Gas Heating

GENERALLY THE CAUSE Make these checks first.	OCCASIONALLY THE CAUSE Make these checks only if the first checks failed to locate the trouble.	RARELY THE CAUSE Make these checks only if other checks failed to locate the trouble.
PROBLEM: Too much heat – Burner cycle too long		
Thermostat heat anticipation incorrect Thermostat out of calibration	Thermostat not level Thermostat location	
PROBLEM: Combustion noise		
Pilot Gas valve Improper burner adjustment Plugged burner orifice Air shutter adjustment (if so equipped)	Inducer motor Primary or safety controls Ignition transformer Flashback to venturi Gas pressure Pressure regulator	Faulty wiring Loose terminals Regulator Dirty or plugged heat exchanger Vent or flue Displaced or damaged baffles Blocked heat exchanger Input too high
PROBLEM: Mechanical noise		
Blower bearings Blower belt cracked or slipping	Inducer motor Blower motor Cabinet	Low voltage Control transformer Control relay or contactor Blower wheel Cracked or ruptured heat exchanger Displaced or damaged baffles Fuel lines
PROBLEM: Air noise		
Blower Cabinet Ductwork small or restricted Air leaks in ductwork	Other air system restrictions	Blower wheel Dirty filters
PROBLEM: Odor		
Vent or flue Flue gas spillage Fuel leaks Cracked or ruptured heat exchanger Blocked heat exchanger	Regulator Low air volume Improper burner adjustment Displaced or damaged baffles Humidifier stagnant water Water or moisture	Faulty wiring Loose terminals Control transformer Dirty filters Dirty or plugged heat exchanger Input too high Input too low Outdoor odors
PROBLEM: Cost of operation		
Combustion motor Blower motor Dirty filters Improper burner adjustment Equipment size	Low air volume Insufficient insulation Excessive infiltration	Low voltage Fan control Orifice Blower belt slipping Dirty blower wheel Ductwork small or restricted Vent or flue Input too high

Appendix B

GAS FURNACE TROUBLESHOOTING FLOW CHART

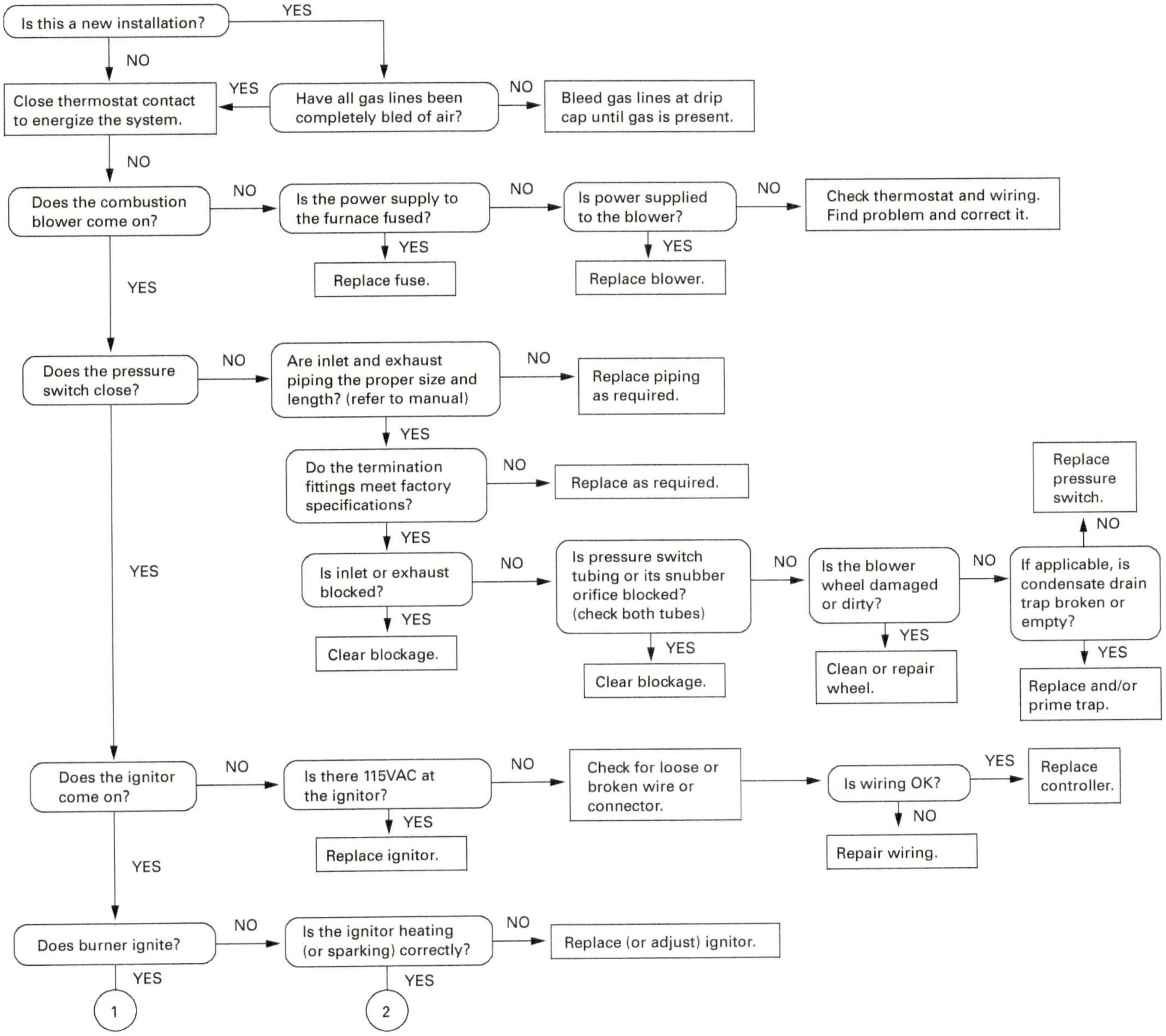

```
Is this a new installation? --YES--> Have all gas lines been completely bled of air? --NO--> Bleed gas lines at drip cap until gas is present.
          |                                      |
          NO                                     YES
          |                                      |
          v                                      v
Close thermostat contact <---YES---------------- 
to energize the system.
          |
          NO
          |
          v
Does the combustion --NO--> Is the power supply to --NO--> Is power supplied --NO--> Check thermostat and wiring.
blower come on?              the furnace fused?             to the blower?            Find problem and correct it.
          |                         |                           |
          YES                       YES                         YES
          |                         v                           v
          |                    Replace fuse.              Replace blower.
          v
Does the pressure --NO--> Are inlet and exhaust piping the proper size and --NO--> Replace piping as required.
switch close?             length? (refer to manual)
          |                         |
          |                         YES
          |                         v
          |              Do the termination fittings meet factory --NO--> Replace as required.
          |              specifications?
          |                         |
          |                         YES
          |                         v
          |              Is inlet or exhaust --NO--> Is pressure switch tubing or its snubber --NO--> Is the blower wheel --NO--> If applicable, is condensate --NO--> Replace pressure switch.
          |              blocked?                    orifice blocked? (check both tubes)              damaged or dirty?          drain trap broken or empty?
          |                         |                         |                                              |                          |
          YES                       YES                       YES                                            YES                        YES
          |                         v                         v                                              v                          v
          |                    Clear blockage.           Clear blockage.                              Clean or repair wheel.    Replace and/or prime trap.
          v
Does the ignitor --NO--> Is there 115VAC at --NO--> Check for loose or broken --> Is wiring OK? --YES--> Replace controller.
come on?                  the ignitor?                 wire or connector.                  |
          |                         |                                                      NO
          YES                       YES                                                    |
          |                         v                                                      v
          |                    Replace ignitor.                                       Repair wiring.
          v
Does burner ignite? --NO--> Is the ignitor heating (or sparking) correctly? --NO--> Replace (or adjust) ignitor.
          |                         |
          YES                       YES
          |                         |
         (1)                       (2)
```

03209 Troubleshooting Gas Heating

Module Five 35

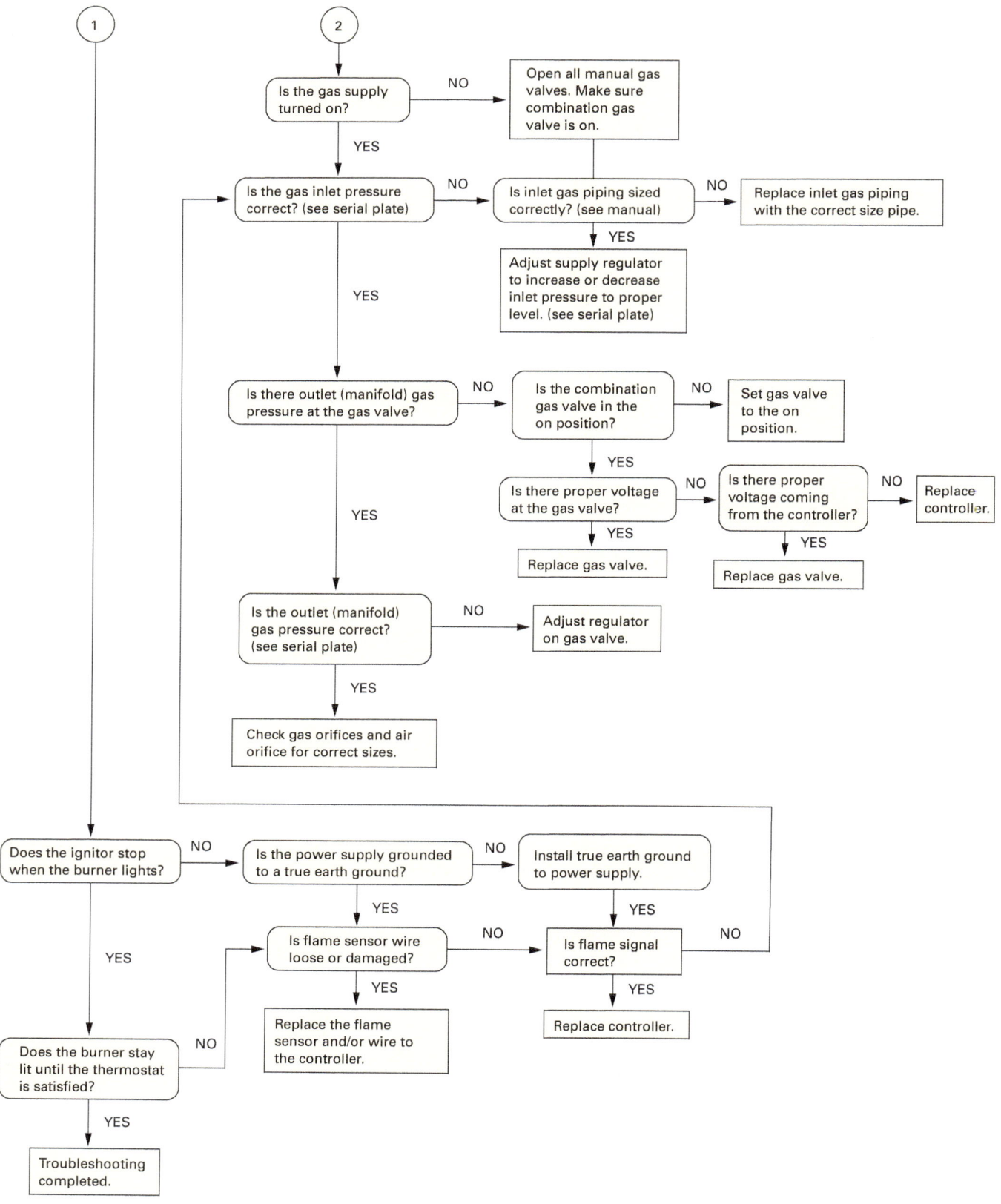

Additional Resources

This module presents thorough resources for task training. The following resource material is suggested for further study.

Refrigeration and Air Conditioning, An Introduction to HVAC/R, Larry Jeffus. Fourth Edition. Air Conditioning and Refrigeration Institute. New York, NY: Pearson.

Figure Credits

© iStock.com/CarolinaSmith, Module opener
BostonHeatingSupply.com, Figure 1A
Courtesy of Emerson Climate Technologies, Figures 1C, 4
Carrier Corporation, Figures 3, 10
Courtesy of Combustion Research Corporation, Figure 15
Enerco Group, Inc., Figure 16
Courtesy of AmbiRad Limited, Figure 17
Courtesy of Superior Radiant Products Ltd., Figure 18
Courtesy of Bacharach, Inc., www.MyBacharach.com, Figures 20, 22

Section Review Answer Key

Answer	Section Reference	Objective
Section One		
1. c	1.1.3	1a
2. a	1.2.2	1b
3. b	1.2.0	1c
4. d	1.4.2	1d
Section Two		
1. c	2.1.2	2a
2. d	2.2.0	2b
Section Three		
1. d	3.1.0	3a
2. b	3.2.0	3b

NCCER CURRICULA — USER UPDATE

NCCER makes every effort to keep its textbooks up-to-date and free of technical errors. We appreciate your help in this process. If you find an error, a typographical mistake, or an inaccuracy in NCCER's curricula, please fill out this form (or a photocopy), or complete the online form at **www.nccer.org/olf**. Be sure to include the exact module ID number, page number, a detailed description, and your recommended correction. Your input will be brought to the attention of the Authoring Team. Thank you for your assistance.

Instructors – If you have an idea for improving this textbook, or have found that additional materials were necessary to teach this module effectively, please let us know so that we may present your suggestions to the Authoring Team.

NCCER Product Development and Revision
13614 Progress Blvd., Alachua, FL 32615

Email: curriculum@nccer.org
Online: www.nccer.org/olf

❏ Trainee Guide ❏ Lesson Plans ❏ Exam ❏ PowerPoints Other _____

Craft / Level: _____ Copyright Date: _____

Module ID Number / Title: _____

Section Number(s): _____

Description: _____

_____

_____

_____

_____

Recommended Correction: _____

_____

_____

_____

_____

Your Name: _____

Address: _____

_____

Email: _____ Phone: _____

Troubleshooting Oil Heating

Overview

In northern sections of the United States, many homes are heated with oil furnaces. There was a time when oil was a very inexpensive way to heat, so oil furnaces became very popular. In other instances, natural gas just was not available to the home as it was being built, so oil was the best alternative. Although the end result is the same as that of a gas furnace, the method of producing heat is significantly different. One of the main differences is that the oil is stored on site and must be pumped to the point of combustion, whereas natural gas is supplied under pressure through pipelines. Although its use as a home heating fuel is diminishing as time goes on, it is still an important fuel that will likely be encountered in the field.

Module 03310

Trainees with successful module completions may be eligible for credentialing through NCCER's National Registry. To learn more, go to **www.nccer.org** or contact us at **1.888.622.3720**. Our website has information on the latest product releases and training, as well as online versions of our *Cornerstone* magazine and Pearson's product catalog.

Your feedback is welcome. You may email your comments to **curriculum@nccer.org**, send general comments and inquiries to **info@nccer.org**, or fill in the User Update form at the back of this module.

This information is general in nature and intended for training purposes only. Actual performance of activities described in this manual requires compliance with all applicable operating, service, maintenance, and safety procedures under the direction of qualified personnel. References in this manual to patented or proprietary devices do not constitute a recommendation of their use.

Copyright © 2018 by NCCER, Alachua, FL 32615, and published by Pearson, New York, NY 10013. All rights reserved. Printed in the United States of America. This publication is protected by Copyright, and permission should be obtained from NCCER prior to any prohibited reproduction, storage in a retrieval system, or transmission in any form or by any means, electronic, mechanical, photocopying, recording, or likewise. To obtain permission(s) to use material from this work, please submit a written request to NCCER Product Development, 13614 Progress Blvd., Alachua, FL 32615.

03310 V5

From *HVAC Level Three, Trainee Guide*. NCCER.
Copyright © 2018 by NCCER. Published by Pearson. All rights reserved.

03310
TROUBLESHOOTING OIL HEATING

Objectives

When you have completed this module, you will be able to do the following:

1. Identify the primary components of an oil-fired furnace and explain its operation.
   a. Describe a basic oil-fired heating system.
   b. Describe the primary components and operation of a pressure-type oil burner.
   c. Describe the safety controls used on oil furnaces.
   d. Describe the fuel supply system used with oil furnaces.
2. Describe how to perform periodic servicing of a typical oil-fired heating system.
   a. Describe the basic servicing procedures performed on an oil-fired system.
   b. Describe how to perform a combustion efficiency test.
3. Describe how to troubleshoot a typical oil-fired heating system.
   a. Describe troubleshooting procedures for typical oil furnace controls.
   b. Describe troubleshooting procedures for common oil heating problems.

Performance Tasks

Under the supervision of your instructor, you should be able to do the following:

Complete two of the following:
1. Remove and reinstall an oil pump in single-pipe and two-pipe systems.
2. Test a cad cell flame detector.
3. Conduct a complete combustion analysis (smoke test and draft included).
4. Remove and replace an oil burner nozzle and set the electrode gap.

Trade Terms

Annual Fuel Utilization Efficiency (AFUE)
Atomize
Cad cell
Coping
Downdraft
Flame-retention burner
Joist
Lockout
Over-firing
Primary control
Refractory
Smoke spot test
Soot
Temperature differential
Weatherized furnace

Industry-Recognized Credentials

If you are training through an NCCER-accredited sponsor, you may be eligible for credentials from NCCER's Registry. The ID number for this module is 03310. Note that this module may have been used in other NCCER curricula and may apply to other level completions. Contact NCCER's Registry at 888.622.3720 or go to **www.nccer.org** for more information.

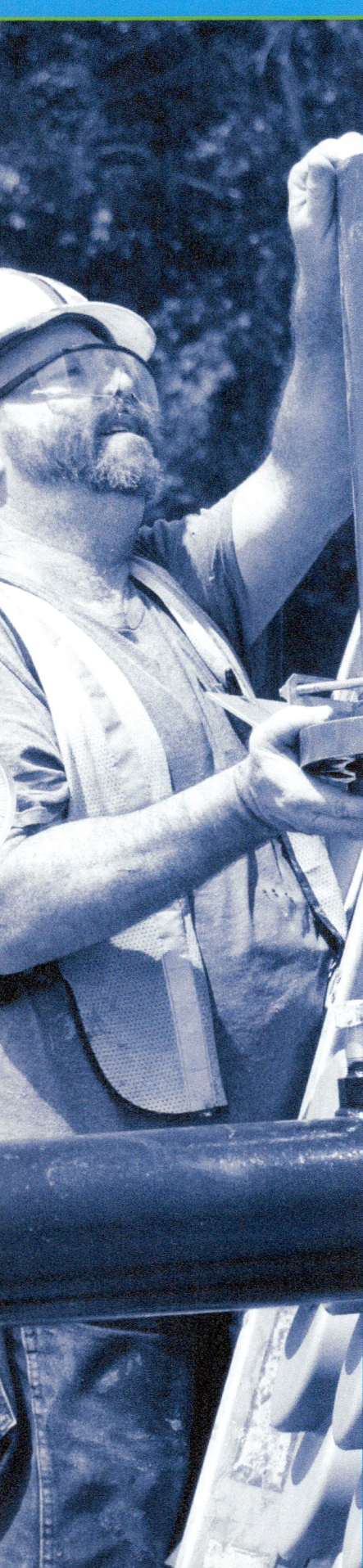

Contents

1.0.0 Oil-Fired Heating Systems 1
  1.1.0 Oil-Fired Heating System Basics 1
  1.2.0 Pressure-Type Oil Burners 2
    1.2.1 Power Assembly 3
    1.2.2 Nozzle Assembly 4
    1.2.3 Ignition System 5
    1.2.4 Flame-Retention Burners 6
  1.3.0 Oil Safety Controls 6
  1.4.0 Oil Supply System 8
    1.4.1 Fuel Oil 8
    1.4.2 Piping Systems 8
2.0.0 Servicing Oil Heating Systems 14
  2.1.0 Servicing Procedures for Oil-Fired Systems 14
    2.1.1 Changing the Oil Filter 16
    2.1.2 Check the Safety Controls 16
    2.1.3 Check the Temperature Rise 16
    2.1.4 Servicing the Oil Burner Assembly 17
  2.2.0 Measuring Combustion Efficiency and Flue Gases 19
    2.2.1 Smoke Spot Testing 21
    2.2.2 Combustion Efficiency and Flue Gas Testing 21
3.0.0 Oil Furnace Troubleshooting 23
  3.1.0 Control Circuit Troubleshooting 23
    3.1.1 Primary Control Troubleshooting 23
    3.1.2 Cad Cell Flame Detector Troubleshooting 26
    3.1.3 Ignition Components Troubleshooting 27
  3.2.0 Troubleshooting Common Oil Heating Problems 28
    3.2.1 Oil Burner Troubleshooting 28
    3.2.2 Fuel Supply Troubleshooting 29
    3.2.3 Odors 29
    3.2.4 Flue and Chimney Exhaust Troubleshooting 30
    3.2.5 Draft Regulator 31
    3.2.6 Solenoid Oil Flow Valves 31
    3.2.7 Electronic Time Delay 32
Appendix A Troubleshooting Summary 37
Appendix B Troubleshooting Chart 41

Figures

Figure 1 Oil furnace 1
Figure 2 An oil-fired heating system 2
Figure 3 Pressure oil burner components 3
Figure 4 Power assembly 4
Figure 5 Nozzle assembly 4
Figure 6 Nozzle patterns 4
Figure 7 Spray angles 5
Figure 8 Ignition transformer 5
Figure 9 Ceramic insulators 6
Figure 10 Electrode position 7
Figure 11 Flame-retention oil burner 7
Figure 12 Primary control for an oil-fired boiler 7
Figure 13 Cad cell 7
Figure 14 Stack switch 8
Figure 15 Single-pipe oil supply system 9
Figure 16 Two-pipe oil supply—burner below tank level 10
Figure 17 Two-pipe oil supply—burner above tank level 11
Figure 18 Oil piping system for rooftop units 12
Figure 19 An oil filter element 16
Figure 20 Hole locations for temperature rise test 17
Figure 21 Oil burner assembly 17
Figure 22 Nozzle wrench 18
Figure 23 Electrode adjustment 18
Figure 24 Electrode setting tool 19
Figure 25 Typical air tube length 19
Figure 26 Electronic gas/oil combustion analyzer 20
Figure 27 Smoke spot tester kit 20
Figure 28 Test hole locations 21
Figure 29 Oil burner control system 23
Figure 30 Operating sequence number one 24
Figure 31 Field wiring diagram number one 25
Figure 32 Primary control and wiring diagram 26
Figure 33 Gun-type burner 28
Figure 34 Improper burner flame 29
Figure 35 Common chimney problems 30
Figure 36 Draft regulator 31
Figure 37 Oil valve circuit 31
Figure 38 Time delay schematic 32

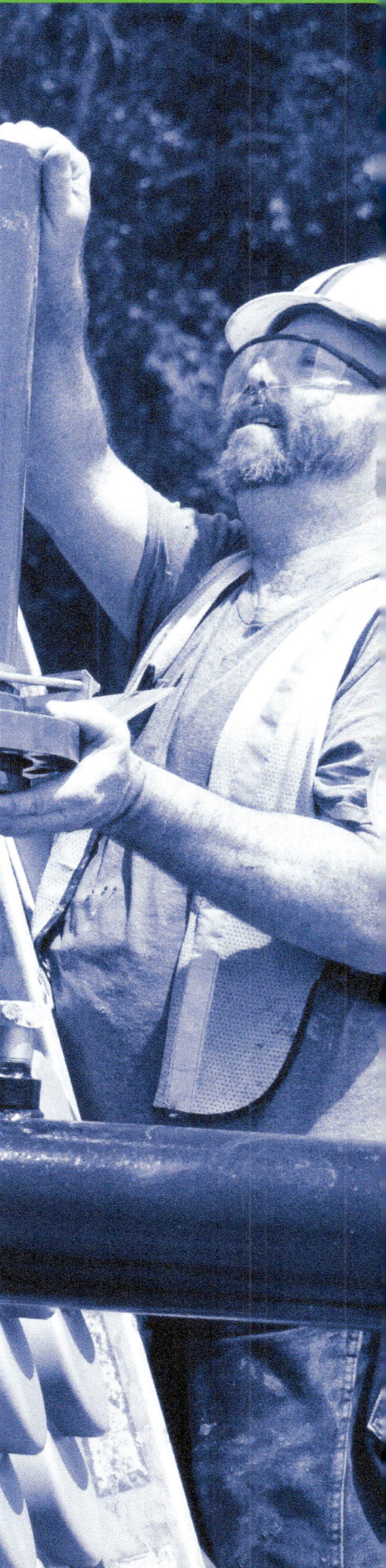

SECTION ONE

1.0.0 OIL-FIRED HEATING SYSTEMS

Objectives

Identify the primary components of an oil-fired furnace and explain its operation.
 a. Describe a basic oil-fired heating system.
 b. Describe the primary components and operation of a pressure-type oil burner.
 c. Describe the safety controls used on oil furnaces.
 d. Describe the fuel supply system used with oil furnaces.

Trade Terms

Annual Fuel Utilization Efficiency (AFUE): An industry standard used to define furnace efficiency.

Atomize: Reduce to a fine spray.

Cad cell: A light-sensitive device (photocell) in which the resistance reacts to changes in the amount of light. In an oil furnace, it acts as a flame detector.

Flame-retention burner: A high-efficiency oil burner.

Lockout: A safety feature that prevents a furnace from automatically cycling if the flame fails.

Primary control: A combination switching and safety control that turns the oil burner on and off in response to signals from the thermostat and safety controls.

Refractory: Materials such as ceramics designed to withstand extremely high temperatures.

Weatherized furnace: A furnace that has been well insulated and has a weather-resistant case as specified by government standards.

Oil furnaces are designed to burn fuel oil, which is mixed with air, then sprayed as a fine mist into the furnace's combustion chamber, where it is ignited by an electric spark. Heat produced in the combustion chamber heats up the furnace heat exchanger. Air returned from the conditioned space is directed over the heat exchanger by a blower in the furnace unit. Heat is then transferred to that air and recirculated by the duct system.

A typical oil furnace is shown in *Figure 1*. Oil furnaces do not achieve the efficiency levels of gas-fired furnaces. Current high-efficiency oil furnaces achieve an Annual Fuel Utilization Efficiency (AFUE) of 86.5 percent in comparison with gas furnaces, which report AFUE ratings up to 98.2 percent. The US Department of Energy has established minimum AFUE ratings for oil furnaces of 83 percent for weatherized furnaces and 78 percent for non-weatherized furnaces. These ratings do not apply to oil furnaces used in mobile homes, which have lower minimum efficiency ratings.

1.1.0 Oil-Fired Heating System Basics

In a typical oil-fired furnace, the oil burner receives oil from a storage tank on the premises (*Figure 2*). The oil burner shoots a spray of oil into the combustion chamber, where it is mixed with combustion air. Electrodes located at the point where the oil spray enters the combustion chamber provide a high-voltage spark that ignites the oil.

The purpose of the combustion chamber is to protect the heat exchanger from flame damage and to provide reflected heat to the burning oil. The reflected heat warms the tips of the flame, enhancing combustion. No part of the flame should touch the combustion chamber surface. If it does, incomplete combustion could result. The chamber must fit the flame and the nozzle must be located at the correct height above the chamber floor.

Three types of materials are commonly used for combustion chamber construction: metal, typically stainless steel; insulating fire brick; and molded ceramic. In small furnaces such as those used in residential applications, the combustion

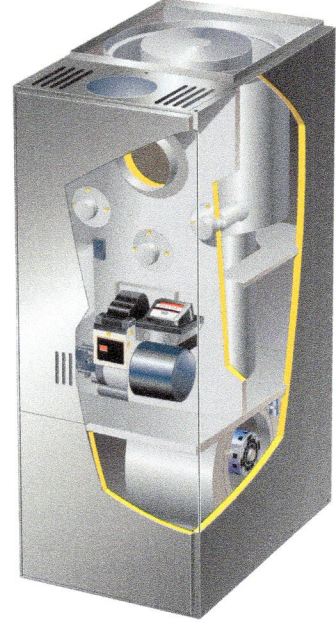

Figure 1 Oil furnace.

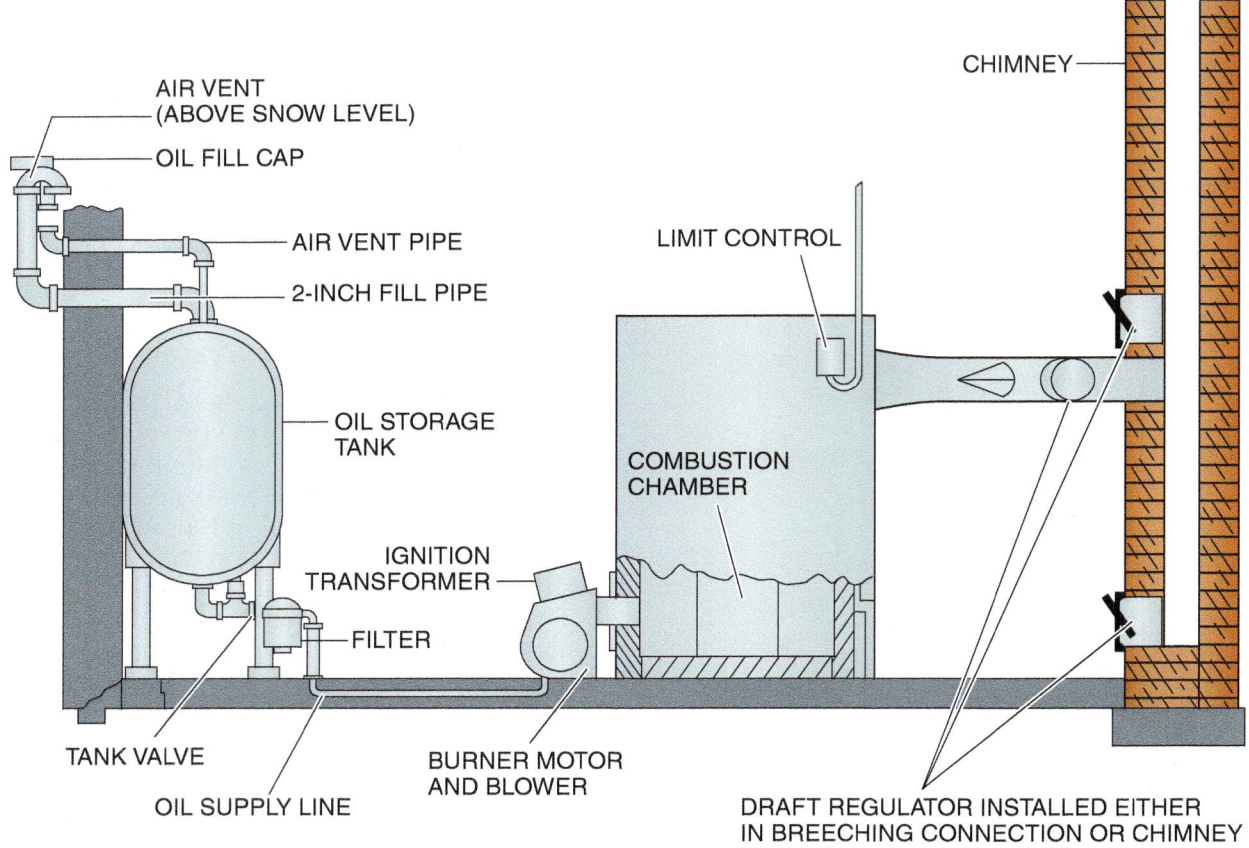

Figure 2 An oil-fired heating system.

chamber is built into the furnace. For large commercial and industrial applications, the combustion chamber may be constructed on site. Such chambers can be made of fire brick or pre-cast ceramic (refractory) material. If pre-cast materials do not fit the job, a combustion chamber may be formed and poured on site using ceramic materials designed for this purpose.

A moldable refractory material, called by the generic name of wet-pack, is used to build or re-line combustion chambers. This material is moist when it is purchased, and is shipped in sealed plastic bags. Because it is moist, it can be easily molded into the desired shape. Continued exposure to air causes the wet-pack to dry into a very hard, rigid structure. The material has very poor heat transfer characteristics, making it a good insulator and chamber construction material.

The heat exchanger of an oil-fired furnace has primary and secondary heating surfaces. The primary heating surface surrounds the flame, whereas the secondary heating surface is often a series of steel sections over which the hot combustion gases pass before leaving the heat exchanger on their way to the chimney flue.

The main elements of an oil-fired system are the following:

- Oil burner assembly
- Controls
- Fuel supply system
- Flue/chimney exhaust system

1.2.0 Pressure-Type Oil Burners

There are two categories of pressure-type oil burners: the low-pressure gun-type and the high-pressure gun-type. In a low-pressure gun-type burner, which may be found in older furnaces, oil and primary air are mixed before going through a nozzle. At a pressure of 1 to 15 psi, the mixture is forced through the nozzle where it atomizes as it exits. Secondary combustion air is drawn into the spray mixture after it is released from the nozzle. An electric spark is used to light the combustible mixture.

A high-pressure gun-type burner forces oil through the nozzle under a pressure of 100 to 150 psi. Some burners may even be designed for operation at pressures up to 200 psi. This breaks the oil into much finer, mist-like droplets. The atomized oil spray leaving the nozzle creates a low-pressure area into which the secondary combustion air is drawn in and mixes with the oil. Combustion air is supplied by a vane fan, which is part of the burner assembly. The high-pressure, gun-type burner is the most popular domestic burner. It is simple in construction and efficient in operation. The parts are mass-produced, readily available, and relatively low in cost.

Oil furnaces in current production provide an atomized oil vapor mixed with the correct proportion of air to the combustion chamber. An ignition transformer provides the spark at a set of electrodes to ignite the fuel-air mixture in the combustion chamber. The primary control locks the system out on a flame failure. The lockout must be manually reset. Cad cells are commonly used to sense the burner light in order to verify the presence of a flame for continued operation.

A pressure-type oil burner assembly (*Figure 3*) consists of a pump, an air tube and nozzle assembly, and a fan. Oil is pumped to the nozzle, which converts it to a fine mist. The oil mist is mixed with air in the air tube (also known as a blast tube) and ignited by an electric spark. The burner cycles on or off in accordance with the demands of the room thermostat. The burner mechanism is outside the furnace or boiler, making it easily accessible for servicing. Grade 2 fuel oil is commonly used with pressure-type oil burners.

1.2.1 Power Assembly

The power assembly of the high-pressure burner unit consists of the motor, fan, and oil pump (*Figure 4*). The motor drives both the fan and oil pump through a single shaft. The fan, mounted on the motor shaft, forces air through a blast tube to provide combustion air for the atomized oil. The inlet to the fan has an adjustable opening so that the air volume can be controlled manually. The oil pump, coupled to the motor shaft, draws oil from the storage tank and delivers it to the nozzle. The fuel/air mixture is ignited by an electric spark that is formed between two electrodes in the nozzle assembly.

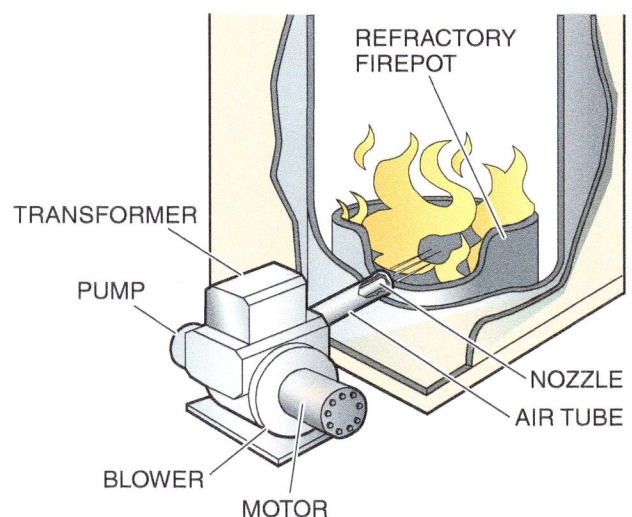

(A) PRESSURE OIL BURNER COMPONENTS

(B) OIL BURNER ASSEMBLY

Figure 3 Pressure oil burner components.

A pressure-regulating valve on the oil pump has an adjustable spring that permits adjustment of the oil pressure. The pump delivers more oil than the system can use. The excess oil is re-

Heating With Biodiesel

Biodiesel is a fuel made from new or recovered vegetable oil or animal fat. For several years, people with diesel-powered cars have been using filtered oil from restaurant deep fryers to power their cars. Experiments with blends of heating oil and biodiesel in oil furnaces have been successful and people using a blend containing 20 percent biodiesel have experienced no problems. In addition, no modifications of the furnaces have been necessary. Further experiments by the US Department of Energy have shown that a 70-30 blend can be used with no change in heating performance. At this writing, if a heating fuel blend contains less than 10 percent of biodiesel, it need not be labeled as being a blend. One of the main benefits of using such a blend is a reduction of carbon emissions. In addition, biodiesel, unlike fossil fuel, is a readily renewable resource.

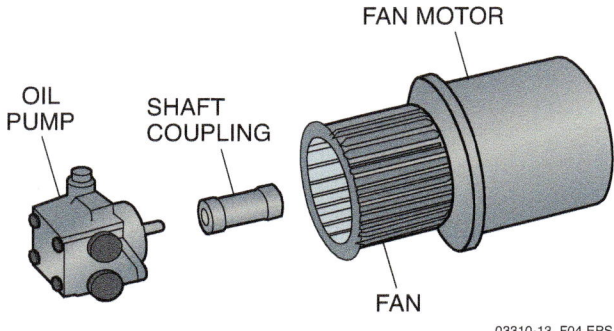

Figure 4 Power assembly.

turned to the tank or bypassed back to the supply line internally. The choice depends on the oil line configuration. A two-pipe system allows for the excess oil to return to the tank, while a single-pipe system does not.

On some burners, a solenoid-controlled oil flow valve stops the flow of oil as soon as power is removed, thereby preventing oil from dripping into the combustion chamber. Oil pumps are designed for single-stage or two-stage operation. The single-stage pump is used where the oil supply is above the level of the burner and the oil is mainly fed to the pump by gravity. The two-stage unit is used where the storage tank is below the burner and the oil must be lifted. The first stage draws the oil to the pump and the second stage provides the pressure required by the nozzle. Pump suction should not exceed 15 inches of vacuum.

1.2.2 Nozzle Assembly

The nozzle assembly (*Figure 5*) consists of the oil feed line, the nozzle, ignition electrodes, and the transformer connections. The nozzle prepares the oil for mixing with the air by atomizing the fuel. Oil passes through a strainer and enters slots

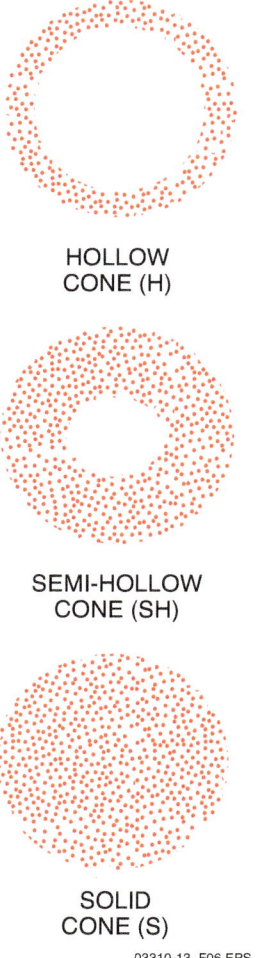

Figure 6 Nozzle patterns.

that direct the oil to the swirl chamber. The swirl chamber gives the oil a rotary motion when it enters the nozzle orifice, shaping the spray pattern. The nozzle orifice increases the velocity of the oil. The oil leaves the nozzle in the form of a mist or spray and mixes with the air from the blast tube.

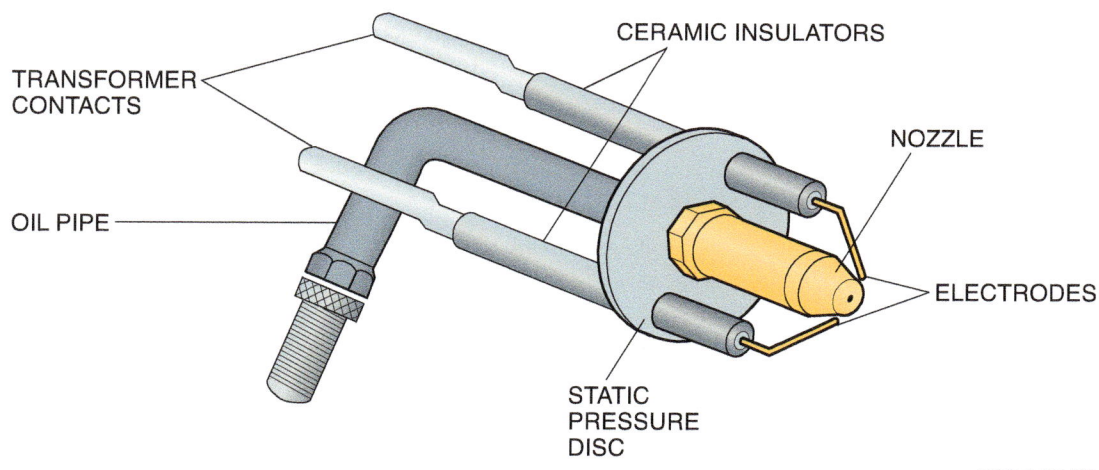

Figure 5 Nozzle assembly.

Nozzles are selected by the manufacturer to provide the correct oil volume and spray pattern for a given oil burner. If it is necessary to replace a nozzle, the replacement should be the same size and type as the original.

Due to the fine tolerances of the nozzle construction, dirty or defective nozzles are usually replaced. Nozzle spray patterns vary with the type of application. The shape of the spray and the angle between the sides of the spray can be varied. There are three basic spray patterns, as shown in *Figure 6*: hollow (H), semi-hollow (SH), and solid (S). The hollow and semi-hollow are most popular on domestic burners as they are more efficient when used with modern combustion chambers.

The angle of the spray must correspond to the type of combustion chamber (*Figure 7*). An angle of 70 to 90 degrees is usually best for square or round chambers; an angle of 30 to 60 degrees is best suited to long, narrow chambers. Again, it is important to use the spray pattern and nozzle angle recommended by the manufacturer for proper performance. Although another spray pattern or nozzle angle may prove functional, it will not provide the best performance and could result in irreparable damage to the furnace.

Oil burner nozzles are sized by the gallons per hour of oil they will pass at the design oil pressure. Information regarding the size and type of an oil burner nozzle is stamped on the nozzle body. Different nozzle manufacturers may use different letters or combinations of letters to identify the spray pattern. When replacing a nozzle, consult the manufacturers catalog information if unfamiliar with the markings.

1.2.3 Ignition System

The ignition system for high-pressure burners consists of a step-up transformer (*Figure 8*) connected to two electrodes. The transformer supplies high voltage that causes a spark to jump between the two electrodes. The force of the air in the blast tube causes the arc to bend into the fuel-air mixture, igniting it.

Ceramic insulators surround the electrodes (*Figure 9*) and serve to position them. The step-up

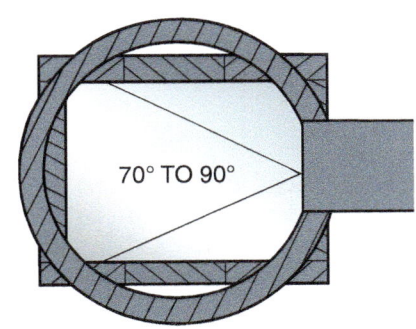

70° TO 90° FOR ROUND OR SQUARE CHAMBERS

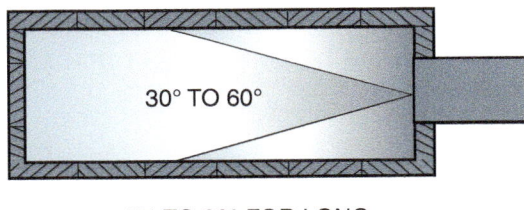

30° TO 60° FOR LONG NARROW CHAMBERS

Figure 7 Spray angles.

Figure 8 Ignition transformer.

Oil Pressure

For many years, it was commonly accepted that the oil-burner pump outlet pressure should be 100 psi. Today, some manufacturers specify higher pressures, and 150 psi has become common. Always check the manufacturer's literature before checking or adjusting the pump outlet pressure.

transformer increases the voltage from 120V to about 10,000V, but reduces the amperage to about 20 milliamps. The ceramic insulators prevent arcing from the electrode to other parts of the burner, since a voltage this high can jump across a significant gap to a grounded surface. The low amperage reduces wear on the electrode tips.

The correct position of the electrodes (*Figure 10*) is important. If the electrodes are not centered on the nozzle orifice, the flame will be one-sided and will cause carbon to form on the nozzle.

The oil burner nozzle flow rate determines burner input. For example, one gallon of Grade 2 fuel oil has a heat content of approximately 140,000 Btus. If using a half-gallon nozzle, the burner input would be 70,000 Btus per hour (Btuh) because it delivers only half as much fuel as the one-gallon nozzle. The nozzle flow rate is stamped on the nozzle body. Note that the flow rate discussed here is related to furnace input and not the output. The output will be lower, determined by the combustion efficiency of the burner.

1.2.4 Flame-Retention Burners

Flame-retention burners (*Figure 11*) are used in modern oil-burning appliances and are designed to produce significant improvements in overall heating efficiency, while simultaneously reducing the number of burner-related service calls. Flame-retention burners use motors that operate at 3,450 rpm instead of the conventional 1,725 rpm. This higher speed is needed to overcome the greater resistance to air flow that is created by the retention ring.

The advantages of a flame-retention burner include the following:

- Better air-oil mixing and flame retention within the air pattern.
- A higher flame temperature with less excess air.
- More usable energy produced for the amount of oil consumed.
- Less effect on flame from stack draft variations.
- Fewer products of incomplete combustion.
- During the off-cycle, the retention head reduces the flow of air through the burner, over the heat exchanger, and out the stack, thereby removing less heated air from the structure.

1.3.0 Oil Safety Controls

All oil-fired furnaces are equipped with a central electrical control known as the primary control (*Figure 12*), which is connected to the room thermostat. The primary control is usually mounted on the burner assembly. It provides both control and safety functions. As a control device, its job is to turn the burner on and off in response to commands from the thermostat. As a safety control, the primary shuts off power to the burner if the burner does not ignite or if the burner flame goes out. Its response is based on one or more simple

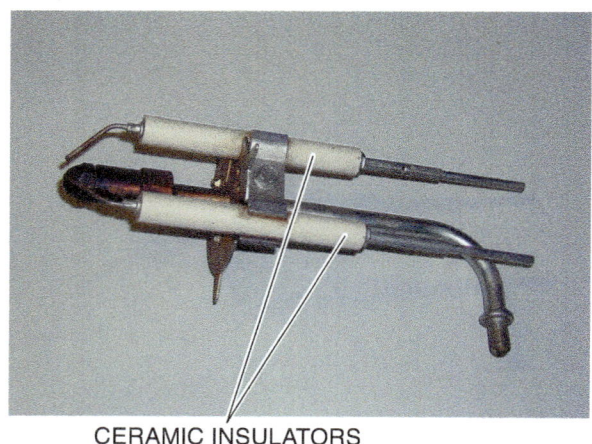

Figure 9 Ceramic insulators.

Nozzle Identification

The specific characteristics of oil burner nozzles are clearly identified. For example, a stamp that reads 0.75, 90H means that the nozzle has a flow rate of 0.75 gallons per hour at the design oil pressure (usually 100 psi), a spray angle of 90 degrees, and a hollow spray pattern.

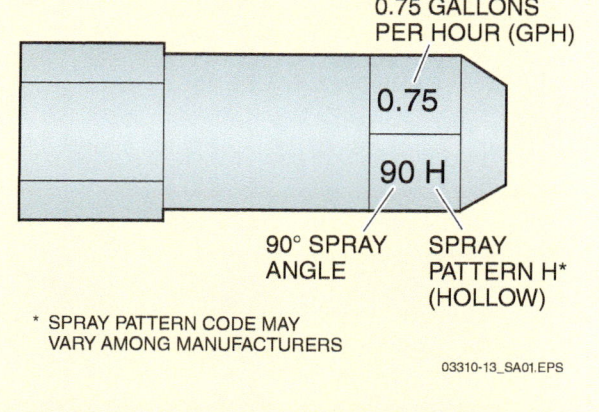

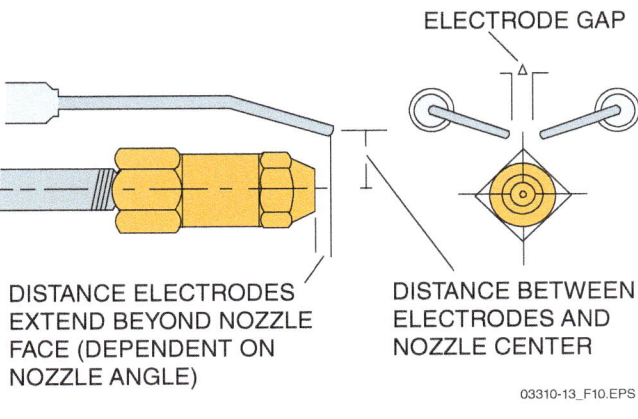

Figure 10 Electrode position.

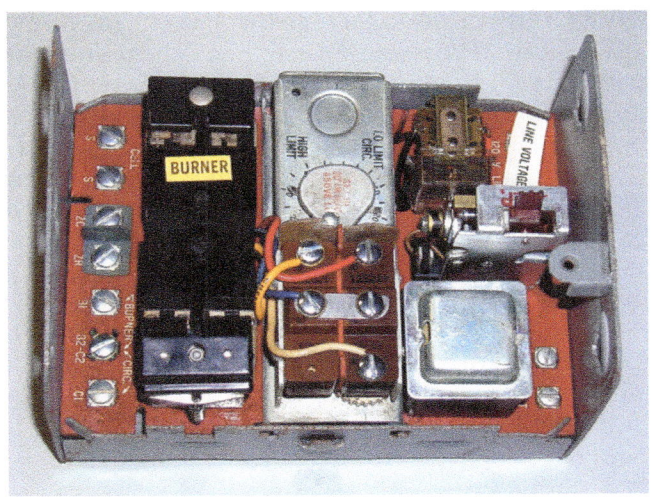

Figure 12 Primary control for an oil-fired boiler.

Figure 11 Flame-retention oil burner.

inputs. The primary control may also control the circulating pump in simple hydronic heating systems.

In modern furnaces, a cadmium sulfide cell, or cad cell (*Figure 13*) is used to sense the presence of a flame. The cad cell performs the same basic function as a gas furnace flame detector, but in a very different way. It is photosensitive, and is installed in the burner assembly on a line of sight with the burner flame. The light of the flame causes the cad cell to generate a small amount of voltage, proving that a flame is established. If the cell does not sense a burner flame, the primary control will not sense any voltage and will turn off the burner.

Another safety device is the limit switch, which is actuated by a bimetal element located in the discharge air plenum of the furnace. It is in series with the power source to the primary control. The limit switch is usually calibrated to

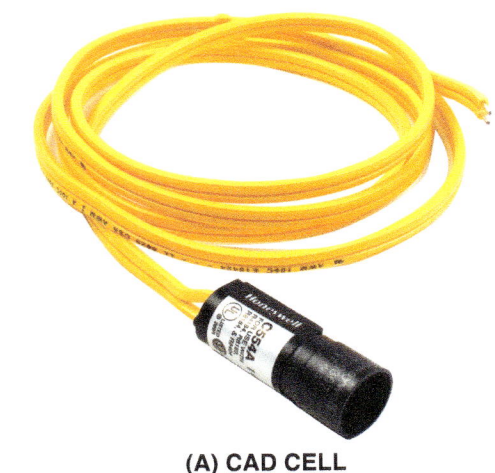

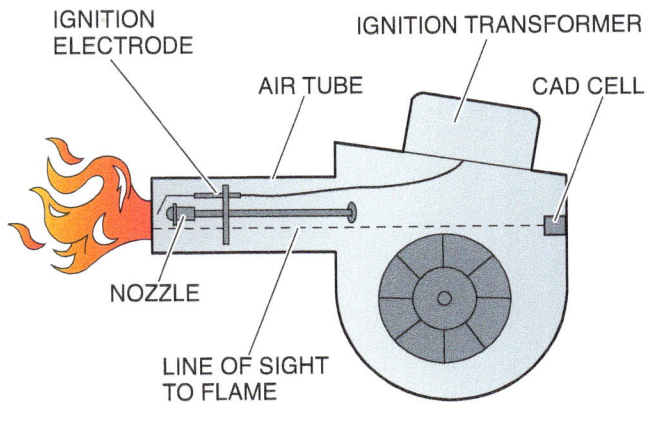

Figure 13 Cad cell.

open if the discharge air temperature reaches or exceeds 200°F. This limit switch will usually reset when the temperature drops 25°F below the cutout point. A secondary (auxiliary) limit switch is an additional safety control used on some down-flow furnaces to protect the conditioned air blower motor against excessively high temperatures. If a limit switch or other safety device trips more than once, the cause should be determined.

Older installations may have a stack switch (*Figure 14*) in place of a cad cell. The stack switch is installed in the flue vent (stack) ahead of the draft damper. It senses heat and shuts down the oil burner if heat is not detected in the stack. Most stack switches combined the function of a primary control and burner ignition proof into a single component. These are more accurately referred to as stack controls or stack relays.

1.4.0 Oil Supply System

Oil furnaces may be supplied from indoor or outdoor oil tanks. Outdoor tanks may be buried. Because many underground oil tanks have leaked, there are strict regulations governing placement, construction, and alarm systems for these tanks. Aboveground tanks over a certain size may require special containment arrangements to ensure the oil from leaks can be captured. No oil tank installation should be undertaken without a clear understanding of local and national codes.

1.4.1 Fuel Oil

Fuel oils are rated according to their Btu per gallon content and the American Petroleum Institute (API) gravity. The API gravity is an index related to the heating values for standard grades of fuel oil. There are six common grades of oil: Nos. 1, 2, 4, 5 (light), 5 (heavy), and 6. The lighter-weight oils have a higher API gravity.

- *Grade 1* – A light-grade distillate for use in vaporizing-type oil burners.
- *Grade 2* – A heavier distillate for use in domestic pressure-type oil burners.
- *Grade 4* – A light residue or heavy distillate used for higher-pressure commercial oil burners.
- *Grade 5 (light)* – A medium-weight, residual-type fuel used in commercial oil burners that are specifically designed for it.
- *Grade 5 (heavy)* – A residual-type fuel for commercial oil burners. Usually requires preheating.
- *Grade 6* – A very heavy residual fuel used for industrial burners. Preheating is necessary in the tank to permit pumping; additional preheating is needed at the burner to convert the fuel into a fine spray or atomize it. Grade 6 is also known as Bunker C.

Grade 1 fuel oil (kerosene) has a slightly lower heat content than Grade 2 fuel oil. In northern climates, Grade 1 may be used if the oil is to be stored in an outside tank. Grade 2 fuel oil has a tendency to thicken when exposed to extreme cold, which makes it hard to pump from the storage tank to the burner. Fuel oil must be properly atomized in order to burn.

1.4.2 Piping Systems

Piping connections to the oil pump are of two configurations: single-pipe and two-pipe. In the single-pipe system, there is only one pipe from the storage tank to the burner. In the two-pipe system, two pipes are routed from the tank to the burner. One carries supply oil; the other carries excess oil returning to the tank from the pump. Oil lines are typically connected using flare fittings; compression fittings should never be used on oil system piping.

Oil piping must be purged of air in order for a solid column of oil to enter the pump. While single-pipe systems generally need to be primed and the air bled from the lines, two-pipe systems prime easily and remain free of air.

Figures 15 through *18* are examples of oil furnace installations that have been used in the past. National, state, and local codes covering oil furnace installations, especially underground tanks, are under constant review and are subject to change. Before beginning any installation, con-

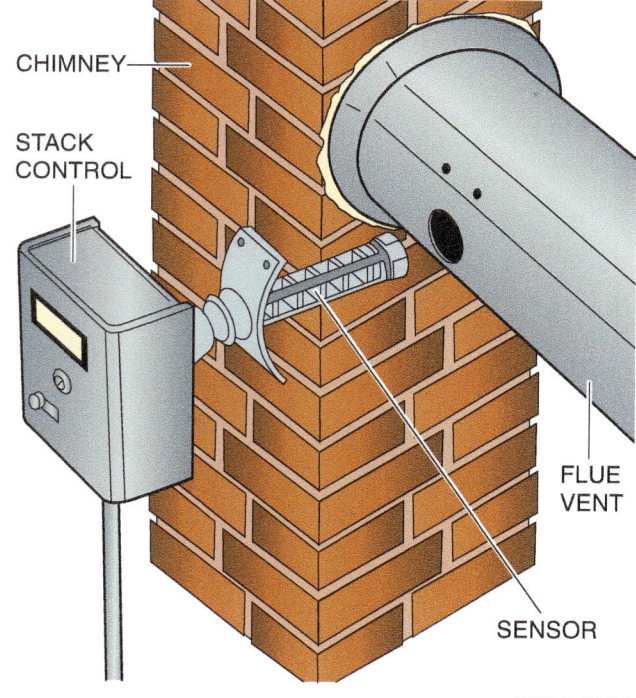

Figure 14 Stack switch.

sult the manufacturer's installation instructions, as well as applicable codes. If an underground tank is used, make sure it meets current construction standards. The US Environmental Protection Agency (EPA) has published special requirements for commercial installations with underground tanks that are different from those for residential installations.

Figure 15 shows a single-pipe system with the supply tank in the same space as the burner. The single-pipe system should only be used when the burner is located on the same level or above the bottom of the oil supply tank. Single-pipe systems often depend primarily on gravity to feed oil to the burner. The pump used on pressure-type burners does create some vacuum to assist oil movement, but not enough to lift the oil very far. In gravity-feed installations, a solenoid oil flow valve at the tank outlet may be appropriate to help prevent oil leakage from a damaged oil line.

The piping on this application must be leak-free. No air should be allowed to enter the oil piping and pump after the supply system has been purged. If air does enter through a leak, the system will likely need to be primed again before the burner will operate.

Figure 16 depicts a two-pipe system with the burner located below or on the same level as the supply tank. However, note that the routing of the oil piping fell below the level of the burner. As a result, a two-pipe system was used.

Note that the storage tank is buried outside the structure. An all-brass check valve should be installed in the supply piping as close to the tank as possible. It is also acceptable to install a globe valve in the supply piping so the oil can be shut off for burner service or removal.

Figure 17 shows the two-pipe system with the burner located above the tank when a lift of less than 15' is required. A check valve and globe shutoff valve should also be included in this piping network.

Figure 18 shows a piping system required where a lift of more than 15' is necessary and multiple units are tied together with a single oil system. A boost pump is often used in this application.

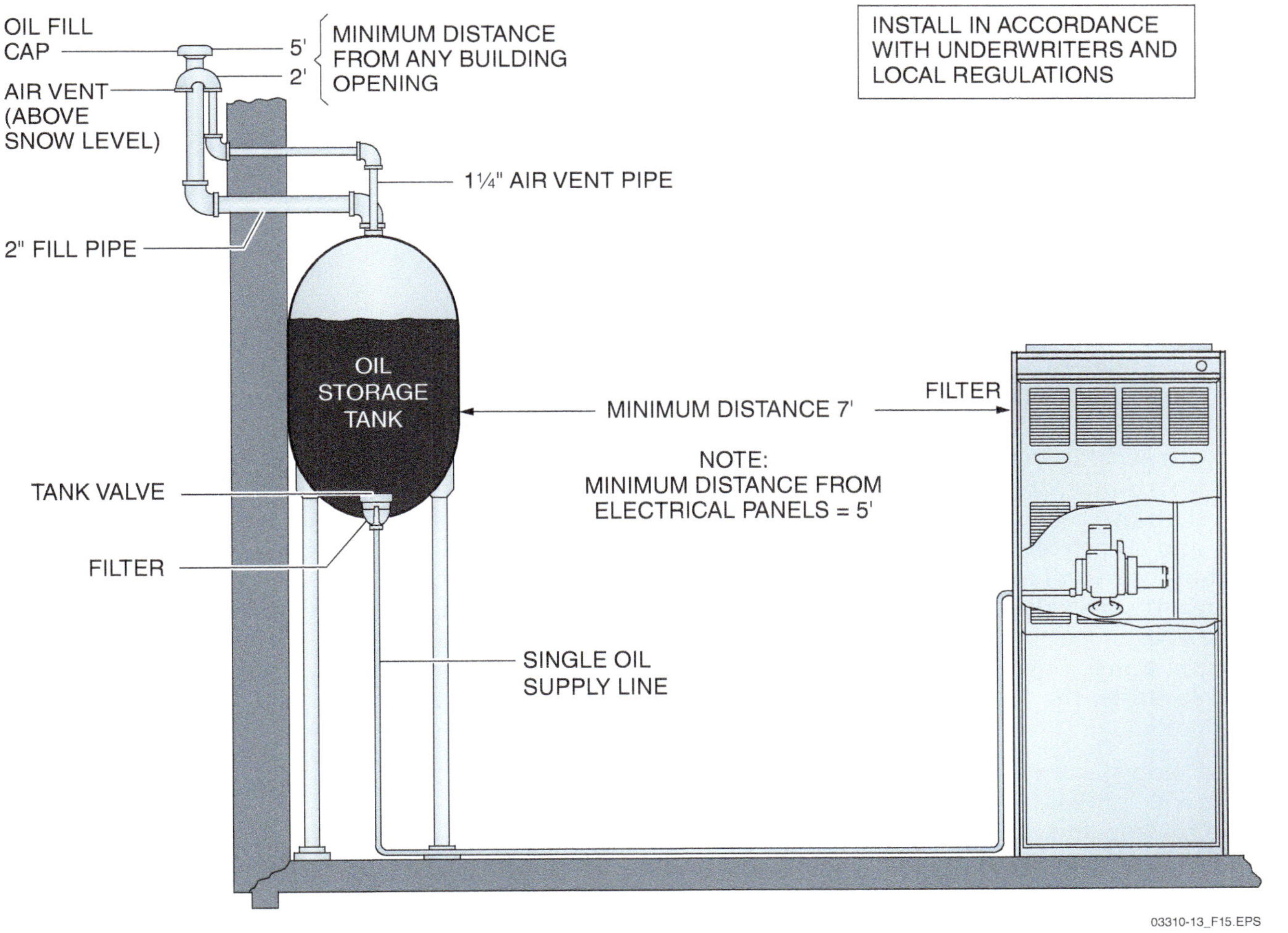

Figure 15 Single-pipe oil supply system.

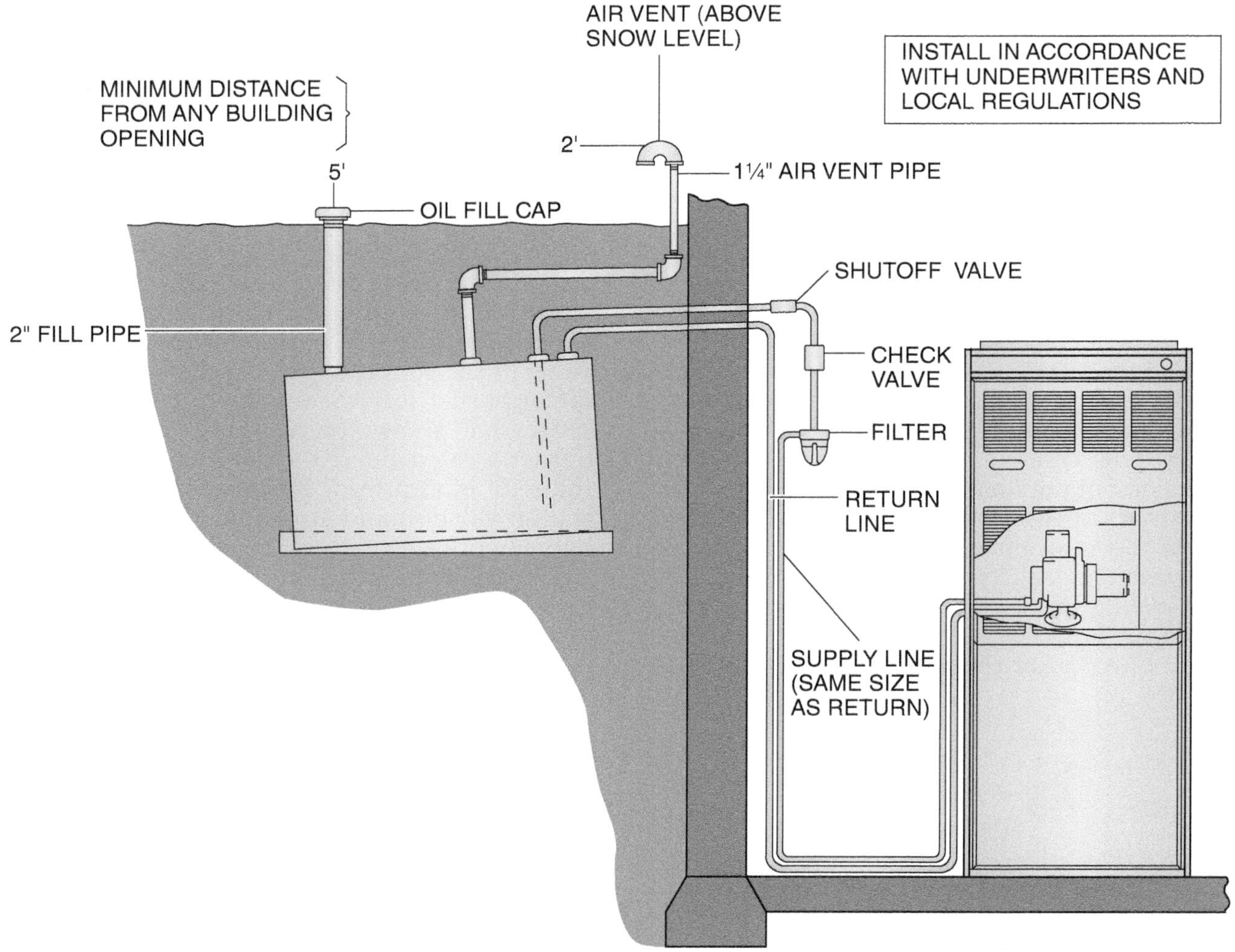

Figure 16 Two-pipe oil supply—burner below tank level.

The schematic in the illustration portrays a series-loop system where all the oil is pumped through individual smaller tanks on each unit in series and then returned directly to the main supply through a separate return line.

Some manufacturers recommend that the piping system use two pipe connections at the boost pump and at each oil rooftop unit. The supply piping from the main tank to the boost pump should include a fusible valve as well as oil filter(s) of adequate size.

To guarantee a trouble-free supply system for multiple units, the following factors should be considered:

- A rooftop system should be limited to a maximum of eight oil-fired units.
- The fittings and lengths of supply and return lines must be accurately calculated to guarantee proper delivery of fuel to all units.
- Supply piping should be run to the farthest burner first and then piped toward the boost pump so that the last unit is the closest to the boost pump. The manufacturer's recommendations as to supply pump pressure must be followed to eliminate seal failure.
- The return piping should never include smaller pipe sizes than the supply piping. One size larger is sometimes recommended on a vertical return line to permit free flow of fuel back to the supply tank.
- A vacuum breaker should be installed at the highest point in the system. It should be located on the last unit in the supply run, closest to the return piping.
- All piping exposed to low ambient outside temperature must be insulated.
- Local and other applicable codes must be followed and will take precedence over any other design and construction considerations for the fuel oil supply system.

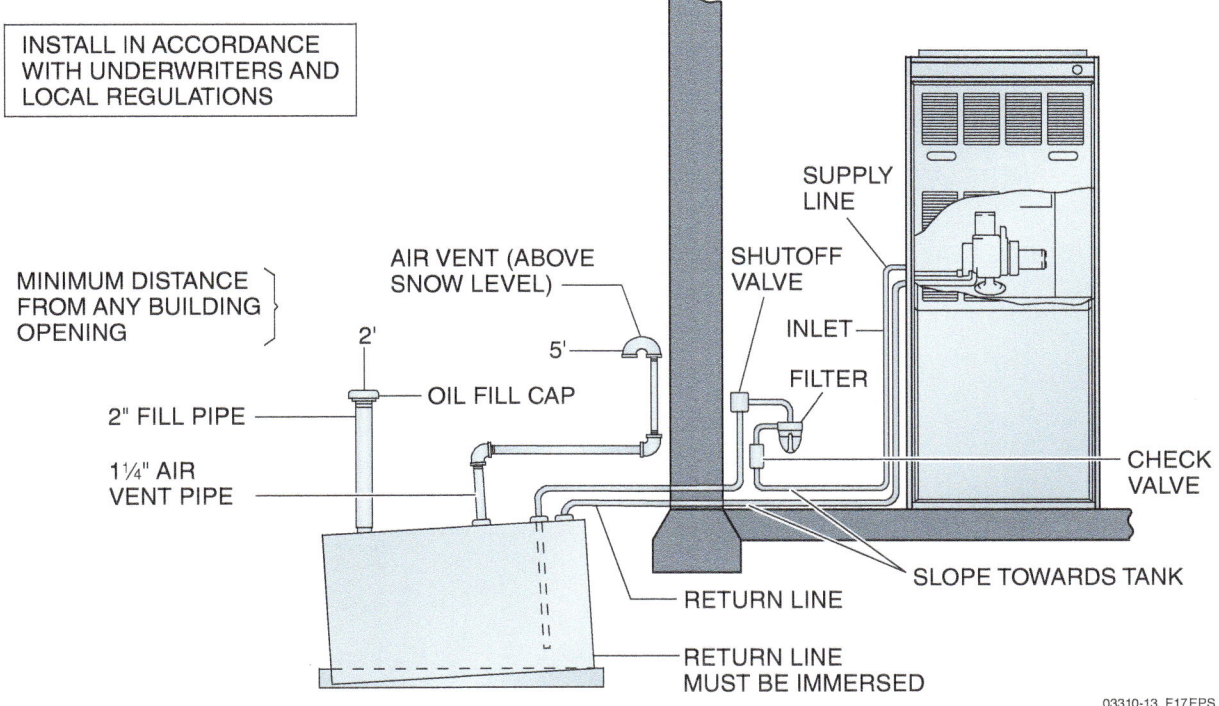

Figure 17 Two-pipe oil supply—burner above tank level.

The Bypass Plug

Most oil pumps are designed to operate on either single-pipe or two-pipe systems. A bypass plug is generally inserted into these pumps for two-pipe applications. For single-pipe applications, the bypass plug must be removed for the pump to function properly. Be sure to follow the manufacturer's instructions carefully for the type of piping system in use.

Moisture in a Fuel Storage Tank

How does water get into a fuel storage tank? Most of it gets in through condensation of water vapor on the interior walls of the tank. Since the tank is vented, moisture-laden air can move into the tank. This can happen with all fuel tanks, but it is a bigger problem in outdoor and buried fuel tanks. Keeping the fuel tank full throughout the year can help prevent air migration and condensation. Any accumulated moisture can be removed by inserting a rigid pipe or hose through the top of the tank until it touches the bottom. The water can then be pumped or siphoned out. Since it is heavier than oil, water settles at the bottom of the tank. Of course, this is also where the pick-up for the burner oil supply is located. As a result, a significant accumulation of water in the tank will be picked up by the pump first.

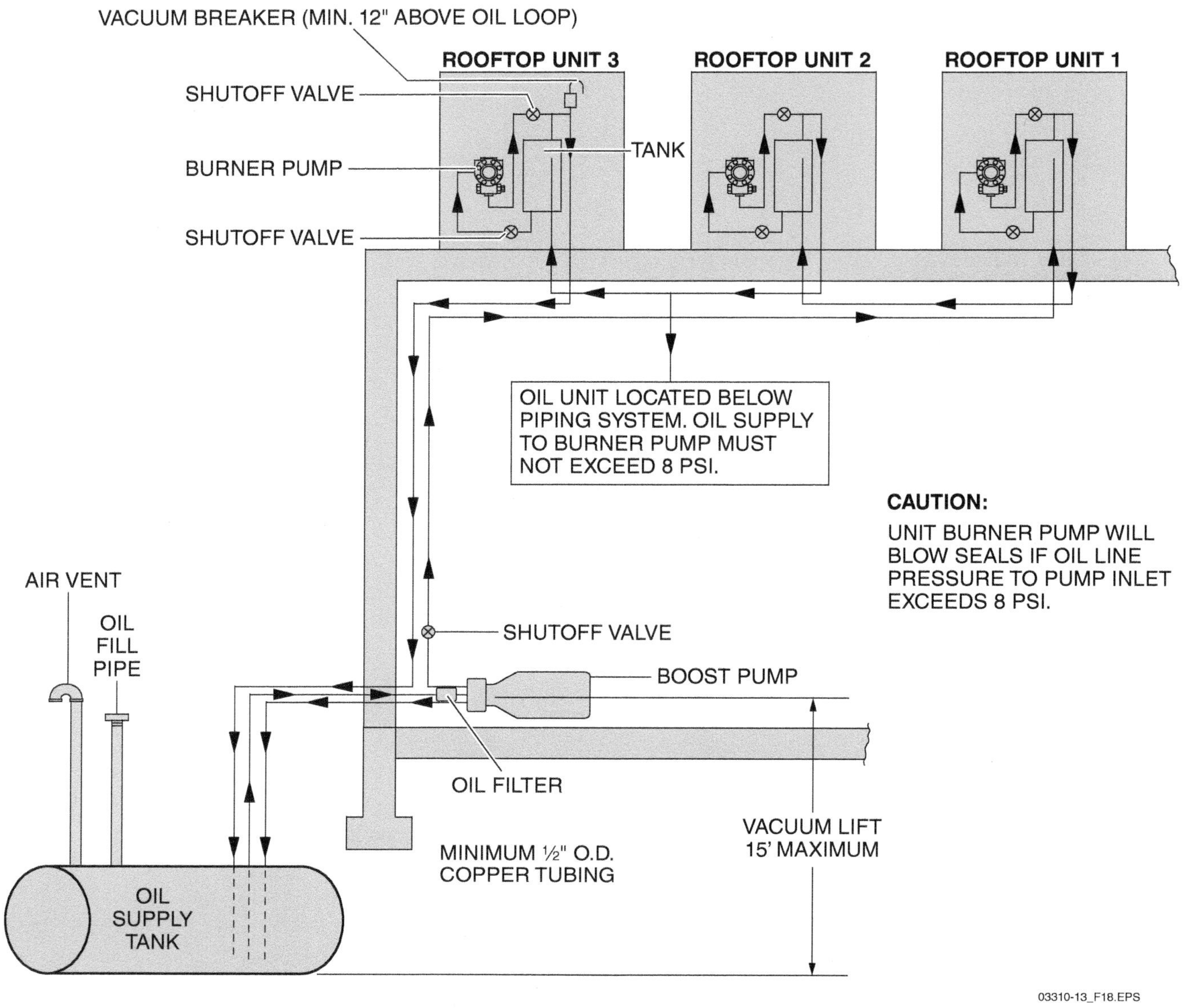

Figure 18 Oil piping system for rooftop units.

Additional Resources

Fundamentals of HVAC/R. Carter Stansfield and David Skaves. Upper Saddle River, NJ: Prentice Hall.

Refrigeration and Air Conditioning, an Introduction to HVAC/R. Larry Jeffus. Air Conditioning and Refrigeration Institute. Upper Saddle River, NJ: Prentice Hall.

1.0.0 Section Review

1. Oil furnaces have the same range of AFUE ratings as gas furnaces.
   a. True
   b. False

2. In an oil furnace, combustion air is supplied by _____.
   a. the vane fan
   b. gravity
   c. the furnace blower
   d. the oil pump

3. The limit switch in an oil furnace is usually found in the _____.
   a. combustion chamber
   b. heat exchanger
   c. discharge plenum
   d. oil burner assembly

4. A two-pipe oil supply system is used when the _____.
   a. oil tank is below the level of the furnace
   b. oil tank is above the level of the furnace
   c. oil tank is located on the roof
   d. oil furnace is located in the basement

Section Two

2.0.0 Servicing Oil Heating Systems

Objectives

Describe how to perform periodic servicing of a typical oil-fired heating system.
   a. Describe the basic servicing procedures performed on an oil-fired system.
   b. Describe how to perform a combustion efficiency test.

Performance Task 1

Remove and reinstall an oil pump in single-pipe and two-pipe systems.

Conduct a complete combustion analysis (smoke test and draft included).

Remove and replace an oil burner nozzle and set the electrode gap.

Trade Terms

Over-firing: A situation caused by an oversized nozzle in which excess oil is pumped into the combustion chamber.

Smoke spot test: A test to determine the level of smoke in the flue gases.

Soot: Fine carbon particles resulting from incomplete combustion. An accumulation of soot on heat exchangers will reduce combustion efficiency and result in increased carbon monoxide production.

Temperature differential: The difference in temperature between the flue gases and the gases above the combustion chamber.

Like the gas furnace, the oil furnace should be checked at the beginning of each heating season. Air filters must be cleaned or replaced, the blower and heat exchangers should be cleaned, and the oil filter should be replaced. Oil burner operation should also be checked. If there is not enough combustion air, the flame may be orange or red instead of yellow. The presence of smoke, soot, or odors could indicate improper oil pressure, poor draft, or an improper mix of oil and air. A pulsating sound could indicate that the flame is touching the combustion chamber. In any of these cases, some cleaning, adjustment, or repair will be needed.

Oil-fired burners typically require more servicing than gas-fired systems because of the continuing need to maintain proper burning efficiency. Inefficient burning will cause soot buildup, which will result in a further loss of efficiency.

The following items are commonly used when servicing oil furnaces:

- AC ammeter
- Bucket
- Carbon dioxide tester (if not part of the combustion test equipment)
- Oil burner combustion test equipment
- Draft gauge, smoke spot tester, CO_2 tester
- Electrode gauge
- Flashlight
- Hand mirror
- Insulated jumper wire
- Multimeter
- Oil pressure gauge kit
- Nozzle wrench (A nozzle tool should always be used to remove the nozzle; pliers and wrenches may cause damage.)
- Stiff brush
- Temperature probes
- Utility knife
- Vacuum cleaner

2.1.0 Servicing Procedures for Oil-Fired Systems

In addition to the specific safety precautions provided in the product literature, the following general safety precautions should be followed when servicing an oil-fired system:

- Unless it is necessary to have the power on for a particular procedure, always turn off and lock out the power when working on the equipment. If it is necessary to check voltage or current, be sure to keep one hand outside the unit and carefully avoid touching bare wires and connections.
- Do not use compression fittings on oil piping. Use flare fittings to prevent air and oil leaks in the fuel lines.
- Wear protective respiratory equipment when cleaning the soot out of a furnace.
- Do not try to clean the oil burner nozzle. It should be replaced.

> **WARNING!**
>
> Never fire an oil furnace if oil has seeped into the combustion chamber and formed a puddle. There is a high risk of explosion or fire. Turn off the oil supply, then pump out or soak up the oil before firing the furnace. Also, be very careful when starting a furnace when the user has used the reset button several times. Each restart attempt sprays oil into the combustion chamber, so a fire or even an explosion can occur when the burner ignites. Also note that the 10,000-15,000 volt ignition transformers used on oil burners are potentially lethal if a defective or damaged high-voltage lead to the electrodes is contacted while the spark is active.

Before performing any service or maintenance procedure, check with the customer and listen to any complaints. Make note of any comment that would indicate a possible source of trouble or equipment malfunction. Then proceed as follows:

- Check the thermostat for proper operation
- With the furnace disconnect turned off, check and clean the blower compartment and blower motor.
- Check the electrical wiring for loose connections, insulation breakdown, and corrosion or dirt on the terminals.
- Inspect the combustion chamber using a hand mirror and flashlight. Look for carbon buildup, holes in the combustion chamber, and erosion of the oil burner head. When the combustion chamber is lined with refractory to isolate the chamber walls from intense heat, look for holes or other damage.
- Check for soot in the heat exchanger:

    Remove the flue or vent pipe.

    Use a flashlight or trouble light to check inside the heat exchanger flue for carbon buildup. It is not necessary to clean the heat exchanger unless there seems to be an excessive amount of carbon or scale on the heat exchanger surfaces.

    If the heat exchanger needs cleaning, refer to the manufacturer's service manual for the proper cleaning procedure.

- Check the oil piping:

    Check all oil supply lines between the supply tank and burner for dents, rust, kinks, loose connections, cracks, or holes. Wetness around fittings, or oil spots on the equipment or the floor, may indicate a leak.

    Check for proper support of oil piping. Check the routing for areas where potential damage could occur.

Soot Removal

There are vacuum cleaners designed specifically for soot removal. While they look something like a shop vacuum, there are significant differences. Soot vacuums have a metal canister that provides a tight seal to prevent soot from escaping. They also use a special fine filter to capture the soot. Soot vacuums are also, in general, more powerful than shop vacuums.

There are several chemical products available that assist in cleaning the soot-coated surfaces of an oil furnace. Always follow both the furnace and chemical manufacturers' instructions and cautions when using soot-removal products.

2.1.1 Changing the Oil Filter

The oil filter (*Figure 19*) should be changed during a preventive maintenance visit. The following is a basic procedure for changing the filter:

- Remove the ring gasket at the top of the filter.
- Check the filter bowl for the name and type of filter element.
- Place a container below the filter to catch any spilled fuel oil.
- Close the fuel line valve and unscrew the bowl mounting screw or attaching hardware. Gently pry the bowl loose from the casting, remove the filter, and place it in the waste bowl container.
- Clean the inside of the filter bowl with filter clean, soft cloth.
- Remove the ring gasket at the top of the filter bowl. It will probably stick to the casting. Scrape away any filter gasket particles with a wide-blade putty knife, and then clean the casting and the top of the bowl with a clean cloth.
- Replace the filter with one of proper size and construction and fill the cartridge with fresh oil.
- Install a new ring gasket, replace the bowl, and firmly tighten the mounting hardware. Be careful not to over-tighten it.
- Open the fuel supply valve and bleed air from the filter bowl through the bleed port at the filter as it fills with oil, if the oil tank is above the level of the burner. This can help speed the priming and air removal process. Then connect a plastic hose to the bleed port of the oil pump and operate the burner. Route the end of the hose into a suitable container to capture the oil. The bleed port is cracked open (roughly ¼ turn) with a small wrench. When oil starts to discharge from the port, close the bleed port.
- Open the oil supply control valve and then check for leaks.

2.1.2 Check the Safety Controls

Check the cad cell flame detector. While the burner is operating, disconnect the cad cell leads at the primary control. The burner should stop in less than 90 seconds. If it does not, the primary control is likely defective. The cad cell should be cleaned during the maintenance procedure. If the burner locks out when combustion appears to be good, the cad cell needs to be tested. Cad cell testing is covered in the next section related to troubleshooting.

Check the safety limit control. Prevent the blower from operating by removing the belt or disconnecting the common wire terminal from the blower relay. Initiate a call for heat at the thermostat and place a suitable thermometer in the heat exchanger plenum. The safety limit control should shut down the furnace at 200°F or after 2½ to 3 minutes of operation. If the furnace continues to operate after reaching the high-limit temperature, shut it off immediately and replace the safety limit control.

2.1.3 Check the Temperature Rise

Temperature rise is the difference in temperature between the air returned from the conditioned space and the air leaving the furnace. It reflects the amount of heat transferred to the air as it passes over the heat exchanger. The temperature range limits are specified by the manufacturer and it found on the furnace label. If the temperature rise is out of tolerance, it can often be corrected by adjusting the furnace blower speed. Note however, that this is also a symptom of a furnace being over- or under-fired by an incorrect nozzle size.

- Initiate a call for heat at the thermostat. Make sure all the furnace access doors are closed and the filters are replaced or clean.
- Drill holes in the supply and return ductwork (*Figure 20*) and insert a temperature probe into each hole. The thermometer in the supply duct must be located out of the line of sight of the furnace heat exchanger to prevent radiant heat from affecting the thermometer.
- If the temperature differential is below the specified range, reduce the blower speed. If it is above the range, make sure that there are no airflow restrictions and that the burner is not over-firing, then increase the blower speed.

Figure 19 An oil filter element.

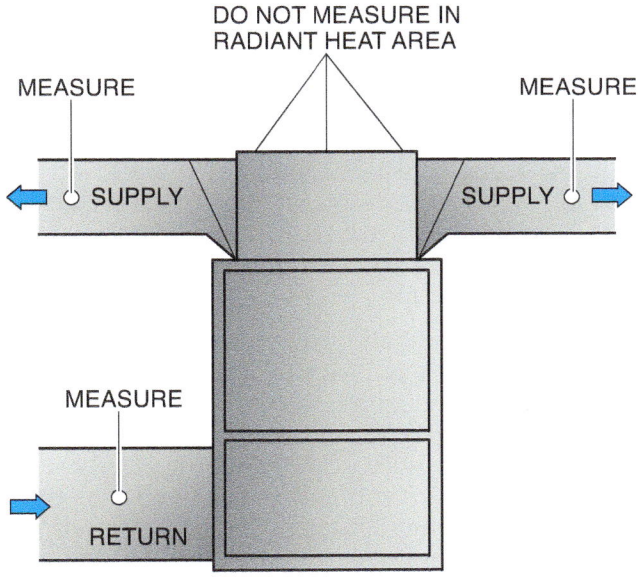

Figure 20 Hole locations for temperature-rise test.

2.1.4 Servicing the Oil Burner Assembly

The oil burner assembly (*Figure 21*) is the primary active component in an oil-fired system. It must be carefully inspected and serviced. The following procedure is a guide to oil burner servicing:

- Turn the power off.
- Disconnect all wiring and piping to the gun and vane fan.
- Remove the assembly, being careful not to damage any of the components of the burner or ignition system and controls. Do not disturb the slip ring for the air gate or lose track of the nozzle adjustment or the insertion dimensions.
- Clean the vane fan wheel with a stiff bristle brush.
- Clean the burner tube with a soft cloth or disposable towels.
- Remove, clean, and inspect the electrodes. Use fine steel wool for cleaning. Do not use sandpaper.
- Remove and replace the oil burner nozzle. Nozzles are inexpensive and should always be replaced as part of any planned maintenance procedure. Make sure the nozzle installed is the correct one for the equipment by checking the furnace or boiler data plate. Do not refer to the data plate on the burner assembly alone, as the burner will have a wide range of nozzle sizes that it can accommodate. The furnace or boiler is the controlling factor for the nozzle size. A nozzle with too large an orifice will result in over-firing, which, in turn, can cause overheating. A nozzle with an orifice that is too small will result in under-firing. This can cause a reduction of flue gas temperature and may result in condensation forming in the flue.
- Replace the nozzle using the nozzle wrench designed for that purpose (*Figure 22*). Do not overtighten it. The nozzle wrench consists of two sockets, with the outer socket larger than the second socket on the inside. The wrench is slipped over the nozzle, where the outer socket seats on the hexagonal section of the oil tube. The inner socket is then pushed over the nozzle, and the nozzle is loosened with minimal effort by rotating the handle attached to the inner socket.
- Reinstall the electrodes and check the position of the electrode tips. Electrode insulators must be clean and free of cracks and carbon buildup. Clean or replace them as necessary. Adjust the gap and spacing from the nozzle (*Figure 23*) as specified by the oil burner manufacturer.

Accurately setting the position of the electrode tips can be very challenging. The angle of the nozzle spray pattern determines how far forward the electrodes are positioned, and this precise measurement must be known to successfully service the burner. Nozzles with narrow angles require that the electrode tips be positioned further forward to ensure that the spark is at the edge of the oil spray. Wider angles require the electrode tips to be moved further back. The distance between the two tips, and the vertical distance of the tips above the nozzle, are also crucial to proper ignition.

Figure 21 Oil burner assembly.

Figure 22 Nozzle wrench.

The simple device pictured in *Figure 24* eliminates a lot of guesswork and simplifies the process. The electrode-setting tool is slipped over the nozzle tip. On the tip of the tool, the various possible spray angles for nozzles are molded into the plastic. Simply select the positioning guide for the angle of the nozzle being adjusted, and then lay the electrodes in the appropriate guide. This will properly position the electrodes in all three dimensions. Once the electrode mount clamp is tightened, the tool is slipped off the end of the nozzle. No additional measuring instruments or nozzle angle charts are required. For technicians servicing oil burners with any regularity, this is an indispensable tool.

- Inspect the cad cell flame detector and clean it if necessary. A photocell flame detector must see light to determine that an oil burner is operating properly. If the face of the cad cell becomes coated with soot or a film of oily dirt, the light is obscured and the cad cell may shut down the burner. During every routine service of an oil burner, clean the face of the cad cell with a soft cloth or swab to remove any soot or oil film. Check for proper contact with the mounting bracket and reinstall the unit.
- Replace the gun assembly and reassemble per the heating plant manufacturer's instructions.

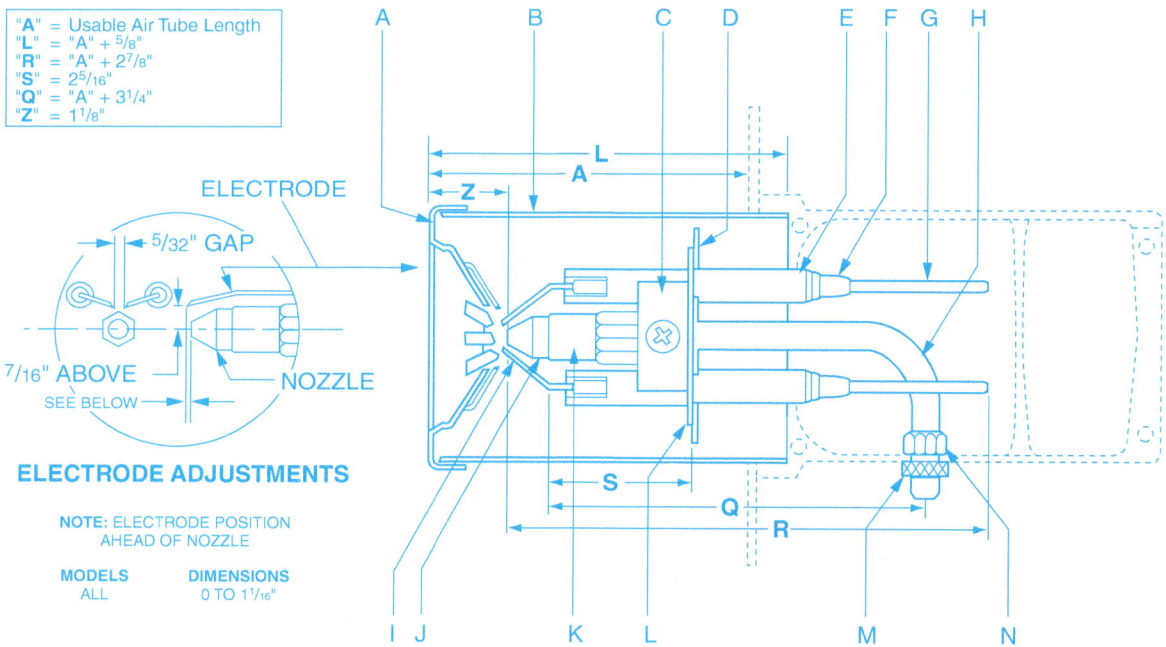

A. BURNER HEAD, SPECIFY TYPE F
B. AIR TUBE
C. ELECTRODE CLAMP, STATIC PLATE, AND NOZZLE LINE SUPPORT ASSEMBLY
D. CENTERING SPIDER
E. PORCELAIN INSULATOR
F. ELECTRODE ROD EXTENSION ADAPTER, AS REQUIRED
G. ELECTRODE ROD EXTENSION, AS REQUIRED
H. NOZZLE LINE AND VENT PLUG
I. ELECTRODE ROD AND TIP
J. NOZZLE
K. NOZZLE ADAPTER – SINGLE
L. STATIC PLATE
M. LOCKNUT BULKHEAD FITTING
N. BULKHEAD FITTING

Figure 23 Electrode adjustment.

Figure 24 Electrode setting tool.

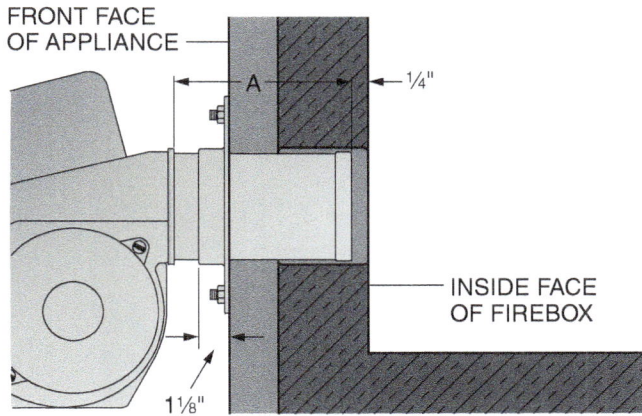

THE AIR TUBE LENGTH (DIMENSION A) IS THE DISTANCE FROM THE FRONT OF THE BURNER HOUSING TO THE DRAIN HOLE IN THE BURNER HEAD

NOTE: ADJUSTABLE FLANGE WIDTH - 1⅛"

Figure 25 Typical air tube length.

Check for air (blast) tube length (*Figure 25*) and reinstall.
- Check the burner motor coupling; replace or service it as necessary.
- Remove and clean the oil pump screen.
- Perform a combustion efficiency check.

2.2.0 Measuring Combustion Efficiency and Flue Gases

Combustion efficiency is a measure of how well an appliance is burning its fuel. The ideal, of course, is 100 percent efficiency, which means all the energy available in the fuel is being used to produce usable heat. Combustion efficiency can therefore be translated into operating cost. That is, the higher the percentage of usable energy extracted from the fuel, the lower the cost of operating the system.

The highest efficiency gas furnaces, with AFUEs exceeding 95 percent, approach this level of efficiency, but oil furnaces do not. The highest AFUE oil furnace at the time of this writing has an AFUE of 86.5 percent, which means that 13.5 percent of the available heat is being wasted in the flue gases.

Combustion efficiency is determined by measuring the temperature of the flue gases and the amount of oxygen and CO_2 in those gases. The lower the flue gas temperature, the higher the combustion efficiency, as it means there is less heat escaping up the chimney. In perfect combustion, all oxygen would be converted to CO_2, so the ratio of oxygen to CO_2 is a measure of combustion efficiency.

Draft is also an important factor in combustion efficiency. Insufficient draft can cause dangerous carbon monoxide (CO) to back up into occupied spaces and can also cause overheating of the combustion chamber. Too much draft will result in lower combustion efficiency.

There are other important factors to consider, however. The most efficient burning produces the least smoke. The smoke produced by burning the fuel contains soot, which will be deposited on the surfaces of the heat exchanger, combustion chamber, and flue gas exhaust system. Heavy soot deposits on the heat exchanger will cause a reduction in heat transfer efficiency and will result in increased operating cost.

The electronic gas/oil combustion analyzer shown in *Figure 26* is a key item of test equipment. It directly measures and displays O_2, CO, stack temperature, draft, differential pressure, combustion air temperature and other optional selected gas emissions. Some testers have optional print-

---

Disposing of Waste Oil

Waste oil cannot simply be thrown away or poured on the ground because it contaminates the environment. Dispose of waste oil at an appropriate recycling facility or service station.

03310 Troubleshooting Oil Heating

Module Six 19

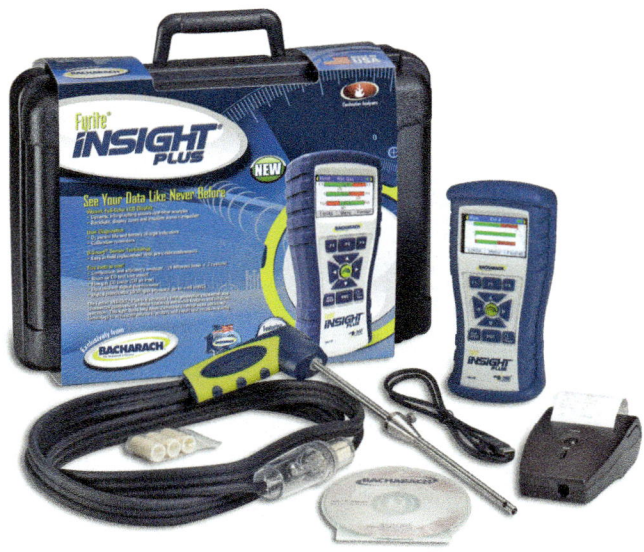

Figure 26 Electronic gas/oil combustion analyzer.

ers that can be connected to the device with a USB cable or by wireless connection.

The instrument simultaneously calculates and displays combustion efficiency, excess air (EA), CO_2, NO_X, and O_2 reference values. The analyzer performs combustion calculations for ten fuels, including natural gas and propane, as well as #2, #4, and #6 fuel oil. The only combustion parameter it does not measure is smoke level. Although there are mechanical instruments available to perform the various tests that are required, the electronic device has a number of advantages:

- It consolidates the various tests performed with a number of different mechanical devices into a single device.

Figure 27 Smoke spot tester kit.

- Electronic instruments allow adjustments to be performed in real time. That is, the adjustments can be made while observing the effect on the instrument readings.
- Electronic instruments are easier to use, more accurate, and reduce the chance of human error.
- Tests can be performed more quickly with electronic instruments.

The combustion analyzer shown in *Figure 26* is equipped with a probe and hose assembly. The probe is inserted into an opening in the flue pipe. The hoses in the hose assembly are terminated in sensors that are plugged into connections on the bottom of the tester.

Going Green

Multi-Fuel System

The system shown here is designed burn coal, wood, or oil. In order to conserve cost in fluctuating oil markets, this furnace can burn wood or coal at any time. At times when there is no one around to stoke the furnace, oil can be used.

2.2.1 Smoke Spot Testing

A smoke spot test should be performed and the air shutter adjusted for proper smoke level before combustion testing is performed. The smoke spot test is performed using a smoke spot tester and comparison chart (*Figure 27*).

- Drill a test hole in the flue vent downstream of the draft diverter and close to the breeching, as shown in *Figure 28*. This opening will also be used for combustion efficiency testing.
- Operate the furnace for 10 to 15 minutes before performing the smoke spot test.
- Before starting the test, insert test paper into the tester in accordance with the manufacturer's instructions.
- Insert the hose of the smoke spot tester into the hole in the vent and draw 10 samples into the tester at 1- to 2-second intervals. The flue gases are drawn in by pulling the handle of the tester back. The gases pass through the filter paper inserted into the chamber, leaving a black circle on the paper. The darkness of the circle indicates the level of smoke in the gases.
- The initial test should show some smoke staining on the paper. If it does not, adjust the air shutter to lower the air until some smoke appears. Then, re-adjust the air shutter for the proper smoke level. For a flame retention head burner, the air shutter should be adjusted for zero smoke. A No. 1 smoke is standard for conventional burners.
- Compare the results to the sample chart provided by the tester manufacturer. For residential equipment that burns No. 2 oil or kerosene, the smoke reading should be between zero and a trace. In a heavy boiler where regular tube cleaning is performed, a smoke reading of two is often acceptable. If it is higher, the air setting at the burner may have to be readjusted.

2.2.2 Combustion Efficiency and Flue Gas Testing

Before starting the test, make sure the inspection door and the blower compartment door are securely closed. The furnace should be operated for at least 10 minutes before testing.

- Set up the combustion analyzer in accordance with the manufacturer's instructions. The hose assembly on the test set typically contains a water trap and filter assembly. Before using the device, make sure the trap is dry and empty, and replace the filter. The test instrument also contains oxygen, CO, and CO_2 sensors, which require periodic replacement. Follow the manufacturer's instructions. It should be started up and allowed to stabilize before inserting the probe into the flue.
- Insert the probe so that the tip is halfway into the vent pipe and allow time for the readings to stabilize.
- Record or print the test results and compare them to the manufacturer's requirements. The values will vary, depending on the type of burner and the AFUE rating of the furnace, so it is important to make this comparison. Leave the probe in the flue pipe. Typical values for a flame retention burner are as follows:
  - Oxygen: 3 to 5 percent
  - CO_2: 10 to 12.5 percent
  - CO: <50 ppm (diluted)
  - Stack temperature: 400°F (60-79 AFUE); 450°F (80-plus AFUE)
  - Overfire draft: 0.02 in wc
  - Stack draft: 0.02 to 0.04 in wc
- If the combustion efficiency is not within the required tolerances, adjust the air shutter on the burner to obtain the correct value. If this value cannot be achieved by air adjustment, troubleshooting will have to be performed to isolate the cause.

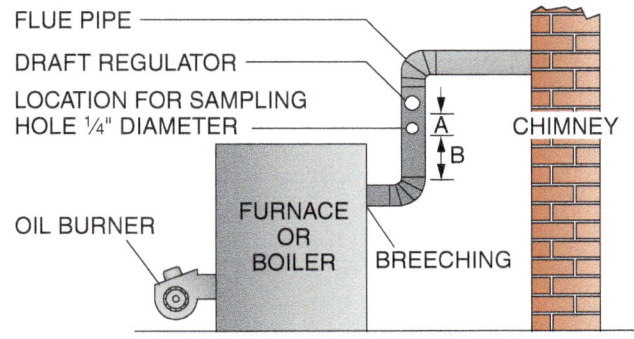
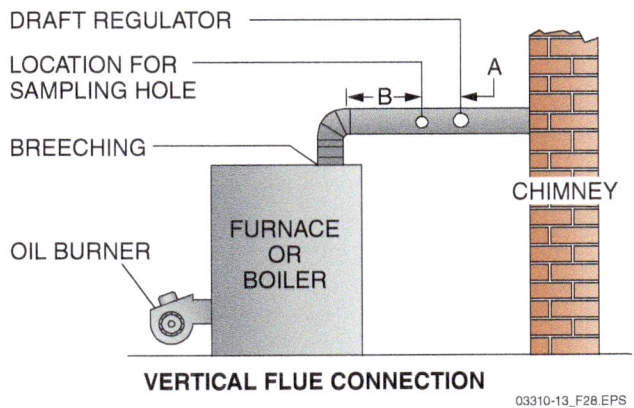

Figure 28 Test hole locations.

If the combustion performance specifications cannot be met, check for air leaks into the combustion chamber and repair the leaks with furnace cement or high-temperature sealant.

To check for dilution by leakage into the combustion chamber, measure the carbon dioxide content at as high a point as possible over the fire. Insert a stainless steel tube through the fire door sample hole and compare the reading with the CO_2 measured in the vent pipe. A differential of more than one percent carbon dioxide between the vent pipe and the over-fire readings will usually indicate air entry into the combustion chamber.

If the required carbon dioxide level cannot be reached without exceeding No. 1 smoke, poor mixing of air and fuel is likely. This could be caused by an improper match between the air pattern and nozzle spray pattern. This problem can frequently be corrected by replacing the nozzle with one having a different spray angle and pattern.

If the proper combustion efficiency cannot be achieved, check for an unbalanced flame. If the flame appears unbalanced, the fuel input may be too low. Check the nozzle size and oil pressure. If the nozzle is partially plugged or defective, replace it with another as recommended by the manufacturer or choose one from an interchange chart.

- Take a negative draft reading through the inspection door (over the flame). A difference between this reading and the reading from the flue of more than 0.002 in wc indicates a dirty or restricted heat exchanger.
- Place the system into operation and observe it through at least one complete heating cycle to make certain all controls are functioning properly.
- With the heating plant in operation for five or more minutes, conduct a CO check in the plenum using an appropriate CO detection device and probe. CO levels of less than 50 parts per million are acceptable.
- Clean up the work area, seal test holes that were drilled, and complete the service call.
- Leave a copy of a complete service and maintenance checklist with the owner.

Additional Resources

Fundamentals of HVAC/R. Carter Stansfield and David Skaves. Upper Saddle River, NJ: Prentice Hall.

Refrigeration and Air Conditioning, an Introduction to HVAC/R. Larry Jeffus. Air Conditioning and Refrigeration Institute. Upper Saddle River, NJ: Prentice Hall.

HVAC Maintenance Procedure. Syracuse, NY: Carrier Corporation.

http://www.bacharach-inc.com/PDF/Brochures/4097percent20Combustion.pdf; Instructions for performing and evaluating a combustion efficiency test using an electronic test instrument.

2.0.0 Section Review

1. Which of the following conditions should occur if the cad cell leads are disconnected while the furnace is running?

    a. The temperature limit switch will open.
    b. The burner will shut off.
    c. The furnace will continue running.
    d. The blower fan will turn off.

2. A sample of flue gases for smoke analysis is drawn into the tester by pumping the handle _____.

    a. once
    b. twice
    c. five times
    d. ten times

SECTION THREE

3.0.0 OIL FURNACE TROUBLESHOOTING

Objectives

Describe how to troubleshoot a typical oil-fired heating system.
a. Describe troubleshooting procedures for typical oil furnace controls.
b. Describe troubleshooting procedures for common oil heating problems.

Performance Task 1

Test a cad cell flame detector.

Trade Terms

Coping: The upper masonry course of a chimney designed to keep water out.

Downdraft: A situation in which outdoor air is forced down a chimney, preventing proper draft. Downdraft can be caused by not extending the top of the chimney above the roof line.

Joist: Framing member used to support floors and ceilings.

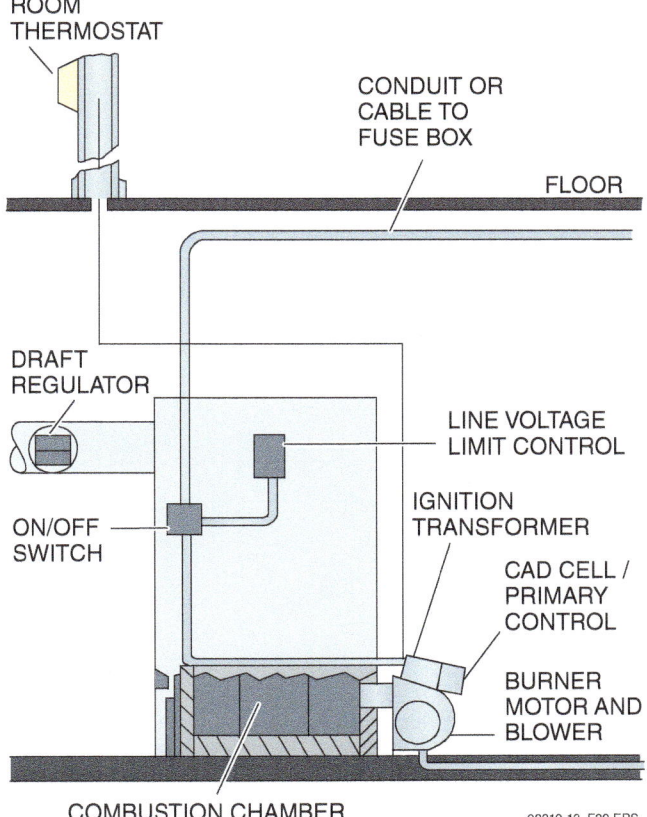

Figure 29 Oil burner control system.

When starting out on a trouble call, it is always a good idea to listen carefully as the customers describe the problem from their perspective. They may not be technically knowledgeable, but what they have experienced can give you a clue to what might be wrong with the oil heating system.

3.1.0 Control Circuit Troubleshooting

The ability to effectively troubleshoot oil furnace problems demands the ability to understand the operating sequence and safety controls for the furnace. This, in turn, requires the ability to interpret the schematic and/or wiring diagrams provided by the manufacturer. The control systems for oil furnaces are generally not as complex as those for gas-fired systems. *Figure 29* provides an overview of an oil furnace control system.

This section begins by providing examples of schematic and wiring diagrams from two furnaces, along with descriptions of their operating sequences.

Figure 30 is the schematic diagram for an oil furnace and add-on air conditioner with a cad cell flame detector. The operating sequence is included on the diagram. *Figure 31* is the wiring diagram for the same furnace. In a troubleshooting situation, the wiring diagram will help determine where to connect the test meter. Although these diagrams represent only one of the many oil-fired systems a technician may encounter, studying them will help gain more insight into the operation of similar systems.

3.1.1 Primary Control Troubleshooting

The primary control shown in *Figure 32* is a combination control and safety device. It contains an electronic switching circuit that turns on the system by transferring line voltage to the burner motor and the ignition transformer when a call for heat is received from the thermostat. Although the low voltage control wiring terminals are visible on top, the line voltage connections enter through the bottom of the control. The primary control is usually mounted on a junction box or similar enclosure that is part of the burner assembly. The safety feature is controlled by the combustion detector, which in this case is a cad

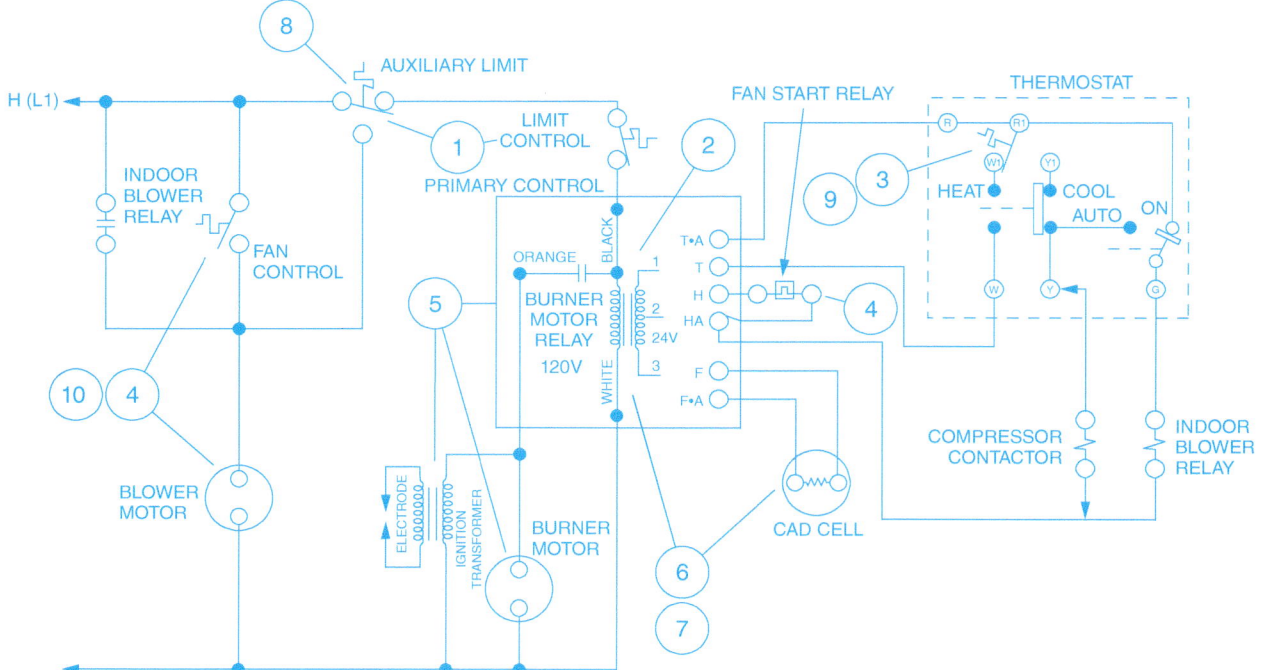

1. Line potential feeds through the secondary limit and limit controls to power primary control.
2. The primary control provides 24-volt control circuit.
3. On a heating demand, the thermostat heating bulb makes.
4. The fan control heater is energized through primary control. After a short period, the heater provides sufficient heat to close the fan contacts. This energizes the blower motor.
5. The primary control simultaneously energizes the burner motor and ignition transformer at the oil burner. The burner motor operates the oil pump and combustion blower to feed air and oil vapor into the combustion chamber. The fuel mixture should ignite with the spark furnished by ignition transformer.
6. If combustion does not take place within the time specified on the primary control, as detected by cad cell, the primary control locks itself out.
7. Should a flame failure occur during the "on" cycle, the primary control locks itself out in response to the cad cell.
8. The secondary limit opens at temperatures above setpoint. This de-energizes the primary control but still allows blower motor operation through the NO contacts.
9. As the heating demand is satisfied, the thermostat heating bulb breaks. This de-energizes the oil burner circuits.
10. The blower motor continues running until furnace temperature drops below fan control setpoint.

Figure 30 Operating sequence number one.

cell flame detector. The flame detector must be matched to the primary control. However, there are very few cad cells to choose from on the market. If the primary control does not receive a continuous signal that combustion is taking place, its relay will be de-energized, turning off the burner motor. The primary control allows a set time to elapse before it shuts off and locks out the burner. The timing of a primary control is not adjustable. Different primary controls have different lockout times. The burner manufacturer selects a primary control that suits the design of the burner. If a flame is established within the prescribed time, burner operation will continue.

The following are operating features of this control circuit:

- The room thermostat senses changes in the space temperature and then signals the primary control to energize or de-energize the burner.
- The primary control acts as the nerve center. It receives signals from the thermostat and flame detector and makes or breaks power to the oil burner.
- The cad cell flame detector has a similar function to that of a thermocouple or flame rod in a gas furnace. In older furnaces, a bimetal detector in the stack was used for this purpose. Either type may be encountered when working on residential and light commercial systems.
- The line voltage limit control de-energizes the burner if overheating occurs.

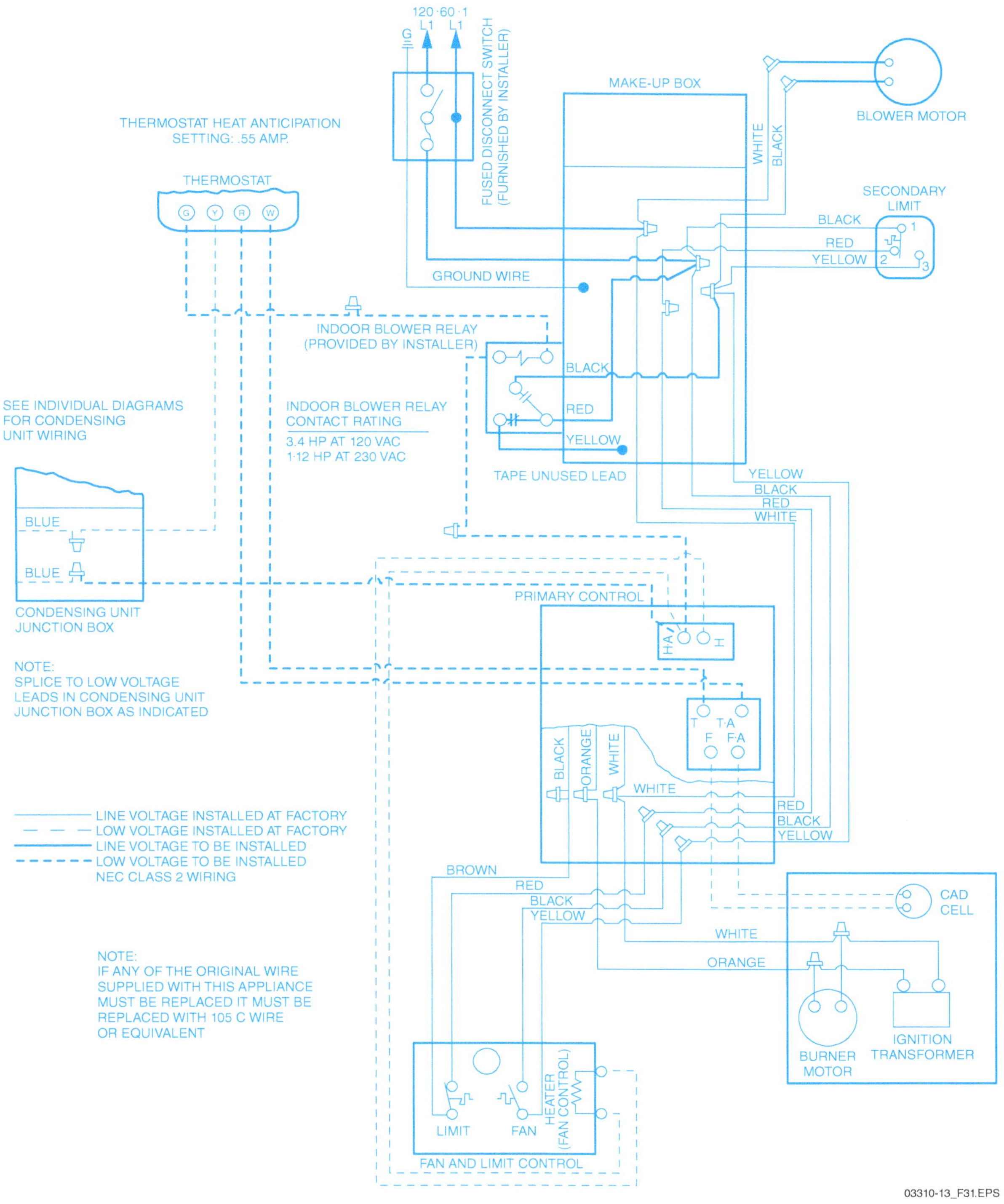

Figure 31 Field wiring diagram number one.

- Auxiliary controls may include devices to control the blower or fan, low-limit controls, zone valve controls, air cleaner controls, humidifier controls, time delay controls, and motor control relays.

Checking the primary control and cad cell must be done with live circuits; therefore, the troubleshooter must observe all the precautions necessary to avoid the danger of electrical shock and equipment damage.

To troubleshoot a primary control if the burner starts and locks out, proceed as follows:

Step 1 Make sure the limit switches are closed and line voltage at the black and white leads of the primary control is 120 volts.

03310 Troubleshooting Oil Heating

Module Six 25

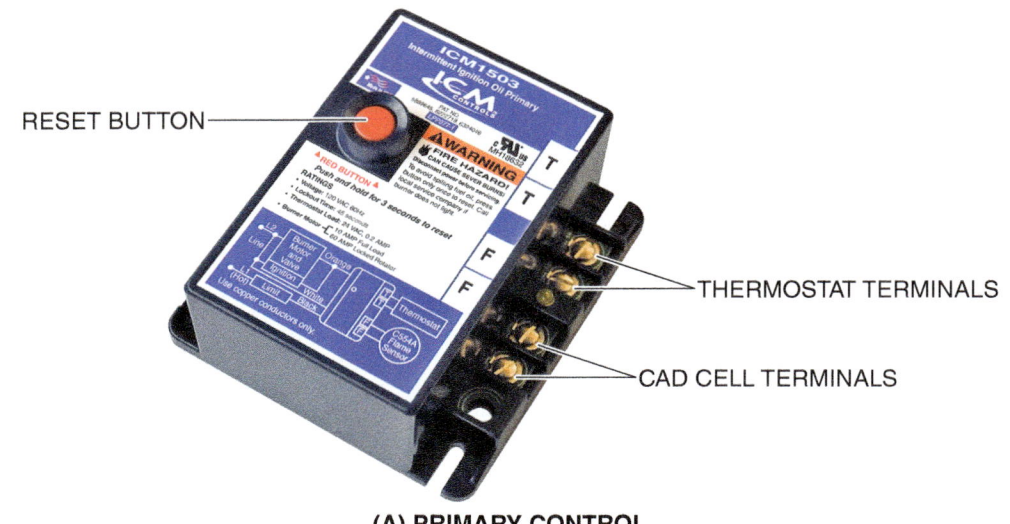

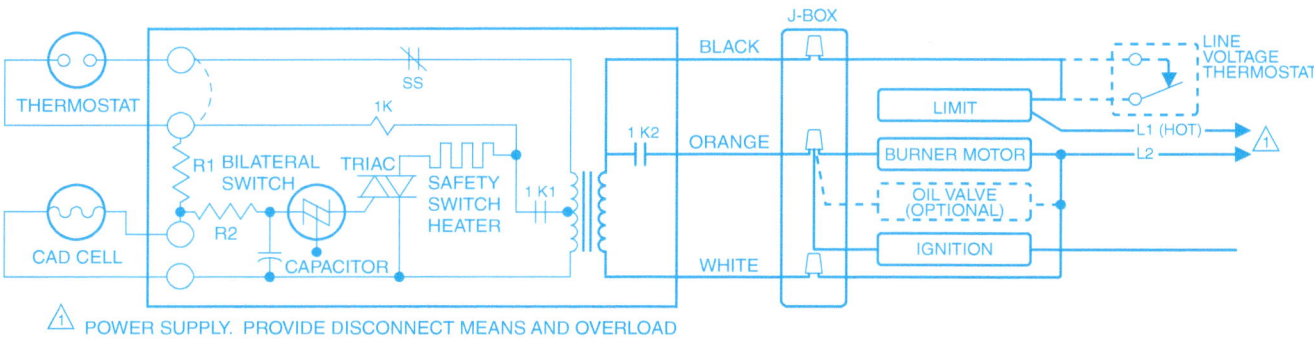

Figure 32 Primary control and wiring diagram.

Switch off the power at the furnace disconnect.

Step 2 Disconnect at least one room thermostat lead from terminals T-T. Place a jumper wire across terminals T-T.

Step 3 Disconnect the two cad cell leads from terminals F-F. Connect one end of a 1,500V resistor to one of the F-F terminals. This resistor will simulate a cad cell that senses light. Make sure that there is enough lead length to quickly connect the other end of the resistor to the other F-F terminal.

Step 4 Close the furnace disconnect, press the reset button on the primary control, and quickly (within 5 to 10 seconds) connect the other end of the 1,500V resistor to the remaining F-F terminal.

Step 5 If the primary control locks out with the resistor across F-F, it indicates that the primary control is defective. When in doubt, wait five minutes and repeat the test, making sure to connect the resistor quickly. If the primary does not lock out with the resistor in place, the control is functioning properly.

3.1.2 Cad Cell Flame Detector Troubleshooting

The cad cell (photocell) flame detector consists of a plug-in, light-sensitive cell and a socket with a mounting bracket and lead wire (one-piece units are also available). It is mounted in the burner unit and wired into the control circuitry as shown in *Figure 32*. The location of the cad cell is critical. It must be positioned so that it can detect the light from the flame in order to function properly.

Before checking the cad cell and primary control, check the following:

- Main power supply (unit disconnect) on

- Electrodes properly positioned and with proper gap
- Contact between ignition transformer and electrodes is good
- Oil pump pressure adequate
- Oil burner nozzle clean and correctly sized
- Oil supply adequate and piping correctly installed

Using a cotton-tipped swab, clean off the soot or oil film that has formed on the face of a cad cell flame detector.

The following is an example of control circuit troubleshooting based on the circuit shown in *Figure 32*.

When checking the cad cell resistance with an ohmmeter, the resistance should range from 300 to 1,000 ohms (Ω) when the burner is operating (light showing). With the cad cell face covered (no light available), the resistance should be roughly 100,000Ω. This is high enough to block any current flow in the circuit. If the cad cell resistance is above 1,600Ω during the burner run cycle, the cad cell may need cleaning or aligning, or the burner flame may need to be adjusted. The equipment required for checking the cad cell and primary control includes the following:

- 0 to 150VAC voltmeter
- 1,500Ω resistor
- Insulated jumper wires
- Ohmmeter

To troubleshoot a cad cell, proceed as follows:

Step 1 Switch off the furnace power. Disconnect at least one lead of the cad cell from primary control terminals and check it using an ohmmeter. Resistance should be around 100,000Ω.

Step 2 Reconnect the cad cell to the terminals, disconnect at least one lead of the room thermostat from the T-T terminals, and place a jumper wire across the T-T terminals.

Step 3 Press the reset button on the primary control and switch on the furnace power to initiate burner operation.

Step 4 After the burner ignites the fuel, disconnect both leads of the cad cell from the terminals. The burner should stop after a short interval (15 to 45 seconds). If the burner does not shut off, the primary control is probably defective.

Step 5 Switch off the furnace power and connect a jumper wire to one of the cad cell terminals.

Step 6 Press the reset button on the primary control and turn on the primary power. The burner should start and run. As soon as the burner starts, connect the other end of the jumper wire to the other cad cell terminal.

Step 7 With the burner operating, measure the resistance across the two disconnected cad cell leads. The resistance should not exceed 1,600Ω. Ideally, it should be between 600Ω and 700Ω.

Step 8 If the resistance is 1,500Ω or less, the cad cell is functioning properly. If the resistance is higher than 1,600Ω, the cad cell face may be dirty, cracked, broken, or misaligned such that it does not completely sense the light from the flame.

3.1.3 Ignition Components Troubleshooting

The ignition system consists of the ignition transformer and the spark electrodes. The ignition transformer can be checked as follows:

> **WARNING!** Follow the manufacturer's instructions and be extremely careful when performing this procedure. The ignition transformer produces a 10,000V pulse.

Perform the following steps to troubleshoot the ignition controls:

Step 1 Turn off power to the unit.

Step 2 Shut off the fuel supply or disconnect the burner motor lead.

Step 3 Connect a voltmeter capable of measuring high voltage across the ignition transformer output terminals, and then turn the power back on. The reading should be at least 10,000V. If this result is not obtained, replace the transformer.

The ignition transformer must make good contact with the contacts of the electrodes. The contacts mounted to each electrode is flexible and can be bent or curved to ensure good contact is made with the transformer high-voltage terminals. If the electrode insulators are cracked or broken, an arc may be established between that electrode and another part of the burner assembly. If the ignition transformer is working properly, yet there is

03310 Troubleshooting Oil Heating   Module Six 27

no arc observed at the end of the electrodes, check their position and gap, and inspect for flaws in the insulators.

3.2.0 Troubleshooting Common Oil Heating Problems

This section provides a troubleshooting guide to common causes of the primary complaints associated with oil heating service calls. The following checks should be made on if the customer complains that the appliance is not producing the desired heat, or is not operating at all:

Step 1 Check for power at the main disconnect switch.

Step 2 Check the burner motor fuse.

Step 3 Check the burner on-off switch.

Step 4 Check the oil supply.

Step 5 Make sure all manual oil valves are open.

Step 6 Check the limit switches. They must be closed.

Step 7 Reset the safety switch, set the thermostat to call for heat, and then perform the following checks:

> **WARNING!** If the switch is off, make sure the combustion chamber is free of oil, or oil vapor, before turning the switch on. Failure to do so could result in an explosion.

- Simulate flame failure by shutting off the oil supply hand valve while the burner is operating. The burner should lock out after the primary control has timed out.
- Simulate ignition or fuel failure by shutting off the power supply while the burner is operating normally. The system should shut down immediately. Wait a few minutes for the stack to cool, and then restore power. The system should restart immediately if the thermostat is calling for heat.

If the system still does not operate properly, note the point at which it fails. Use the troubleshooting chart in *Appendix A* to determine which parts of the oil heating system are most likely to be at fault. *Appendix B* shows the most likely causes of common oil heating system failures.

3.2.1 Oil Burner Troubleshooting

The components of the burner system include the oil supply components and the burner. The gun-type atomizing burner, shown in *Figure 33*, is the most commonly used unit. This type of burner uses a fuel pump to deliver oil under pressure (typically 100 psig, but sometimes up to 150 psig) to a calibrated nozzle where it is broken into a fine, cone-shaped mist. An electric spark provided by an ignition transformer and electrode ignites the mixture.

The oil burner nozzle assembly is a common source of problems. The following conditions can be traced to a faulty nozzle:

- No flame (partial or complete blockage)
- Pulsating pressure
- Flame changing in size and shape
- Flame impinging on (touching) the sides of the combustion chamber (*Figure 34*)
- Low carbon dioxide reading in flue gases (less than 8 percent)
- Delayed ignition
- Oil odors

If the nozzle appears to be faulty, it should be replaced. In general, it is a good idea to replace it annually. If the nozzle strainer is clogged, determine the cause. The fuel oil line or the oil pump

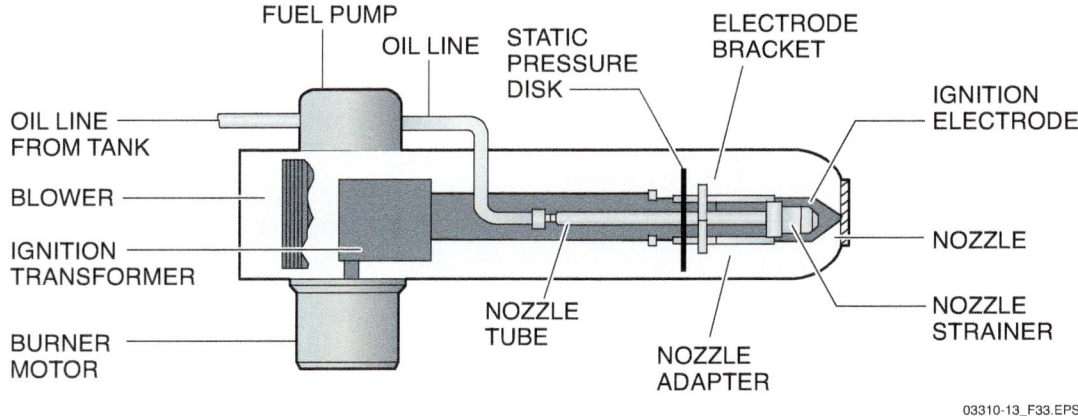

Figure 33 Gun-type burner.

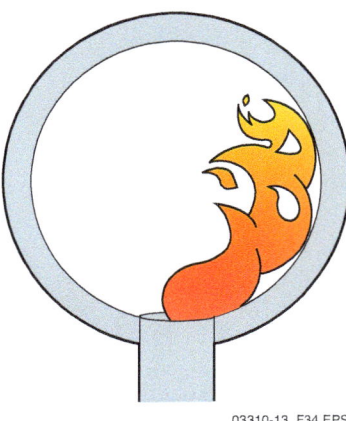

Figure 34 Improper burner flame.

strainer may be plugged, or there may be water or other contaminants in the fuel tank.

If a varnish-like substance has formed on the nozzle, it indicates overheating. There are four major causes of overheating:

- An over-fired burner (nozzle too large)
- The fire burning too close to the nozzle
- Too small a firebox
- The nozzle being positioned too far forward

If the fire is too close to the nozzle, a small increase in oil pump pressure may correct the problem. When the nozzle is too far forward, it may be moved back by using a short adapter or by shortening the oil line.

3.2.2 Fuel Supply Troubleshooting

If an oil pump is working properly, the needle on its pressure gauge should be steady. If the needle pulsates, check for a partially clogged fuel filter or pump screen and/or air leaks in the fuel supply lines. A vacuum gauge connected to the inlet side of a single-stage pump should read less than 10 inches of vacuum. The vacuum gauge on a two-stage pump should read less than 15 inches of vacuum. If a higher vacuum is detected, check for kinks or restrictions in the fuel supply lines and/or a plugged fuel filter. Also listen for a whine at the pump, which can indicate a high suction and a possible supply restriction. Before checking the pump, valves, and controls, check the fuel supply for proper installation and piping.

Oil supply lines in residential installations generally use ⅜" (9.53mm) OD soft copper tubing. To prevent oil and air leaks in fuel lines, always use flare fittings on copper tubing connections. Compression fittings should never be used on copper fuel supply lines.

Water can form in the fuel tank due to condensation. If it collects to a sufficient depth in the bottom of the tank, it can make its way to the pump. Here it can cause corrosion and eventually ruin the pump, in addition to preventing ignition. Fuel tanks should be periodically checked for the presence of water. If water is found, it must be removed.

3.2.3 Odors

To find the source of a fuel oil odor, proceed as follows:

Step 1 Check for leaks in the oil supply and return piping systems. Tighten or replace connections or ruptured lines as necessary.

Step 2 Check the fuel tank for proper venting to the outdoors if it is a basement tank. Correct venting problems where necessary and securely tighten or seal the vent connections.

> **GOING GREEN**
>
> **Disposing of Oil Storage Tanks**
>
> Oil storage tanks for heating systems, particularly those for residential systems, may not be covered by the same federal regulations as industrial underground storage tanks, but they still contain potentially hazardous fuel oil residue. Always check and abide by all applicable federal, state, and local codes when disposing of oil storage tanks.

Over-Fired Oil Burners

How can you tell if an oil burner is over-fired? First, check the furnace information plate to find the input rate. Then, inspect the oil burner nozzle and find the flow rate stamped on the nozzle body. The nozzle flow rate determines burner input. For example, one gallon of No. 2 fuel oil has a heat content of approximately 144,000 Btus. A one-gallon-per-hour nozzle would therefore deliver 144,000 Btuh of input. If a half-gallon-per-hour nozzle is used, the burner input is 72,000 Btuh. A nozzle that produces an input above the computed rate, as determined using the nozzle body stamp and the furnace information plate, is over-fired.

Step 3 Check for seepage around the fill or vent connections if there is an oil film on the tank.

Step 4 Check for seepage through the oil pump cover gasket or the pump shaft bushings. If necessary, replace the pump.

Step 5 Check for loose high-pressure oil line connections to the gun assembly, and tighten the connections or replace the oil line as necessary.

Step 6 Check for fuel filter leakage, correct the cause, and replace the filter.

3.2.4 Flue and Chimney Exhaust Troubleshooting

The flue pipe and chimney exhaust can also become a source of odor or improper burning operation. As with any other HVAC systems, the proper design and installation of the flue pipe and chimney exhaust system eliminate problems that could be encountered by the HVAC troubleshooter.

Refer to *Figure 35* and the numbered list in the text below for an explanation of some of the areas that might lead to chimney problems:

1. Downdraft might be caused by the top of the chimney being lower than surrounding objects. The correction requires extending the chimney above all objects within 30' or as recommended by local building codes.

2. A restricted opening may be caused by problems with the chimney cap, ventilator, or coping.

3. Obstructions in the chimney can be located by using a light and a mirror.

4. A joist projecting into the chimney can cause problems.

5. A break in the chimney lining or tile may be present.

6. A collection of soot at a narrow opening or turn may be present.

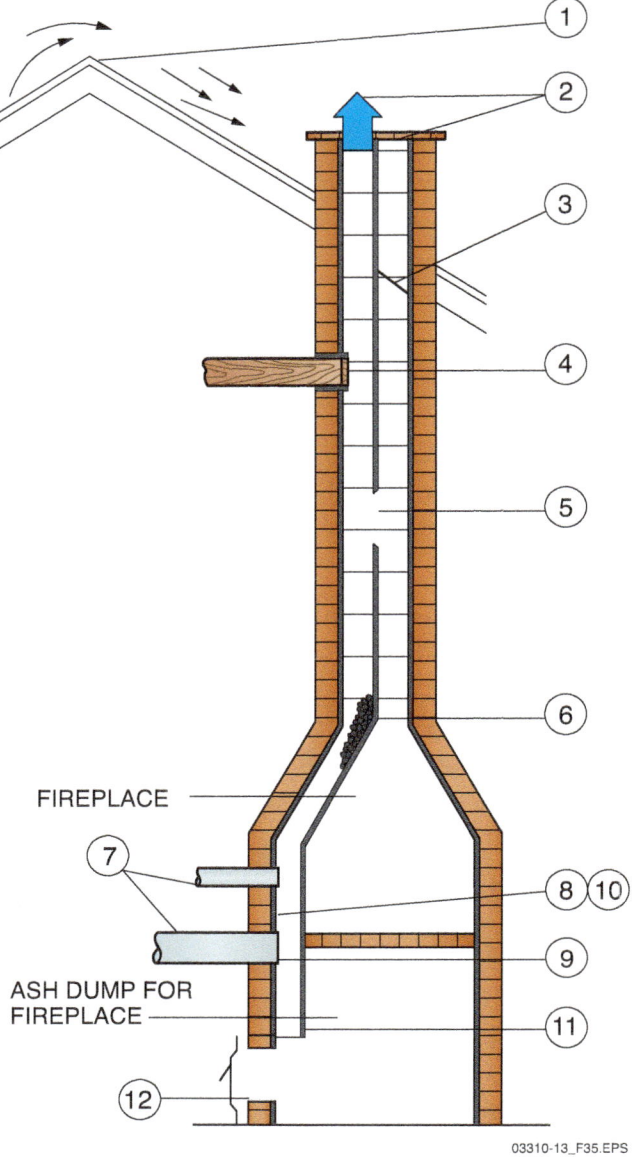

Figure 35 Common chimney problems.

7. Two or more openings may exist in the same chimney. Refer to local codes for recommendations.

8. Loose or open connections at joints in the vent pipe may be present.

9. A burner smoke pipe that extends into the chimney can cause problems.

Oil Burner Nozzle Wear

Oil burner nozzles can wear out over time. The abrasive action of microscopic particles in the oil, which is forced at high pressures through a relatively small orifice, can erode the inside of the orifice. In time, this abrasion will distort the nozzle spray pattern.

10. A loosely seated vent pipe in the chimney can cause problems.
11. Failure to extend the flue pipe into the flue opening can cause problems.
12. An oversized chimney might not be able to establish an adequate draft.

3.2.5 Draft Regulator

A draft regulator (*Figure 36*) consists of a small door in the side of the flue pipe. It is hinged near the center and controlled by adjustable weights. A draft regulator is necessary to control heat and combustion. If too high a draft exists, it causes undue loss of heat through the chimney. If too little draft is available, incomplete combustion could result. The draft regulator should maintain a constant draft over the fire, usually 0.01 to 0.03 in wc.

3.2.6 Solenoid Oil Flow Valves

Solenoid on-off flow valves can be mounted directly in the pipeline or on support brackets.

These magnetic valves provide fuel oil to the burner only when the burner operates. They close immediately upon a loss of power. Some models have an integral thermistor that delays opening until the burner motor reaches full speed. They are wired in parallel to the line voltage supply to the burner motor (*Figure 37*).

Under normal operation, the valve opens when the thermostat calls for heat and closes immediately when the call for heat is ended. These valves usually offer reliable service and should not be replaced until all other sources of trouble have been eliminated. A failure usually results in the valve not opening, resulting in no fuel supply to the burner. To check solenoid oil flow valve operation, proceed as follows:

Step 1 Check the valve operation by listening for an audible click when it opens and closes.

Step 2 If the valve does not open when the thermostat calls for heat:
- Check for normal fuel pressure.
- Make sure that the bleed line is not obstructed.
- Check the power supply at the valve. If no voltage is available, check the power source and circuit controls (relays and primary control). If the proper voltage is available, replace the coil or the entire valve.

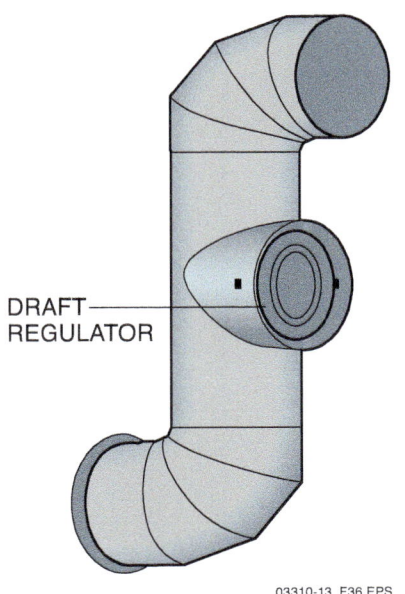

Figure 36 Draft regulator.

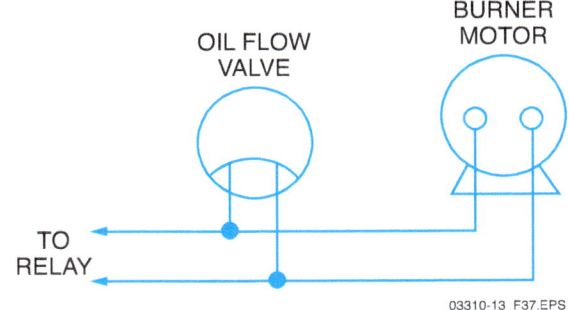

Figure 37 Oil valve circuit.

Solenoid Oil Flow Valve

A solenoid oil flow valve can help maintain the prime in the oil piping when the tank is below the burner. It also provides a safety feature by preventing oil flow if there are leaks or breaks in the oil line. This is especially true if the valve is used in a gravity-feed unit, where the tank or oil level is higher than the burner.

3.2.7 Electronic Time Delay

A few burners may be fitted with an electronic time delay external to the primary control. The electronic time delay allows the burner motor time to establish draft. It also can be used to establish first-stage flame before the opening of a second-stage fuel valve on an oil burner. It is connected in the control circuit as illustrated in the schematic diagram in *Figure 38*. On a call for heat, the electronic circuit in the time delay device (ST 70) delays the opening of the oil valve for approximately five seconds. Troubleshooting the electronic time delay consists of checking for supply voltage. If the input voltage is present and the delay does not function by passing the voltage through to the output, it must be replaced.

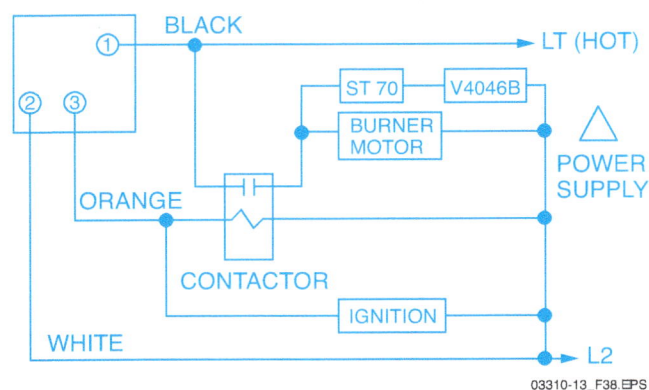

Figure 38 Time delay schematic.

Chimney Liners

Correcting masonry-related chimney problems is much easier today due to advances in technology. If a chimney is oversized or is improperly lined, a flexible chimney liner kit rated for use with fuel oils can be installed to correct the problem.

Additional Resources

Fundamentals of HVAC/R. Carter Stansfield and David Skaves. Upper Saddle River, NJ: Prentice Hall.

Refrigeration and Air Conditioning, an Introduction to HVAC/R. Larry Jeffus. Air Conditioning and Refrigeration Institute. Upper Saddle River, NJ: Prentice Hall.

3.0.0 Section Review

1. In *Figure 30*, the indoor blower will energize when the _____.
   a. thermostat calls for heat
   b. limit control closes
   c. thermostat fan switch is set to ON
   d. cad cell senses burner firing

2. Flare fittings should always be used to connect oil supply piping.
   a. True
   b. False

Summary

Oil furnaces are designed to burn fuel oil, which is mixed with air, then sprayed as a fine mist into the furnace's combustion chamber, where it is ignited. An oil burner system consists of the oil supply and the burner. Gun-type atomizing units are the most common burners used. Air returned from the conditioned space is directed over the heat exchanger by a blower in the furnace unit. Heat is transferred to the air and recirculated by the duct system. The primary component of an oil furnace is the pressure burner, which contains the fuel pump and the nozzle.

Periodic servicing, including inspection, cleaning, and combustion testing are essential for an oil-fire furnace. A critical part of the servicing is measuring combustion efficiency. Oil furnace troubleshooting usually starts with an owner's complaint, which must be analyzed in a logical manner to arrive at the solution to the problem. One of the critical tasks of the HVAC technician is to check a system thoroughly before condemning any control or component. Oil burner nozzle assemblies are a common source of problems. Problems may also occur in oil furnace control systems. Components to check include thermostats, primary controls, flame detectors, limit controls, auxiliary controls, and ignition components. Auxiliary components such as solenoid oil flow valves and electronic time delays, may also cause problems.

At the time of the troubleshooting call, it is a good idea to replace filters and similar devices that will contribute to the efficient operation of the system. The responsible troubleshooter will not leave the site until the reason for the system failure has been corrected. To make sure of this, the system should be cycled several times and then monitored for a reasonable amount of time after startup.

Review Questions

1. The combustion efficiency of a non-weatherized oil furnace must be at least _____.
   a. 65 percent
   b. 78 percent
   c. 92 percent
   d. 100 percent

2. The vane fan in a high-pressure oil burner is used to _____.
   a. prevent overheating
   b. cool the oil
   c. atomize the oil
   d. provide combustion air

3. The burner input, in Btus, is determined by the _____.
   a. type of fuel
   b. nozzle flow rate
   c. position of the electrodes
   d. size of the ignition transformer

4. An advantage of a flame-retention oil burner is _____.
   a. it provides more oil to the flame
   b. with a flame-retention burner, combustion occurs inside the burner, rather than in a combustion chamber
   c. the flame-retention burner operates at a much lower speed
   d. the flame-retention burner produces a higher temperature flame

5. Which of the following receives a signal from the thermostat and switches power to the burner and other components?
   a. Cad cell
   b. Limit switch
   c. Primary control
   d. Master cylinder

6. The device most commonly used in oil furnaces to verify a continuous flame is the _____.
   a. flame rectifier
   b. thermopile
   c. cad cell
   d. hot surface igniter

7. The return lines in a two-pipe system should _____.
   a. never be smaller than the supply lines
   b. be the same size or larger than the supply lines
   c. always be smaller than the supply lines
   d. always be larger than the supply lines

8. A single-pipe oil system is used when the _____.
   a. oil supply level is above the burner
   b. oil supply level is below the burner
   c. the vertical distance from the tank to the furnace is more than 15'
   d. the vertical distance from the tank to the furnace is more than 30'

9. An oil burner flame should be _____.
   a. blue with a yellow tip
   b. red with a yellow tip
   c. yellow
   d. orange

10. In a residential oil furnace burning No. 2 oil, the smoke reading should be _____.
    a. zero
    b. between zero and a trace
    c. between 1 and 2
    d. no greater than 2

11. The test hole for smoke spot and combustion efficiency testing should be drilled _____.
    a. as close as possible to the furnace
    b. between the draft regulator and the chimney
    c. at least two flue pipe diameters from the furnace breeching
    d. at least one flue pipe diameter from the furnace breeching

12. If the burner flame is burning off to one side, striking the combustion chamber wall, the most likely cause is _____.
    a. electrode spark gap too wide
    b. a damaged or worn nozzle
    c. weak ignition transformer
    d. loose oil pump coupling

13. A pulsating needle on an oil pump pressure gauge can indicate _____.
    a. a defective cad cell
    b. clogged fuel filter
    c. cracked heat exchanger
    d. defective nozzle

14. A downdraft in a chimney could be caused by _____.
    a. the top of the chimney being too high
    b. the chimney being below surrounding surfaces
    c. an obstruction in the chimney
    d. a crack in the chimney lining

15. An electronic time delay mounted on an oil burner external to the primary control is sometimes used to _____.
    a. control the second stage of a two-stage burner
    b. delay burner ignition until oil flow is established
    c. delay burner ignition until airflow is established
    d. limit the amount of time the burner runs on each cycle

Trade Terms Introduced in This Module

Annual Fuel Utilization Efficiency (AFUE): An industry standard used to define furnace efficiency.

Atomize: Reduce to a fine spray.

Cad cell: A light-sensitive device (photocell) in which the resistance reacts to changes in the amount of light. In an oil furnace, it acts as a flame detector.

Coping: The upper masonry course of a chimney designed to keep water out.

Downdraft: A situation in which outdoor air is forced down a chimney, preventing proper draft. Downdraft can be caused by not extending the top of the chimney above the roof line.

Flame-retention burner: A high-efficiency oil burner.

Joist: Framing member used to support floors and ceilings.

Lockout: A safety feature that prevents a furnace from automatically cycling if the flame fails.

Over-firing: A situation caused by an oversized nozzle in which excess oil is pumped into the combustion chamber.

Primary control: A combination switching and safety control that turns the oil burner on and off in response to signals from the thermostat and safety controls.

Refractory: Materials such as ceramics designed to withstand extremely high temperatures.

Smoke spot test: A test to determine the level of smoke in the flue gases.

Soot: Fine carbon particles resulting from incomplete combustion. An accumulation of soot on heat exchangers will reduce combustion efficiency and result in increased carbon monoxide production.

Temperature differential: The difference in temperature between the flue gases and the gases above the combustion chamber.

Weatherized furnace: A furnace that has been well insulated and has a weather-resistant case as specified by government standards.

Appendix A

TROUBLESHOOTING SUMMARY

PROBLEM	POSSIBLE CAUSE	CORRECTIVE ACTION
Burner motor does not start		
Trouble in primary control	Check cad cell	Replace if resistance is incorrect
Trouble in thermostat	Broken wires or loose connections Defective thermostat	Replace wires or tighten connections Replace thermostat
Faulty burner components	Broken wires or loose connections Motor start switch or thermal overload switch open Defective motor Defective pump	See manufacturer's instructions
Burner motor starts – no flame established		
Trouble in primary control	Check cad cell or stack relay Loose connections or broken wires between ignition transformer and primary control	Replace wires or tighten connections
Trouble in ignition system	Defective transformer Ignition electrodes: • Improperly positioned • Spaced too far apart • Loose • Dirty Dirty or damaged ceramic insulators	Replace transformer Check manufacturer's instructions Replace or clean electrode assembly
Faulty burner components	Dirty nozzle Loose, misaligned, or worn nozzle Clogged oil pump strainer Clogged oil line Air leak in suction line Defective pressure regulator valve Defective pump Improper draft Water in oil Oil too heavy	Correct per manufacturer's instructions
Burner motor starts – flame goes on and off after start-up		
Trouble in primary control	Cad cell problems	Check and/or clean cad cell
Trouble in ignition system	Limit settings too low Dirty oil filter	Readjust setting or differential Replace filter
Trouble in thermostat	Broken wires or loose connections Defective anticipator Defective thermostat	Replace broken wires or tighten connections Adjust anticipator Replace thermostat
Faulty burner components	Low oil pressure Defective pump, clogged or dirty oil lines Fluctuating water level (hydronic systems) Weak transformer Cracked electrodes	See manufacturer's instructions Replace transformer Replace porcelain electrodes

PROBLEM	POSSIBLE CAUSE	CORRECTIVE ACTION
System overheats house		
Trouble in primary control	Defective primary	With burner running, disconnect one low-voltage thermostat lead from primary control terminal; if relay does not drop out, replace primary
Trouble at thermostat	Wiring shorted	Repair or replace wiring
	Thermostat out of calibration	Recalibrate thermostat with an accurate thermometer
	Defective anticipator or anticipator improperly set	Replace or readjust anticipator; replace thermostat, if necessary
	Thermostat stuck in call-for-heat	Repair or replace thermostat
	Thermostat improperly located in area under control, or draft on stat through wall	Relocate thermostat: must be out of drafts, away from radiating surfaces, ducts, steam pipes, and sunny locations
	Thermostat not level	Level thermostat
Trouble in distribution system	Ductwork close to thermostat	Relocate thermostat
	Flow control valve (hot water) stuck in ON position	Repair or replace valve
	Circulator does not stop running	Check circulator circuits, replace switching device
Faulty burner components	Extremely oversized system	Replace system
	System control valves stuck in ON position	Repair or replace control valve
System underheats house		
Trouble at thermostat	Open or loose wiring	Repair or replace wiring
	Thermostat out of calibration	Recalibrate thermostat with an accurate thermometer
	Defective anticipator or anticipator improperly set	Replace or readjust anticipator; replace thermostat, if necessary
	Thermostat stuck in not calling for heat	Repair or replace thermostat
	Thermostat improperly located	Relocate to sense more accurately
	Thermostat not level	Level thermostat
Trouble at high- or low-limit controller	Controller set too low	Raise setpoint
	Slow to return to ON position	Adjust differential
	Controller defective	Replace controller
Trouble at blower (for forced-air system)	Cutting out or cycling on overload	Check air filters
	Burned out blower motor	Replace motor
	Running too slow or inadequate capacity	Check/replace motor
	Fan belt broken or loose	Replace or tighten belt
System defects	Undersized furnace or boiler system (possible if addition has been made to house)	Replace furnace or boiler system
	Distribution system inadequate	Correct problem
	Sooted heat exchanger	Clean heat exchanger
	Dirty warm-air filters	Replace filters
	Dirty boiler water (steam system)	Drain/refill boiler

03310 Troubleshooting Oil Heating

PROBLEM	POSSIBLE CAUSE	CORRECTIVE ACTION
System underheats house (continued)		
Faulty burner components	Fuel line clogged or strainer plugged Air in fuel line or excess air in system Dirty or improperly sized nozzle Low oil pressure or defective pump Motor does not come up to proper speed Inoperative or inadequate circulator (hydronic system)	Clear clog Check fuel lines for leaks Replace nozzle Adjust or replace pump Replace motor Replace circulator
System cycles too frequently (this problem is outside the primary control; the most common problems are listed below)		
Trouble at thermostat	Poor location Poorly adjusted anticipator Loose connections or defective wiring	Relocate thermostat Readjust anticipator; replace thermostat, if necessary Tighten connections or replace broken wires
Trouble at high limit	Limit set too low Differential too narrow	Reset limit Reset differential; replace control, if necessary
Faulty burner system	Dirty or clogged air filter Loose connections Fluctuating voltage Intermittent shorts	Replace filter Repair connections Contact utility Find short circuit
Relay chatters after pulling in		
Trouble at thermostat	Open contacts	Replace thermostat
Trouble at primary control	Supply voltage too low Defective load relay	Check for loose connections Replace primary

03310-13_A01C.EPS

Appendix B

Troubleshooting Chart

GENERALLY THE CAUSE Make these checks first.	OCCASIONALLY THE CAUSE Make these checks only if the first checks failed to locate the trouble.	RARELY THE CAUSE Make these checks only if other checks failed to locate the trouble.
PROBLEM: No heat – burner fails to start		
Power failure Blown fuses or tripped circuit breaker Open disconnect switch or blown fuse Thermostat set too low Thermostat switch in improper position Primary or safety controls Flame detector	Control relay or contactor Limit controls Thermostat not level Fuel pump Burner motor	Faulty wiring Loose terminals Low voltage Thermostat faulty
PROBLEM: No heat – burner starts but fails to ignite		
Electrodes Ignition transformer Improper burner adjustment Oil supply and oil pressure Manual fuel valve	Fuel pump Nozzle Burner motor Oil line filter Fuel lines	Low voltage Water in oil
PROBLEM: No heat – burner starts and fires, then loses flame before locking out on safety		
Fuel pump Nozzle Oil supply and oil pressure Oil line filters Water in oil Primary or safety controls Flame detector Improper burner adjustment	Fuel lines Nozzle Displaced or damaged baffles	Faulty wiring Loose terminals Vent or flue problem
PROBLEM: Not enough heat – burner cycle too short		
Control relay or contactor Thermostat heat anticipator setting Thermostat not level Thermostat location Dirty filters	Limit controls Fan controls Blower bearings Blower motor Defective blower wheel Dirty blower wheel	Faulty wiring Loose terminals Low voltage Low air volume Ductwork small or restricted
PROBLEM: Not enough heat – burner cycle too long		
Blower belt broken or slipping Dirty filters Dirty or plugged heat exchanger	Fuel pump Low air volume Dirty blower wheel Oil line filter	Ductwork small or restricted Oil supply and oil pressure

03310-13_A02A.EPS

GENERALLY THE CAUSE Make these checks first.	OCCASIONALLY THE CAUSE Make these checks only if the first checks failed to locate the trouble.	RARELY THE CAUSE Make these checks only if other checks failed to locate the trouble.
PROBLEM: Too much heat – burner cycles too long		
Thermostat heat anticipator setting too high Thermostat out of calibration	Thermostat not level Thermostat location	
PROBLEM: Too much heat – burner runs continuously		
Faulty wiring Control relay or contactor	Thermostat faulty	Thermostat not level Thermostat location
PROBLEM: Combustion noise		
Nozzle Electrode Dirty or plugged heat exchanger Fuel pump Burner air adjustment	Burner motor Ignition transformer	Faulty wiring Loose terminals Vent or flue problem Oil supply
PROBLEM: Air noise		
Blower Ductwork small or restricted	Cabinet	Blower wheel Dirty filters
PROBLEM: Cost of operation		
Burner motor Barometric control Blower motor Dirty filters Improper burner adjustment Inefficient burner	Low air volume Insufficient insulation Excessive infiltration	Low voltage Fan control Nozzle Ignition transformer Blower belt broken or slipping Dirty blower wheel Ductwork small or restricted Vent or flue problem Input too low
PROBLEM: Mechanical noise		
Blower bearings Blower motor Blower belt broken or slipping Blower wheel out of balance	Cabinet	Low voltage Control relay or contactor Fuel pump Burner motor Oil valve Ruptured heat exchanger Displaced or damaged baffles Fuel lines
PROBLEM: Odor		
Barometric control Vent or flue problems Flue gas spillage Fuel lines Oil supply Oil pump shutoff valve	Low air volume Improper burner adjustment Ruptured heat exchanger Displaced or damaged baffles Blocked heat exchanger Stagnant water in humidifier Water or moisture in air system	Faulty wiring or loose terminals Control transformer Fuel pump Dirty filters Dirty heat exchanger Input too high Input too low Outdoor odors

03310-13_A02B.EPS

Additional Resources

This module presents thorough resources for task training. The following resource material is suggested for further study.

Fundamentals of HVAC/R. Carter Stansfield and David Skaves. Upper Saddle River, NJ: Prentice Hall.
Refrigeration and Air Conditioning, an Introduction to HVAC/R. Larry Jeffus. Air Conditioning and Refrigeration Institute. Upper Saddle River, NJ: Prentice Hall.
HVAC Maintenance Procedure. Syracuse, NY: Carrier Corporation.
http://www.bacharach-inc.com/PDF/Brochures/4097%20Combustion.pdf; Instructions for performing and evaluating a combustion efficiency test using an electronic test instrument.

Figure Credits

wCourtesy of Lennox Industries, Inc., Figure 1
France/A Scott Fetzer Company, Figure 8
RW Beckett Corporation, Figure 11
Courtesy of Honeywell International, Figure 13 (photo)
Courtesy of RectorSeal®, SA02
Carlin Combustion Technology, Inc., Figure 21

Courtesy of Alpha American Company - Yukon Furnaces, SA03
Courtesy of Bacharach, Inc./www.MyBacharach.com, Figure 26
Courtesy of Dwyer Instruments, Inc., Figure 27
Courtesy of ICM Controls, Figure 32 (photo)

Section Review Answers

Answer	Section Reference	Objective
Section One		
1.b	1.0.0	1a
2.a	1.2.0	1b
3.c	1.3.0	1c
4.a	1.4.2	1d
Section Two		
1.b	2.1.2	2a
2.d	2.2.1	2b
Section Three		
1.c	3.1.0	3a
2.a	3.2.2	3b

NCCER CURRICULA — USER UPDATE

NCCER makes every effort to keep its textbooks up-to-date and free of technical errors. We appreciate your help in this process. If you find an error, a typographical mistake, or an inaccuracy in NCCER's curricula, please fill out this form (or a photocopy), or complete the online form at **www.nccer.org/olf**. Be sure to include the exact module ID number, page number, a detailed description, and your recommended correction. Your input will be brought to the attention of the Authoring Team. Thank you for your assistance.

Instructors – If you have an idea for improving this textbook, or have found that additional materials were necessary to teach this module effectively, please let us know so that we may present your suggestions to the Authoring Team.

NCCER Product Development and Revision
13614 Progress Blvd., Alachua, FL 32615

Email: curriculum@nccer.org
Online: www.nccer.org/olf

❏ Trainee Guide   ❏ Lesson Plans   ❏ Exam   ❏ PowerPoints   Other _____

Craft / Level: _____   Copyright Date: _____

Module ID Number / Title: _____

Section Number(s): _____

Description: _____

_____

_____

_____

Recommended Correction: _____

_____

_____

_____

Your Name: _____

Address: _____

_____

Email: _____   Phone: _____

Troubleshooting Accessories

OVERVIEW

Residential furnaces, especially in cold, dry northern climates, are often equipped with humidifiers to maintain a healthy moisture level in the indoor air. Electronic air cleaners may also be installed in both residential and commercial environments. Economizers are common in commercial systems. Energy and heat recovery ventilators are another accessory that can be applied in both residential and commercial systems. Servicing these types of devices requires some specialized knowledge. This is especially true of economizers, because there are several types, each with its own specific operating characteristics.

Module 03312

Trainees with successful module completions may be eligible for credentialing through NCCER's National Registry. To learn more, go to **www.nccer.org** or contact us at **1.888.622.3720**. Our website has information on the latest product releases and training, as well as online versions of our *Cornerstone* magazine and Pearson's product catalog.

Your feedback is welcome. You may email your comments to **curriculum@nccer.org**, send general comments and inquiries to **info@nccer.org**, or fill in the User Update form at the back of this module.

This information is general in nature and intended for training purposes only. Actual performance of activities described in this manual requires compliance with all applicable operating, service, maintenance, and safety procedures under the direction of qualified personnel. References in this manual to patented or proprietary devices do not constitute a recommendation of their use.

Copyright © 2018 by NCCER, Alachua, FL 32615, and published by Pearson, New York, NY 10013. All rights reserved. Printed in the United States of America. This publication is protected by Copyright, and permission should be obtained from NCCER prior to any prohibited reproduction, storage in a retrieval system, or transmission in any form or by any means, electronic, mechanical, photocopying, recording, or likewise. To obtain permission(s) to use material from this work, please submit a written request to NCCER Product Development, 13614 Progress Blvd., Alachua, FL 32615.

03312 V5

From *HVAC Level Three, Trainee Guide*. NCCER.
Copyright © 2018 by NCCER. Published by Pearson. All rights reserved.

03312
TROUBLESHOOTING ACCESSORIES

Objectives

When you have completed this module, you will be able to do the following:

1. Describe how to troubleshoot various HVAC system accessories.
   a. Describe how to approach the troubleshooting process.
   b. Describe how to troubleshoot humidifiers.
   c. Describe how to troubleshoot electronic air cleaners.
   d. Describe how to troubleshoot UV lighting devices.
2. Describe how to troubleshoot accessories related to the introduction of outside air.
   a. Describe how to troubleshoot economizers.
   b. Describe how to troubleshoot recovery ventilators.

Performance Task

Under the supervision of your instructor, you should be able to do the following:

1. Using the correct tools and circuit diagrams, isolate and correct malfunctions in selected accessories.

Trade Terms

Free cooling
Mechanical cooling
Water hammer

Industry-Recognized Credentials

If you are training through an NCCER-accredited sponsor, you may be eligible for credentials from NCCER's Registry. The ID number for this module is 03312. Note that this module may have been used in other NCCER curricula and may apply to other level completions. Contact NCCER's Registry at 888.622.3720 or go to **www.nccer.org** for more information.

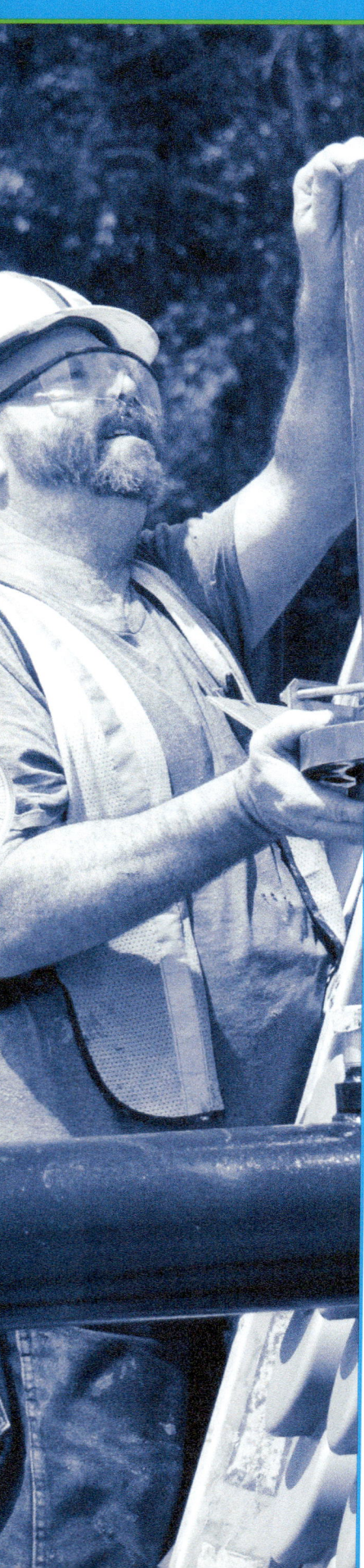

Contents

1.0.0 Accessory Troubleshooting 1
  1.1.0 Troubleshooting Approach 1
    1.1.1 Electrical Troubleshooting 1
    1.1.2 Troubleshooting Aids 2
    1.1.3 Sequence of Operation 2
  1.2.0 Troubleshooting Humidifiers 2
  1.3.0 Troubleshooting Electronic Air Cleaners 8
  1.4.0 Troubleshooting Ultraviolet Lamps 11
2.0.0 Outside Air Accessory Troubleshooting 15
  2.1.0 Economizer Operation 15
    2.1.1 Economizer Types 16
    2.1.2 Troubleshooting Economizers 18
  2.2.0 Recovery Ventilators 19

Figures and Tables

Figure 1   Technician troubleshooting with a multimeter 2
Figure 2   Sequential operating sequence.. 3
Figure 3   Graphical presentation of an operating sequence 4
Figure 4   Control circuit schematic 4
Figure 5   Schematic of a typical fan-powered humidifier circuit................... 5
Figure 6   Infrared humidifier schematic 5
Figure 7   Humidistat.. 6
Figure 8   Recommended indoor relative humidity in the winter 6
Figure 9   Digital humidistat 7
Figure 10 Humidifier capacity chart... 8
Figure 11 Electronic air cleaner (EAC) 8
Figure 12 Electronic air cleaner operation 9
Figure 13 Electronic air cleaner schematic... 10
Figure 14 Electronic air cleaner manufacturer's troubleshooting
           flow chart... 12
Figure 15 Ultraviolet lamp.. 12
Figure 16 Rooftop unit with an economizer installed..................................... 15
Figure 17 Basic economizer system 16
Figure 18 Simplified schematic of an economizer.. 17
Figure 19 Dry-bulb economizer 18
Figure 20 Enthalpy economizer.. 18
Figure 21 Integrated economizer 18
Figure 22 Motorized rectangular damper.. 19
Figure 23 Energy recovery ventilator 20
Figure 24 Recovery ventilator troubleshooting chart..................................... 22

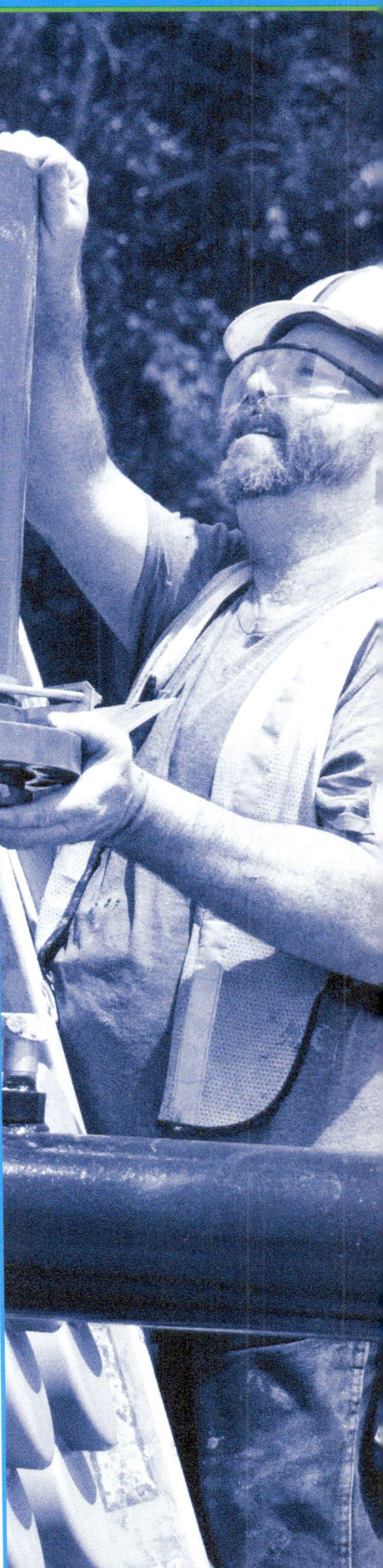

SECTION ONE

1.0.0 ACCESSORY TROUBLESHOOTING

Objectives

Describe how to troubleshoot various HVAC system accessories.
   a. Describe how to approach the troubleshooting process.
   b. Describe how to troubleshoot humidifiers.
   c. Describe how to troubleshoot electronic air cleaners.
   d. Describe how to troubleshoot UV lighting devices.

Performance Task

1. Using the correct tools and circuit diagrams, isolate and correct malfunctions in selected accessories.

Trade Terms

Water hammer: A banging noise in water pipes that occurs when the water supply solenoid valve on a humidifier abruptly shuts off water to the humidifier.

Troubleshooting the accessories and optional equipment used with HVAC systems is done in the same way as troubleshooting basic heating, cooling, or air distribution equipment. It involves close examination of the equipment to detect telltale signs of what may be wrong. Following standard procedures makes troubleshooting accessories easier than a trial-and-error method.

By using the manufacturer's wiring diagrams, specifications, and troubleshooting aids, the cause of almost any problem can be quickly isolated. The HVAC technician must be able to listen to the owner's complaint, observe the system in operation, then identify and pursue the most probable cause of the malfunction. Accurate diagnosis and repair of HVAC accessories also requires the proper selection and safe use of test equipment and instruments.

Many kinds of accessories can be used with HVAC equipment. In addition, accessories used with residential equipment are often different from those used with commercial equipment. This module will cover troubleshooting of the common accessories and optional equipment introduced in the *HVAC Level Two* module, *Air Quality Equipment*. They include the following:

- Humidifiers
- Electronic air cleaners
- Economizers
- Heat recovery ventilators

1.1.0 Troubleshooting Approach

For the purpose of this module, problems are divided into two categories: electrical and mechanical. This does not mean that all malfunctions fit easily into these two categories. Obviously, there are problems in which one affects the other. Whether troubleshooting an electrical or mechanical problem, using a logical, structured approach to troubleshooting is essential. The technician must know how to diagnose specific control devices and components used with accessories. A good set of service tools and the ability to read electrical wiring diagrams are a must. Do not guess at the cause of a problem. Also, review the material presented in the *HVAC Level Three* module, *Control Circuit and Motor Troubleshooting*. It describes the elements necessary to perform a systematic approach to control troubleshooting. Many of the problems related to accessories are control-related. Even when this is not the case, some testing and manipulating the controls is often required to pinpoint the real problem area. The module also discusses the different kinds of troubleshooting aids provided by equipment manufacturers, both in their equipment and in their service literature.

1.1.1 Electrical Troubleshooting

The electrical troubleshooting information covered in this module is limited to those electrical devices specific to certain accessories. Troubleshooting most of the electrical problems and components in HVAC equipment is performed in the same way, regardless of the type of equipment. Troubleshooting also requires the use of various common tools and test instruments (*Figure 1*).

For quick reference, troubleshooting topics that are relevant to accessories are listed here:

- Isolating to a faulty circuit first via the process of elimination.
- Isolating to a faulty component.
- Single-phase and three-phase input voltage measurements.
- Fuse and circuit breaker checks.

Figure 1 Technician troubleshooting with a multimeter.

- Resistive and inductive load checks.
- Switch and relay/contactor checks.
- Control transformer checks.

When troubleshooting an accessory that uses an electronic control, do not arbitrarily replace the board. First make sure the module or board has failed. In some equipment, this means using a built-in diagnostic system to indicate that a board has failed. If the equipment does not have a built-in diagnostic capability, make sure to isolate the board as the cause of a problem by using the troubleshooting information in the product literature. Often this means checking that the correct input signals are available to the control. If the correct inputs are available, and the control is providing the correct output signals, the problem is probably not with the control board, but elsewhere in the system.

1.1.2 Troubleshooting Aids

The first action, when troubleshooting an accessory, is to consult the product literature. This literature usually describes the mechanical and electrical operation of the equipment, including sequence of operation. Voltage or signal inputs and outputs related to the operation of the accessory are also defined.

Service literature almost always includes troubleshooting aids such as wiring diagrams and troubleshooting charts. These aids will help to isolate an equipment problem to the specific failed component.

1.1.3 Sequence of Operation

Knowing the sequence of operation of any accessory is absolutely necessary to successfully troubleshoot a problem. While each manufacturer may have very specific sequences for their equipment, most follow a common sequence, regardless of the brand. For example, in gas forced-air furnaces, the burner ignition sequence is nearly universal: on a call-for-heat, the inducer energizes. When inducer airflow is proven, the igniter warms up, the gas valve opens, flame is proved, and the blower operates. Similarly, common sequences occur with accessories.

The purpose of a humidifier used with a forced-air furnace is to introduce moisture into the structure through the air distribution system. Without the blower operating, no moisture will be distributed. For that reason, many humidifiers have their operation tied directly to blower operation. If the blower is not running, the humidifier will not operate. Electronic air cleaners (EACs) also require air movement through them for proper operation. If airflow (blower operation) is missing from the sequence of operation of these accessories, they will not operate or will function poorly. With either of these accessories, there may be some component that proves airflow to allow normal operation.

The sequence of operation can be presented in several ways in the product literature. One way is a list of steps in a normally occurring sequence. Step one is a certain event that must occur before step two (another event) can occur. Sometimes the operating sequence is presented graphically, using a flow chart or a spreadsheet to show events and their relationship to other events over time. *Figure 2* shows an operating sequence arranged as the steps occur. *Figure 3* is a graphical representation of an operating sequence. Both sequences of operation are for steam humidifiers.

1.2.0 Troubleshooting Humidifiers

A humidifier is a device used to add humidity. All humidifiers introduce water vapor into a building's conditioned air at a predetermined rate; therefore, they must be connected to a water supply. Some humidifiers also require a drain connection. Humidifiers must be carefully installed and mounted as directed in the manufacturer's installation instructions. Some are motor-driven while others are not. They are controlled by a humidistat in almost all applications.

Most humidifier problems can be prevented with periodic maintenance. Evaporative humidifiers leave behind minerals contained in the water that has evaporated. These minerals are left on the media and in the water reservoir. The float, float valve orifice, and valve seat for most humidifiers need periodic cleaning to remove mineral

OPERATION	UNIT ACTIVITY
Power is provided or restored to the unit	• HPC-450i controller is energized. • LEDs three times; On LED remains lit. • All sensors activated.
Water level sensor determines the steam chamber is empty *Unit is prepared for a humidification cycle*	• Following short delay, the fill valve opens and the Fill LED illuminates in unison. • Water flows into the steam chamber for up to 14 minutes, depending upon chamber size. • Water level sensor determines chamber is full. The HPC-450i de-energizes the fill valve and related LED.
On a call for humidification by the control device	• A DC signal is completed at the control, applying DC power to terminal H1 on humidifier. The signal will range from 1 to 10 VDC based on the ⌧RH between the new setpoint and the measured space RH. • The fan relay will energize in 5-7 seconds. • The HPC-450i will monitor for closed circuits in the fan proving switch and the high limit switch. • If circuits remain closed, the STEAM LED will illuminate and remain on as long as the demand for humidity is present. • The STEAM LED will de-energize if the monitored circuits open or the DC demand signal falls below 1 VDC. • After a 20-second delay, the contactor will close to energize the electric elements. • In roughly 9 minutes, the water in the steam chamber will boil and steam will be generated. Steam enters the dispersion hose and enters the manifold. **WARNING:** Steam chamber and hose will be very hot!
As humidification continues, the water level in the steam chamber will fall	• Fill sequence as noted above begins.
The demand for humidity ceases	• Steam LED cycles off. • The electric elements are now de-energized. • The fan continues to run for 4 minutes, and then the fan relay is de-energized by the HPC-450i.

Figure 2 Operating sequence.

deposits. If so equipped, strainers should also be cleaned. In addition, algae, bacteria, and virus growth can cause problems. Algaecide can be used to help neutralize algae growth.

Most manufacturers recommend that the water reservoir be drained and humidifier components cleaned about every two months. In hard water areas, more frequent cleaning may be necessary.

POWER ON	DRAINING CYCLE			FILLING CYCLE	BOILING CYCLE		
1	2	3	4	5	6	7	8
Up to 5 sec	Up to 3 min	Up to 4 min	Up to 2 min	Up to 2 min	Up to 5 sec	Up to 6 min	Up to 20 sec
	Up to 9 minutes total for drain cycle				1 steam cycle		

Figure 3 Graphical presentation of an operating sequence.

Figures 4, 5, and *6* show typical control circuits for three different humidifiers.

Figure 4 is a fairly simple control circuit. The furnace blower motor and a 24V transformer are connected in parallel and are simultaneously energized when the furnace fan control closes. Note that this transformer is dedicated to the humidifier only; a separate transformer serves the primary equipment. A humidistat in the low-voltage circuit is in series with a 24V motor in the humidifier. In this control circuit scheme, the humidifier will operate only if the blower motor and transformer are energized and the humidistat is calling for moisture. Sometimes a 24V solenoid will be connected in parallel with the 24V motor, allowing water to flow to the humidifier.

Figure 5 is a variation of the circuit shown in *Figure 4.* In *Figure 5,* the humidifier motor and water solenoid valve are powered by 120VAC supplied through a normally open set of relay (R) contacts. In this circuit, the humidistat contacts close to power the coil of the relay. When energized, the relay contacts provide a circuit through which the humidifier fan and solenoid valve are energized.

Figure 6 is the control circuit for an infrared humidifier. Unlike other humidifiers, an infrared humidifier does not depend on heat supplied by the furnace to evaporate water. Instead, infrared heaters heat a pan of water and evaporate it in much the same way that sunlight evaporates a puddle of rainwater. The infrared heaters are powered by line voltage and energized by a contactor that is controlled by a timer. Devices in the control circuit are connected in series. For example, the humidistat may be calling for moisture, but if the sail switch is not sensing airflow in the duct and/or the high-limit aquastat is open, the humidifier will not operate.

Troubleshooting all three of these different circuits involves checking for the presence of power at key locations; testing circuits for open and closed switches, either by continuity or voltage testing; testing the electrical integrity of various components; and checking the mechanical integrity of various load components.

Common problems encountered with humidifiers follow:

- *Humidifier not running/not providing moisture* – This is usually caused by an electrical control circuit problem. As applicable, check the overload protection, humidistat, and control circuit components. Also check to see if the humidifier motor has failed. Electrical failures such as a tripped overload or failed motor can result from the rotating media being bound as a result of a mineral buildup or mechanical failure. A solenoid valve that is stuck closed or has an open coil will not supply water to the humidi-

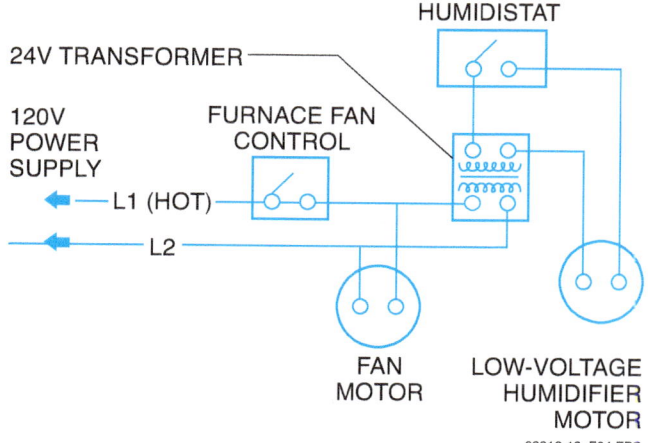

Figure 4 Control circuit schematic.

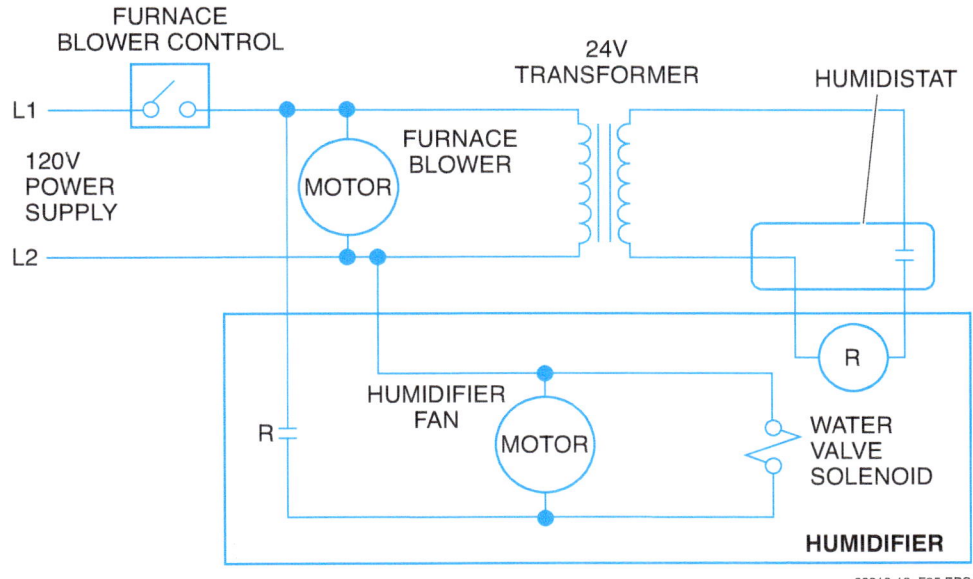

Figure 5 Schematic of a typical fan-powered humidifier circuit.

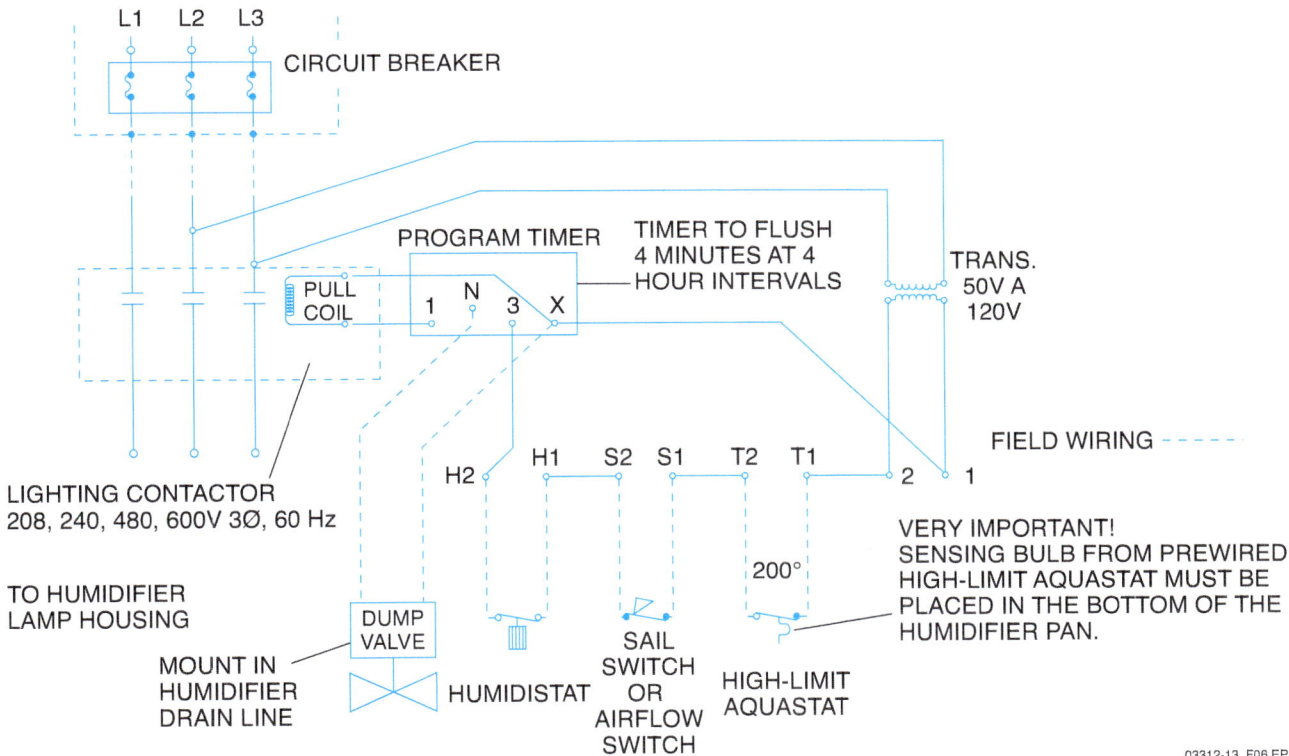

Figure 6 Infrared humidifier schematic.

Servicing Accessories

When servicing accessories, it is often necessary to consult the service literature for the main HVAC unit with which the accessory is being used. Also, consult the service literature for some individual components, such as the humidistat or thermostat, used in the accessory control system.

fier and no moisture will be added to the structure, even if the humidifier motor is running.

- *Water overflow* – Check for a dirty, misadjusted, or faulty float valve. If the humidifier is equipped with a solenoid valve, check for dirt or scale lodged between the valve plunger and seat that may cause the valve to stick open, even when de-energized. The valve may need cleaning, adjusting, or replacement. Also, check for water backing up the drain line as a result of the line being plugged or restricted.

> **CAUTION**
> Water overflowing from a humidifier can corrode metal parts, including furnace heat exchangers, and lead to premature failure of the part. Water can also damage electrical and electronic components.

- *Noise* – If the unit is equipped with a solenoid valve, the clicking noise it makes when it opens and closes can be amplified by sheet metal ductwork. To help reduce this problem, check to make sure the unit is snug against the plenum. The solenoid valve is powered by alternating current that operates at a frequency of 60Hz. That frequency can sometimes be heard as a humming sound that can also be amplified by the ductwork. One manufacturer supplies a rectifier kit that can be installed on the terminals of the solenoid. The rectifier converts AC to DC to help eliminate the hum. High pressure in the humidifier's water supply line may cause **water hammer** when the solenoid opens and closes. It may be necessary to place a length of high-pressure hose before the valve, or install a pressure-reducing device in the supply line to reduce this problem.
- *Low or high levels of humidity* – The recommended level of relative humidity (RH) that should be maintained in a building during the winter is dependent on several factors. Condensation on windows and inside exterior walls are signs of a relative humidity value that is too high. Results of low RH can include the following:

  – Dry, itchy skin
  – Static electricity shocks
  – Clothing static cling
  – Sinus problems
  – Chilly feeling
  – Sickly pets and plants
  – Loose furniture joints

When the humidity is too high or too low, first check the humidistat setpoint (*Figure 7*). Electromechanical humidistats like the one shown in *Figure 7* have been used for many years. They contain a material that expands and contracts relative to the moisture level in the air. That material is connected to a switch in the humidistat, which opens and closes depending on the moisture levels in the room and the setting of the humidistat. Because these parts are electromechanical, their calibration can drift over time.

To check calibration, compare the humidistat dial setting to the relative humidity level in the conditioned space, as measured with a sling or digital psychrometer. First, turn the humidistat dial to its lowest setting, and then slowly increase the setting until a light click is heard. The relative humidity setting indicated by the dial should be close to the relative humidity measured with the sling psychrometer. If the humidistat is out of calibration, adjust it, if possible. *Figure 8* lists typical recommended levels of indoor relative humidity that should be maintained at various outdoor winter temperatures.

Figure 9 shows a digital humidistat that electronically senses moisture levels in the indoor air.

Figure 7 Humidistat.

AT OUTDOOR TEMPERATURE (°F)	RECOMMENDED (SAFE) INDOOR RH (%)
−20	15
−10	20
0	25
10	30
20	35
30	40

BASED ON AN INDOOR TEMPERATURE OF 72°F

Figure 8 Recommended indoor relative humidity in the winter.

Electronic humidistats often will use an optional thermistor to sense outdoor air dry-bulb temperature. With outdoor temperature known, the control's software can automatically maintain indoor relative humidity at a level that avoids indoor condensation on windows. Electronic humidistats offer several advantages over older electromechanical types, including accuracy and more reliable operation.

Modern room thermostats can contain a built-in humidistat that measures and controls humidity electronically. One manufacturer refers to this control as a Thermidistat™. The desired humidity level is entered into the control by pressing a button and selecting the desired humidity level on a digital display. The control then operates the humidifier to maintain the desired humidity level during heating operation as with a conventional humidistat. If an outdoor temperature sensor is used with the combination thermostat/humidistat, the control automatically adjusts the indoor humidity setpoint according to outdoor temperature.

This type of humidistat can also control the cooling system to dehumidify. Normal air conditioning operation removes humidity from the air. However, there may be times when humidity is high, but the temperature in the space is not high enough to initiate a call for cooling. In these cases, the control calls for compressor operation. Typically this arrangement is used with a variable-speed blower and an electronic control in the air handler. When a call for dehumidification is present without a call for cooling, the blower operates at a much lower speed to maximize moisture removal. The software in the control and in the air handler allows dehumidification without overcooling the space. To enable this feature, the room control and/or the air handler must be properly configured according to the manufacturer's instructions.

Some reasons for low humidity not directly related to the humidifier itself are as follows:

- The water supply feed valve/saddle supply valve is closed or not fully open.
- Excessive air infiltration from permanently opened windows, exhaust fans, or open fireplace dampers can cause low humidity.
- An oversized furnace can cause low humidity because burner cycles are too short for the humidifier to do its job. Remember that many humidifiers can only add moisture to the air when the blower is also functioning. With this condition, it may be necessary to set the blower for continuous operation so that the humidifier can run as needed. However, it is important to note that the capacity of this type of humidifier (evaporative) is substantially reduced when only room air is being used, as opposed to using heated air for evaporation. The capacity of steam humidifiers would be unaffected.

Excessive humidity may be caused by continuous operation of the humidifier as a result of a failed humidistat or a defective or sticking contact on the humidifier relay. Either low or high humidity levels can also be caused by using a humidifier with the wrong capacity. Humidifiers are typically rated in gallons of water per day. Humidifier selection depends on the volume of the building or area in square feet (ft^2). It also depends on the building's air tightness. *Figure 10* shows a typical graph used for the selection of residential humid-

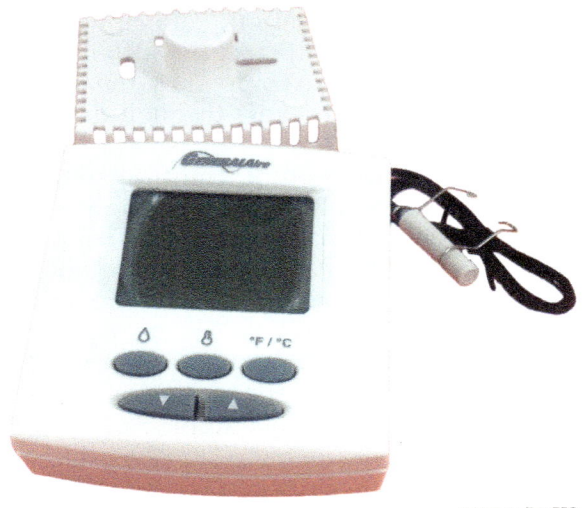

Figure 9 Digital humidistat.

Humidifier Selection

To reduce mineral deposit buildup or algae growth, select a humidifier design that minimizes these problems. For example, many bypass and fan-powered humidifiers lack a water reservoir. Water flows through the humidifier only when needed; excess water is drained away. Eliminating the reservoir reduces algae growth, while draining the excess water helps flush out mineral deposits.

ifiers. Similar graphs and/or charts are available for commercial and industrial humidifiers.

The terms *loose*, *average*, and *tight* are defined as follows:

- *Loose* – The building has little insulation, no vapor barriers, and no storm doors or windows. In homes, it can also mean an undampered fireplace. The air exchange rate is about 1.5 changes per hour.
- *Average* – The building is insulated, has vapor barriers, and has loose storm doors or windows. In homes, it can also mean a dampered fireplace. The air exchange rate is about 1.0 change per hour.
- *Tight* – The building is well insulated, has vapor barriers, and tight storm doors or windows. In homes, it can also mean a dampered fireplace. The air exchange rate is about 0.5 change per hour.

1.3.0 Troubleshooting Electronic Air Cleaners

Electronic air cleaners (EACs) (*Figure 11*) remove airborne particles and odors from the air circulated to the conditioned space. They can be stand-alone units or may be mounted in the return-air side of a duct system. EACs create an electrostatic field that produces a positive or negative charge on the particles being carried along in the air stream. The EAC then removes the charged particles from the air stream by drawing them to oppositely charged collector plates.

EACs have a high-voltage, solid-state power supply. The high voltage produced by the power supply, which can range from 6,000 to 10,000 volts DC, is used to electrically charge (ionize) all particles in the air that pass through the filter. After the particles are ionized, the high voltage is then used to attract, collect, and hold the particles in the filter. *Figure 12* shows typical EAC operation.

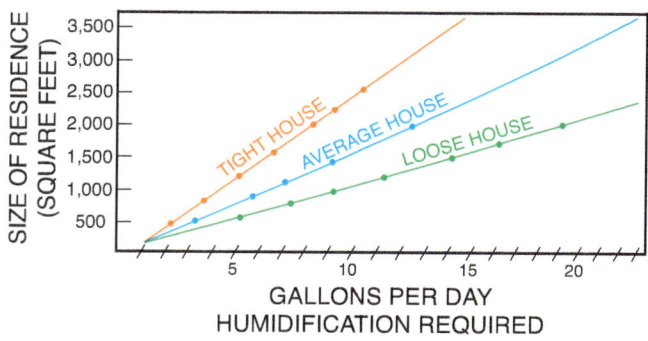

Figure 10 Humidifier capacity chart.

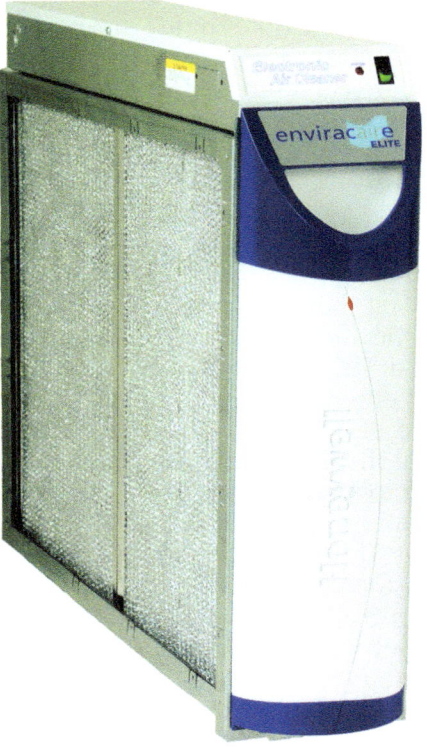

Figure 11 Electronic air cleaner (EAC).

As shown, the filter portion of an EAC consists of a prefilter, ionizer section, collector section, and charcoal filter section. As the air enters the EAC, larger particles are trapped by the prefilter section. Smaller particles pass through the prefilter to the ionizing section.

This section consists of a fine tungsten wire grid (ionizer grid) connected to the high-voltage DC power supply. An electrostatic ionizing field created by the high voltage on the wire grid gives particles a positive or negative charge as they pass through the grid. These charged particles are then drawn into the collector section.

The collector section consists of a series of equally spaced, parallel collector plates connected to the high-voltage DC power supply; even- and odd-numbered plates are at a positive and negative DC voltage, respectively. As ionized particles flow between the plates, they are attracted to and held on oppositely charged collector plates. The air, cleaned of pollutants, then passes through the charcoal filter section where odors are absorbed. From there, air passes to the conditioned space. Pollutants remain held in the collector section until the filter is cleaned.

Figure 13 shows the schematic diagram for a typical EAC. In the unit shown, the high-voltage power supply is made from discrete components such as diodes, capacitors, and resistors. In newer

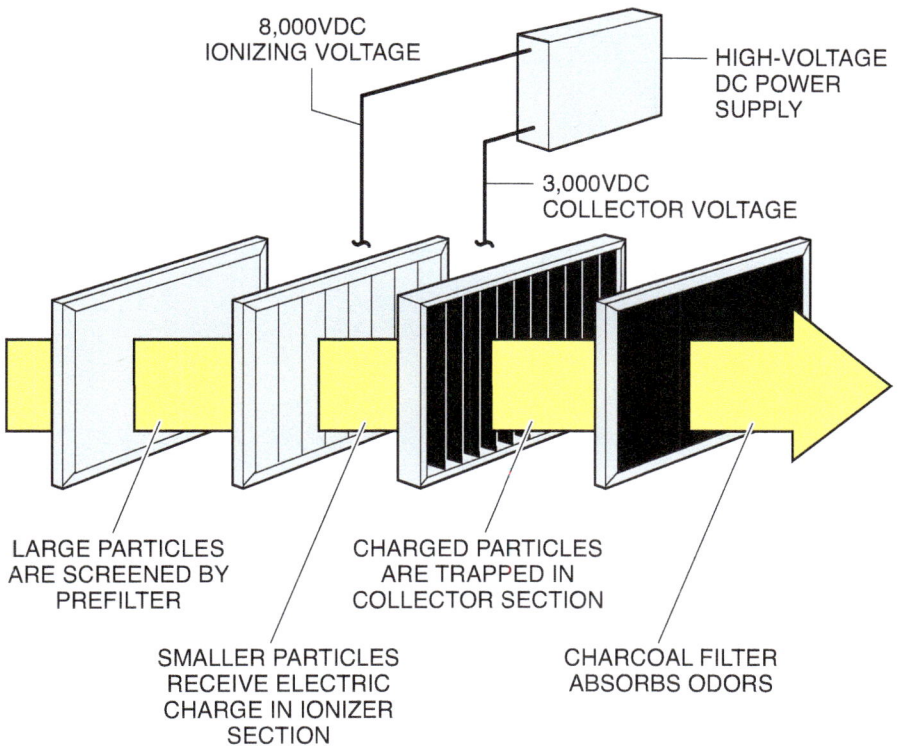

Figure 12 Electronic air cleaner operation.

Bypass Humidifiers

A bypass humidifier is normally installed on the supply duct with the bypass duct connecting to the return. The humidifier can operate just as effectively if installed on the return duct with the bypass connecting to the supply plenum. This makes bypass humidifiers more versatile than other types of humidifiers, which must always be installed on the supply plenum.

modules, the power supply may be a solid-state circuit that is fully enclosed in a non-repairable, sealed assembly.

The EAC is normally electrically connected to the furnace fan circuit and operates only when the furnace blower motor is energized. On some units, this control is through contacts of a sail

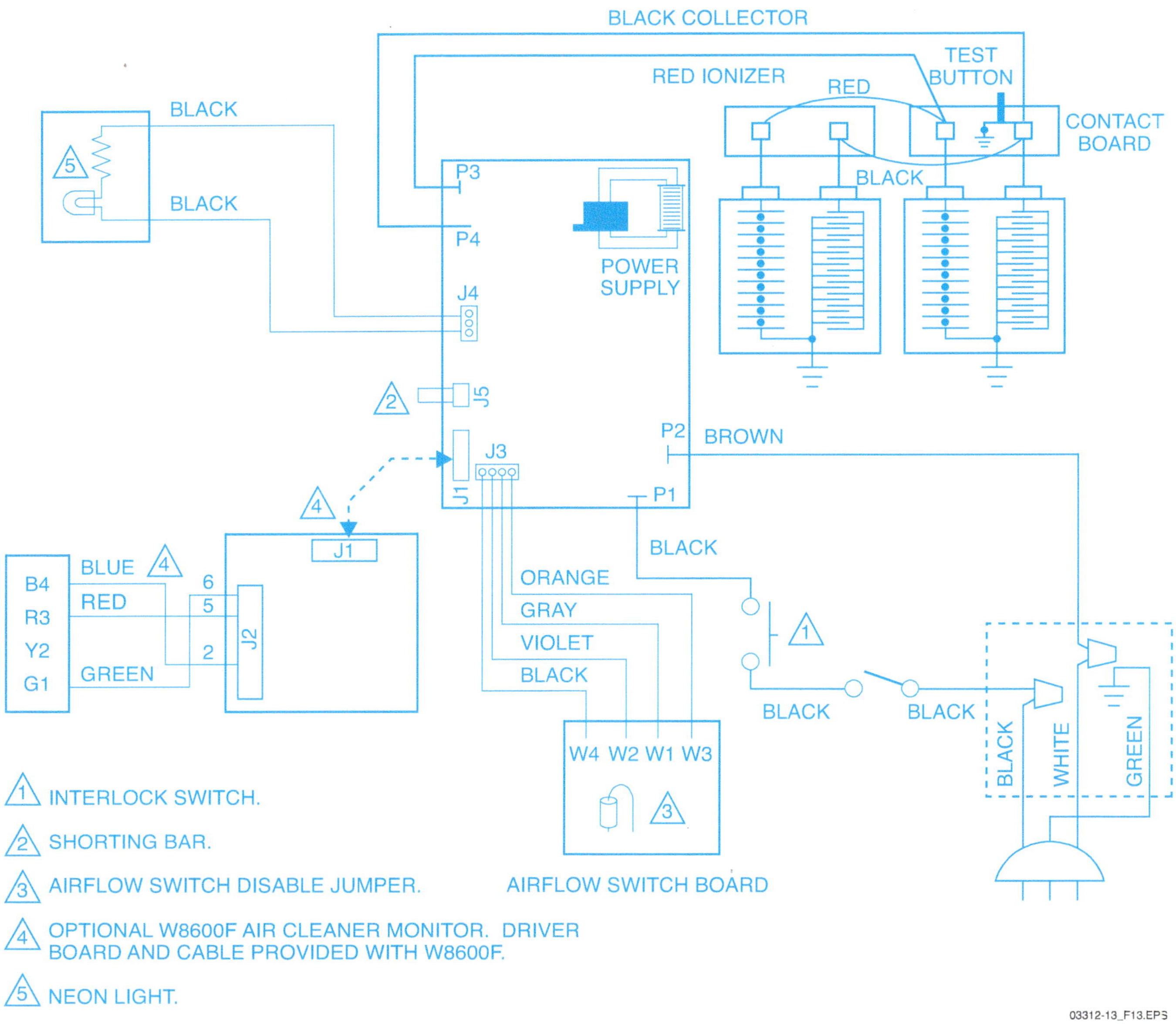

Figure 13 Electronic air cleaner schematic.

switch installed in the air stream. Others may use contacts of the fan control relay.

Some EACs are electrically interlocked so that the unit shuts off automatically when the service doors are opened to gain access to the unit.

EACs do not clean air effectively when dirty. Electronic cells and protective screens are generally cleaned every two to three months. The electronic cells of most residential models can be safely placed in a dishwasher. Large commercial

Humidifiers and Hot Water

Some bypass and fan-powered humidifier installers tap into a hot water line instead of a cold water line to supply water to the humidifier. However, some manufacturers prohibit this practice. Hotter water vaporizes much more quickly in the humidifier, but the energy used to heat the water increases its cost. If hot water is used to supply a humidifier, the flow rate must be carefully controlled to prevent waste.

models may be equipped with built-in washing apparatus and automated washing controls.

Some manufacturers have designed their equipment so that the power indicator flashes when the EAC needs cleaning. If cleaned regularly, many EAC problems can be avoided. However, some conditions can cause an EAC to become dirty and overloaded in an abnormally short time. These conditions include the following:

- New carpeting has been installed.
- Untreated concrete floors exist.
- Dusty construction work is being performed in the area.
- Extreme amounts of pollen enter the building during the spring.

EAC operation is sensitive to airflow. When operated at airflow rates above those recommended by the manufacturer, they can become quite inefficient. When the airflow is reduced below the manufacturer's recommended minimum, enough ozone can be generated to cause an annoying odor. If the EAC is operating and is clean, but is not cleaning the air, or if the smell of ozone is present, check to make sure that the system airflow matches the EAC specifications.

The recommended procedure for troubleshooting problems in an EAC is to first perform a visual inspection to identify any mechanical damage that might be the cause of the problem. For electrical problems, use the troubleshooting aids in the manufacturer's service literature to fault-isolate the unit. *Figure 14* shows a typical manufacturer's troubleshooting flow chart. Note that the chart calls for a reduced ionizer voltage (7,500 to 8,500VDC) when the cells are in place, compared to the value with the cells removed. This is due to the voltage drop that occurs when a load is imposed on the high-voltage transformer. This is also true of the collector voltage, at roughly the same percentage of reduction.

> **CAUTION**
> If measuring high voltage while troubleshooting an electronic air cleaner, use a voltmeter that is designed and rated for measuring the high voltages encountered in the unit.

1.4.0 Troubleshooting Ultraviolet Lamps

Ultraviolet (UV) lamps (*Figure 15*) are used for microbe control on evaporative coils, drain pans, and duct insulation. They are also used to kill airborne pathogens and may be used in conjunction with an electronic air cleaner. Two different methods are used for air purification. One highly effective method constantly bathes a fixed object such as an evaporator coil with UV light. The other method purifies the air with UV lamps positioned in a moving airstream. Since microbes in a moving airstream can move past the lamps fairly fast, their exposure to UV rays is limited, making this method of air purification less effective.

Saddle Valves

To obtain water for a humidifier, a nearby water line is often tapped using a saddle valve. This valve is clamped around a copper water line and contains a piece of rubber that acts as a gasket to prevent water leakage. The water line is tapped, or penetrated, by turning in the handle of the valve. This causes a steel point to penetrate the wall of the copper tube. When the tube is pierced, the steel point seats itself against the copper to shut off or allow water flow. One manufacturer states that adequate humidifier water flow can be obtained by opening the valve no more than one turn. Some local codes do not allow the use of saddle valves. They do have a tendency to leak over time, as the rubber seal dries out and hardens.

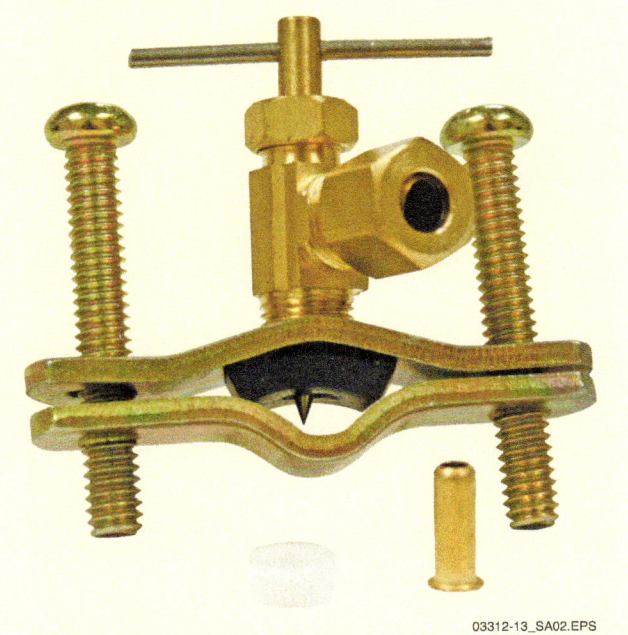

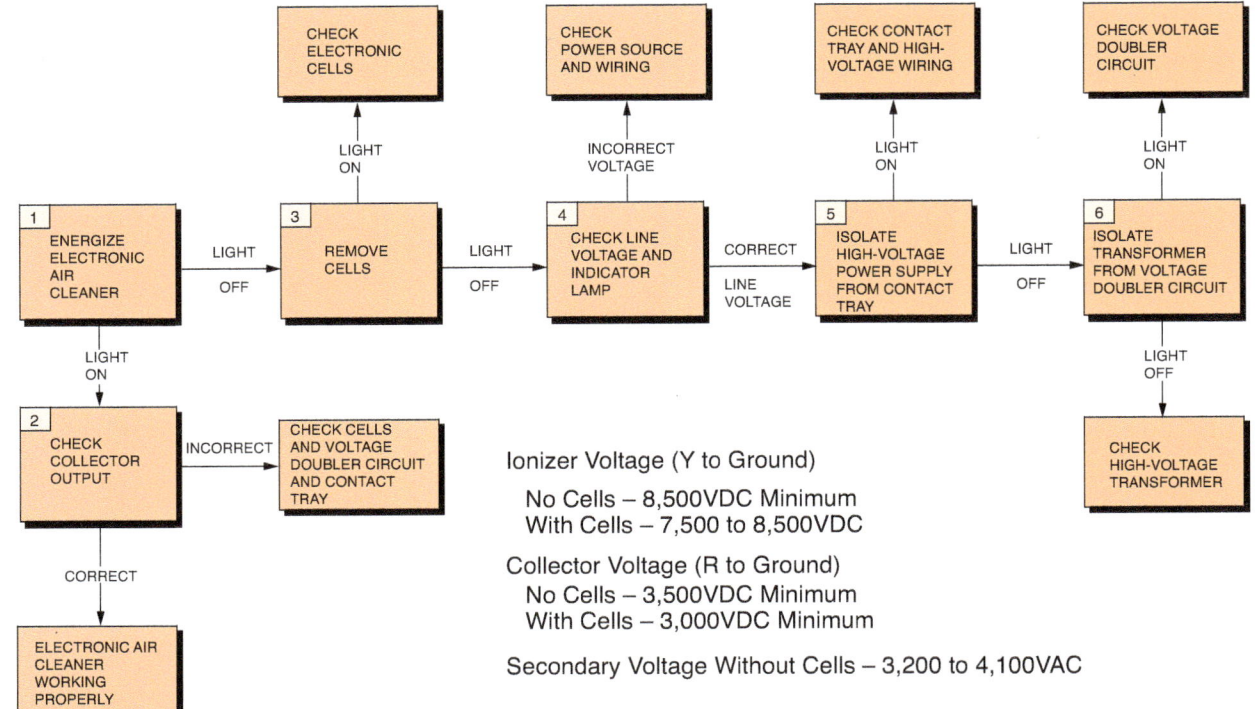

Figure 14 Electronic air cleaner manufacturer's troubleshooting flow chart.

Figure 15 Ultraviolet lamp.

WARNING! The light from ultraviolet lamps can damage eyes. When servicing these lamps, follow all manufacturer safety warnings and make sure the lamp is switched off and disconnected from the power supply.

There is not much to troubleshooting most ultraviolet lamps. If a lamp is not working, check its power supply. Check the lamp itself for physical damage and make sure no debris is in the lamp socket.

Ultraviolet lamps must be replaced after a certain number of hours of operation. Consult the manufacturer's instructions for information about the life span of the lamps and at what intervals they should be changed.

WARNING! Ultraviolet lamps contain a small quantity of mercury, a hazardous substance. If a lamp breaks during handling, use caution when cleaning up the debris and dispose according to local regulations.

Connecting Accessory Devices

Many HVAC equipment manufacturers pre-wire their products with sets of terminals to which accessory devices such as humidifiers and air cleaner may be connected. This eliminates wiring errors and field modification of their product to accommodate the accessory device. The manufacturer of this furnace has supplied line-voltage terminals to which an electronic air cleaner can be connected and 24V terminals for connecting a humidifier.

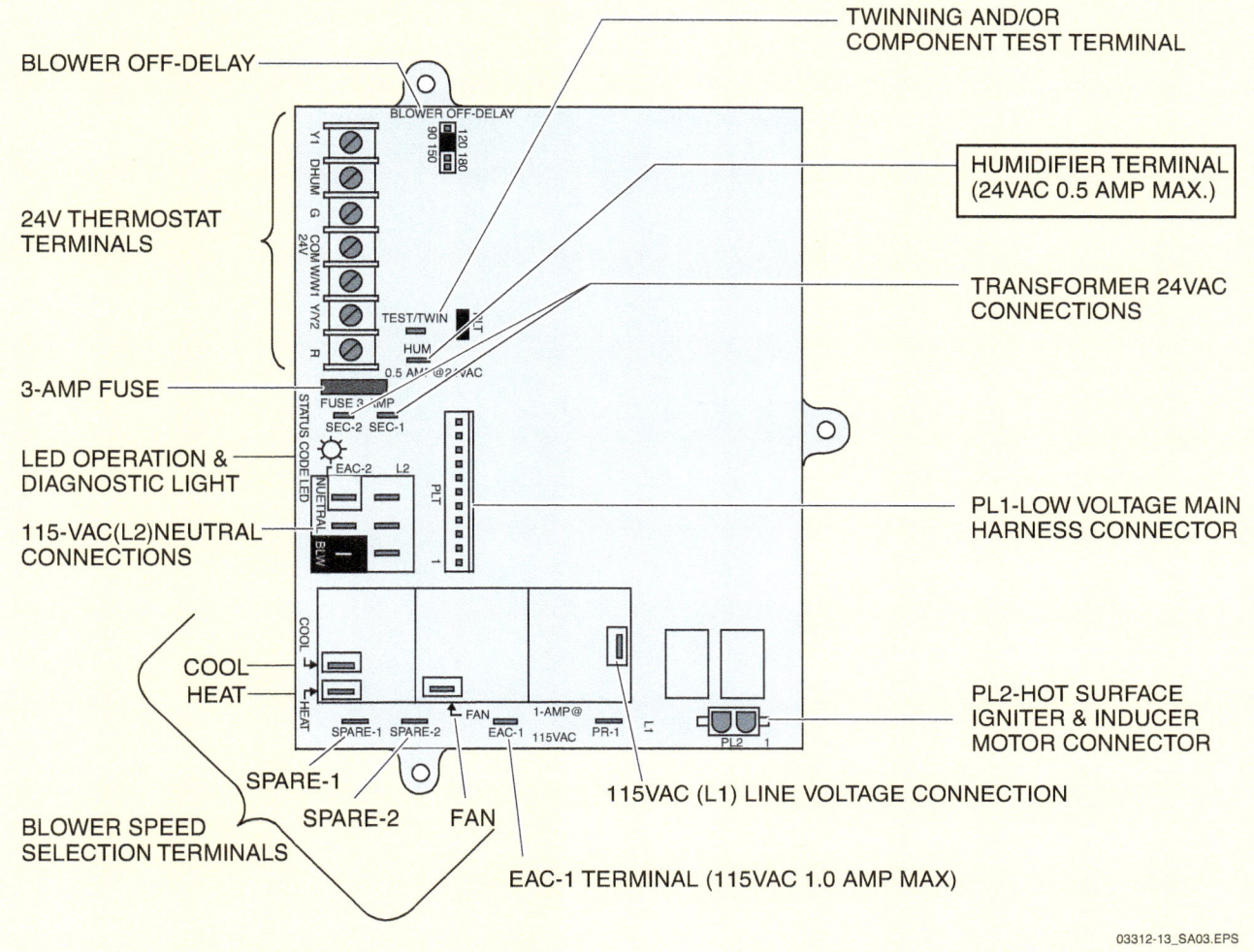

Additional Resources

Refrigeration and Air Conditioning, An Introduction to HVAC/R, Fourth Edition. Larry Jeffus. Air Conditioning and Refrigeration Institute. Prentice Hall.

1.0.0 Section Review

1. A humidifier that works with a forced-air furnace system will operate if the blower is not running.
   a. True
   b. False

2. In the control circuit of an infrared humidifier, devices are typically connected _____.
   a. in parallel with the humidistat
   b. next to the infrared lamps
   c. in series with the humidistat
   d. in parallel with the blower motor

3. What does the presence of an ozone odor during electronic air cleaner operation indicate?
   a. Low airflow
   b. Normal airflow
   c. High airflow
   d. Low voltage

4. Why are microbes in a moving airstream less likely to be killed by UV light?
   a. They are moving too fast.
   b. They are immune to UV light.
   c. The light is not concentrated.
   d. The air acts as an insulator.

SECTION TWO

2.0.0 OUTSIDE AIR ACCESSORY TROUBLESHOOTING

Objective

Describe how to troubleshoot accessories related to the introduction of outside air.
  a. Describe how to troubleshoot economizers.
  b. Describe how to troubleshoot recovery ventilators.

Performance Task

1. Using the correct tools and circuit diagrams, isolate and correct malfunctions in selected accessories.

Trade Terms

Free cooling: A mode of economizer operation. During cooling operation, the system compressor is turned off and the indoor fan is used to bring outside air into the building to provide cooling.

Mechanical cooling: The cooling provided in the conventional manner by the compressor and mechanical refrigeration system.

Before HVAC equipment was readily available, staying comfortable was a much different task. If it became too warm in a building or the air seemed a little stale, the solution was simple: open a window and let in cool, fresh air. The result was not often effective; the biggest drawback was no control over the quality and temperature of the air entering the building. In addition, dust and pollen could easily penetrate many window screens. Modern HVAC equipment also has the ability to bring in outside air, but the difference is that equipment is able to control the quantity, temperature, and quality of the air that enters. Two devices commonly used for this purpose are economizers and recovery ventilators.

2.1.0 Economizer Operation

An economizer (*Figure 16*) is an accessory for commercial units that reduce operating cost during cooling by using outdoor air for cooling (free cooling) and, whenever possible, reducing compressor (mechanical cooling) operation. It does this by controlling the amount of outside air that is brought into a conditioned space. *Figure 17* shows a basic economizer system. It consists of a damper actuator assembly and related economizer control module.

Control signals applied to the economizer control module come from three sources: the thermostat located in the conditioned space, an enthalpy (total heat) sensor located in the outdoor air duct, and the discharge air sensor located on the discharge side of the system evaporator coil.

The setpoint controls the changeover from mechanical to free cooling. The outdoor air enthalpy sensor monitors the air dry-bulb temperature and humidity, allowing the use of outdoor air at higher temperatures for free cooling, as long as the humidity is low. The mixed, or discharge, air sensor monitors the average air temperature on the leaving side of the indoor coil and compares this to the predetermined setpoint. It also monitors the temperature of the return air coming from the conditioned space.

There are many types of economizers. A simplified schematic for an economizer is shown in *Figure 18*. The operation of this unit, as briefly described here, is similar to that of many economizers found in the field.

On a call for cooling by the space thermostat, the system operates as follows:

When the enthalpy of the outdoor air is below the setpoint, the outdoor air damper is driven open and the return air damper is driven closed to maintain a temperature selected by the setpoint of the mixed/discharge air sensor. Typically, this is between 50°F and 56°F.

If the enthalpy of the outdoor air is above the setpoint, the outdoor air damper closes to its preset minimum position. A call for cooling from the

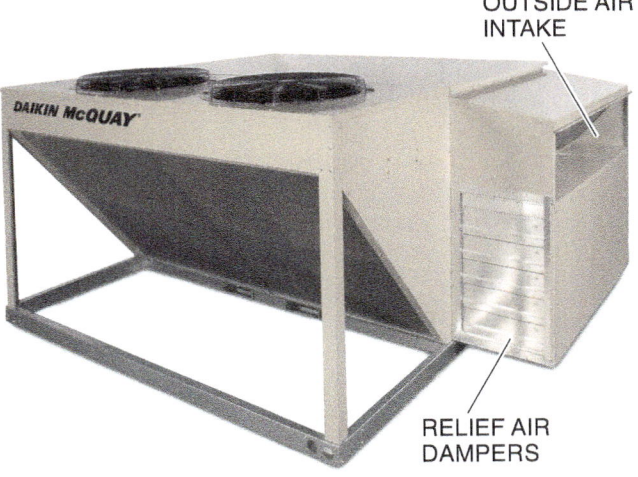

Figure 16 Rooftop unit with an economizer installed.

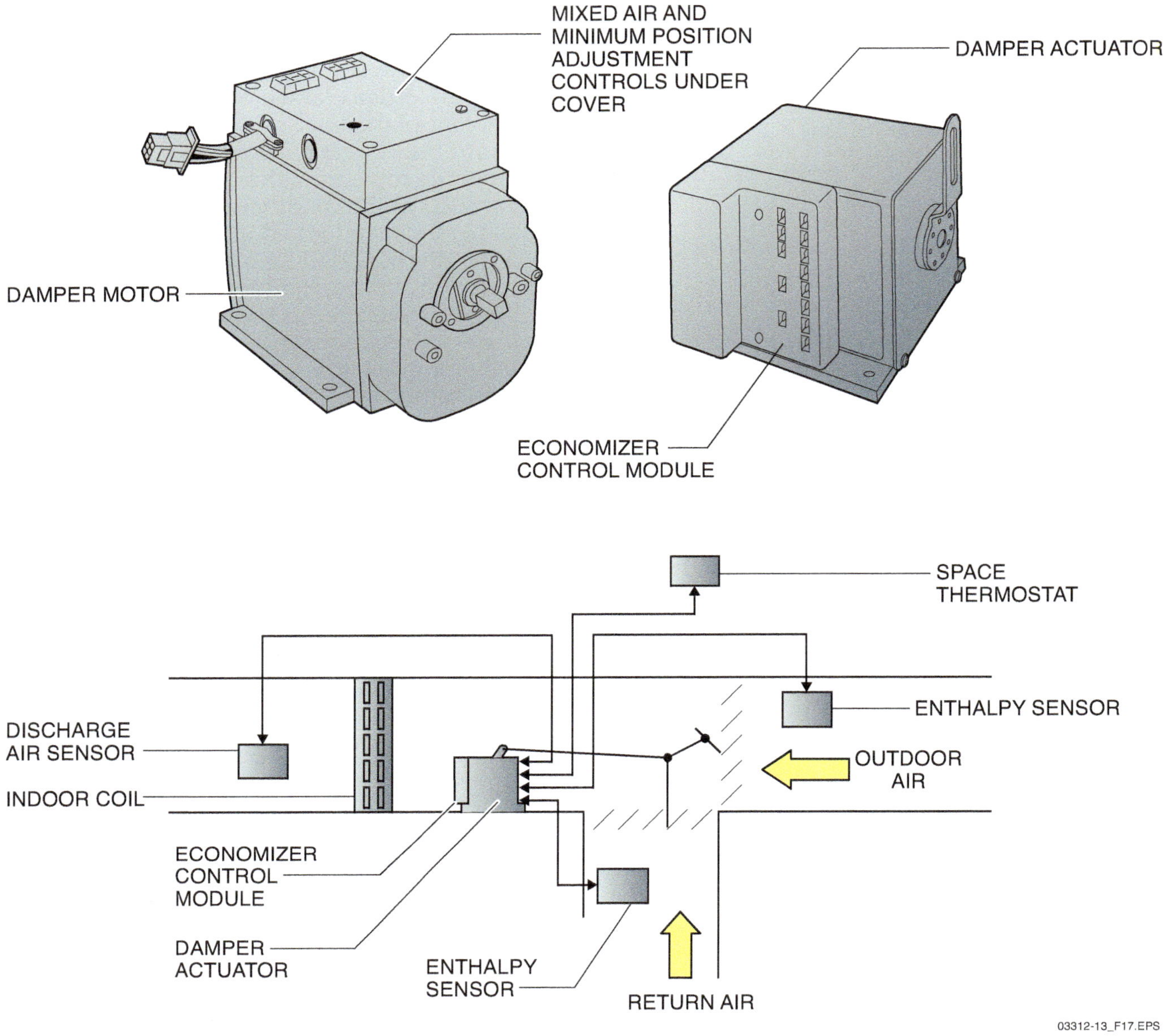

Figure 17 Basic economizer system.

space thermostat turns on mechanical cooling. During the unoccupied period, the damper actuator spring returns the outdoor damper to the fully closed position. A minimum position for outside air is typically not required when the building is unoccupied. Closing the outside air damper during this period saves energy.

With a two-stage cooling thermostat, the first stage will always be satisfied by economizer operation, assuming outdoor temperature conditions are suitable. The mechanical cooling can be operated by the second stage of a cooling thermostat.

Some economizer systems use a second enthalpy sensor located in the return duct. This control method is called differential enthalpy. The economizer control selects the air (outside or return) with the lower enthalpy for cooling. If the outdoor air has lower enthalpy than the return air, then the outdoor air damper will be opened to bring in outdoor air for free cooling.

2.1.1 Economizer Types

When troubleshooting economizers, three basic types will be encountered: dry-bulb economizers, enthalpy economizers, and integrated economizers. The dry-bulb economizer (*Figure 19*) uses an outdoor air temperature sensor (O_{AT}) to activate a control relay whenever the outdoor dry-bulb temperature falls below a user-specified value (typically 55°F). The control relay (CR) locks out the compressor. Space cooling is accomplished by

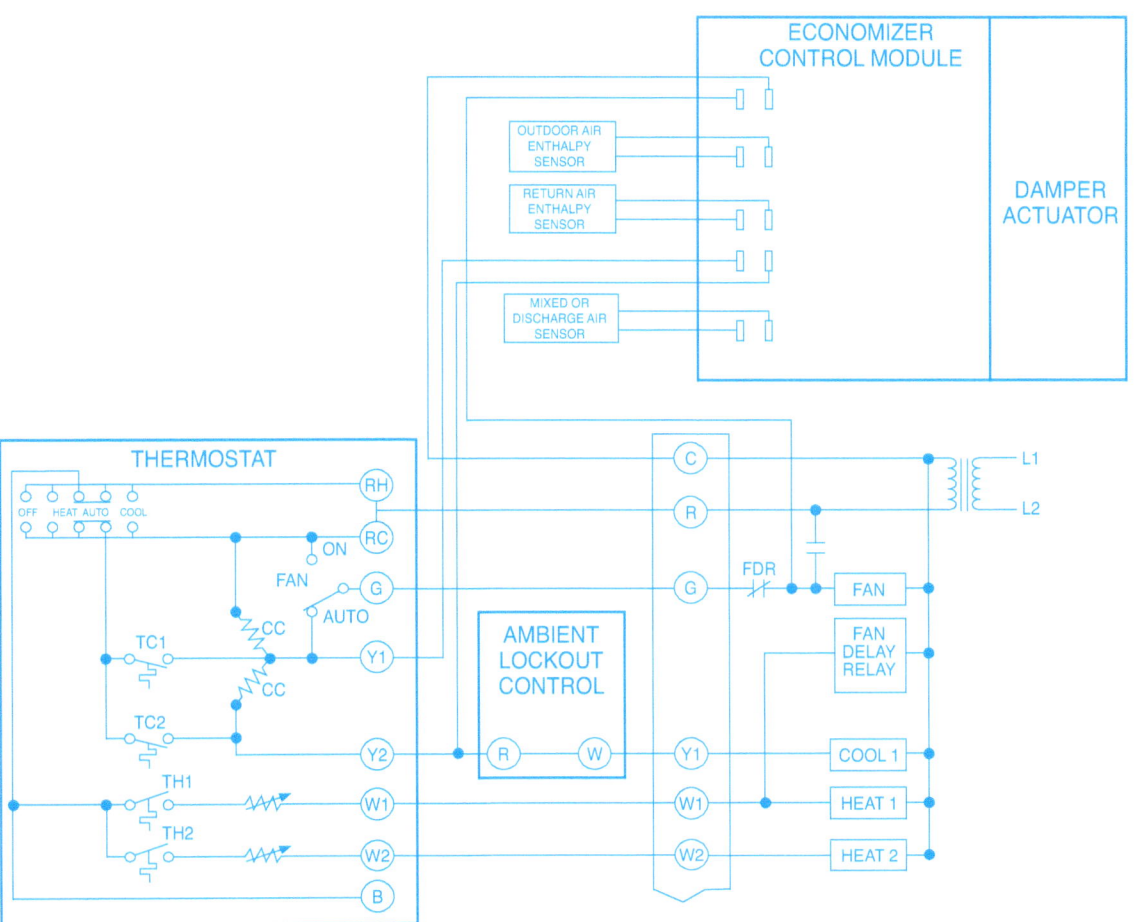

Figure 18 Simplified schematic of an economizer.

modulating the amount of outdoor air entering the return side of the supply air fan. As the outdoor air temperature falls below 55°F, more return air may be mixed with outdoor air to maintain a 55°F supply air temperature. This is the simplest and least costly type of economizer control.

The enthalpy economizer (*Figure 20*) uses an outdoor air temperature sensor (O_{AT}) wired in series with an enthalpy sensor (hOA) and a control relay (CR). When the outdoor air temperature is below 60°F and the outdoor enthalpy is below 24.7 Btu/lb (both common setpoints), the control relay activates to prevent compressor operation.

Economizers

In addition to energy savings and reduced long-term maintenance costs, economizers provide the natural advantage of good building ventilation. While in the free-cooling mode, the system is using anywhere from 50 to 100 percent outside air. This level typically exceeds the code-required ventilation rates for most building applications. Economizers also provide the option of ventilating a building at the needed rate based on occupancy. This is normally achieved by using CO_2 sensors mounted in rooms that may have high occupancy rates; conference rooms, assembly halls, restaurants, training rooms, and meeting centers are examples. These rooms are often used on a sporadic scheduling basis and therefore less than fully occupied much of the time. By interlocking a CO_2 sensor with the economizer, the ventilation rate can be controlled when the system is either in a design cooling or design heating mode, when 100 percent outside air is not desirable. The CO_2 sensor only increases the rate of ventilation to match the human occupancy load, thus maximizing both energy savings and human comfort with superior ventilation. This control sequence is called demand control ventilation (DCV) and is becoming an expected feature in medium-to heavy-commercial applications.

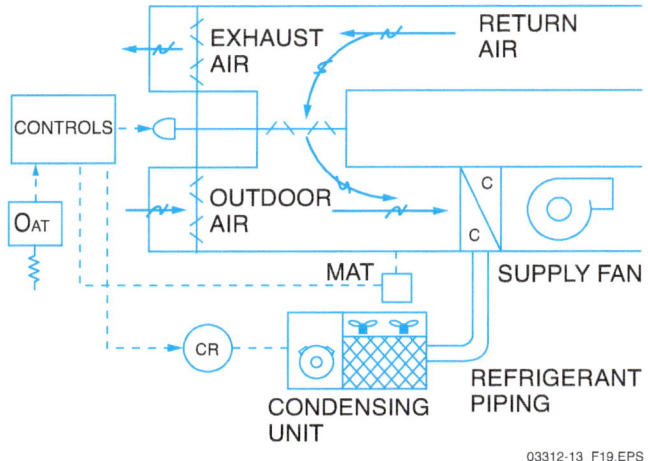

Figure 19 Dry-bulb economizer.

controls of the central unit. The controller calculates outdoor air and return air enthalpy values. Below a pre-determined outdoor air temperature (55°F in this example), the economizer provides cooling with the compressor off, just like a dry-bulb economizer.

For outdoor air temperatures above 55°F, when the outdoor enthalpy is less than the return air enthalpy, 100 percent outdoor air is used to assist compression cooling. When the outdoor air enthalpy is above the return air enthalpy, outdoor air dampers are set to the minimum ventilation position and the compressor provides the needed cooling capacity.

2.1.2 Troubleshooting Economizers

Like many HVAC products, economizers operate basically the same regardless of the manufacturer. For that reason, basic troubleshooting steps are the same. However, before troubleshooting any economizer, always consult the specific manufacturer's service literature before proceeding. Although the functional concepts may remain the same, the controls used can differ significantly.

Economizers depend on dampers to control and mix air. Devices which control the dampers are electronic modules and electronic sensors that provide inputs to the modules. The electronic controls and their sensors tend to be very reliable and are not as prone to problems or failure as the elec-

Zone cooling is accomplished by modulating the amount of outdoor air entering the return side of the supply air fan. An enthalpy sensor is able to sense the total heat in the air (sensible heat and latent heat). The outdoor air temperature sensor used with dry-bulb economizers can only detect sensible heat. An enthalpy economizer will prevent the entry of overly-humid outdoor air into a structure, even if the dry-bulb temperature of the air is fairly low.

The integrated economizer (*Figure 21*) uses two sensors to measure enthalpy: a temperature sensor and humidity sensor. An outdoor air temperature sensor (O_{AT}), outdoor air humidity sensor (hOA), return air temperature sensor (R_{AT}), and return air humidity sensor (hOA) are wired to the

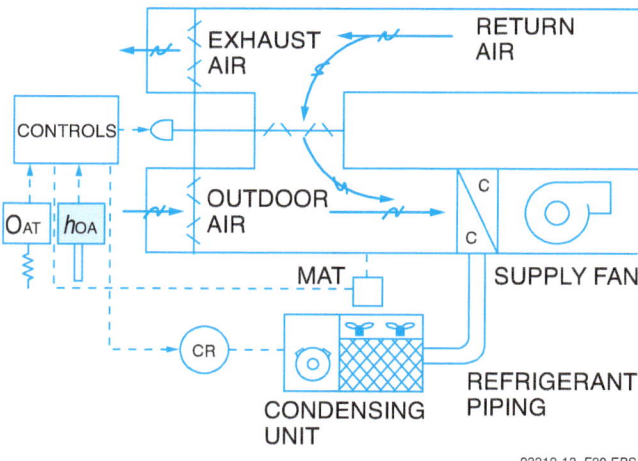

Figure 20 Enthalpy economizer.

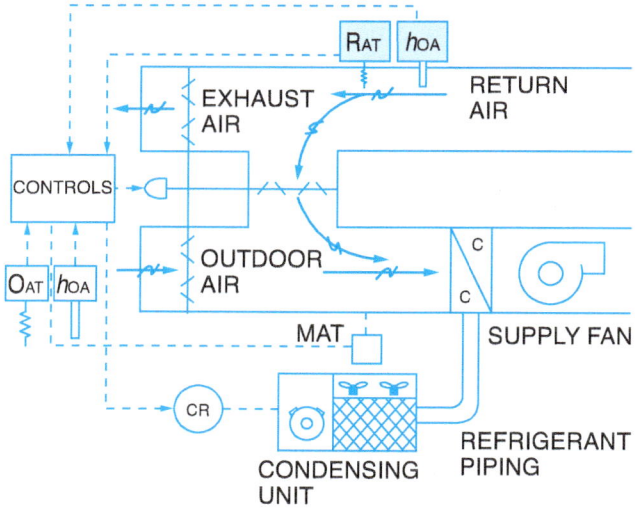

Figure 21 Integrated economizer.

Economizer Control Modules

Older economizer control modules used conventional relay circuitry, while most new ones are solid-state electronic modules.

tromechanical components in the system, such as the dampers and their drive motors (actuators). When first approaching a system equipped with an economizer, it is important to check for normal operation. If any problems are found, proceed to find the root cause.

The following procedure should be used to confirm normal operation:

- *Check sensors* – Temperature sensors used with modern economizers tend to be electronic and very reliable. However, their placement and connections to the control module can affect economizer operation. Check that all sensors are placed in locations according to the manufacturer's instructions. Check the sensors for cut or pinched leads, or loose and/or corroded connections.
- *Check outdoor air damper operation* – The outdoor air damper can be seen under the rain hood of the economizer, or from the return air plenum. Under certain occupancy conditions, the damper may be fully closed. Shut off power to the rooftop unit at the disconnect. With power removed, the outdoor air damper will slowly close. Restoring power should return the damper to its minimum position. With the damper in its minimum position, vary the potentiometer on the control module that controls the minimum position. If determined that varying the potentiometer changes damper position, return the potentiometer to its original position. It is also important to ensure that the mechanical operation of the damper and linkages is smooth and does not bind.
- *Check economizer operation* – Disconnect the room thermostat from the economizer and use a jumper wire to simulate a first-stage call-for-cooling. With a two-stage cooling thermostat, the first stage will always be satisfied by economizer operation, assuming outdoor temperature conditions are suitable. If outdoor temperature is low enough, the outdoor air damper and return air damper will modulate and the blower in the rooftop unit will mix the air supplied to the conditioned space. A low outdoor air temperature can be simulated by cooling the outdoor air temperature sensor with ice in a plastic bag for testing purposes.

The dampers and damper motors must function properly for an economizer to be effective. Leaky or inoperative dampers waste energy. To troubleshoot dampers (*Figure 22*), make the following checks:

- Check that all voltages and electrical signals used to actuate the damper are within specifi-

Figure 22 Motorized rectangular damper.

cations. Ensure that all sensors are positioned correctly for proper sensing.
- Observe damper motors and actuators through an operating cycle to check for defects or binding. Make sure that all mounting bolts are securely fastened.
- Check that the linkage from the actuators is adjusted to ensure that the damper blades fully open or close within the stroke or travel of the actuator arm.
- Check that all blades close tightly in the closed position. If necessary, adjustments should be made to damper linkage(s) to close any partially open blades. Remember that dampers will not always be commanded to close fully, especially in an occupied mode of operation. Do not make the error of adjusting dampers to a fully closed position when the controls are not calling for this setting.
- Make sure that all damaged blades are replaced. Dirt, soot, and lint should be removed, especially around moving parts.
- Check the blade edge and side seals. Replace where necessary.
- Check the pins, straps, and bushings/bearings for wear, rust, or corrosion and replace any as required.
- Inspect the condition of any caulking used to make damper frames tight to the unit.

2.2.0 Recovery Ventilators

Energy recovery ventilators (ERVs) and heat recovery ventilators (HRVs) draw fresh outdoor air into the home. At the same time, they expel stale indoor air to the outside (*Figure 23*). Expelled air carries odors and other unwanted contaminants with it.

While removing stale air from the home, recovery ventilators also recover much of the heat and/or cooling energy from the outgoing indoor air and transfer it to the fresh incoming air.

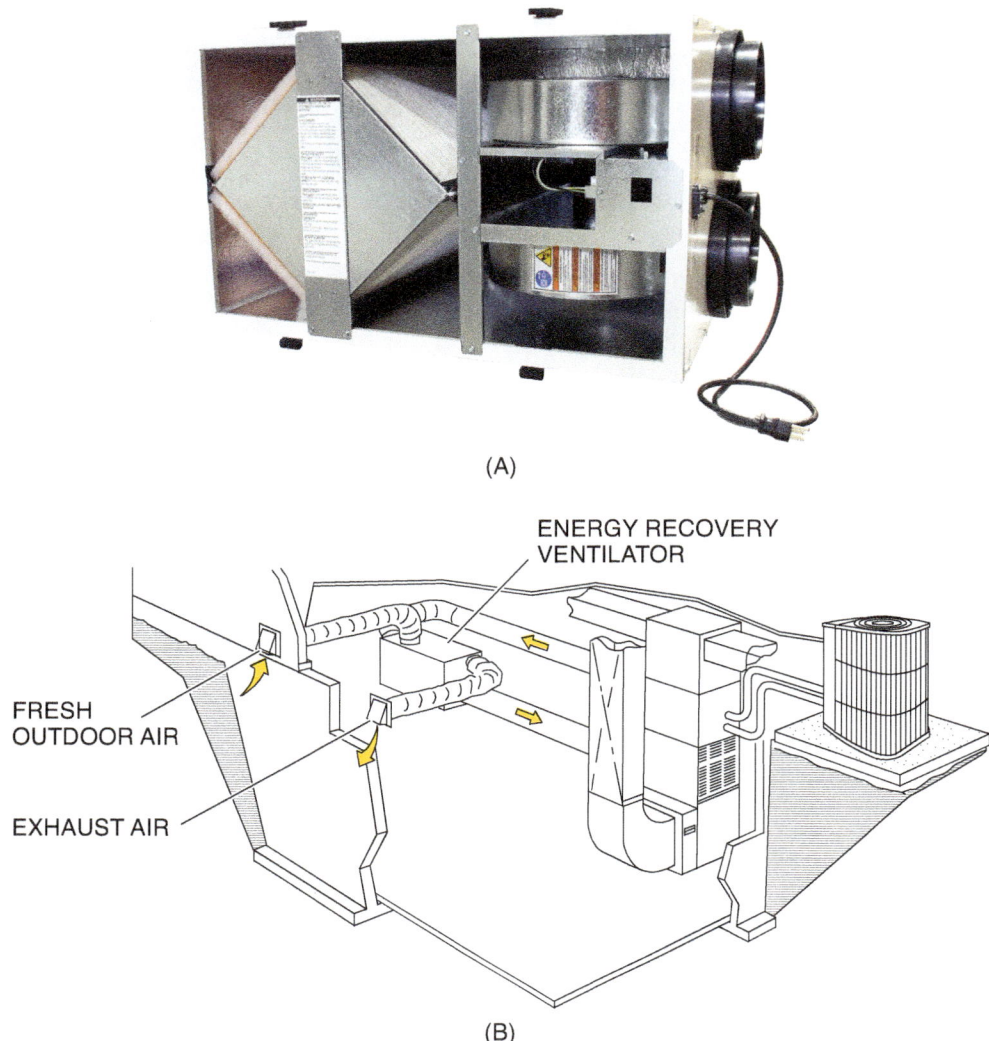

Figure 23 Energy recovery ventilator.

ERV and HRV units are similar in construction and in operation. The heart of both units is a special heat exchanger. In the ERV, this heat exchanger may be referred to as an energy recovery core, since it transfers both sensible and latent heat between the fresh incoming air and the stale air, allowing it to remove moisture from the air during the summer cooling season. In the HRV, the heat exchanger operates in a similar manner, except that it is designed to transfer only sensible heat between the entering and leaving air streams.

ERVs provide more efficient year-round operation than HRVs. For this reason, they are used in most areas of the United States and some parts of Canada. HRVs are used to supply fresh air, quickly remove excessive moisture, and recover heat energy during the heating season. They are more popular in areas with colder climates and extended heating seasons.

An ERV/HRV can be installed as a standalone unit with its own ductwork. However, many are installed as part of a forced-air heating/cooling system. This can influence the load imposed on the HVAC system blower. Also, for the ventilator

Combustion Air

A recovery ventilator does more than just refresh indoor air. In a tightly built structure with a fossil fuel furnace, the system can be designed so that the ventilator brings fresh combustion air into the structure.

to operate properly in these cases, the system capacity must be able to handle the additional heating and/or cooling load from the introduction of outside air. Although the heat exchange process is significant, these units are not capable of total energy transfer. As a result, air that is colder or warmer than the desired temperature indoors is almost always being introduced.

The ERV/HRV can sometimes be controlled independently by a simple wall-mounted control. A blower interlock relay incorporated in the furnace/fan coil control circuit ties the operation of the ERV/HRV unit to the furnace or fan coil blower.

When the ERV/HRV is in operation, the interlock relay is energized, causing the furnace or fan coil blower to operate. To maintain a good transfer of energy and proper airflow, the air system must be properly balanced and the ERV/HRV components clean of dust and dirt. The system must be balanced so intake and exhaust airflows are about equal. Typically, the ventilator is set up to bring in slightly more air than it exhausts. This results in a slightly positive pressure in the structure. If the structure is at a negative pressure, outdoor air is drawn into the space uncontrolled around windows and doors. Balancing is done with a balancing kit at installation and any time an airflow problem is suspected.

The ERV/HRV air filters, screens, and recovery core must be cleaned to remove dust and dirt. These tasks should be performed at intervals recommended by the ventilator manufacturer. Typically, the recovery core of an ERV should be cleaned about every three months. The core of an HRV should be cleaned about once each year at the end of the heating season.

> **NOTE**
> Some cores are made of cardboard and cannot be washed with water. Follow the manufacturer's instructions for the specific model being serviced carefully.

Troubleshooting an inoperative or malfunctioning ERV/HRV involves isolating the problem to one of its three main parts: the wall control, electronic control board, and fan motors. Normally, the wall control is either functional or defective, which can be determined easily enough with a VOM. In many cases, it is simply one or two switches. To troubleshoot the electronic control board, the wall control must be electrically connected before the unit will operate properly. Any configuration jumpers located on the board must match the configuration specified in the service literature. The easiest way to check their operation is to use the wall control to initiate low-speed and high-speed operation.

Troubleshooting electrical problems in a ventilator must always be done in accordance with the detailed troubleshooting procedures and diagrams provided in the service literature supplied with the unit.

Typical recovery ventilator problems and possible causes are as follows:

- *Unit not operating* – Check power supply. Check that operating controls are properly set.
- *Inadequate airflow* – Check outside air intake and connecting duct for blockage. Ensure that any dampers are open and no duct air leaks or disconnected ducts are present. Check that the heat exchanger is not dirty or frozen.
- *Frozen coil* – Some frost on the heat exchanger is normal. Check the defrost circuits. Check system airflow balance.
- *Excess humidity* – Adjust dehumidifier setpoint. Remove sources of excess humidity. Check system airflow balance.
- *Inadequate humidity* – Adjust dehumidifier setpoint. Check/adjust recovery ventilator blower speed.
- *Water in unit* – Check for plugged drain pan/drain lines. Check if unit is level.
- *Noisy operation* – Check fans and fan motors. Check recovery core for cleanliness.

Figure 24 is an example of troubleshooting information provided by one HRV manufacturer.

Some form of air filtration is usually a part of an ERV/HRV unit as well. The filter must be serviced periodically. If the unit is operated without a filter, costly and possibly permanent damage can occur to the heat exchanger of the unit.

TROUBLESHOOTING YOUR HRV SYSTEM

SYMPTOM	CAUSE	SOLUTION
Poor air flows	• ¼" (6 mm) mesh on the outside hoods is plugged • filters plugged • core obstructed • house grilles closed or blocked • dampers are closed if installed • poor power supply at site • ductwork is restricting HRV • improper speed control setting • HRV airflow improperly balanced	• clean exterior hoods or vents • remove and clean filter • remove and clean core • check and open grilles • open and adjust dampers • have electrician check supply voltage at house • check duct installation • increase the speed of the HRV • have contractor balance HRV
Supply air feels cold	• poor location of supply grilles, the airflow may irritate the occupant • outdoor temperature extremely cold	• locate the grilles high on the walls or under the baseboards, install ceiling mounted diffuser or grilles so as not to directly spill the supply air on the occupant (eg, over a sofa) • turn down the HRV supply speed. A small duct heater (1 kw) could be used to temper the supply air • placement of furniture or closed doors is restricting the movement of air in the home • if supply air is ducted into furnace return, the furnace fan may need to run continuously to distribute ventilation air comfortably
Dehumidistat is not operating	• outdoor temperature is above 15°C (59°F) • improper low voltage connection • external low voltage is shorted out by a staple or nail • check dehumidistat setting; it may be on OFF	• dehumidistat is functioning normally • check that the correct terminals have been used • check external wiring for a short • set the dehumidistat at the desired setting
Humidity levels are too high. Condensation is appearing on the windows	• dehumidistat is set too high • HRV is undersized to handle a hot tub, indoor pool, etc. • lifestyle of the occupants • moisture coming into the home from an unvented or unheated crawl space • moisture is remaining in the washroom and kitchen areas • condensation seems to form in the spring and fall • HRV is set at too low a speed	• set dehumidistat lower • cover pools, hot tubs when they are not in use • avoid hanging clothes to dry, storing wood and venting clothes dryer inside. Heating wood may have to be moved outside • vent crawl space and place a vapor barrier on the floor of the crawl space • ducts from the washroom should be sized to remove moist air as effectively as possible, use of a bathroom fan for short periods will remove additional moisture • on humid days, as the seasons change, some condensation may appear, but the home's air quality will remain high with some HRV use • increase speed of the HRV
Humidity levels are too low	• dehumidistat control is set too low • blower speed of HRV is too high • lifestyle of occupants • HRV air flows may be improperly balanced	• set dehumidistat higher • decrease HRV blower speed • humidity may have to be added through the use of humidifiers • have a contractor balance HRV airflows
HRV and / or ducts frosting up	• HRV air flows are improperly balanced • malfunction of the HRV defrost system	• Note: minimal frost build-up is expected on cores before unit initiates defrost cycle functions • have HVAC contractor balance the HRV • ensure damper defrost is operating during self-test
Condensation or ice build-up in insulated duct to the outside	• incomplete vapor barrier around insulated duct • a hole or tear in outer duct covering	• tape and seal all joints • tape any holes or tears made in the outer duct covering • ensure that the vapor barrier is completely sealed
Water in the bottom of the HRV	• drain pans plugged • improper connection of HRV's drain lines • HRV is not level • drain lines are obstructed • HRV heat exchange core is not properly installed	• ensure O-ring on drain nozzle sits properly • look for kinks in line • check water drain connections • make sure water drains properly from pan

Figure 24 Recovery ventilator troubleshooting chart.

Make-Up Air Devices

Recovery ventilators refresh the indoor air and can also replace and supply air used for combustion in tightly-built structures. Combustion air can be supplied to a fuel-burning appliance in other ways without the expense of installing a recovery ventilator. Make-up air devices simply supply combustion air without the complexity found in an energy-recovery device. The simplest make-up air device is an air duct that brings in outside air to the vicinity of the furnace or boiler. A screen in the intake stops debris from entering the duct. One manufacturer offers a device that installs on the return duct of a forced air furnace. When the blower runs, the negative pressure created in the return duct draws in outside air through a duct connected to the device. A barometric damper in the device opens under negative pressure and helps keep air from entering the return duct when the blower is not running. Several other manufacturers offer motorized make-up air controls that provide air to the vicinity of a furnace or boiler. Combustion air is only supplied when there is a call-for-heat signal from the room thermostat. On a call-for-heat, the blower in the make-up air control starts, drawing in outdoor air through a duct. When combustion airflow is proven by a flow-sensing switch, the make-up air control allows the normal burner ignition sequence to occur. Some motorized make-up air devices temper the cold incoming air by mixing it with indoor air. In cold climates, the duct that brings in outside air must be insulated to prevent condensation and/or frost from forming on the duct.

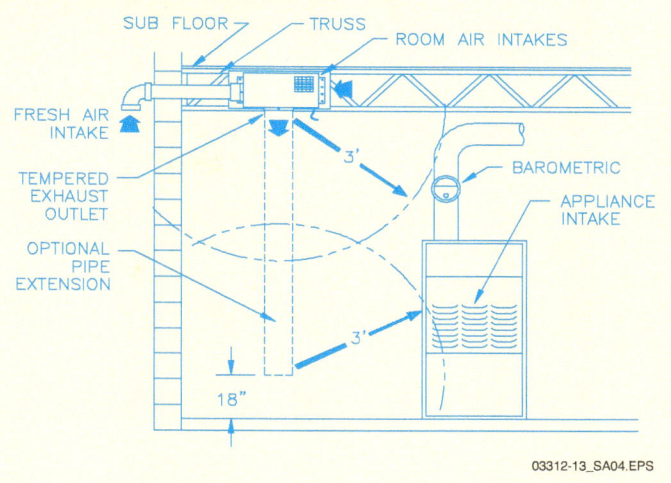

Additional Resources

Refrigeration and Air Conditioning, An Introduction to HVAC/R, Fourth Edition. Larry Jeffus. Air Conditioning and Refrigeration Institute. Prentice Hall.

2.0.0 Section Review

1. What about an enthalpy economizer prevents overly humid air from entering a structure?
    a. It contains a de-humidifier.
    b. It can sense latent heat in the air.
    c. It contains tightly fitted dampers.
    d. It contains multiple enthalpy sensors.

2. The recovery core of a heat recovery ventilator (HRV) recovers _____.
    a. all indoor heat
    b. latent heat only
    c. sensible and latent heat
    d. sensible heat only

Summary

To effectively troubleshoot HVAC accessories, the HVAC technician listens to a customer's complaint, performs an independent analysis of a problem, then initiates and performs a systematic, step-by-step troubleshooting method that results in the correction of the problem. The technician must understand the purpose and principles of operation of each component in the accessory being serviced. Also, the relationship between the operation of the accessory and the main HVAC equipment with which it is used must also be understood. The technician must be able to tell whether a given device is functioning properly and be able to recognize the symptoms arising from the improper operation of any part of the equipment. On the basis of symptoms revealed through analysis of a system's operation, the technician should use the troubleshooting aids provided by the system manufacturer to identify and repair the cause of the problem.

Review Questions

1. Some humidifiers are controlled by a(n) _____.
   a. check valve
   b. economizer
   c. ventilator
   d. humidistat

2. If a humidifier is not running, the problem is usually caused by a(n) _____.
   a. faulty control circuit
   b. poor water quality
   c. high RH condition
   d. oversized furnace

3. When a humidifier experiences water overflow, the problem is often caused by a(n) _____.
   a. faulty control circuit
   b. dirty float valve
   c. high RH condition
   d. open solenoid coil

4. Signs of high relative humidity include _____.
   a. dry, itchy skin
   b. loose furniture joints
   c. static electricity shocks
   d. condensation on windows

5. One possible cause of a low humidity complaint could be the _____.
   a. float valve is set too high
   b. water supply feed valve closed
   c. furnace blower running in high speed
   d. condensing unit is unable to start

6. A tight building is characterized by the _____.
   a. lack of a vapor barrier
   b. presence of a fireplace
   c. lack of insulation
   d. presence of tight windows

7. In an electronic air cleaner, the output of the high-voltage power supply is applied to _____.
   a. the ionizer section
   b. the collector section
   c. both the collector and ionizer
   d. the prefilter

8. When an electronic air cleaner is operated at an airflow above that recommended by the manufacturer, the filter's cleaning ability is _____.
   a. slightly better
   b. worse
   c. the same
   d. doubled

9. The enthalpy sensor(s) used with an economizer control _____.
   a. room temperature enthalpy only
   b. outdoor air humidity enthalpy only
   c. changeover from free cooling to mechanical cooling
   d. both room temperature and humidity enthalpy

10. An economizer uses a mixed/discharge air sensor to control the open or closed position of _____.
    a. both the outdoor and return dampers during free cooling
    b. the outdoor damper during free cooling
    c. the outdoor damper during mechanical cooling
    d. both the outdoor and return dampers during both free and mechanical cooling

11. In an economizer that uses differential enthalpy sensing, what type of air does the economizer control module select for cooling?
    a. only outside
    b. only return
    c. either outside or return
    d. mixed/discharge

12. How many air temperature sensors does a dry-bulb economizer use?
    a. Multiple
    b. One
    c. Two
    d. Three

03312 Troubleshooting Accessories

13. When troubleshooting a problem suspected to be caused by a malfunctioning economizer damper, one of the first things to do is _____.
    a. check the control voltage applied to the compressor
    b. check an operating damper for defects and binding
    c. jumper the Y-C contacts on the room thermostat
    d. replace the damper with a known good one

14. A device that uses a heat exchanger to transfer heat from warmed indoor air to ventilation air entering a building is a(n) _____.
    a. economizer
    b. enthalpy control
    c. recovery ventilator
    d. mixed air thermostat

15. Energy recovery ventilators are more likely to be found _____.
    a. only in cold climates
    b. throughout the United States
    c. installed with heat pumps
    d. installed with hydronic systems

Trade Terms Introduced in This Module

Free cooling: A mode of economizer operation. During cooling operation, the system compressor is turned off and the indoor fan is used to bring outside air into the building to provide cooling.

Mechanical cooling: The cooling provided in the conventional manner by the compressor and mechanical refrigeration system.

Water hammer: A banging noise in water pipes that occurs when the water supply solenoid valve on a humidifier abruptly shuts off water to the humidifier.

Additional Resources

This module presents thorough resources for task training. The following resource material is suggested for further study.

Refrigeration and Air Conditioning, An Introduction to HVAC/R, Fourth Edition. Larry Jeffus. Air Conditioning and Refrigeration Institute. Prentice Hall.

Figure Credits

Courtesy of Arnold Service Co., Inc., Module opener
Tim Dean, Figure 1
Courtesy of Thermolec Ltd., Figure 3
Courtesy of General Filters, Inc., Figure 9
Courtesy of PlumbersStock.com, SA02
Courtesy of Honeywell International, Figures 11, 13
Carrier Corporation, SA03

American Air & Water, Inc., Figure 15
Courtesy of Daikin McQuay, Figure 16
Jackson Systems LLC-- Controls Done Right™, Figure 22
Courtesy of RenewAire LLC, Figure 23 (photo)
Courtesy of Airia Brands, Inc., Figure 24
Courtesy of Tjernlund Products, Inc., SA04

Section Review Answer Key

Answer	Section Reference	Objective
Section One		
1. b	1.1.3	1a
2. c	1.2.0	1b
3. a	1.3.0	1c
4. a	1.4.0	1d
Section Two		
1. b	2.1.1	2a
2. d	2.2.0	2b

NCCER CURRICULA — USER UPDATE

NCCER makes every effort to keep its textbooks up-to-date and free of technical errors. We appreciate your help in this process. If you find an error, a typographical mistake, or an inaccuracy in NCCER's curricula, please fill out this form (or a photocopy), or complete the online form at **www.nccer.org/olf**. Be sure to include the exact module ID number, page number, a detailed description, and your recommended correction. Your input will be brought to the attention of the Authoring Team. Thank you for your assistance.

Instructors – If you have an idea for improving this textbook, or have found that additional materials were necessary to teach this module effectively, please let us know so that we may present your suggestions to the Authoring Team.

NCCER Product Development and Revision
13614 Progress Blvd., Alachua, FL 32615

Email: curriculum@nccer.org
Online: www.nccer.org/olf

❏ Trainee Guide ❏ Lesson Plans ❏ Exam ❏ PowerPoints Other _____

Craft / Level: _____ Copyright Date: _____

Module ID Number / Title: _____

Section Number(s): _____

Description: _____

Recommended Correction: _____

Your Name: _____

Address: _____

Email: _____ Phone: _____

Zoning, Ductless, and Variable Refrigerant Flow Systems

OVERVIEW

The implementation of a zoning system that uses a single air source but distributes the air in a highly controlled manner makes a standard constant-volume system more effective. Ductless systems may be single zone or multi-zone, offering small-capacity heating and cooling to one or more areas. Advanced variable refrigerant flow systems take this approach a step further with increased capacity. This module describes the operation and application of these innovative approaches to comfort cooling and heating.

Module 03315

Trainees with successful module completions may be eligible for credentialing through the NCCER Registry. To learn more, go to **www.nccer.org** or contact us at 1.888.622.3720. Our website has information on the latest product releases and training, as well as online versions of our *Cornerstone* magazine and Pearson's product catalog.

Your feedback is welcome. You may email your comments to **curriculum@nccer.org**, send general comments and inquiries to **info@nccer.org**, or fill in the User Update form at the back of this module.

This information is general in nature and intended for training purposes only. Actual performance of activities described in this manual requires compliance with all applicable operating, service, maintenance, and safety procedures under the direction of qualified personnel. References in this manual to patented or proprietary devices do not constitute a recommendation of their use.

Copyright © 2018 by NCCER, Alachua, FL 32615, and published by Pearson, New York, NY 10013. All rights reserved. Printed in the United States of America. This publication is protected by copyright, and permission should be obtained from NCCER prior to any prohibited reproduction, storage in a retrieval system, or transmission in any form or by any means, electronic, mechanical, photocopying, recording, or likewise. To obtain permission(s) to use material from this work, please submit a written request to NCCER Product Development, 13614 Progress Blvd., Alachua, FL 32615.

03315 V5

From *HVAC Level Three, Trainee Guide*. NCCER.
Copyright © 2018 by NCCER. Published by Pearson. All rights reserved.

03315
Zoning, Ductless, and Variable Refrigerant Flow Systems

Objectives

When you have completed this module, you will be able to do the following:

1. Identify common zoning systems and describe the basic approach to troubleshooting.
   a. Describe common zoning system components.
   b. Describe the sequence of operation for common zoning systems.
   c. Describe the basic approach to troubleshooting zoning systems.
   d. Describe common VVT-system zoning components.
   e. Describe the operating characteristics of VVT control systems.
2. Identify ductless and variable refrigerant flow systems and describe the basic approach to troubleshooting.
   a. Identify and describe the operation of ductless split-system equipment.
   b. Describe the installation and troubleshooting of typical ductless systems.
   c. Identify and describe the operation of variable refrigerant flow systems.
   d. Describe the installation and troubleshooting of variable refrigerant flow systems.

Performance Tasks

Under the supervision of your instructor, you should be able to do the following:

1. Troubleshoot and repair a typical ductless or variable refrigerant flow system.
2. Program the controller for a ductless or variable refrigerant flow system.

Trade Terms

Dump zone
Polling
Pressure-dependent operation
Pressure-independent operation
Refnet joints

Industry Recognized Credentials

If you are training through an NCCER-accredited sponsor, you may be eligible for credentials from NCCER's Registry. The ID number for this module is 03315. Note that this module may have been used in other NCCER curricula and may apply to other level completions. Contact NCCER's Registry at 888.622.3720 or go to **www.nccer.org** for more information.

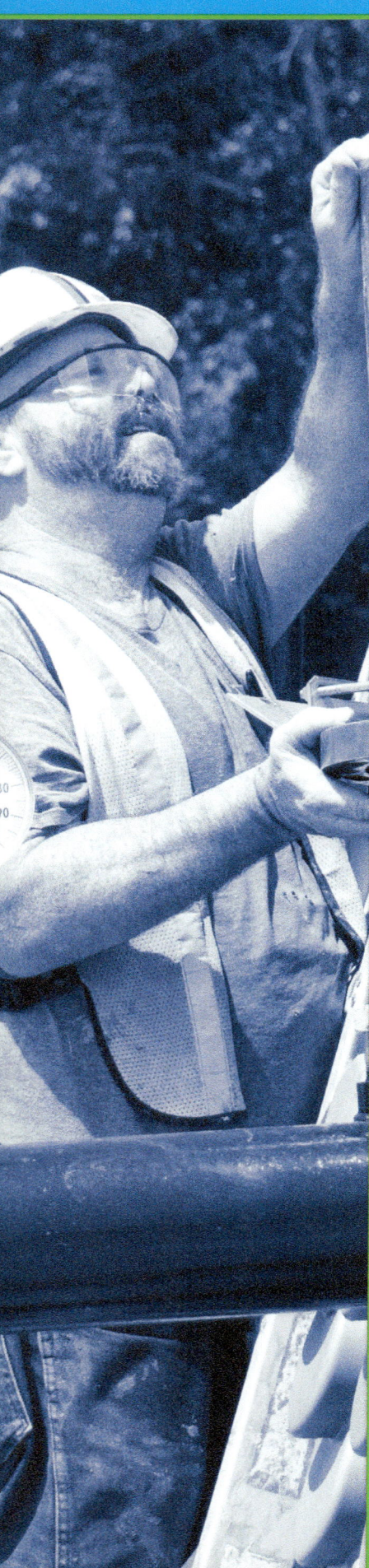

Contents

1.0.0 Common Zoning Systems 1
  1.1.0 Zoning System Components 1
    1.1.1 Zone Thermostats 2
    1.1.2 Zone Dampers 2
    1.1.3 Bypass Dampers 4
    1.1.4 Zoning Control Panels 4
    1.1.5 Additional Devices 9
  1.2.0 Zoning System Operation 10
    1.2.1 Ventilation Modes 11
    1.2.2 Heating and Cooling 11
    1.2.3 Multistage Systems 12
    1.2.4 High- and Low-Limit Operation 12
  1.3.0 Zoning System Troubleshooting 13
    1.3.1 Zone Panels 13
    1.3.2 Zone Thermostats and Dampers 15
  1.4.0 VVT System Components 16
    1.4.1 Bypass Damper 17
    1.4.2 VVT Zone Dampers 18
    1.4.3 Zone Controllers 18
    1.4.4 Relay Packs and Unit Controllers 19
  1.5.0 VVT System Operation 19
    1.5.1 Communication 19
    1.5.2 Pressure-Dependent vs. Pressure-Independent Operation 20
    1.5.3 Occupied Cooling 22
    1.5.4 Occupied Heating 23
    1.5.5 Unoccupied Period 23
    1.5.6 Simultaneous Cooling and Heating 23
2.0.0 Ductless and Variable Refrigerant Flow (VRF) Systems 25
  2.1.0 Ductless Split Systems 25
    2.1.1 Ductless Split-System Condensing Units 27
    2.1.2 Ductless Split-System Indoor Units 27
    2.1.3 Multiple Indoor Units 29
  2.2.0 Ductless Split-System Installation and Troubleshooting 30
    2.2.1 Installing Ductless Systems 30
    2.2.2 Troubleshooting Ductless Systems 34
  2.3.0 Variable Refrigerant Flow Systems 34
    2.3.1 VRF Equipment 38
  2.4.0 VRF System Installation and Troubleshooting 41
    2.4.1 VRF System Installation 41
    2.4.2 VRF System Troubleshooting 46

Figures

Figure 1   Zoned system schematic 2
Figure 2   Programmable zone thermostat 2
Figure 3   Master thermostat and matching slave thermostat 3
Figure 4   Zone dampers 3
Figure 5   Diffuser/zone damper assembly 4
Figure 6   Bypass dampers 4
Figure 7   Dump zone and direct return bypass zones 5
Figure 8   BZD Barometric Zone Damper™ 6
Figure 9A Zoning system control panel 7
Figure 9B Zoning system control panel 8
Figure 10 Wireless zone damper and thermostat 9
Figure 11 Wireless zone control panel 9
Figure 12 Duct static-pressure sensor 10
Figure 13 Zone damper in the open position 11
Figure 14 High- and low-limit setpoint adjustment 13
Figure 15 Panel power indicator light 14
Figure 16 Panel configuration switches 14
Figure 17 Zoning system status lights 15
Figure 18 Zone thermostat terminal blocks 15
Figure 19 Three-wire zone damper terminal block 16
Figure 20 Variable volume/temperature (VVT) system 17
Figure 21 A VVT terminal awaiting installation 18
Figure 22 Zone controller wiring 20
Figure 23 VVT control communication schematic 21
Figure 24 Polling network 21
Figure 25 Pressure-dependent zone damper modulation 22
Figure 26 Common ductless split-system installation 26
Figure 27 Handheld remote control for a ductless system 27
Figure 28 Ductless split-system condensing unit 27
Figure 29 High sidewall air handler 28
Figure 30 Ductless system condensate pumps 28
Figure 31 In-ceiling cassette air handler 29
Figure 32 Cassette piping and wiring 29
Figure 33 Floor-mounted indoor unit 29
Figure 34 Direct-expansion indoor unit hookup 30
Figure 35 Ductless condensing units serving a residential building
           in Shanghai, China 30
Figure 36 Wall-mounted condensing unit 31
Figure 37 Ductless indoor unit mounting plate 32
Figure 38 Wall sleeve and opening detail 32
Figure 39 Line set cover 32
Figure 40 Remote control button layout 33
Figure 41 Ductless system fault codes 35
Figure 42 Ductless system troubleshooting chart 36
Figure 43 Fault code guide 37

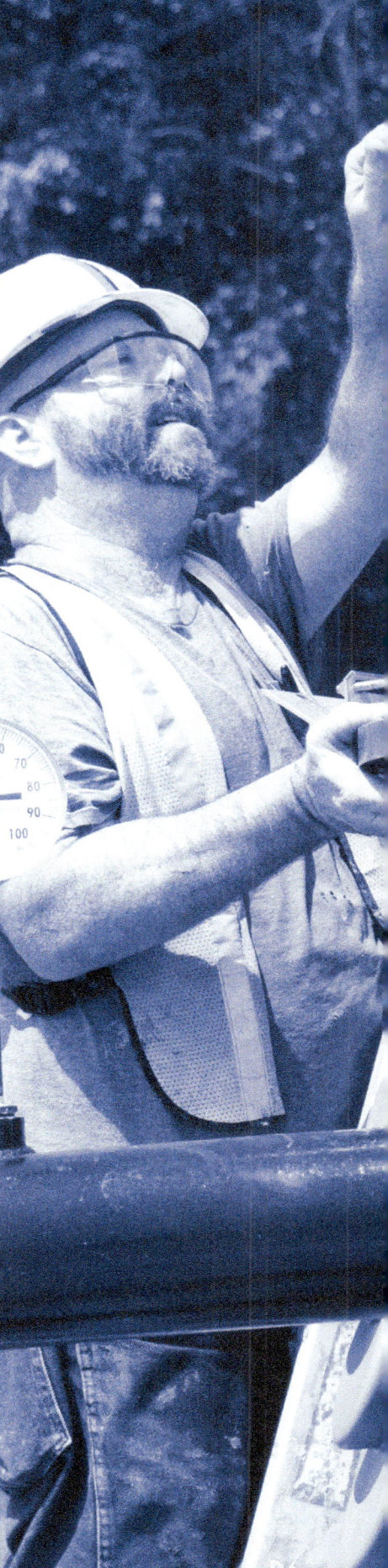

Figure 44	Problem description	38
Figure 45	Troubleshooting flow chart	39
Figure 46	VRF-system refrigerant flow schematic	40
Figure 47	VRF condensing unit	40
Figure 48	VRF fan-coil unit designed for concealment	40
Figure 49	Navigation Remote Controller™	41
Figure 50	VRF-system piping and wiring overview	42
Figure 51	Refnet joints and headers	43
Figure 52	Refnet joint positioning	44
Figure 53	Branch selector boxes	44
Figure 54	VRF-system communication wiring terminals	45
Figure 55	Daisy-chain wiring	46
Figure 56	Partial symptom-based troubleshooting guide	47
Figure 57	Partial remote control maintenance menu	49
Figure 58	Example error code chart	50
Figure 59	Error code E7 explained	51
Figure 60A	Flow chart to troubleshoot the condenser fan motor	52
Figure 60B	Flow chart to troubleshoot the condenser fan motor	53
Figure 61	The Check 6 troubleshooting procedure	54
Figure 62	Pressure-sensor test chart	55
Figure 63	Daikin Service Checker	56

Section One

1.0.0 Common Zoning Systems

Objective

Identify common zoning systems and describe the basic approach to troubleshooting.
a. Describe common zoning system components.
b. Describe the sequence of operation for common zoning systems.
c. Describe the basic approach to troubleshooting zoning systems.
d. Describe common VVT-system zoning components.
e. Describe the operating characteristics of VVT control systems.

Trade Terms

Dump zone: An area used as a point to discharge unneeded supply air from a zoning system. The dump zone is typically in the conditioned space, but not an area where maintaining a specific space temperature is essential. A dump zone generally has no supply air grilles, but a return grille is often installed to relieve room pressure and return mixed air back to the system.

Polling: A network-communications arrangement where a VVT system seeks input from individual zones, determines what demands have priority, and then responds accordingly. Electronic participants in a polling system are referred to as voters.

Pressure-dependent operation: A VVT-system approach where dampers are positioned based on demand only, or operate in two positions. The result is that the design volume of air may or may not be flowing through the damper at any time. The volume of air flow through each damper is dependent on the duct pressure; it fluctuates as the pressure fluctuates. A minimum damper position is set, but air volume is not a controlling factor, and so the minimum damper position is a fixed point.

Pressure-independent operation: A VVT-system approach where dampers are positioned by demand and by the actual volume of air flowing through the damper. To achieve a specific amount of airflow, the damper positions itself as necessary. Maximum and minimum airflow setpoints are in effect, and the damper will modulate as necessary to remain within these boundaries.

Traditional heating and cooling systems maintain the desired conditions throughout a home or small business by answering to a single thermostat. The system has no idea of the conditions in every room or area; it responds only to the conditions at the thermostat location. This often results in areas that are too hot or too cold, even with a properly balanced system. Throughout a typical day and night, the load in an area can change dramatically as the sun moves across the sky, or as internal loads cycle on and off.

A zoning system allows a single heating and cooling system to respond to two or more thermostats, providing conditioned air only to those areas (zones) that presently need heating or cooling. This enables a single system to accommodate the needs of the structure in a more precise way, and it also saves energy. For example, the bedrooms of a home may be considered one zone, with the living area another zone. If the bedrooms are unoccupied all day, the temperature can be set up or set back. The system can then focus on providing conditioned air only to the living area. At night, the zone of the living area can be set up or set back while the system provides comfort to the bedroom areas. The system is also able to heat or cool both zones at the same time.

This section examines basic zoning systems and explains how to test and troubleshoot them when problems arise.

1.1.0 Zoning System Components

Zone control establishes separately controlled spaces, or zones, where different temperatures can be maintained within a building. Each of these areas may be referred to as a control zone. Zoning is used to overcome variations in cooling or heating loads that occur in specific areas of the building at different times of the day. It provides occupants with a means to maintain different temperatures in each of these zones.

A typical zoned system (*Figure 1*) consists of an electronic control panel that is connected to several zone thermostats and motorized dampers. Each thermostat reports its needs, if any, to the control panel in much the same way that all thermostats operate. They are normally powered by a 24VAC control circuit, and contacts open or close in the thermostats when heating or cooling is needed. The zone damper, which is installed in the supply duct to each zone, controls the flow of conditioned air to the zone. The control panel is also connected to the control circuits of the building's heating/cooling equipment.

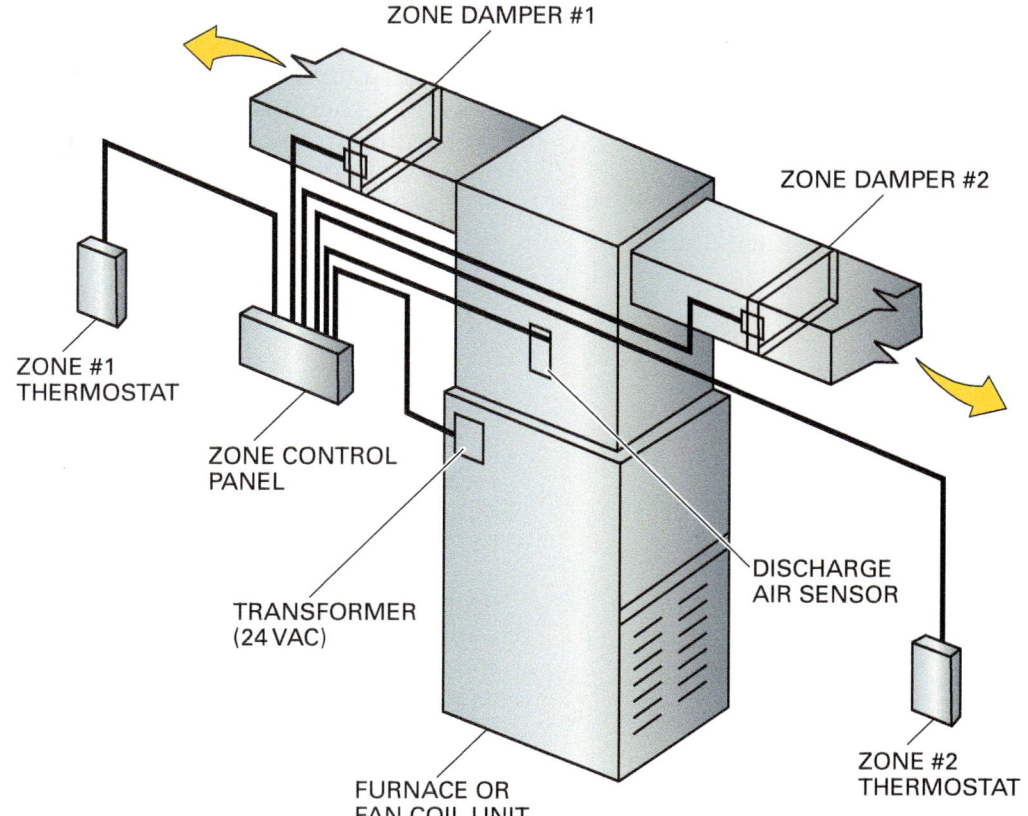

Figure 1 Zoned system schematic.

1.1.1 Zone Thermostats

Zone thermostats (*Figure 2*) look like any other thermostat. They usually have buttons or a touch screen that allow the user to change the setpoint(s) and the mode of operation. Programmable models allow the user to set schedules for setpoint changes in the zone. For some zoning systems, the thermostat selected must be compatible with the HVAC equipment. For example, if a heat pump is used, the zone thermostats must be able to accommodate heat pumps. However, this is not true for every system.

In some simple older systems, the means to change the mode of operation is not given to each zone thermostat. Rather, one thermostat is designated as the master (*Figure 3*). Only the master thermostat provides mode-switching capability. The system will operate in the heating or cooling mode, or it can be set for auto-changeover if the thermostat is equipped with this function. All the other zone thermostats, referred to as slaves (like the one shown in *Figure 3*), are programmed with setpoints. But as the name implies, they operate at the discretion of the master thermostat. If a zone calls for heating, heating will only be provided if the master thermostat is set to heating. Otherwise, there is no system response to the zone's request. Again, this is a feature found in older zoning systems. It may still be encountered by technicians, but is rarely installed any more.

Figure 2 Programmable zone thermostat.

1.1.2 Zone Dampers

Zone dampers are rectangular or round (*Figure 4*) and are available in many sizes. Most zone dampers are two-position models. They are usually powered to their closed position and are

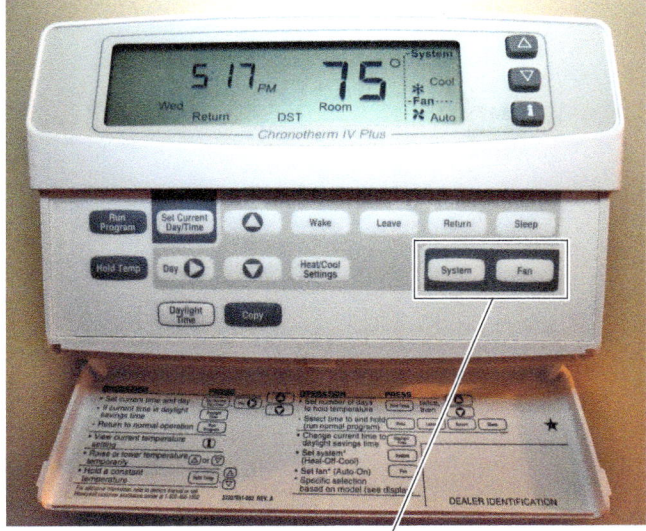
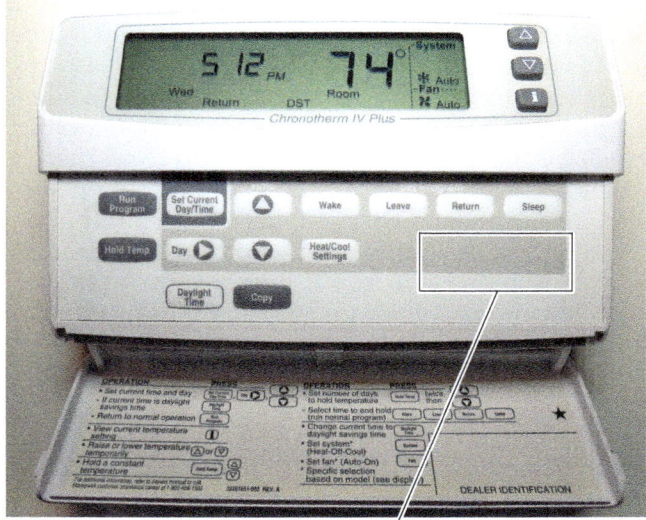

(A) MASTER THERMOSTAT

(B) ZONE (SLAVE) THERMOSTAT

Figure 3 Master thermostat and matching slave thermostat.

(A) RECTANGULAR ZONE DAMPER

(B) ROUND ZONE DAMPER

Figure 4 Zone dampers.

equipped with springs that return the damper to its open position when the motor is de-energized. This may seem backwards, as one might expect the damper to be powered open when heating or cooling is called for by the thermostat. However, by maintaining power on the damper when it is closed and using a spring to open it when it is de-energized, the damper becomes fail-safe. If the motor or its power source fails, the damper will open and remain in this position. This ensures that a zone can still be conditioned in exceptionally cold or hot weather despite a damper failure.

Some dampers can modulate their position, stopping at any point between open and closed. Rather than being equipped with a spring, they are powered by a motor in both directions. This is a function of the motor rather than the damper. Either spring-assisted motors or power-driven motors can generally be attached to the same damper assembly.

Even two-position dampers can be adjusted so that they do not fully close when powered in that direction. This is referred to as a minimum position setting. Preventing a zone damper from fully closing ensures that the zone will always receive a small amount of air for ventilation any time the system is running. It can also reduce the amount of air that must be bypassed when some zones are satisfied.

Although zone dampers are most often placed in trunk and/or branch ducts, they can also be attached to a single diffuser (*Figure 5*). This setup can be used to help improve the conditions in a single room or area that is consistently over-cooled or over-heated by a constant-volume system, without installing an entire zoning system. The assembly is equipped with a temperature sensor that determines if the system is in the heating or cooling mode by the supply duct temperature, and then it reacts accordingly. For example, if the cooling system is running and the wall thermostat falls below the setpoint, the damper will close to stop the airflow to this one diffuser. If it later senses that the heat is running, based on the duct temperature, it will open and allow the cold zone to warm up.

1.1.3 Bypass Dampers

Airflow in a heating and cooling system is important. With a furnace, a certain amount of airflow must be maintained across the heat exchanger to keep the furnace from overheating. In cooling systems, a certain amount of airflow must be maintained to keep the coil from freezing, and to ensure there is enough load to vaporize the refrigerant in the coil and prevent liquid floodback. This is especially true of systems using a fixed restrictor. For heat pumps operating in the heating mode, a significant reduction in indoor airflow will cause excessive head pressure. The problem of airflow across coils and/or heat exchangers is a significant concern in zoning systems.

An increasing number of systems are equipped with ECMs that modulate their speed and therefore the air volume. As zone dampers close, the motor can simply slow down to maintain a consistent static pressure in the duct. However, the furnace heat exchanger or indoor coil still require a minimum air volume for the system to function properly.

Since one or more zone dampers may be closed at any time while a system is in operation, the total amount of airflow into the conditioned space changes. There must be a method of ensuring that the airflow through the furnace or air handling unit remains nearly constant, without overheating or overcooling satisfied zones. Small fluctuations may be acceptable, especially if a thermostatic expansion valve is used as a metering device. Furnaces generally have at least a 30°F (17°C) range of acceptable temperatures in their temperature rise, but large reductions in airflow cannot be tolerated.

In typical zoning systems, this function is provided by the bypass damper (*Figure 6*). The bypass damper is the most important damper in a zoning system. As zone dampers close, the air volume through the air handling unit or furnace drops. To maintain a consistent volume of flow through the unit, air must bypass the closed zone

Figure 5 Diffuser/zone damper assembly.

(A) MOTORIZED, MODULATING BYPASS DAMPER

(B) BAROMETRIC BYPASS DAMPER

Figure 6 Bypass dampers.

dampers and go elsewhere. Since the number of closed zone dampers is unpredictable moment to moment, the amount of air to be bypassed also changes. While zone dampers are often two-position models (fully open or fully closed), bypass dampers modulate their position. This means that they can stop at any point between fully open and fully closed to satisfy the needs of the system.

Two types of bypass dampers are in common use: barometric dampers and motorized dampers. Both types are shown in *Figure 6*. The barometric damper is sometimes chosen as a cost-cutting measure, because it does not require electrical connections and control wiring; it is not motor-driven. The barometric damper's balancing mechanism is adjusted to open the damper when the system pressure reaches a predetermined pressure. Due to the action of the barometric damper, it cannot be installed in a vertical position. The motorized, modulating damper is also positioned based on system pressure, but it uses a sensor and damper motor.

There are two primary ways to apply a bypass damper, as shown in *Figure 7*. Air can be bypassed to a dump zone, or it can be bypassed directly from the supply air duct into the return air duct. In either case, the amount of air moving across the coil or heat exchanger remains consistent. A dump zone is an area located within a zoned building that does not require precise temperature control, such as a basement, large foyer, or hallway. In a commercial building, especially one that uses the ceiling plenum as a return air plenum, the air can be dumped into the ceiling area.

The farther away from the air handling unit it is dumped, the better the system functions. As the zone thermostats in the controlled areas are satisfied and the related zone dampers close, any excess system airflow is routed through the bypass damper into the dump zone. The bypass damper acts as a pressure regulator to maintain the design pressure in the supply trunk for the operating zones.

One benefit of using a dump zone bypass is that the heating or cooling capacity supplied by the bypassed air will pick up part of the building load. Another advantage is that the return air temperature to the furnace or coil does not change dramatically when the dump zone opens. The air that is dumped continues to mix with room air before returning to the unit. A return air grille is normally located in a dump zone area to avoid pressure buildup in the space.

The disadvantage is that the area where the air is discharged is not controlled, and the temperature may become too hot or too cold. Some installations using a dump zone may waste conditioned air, if it is an area where the occupants do not care about the temperature at all. If a barometric damper is used as a bypass, it is best used with a dump zone. When a direct bypass to the return duct is installed, barometric dampers do not perform as well.

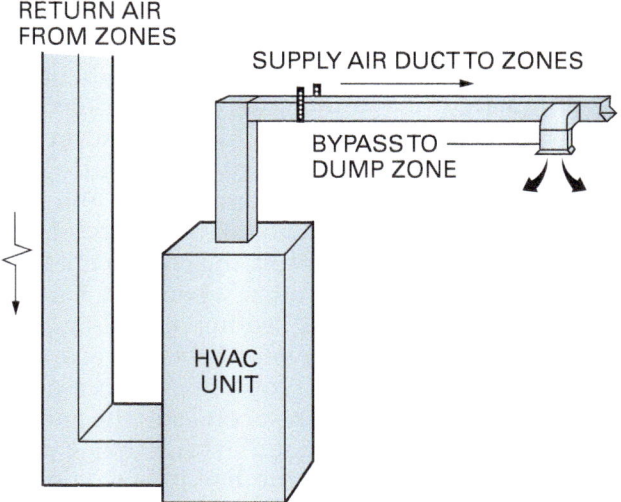

DUMP ZONE BYPASS IN BASEMENT

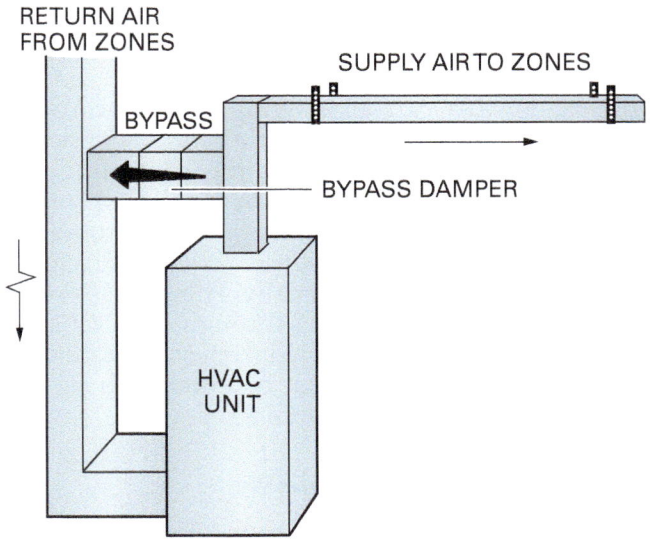

DIRECT RETURN BYPASS

Figure 7 Dump zone and direct return bypass zones.

The direct return bypass uses a damper installed in a bypass duct near the furnace or fan coil. Since the air discharged by the damper is bypassed directly to the return inlet of the furnace or air handler, it has no chance to pick up any part of the building heating or cooling load. This is a disadvantage, because when the bypass damper is wide open the air temperature returning to the equipment will significantly increase during heating and decrease during cooling. Bypassed air that is too hot can cause the furnace to shut off on its high limit. Air that is too cold can cause the air conditioning coil to freeze up.

To avoid excessive temperature swings in the heating or cooling equipment, a direct return bypass is usually sized to handle only about 25 percent of the total system air volume when the bypass damper is fully open. In addition, protection is provided by using a duct temperature sensor. The sensor monitors the temperature and will cycle the equipment off when it reaches preset low and high setpoints. As a result, the equipment may cycle on and off far more often when only a small part of the system requires heating or cooling.

The use of bypass dampers is being reconsidered in some areas of the country. When a great deal of air is being bypassed consistently, energy efficiency suffers. One manufacturer has a new approach that eliminates the use of a single bypass damper. In the process, it also eliminates the problem of hot or cold air being dumped directly into the return duct of the systems. A BZD Barometric Zone Damper™ (*Figure 8*) is a new type of zone damper that combines a motorized damper blade with barometric operation. Although the damper blade is motorized, half of the blade area can also swing open barometrically. A small weight is provided for adjustment.

When a zone calls for heating or cooling, the entire blade moves to the open position. When it is closed, and pressure begins to build throughout the duct system, the barometric blade of each closed zone damper can open partially. This relieves some of the pressure in the duct through each satisfied zone. Although the satisfied zones are receiving unwanted air, it should not be enough to significantly raise or lower the room temperature.

1.1.4 Zoning Control Panels

The zoning control panel (*Figure 9A* and *Figure 9B*) provides overall system management. It serves as a centralized point of connection for the thermostats, dampers, sensors, and HVAC equipment it

Figure 8 BZD Barometric Zone Damper™.

controls. Although zoning control panels can vary significantly in layout, most provide the same basic type of connections shown in *Figure 9B*.

Zoning control panels must be selected according to the application requirements. All manufacturers offer panels that can control some finite number of zones. For example, a two-zone panel can handle only two zones and may be appropriate for a simple two-story structure. Another panel may handle six zones. Panels that can accommodate as many as 20 zones are available. Beyond that, more sophisticated approaches to the overall system design are required.

The number of zones that the designer anticipates using is generally the first concern in panel selection. Note that panels capable of handling some fixed number of zones do not need connections made to every zone output. A six-zone panel, for example, can function just fine with only four or five zones connected.

The type of HVAC equipment to be controlled is the next consideration. Standard, single-stage heating and cooling equipment is fairly simple. Multistage systems may require a different panel. The panel shown in *Figure 9B* can control two stages of cooling (through the Y1 and Y2 terminals) and two stages of heating (through the W1 and W2 terminals). If a heat pump or dual-fuel

HVAC Equipment for Zoned Systems

A single-speed compressor and a fixed-speed indoor blower deliver a fixed amount of capacity, regardless of the number of zones calling for conditioned air. If all the zones are calling for air, the system works properly. If fewer than all zones call for air, the excess system capacity must be handled in some way, normally with a bypass arrangement.

Two-speed or modulating-capacity compressors, two-stage or modulating-capacity gas furnaces, and variable-speed ECM blower motors can respond much better to a reduced number of zones calling for conditioned air. When fewer than all zones call for conditioned air, the system responds by operating at a lower compressor capacity, a lower heating capacity, and/or a lower fan speed. The result is better indoor comfort, increased energy efficiency, and a quieter system. With the ability to stage system capacity, it is also likely that a bypass damper is unnecessary, further adding to system efficiency.

Unique Zoning Damper

Zoning dampers are available to meet nearly every possible installation situation. The model shown here replaces the take-off at the main trunk. As a result, it saves on the cost of buying a take-off, and the installation labor for the damper itself is no more than that of the take-off.

Figure Credit: Courtesy of ZONEFIRST

system is in use, this must also be factored into the selection. The control panel shown can be switched between heat pump or conventional fossil-fuel furnace service. Not every panel can properly control every type of system. There are many unique equipment capabilities to consider, such as variable-speed blowers, variable-speed compressors, and humidity control.

Most HVAC equipment manufacturers have their own zoning system products, some of which are specifically designed to work with their more advanced products. Using the proper control panel for a given type of equipment allows the user to take advantage of their advanced system features.

The control panel must also be selected to coordinate with the type of zone and bypass dampers that will be used. Zone dampers that are powered closed and spring open when de-energized (two-positions) are often referred to as two-wire dampers; only two wires are required to operate them. Zone dampers that modulate their position require three wires and are referred to as three-wire dampers. The panel shown in *Figure 9B* can accommodate three zones controlled by three-wire dampers.

Although some panels may offer the flexibility of working with either type of damper, others do not. This is a cost-saving measure, rather than a lack of technology. While programmable panels can be built that can work with almost every conceivable design and equipment, it may not be a cost-effective product.

One other very significant option available in zoning systems is wireless communication. Dampers (*Figure 10*) are normally wired back to the zone panel, or to a centrally located 24VAC transformer, since they require a source of power for the damper motor. A small 120VAC/24VAC control transformer located adjacent to the damper is also an option. A single centrally located transformer is more common. The thermostats are wireless, as is panel-to-damper communication. The panel shown in *Figure 11* has no terminals for thermostats or zone dampers. The system temperature sensors (not to be confused with the thermostats) are wired to the panel and do not communicate wirelessly in this product.

Wireless zoning systems are gaining in popularity, due to the significant labor savings in their installation. In addition to complete systems, there are also wireless dampers and thermostats that can operate independently and handle problem areas that overheat or overcool.

Figure 9A Zoning system control panel.

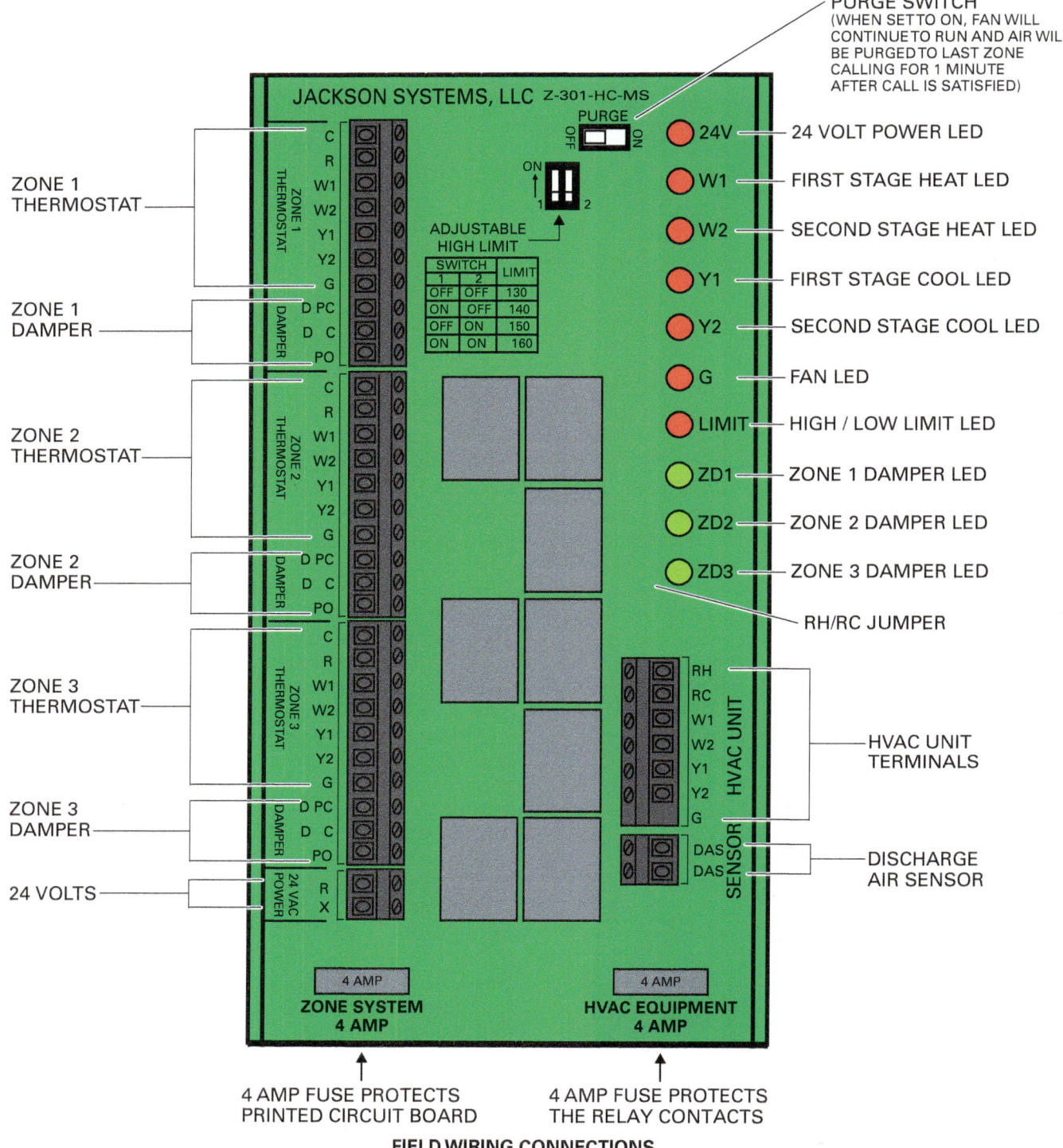

Figure 9B Zoning system control panel.

Figure 10 Wireless zone damper and thermostat.

The damper shown in *Figure 10* also has a sensor to monitor the supply air temperature. This determines whether the thermostat operates in a heating or cooling mode. For example, if the sensor indicates that there is 57°F (14°C) air in the supply duct, the system is obviously in the cooling mode. If the thermostat becomes satisfied and requires no more cooling, it will close the damper. The opposite action takes place when the sensor indicates the main system is in the heating mode.

Some zoning panels require the installer to program a variety of operating parameters. However, most of today's systems use a series of small switches on the panel for configuration instead. Most of these switches are two-position, although some may be three-position. Choosing a specific position for each switch configures the panel as needed for the application. For example, one switch may allow the installer to choose between a heat pump or a gas furnace. Another may allow the installer to choose whether the heat pump reversing valve is energized in the heating mode or the cooling mode.

1.1.5 Additional Devices

Most systems, especially those that are equipped with a bypass damper, require some additional controls. One such control is the device that provides the information needed to position the bypass damper. A static-pressure sensor (*Figure 12*) monitors the pressure in the duct and generates a signal that is used to position a motorized bypass damper. Barometric dampers are manually adjusted and require no further control.

High- and low-limit switches or sensors are required for systems with bypass dampers as well. They may also be needed in two-zone systems without a bypass damper for maximum protection. The high-limit is used to protect the heating unit, and typically monitors the discharge air temperature of the system. It opens when a preset limit is reached. This does not lock out the equipment; it simply cycles the heating unit off until the temperature is reduced to an acceptable level.

A low-limit protects the cooling unit from frozen coils and possible liquid refrigerant floodback. If the discharge air temperature falls below a preset level, the switch opens and the compressor is cycled off until the air temperature rises

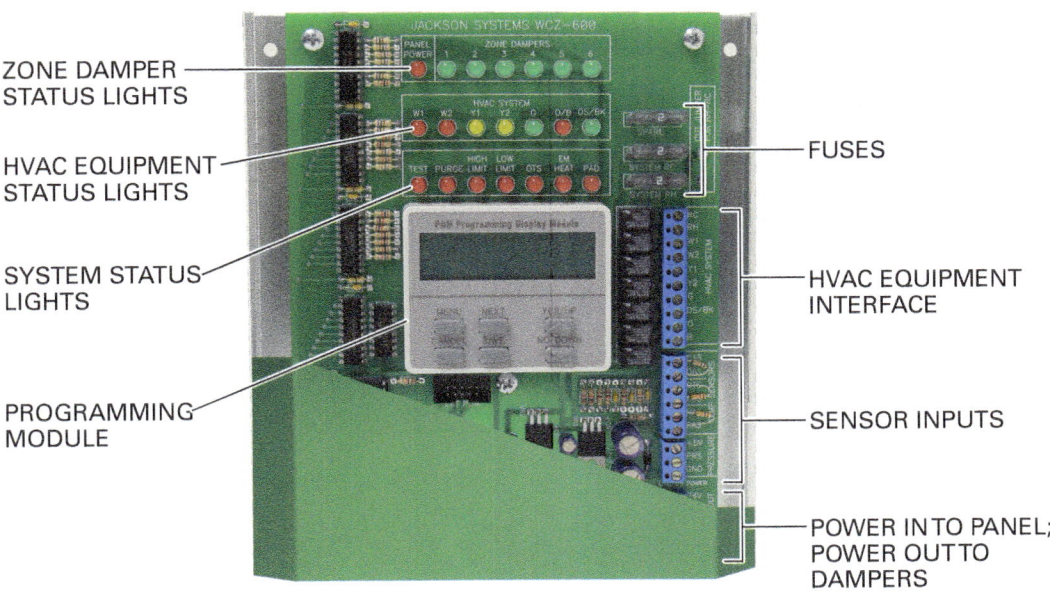

Figure 11 Wireless zone control panel.

Figure 12 Duct static-pressure sensor.

significantly. Since most systems are equipped with short-cycle protection, the compressor will generally be off for at least five minutes. Instead of monitoring air temperatures, a low-limit switch may be placed on the suction line of the compressor instead. Refer to the manufacturer's literature for the recommended approach.

Instead of switches, a single sensor can provide a discharge temperature indication to the control panel. The high- and low-limit switching is then accomplished by the panel. The logic of the panel determines the proper course of action when temperature limits are reached. Some systems may use an outdoor air sensor to provide a temperature indication that the panel can use for tasks such as controlling the supplementary heat of a heat pump system. If the temperature outdoors is above a preset balance point, the supplemental heat will remain locked out (except in the defrost mode) until the temperature falls below the setpoint.

The interconnecting wiring for most zone system components uses standard 18 AWG thermostat wire, because the current loads are quite low. Damper motors draw very little current (10VA, or roughly 0.4 amps), and they typically represent the highest electrical loads in the zoning system.

1.2.0 Zoning System Operation

There are many variations in zoned systems, but the principles of operation are basically the same. Each zone is controlled by its own thermostat that operates the zone damper (through the control panel). When all zones are satisfied and the HVAC equipment is idle, the zone dampers are all in their open position. This usually means that the damper motor is de-energized and the spring-return mechanism has opened the damper. Most dampers provide a means of visually determining the damper position, as shown in *Figure 13*. In addition to lights, the damper shaft may have a tab or visible line on it, showing the position of the blade.

When a zone thermostat calls for heating or cooling, the control panel starts the equipment and closes the dampers to all the satisfied zones. Some control panels provide for automatic changeover of the system from the heating mode to the cooling mode. When all the zone thermostats are again satisfied, the zone relays are de-energized at the panel. This causes all the zone dampers to return to the open position.

Indoor fan operation may be allowed to continue for a preset period to provide continuous air circulation in any or all the zones. In addition, it makes use of remaining boiling refrigerant in the evaporator coil, or heat that remains in the heat exchanger of a furnace. This is often referred to as a purge cycle (not to be confused with the purge cycle of a fossil-fuel burner).

There are a number of variations on this basic operating sequence. Some of these variations are discussed in the following sections. It is very important to read the manufacturer's literature for

Zone Dampers for Existing Systems

Existing systems can often be retrofitted for zoning but installing common zone damper assemblies in these systems can be labor-intensive. A damper has been developed specifically for this purpose. The damper shown here can be installed directly into an existing round duct. Since the assembly consists only of a blade and a motor, they save on material as well as labor costs.

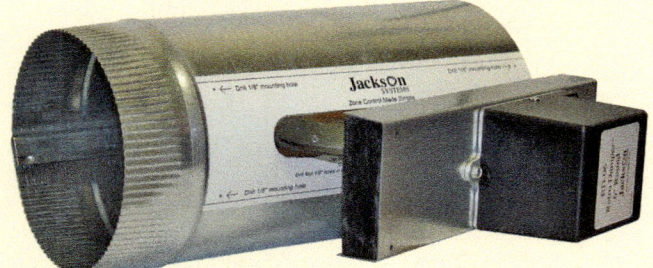

Figure Credit: Jackson Systems, LLC – Controls Done Right™

Figure 13 Zone damper in the open position.

each system thoroughly before installing a zoned system or troubleshooting one. Never assume that one system works the same way as another. Even minor differences in the sequence of operation can lead a technician to the wrong conclusion, wasting both time and money.

1.2.1 Ventilation Modes

In most zoning systems, each thermostat is equipped with both fan and system mode switches. Any of the zone thermostats can start the system fan by turning the fan switch from Auto to On. The zone damper for any thermostat fan switch set to On will remain open and allow ventilation air to enter. Any zone thermostat with the fan switch remaining in Auto will power the zone damper closed during the ventilation mode. If any of the zones call for cooling or heating, the system will then respond normally to the call by closing dampers for zones that do not need conditioned air. In other words, even with the fan switch set to On for ventilation, the damper will close if unwanted conditioned air is being provided. Once the heating or cooling operation stops, the damper will reopen to restore ventilation airflow.

1.2.2 Heating and Cooling

There are several possible variations in the needs of different zones. Sometimes one zone will request cooling while another requests heating.

Additional Heat

In many homes, there is one small room or area that always seems to be a bit too cold or hot when the rest of the home is comfortable. The Hot Pod™ is one way this problem can be solved. Equipped with a small fan and connected to its own wall thermostat, the fan can help boost the flow of air into a single room or branch duct. It is also fitted with an electric heating element. When the thermostat calls for additional heat, both the integral fan and the electric heater are energized to warm the area.

The Hot Pod™ is not connected to a zoning system in any way; it is a standalone device.

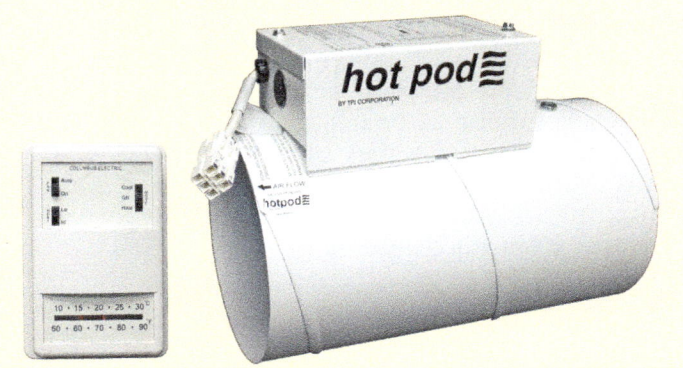

Figure Credit: Jackson Systems, LLC – Controls Done Right™

In fact, using it with a zone system could create some problems. If the zone damper for the area the Hot Pod™ serves is closed, then its source of air is cut off. Although there would be little harm in the cooling mode, the electric heater would overheat without airflow across it, and would then cycle off and on through its limit switch. Although it is protected, there would be no benefit from its operation.

Generally, the zone calling first takes priority. The zone panel will start the appropriate equipment to satisfy the needs of that zone first. Once that zone is satisfied, the system will switch and satisfy the other zone.

Some systems have programming that allows for mode control to be shared without conflict. For example, assume a call for both heating and cooling from different zones is made at nearly the same time, and that the cooling request was made first. The cooling equipment will start, and the damper will close on all systems that do not need cooling. Without the ability to share operating time, this status could continue indefinitely—the zone needing cooling could continue to dominate until it is satisfied.

With programming that allows sharing, the dominating call will only be served for a fixed period; a 20-minute maximum is common. At that point, the system will stop cooling and then start the heating equipment to answer the heating call. If the call for heat is satisfied in less than 20 minutes, the system will switch back to cooling again. If not, it will switch back to cooling again when the maximum time limit is reached.

Some systems have both minimum-on and minimum-off times for the cooling equipment. Short-cycling HVAC equipment increases wear and tear, especially on the compressor. Compressors experience more electrical stress during startup than any other time. Cooling systems often have a 3- to 5-minute minimum-off time before they are restarted. Technicians must pay attention to the setup of such systems. Thermostats often have built-in time delays for this purpose. Zone panels may also have short-cycle protection enabled, and a delay-on-break short-cycle timer may be installed in the condensing unit. These times may be compounded as a result. Minimum-on times may be shorter than minimum-off times. At least two minutes of on time after a start is common.

1.2.3 Multistage Systems

Ideally, a zone's need for heating or cooling is taken care of by the first stage of heating or cooling in a reasonable amount of time. This is the energy-efficient point of having a multistage system. However, zone control panels must have some sort of programming that determines when the second stage of heating or cooling is called upon. This eliminates the need to use multistage thermostats in each zone.

A common method is based on the amount of time that a call for heating or cooling remains. The timing begins when the zone thermostat first calls for cooling or heating. After a preset period, if the zone thermostat has not been satisfied, the zone panel will start the second stage of cooling or heating. The timing is adjustable on most systems.

More sophisticated systems allow for each zone to have different timing before the second stages of cooling or heating are called upon. For example, a living area zone could be set to allow a 5-minute run time before a second stage of capacity can start. A separate zone may then be set to allow 11 minutes of run time before a second stage can start. Note that these are not suggestions, but examples only; the specific installation and the habits of the occupants must be considered.

Sophisticated systems may also allow for a minimum number of zones to be calling for conditioned air before a second stage can start. This intelligent feature allows the installer to prevent the second stage of heating or cooling to start when only one or two zones in a large system are calling, regardless of demand. In a well-designed system, all stages of heating or cooling would not normally be required to satisfy the needs of a single zone. There are occasional exceptions, however.

1.2.4 High- and Low-Limit Operation

Many two-zone systems do not require bypass dampers or high- and low-limits for equipment protection. However, as the number of zones increases and become smaller, both bypass dampers and temperature-limit devices are usually required.

High- and low-limit switches or sensors cycle the cooling and heating equipment to prevent overheating, coil freezing, and refrigerant floodback. The primary idea is to prevent equipment safety controls, such as a furnace high-limit switch or a condensing unit low-pressure safety switch, from activating and shutting the equipment down completely. Some of these equipment safety devices require a manual reset. In addition, they are generally designed to activate near the performance limits of the equipment. When applying a zoning system, there is no desire to push the equipment performance to this level. By using high- and low-limit switches or sensors, the zoning system can cycle off the equipment before equipment safety devices are activated.

When a bypass damper is partially or fully open, excessively cool or warm air can be delivered from the heat exchanger or coil. Monitoring the discharge air temperature allows the system to cycle the heating or cooling equipment off until the air temperatures return to a safe level.

If the equipment is cycled off due to a high- or low- temperature condition, the rest of the zoning system continues to operate normally. The zone dampers of zones that are calling for conditioned air remain open and the blower continues to circulate air. Once the air temperature changes sufficiently, the equipment will be cycled back on. However, the start time will be subject to other limits, such as the minimum-off time programmed into the panel. In other words, both time and air temperature will likely play a part in how long the equipment remains off.

Low limits are sometimes fixed and cannot be changed by the installer or user. A common setpoint is 45°F (7°C). Other panels may allow an adjustment within a small range, as shown in *Figure 14*. The high limit is usually adjustable within a range, such as 110°–160°F (43°–71°C).

1.3.0 Zoning System Troubleshooting

Troubleshooting a zoning system is not difficult if you understand basic control troubleshooting. However, like thermostatic expansion valves, zoning controls are often suspected or blamed for a problem caused by other components. This is often due to a lack of understanding of what electronic controls are supposed to do. To successfully troubleshoot any zoning system, it is essential that the technician study the literature for that specific system. Without a clear understanding of the sequence of operation, troubleshooting cannot be accomplished effectively.

The troubleshooting process for a non-zoned system often starts as a no-heat or no-cooling service call. In these cases, technicians begin the process by determining if the system has power applied to it. If the equipment is powered, the process moves to determining if a control voltage is provided. When line voltage and control power are both present, the next step is to determine if a call for cooling or heating is coming from the thermostat.

With a zoning system, this initial troubleshooting process should be the same. The existence of a zoning system should not dramatically change that process. When there is a lack of cooling or heating throughout the facility, start with these simple tests to verify power. If the heating or cooling equipment is operating, verify that it is operating correctly before proceeding to the zoning system controls. Check the refrigerant charge, air filter condition, and other common causes for comfort complaints. It is best to make these tests with all zone thermostats calling for cooling or heating. This ensures that the maximum available load is being applied to the equipment.

The real change in the troubleshooting process begins when a call for the cooling or heating equipment to operate is being checked. The zoning panel typically takes the place of the single wall thermostat. Control power to energize a furnace, blower, and/or cooling unit comes from the zoning panel connections instead of directly from a thermostat.

1.3.1 Zone Panels

Troubleshooting of a zone system typically begins when the heating or cooling equipment is not running at all, and power has been confirmed. Then, if the complaint is related to a single zone, the troubleshooting process begins directly at the zoning system. This type of complaint will be discussed later in this section.

After reviewing the literature for the zoning system, use a voltmeter and check to ensure the panel is being powered. Zone panels are usually powered by a 24VAC source. The voltage applied should generally be within 10 percent, but check the manufacturer's literature to be sure of the range. Most panels have light-emitting diode (LED) lights to indicate a variety of conditions. If no lights are illuminated at all, this may indicate a lack of power. A panel power lamp is very common (*Figure 15*).

Once it is certain the panel is powered, it is a good practice to check any available programming next. There are many different panels in use, each with different capabilities and characteristics. If the programming or setup is incorrect for the application, problems will result. This step also provides valuable information to the technician. During the rest of the process, it is helpful to know precisely what the time and temperature settings are for various functions. Use the manufacturer's literature to review the panel programming carefully.

Figure 14 High- and low-limit setpoint adjustment.

Figure 15 Panel power indicator light.

cooling is expected, check for voltage at the panel terminal board.

Jumper wires can also be used at the panel outputs to check the integrity of controls downstream. For example, placing a jumper wire from R/RC to Y at the panel should start the condensing unit. If it does, all wiring and controls downstream are functioning. If outputs to the HVAC equipment are not present at the panel, and it has been confirmed that other conditions are correct, the panel is likely defective.

Remember that temperature sensors, generally thermistors, can also prevent heating and cooling equipment from operating. If a failed sensor indicates to the panel that the supply air temperature is too high or too low, outputs to the equipment will be prevented. Before condemning a panel that does not provide outputs when all other factors are correct, first check the accuracy of the temperature sensors. Testing and troubleshooting thermistors was covered in another module of this curriculum.

Also check the position of any dual-inline package (DIP) switches located on the panel (*Figure 16*). Most panels are configured using switches of this type. For example, if a panel is set up to require at least two zones calling for cooling to start the equipment, then a single zone requesting cooling will not result in equipment operation. A technician could mistakenly assume that the panel is defective, since there will be no output to the blower or condensing unit when the single zone calls for operation. In reality, the panel is doing exactly what it was programmed to do. Short-cycle timer operation on the board can also fool an impatient technician by delaying the start of the equipment.

Assuming that a sufficient number of zones are calling for cooling, the equipment outputs at the panel can be checked in the same way as any thermostat. R is the leg of control power being broken through the panel relays; it is the source of power that will feed back to the equipment via the W, Y, and G wires (as well as other wires, if required). If an output for

Figure 16 Panel configuration switches.

Check That Panel

The thermostat brand is not an accurate indication of the brand of zoning system that has been installed. Unlike many older systems, many of today's zoning systems can function with any common thermostat used for the zones. Therefore, a Honeywell thermostat may be found operating with a Jackson Systems zone panel. Always check the zone panel to confirm the manufacturer. Securing the correct installation and troubleshooting data for the panel is extremely important when troubleshooting zone controls.

Most panels have a number of LEDs to indicate the status of the system and various conditions. They can be a valuable aid. Pay attention to them while troubleshooting and observing system operation. As shown in *Figure 17*, some panels have lights that indicate that the high- or low-limit programming is active. If these indicators are lit but the discharge air temperature is normal, then either the sensor or the panel is likely at fault.

1.3.2 Zone Thermostats and Dampers

Except in wireless systems, zone thermostats are wired back to the zoning control panel (*Figure 18*). The zone thermostats can be tested for proper outputs like any other thermostat. From the panel location, a technician can test to see if thermostat circuits are open or closed by measuring voltage at the appropriate terminals. If a thermostat is suspected of being faulty, it is always best to confirm this at the thermostat before replacing it. This eliminates the possibility of a problem in the interconnecting wiring.

Remember that individual thermostats may be programmable. Sometimes a problem can be as simple as a user not realizing that a program has been invoked to change the setpoint at some time of day, causing discomfort. Also, keep in mind that if a setpoint is scheduled to change during the time a technician is involved in the troubleshooting process, it can create confusion.

A good way to start the process of checking thermostats and dampers is to begin with setting all zones to call for either heating or cooling. During this period, depending on the conditions, technicians may wish to cycle the power to the heating and cooling equipment off. This will prevent the equipment from running during a test of zone thermostats and dampers. With all zones calling, evaluate the system airflow and visually confirm that every zone damper is open, as it should be. The bypass damper should be completely closed when all the zone dampers are open. This should be confirmed visually as well. Remember that barometric bypass dampers are controlled only by weight and balance. If they are not fully closed, make the necessary adjustments.

Figure 17 Zoning system status lights.

Figure 18 Zone thermostat terminal blocks.

Once proper operation has been verified with all zones calling, begin adjusting one zone thermostat at a time. Set it to a low or high setting that will satisfy the thermostat. Confirm that this damper closes fully. When possible, observe the damper shaft or blade as it travels. More often than not, dampers fail to open or close properly due to binding or stress placed on the housing during or after installation. Often, there will be some distinguishable increase in airflow from the other zones.

The bypass damper may or may not move slightly. This depends primarily on the size of the zone compared to the total system and the number of zones. Remember that systems using a variable-speed blower motor will automatically reduce the system airflow; opening the bypass damper is likely unnecessary for these systems. In fact, they may have no bypass damper at all.

Continue this process for each of the connected zones. Although it may be tempting to control the entire process from the zoning control panel, doing so ignores too many variables that need to be checked. Control each zone from its thermostat and, whenever possible, observe the damper motor as it travels.

When zone dampers do not respond as expected, especially when commanded to close, begin by checking to ensure that an output is being sent from the zoning control panel. If not, make sure that the thermostat is signaling the panel appropriately. Because two-wire damper motors use a spring to open the damper, a two-wire damper motor with no input power to close the damper should always open on its own. If it doesn't, binding or stress on the damper blade may be the culprit. Three-wire dampers (*Figure 19*) must be powered to both the open and closed positions. These dampers will not move at all unless powered in one direction or the other.

Note the markings on the terminal board in *Figure 19*. The COM wire is the Common wire from the 24VAC power source at the panel. The center terminal, marked PO, provides the second leg of power to open the damper. The PC terminal powers the damper closed. To determine if the damper is being commanded to move in either direction, check for 24VAC between COM and either of the other two terminals. If 24VAC is present, that command is being sent to the damper. The PO and PC terminals should never be energized at the same time. That would likely indicate a significant problem with the panel itself, or possibly wiring damage that is allowing two conductors to make contact (a short).

Note the words Manual Gear Release above the terminal board. Many dampers have a means of disengaging the motor from the damper blade.

Figure 19 Three-wire zone damper terminal block.

This allows the damper to be positioned manually. If a zoning panel fails, a technician can manually disengage and open all the dampers. This allows the system to operate as a non-zoned system until repairs can be made.

When in doubt about the operation of a system or an individual component, contact the manufacturer directly. Many companies offer technical assistance by phone for challenging problems. They prefer to do so while the technician is on site, so they can suggest specific troubleshooting steps to implement. However, technicians should never rely on the manufacturer to replace self-study of the system and a sincere effort to first determine the problem on their own.

1.4.0 VVT System Components

Variable volume and temperature (VVT) systems have a lot in common with standard zoning systems, but they are more suited to the commercial environment. Although there are many similarities, there are also significant differences. The components of a VVT system communicate in a more sophisticated way.

VVT systems have changed in many ways over the years. Each generation of VVTs offers different capabilities and may communicate differently. The task of programming most VVT systems is as simple as following a checklist and understanding the details of the design well enough to make good decisions. Even though a task may be fairly consistent from one product or model to another, terminology and features can differ dramatically. Whenever working with a VVT system, it is crucial to obtain and review the technical data for that particular system and its specific components.

Figure 20 shows the layout of a typical VVT system. VVT systems are best used in commercial buildings where they control systems up to 25 tons. They are used on larger systems than most common zoning systems, and usually have more zones as a result. The arrangement shown here allows a single constant-volume packaged unit to meet individual zone cooling and heating needs. Each room is a control zone with its own communicating zone damper controller and room sensor.

Many of these components will look familiar, as they are similar to those in common residential zoning systems. Zone dampers are in place and there is a wall-mounted sensor in each zone. In a common zoning system, this would be a thermostat instead of a sensor alone. In a VVT system, the zone thermostat or sensor communicates its needs and status in a completely different way.

Another notable difference is found in the interconnecting wiring. In the VVT system, the zone dampers are communicating with each other through a bus. In addition, they are communicating with a device in the unit and with the bypass damper controller.

1.4.1 Bypass Damper

Because the central HVAC system is usually a constant-volume system, a bypass damper must be provided. As zone dampers throttle, the bypass damper is opened to keep the airflow through the unit consistent. They are controlled based on the static pressure of the supply air duct. The damper bypasses supply air directly to the return duct or to a ceiling return plenum.

This same concept applies to common zoning systems. However, there is often one distinct advantage to applying the system in a commercial building rather than a residence. One of the consistent complications of any zoning system is how to handle the bypass air. Bypassing it directly into the return duct works, but it is not a perfect solution and the equipment must be cycled off at times for its own protection. A dump zone also works but is highly wasteful.

In many commercial buildings, the plenum area above a suspended ceiling often serves as a return air plenum. This eliminates the need to install an elaborate return air system; the unit

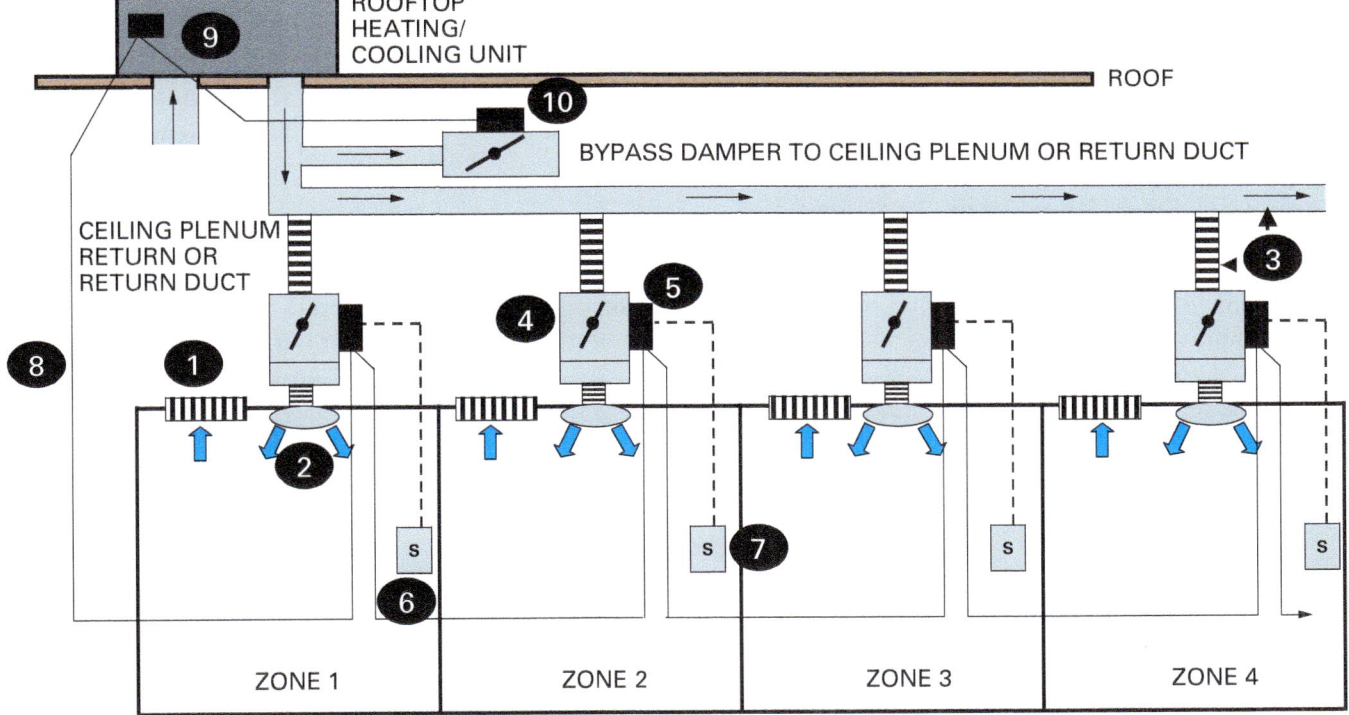

Figure 20 Variable volume/temperature (VVT) system.

simply draws its return air from the plenum through a single duct opening. One or more ceiling tiles are replaced with grilles to allow room air to easily migrate into the return air plenum.

This plenum approach can work in favor of a common zoning or VVT system, but it is extremely rare in residential construction. The bypass air can be dumped into the ceiling plenum a considerable distance away from the unit return air duct. This concept is illustrated in *Figure 20*. As it naturally travels back toward the return duct, there is plenty of time for the cooled or heated bypass air to mix with air in the plenum before reaching the unit. This helps to stabilize the return air temperature and prevent it from reaching extremes. In addition, this conditioned air is still being used in a positive way to condition the building, rather than being wasted into a dump zone. Allowing bypassed air to flow into a remote part of a ceiling return plenum offers the best performance in both VVT and zoned systems.

1.4.2 VVT Zone Dampers

Zone dampers, commonly referred to as VVT terminals (*Figure 21*), are available for VVT systems in both round and rectangular types. The most economical installation cost is achieved when the terminal size and shape match those of the branch duct in which it is installed, and when the number of dampers is minimized. However, it is important to properly size a damper for the volume of air to be passed. It should not be assumed that an existing branch duct size is correct. Avoid choosing the zone damper based only on the duct size.

Round air terminals are made in several sizes, typically ranging from 160 to 1,125 cfm (0.4 to 2.8 tons of cooling capacity). They have dampers that range from 6 to 16 inches (15 to 40 cm) in diameter to match typical branch duct sizes. Round dampers generally come without any insulation; insulation and a vapor barrier are applied externally during installation. They are normally installed on jobs where the duct pressure at the terminal inlet is 1.0 in. w.c. or less.

Rectangular air terminals are typically made in sizes ranging from 8" × 10" to 8" × 24" (20 × 25 cm to 20 × 60 cm), where 8" (20 cm) represents the height of the terminal. Rectangular air terminals are available from the factory with an insulated duct sleeve designed for new duct systems. The sheet metal sleeve has internal insulation that complies with fire protection codes. The internal damper is made of multiple opposing blades and is powered by a high-torque actuator. The actuator delivers sufficient torque to handle higher velocity, higher static inlet pressures, and more turbulence than can be handled by round dampers. The actuators for VVT systems are significantly more powerful than those used in standard residential zoning systems. This is necessary due to the higher pressures, velocities, and air volumes found in some commercial duct systems.

Normally in a VVT system, when a zone needs heat and the central unit is in the cooling mode, the zone needing heat must wait. However, VVT system terminals may be equipped with an auxiliary heat source. This is usually electric heat, but a hydronic heating coil can also be used. If a rectangular air terminal is used, the hot water coil or electric duct heater is mounted directly on the discharge of the terminal. The auxiliary heat source is used to provide heat to a needy zone while other zones need cooling.

1.4.3 Zone Controllers

In older VVT systems, there is one primary zone controller. It may be referred to as a linkage coordinator, master zone controller, or monitor thermo-

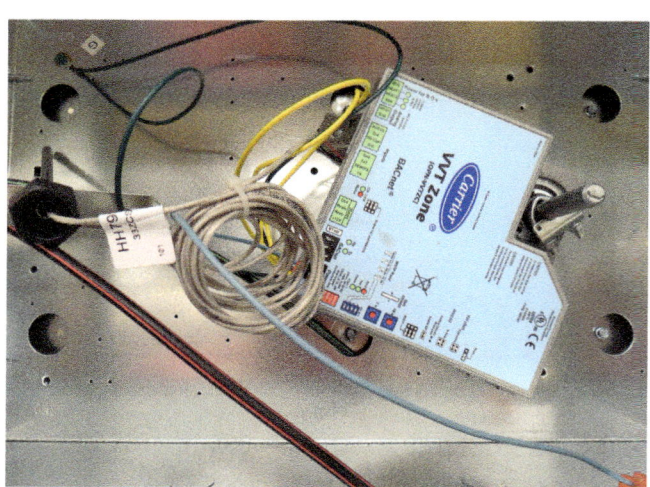

Figure 21 A VVT terminal awaiting installation.

stat. There are also secondary, or zone, controllers. The primary zone controller sends time of day, system mode of operation, and outdoor temperature information to each of the zone controllers. It also monitors the status of each of the zone controllers on its communications bus and selects the mode of operation for the central unit.

While providing these functions in most installations, it is also responsible for managing its own zone damper, like the rest of the zones. When a sufficient number of zones call for heat, the master controller switches the central unit into heat. When all heating zones have been satisfied, it switches the unit back into cooling. In this manner, the system provides cooling and heating to a building that has varying needs.

However, the system cannot provide simultaneous heating and cooling. In this respect, VVT systems suffer the same limitation as a standard zoning system. Recall that some terminals can be fitted with auxiliary heat. Although they help maintain comfort, auxiliary heaters are generally energy-wasteful devices, since the main unit is operating in the cooling mode. This duplicates energy consumption as the auxiliary heat and main unit work against each other.

The monitor thermostat is where most system programming is accomplished. It can be equipped with or without a time clock. The zone thermostats do not have their own time clock; instead, they share this information from the monitor thermostat via the communication bus. The monitor thermostat also has system switches, while the zone models do not. Time-of-day scheduling can often be accomplished either at the zone thermostat or at the monitor. Each zone can operate on its own schedule.

Newer VVT control systems have done away with the concept of monitor and zone thermostats with independent programming. The monitor thermostat has been replaced by computer

Monitor-Only Usage

When a VVT system is first installed in a new building, comfort conditions may need to be provided before the complete control system has been installed and set up. Monitor thermostats typically have a monitor-only mode. In this mode, it does not operate its own zone damper. Instead, it merely controls the HVAC equipment. A monitor thermostat can be used this way to operate a unit temporarily while the rest of the system is under construction.

The Popularity of VVT

You are likely to find some version of a VVT system in many commercial buildings. It has gained widespread acceptance and offers building owners comfort at an affordable price. In addition, 80 percent of all the buildings in the United States are 20 tons or smaller in size—the type of building best served by this system.

software that interfaces with zone sensors, zone dampers, and the HVAC equipment through a network. Programming for each zone and the system as a whole is done at the PC keyboard. Each zone damper control module stores the information pertaining to its zone once programmed through the PC and network. It can then operate on its own regardless of whether the controlling PC is connected or not.

1.4.4 Relay Packs and Unit Controllers

Since VVT control systems communicate differently, a method of communicating with the HVAC equipment is required. For most systems using a monitor thermostat, a relay pack installed in the equipment provides this service. Signals from the monitor thermostat are accepted, and the relays open and close in response to start and stop equipment operations.

Newer systems that are driven through computer software use a digital controller inside the equipment to provide this function. Like newer zone damper controllers, they also store their own programming information.

1.5.0 VVT System Operation

VVT systems communicate differently than common zoning systems. Most zoning systems operate by closing and opening 24VAC circuits. The components respond to the addition of control power, or the lack of it. Each zone damper opens or closes; it does not provide or consume any information. VVT system components communicate important information and share actual values with each other, working as a team at all times.

1.5.1 Communication

Figure 22 shows the wiring arrangement for a zone controller on a system equipped with a monitor thermostat. The zone controller electronics are housed in the front portion of the device,

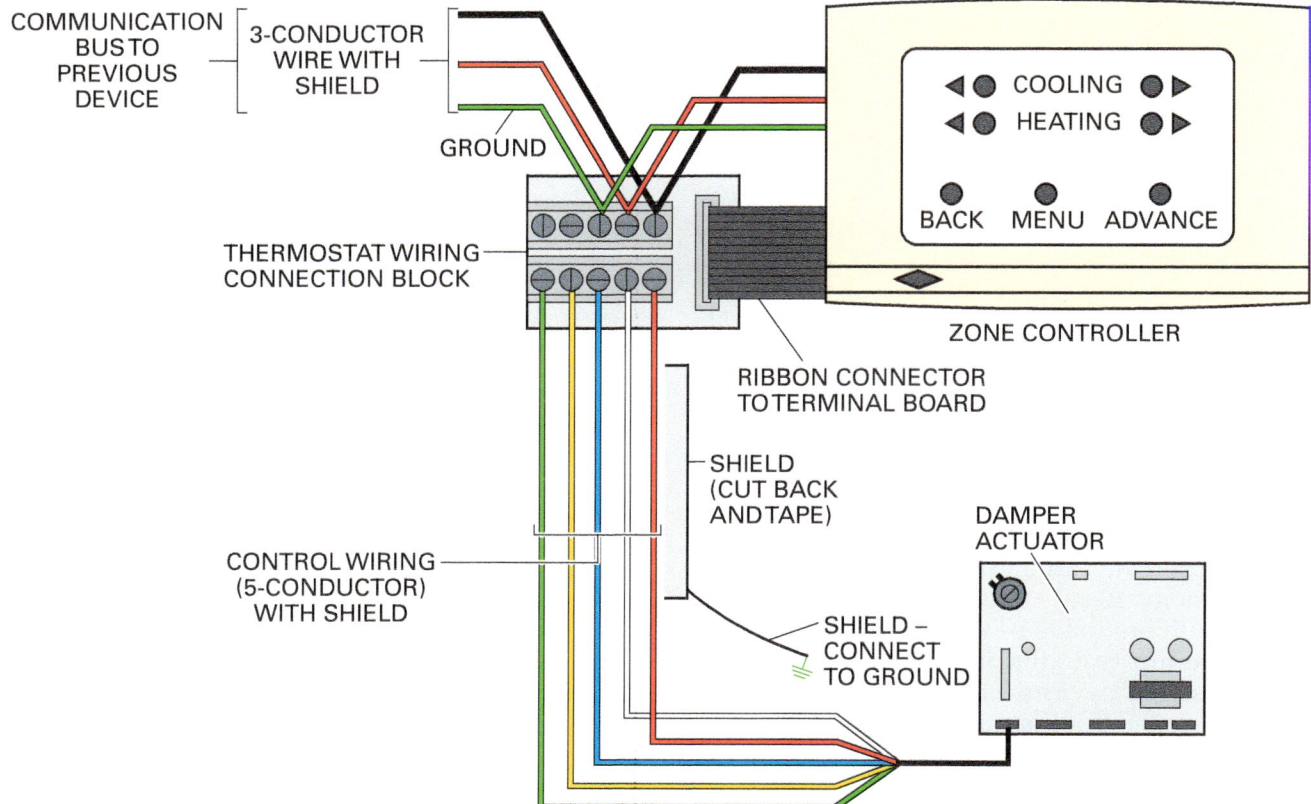

Figure 22 Zone controller wiring.

which is then connected to its wiring subbase by a ribbon cable. At the top of the diagram, the red, black, and green wires provide the path for communications between the devices. This is the communication bus. Signals are transferred through the red and black wires along the bus from each device, and all other devices share the information. Shielded wire is used for communication to prevent electrical noise generated by high-voltage and/or high-current load wiring nearby. Each device in the VVT control system is connected to the bus the same way.

Figure 23 shows a VVT system control communication schematic. The key to VVT system operation and control is the use of zone controllers that can communicate with each other through the bus. As shown, each zone has its own damper controller mounted on a VVT air terminal. A zone controller monitors temperature and allows the zone occupants to modify their own cooling and heating setpoints.

This type of VVT system uses a polling approach to determine the needs of the individual zones and to gather information from other system devices. In a polling network (*Figure 24*), the monitor thermostat requests, or polls, information from each of the other controllers, one at a time, and in a certain order. Information from other accessory devices, such as an outdoor air temperature sensor, will also be polled. The polling process is executed in the same order continuously.

In this network, the response time may be slow due to the time required to complete the polling sequence. This is particularly true when considering alarms. For example, consider that Controller 4 is being polled in a network consisting of 20 devices. If Controller 3 goes into alarm at this moment, the alarm cannot be communicated until the polling device has communicated with the other controllers and returns to polling controller 3 again.

The single monitor thermostat in the system does the polling. It polls the other zone controllers and devices at regular intervals. The monitor thermostat then chooses the appropriate mode of operation and communicates that to the HVAC unit. Polling is a continuous process that occurs as long as the system is powered.

1.5.2 Pressure-Dependent vs. Pressure-Independent Operation

Most simple VVT systems work in a pressure-dependent operation mode. This is a simpler approach and requires fewer and less-sophisticated

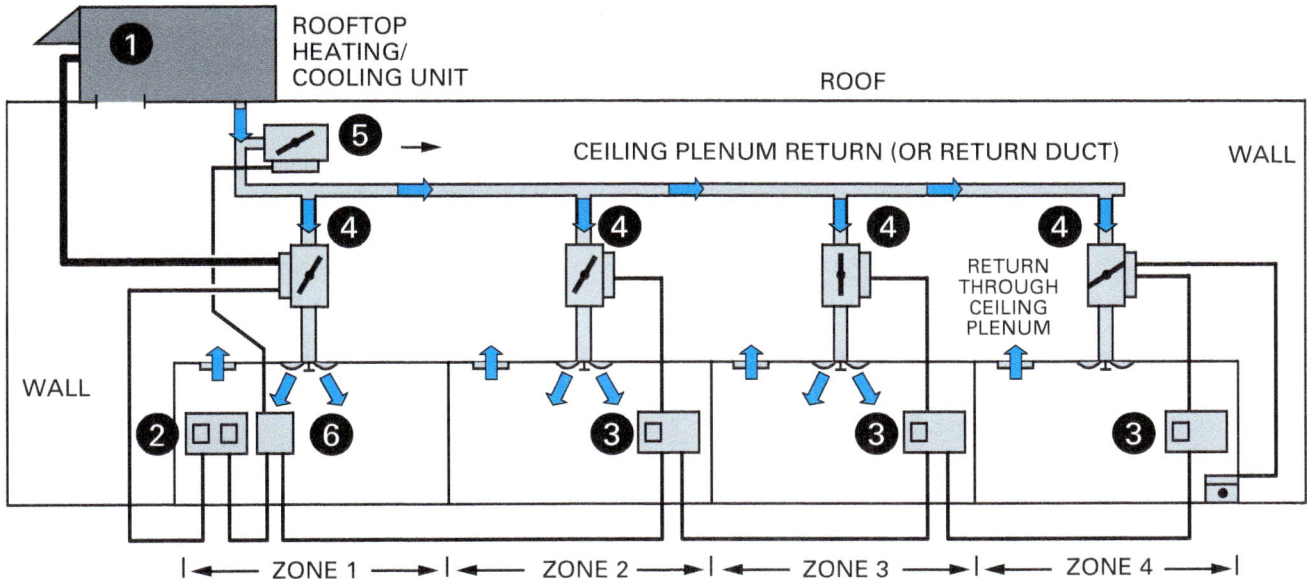

1. PACKAGED AIR CONDITIONER OR SPLIT SYSTEM
2. MONITOR THERMOSTAT WITH TIME CLOCK
3. ZONE CONTROLLER
4. ZONE DAMPER WITH ACTUATOR
5. BYPASS DAMPER WITH ACTUATOR AND STATIC PRESSURE SENSOR
6. BYPASS CONTROLLER

Figure 23 VVT control communication schematic.

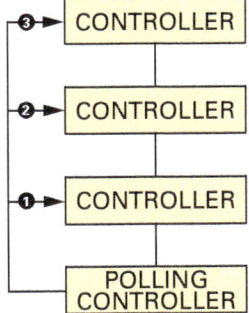

Figure 24 Polling network.

controls. In a pressure-dependent system, the actual volume of airflow through a given zone damper is not monitored. The damper is positioned between the open and closed position based on the demand for the zone. *Figure 25* shows an example of this. The damper position is changed incrementally as the demand increases or decreases. Usually, when the setpoint and room temperature are more than 1.5°F apart, the damper is fully opened. However, this is generally a changeable value. A pressure-dependent system has no idea how much air is flowing through a given damper and has no means of control over the volume. The actual volume will depend on the pressure present in the duct at any given moment.

Systems that use **pressure-independent operation** are more sophisticated. In these systems, the airflow through a zone damper is monitored using a velocity sensor. Based on the measured velocity and the area of the duct in which the damper is installed, the system calculates the actual air volume flowing through. Programmed values for the minimum and maximum airflow in a zone are then used to determine damper position. The basic operation remains the same: more demand for a given mode of operation (heating or cooling) results in the damper opening further. However, with the air volume constantly being monitored, the damper will only open until the maximum airflow setpoint is reached. Regardless of how high the demand may be, it will not open further.

The airflow calculation is also used to set the damper at its minimum position for ventilation. In a pressure-dependent system, the only accurate way of determining the minimum position for the damper is to measure the airflow manually and set the minimum position of the damper accordingly. The minimum position of the damper will always be the same, regardless of pressure changes in the duct. In a pressure-independent system, the actual minimum damper position will be different as the duct pressure changes. In a manner of speaking, pressure-independent systems are self-balancing.

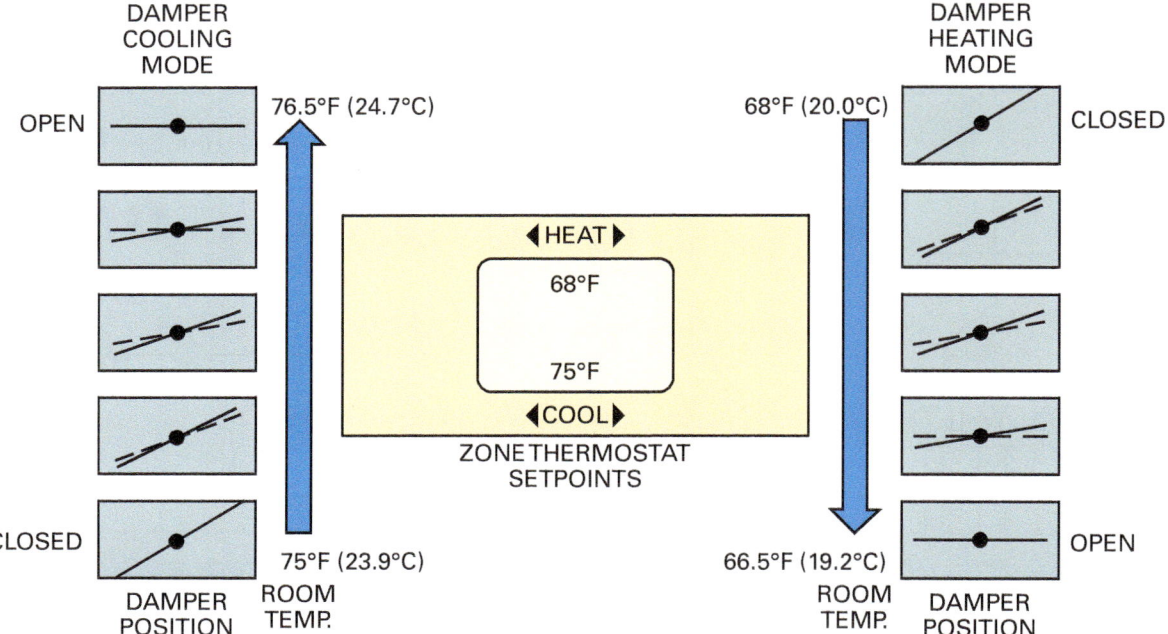

Figure 25 Pressure-dependent zone damper modulation.

1.5.3 Occupied Cooling

When one or more zones become occupied, the monitor thermostat determines that the building is occupied and may tell the system indoor fan to start. In many VVT systems used in the commercial market, the indoor fan runs continuously when the building is occupied. At the same time, the system's outside air dampers open to their minimum ventilation position. Running the fan continuously is the only way to ensure that each area gets the required amount of fresh air, regardless of whether heating or cooling is needed. Some systems can also respond to the input of a CO_2 sensor, and position the outside air dampers to meet the demand for fresh air based on actual conditions. Dehumidification modes are yet another feature in advanced systems.

Zone damper assemblies are rarely programmed to go to a completely closed position, especially during occupied hours. Even when a zone is not calling for heating or cooling, fresh air needs to be provided in a commercial setting. Therefore, each zone damper is generally set to close down to a minimum position setpoint. This leaves the damper open slightly. Ventilation air can then enter the space during all occupied hours.

Each zone controller has its own setpoints. There is usually a separate setpoint for heating and cooling. The zone controller works with the monitor thermostat to determine how far away the actual zone temperature is from the setpoint. This is called the demand. The monitor thermostat regularly polls the individual zone controllers, and some calculate the total average cooling demand and total average heating demand for the group. These demands may be displayed at the monitor thermostat or another user interface. The average demand is weighted based on the design airflow to be delivered to a zone. An 800-cfm zone will count twice as much in the demand average calculation as a 400-cfm zone.

Demand may also be based solely on individual zone votes. Each zone, regardless of size, gets one vote. Zones vote for heating or cooling, and the majority wins. Cooling is usually the priority. If the average cooling demand exceeds the average heating demand and is greater than the minimum level of demand required for the cooling mode, the system enters the cooling mode of operation and communicates that information to each zone controller.

More than one vote, and sometimes as many as three or four votes, is typically required to get the system to operate. The minimum level of demand required is a programmable feature of the system. This is true for systems that calculate weighted demand and for those that are simply counting individual votes. This prevents the system from operating a large HVAC unit and system to satisfy a small demand, which often results in significant energy use. A great deal of air would need to be bypassed to serve a single small zone demand.

A mode timer starts when the cooling mode begins. VVT systems can be programmed to operate in the cooling mode for a maximum specified period before switching to the heat mode to satisfy rising heating demand from other zones. This prevents a few zones from totally dominating the system. Most systems have minimum operating times as well. Once the cooling mode starts, it will run for some preset period, at least, before shutting down or switching to the heating mode.

Since this is the occupied mode, the outside air damper should already be open to provide ventilation air when cooling begins. Compressor stages are added as necessary. If the system is equipped with an economizer, it may be activated as the first stage of cooling if the outdoor conditions are favorable. With cool air being supplied, each zone controller modulates its individual air terminal damper as needed to maintain the zone cooling setpoint. The farther the room temperature is from the setpoint, the more the damper will open. As the temperature closes in on the setpoint, the damper will begin to modulate toward the closed position.

As the air terminals throttle airflow, the supply duct pressure rises and the airflow through the central packaged unit decreases. A bypass controller monitors the rise in duct pressure and opens the bypass damper accordingly. This allows supply air to be recirculated back to the return side of the central unit and maintains a constant airflow across the DX cooling coil.

Like common zoning systems, temperature sensors are used to monitor the discharge air temperature. Minimum and maximum values are programmed, and the heating and cooling equipment is cycled off as necessary to prevent problems.

If the average cooling demand continues to rise, the central unit controller adds cooling capacity stages as necessary. Capacity stages are then cycled on and off to satisfy the average demand setpoint for cooling. The cooling mode ends when the average cooling demand has dropped below the field-configured minimum average cooling demand, a greater average heating demand is encountered, or a heating demand has existed for a specified period of time while the unit was in the cooling mode.

1.5.4 Occupied Heating

Should any zone require heating while the central unit is in cooling, the zone controller closes that air terminal's damper to the minimum position and sends the zone heating demand to the monitor thermostat or linkage coordinator. Since the damper must remain at a minimum position for ventilation and is not fully closed, some continued cooling can and does occur. This drives the space temperature even further from the setpoint, increasing the demand. If the average demand for heating throughout the system becomes greater than the average demand for cooling, and the cooling mode minimum timer has elapsed, the air source controller switches the central unit from cooling to heating. It then communicates to the zone controllers that the heating system mode is active. The mode timer is also restarted.

Those zones needing heat modulate their air terminal dampers open to meet the zone heating setpoint. All other air terminal dampers are set to their minimum ventilation position, even if any zone starts to need cooling. The monitor thermostat or linkage coordinator continues to poll the zones and accessory devices and responds when necessary.

If the average heating demand continues to rise, the central unit controller adds heating stages (if available) as necessary. The stages are then cycled on and off to satisfy the average demand setpoint for heating. The heating mode ends when the average heating demand has dropped below the field-configured minimum average heating demand, when a greater average cooling demand is encountered, or cooling demands have gone unanswered for a preset period of time.

During the heating mode, the bypass controller continues to function and position the bypass damper, ensuring that a constant air volume flows across the heat exchanger and/or coil.

1.5.5 Unoccupied Period

When the monitor thermostat determines that all zones are unoccupied, it requests the central unit to stop and all zone dampers to be returned to their minimum positions. The central unit compressors and supply air fan stop, and the outside air damper closes. If the zone controllers are equipped with setback temperatures, the central unit operates just as it did during the occupied period to maintain the new setpoints, except that the outside air dampers remain closed.

1.5.6 Simultaneous Cooling and Heating

When the central unit is in cooling, simultaneous heating at the zone level can be provided if zone air terminals are equipped with a heat source. If the system is so configured, a zone heater can be activated by its controller. The zone damper will remain at its minimum open position, since ventilation rates must still be maintained. The zone heater, whether it be electric or hydronic, reheats the supply air to warm the zone. When the central unit is in heating, however, simultaneous cooling cannot be provided by a VVT system.

Normal operation of the VVT system does not involve the use of zone heaters. It comes close to satisfying a building's simultaneous need for cooling and heating by time-sharing central equipment heating and cooling modes. In fact, many VVT systems are not equipped with zone heaters at all. Their installation increases expenses significantly due to the added costs of equipment and electrical wiring.

The simultaneous need for cooling and heating in small buildings tends to be seasonal, occurring in the spring and fall when the building loads are significantly less than the peak design loads. However, internal loads are also a factor. Because VVT uses a central unit designed for the peak cooling and heating loads, the unit can quickly switch from one mode to another and satisfy the building's smaller part-load conditions.

VVT, with its time-cycling of capacity, works well in building sizes up to 25 tons. They are best suited for buildings with zones that have fairly equal loads. If one or two of many zones have an intense amount of heat gain compared to the others, comfort problems will usually result. These unique situations often need to be addressed with unique solutions, such as a dedicated ductless system.

Additional Resources

ANSI/ACCA Manual Zr, Residential Zoning. Current edition. Arlington, VA: Air Conditioning Contractors of America.

1.0.0 Section Review

1. In a system with a master thermostat, _____.
   a. the system mode switches on the master and the zone thermostat must be set to the same position
   b. the master thermostat alone determines the mode
   c. each individual zone can determine the operating mode of the system
   d. an auto-changeover feature cannot be used

2. High- and low-temperature limit devices associated with zoning systems are designed to _____.
   a. cycle-off the equipment before the equipment safety devices are activated
   b. act in place of the safety devices installed in the equipment
   c. protect the zone dampers from thermal damage
   d. protect against overheating or overcooling a zone

3. When a user with a zoning system complains they have no cooling at all, the best way to begin is to _____.
   a. check the individual zone dampers
   b. test the individual zone thermostats
   c. check for power at the HVAC equipment
   d. check the zoning panel programming

4. An older VVT system will have _____.
   a. only one primary zone controller
   b. two primary controllers
   c. three primary controllers
   d. at least four primary controllers

5. In which VVT system operating mode would the outside air damper most likely remain closed?
   a. Occupied cooling
   b. Occupied heating
   c. Ventilation
   d. Unoccupied

Section Two

2.0.0 Ductless and Variable Refrigerant Flow (VRF) Systems

Objective

Identify ductless and variable refrigerant flow systems and describe the basic approach to troubleshooting.
  a. Identify and describe the operation of ductless split-system equipment.
  b. Describe the installation and troubleshooting of typical ductless systems.
  c. Identify and describe the operation of variable refrigerant flow systems.
  d. Describe the installation and troubleshooting of variable refrigerant flow systems.

Performance Tasks

1. Troubleshoot and repair a typical ductless or variable refrigerant flow system.
2. Program the controller for a ductless or variable refrigerant flow system.

Trade Terms

Refnet joints: Engineered joints used as connection points in a VRF-system refrigerant network. Refnet joints connect the main refrigerant lines to branch runs or to indoor units.

Ductless split systems have been extremely popular in Europe and Asia for many years, dominating the housing industry in many locations. They have even been used extensively in high-rise residential buildings overseas. The popularity of ductless systems has greatly increased in the United States after a rather slow start.

Variable refrigerant flow (VRF) systems are a newer product. Although they share some characteristics with ductless systems, VRFs represent a significantly different approach to zoning, utilizing equipment instead of an air distribution system. VRF systems are not necessarily ductless.

2.1.0 Ductless Split Systems

In some heating and cooling applications, it is difficult or impractical to install a ducted air distribution system. Examples include homes heated with hydronic systems, or a historic building that does not have the space to accommodate ducts without undesirable structural modifications. Other unique and challenging applications include highway tollbooths, guard shacks, stand-alone ATMs outside of banks, fast-food restaurant drive-through windows, and add-on sunrooms. The options for cooling in these applications include the use of ductless split systems.

Figure 26 shows one example of a ductless split system installation. Ductless split systems are available in both straight-cooling and heat pump configurations.

As the name implies, these are split systems that use a ductless air handling unit. The narrow profile and horizontal discharge of air from the air handler make it ideal for installations where space is limited. The air handlers used in ductless split systems can be mounted on the floor, high on a wall, and either on or in the ceiling. Although ductless systems, also referred to as mini-splits, are normally installed without any duct at all, some manufacturers offer indoor units that can be concealed. They can usually accommodate a small amount of duct routed to one or two grilles. However, the blowers used cannot overcome a substantial amount of external static pressure, so the amount of ductwork must be limited.

Refrigerant tubing and wiring connect the indoor and outdoor components of the system. The system is typically controlled with a handheld remote similar to a TV remote (*Figure 27*). Some systems also allow the use of an optional wall-mounted room thermostat that is hardwired.

Ductless Hydronic Systems

Ductless hydronic split systems are also available. The indoor units look much the same as units that use refrigerant, but hydronic split systems are equipped with water coils instead of refrigerant coils. Ductless hydronic systems do not heat. However, they are advantageous to use in buildings that already have hydronic or steam heating systems, which are also ductless, or in heated buildings so old that retrofitting with ductwork would be difficult. As a result, ductless hydronic systems are more popular in the North, where these systems are more common.

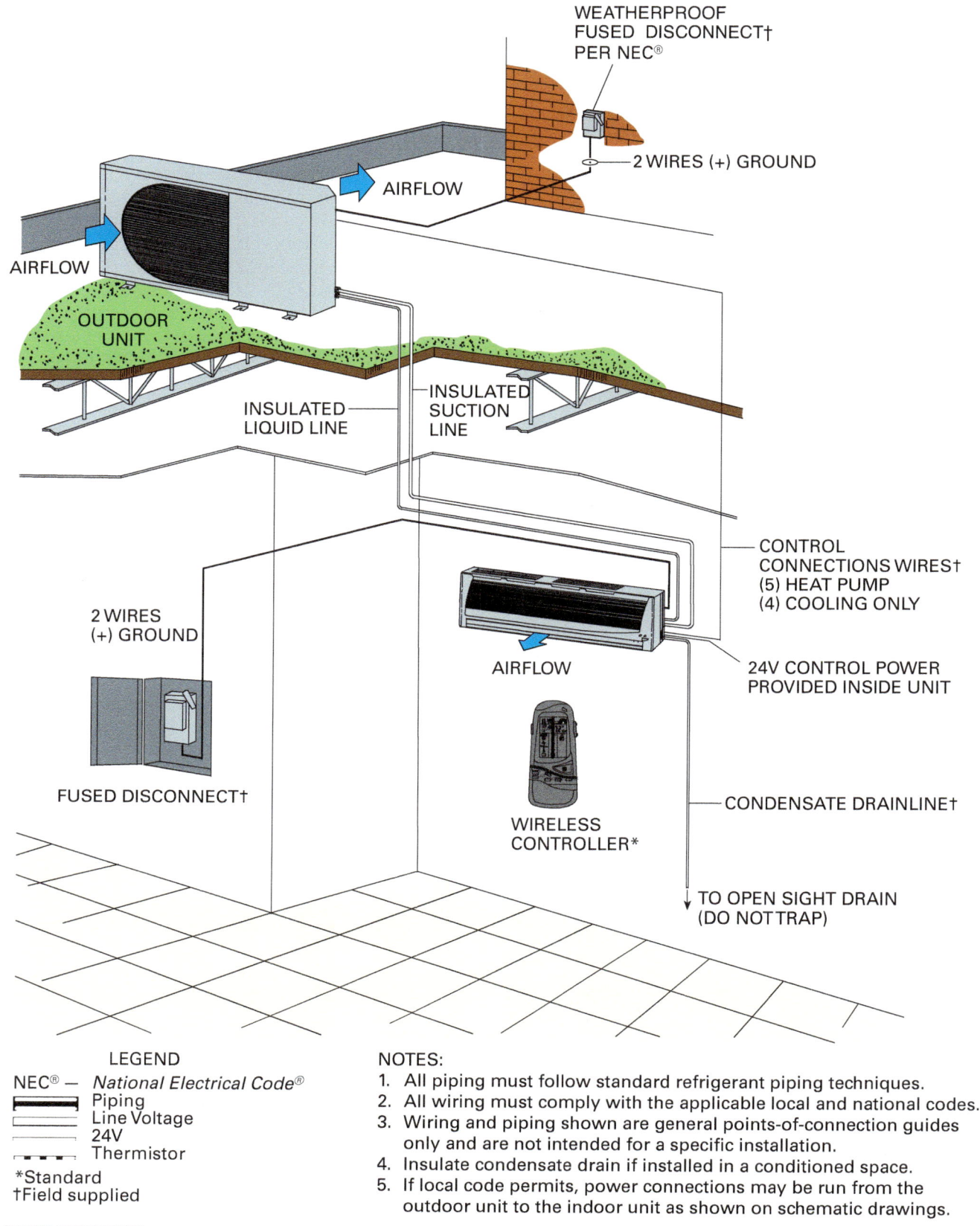

Figure 26 Common ductless split-system installation.

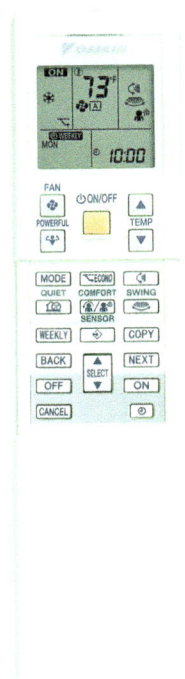

Figure 27 Handheld remote control for a ductless system.

Figure 28 Ductless split-system condensing unit.

2.1.1 Ductless Split-System Condensing Units

The outdoor unit (*Figure 28*) contains the compressor, condenser coil, condenser fan, and controls like any other split-system condenser. However, other aspects of their construction differ considerably. They are generally designed as rectangular slabs that resemble a large suitcase. The condenser coil is mounted horizontally in the cabinet and air is either blown or drawn through the coil and discharged horizontally. Higher-capacity units may have the appearance of two units stacked on top of each other, with two condenser fans.

Units are typically powered by single-phase 115VAC or 230VAC power. Power for the indoor fan coil is typically supplied through the outdoor unit. The metering device is often located in the outdoor unit and not inside the fan coil as in traditional systems. An electronic expansion valve is usually required for best performance. The result is that the tube that carries refrigerant to the indoor fan coil in the cooling mode does not contain subcooled liquid refrigerant, as is the case with a traditional split system. Instead, it carries a low-pressure, low-temperature liquid/gas refrigerant mix. For this reason, both refrigerant lines must be insulated to prevent capacity loss and condensation on the lines. Electronic controls are widely used in ductless split systems.

Many ductless systems use variable-speed compressors powered by inverters that change the provided AC power to DC. Others use technology that converts the single-phase power to a three-phase AC power source, using a changing frequency to vary the speed. This provides the precise amount of capacity in demand at any time. Since many ductless systems can support more than one indoor unit, the required capacity can vary widely as indoor units start and stop in response to their individual temperature controls. Variable-speed compressors provide a major portion of the energy savings ductless systems offer, as well as precision capacity control.

Because the indoor unit(s) is typically powered from the outdoor unit, the interconnecting wiring between the two includes both line-voltage power as well as control inputs.

2.1.2 Ductless Split-System Indoor Units

The indoor unit contains the evaporator coil, indoor blower, and a control board. A washable air filter is included. The blower wheel is the centrifugal type, small in diameter but elongated. The metering device is typically in the outdoor unit. The refrigerant piping and wiring can enter through the back of the unit, where it remains unseen on the inside. It can also enter from the side and be exposed on the inside wall of the room where appearances are not important.

The variety of different air handlers available for ductless split systems greatly enhances their versatility. Air handler types include but are not limited to the following:

- High sidewall
- Ceiling mount
- In-ceiling cassette
- Floor mount

High sidewall air handlers (*Figure 29*) are usually rectangular and shallow in depth. When mounted high on a wall, the fan in the air handler can more

Figure 29 High sidewall air handler.

effectively move the air into and through the room. The forward-inclined centrifugal blower in this type of air handler is small in diameter and almost as long as the air handler is wide. The moving motorized louvers on the air outlet help distribute the air more evenly throughout the room, if desired.

If the air handler is mounted on an outside wall, the condensate can be drained through the wall by gravity. If mounted on an inside wall, an accessory condensate pump can be used for condensate disposal (*Figure 30*). Some condensate pumps fit inside the unit cabinet, while others must be mounted exposed on the wall. These small pumps cannot raise the condensate very high, so drainage must be carefully considered in every installation.

High sidewall air handlers are typically attached to the wall on a mounting plate. The mounting plate also acts as a template for drilling holes through the wall for the refrigerant piping, wiring, and condensate drain. Mounting instructions are often provided on the plate as well as in the installation manual. Control of the system is accomplished with a handheld remote. Some systems offer the option of a traditional hard-wired, wall-mounted controller. Regardless of which type is used, room temperature sensing is done at the indoor unit return air inlet.

Ceiling-mounted air handlers, often called under-ceiling air handlers, have many of the same characteristics as high sidewall air handlers. The main difference is that they are suspended beneath a ceiling. They are often located adjacent to an outside wall so that refrigerant piping, wiring, and condensate piping can be routed to the unit without being exposed. If located away from an outside wall, a condensate pump is needed to dispose of condensate. Control of the unit is the same as that of a high sidewall unit.

In-ceiling cassettes (*Figure 31*) are usually mounted in drop ceilings with the face of the unit flush with the ceiling. All that may protrude slightly below the ceiling line is the return air intake (located in the center) and the supply air distribution louvers (located around the perimeter). When installed, the unit has the appearance of a concentric ceiling diffuser. The blower draws air in through the center of the diffuser and distributes it out through the four supply louvers. Refrigerant lines, wiring, and condensate lines

(A) UNIT-MOUNTED

(B) EXPOSED AFTERMARKET INSTALLATION

Figure 30 Ductless system condensate pumps.

are all located above the drop ceiling (*Figure 32*). A condensate pump is often required in this application. Control is similar to that of other ductless split-system air handlers.

In-ceiling cassettes are different from the other types of ductless split-system air handlers in that a limited length and size of ductwork can sometimes be attached to the air handler to supply an adjoining space. A knockout, visible on the side of the cassette, is provided for attaching the duct. If the ducted air is supplied to an adjoining room, air return must be accomplished using an undercut door or other method. Some cassettes also have provisions for the introduction of outside air for ventilation. A power ventilation accessory fan is used to overcome the friction loss of the interconnecting duct.

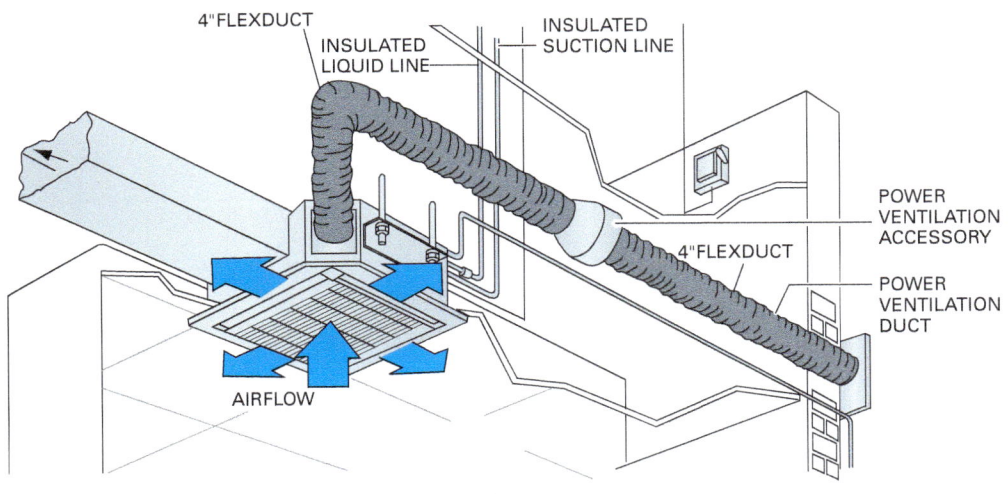

Figure 31 In-ceiling cassette air handler.

Figure 32 Cassette piping and wiring.

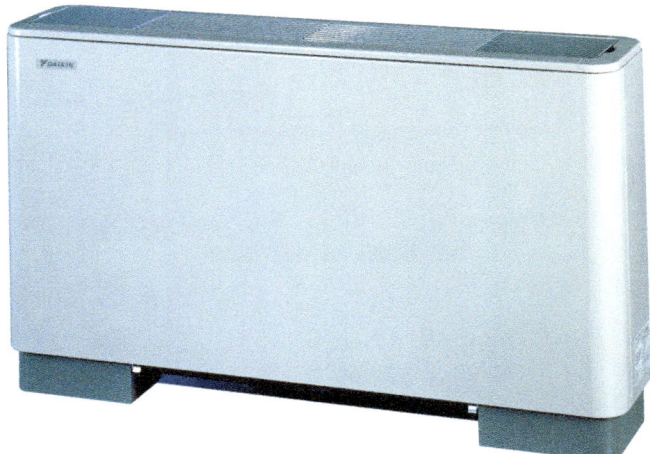

Figure 33 Floor-mounted indoor unit.

On floor-mounted or console air handlers (*Figure 33*), air louvers on top direct airflow into the room. Return air enters and is filtered through the bottom. These units are typically mounted against a wall for a through-the-wall connection to the outdoor unit. Condensate disposal and control options are similar to those of the other types of air handlers.

2.1.3 Multiple Indoor Units

Ductless split systems that allow several air handlers to be attached to a single outdoor unit are also available (*Figure 34*). The different areas where the air handlers are located become, in effect, control zones. Each zone has independent temperature control. The outdoor unit has multiple service valves or other connection points where the refrigerant lines supplying each fan coil are connected. The larger the capacity of the condensing unit, the more fan coils it can likely supply with refrigerant. Again, the metering device(s), typically an electronic expansion valve, is in the outdoor unit.

Due to the zoned nature of multiple configurations, there will be times when the full capacity of the outdoor unit will not be required. For example, if only one zone of a three-zone system requires cooling, the compressor will have excess capacity. Each manufacturer addresses compressor capacity control differently. Solutions may include multiple smaller compressors, two-stage compressors, or variable-speed compressors.

Increased versatility can be realized when different indoor units are needed for different zones. For example, in a four-zone system, the first zone may use a ceiling-mounted unit, the second and

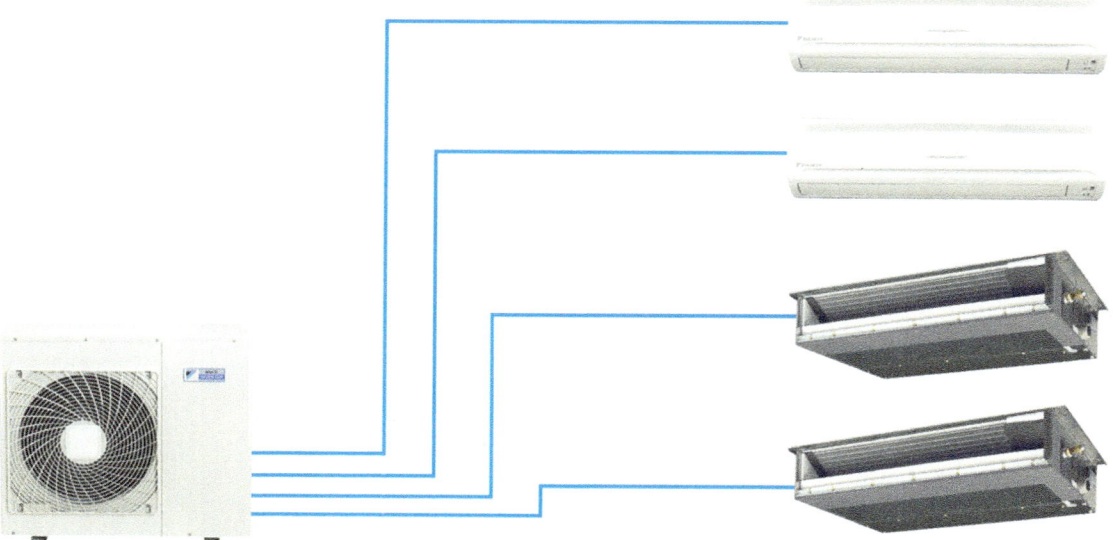

Figure 34 Direct-expansion indoor unit hookup.

third zones may use high sidewall units, and the fourth zone may use an in-ceiling cassette unit. Each indoor unit has its own separate set of refrigerant lines that are routed to the condensing unit.

2.2.0 Ductless Split-System Installation and Troubleshooting

Although some generic installation issues are presented here, always check the manufacturer's installation manual for the exact model being installed. The installation instructions vary from one manufacturer to another and from one model to another. The proper information must be acquired in order to perform troubleshooting and repair tasks successfully as well. In some cases, manufacturers require factory training in their product line. Regardless of whether or not it is required, factory training is essential to effectively install and troubleshoot these unique systems.

2.2.1 Installing Ductless Systems

The installation of ductless split systems is similar to conventional split-system air conditioners and heat pumps. Outdoor units should be installed to allow proper airflow and access for service. Remember that, unlike standard condensing units, the condenser air is discharged horizontally. This allows these units to be installed underneath a structure, such as a deck.

When locating the outdoor unit, installers must ensure that the condenser air intake and discharge areas are not obstructed. As with all units, they should not be located where flammable gases could be present. The potential for snowfall and the prevailing wind direction must also be considered. Due to the shape, size, and weight of ductless condensing units, it is always recommended that they be bolted securely to their foundation. Otherwise, the unit is likely to be blown over by strong winds or quickly removed by vandals. Before securing the unit in its final location, ensure that all clearances required per the installation manual have been addressed. Some heat pump units may require a short section of drain line to be attached to the pan of the outdoor unit. This is best done before the unit is secured.

In many countries, ductless condensing units can be seen mounted on the outside wall of residential high-rise buildings, from top to bottom (*Figure 35*). Servicing the equipment on these buildings is not without some level of risk. However, thousands of buildings like this in Europe and Asia are heated

Figure 35 Ductless condensing units serving a residential building in Shanghai, China.

Figure 36 Wall-mounted condensing unit.

and cooled by ductless systems. Since the condenser air can be drawn in through the sides as well as the back, the unit can be placed close to a wall. This approach can also be used at a residence where heavy snowfall is expected (*Figure 36*).

The refrigerant piping is most often connected to the indoor and outdoor units by flare fittings, although several different connection types may be encountered. Where flare fittings are used, it is common for the installation guidelines to provide a specific torque to be applied to the two different flare nut sizes. Larger systems are more likely to have brazed connections, due to the line size. Joints in piping between the indoor and outdoor units of all systems are commonly brazed.

The installation instructions will provide other refrigerant piping guidelines, such as the following:

- The proper line size
- Maximum length of the lines
- Maximum height of the refrigerant piping
- The amount of refrigerant to be added for each foot of refrigerant piping over a certain length

Both refrigerant lines are usually insulated, as both can produce condensation and suffer from heat transfer to the surrounding environment. Following installation, the refrigerant lines must be properly leak-tested and evacuated before the refrigerant contained in the outdoor unit is released into the system.

Note that some manufacturers prohibit adding a filter-drier. Therefore, any refrigerant piping work must be done using best practices for cleanliness and joint quality. A nitrogen purge is required for all brazing work on the system. Triple-evacuations are a must to ensure that all moisture has been removed.

One reason for prohibiting a filter-dryer on some units is that many ductless and related systems from Asian manufacturers use polyvinylether (PVE) oil instead of polyolester (POE) oil. Both oils are very hygroscopic, taking in water at every opportunity. POE oils, however, tend to create molecular bonds with water that makes it difficult to remove the moisture contamination through an evacuation process alone. PVE oils do not bond with water in the same way, and therefore moisture can be removed. There are some additional benefits to PVE oils, but the fact that water can indeed be removed through evacuation is one of the most significant advantages to technicians.

The installation procedure for the indoor unit depends on the chosen style. The typical wall-mounted indoor unit simply hangs from a metal plate mounted securely to the wall (*Figure 37*). Note the torpedo level across the two tabs, used to level the mounting plate. The unit must be level to avoid condensate drainage problems and to present a neat appearance.

Per the manufacturer's instructions, a hole is bored through the wall to accommodate the piping and wiring. Piping and wiring can also be routed within the wall itself. The hole through the wall is bored at a slight angle down towards

Figure 37 Ductless indoor unit mounting plate.

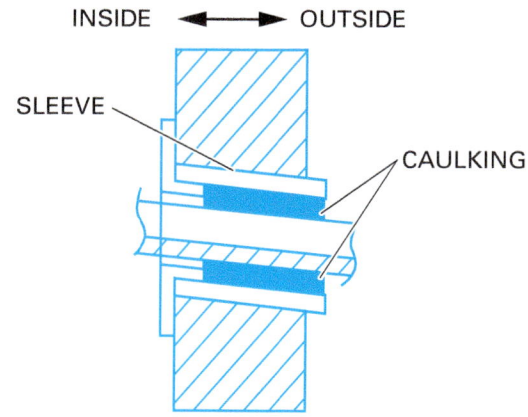

Figure 38 Wall sleeve and opening detail.

the outside (*Figure 38*). This prevents rain from entering the opening and encourages drainage. In addition to a structural sleeve that is often required, such as a piece of SCH 40 or SCH 80 pipe, a plastic sleeve is often used to line the opening. There are two approaches to making refrigerant line connections. One method uses factory-connected flexible refrigerant lines and a drain line on the indoor unit. They are long enough to pass through the wall, where the connections to the lines can be made outdoors with the indoor unit installed on the mounting plate. The installation instructions for most units include options for connections as the unit is approached from different directions, such as right-side, right-back, or right-bottom.

Another approach is to route the refrigerant lines through the wall from the outside and make the connection at the indoor unit. These units are designed to hang from the top of the mounting plate with a spacer placed behind it to hold it away from the wall. Regardless of which approach is used to make refrigerant line connections, condensate drain lines are never trapped.

The wiring from the outdoor unit powers the indoor unit and provides communication between the two pieces. All applicable codes for wiring must be followed. Marked terminal blocks are provided on both units to make the wiring connections straightforward. For a neat appearance, the refrigerant piping and wiring can be concealed in special line set covers. A variety of fittings helps to accommodate almost any situation. The material from some manufacturers can also be painted to blend in with the environment (*Figure 39*).

Unlike conventional systems, most ductless split systems do not use a 24VAC control circuit. Instead, DC communication signals between the outdoor unit and indoor unit(s) control all functions of the system. Such electronic controls are widely used in ductless split systems. These controls may have to be set up or configured during installation

Figure 39 Line set cover.

to ensure that the outdoor unit is communicating with its various air handlers. Configuring the unit may include placing jumpers on the PC board across specific pins or setting a series of DIP switches. Since each manufacturer has their own electronic controls, it is important to carefully read the installation instructions for the initial setup.

The remote control must be set prior to startup. They are battery-operated, usually running on two or three AAA batteries. Once power has been applied to the outdoor unit, the remote control must be programmed for the present day of the week and time of day. The system controlled by the remote shown in *Figure 40* has five modes of operation:

- *Heat* – The unit operates in the heating mode as a heat pump. Most do not have supplementary heat; therefore, when they enter the defrost mode, the indoor blower remains off.
- *Cool* – The unit operates in its normal cooling mode.
- *Dry* – When placed in the Dry mode, the ability to manually control the temperature setpoint and the airflow is locked out. The unit attempts to dehumidify by slowing the fan speed significantly while also attempting to keep the room temperature the same.
- *Auto* – The unit switches automatically between Cool and Heat as needed to maintain the desired settings.
- *Fan* – This mode simply provides air movement without conditioning.

There are other settings that are user-controlled. The terminology and features vary between companies and products. Always follow the manufacturer's operating guidelines for the specific unit. Moving from the top of *Figure 40* clockwise, the features of this remote are described as follows:

- The Home Leave feature allows the user to press a single button to force the system to respond to settings preferred when the space is vacant. Both a temperature and an airflow rate can be preset.
- The Vertical/Horizontal Louver or Swing buttons allow the user to set the position of the discharge blades on the supply air outlet. This affects the direction of the airflow.
- The Intelligent Eye or Sensor button controls an energy-saving feature. The sensor can detect human movement. After a period of time passes without movement being sensed, the unit will change the setpoint to save energy. This particular unit changes the temperature up or down 3.6°F (2°C), depending on the operating mode. The angle of the sensing eye on the face of the unit can be adjusted to face the appropriate direction for movement.
- The Timer On (and Timer Off) buttons, along with the Select button, allow the user to set times of day for the system to start and stop.

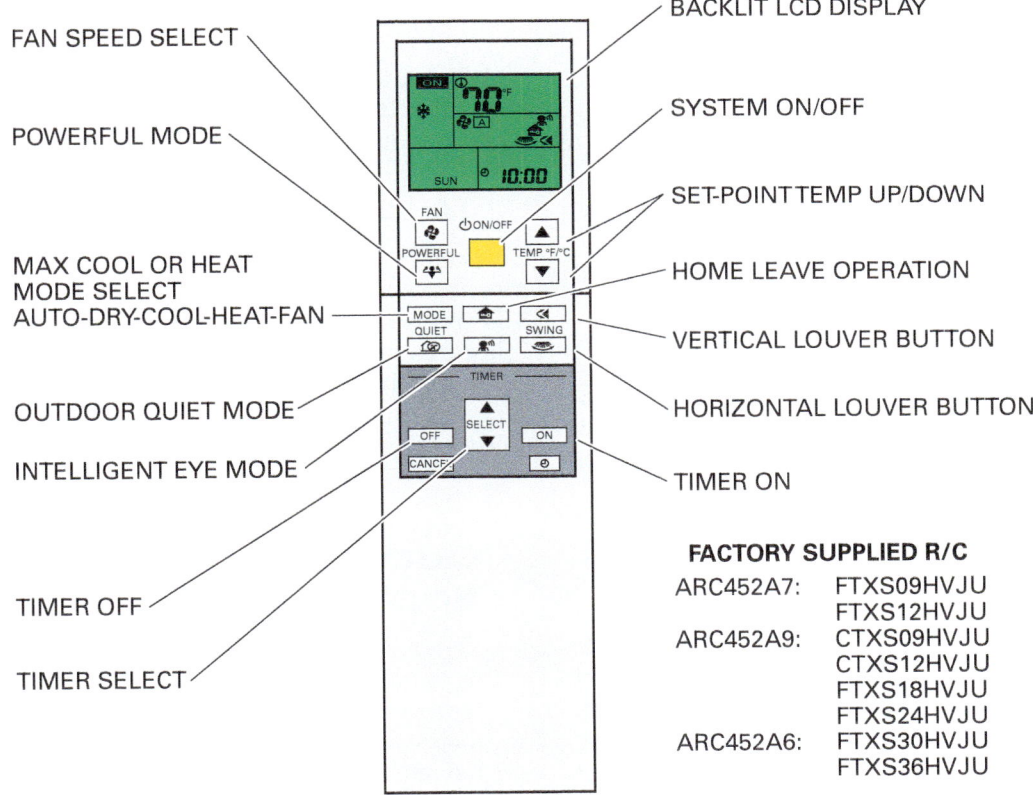

Figure 40 Remote control button layout.

- The Outdoor Quiet button affects the outdoor unit operation. At night or other times when the sound of the outdoor unit may be a nuisance, the button can be used to change the speed and frequency of the condenser fan. This feature is not available in Powerful mode.
- The Powerful setting disregards any previous settings for airflow control and maximizes the cooling capacity for 20 minutes.

Some systems may offer full programming with different settings based on time of day. During startup, all the functions and settings of the remote should be checked and proper operation confirmed. The remote control described here has a Trial mode that is available to the operator when the system is first started. In Trial mode, the system operates on its own for roughly 30 minutes, performing various operational tests. Any problems are reported through fault codes displayed on the remote or at the unit control board outdoors.

Training the homeowner how to use the remote control is essential. Most of the button labels are non-technical in nature and user-friendly. For example, the Quiet button slows the condenser fan motor. The average homeowner may not know what a condenser fan motor is but will easily understand that pushing this button results in quieter operation.

2.2.2 Troubleshooting Ductless Systems

While ductless systems have proven to be very reliable, troubleshooting and repair will likely be required eventually. Troubleshooting the refrigerant circuit is not significantly different from common split systems. However, the metering device is usually located in the outdoor unit. The service valves located at the outdoor unit provide the only refrigerant circuit access. Since the refrigerant lines are virtually inaccessible at the indoor unit, all pressure and temperature readings are taken at the condensing unit. Procedures such as refrigerant charging should be done according to the manufacturer's instructions using commonly accepted service practices.

The controls on ductless systems are significantly different from common systems. Electronic controls often have built-in diagnostics in the form of fault code LEDs. The LEDs flash in a particular sequence or for varying periods of time to indicate the fault that was sensed. Labels on the equipment and the service literature can help the technician decipher any fault codes, and manufacturers provide hotlines for technicians to call if further assistance is required. Understanding the fault codes is essential to the troubleshooting process.

The remote on some systems can also query the system for faults. An astute user can view the trouble codes and report them to the service agency when the service call is placed. *Figure 41* shows a typical list of fault codes for one specific system. As noted at the top of the chart, the fault codes can be viewed by repeatedly pressing a specific key.

Ductless system controls can be very complex. Factory-supported, equipment-specific training provides the best possible preparation for effectively troubleshooting and repairing these systems. Always use the manufacturer's instruction manual for the specific unit being serviced. This literature usually provides troubleshooting charts, such as the one shown in *Figure 42*, that help to point technicians in the right direction. As shown here, the chart references a particular page of the manual for additional information. A chart showing all the possible error codes (*Figure 43*) also guides technicians to a specific page that provides additional instructions to help resolve the problem.

Once an error code is defined and a referenced page number is accessed, a flow chart is often found to guide the technician through the appropriate actions to take. As an example, note the error code EA in *Figure 43*, which refers to a four-way (reversing) valve abnormality. The code description references page 119 of the service manual. A more detailed description of the problem is found on this page (*Figure 44*). Next, a flow chart is provided to help resolve the problem (*Figure 45*). Due to the complexity of the system controls, these troubleshooting aids are indispensable.

2.3.0 Variable Refrigerant Flow Systems

Variable refrigerant flow (VRF) systems represent a completely different approach to zoning. Although they share some similarities with ductless systems, a larger assortment of ducted indoor units is available. In fact, ducted indoor units represent roughly half of the equipment connected to VRF systems worldwide.

Ductless System Sophistication

Here's a clue to the level of sophistication that can be found in ductless systems. The service manual for a relatively standard indoor and outdoor matched combination, portions of which were referenced in this module, is 414 pages in length!

2. **Press the TIMER CANCEL button repeatedly until a continuous beep is produced.**
- The code indication changes as shown below, and notifies with a long beep.

	CODE	MEANING
SYSTEM	00	NORMAL
	U0	REFRIGERANT SHORTAGE
	U2	DROP VOLTAGE OR MAIN CIRCUIT OVERVOLTAGE
	U4	FAILURE OF TRANSMISSION (BETWEEN INDOOR UNIT AND OUTDOOR UNIT)
INDOOR UNIT	A1	INDOOR PCB DEFECTIVENESS
	A5	HIGH PRESSURE CONTROL OR FREEZE-UP PROTECTOR
	A6	FAN MOTOR FAULT
	C4	FAULTY HEAT EXCHANGER TEMPERATURE SENSOR
	C9	FAULTY SUCTION AIR TEMPERATURE SENSOR
OUTDOOR UNIT	EA	COOLING-HEATING SWITCHING ERROR
	E5	OL STARTED
	E6	FAULTY COMPRESSOR START UP
	E7	DC FAN MOTOR FAULT
	E8	OPERATION HALT DUE TO DETECTION OF INPUT OVER CURRENT
	F3	HIGH TEMPERATURE DISCHARGE PIPE CONTROL
	F6	HIGH PRESSURE CONTROL (IN COOLING)
	H6	OPERATION HALT DUE TO FAULTY POSITION DETECTION SENSOR
	H8	CT ABNORMALITY
	H9	FAULTY SUCTION AIR TEMPERATURE SENSOR
	J3	FAULTY DISCHARGE PIPE TEMPERATURE SENSOR
	J6	FAULTY HEAT EXCHANGER TEMPERATURE SENSOR
	L4	HIGH TEMPERATURE AT INVERTER CIRCUIT HEATSINK
	L5	OUTPUT OVERCURRENT
	P4	FAULTY INVERTER CIRCUIT HEATSINK TEMPERATURE SENSOR

NOTE

1. A short beep and two consecutive beeps indicate non-corresponding codes.
2. To cancel the code display, hold the TIMER CANCEL button down for 5 seconds. The code display also cancels itself if the button is not pressed for 1 minute.

Figure 41 Ductless system fault codes.

Like ductless systems, a single VRF outdoor unit can support a number of indoor units. With ductless systems, each indoor unit has an independent set of refrigerant lines that must be connected directly to the outdoor unit. A VRF system uses only one set of refrigerant lines and distributes the refrigerant to each indoor unit through the same circuit (*Figure 46*).

VRF systems control the flow of refrigerant to each connected indoor unit as necessary. The primary task of the outdoor unit is to produce a varying amount of refrigerant flow at a preset subcooling value on demand. Daikin, the industry leader in VRF systems and the original developer of the technology, uses a subcooling value of 9°F (5°C). Rather than using information gathered from the indoor units, the condensing unit modulates the refrigerant flow volume to maintain the 9°F subcooling setpoint.

Instead of using the generic term *variable refrigerant flow*, Daikin has trademarked the term *variable refrigerant volume (VRV™)*. The indoor units are equipped with electronic expansion valves (EEVs) and are designed to maintain a 9°F superheat value when they are in operation. The subcooling setpoint for other systems may vary.

As individual indoor units start and stop, the refrigerant flow in the circuit changes. This change in the flow rate results in a change in the subcooling value of refrigerant leaving the condenser. For example, assume two indoor units at 9,000 Btuh each are in steady operation. A third unit at 12,000 Btuh capacity is then brought online to cool a third area. Proportional-Integral-Derivative (PID) control programmed in the

Metering Devices

Ductless systems place the system metering device, which is usually an EEV, in the outdoor unit. VRF systems use EEVs exclusively, but those responsible for metering refrigerant for the indoor units are located there, not in the outdoor unit. However, EEVs may also be found in the outdoor unit to control the flow of refrigerant within the unit itself for various purposes.

Symptom	Check Item	Details of Measure	Reference Page
The unit does not operate.	Check the power supply.	Check if the rated voltage is supplied.	—
	Check the type of the indoor unit.	Check if the indoor unit type is compatible with the outdoor unit.	—
	Check the outdoor temperature.	Heating operation cannot be used when the outdoor temperature is 24°C (75.2°F) or higher, and cooling operation cannot be used when the outdoor temperature is below 10°C (50°F).	—
	Diagnose with remote controller indication.	—	94
	Check the remote controller addresses.	Check if address settings for the remote controller and indoor unit are correct.	388
Operation sometimes stops.	Check the power supply.	A power failure of 2 to 10 cycles stops air conditioner operation. (Operation lamp OFF)	—
	Check the outdoor temperature.	Heating operation cannot be used when the outdoor temperature is 24°C (75.2°F) or higher, and cooling operation cannot be used when the outdoor temperature is below 10°C (50°F).	—
	Diagnose with remote controller indication.	—	94
The unit operates but does not cool, or does not heat.	Check for wiring and piping errors in the connection between the indoor unit and outdoor unit.	—	—
	Check for thermistor detection errors.	Check if the thermistor is mounted securely.	—
	Check for faulty operation of the electronic expansion valve.	Set the unit to cooling operation, and check the temperature of the liquid pipe to see if the electronic expansion valve works.	—
	Diagnose with remote controller indication.	—	94
	Diagnose by service port pressure and operating current.	Check for refrigerant shortage.	104
Operating noise and vibrations	Check the output voltage of the power module.	—	153
	Check the power module.	—	—
	Check the installation condition.	Check if the required spaces for installation (specified in the installation manual) are provided.	—

Figure 42 Ductless system troubleshooting chart.

	Error Codes	Description		Reference Page
System	00	Normal		—
	U0★	Refrigerant shortage		104
	U2	Low-voltage detection or over-voltage detection		107
	U4	Signal transmission error (between indoor unit and outdoor unit)		109
	UA	Unspecified voltage (between indoor unit and outdoor unit)		112
Indoor Unit	A1	Indoor unit PCB abnormality		95
	A5	Freeze-up protection control or heating peak-cut contro		97
	A6	Fan motor or related abnormality	DC motor (FTXS series)	99
			AC motor (FDXS series)	101
	C4	Indoor heat exchanger thermistor or related abnormality		103
	C9	Room temperature thermistor or related abnormality		103
Outdoor Unit	E1	Outdoor unit PCB abnormality		113
	E5★	OL activation (compressor overload)		115
	E6★	Compressor lock		116
	E7★	DC fan lock		117
	E8	Input overcurrent detection		118
	EA	Four-Way valve abnormality		119
	F3	Discharge pipe temperature control		121
	F6	High pressure control in cooling		123
	H0	Compressor system sensor abnormality		125
	H6	Position sensor abnormality		128
	H8	DC voltage / current sensor abnormality (09/12 class only)		131
		CT or related abnormality (24/30/36 class only)		132
	H9	Outdoor temperature thermistor or related abnormality		134
	J3★	Discharge pipe thermistor or related abnormality		134
	J6	Outdoor heat exchanger thermistor or related abnormality		134
	L3	Electrical box temperature rise		136
	L4	Radiation fin temperature rise		138
	L5★	Output overcurrent detection		140
	P4	Radiation fin thermistor or related abnormality		134
	U7	Signal transmission error on outdoor unit PCB (24/30/36 class only)		111

★: Displayed only when system-down occurs.

Figure 43 Fault code guide.

4.16 FourWay Valve Abnormality

Remote Controller Display	EA
Method of Malfunction Detection	The room temperature thermistor and the indoor heat exchanger thermistor are checked if they function within their normal ranges in each opeartion mode.
Malfunction Decision Conditions	A following condition continues over 1 ~ 10 minutes after operating for 5 ~ 10 minutes. ■ Cooling / Dry (room thermistor temp. – indoor heat exchanger temp.) < –5°C (–9°F) ■ Heating (indoor heat exchanger temp. – room thermistor temp.) < –5°C (–9°F) ■ If the error repeats, the system is shut down. ■ Reset condition: Continuous run for about 60 minutes without any other error
Supposed Causes	■ Disconnection of fourway valve coil ■ Defective fourway valve, coil, or harness ■ Defective outdoor unit PCB ■ Defective thermistor ■ Refrigerant shortage ■ Water mixed in refrigerant ■ Defective stop valve

Figure 44 Problem description.

indoor units allows each to adjust its expansion device based on data gathered from the zone.

As the indoor unit EEV opens to admit refrigerant, the subcooling value at the condensing unit drops. This is because the refrigerant is now moving through the condenser more rapidly, with less time exposed to condenser airflow. The condensing unit responds by increasing the speed of the compressor(s) to move more refrigerant, settling at a speed that again results in a 9°F subcooling value.

The outdoor unit contains temperature and pressure sensors and adjusts compressor speed and other variables to maintain the target superheat and subcooling for the combined load of all attached indoor units. The control programming for the outdoor unit determines compressor speed as well as the ideal subcooling based on the superheat values of the combined load. Subcooling is aided by a subcooling heat exchanger positioned downstream of the condenser coil. If subcooling is lower than desired, the heat exchanger valve is actuated to provide additional cooling.

VRF systems are available in straight cooling, heat pump, and heat-recovery models. Like conventional heat pumps, VRF heat pumps can provide both heating and cooling, but not at the same time. Heat-recovery systems, however, can provide both simultaneously. With heat-recovery VRF systems, while one or more indoor units are operating in the cooling mode, other units can operate in the heating mode. This is done by routing a hot-gas line to each indoor unit, along with the liquid and suction lines. Heat that is absorbed by the system from units in the cooling mode is then transferred to areas that require heating. This feature is unique among heating and cooling systems, other than in specialized swimming pool systems that transfer collected heat to the pool water.

2.3.1 VRF Equipment

VRF condensing units (*Figure 47*) are typically available up to 14 tons in size. They can be connected together by a piping manifold to achieve up to 38 tons of capacity in a single system. Depending on the number of outdoor units coupled together, a single system can support up to 64 indoor units (this may vary by manufacturer).

Units with 8 tons of capacity or more usually have multiple compressors. When two compressors are used, one may be a fixed-speed model, while the other is a variable-speed model powered by an inverter. Both compressors may be variable-speed in other models or brands. The inverter controls the frequency of power to the compressor(s), thereby controlling the speed.

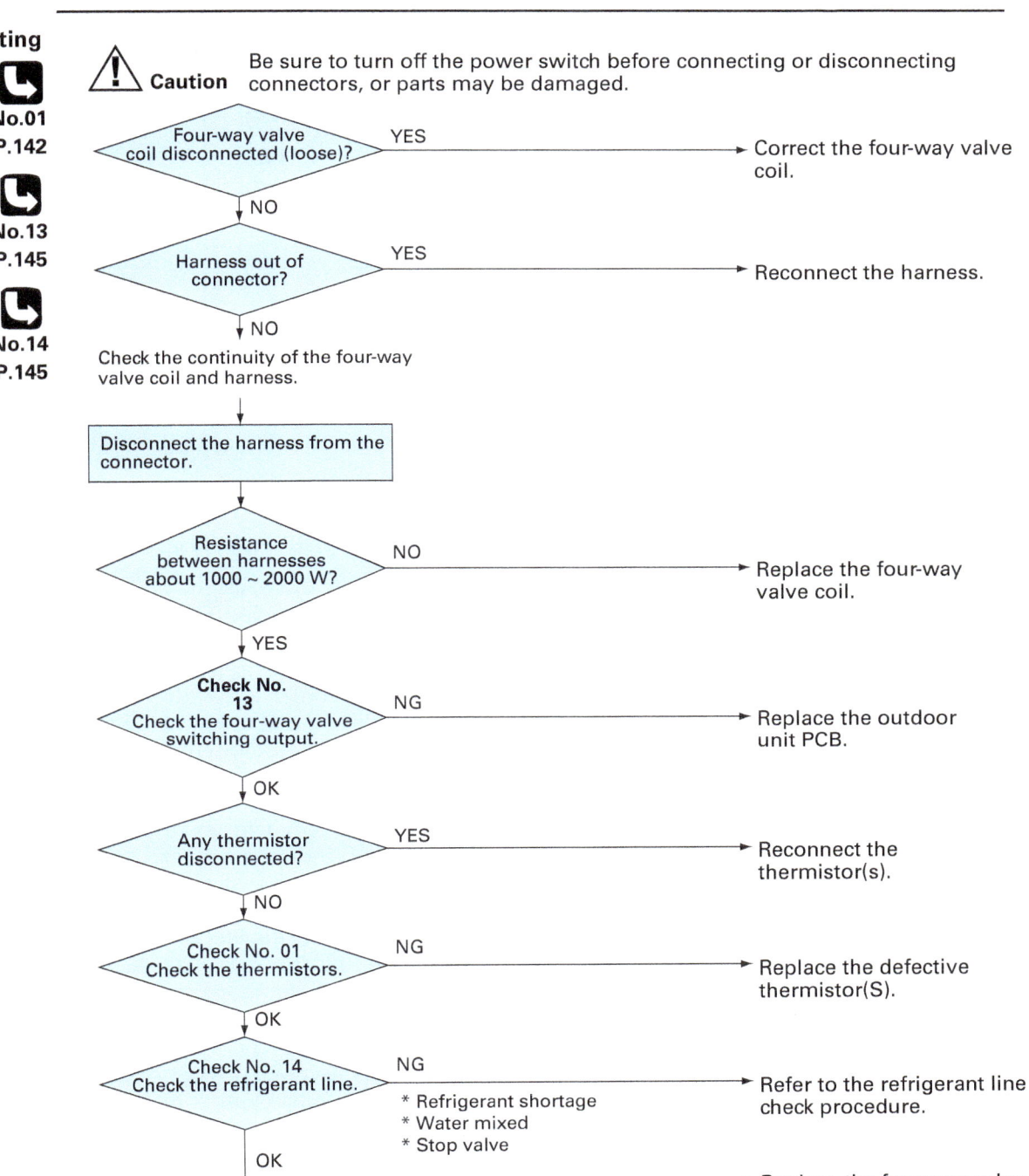

Figure 45 Troubleshooting flow chart.

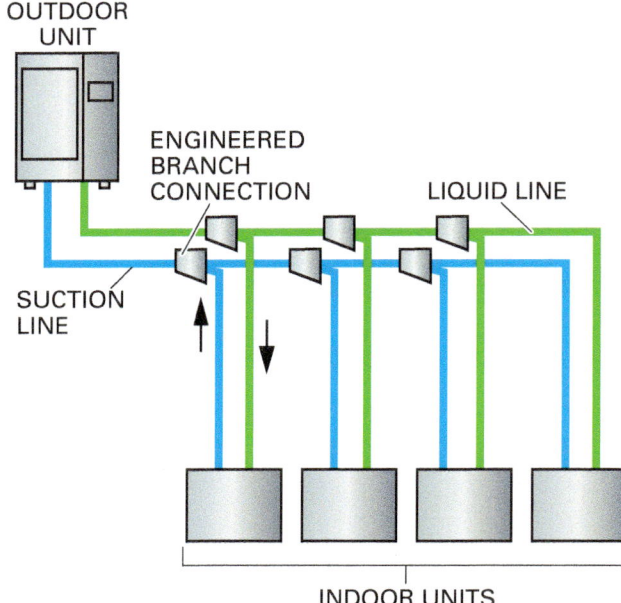

Figure 46 VRF-system refrigerant flow schematic.

Figure 47 VRF condensing unit.

Compressors using this technology can typically reduce their capacity to roughly 15 percent of maximum and operate at virtually any point between 15 and 100 percent.

Other benefits of variable-frequency driven compressors include the following:

- Very low starting currents
- Soft, quiet starts that do not cause lights to flicker due to the starting current
- System pressures that build slowly, reducing stress
- Fewer start and stop cycles, which are stressful to common compressors
- Enhanced ability to dehumidify without overcooling

In a system where the refrigerant flow rate is constantly changing, the return of refrigeration oil to the compressor is a concern. If the refrigerant lines were sized to maintain a proper velocity for oil flow at the lowest possible flow rate, they would be too small to support the flow needed at maximum capacity. Therefore, VRF systems incorporate oil separators at the outdoor unit to minimize the amount of oil in circulation.

In addition, some manufacturers apply what may be referred to as an oil recovery mode. After a predetermined period of operation, the system positions all the EEVs to maximize the flow rate, and then maximizes the speed of the compressor(s). This increases the refrigerant velocity to its maximum point throughout the refrigerant circuit, giving any oil remaining in the circuit an opportunity to move again. Oil recovery cycles typically last less than 10 minutes.

VRF condensing units are typically air-cooled, but water-cooled units are also available. Like commercial water-cooled heat pump systems, a means of heat rejection such as a cooling tower or dry cooler is required. In addition, a boiler or similar source of heat is likely needed to add heat to the loop in the winter. The two parts of the system work to maintain the loop water temperature in the appropriate range—such as 67°F to 95°F (19°C to 35°C), for example.

VRF indoor units are available in a variety of configurations. Models that are ductless and look like other ductless units are quite popular. Ducted units are available as well. Indoor units that are virtually the same as traditional fan-coil units used with standard split systems are available, as are units designed to be concealed in small spaces (*Figure 48*). Some units can be fitted with electric heat if necessary. Most ducted models can accommodate outside-air connections for improved IAQ.

VRF-system indoor and outdoor units must constantly communicate with each other. Most use a DC control protocol that facilitates communication between the units. Information is transmitted to identify mode of operation, calls for function, part-load operation, and malfunctions. Due

Figure 48 VRF fan-coil unit designed for concealment.

to the collective nature of the programming, data is available for monitoring system operation and identifying malfunctions in all connected components. Most VRF manufacturers offer a data-interface product that allows the technician to connect and analyze system operation and performance and also troubleshoot problems. Tablets and laptops are common ways to access the data.

If a specific indoor unit is having problems, you can see the readings from each thermistor, the calculated superheat, the precise position of the EEV, and information related to the blower. Again, the indoor and outdoor units share a great deal of information, but their function is primarily based on information collected within their own cabinet.

The VRF-system indoor units can be controlled several different ways. One way is through a wireless remote, much like those used with ductless systems. A simple wall-mounted, wired controller is also available. Depending on the manufacturer, even the simple control devices for VRF systems can control up to 16 indoor units from a single device. The wired control shown in *Figure 49* is the Navigation Remote Controller™, a trademarked device from Daikin. This controller provides advanced scheduling in addition to controlling up to 16 units.

2.4.0 VRF System Installation and Troubleshooting

VRF systems have unique refrigerant piping requirements compared to common systems. Failing to follow the specific installation guidelines for the piping will undoubtedly result in functional problems. Sophisticated piping systems present many opportunities for errors and poorly constructed joints. The quality of the overall installation is of the utmost importance to the operation and longevity of the system.

Twinned/Linked VRF Outdoor Units

Manufacturers are beginning to introduce additional options to the VRF market. To increase the overall capacity for a system, two and possibly more VRF outdoor units can be connected to a single refrigerant circuit serving many indoor units. The advantages to twinning or linking VRF condensing units on one circuit include the following:

- Reduced piping, which also results in less refrigerant volume required
- Higher efficiencies at partial loads due to adding additional heat-transfer surfaces
- Improved heat recovery (when used) from having a wider variety of possible sources on the circuit

The proper design of the refrigerant piping system is critical. The manufacturer must be included in any design process involving multiple outdoor units connected to a single refrigerant circuit.

The interconnecting communications wiring must also be done with great care and attention to the details provided by the system manufacturer. Due to the level of sophistication of the controls, troubleshooting can often be done with very few tools in hand; the system itself is able to provide much of the information needed to make an informed repair decision.

> **NOTE**
> Manufacturers of VRF and similar systems often require that the installers as well as the technicians servicing the systems be certified through a corporate training program. This is an important warranty consideration, as the manufacturer may deny warranty claims on systems installed and serviced by technicians who are not certified on that specific product line.

2.4.1 VRF System Installation

System designers often work with equipment manufacturers in the design process by providing the building plans along with a preferred equipment layout that suits the air distribution needs. The manufacturer then provides a piping and control layout, identifying the proper joints and other components that will ensure optimal performance.

Engaging the manufacturer in the process helps to shift some of the contractor's burden for

Figure 49 Navigation Remote Controller™.

performance to the manufacturer. If the manufacturer's guidance and plans are carefully followed, fewer problems are likely to occur; and those that may occur can be resolved in partnership with the manufacturer. VRF system installations must be planned with great care and the proper knowledge. As one industry professional puts it, "You cannot properly design a VRF system on the tailgate of a pickup truck."

The unique installation requirements for VRF systems are generally associated with the refrigerant piping and control wiring. *Figure 50* provides an overview of the refrigerant piping and control wiring for a typical VRF system.

The same refrigerant circuit serves all the indoor units. This requires careful management of the refrigerant flow and great attention to detail during the installation process. Special fittings and piping assemblies are engineered for VRF systems and must be utilized. These special piping accessories are available through the product manufacturer.

All refrigerant lines, including the hot-gas line used for systems capable of heat recovery and simultaneous heating and cooling, must be insulated individually. The insulated lines are sometimes wrapped together as a group with a special tape, with communication wiring also included in the bundle.

Refrigerant filter-driers are not recommended during the installation. Although a nitrogen purge during brazing is important in every installation, it is even more important with VRF systems. Since these systems have multiple screens, check valves, capillary tubes, and orifices, solid contaminants must be avoided. It is also important that traps in the refrigerant line be avoided. Remember that the refrigerant flow rate may sometimes be less than ideal under light loads, which hinders oil from returning to the compressor. Traps will further hinder oil return.

Engineered branch-line connections and headers (*Figure 51*) are either made, recommended, or approved by the manufacturer. They must be used to make primary connections where directed. These components are often referred to as **refnetjoints**. The refnet (refrigerant network) joints include a special distributor to help ensure the balanced distribution of refrigerant in both directions.

Note the stepped-down construction of the tubing for the connections. The components have been designed so that a single style of refnet joint can be used with multiple pipe sizes. The installer simply uses the tube size required for the installation, cutting off any portion of the connection that is smaller than the desired size. This reduces the required inventory of installation parts. Manufacturers also provide preformed insulation that fits these special piping components properly; field-insulating these parts properly would be challenging at best.

The refnet joints must be installed according to the manufacturer's instructions. The system designer must size the piping as specified by

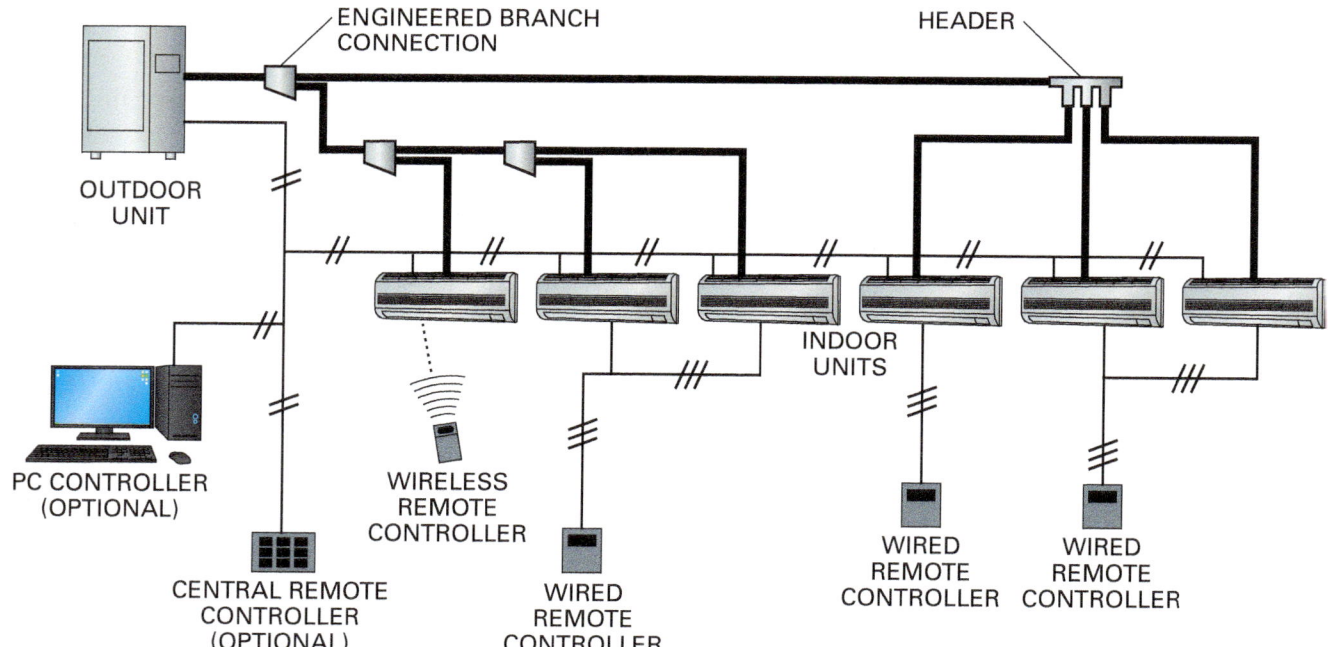

NOTE: ONLY ONE OF TWO REFRIGERANT LINES SHOWN.

Figure 50 VRF-system piping and wiring overview.

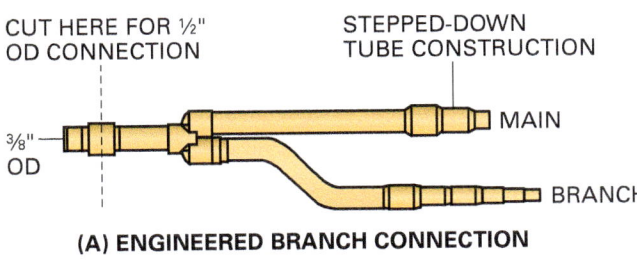

(A) ENGINEERED BRANCH CONNECTION

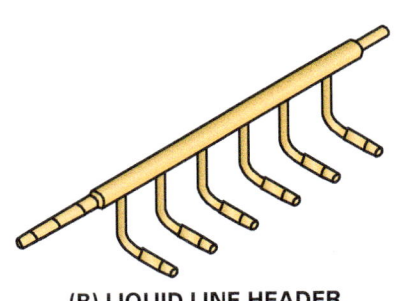

(B) LIQUID LINE HEADER

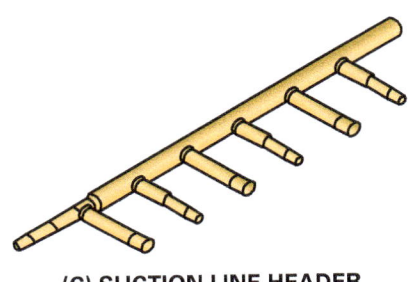

(C) SUCTION LINE HEADER

Figure 51 Refnet joints and headers.

the manufacturer and choose the location of the joints wisely. When installing refnet joints and headers, consider the following:

- The refnet joint can be oriented straight up, straight down, or horizontally with the main and branch lines side-by-side and level with each other. It should never be installed in a manner that places the main line and branch line above or below each other.
- Headers must always be installed so that the branch lines leaving it are horizontal, never vertical. Remember that the liquid and suction headers are constructed differently. Both headers are shown properly oriented for installation in *Figure 51*. Unused ports on a header must be closed by brazing or soldering.
- A distance of 20" (51 cm) is generally required after an elbow or similar fitting before a refnet joint is placed. The next refnet joint must then be at least 40" (102 cm) downstream. Refer to *Figure 52*.

Manufacturers also have requirements for the size and length of each segment of the refrigerant network. The requirements will differ from one brand to another, and possibly from one model series to another. It is essential to follow the manufacturer's specifications for the particular system being installed. In most cases, installation technicians are not responsible for sizing or

Special Tool Kits

Since ductless and VRF system manufacturers specify torque values for each size of flare nut on their systems, special tool kits have been developed for these installations. Tool manufacturers and equipment manufacturers alike offer kits specifically to support the installation of these systems. The kits typically include a refrigerant gauge manifold with hoses; flaring tools; gauges to check the quality of a formed flare; a torque wrench with various sizes of open-end attachments to fit the flare nuts; metric and/or imperial hex wrenches; and various brass adapters and valve core tools.

These systems, primarily from Asian manufacturers, use PVE oils. Therefore, it is preferred that a refrigerant gauge manifold set be used exclusively on their equipment to prevent cross-contamination of oil charges.

Figure Credit: Courtesy of CPS Products

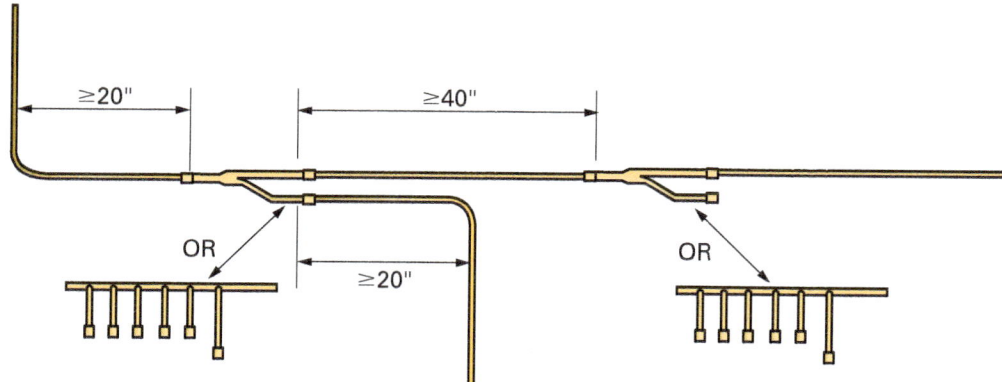

Figure 52 Refnet joint positioning.

designing the layout of the refrigerant piping, but they should understand the requirements behind the design.

The piping layout may include expansion loops or flexible sections of refrigerant line in key locations that may be subjected to significant levels of expansion and contraction. Otherwise, the piping may be damaged by these powerful forces.

For most VRF systems, maximum and/or minimum values will be provided for the following portions of the refrigerant network:

- The maximum distance from the first refnet joint to the farthest indoor unit.
- The maximum distance from the outdoor unit to the nearest refnet joint.
- The maximum distance from a refnet joint to the indoor unit it serves.
- The minimum distance between refnet joints.
- The maximum difference in length between the longest and shortest run connecting refnet joints to an indoor unit.
- The maximum difference in height from top to bottom

The refrigerant piping between condensing units that are connected to a single circuit requires great attention to detail. If multiple units are connected improperly, oil return problems as well as functional difficulties will likely result.

For systems that provide heat recovery, which allows the system to provide simultaneous heating and cooling, an additional device is required. A branch selector box (*Figure 53*) or similar apparatus is required to manage the flow of refrigerant, facilitating the changeover from one mode to the other. A heat-recovery system has three refrigerant lines, including the necessary hot-gas line. However, indoor units have only two line connections—one for liquid and one for vapor.

The branch selector box manages the refrigerant flow as needed between the outdoor and indoor units. To do so, it must be made a part of the communications network. While branch selector boxes often support a single indoor unit, models are available that support up to twelve indoor units. This style can take the place of refnet headers in the refrigerant piping, taking the three refrigerant lines from the outdoor systems and distributing refrigerant to multiple indoor units from a single enclosure.

The communication and control wiring between the indoor units, outdoor unit(s), and

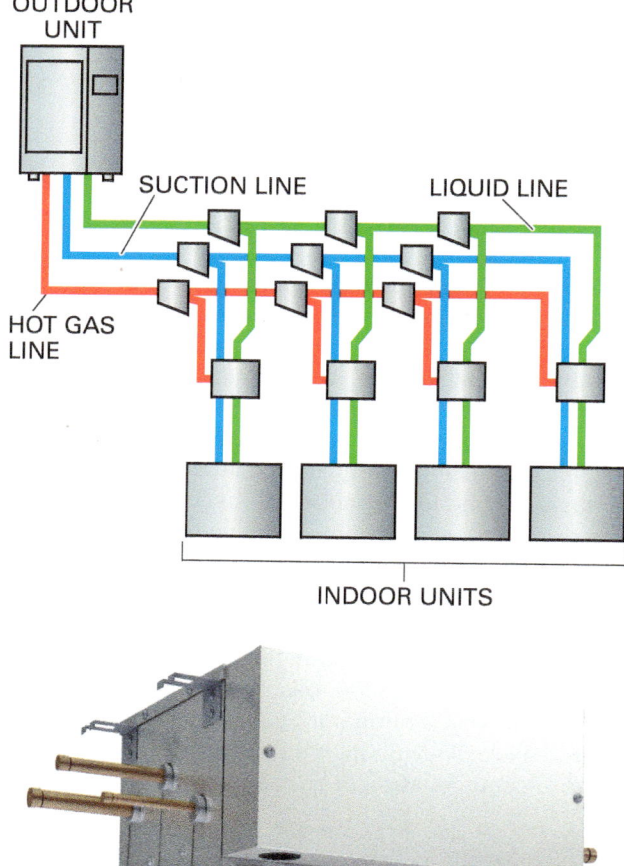

Figure 53 Branch selector boxes.

wired controllers is also critical to proper system operation. Incorrect wiring methods will result in lost communications. Terminals are provided on each unit for the interconnecting wiring. *Figure 54* shows the terminal designations for one line of equipment. Note that the communication wiring terminals are designated as transmission terminals by this manufacturer.

The terminals designated Forced Off are used to interlock the equipment with safety devices, such as the float switch of a condensate pump. On the outdoor unit, the terminals marked Q1 and Q2 are used when multiple condensing units are connected together on the same system. The outdoor units would communicate through this circuit. The P1 and P2 terminals are used to connect a wired controller. As the wiring travels between one unit and another, the polarity must remain the same; the same wire must be connected to all F1 terminals throughout the system, for example. Remember that these terminal designations can differ from manufacturer to manufacturer.

The wire typically used for communications is two-conductor 18 AWG stranded, non-shielded wire. If the length is excessive, 16 AWG may be required. Always follow the manufacturer's requirements carefully, as the requirements can differ. Three conductors may be required for some applications. This type of wire rarely needs to be in conduit, so it is often routed through ceiling cavities by following other runs of piping or building structural components. However, to avoid interference, the communications wiring should never be strapped along the outside of conduit carrying AC-powered wiring.

Shielded wire is not usually required. If it is used, the shield must be grounded at every component to which the conductors are connected. The wire should not be spliced between components; a continuous piece of wire should be routed from component to component. Follow the manufacturer's requirements carefully.

Communication wiring is routed in a daisy-chain configuration (*Figure 55*). Star connection wiring configurations should not be used. All terminations and connections occur at the equipment. There should be no connections or junctions in the wiring elsewhere. Although other strategies may be fine for distributing power, they do not allow for proper communication signals in this type of system.

Once the installation of a system is complete, the system must be charged with refrigerant. Most standard HVAC units of 5 tons capacity or less come with a sufficient amount of refrigerant charge to handle a specific length of piping. When charging VRF systems, the entire piping system

(A) INDOOR UNIT TERMINALS

(B) BRANCH SELECTOR BOX TERMINALS

(C) OUTDOOR UNIT TERMINALS

Figure 54 VRF-system communication wiring terminals.

Thermostat? What Thermostat?

Throughout this section, the word *thermostat* has seldom been used. With standard heating and cooling systems, a wall thermostat is typically in control of the system. However, with ductless and VRF systems, the device used to control the systems is referred to as a remote. We tend to think of remote controls as portable devices, like the ones used to control TVs. However, with these systems even wall-mounted devices are referred to as remote controls, or simply remotes.

This term is actually more accurate than "thermostat", as the remote control for ductless and VRF systems does not monitor room temperature. Room temperature is usually monitored by a thermistor at the return air inlet of the indoor unit(s). The remote is used to change settings and acquire information, but it does not stop and start the unit in response to room temperature.

Figure 55 Daisy-chain wiring.

all the way back to the condensing unit must be considered. The required weight of refrigerant is calculated based on the length of the lines and system components, and then weighed into the system. Each foot of liquid line of a given size can hold a specific volume of refrigerant. Therefore, the installed length and internal volume for each line size must be calculated. Each specific indoor unit model and added piping component requires a certain amount of refrigerant to be added.

Most manufacturers require charging to be done exclusively by weight. When the charge becomes incorrect due to leaks or other issues, the remaining charge must be recovered and the system recharged from scratch by weight. Attempting to charge a system using traditional superheat and subcooling methods is not effective. This significant task is yet another reason why the quality of piping and joints must be at its highest for VRF systems.

2.4.2 VRF System Troubleshooting

Proper training is essential in order to successfully troubleshoot VRF systems. Although they may seem complex in many ways, the sophisticated controls and communications can actually do most of the troubleshooting for you. With standard systems, technicians often reach for the gauge manifold first in every troubleshooting situation. With VRF systems, troubleshooting should never begin with connecting a set of gauges.

Troubleshooting some common problems in VRF systems can begin with the symptoms at hand. Charts and tables that provide guidance based on symptoms (*Figure 56*) are offered by many manufacturers. In some cases, the suggested solution may seem overly simplified, but many problems are caused by simple errors or a misunderstanding of how the system operates.

For more complex problems or those that do not have a simple solution (such as moving the curtains out of the way of the indoor unit discharge), troubleshooting often begins at the remote control. This is where the real value of the communications network shines through. In many VRF systems, the operating parameters of each piece of equipment can be viewed from a single point, regardless of how large or expansive the system may be.

Each manufacturer likely has a different approach to accessing the needed information from a remote control. This is done by using menus to access test functions where error codes related to the problem will be displayed, or by using a

HFC-410A Flare Joints

Due to the higher pressures associated with HFC-410A, the requirements for forming reliable flares are slightly different. The outer diameter of the flare should generally be larger, and the flare deeper. With most flaring tools, the user must take care to set the tube in the block at the proper depth. The 45-degree flaring tool shown here takes the guesswork out of the task, automatically setting the tube at the appropriate depth for HFC-410A flares.

For example, the SAE specification for the diameter of a ⅜" flare cone is 0.488". The diameter specified by most ductless-system manufacturers for their HFC-410A equipment is 0.520". This tool from Yellow Jacket®, a division of Ritchie Engineering, creates flares to that specification. Ductless system manufacturers may also offer flaring tools that are specifically designed to create flares meeting their specifications.

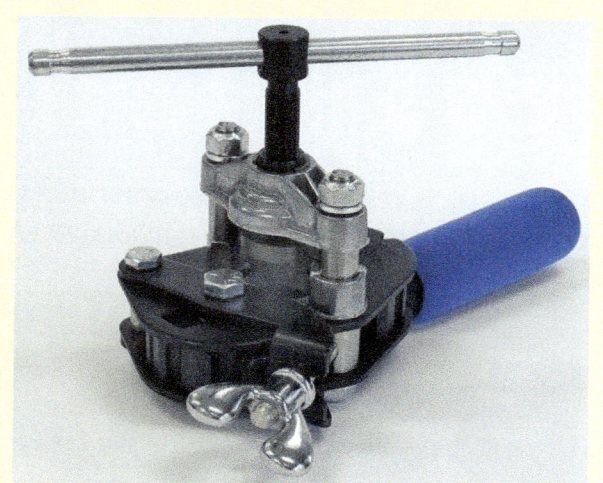

Figure Credit: Courtesy of Ritchie Engineering Company, Inc., YELLOW JACKET Products Division

	Symptom		Supposed Cause	Countermeasure
1	The system does not start operation at all.		Blowout of fuse(s)	Turn Off the power supply and then replace the fuse(s).
			Cutout of breaker(s)	• If the knob of any breaker is in its OFF position, turn ON the power supply. • If the knob of any circuit breaker is in its tripped position, do not turn ON the power supply. 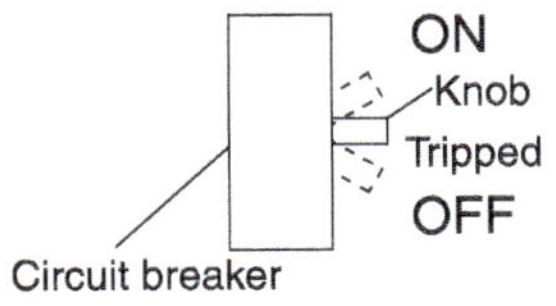
			Power failure	After the power failure is reset, restart the system.
2	The system starts operation but makes an immediate stop.		Blocked air inlet or outlet of indoor or outdoor unit	Remove obstacle(s).
			Clogged air filter(s)	Clean the air filter(s).
3	The system does not cool or heat air well.		Blocked air inlet or outlet of indoor or outdoor unit	Remove obstacle(s).
			Clogged air filter(s)	Clean the air filter(s).
			Enclosed outdoor unit(s)	Remove the enclosure.
			Improper set temperature	Set the temperature to a proper degree.
			Airflow rate set to **LOW**	Set it to a proper airflow rate.
			Improper direction of air diffusion	Set it to a proper direction.
			Open window(s) or door(s)	Shut it tightly.
		[In cooling]	Direct sunlight received	Hang curtains or shades on windows.
		[In cooling]	Too many persons staying in a room	
		[In cooling]	Too many heat sources (e.g. OA equipment) located in a room	
4	The system does not operate.	The system stops and immediately restarts operation.	If the **OPERATION** lamp on the remote controller turns ON, the system will be normal. These symptoms indicate that the system is controlled so as not to put unreasonable loads on the system.	Normal operation. The system will automatically start operation after a lapse of five minutes.
		Pressing the **TEMP ADJUST** button immediately resets the system.		
		The remote controller displays **UNDER CENTRALIZED CONTROL**, which blinks for a period of several seconds when the **OPERATION** button is depressed.	The system is controlled with centralized controller. Blinking display indicates that the system cannot be operated using the remote controller.	Operate the system using the **COOL/HEAT** centralized remote controller.
		The system stops immediately after turning **ON** the power supply.	The system is in preparation mode of micro computer operation.	Wait for a period of approximately one minute.
5	The system... displays		The system stops due to an interruption in communication caused by electrical...	Remove causes of electrical noises. If these causes are removed, the system will automatically restart.

Figure 56 Partial symptom-based troubleshooting guide.

unique sequence of key-presses to access the desired information. In some cases, error codes may be displayed on the user's main display. Unique methods of accessing service and maintenance menus provide a layer of protection. Many settings are not understood by users but can lead to serious problems if they are casually or accidently changed.

The first menus encountered can provide operating and maintenance information for one or more units on the network (*Figure 57*). All communicating components in the network have a unique address, allowing the technician to access the information for a specific unit. Problems that affect the entire system are likely related to the outdoor unit or the refrigeration cycle. When only one area experiences a performance issue, the indoor unit serving that area is likely where the problem lies. In most systems, the information for a specific piece of equipment can be accessed from any single remote. When a unit number or address is selected, the information for that unit is displayed. As shown here, even the operating hours for each piece of equipment can be viewed.

Some distinction is usually made between problems that cause system shutdown and those that merely represent a reason for concern. The latter are usually referred to as warnings. In this case, a light on the remote and/or indoor unit may remain on, but a warning message is displayed on the remote. For problems that cause system or unit shutdown, this same light may blink rapidly and an error code will be displayed on the remote. Problem indicators will be different from system to system.

Once the correct menu is accessed, one or more error codes will be displayed. Each individual code should be carefully considered. In some cases, one problem causes other problems that may result in additional error codes. For example, on the error code chart shown in *Figure 58*, there are several error codes that are related to the refrigerant condensing section. Error code E3 indicates the activation of a high-pressure switch; code E7 is related to a malfunction of the condenser fan motor; and code H3 is related to the failure of the high-pressure switch. Only one problem may truly exist, but it may cause more than one error code to be displayed.

Once the error code is determined, the technician is guided to specific pages in the service manual that provide more detailed information. An explanation of the actual circumstance that caused the error code to be generated is usually found here. A flow chart to guide the technician's troubleshooting activities is often included as well.

The earlier example of error code E7 will be used here to demonstrate the use of the troubleshooting guide. Code E7 refers to a malfunction of the outdoor unit fan motor. The technician is directed to page 138 of the service manual where an explanation of the error code is provided (shown in *Figure 59*). The text explains that this malfunction is detected when there are abnormal currents from the main inverter or the fan motor inverter or when there is a lower-than-expected fan rpm.

Maintenance Menu	Item 2	Remarks	
1. Model Name	1. Unit No.	Select the Unit No. you want to check.	
	2. Indoor unit		
	3. Outdoor unit		
2. Operating Hours	1. Unit No.	Select the Unit No. you want to check.	
	2. Indoor unit operating time	All of these are displayed in hours.	
	3. Indoor fan operation		
	4. Indoor unit energized time		
	5. Outdoor operating time		
	6. Outdoor fan 1 operation		
	7. Outdoor fan 2 operation		
	8. Outdoor comp. 1 operation		
	9. Outdoor comp. 2 operation		
3. Indoor Unit Status	1. Unit No.	Select the Unit No. you want to check.	
	2. FAN	Fan tap	
	3. FLAP	Swing, fixed	
	4. Speed	Fan speed (rpm)	
	5. EV	Degree that electronic expansion valve is open (pls)	
	6. MP	Drain pump ON/OFF	
	7. EH	Electric heater ON/OFF	
	8. Hu	Humidifier ON/OFF	
	9. TBF	Anti-freezing control ON/OFF	
	10. FLOAT		
	11. T1/T2		
	12. Unit No.	Select the Unit No. you want to check.	
		SkyAir	VRV
	13. Th1	Suction air thermistor	Suction air thermistor
	14. Th2	Heat exchanger thermistor	Heat exchanger liquid pipe thermistor
	15. Th3	—	Heat exchanger gas pipe thermistor
	16. Th4	Discharge air thermistor	Discharge air thermistor
	17. Th5	—	—
	18. Th6	—	—
4. Outdoor Unit Status	1. Unit No.	Select the Unit No. you want to check.	
	2. FAN step	Fan tap	
	3. COMP	Compressor power supply frequency (Hz)	
	4. EV1	Degree that electronic expansion valve is open (pls)	
	5. SV1	Solenoid valve ON/OFF	
		SkyAir	VRV
	6. Th1	Outdoor air thermistor	—
	7. Th2	Heat exchanger thermistor	—
	8. Th3	Discharge pipe thermistor	—
	9. Th4	Heat exchanger deicer thermistor	—
	10. Th5	Heat exchanger gas pipe thermistor	—
	11. Th5	Liquid pipe thermistor	—
5. Forced Defrost (SkyAir only)	1. Forced defrost ON	Enables the forced defrost operation.	
	2. Forced defrost OFF	Disables the forced defrost operation.	

Figure 57 Partial remote control maintenance menu.

 03315 Zoning, Ductless, and Variable Refrigerant Flow Systems   Module Eight   49

	Malfunction code	Operation lamp	Malfunction contents	Page Referred
Indoor Unit	A0	l	Error of External Protection Device	109
	A1	l	PCB Defect	110
	A3	l	Malfunction of Drain Level Control System (S1L)	111
	A6	l	Fan Motor (M1F) Lock, Overload	113 115 116
	A7	k	Malfunction of Swing Flap Motor (M1S)	119
	A9	l	Malfunction of Moving Part of Electronic Expansion Valve (Y1E)	121
	AF	k	Drain Level above Limit	123
	AJ	l	Malfunction of Capacity Determination Device	124
	C4	l	Malfunction of Thermistor (R2T) for Heat Exchanger	125
	C5	l	Malfunction of Thermistor (R3T) for Gas Pipes	126
	C9	l	Malfunction of Thermistor (R1T) for Suction Air	127
	CJ	k	Malfunction of Thermostat Sensor in Remote Controller	128
Outdoor Unit	E1	l	PCB Defect	129
	E3	l	Actuation of High Pressure Switch	130
	E4	l	Actuation of Low Pressure Sensor	133
	E5	l	Inverter Compressor Motor Lock	135
	E6	l	STD Compressor Motor Overcurrent/Lock	—
	E7	l	Malfunction of Outdoor Unit Fan Motor	138
	E9	l	Malfunction of Moving Part of Electronic Expansion Valve (Y1E, Y2E)	141
	F3	l	Abnormal Discharge Pipe Temperature	143
	F6	l	Refrigerant Overcharged	145
	H3	k	Failure of High Pressure Switch	—
	H4	l	Actuation of Low Pressure Switch	—
	H7	l	Abnormal Outdoor Fan Motor Signal	—
	H9	l	Malfunction of Thermistor (R1T) for Outdoor Air	146
	J2	l	Current Sensor Malfunction	—
	J3	l	Malfunction of Discharge Pipe Thermistor (R2T)	147
	J5	l	Malfunction of Thermistor (R3T, R5T) for Suction Pipe	148
	J6	l	Malfunction of Thermistor (R4T) for Outdoor Unit Heat Exchanger	149
	J7	l	Malfunction of Liquid Pipe Thermistor (R7T)	150
	J9	l	Malfunction of Subcooling Heat Exchanger Gas Pipe Thermistor (R6T)	151
	JA	l	Malfunction of High Pressure Sensor	152
	JC	l	Malfunction of Low Pressure Sensor	154
	L0	l	Inverter System Error	—
	L1	l	Malfunction of PCB	157
	L4	l	Malfunction of Inverter Radiating Fin Temperature Rise	159
	L5	l	Inverter Compressor Abnormal	162
	L8	l	Inverter Current Abnormal	164
			Start up Error	166

Figure 58 Example error code chart.

3.17 "E7" Malfunction of Outdoor Unit Fan Motor

Remote Controller Display	E7
Applicable Models	RXYMQ36 · 48PVJU
Method of Malfunction Detection	Detect a malfunction based on the current value in the inverter PCB (as for motor 2, current value in the fan PCB). Detect a malfunction for the fan motor circuit based on the number of rotation detected by hole IC during the fan motor operation.
Malfunction Decision Conditions	■ Overcurrent is detected for inverter PCB (A2P) or fan inverter PCB (A3P) (System down is caused by 4 times of detection.) ■ In the condition of fan motor rotation, the number of rotation is below the fixed number for more than 6 seconds. (System down is caused by 4 times of detection.)
Supposed Causes	■ Failure of fan motor ■ Defect or connection error of the connectors/ harness between the fan motor and PCB ■ The fan cannot rotate due to any foreign substances entangled. ■ Clear condition: Continue normal operation for 5 minutes

Figure 59 Error code E7 explained.

Notice that if this malfunction is detected four times, it will cause unit shutdown. Otherwise, only a warning message will be displayed on the remote and the unit will continue to operate. Possible causes of the problem listed here include a fan motor failure; a defect in the wiring or connections to the motor; or a foreign object interfering with the rotation of the motor. The final note indicates that the warning message will go away if normal operation is detected for a full five minutes.

A flow chart is provided on the next page of the manual to further guide the technician (*Figure 60A* and *Figure 60B*). Specific steps to follow are shown, with a resolution for each step based on the result. In some cases, the technician may be guided to another page for a procedure. For example, the final step in the flow chart, tagged as Check 6, directs the technician to page 215 for guidance on checking the fan motor connector. The procedure from that page is shown in *Figure 61*. The technician is instructed to use a meter to measure the resistance from terminal Vcc to the other three wires, and to check the resistance from the Ground wire to the other three terminals. Returning to the flow chart, the technician reads the statement, answers yes or no based on the test results, and then completes the repair as prescribed.

Caution Be sure to turn off the power switch before connecting or disconnecting connector, or parts may be damaged.

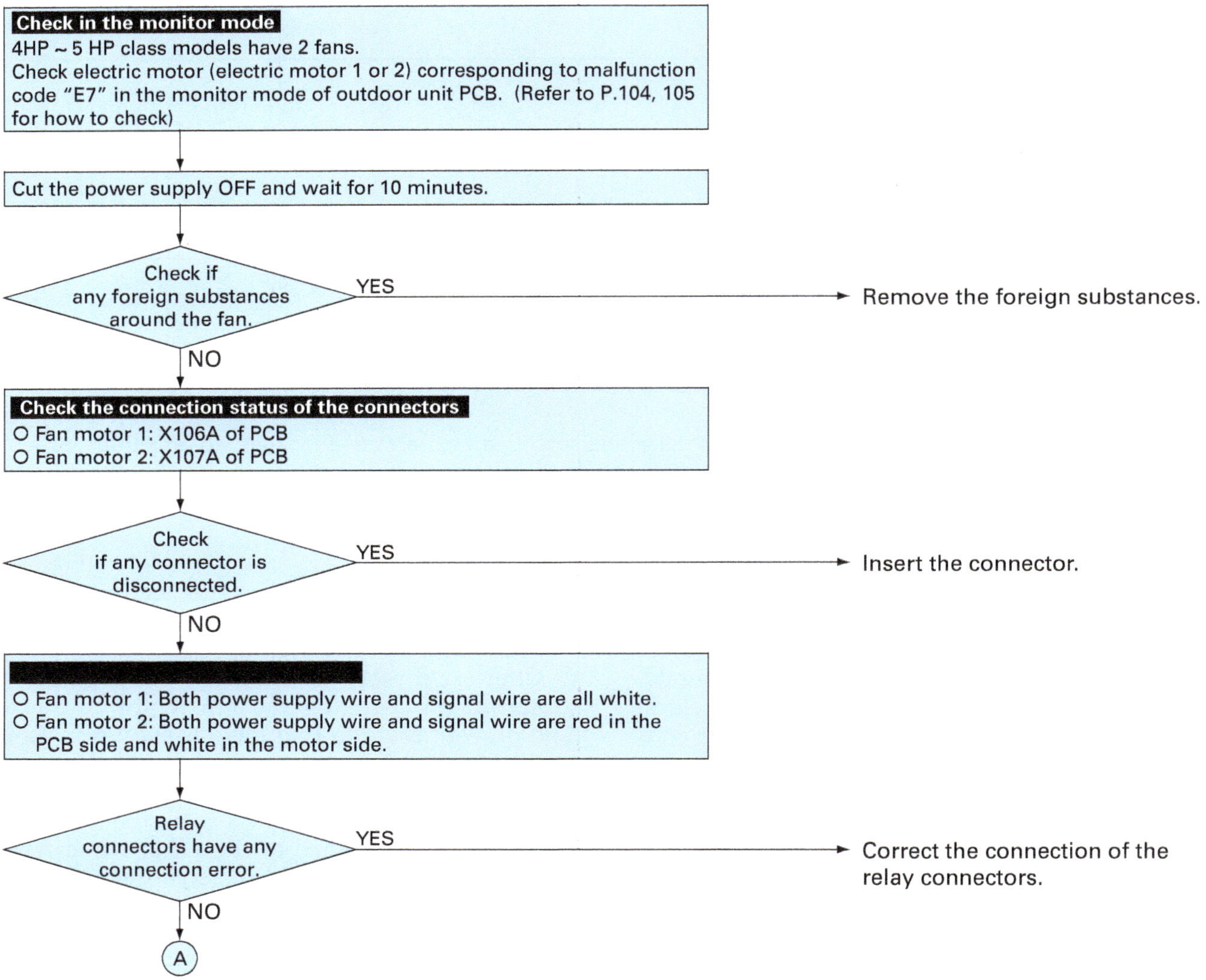

Check in the monitor mode
4HP ~ 5 HP class models have 2 fans.
Check electric motor (electric motor 1 or 2) corresponding to malfunction code "E7" in the monitor mode of outdoor unit PCB. (Refer to P.104, 105 for how to check)

↓

Cut the power supply OFF and wait for 10 minutes.

↓

Check if any foreign substances around the fan. — YES → Remove the foreign substances.

↓ NO

Check the connection status of the connectors
○ Fan motor 1: X106A of PCB
○ Fan motor 2: X107A of PCB

↓

Check if any connector is disconnected. — YES → Insert the connector.

↓ NO

○ Fan motor 1: Both power supply wire and signal wire are all white.
○ Fan motor 2: Both power supply wire and signal wire are red in the PCB side and white in the motor side.

↓

Relay connectors have any connection error. — YES → Correct the connection of the relay connectors.

↓ NO

(A)

Figure 60A Flow chart to troubleshoot the condenser fan motor.

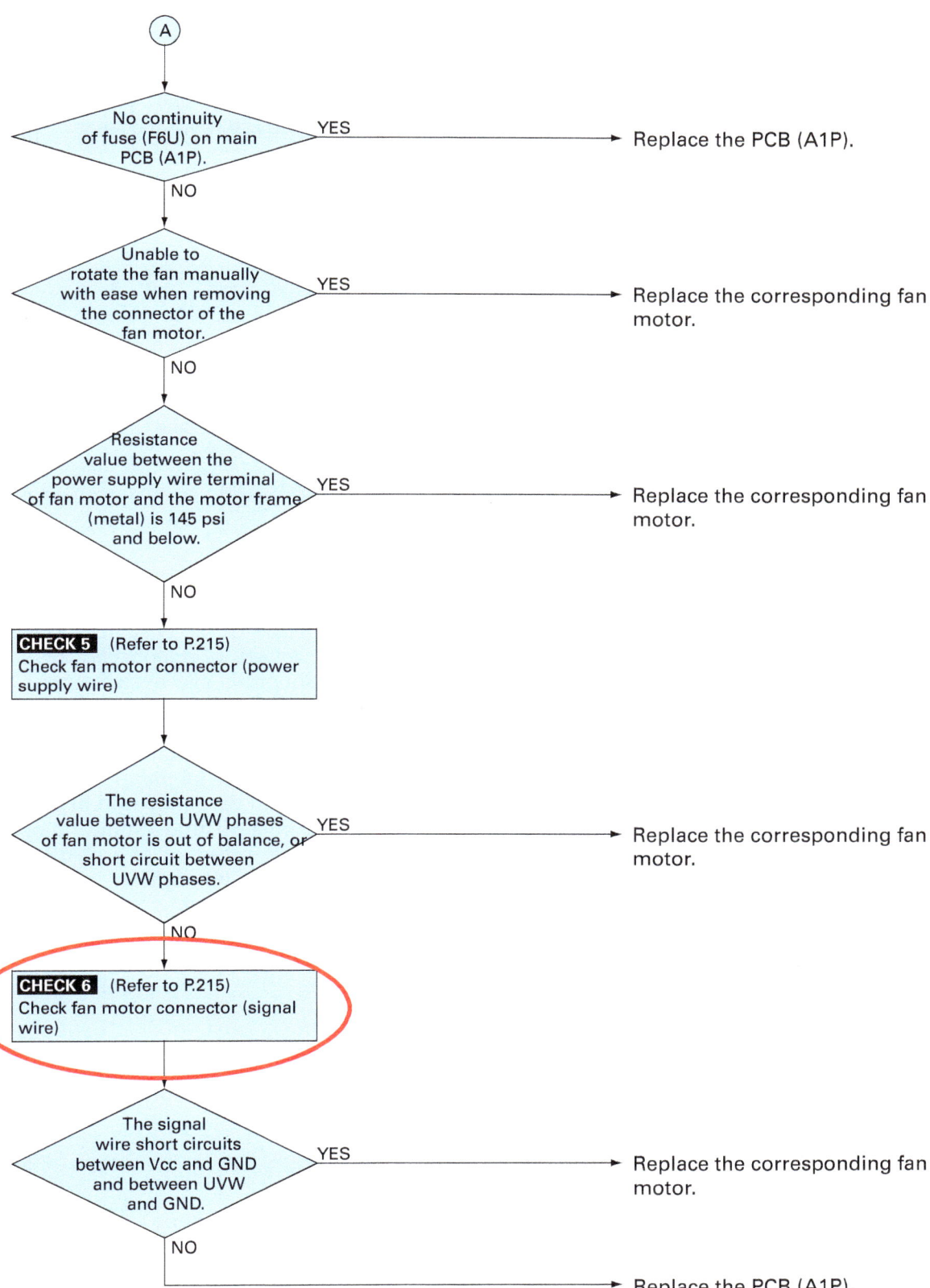

Figure 60B Flow chart to troubleshoot the condenser fan motor.

CHECK 6

(1) Turn off the power supply. (Signal wire)
(2) Measure the resistance between Vcc and each phase of U,V,W, and GND and each phase at the motor side connectors (five-core wire) to check that the values are balanced within the range of ± 20 %, while connector or relay connector is disconnected.
Furthermore, to use a multi-meter for measurement, connect the probe of negative pole to Vcc and that of positive pole to GND.

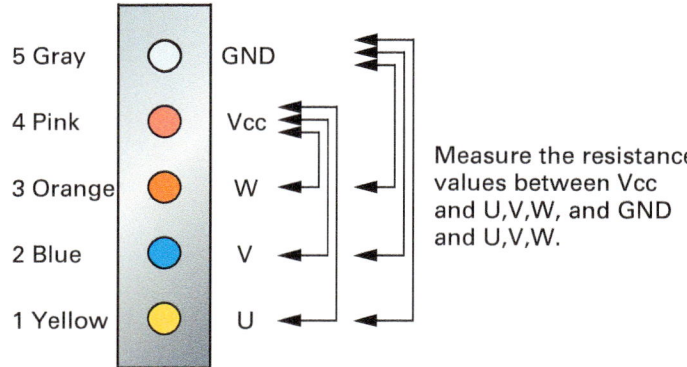

Figure 61 The Check 6 troubleshooting procedure.

For testing thermistors, charts are provided that indicate the resistance reading for a given temperature. Charts are also provided for pressure sensors to check their accuracy (*Figure 62*). The high- and low-pressure sensors have different characteristics, as shown on the chart.

There are even more powerful ways to view and troubleshoot many VRF systems. Some manufacturers have testing equipment that serves as an interface between the system and a laptop computer. The device shown in *Figure 63* is known as the Service Checker. It allows a technician to communicate with the entire system using a laptop instead of the very limited displays and keys characteristic of remote controls.

Troubleshooting VRF systems can be challenging. Proper training in the product line being serviced is essential to success with VRF systems. However, properly installed VRF systems have a high degree of reliability, and many problems are simpler than they may initially appear. Issues can arise from a misunderstanding of the system operation by the user, or as a result of untrained technicians who attempt to troubleshoot these systems using a conventional approach.

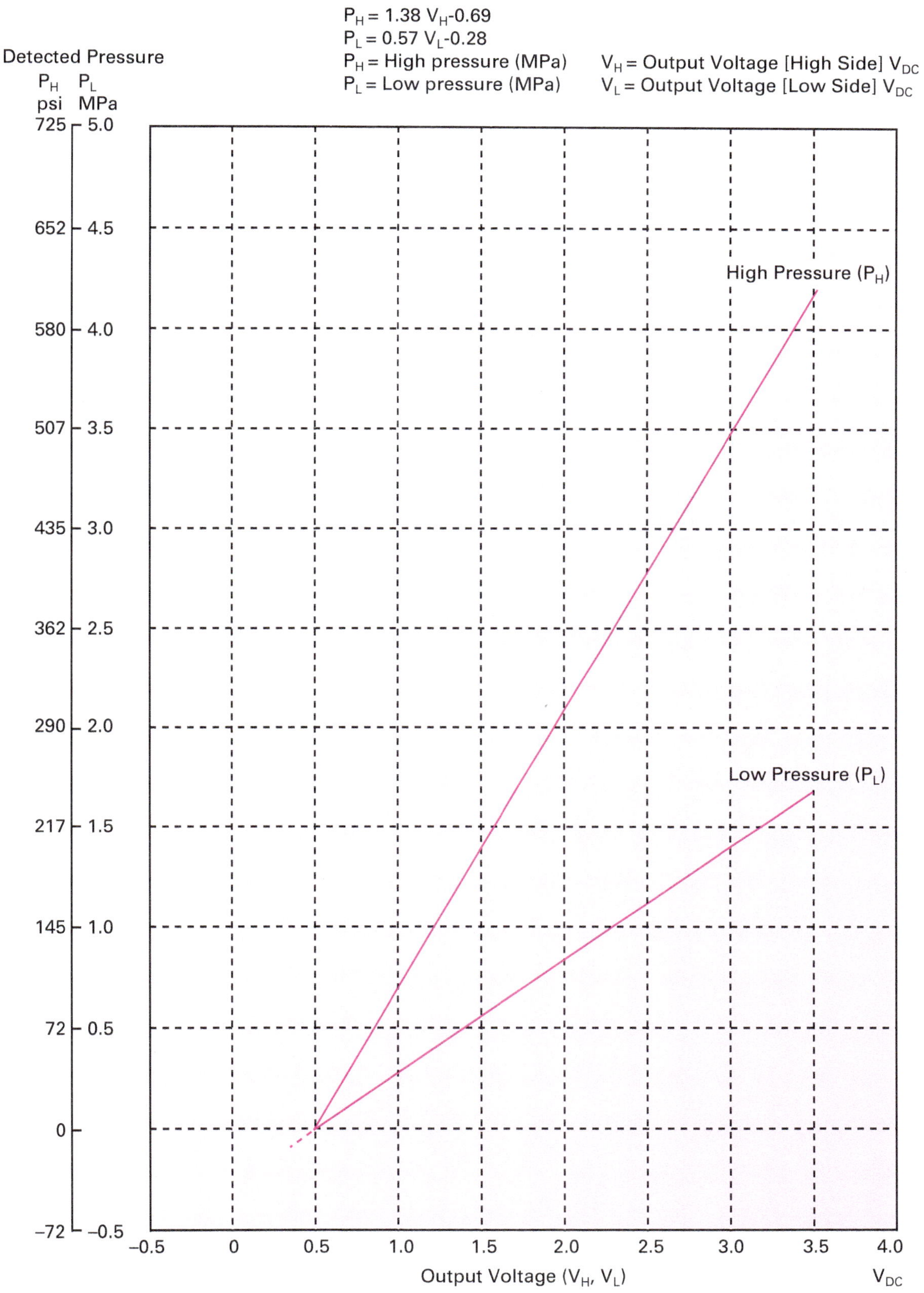

P_H = 1.38 V_H-0.69
P_L = 0.57 V_L-0.28
P_H = High pressure (MPa)
P_L = Low pressure (MPa)
V_H = Output Voltage [High Side] V_{DC}
V_L = Output Voltage [Low Side] V_{DC}
Detected Pressure
P_H psi
P_L MPa
725 5.0
652 4.5
580 4.0
507 3.5
435 3.0
362 2.5
290 2.0
217 1.5
145 1.0
72 0.5
0
–72 –0.5
High Pressure (P_H)
Low Pressure (P_L)
–0.5
0
0.5
1.0
1.5
2.0
2.5
3.0
3.5
4.0
Output Voltage (V_H, V_L)
V_{DC}

Figure 62 Pressure-sensor test chart.

Connection Method Checker Type III

Connection methods for SkyAir, VRV-S & VRV:
- "DIII-net" F1-F2 connection VRV
- OUTDOOR F1-F2

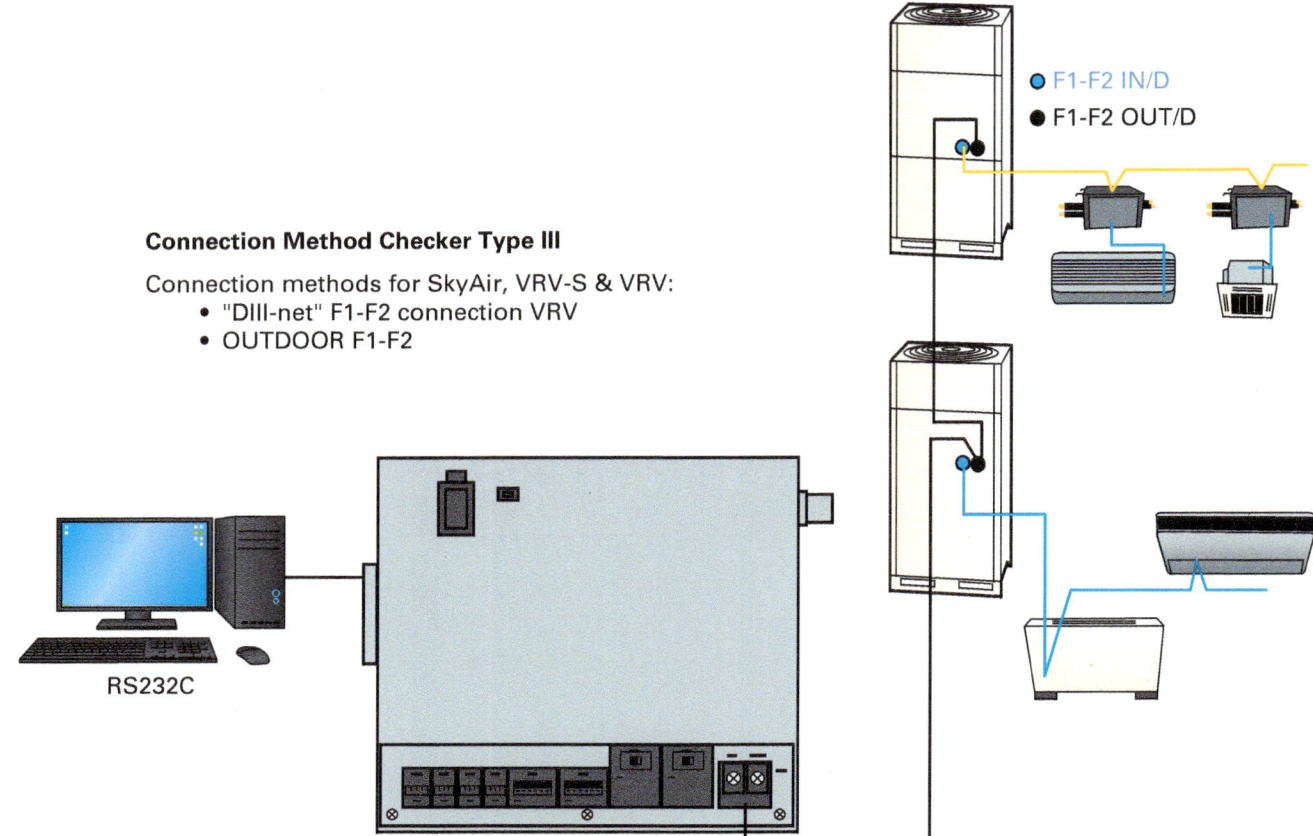

Figure 63 Daikin Service Checker.

Additional Resources

ASHRAE Handbook—HVAC Systems and Equipment. Current edition. Atlanta, GA: The American Society of Heating, Air Conditioning, and Refrigeration Engineers.

2.0.0 Section Review

1. The metering device used in ductless split systems is typically a(n) _____.
   a. thermostatic expansion valve
   b. automatic expansion valve
   c. electronic expansion valve
   d. capillary tube

2. Refrigerant piping connections at the indoor and outdoor unit of most ductless systems are made _____.
   a. by brazing
   b. with flare fittings
   c. using pressed fittings
   d. with unique screwed couplings

3. At least one manufacturer designs the outdoor units to control refrigerant flow in direct response to the _____.
   a. subcooling value at the outdoor unit
   b. superheat value at the outdoor unit
   c. number of indoor units in operation
   d. capacity of the largest indoor unit

4. The wiring used for communications between the system components is most often _____.
   a. shielded
   b. Class 1
   c. solid
   d. stranded

SUMMARY

Zoning, ductless, and VRF systems offer different approaches to making indoor comfort more personalized and accurate. Although they are generally thought of as being used for small applications, this is not necessarily the case. Light and even heavy commercial systems of almost any size can employ zoning to increase their effectiveness. VRF systems can provide significant capacity, typically up to 30 tons per system. As the capacity requirements increase, VAV systems become more cost-effective and practical.

Zoning systems use motorized dampers and related controls to manage system airflow with more precision and provide users with multiple thermostats to control the conditions in a group of smaller areas. Ductless systems offer the same basic user flexibility but do so using multiple sources of airflow from small-capacity, compact indoor units. VRF systems use refrigerant management techniques and sophisticated control and communication circuits to serve multiple air sources from a single refrigeration circuit. In all three systems, both heating and cooling can be provided.

Troubleshooting these systems requires excellent electrical skills and the ability to read and follow technical information and guidance. Without the proper information for the specific system in use, troubleshooting can be difficult and may even result in additional problems. However, a properly trained technician who uses the manufacturer's instructions and troubleshooting guides will be able to efficiently service these systems. Developing the advanced skills needed to understand these systems and successfully install or repair them is crucial to technicians who want to do well in the trade.

Review Questions

1. When a single zone damper is installed in a constant-volume system to control the temperature in one space, it is aware of the system operating mode through _____.
   a. communication with the main control system
   b. a direct connection to the system thermostat
   c. a temperature sensor placed in the supply air duct
   d. a temperature sensor placed in the return air duct

2. To avoid air pressure buildup in a dump zone, _____.
   a. a return air grille can be installed there
   b. another zone damper is used to control pressure
   c. a pressure transmitter in the space controls bypass damper position
   d. most of the zone dampers must remain open at all times

3. The status of the cooling control wiring between a zone control panel and the condensing unit can be quickly checked from the panel by placing a jumper across _____.
   a. R/RC and G
   b. R/RC and W
   c. R/RC and Y
   d. R/RC and O

4. In some VVT systems, the individual zone thermostats receive time of day and system operating information from _____.
   a. a monitor thermostat
   b. their own internal programs
   c. the HVAC equipment
   d. no source; the information is not necessary

5. Pressure-dependent VVT systems _____.
   a. do *not* have modulating zone dampers
   b. will cease to function if the duct pressure is more than 0.5 in. w.c.
   c. monitor the actual volume of air flowing through a zone damper
   d. do *not* monitor the actual volume of air flowing through the zone damper

6. In a typical ductless system operating in the cooling mode, _____.
   a. both refrigerant lines are at a low pressure
   b. both refrigerant lines are at a high pressure
   c. the vapor line is always at a high pressure
   d. the liquid line is always at a high pressure

7. Which of the following types of oil are more commonly found in ductless systems from Asian manufacturers?
   a. Mineral oil
   b. POE oil
   c. PVE oil
   d. PAG oil

8. On a ductless system remote control, what function would a Quiet setting provide?
   a. It reduces the speed and capacity of the system compressor.
   b. It causes the indoor air supply to be directed away from occupants.
   c. It slows the blower speed of the indoor unit.
   d. It slows the condenser fan speed of the outdoor unit.

9. The wiring configuration required for VRF system communication is the _____.
   a. homerun
   b. star
   c. synchronous
   d. daisy-chain

10. Which of the following statements about VRF system installation is correct?
    a. Refnet joints can only be installed in specific positions.
    b. Stranded or solid wire can be used for communication between components.
    c. Refrigerant charging is best done using the subcooling method.
    d. Only the vapor line needs to be insulated.

Trade Terms Introduced in This Module

Dump zone: An area used as a point to discharge unneeded supply air from a zoning system. The dump zone is typically in the conditioned space, but not an area where maintaining a specific space temperature is essential. A dump zone generally has no supply air grilles, but a return grille is often installed to relieve room pressure and return mixed air back to the system.

Polling: A network-communications arrangement where a VVT system seeks input from individual zones, determines what demands have priority, and then responds accordingly. Electronic participants in a polling system are referred to as voters.

Pressure-dependent operation: A VVT-system approach where dampers are positioned based on demand only, or operate in two positions. The result is that the design volume of air may or may not be flowing through the damper at any time. The volume of air flow through each damper is dependent on the duct pressure; it fluctuates as the pressure fluctuates. A minimum damper position is set, but air volume is not a controlling factor, and so the minimum damper position is a fixed point.

Pressure-independent operation: A VVT-system approach where dampers are positioned by demand and by the actual volume of air flowing through the damper. To achieve a specific amount of airflow, the damper positions itself as necessary. Maximum and minimum airflow setpoints are in effect, and the damper will modulate as necessary to remain within these boundaries.

Refnet joints: Engineered joints used as connection points in a VRF-system refrigerant network. Refnet joints connect the main refrigerant lines to branch runs or to indoor units.

Additional Resources

This module presents thorough resources for task training. The following reference material is recommended for further study.

ANSI/ACCA Manual Zr, Residential Zoning. Current edition. Arlington, VA: Air Conditioning Contractors of America.

ASHRAE Handbook—HVAC Systems and Equipment. Current edition. Atlanta, GA: The American Society of Heating, Air Conditioning, and Refrigeration Engineers.

Figure Credits

© iStock.com/THANATASDcom, Module opener
Jackson Systems, LLC - Controls Done Right™, Figures 2, 4–6, 8–11, SA02, SA03, 15, 17
Courtesy of ZONEFIRST, SA01, Figure 12
Carrier Corporation, Figures 26, 31
Daikin Industries, Ltd., Figures 27–29, 33, 34, 37, 40–45, 47–49, 52, 53 (photo), 54, 56–63
Courtesy of Sauermann N.A. Corp., Figure 30A
Courtesy of DiversiTech, Figures 30B, 38 (photo), 39
Courtesy of Michael Chasse, the quad in Quadomated.com, Figure 36
Courtesy of CPS Products, SA04
Courtesy of Ritchie Engineering Company, Inc., YELLOW JACKET Products Division, SA05

Section Review Answer Key

Answer	Section Reference	Objective
Section One		
1. b	1.1.1	1a
2. a	1.2.4	1b
3. c	1.3.0; 1.3.1	1c
4. a	1.4.3	1d
5. d	1.5.5	1e
Section Two		
1. c	2.1.1	2a
2. b	2.2.1	2b
3. a	2.3.0	2c
4. d	2.4.1	2d

NCCER CURRICULA — USER UPDATE

NCCER makes every effort to keep its textbooks up-to-date and free of technical errors. We appreciate your help in this process. If you find an error, a typographical mistake, or an inaccuracy in NCCER's curricula, please fill out this form (or a photocopy), or complete the online form at **www.nccer.org/olf**. Be sure to include the exact module ID number, page number, a detailed description, and your recommended correction. Your input will be brought to the attention of the Authoring Team. Thank you for your assistance.

Instructors – If you have an idea for improving this textbook, or have found that additional materials were necessary to teach this module effectively, please let us know so that we may present your suggestions to the Authoring Team.

NCCER Product Development and Revision
13614 Progress Blvd., Alachua, FL 32615

Email: curriculum@nccer.org
Online: www.nccer.org/olf

❏ Trainee Guide ❏ Lesson Plans ❏ Exam ❏ PowerPoints Other _____

Craft / Level: _____ Copyright Date: _____

Module ID Number / Title: _____

Section Number(s): _____

Description: _____

_____

_____

_____

Recommended Correction: _____

_____

_____

_____

Your Name: _____

Address: _____

_____

Email: _____ Phone: _____

Commercial Hydronic Systems

Overview

In commercial and industrial hydronic systems, water is used as the heat exchange medium for both heating and cooling. Chillers provide the cold water for cooling, using different types of compressors. Hot water for heating is provided by various types of boilers. Commercial hydronic systems have unique servicing and maintenance requirements, including maintaining proper chemical balance of the water to prevent fouling, corrosion, and similar problems. The layout of water distribution systems varies with the size and type of building.

Module 03305

Trainees with successful module completions may be eligible for credentialing through the NCCER Registry. To learn more, go to **www.nccer.org** or contact us at 1.888.622.3720. Our website has information on the latest product releases and training, as well as online versions of our *Cornerstone* magazine and Pearson's product catalog.

Your feedback is welcome. You may email your comments to **curriculum@nccer.org**, send general comments and inquiries to **info@nccer.org**, or fill in the User Update form at the back of this module.

This information is general in nature and intended for training purposes only. Actual performance of activities described in this manual requires compliance with all applicable operating, service, maintenance, and safety procedures under the direction of qualified personnel. References in this manual to patented or proprietary devices do not constitute a recommendation of their use.

Copyright © 2018 by NCCER, Alachua, FL 32615, and published by Pearson, New York, NY 10013. All rights reserved. Printed in the United States of America. This publication is protected by Copyright, and permission should be obtained from NCCER prior to any prohibited reproduction, storage in a retrieval system, or transmission in any form or by any means, electronic, mechanical, photocopying, recording, or likewise. To obtain permission(s) to use material from this work, please submit a written request to NCCER Product Development, 13614 Progress Blvd., Alachua, FL 32615.

03305 V5

From *HVAC Level Three, Trainee Guide.* NCCER.
Copyright © 2018 by NCCER. Published by Pearson. All rights reserved.

03305
COMMERCIAL HYDRONIC SYSTEMS

Objectives

When you have completed this module, you will be able to do the following:

1. Describe basic concepts related to water as a substance and its movement.
   a. Describe the basic properties of water and water pressure.
   b. Explain how pressure drop, head pressure, and static pressure are related to hydronics.
2. Describe various commercial hot-water heating system components and subsystems.
   a. Identify various types of hot-water boilers and their common controls.
   b. Describe the construction of common centrifugal pumps.
   c. Identify various types of valves and other commercial hydronic system components.
   d. Identify common commercial piping systems and their characteristics.
   e. Describe how a typical hydronic piping system is balanced.
3. Explain the basic concepts of chilled-water systems and the related components.
   a. Explain the basic concepts of chilled-water cooling systems.
   b. Identify various types of chillers and their common controls.
   c. Identify various types of cooling towers and evaporative condensers.

Performance Tasks

Under the supervision of your instructor, you should be able to do the following:

1. Identify the major components of commercial hot-water heating and chilled-water cooling hydronic systems.
2. Identify the types of common piping configurations used with commercial hot-water and chilled-water hydronic systems.

Trade Terms

Boiler horsepower (BoHP)
Firetube boiler
Flash economizer
Lithium bromide
Oxygen scavengers
Secondary coolant
Superheated steam
Thermal economizer
Venturi tube
Watertube boiler

Industry-Recognized Credentials

If you are training through an NCCER-accredited sponsor, you may be eligible for credentials from NCCER's Registry. The ID number for this module is 03305. Note that this module may have been used in other NCCER curricula and may apply to other level completions. Contact NCCER's Registry at 888.622.3720 or go to **www.nccer.org** for more information.

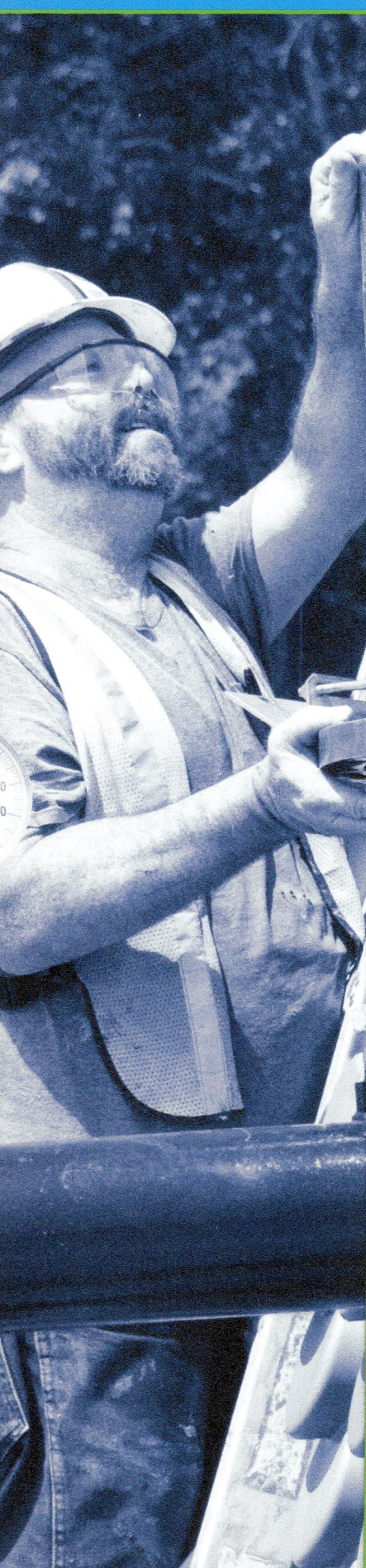

Contents

1.0.0 Water and Movement 1
   1.1.0 Properties of Water 1
      1.1.1 Water Pressure 2
   1.2.0 Movement of Water Through Piping Systems 2
      1.2.1 Pressure Drop 2
      1.2.2 Head Pressure 3
      1.2.3 Static Pressure 3
2.0.0 Commercial Hot-Water Heating System Components 6
   2.1.0 Commercial Hot-Water Boilers 6
      2.1.1 Copper-Finned Tube Boilers 7
      2.1.2 Cast-Iron Sectional Boilers 8
      2.1.3 Steel Firetube and Watertube Boilers 9
      2.1.4 Steel Vertical Tubeless Boilers 10
      2.1.5 Electric Boilers 11
      2.1.6 Commercial Condensing Boilers 12
      2.1.7 Boiler Operating/Safety Controls 12
   2.2.0 Centrifugal Pumps 13
      2.2.1 Centrifugal Pump Construction 13
      2.2.2 Double-Suction Centrifugal Pumps 14
      2.2.3 Multistage Centrifugal Pumps 17
   2.3.0 Valves and Components 17
      2.3.1 Multi-Purpose Valves 17
      2.3.2 Balancing Valves and Flow Meters 18
      2.3.3 Butterfly Valves 19
      2.3.4 Expansion and Compression Tanks 21
      2.3.5 Air Management 22
   2.4.0 Common Commercial Piping Systems 23
      2.4.1 Four-Pipe Systems 23
      2.4.2 Primary-Secondary Piping Systems 24
   2.5.0 Water System Balancing 25
      2.5.1 Pre-Balance Checks 25
      2.5.2 Water System Balancing Procedure 26
3.0.0 Chilled-Water Systems 29
   3.1.0 Basic Chilled-Water System Concepts 29
   3.2.0 Chillers 30
      3.2.1 Reciprocating and Scroll Chillers 31
      3.2.2 Centrifugal Chillers 32
      3.2.3 Screw Chillers 33
      3.2.4 Absorption Chillers 34
      3.2.5 Chiller Operating/Safety Controls 35
   3.3.0 Cooling Towers and Evaporative Condensers 36
      3.3.1 Cooling Towers 37
      3.3.2 Evaporative Condensers 38

Figures

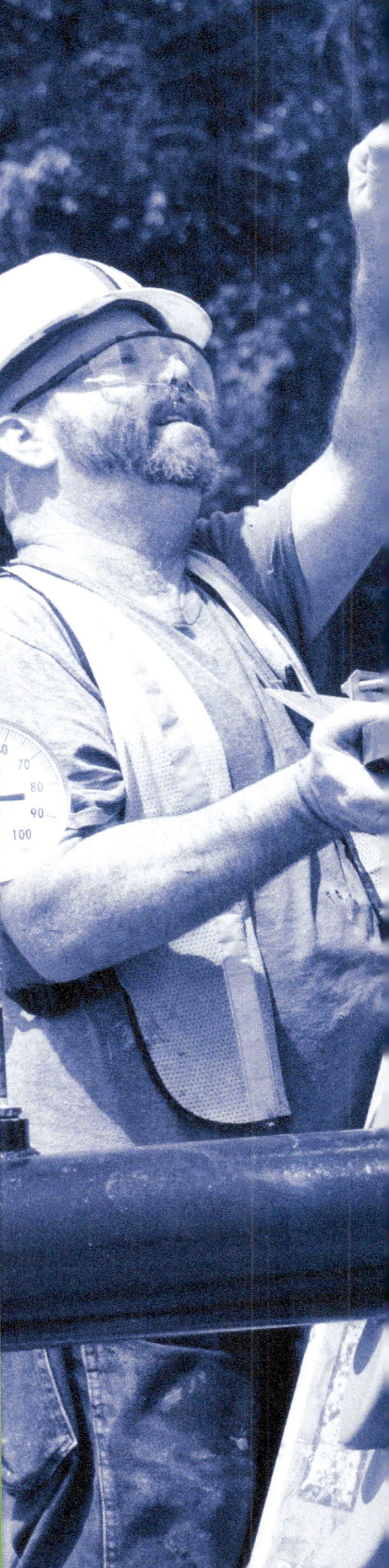

Figure 1	Change in water pressure with depth	2
Figure 2	Pressure drop.	3
Figure 3	Relationship of head pressure to height.	3
Figure 4	Static pressure.	4
Figure 5	Commercial copper-finned tube boiler cutaway.	8
Figure 6	Boiler with sealed combustion chamber.	8
Figure 7	Cast-iron sections in a packaged boiler.	8
Figure 8	Cast-iron packaged boiler operation.	9
Figure 9	Firetube and watertube boilers.	10
Figure 10	Scotch Marine boiler.	10
Figure 11	Vertical tubeless boiler.	11
Figure 12	Electric hot-water boiler.	11
Figure 13	Cutaway of a modern commercial condensing boiler.	12
Figure 14	Float-operated, low-water cutoff control.	13
Figure 15	Flow switch.	13
Figure 16	Centrifugal pump.	15
Figure 17	Centrifugal pump components.	16
Figure 18	Impeller types.	17
Figure 19	Double-suction centrifugal pump.	17
Figure 20	Multistage centrifugal feedwater pump.	17
Figure 21	Multi-purpose valve.	18
Figure 22	Balancing valve.	19
Figure 23	Venturi tube and orifice plate flow-metering device.	20
Figure 24	Butterfly valve.	20
Figure 25	Wafer butterfly valve.	21
Figure 26	Wafer lug butterfly valve.	21
Figure 27	Two-flange butterfly valve.	21
Figure 28	Pressurized expansion tanks.	22
Figure 29	Commercial air separator.	22
Figure 30	Air-control component arrangement.	23
Figure 31	Air-elimination component arrangement.	23
Figure 32	Four-pipe water system with one-coil and two-coil terminals.	24
Figure 33	Simplified primary-secondary chilled-water system.	25
Figure 34	Packaged air-cooled chiller.	29
Figure 35	Basic chilled-water system.	30
Figure 36	Chilled-water system with a packaged air-cooled chiller.	31
Figure 37	Packaged air-cooled scroll chiller.	31
Figure 38	Centrifugal chiller.	32
Figure 39	Air-cooled screw chiller.	34
Figure 40	Chiller operator interface.	35
Figure 41	Direct-contact natural-draft cooling tower.	37
Figure 42	Cooling tower float-controlled makeup water system.	38
Figure 43	Indirect-contact natural-draft cooling tower.	38
Figure 44	Indirect-contact induced-draft cooling tower.	39
Figure 45	Evaporative condenser.	40

Section One

1.0.0 Water and Movement

Objective

Describe basic concepts related to water as a substance and its movement.
a. Describe the basic properties of water and water pressure.
b. Explain how pressure drop, head pressure, and static pressure are related to hydronics.

Trade Terms

Oxygen scavengers: Chemicals added to boiler feed water that chemically combine with dissolved oxygen to prevent it from causing corrosion within the system. Typically creates a sludge that must be purged with boiler blowdowns or skimming.

Before addressing more complex hydronic subjects, a review of basic terms and concepts related to water will prove helpful. Many problems with hydronic system operation can be identified or resolved if the technician has a thorough understanding of these basic concepts.

1.1.0 Properties of Water

Water (H_2O) is a chemical compound of two chemical elements: hydrogen (H) and oxygen (O). It can exist as ice, water, or steam (solid, liquid, or vapor) due to changes in temperature and/or pressure. These three forms of water are referred to as *states*, or *phases*. At the atmospheric pressure found at sea level, water freezes (becomes solid) at 32°F (0°C), and it boils (becomes vapor) at 212°F (100°C).

Liquid water changes density with a change in temperature; the higher the temperature of water at a given pressure, the lower its density. This change in weight per unit volume is due to the expansion in water volume as the water molecules move farther apart with temperature. For example, one cubic foot of liquid water weighs 62.41 pounds (999.9 kg/m^3) at 32°F (0°C). At 200°F (93°C), one cubic foot weighs 59.45 pounds (952.6 kg/m), a density difference of roughly 3.0 lb/ft^3 (47.3 kg/m^3).

Different water densities resulting from locations at different temperatures can cause natural circulation of water in a closed system due to gravity alone. This natural circulation is also referred to as *thermal circulation*. Water itself is noncorrosive to most materials used in hydronic systems. The oxygen within the water molecule is basically inert. However, a variable amount of air, which is mainly a mixture of nitrogen and oxygen, is also dissolved in water. Although nitrogen is also relatively inert, the free, dissolved oxygen will cause corrosion. For example, common rust is a compound formed from a chemical reaction between iron and oxygen in the presence of water.

For this reason, only minimal amounts of fresh water should be admitted to the system after the initial fill, since it contains some dissolved oxygen. This minimizes the amount of oxygen available. Corrosion stops when the available free oxygen has been consumed. Open systems (systems that are open to the atmosphere at one or more points in the system) allow the water to continuously absorb additional oxygen from the air. Open systems are not used for hydronic heating and cooling systems that circulate throughout a building. However, they are still used for water-cooled condensers and similar applications.

Each time fresh water enters a closed system, when filling following a leak repair for example, additional oxygen is admitted and fresh corrosion begins until all free oxygen is consumed. If the admission of fresh water continues, components in the system will suffer additional corrosion and possible failure.

Because most systems use automatic valves to admit fresh water, hydronic systems should be inspected regularly for leaks. A small leak can result in the constant and unmonitored addition of fresh, oxygenated water. On larger systems, water meters may be installed to monitor the volume of fresh water being added over a given period of time. Checking the meter provides a way to see if a significant amount of water is being added.

As water is heated in a boiler, air dissolved in the water is liberated in the form of bubbles. Eventually, most of the dissolved air collects as gas bubbles or pockets in the system. That air must be removed from the system for it to function properly.

Because public and well water contains various amounts and types of minerals, scale can form in the piping. Scale reduces both heat transfer and water flow, another reason to minimize the admittance of fresh, untreated water. Special chemicals are often added to systems to reduce scaling. Certain chemicals called **oxygen scavengers** work to reduce dissolved oxygen by chemically combining with the oxygen. This reaction produces chemically harmless materials and prevents the oxygen from bonding with iron to form rust. Always check with the boiler manufacturer before adding any water treatment chemical to boiler water, however. The wrong chemical may damage the boiler and void any warranties. The subject of water treatment is covered in depth in NCCER Module 03308, *Water Treatment*.

1.1.1 Water Pressure

The change in water pressure below any point in a water system can be calculated using basic arithmetic. This approach assumes the water is not moving—it is the change in static water pressure. If the reference point is the surface of the water, as shown in *Figure 1*, then the calculated change in pressure is the actual static pressure at the depth of interest.

To conceptualize this principle, consider a cubic foot of water. Reference tables tell you that one cubic foot of water weighs 62.4 pounds at 39.2°F (4°C). The bottom surface of the cubic foot is 1 square foot. Thus, the pressure exerted by water on the bottom of the 1 ft^3 volume is 62.4 pounds per square foot (psf). You could calculate this pressure using the following formula:

$$P = \text{pressure change per foot of depth} \times D$$

Where:

P = pressure
D = depth in feet

In this example:

$$P = 62.4 \text{ psf/ft} \times 1 \text{ ft}$$
$$P = 62.4 \text{ psf}$$

Similarly, the pressure at two feet from the top of a water system is:

$$P = 62.4 \text{ psf/ft} \times 2 \text{ ft}$$
$$P = 124.8 \text{ psf}$$

Technicians usually measure system pressures in pounds per square inch (psi). To convert pressures in psf to psi, divide the result by the factor 144 in^2/ft^2, since there are 144 square inches in a square foot:

Pressure at one-foot depth:
$$P = 62.4 \text{ psf/ft} \div 144 \text{ in}^2/\text{ft}^2 = 0.433 \text{ psi}$$

Pressure at two-foot depth:
$$P = 124.8 \text{ psf/ft} \div 144 \text{ in}^2/\text{ft}^2 = 0.867 \text{ psi}$$

The same process works for metric units. One cubic meter of water weighs 1,000 kg at 4°C (39°F). The pressure at one-meter depth is 1,000 kg per square meter. In metric units, it is common to measure pressure in kilopascals (kPa). This pressure converts to 0.981 kPa at one-meter depth. To calculate the metric pressure at any depth in meters, substitute this value into the formula for pressure, as follows:

$$P \text{ (in kPa)} = 0.981 \text{ kPa/m} \times D \text{ (in meters)}$$

The key thing to keep in mind is that the static pressure of water increases below a point where it is measured in a system and decreases above that point at a rate of about 0.43 psi/ft or 0.98 kPa/m.

1.2.0 Movement of Water Through Piping Systems

Water movement in a piping system is affected by several factors. Examples include the resistance to water flow within the piping and the pressure available to force water past that resistance.

1.2.1 Pressure Drop

Pressure drop is the difference in pressure between two points in a hydronic system due to fluid flow and changes in elevation. It is an important factor in all hydronic systems. In *Figure 2*, there is a 5 psi (34.5 kPa) drop in pressure between two points when water is flowing at a rate of 5 gallons per minute (gpm) (19 L/min). Because of the length and volume of flow required in many commercial or industrial applications, as opposed to smaller residential systems, pressure drop in

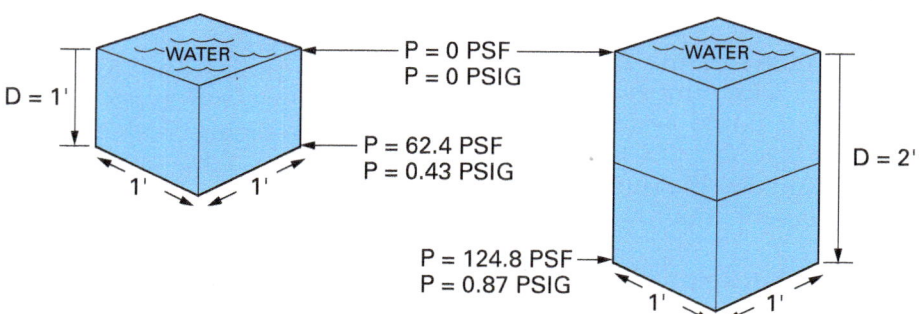

Figure 1 Change in water pressure with depth.

piping must be fully understood and given careful consideration. Where there is no flow or water movement in horizontal pipe, no pressure drop is encountered.

> **NOTE**
> Without a pressure drop in a system, there would be no fluid flow. This is because fluids only flow from higher to lower pressure. For this reason, hydronic systems require pumps to create a pressure difference.

When flow is present, fluid drag (a type of friction) occurs and increases roughly in direct proportion to the square of the change in flow velocity. The change in pressure drop resulting from a change in flow rate can be estimated as follows:

Final pressure drop =
$$\left(\frac{\text{Final gpm flow rate}}{\text{Initial gpm flow rate}}\right)^2 \times \text{initial pressure drop}$$

> **NOTE**
> This formula is reasonably accurate for relatively low flow rates through a large-diameter, horizontal pipe with no change in fluid temperature between the points. It is also valid when replacing Imperial units with equivalent metric units.

Hydronic system pumps must provide sufficient power to overcome fluid drag encountered in the piping system and all components. They accomplish this by creating a sufficiently large pressure difference. Because piping systems differ, the pressure drop for piping must be calculated for each system. The manufacturer's specifications for other hydronic system components, such as coils and heat exchangers, provide pressure drop data for various flow rates. This information may be expressed in a variety of units, so the following conversions may be helpful:

- 1 pound per square inch = 2.31 feet of water
- 1 foot of water = 0.433 pounds per square inch
- 1 foot of water = 12,000 milli-inches water
- 1 inch of water = 1,000 milli-inches water

1.2.2 Head Pressure

The quantity head pressure is another measure of pressure, often expressed in feet of water. It is most often used to describe the capacity of a circulating pump, indicating the height of a column of water that the pump can lift, without considering pressure drop from friction losses. A head

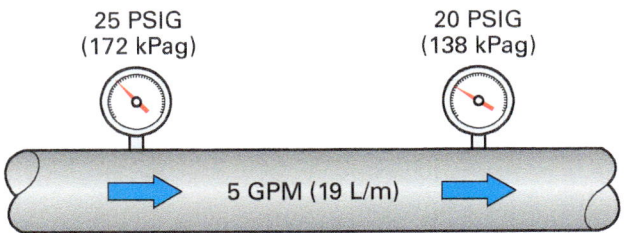

Figure 2 Pressure drop.

pressure value is, therefore, often referred to as *feet of head*.

Since a column of water 1' in height produces a static pressure of 0.433 psig (0.30 kPag), simple calculations prove that lifting a column of water 2.31' (0.70 m) in height requires a pressure of 1 psig (6.9 kPag). *Figure 3* illustrates these conversions.

1.2.3 Static Pressure

Static pressure (*Figure 4*) is created by the height of the water above the point of measurement in the system. Recall that a column of water 1' (0.305 m) high above a pressure gauge produces a pressure of 0.43 psi (3.0 kPa) at the gauge. Therefore, static pressure is equal to 0.43 psi for each additional foot (or 9.7 kPa/m) of height above the system gauge.

Consider a simple water system to be an upright loop of water confined in a pipe. As such, the static pressure in one of the vertical pipes of the loop is identical to the static pressure at the same level in the opposite vertical pipe. Static pressure levels in a system decrease as one moves from the lowest point in the system toward the highest point in the system. At the highest point, the static pressure is 0 psig (0 kPag) if the system is vented to the atmosphere. Remember that this

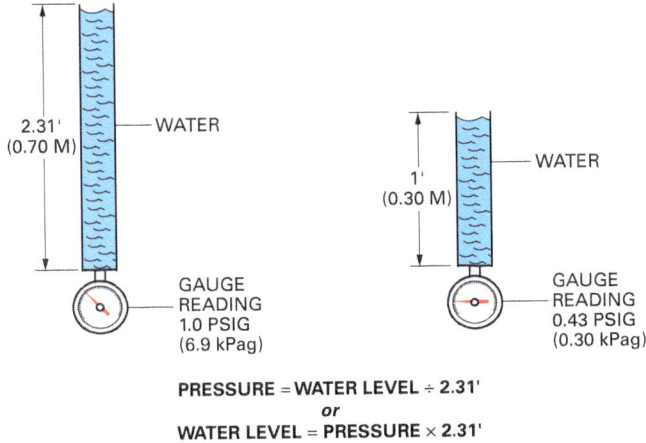

Figure 3 Relationship of head pressure to height.

measured value must be stated in gauge-pressure units, since the weight of the atmosphere is also pushing down on the water column.

For example, if the highest point holding water in a hydronic system is 30' (9.2 m) above the hot-water boiler pressure gauge, the static pressures within the system are:

- 30' (highest point in system) = 0 psig, or 0 kPag
- 20' (10' below highest point) = 4.3 psig (10' × 0.43 psig/ft), or 3.0 kPag
- 10' (20' below highest point) = 8.6 psig (20' × 0.43 psig/ft), or 59.3 kPag
- 0' (30' below highest point) = 12.9 psig (30' × 0.43 psig/ft), or 88.9 kPag

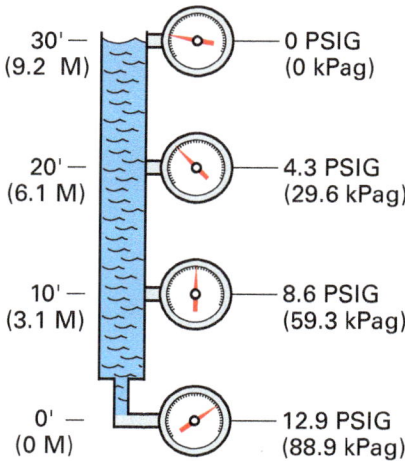

Figure 4 Static pressure.

Think About It
Head Pressure Relationships

Use the conversions shown in *Figure 3* to answer the following questions:

1. The water level in a water system reservoir is maintained at 150' (46 m) above its use level. What is the pressure resulting from this head indicated by a gauge installed at the system use wlevel?
2. A gauge installed at the use level of a water system reads 64.5 psig (444.7 kPag). What is the water level height (head pressure) in the water system reservoir that causes this reading?

Additional Resources

Air-Cooled Chillers, TDP-622. Current edition. Syracuse, NY: Carrier Corporation.

Water-Cooled Chillers, TDP-623. Current edition. Syracuse NY: Carrier Corporation.

Refrigeration and Air Conditioning: An Introduction to HVAC/R, Larry Jeffus. Current edition. New York, NY: Pearson.

1.0.0 Section Review

1. What substance in water causes it to be corrosive to hydronic system components?
   a. Dissolved oxygen
   b. Suspended particles
   c. Dissolved nitrogen
   d. Oxides of iron

2. If there is no flow or water movement in horizontal pipe, _____.
   a. there are no elbows or fittings in the pipe
   b. there is no pressure drop
   c. the pipe interior is coated with Teflon®
   d. the water is chemically treated

Section Two

2.0.0 Commercial Hot-Water Heating System Components

Objective

Describe various commercial hot-water heating system components and subsystems.
  a. Identify various types of hot-water boilers and their common controls.
  b. Describe the construction of common centrifugal pumps.
  c. Identify various types of valves and other commercial hydronic system components.
  d. Identify common commercial piping systems and their characteristics.
  e. Describe how a typical hydronic piping system is balanced.

Performance Tasks

1. Identify the major components of commercial hot-water heating and chilled-water cooling hydronic systems.
2. Identify the types of common piping configurations used with commercial hot-water and chilled-water hydronic systems.

Trade Terms

Boiler horsepower (BoHP): The heat necessary to vaporize 34.5 pounds (15.7 kg) of water per hour at 212°F (100°C) and 0 psig (0 kPag). This is equal to a heat output of 33,475 Btus per hour or about 140 square feet (13 m^2) of steam radiation. 1 BoHP is also equal to 9.803 kilowatts.

Firetube boiler: A boiler design where the hot flue gases flow inside the tubes, which are surrounded by the water to be heated.

Superheated steam: Steam at a temperature above the saturation temperature for its pressure.

Venturi tube: A short tube with flaring ends and a constricted throat that is used for measuring flow velocity by comparing the throat pressure to the inlet pressure. (Throat pressure decreases as the fluid's velocity increases.)

Watertube boiler: A boiler design where the water to be heated flows inside the tubes, with the hot flue gases flowing around the outside of the tubes.

A number of common hydronic devices and components apply to commercial and industrial systems. There are also some hydraulic devices, components, and configurations specific to larger systems. Primary commercial hot-water system components include the following:

- Hot-water boilers
- Operating and safety controls associated with commercial boilers
- Expansion and compression tanks
- Air control devices
- Circulating pumps
- Valves

2.1.0 Commercial Hot-Water Boilers

A boiler heats water using different fuels or electricity as the heat source. Some boilers can be fired with more than one fuel, using either burner conversion or dual-fuel burners. Gaseous fuels include natural gas, manufactured gas, and liquefied petroleum (propane and butane). Oil fuels include both lightweight and heavy oils. Solid fuels include coal, wood, and flammable solid waste.

Commercial and industrial markets are dominated by fuel oil and natural gas, due to their availability and high energy content. There has been a resurgence of natural gas use, as new and plentiful domestic sources have been discovered and exploited. Natural gas is delivered to a given facility via pipeline, eliminating dependence upon ground transportation and requiring no on-site storage.

Although rare, large electric boilers are also used in some applications. They are often used by electric power utilities in company-owned facilities as a means of proving their commitment to the electric power they provide. Geothermal heat pumps can also be used to produce hot water for comfort heating.

Combustion in a fired boiler occurs by combining the fuel with oxygen and igniting the mixture. The methods used for combustion in gas-fired and oil-fired boilers are basically the same as those used with furnaces. Fuel ignition can be achieved by use of a standing gas pilot, an intermittent igniter of the electric-spark variety, or a hot surface igniter. The presence of a pilot or flame is proven by a protective device before the main gas or oil supply valve is allowed to open.

Most boiler combustion systems have a pre-purge and/or purge sequence to make sure that

> ### Clean Natural Gas
>
> With increasing concerns about global warming, people are looking at natural gas as the best fuel to burn to reduce carbon emissions. Natural gas produces less carbon emissions than both oil and coal. In addition, natural gas does not produce ash particles like coal and oil. Due to recent advances in drilling technology, the United States now has about a 100-year supply of this very clean fuel. Furnaces and boilers fired by natural gas are popular for heating applications ranging from small water heaters to massive industrial steam boilers.

there is not a combustible mixture already in the combustion chamber that could cause an explosion during startup. The methods used to vent boilers are similar to those used with furnaces.

The construction and operation of hot-water and steam boilers are similar, with two exceptions. The operating and safety controls used with hot-water boilers are different from those used with steam boilers. Also, hot-water boilers are entirely filled with water, while steam boilers require a steam space to operate properly.

Low-pressure boilers are the most widely used type of boiler. Low-pressure hot-water boilers can be built to have working pressures of up to 160 psig (1,103 kPag). However, they are typically designed for a 30 psig (207 kPag) maximum working pressure. They are generally operated below that maximum pressure level, with 12–15 psig (83–103 kPag) being common. Low-pressure hot-water boilers are generally limited to a maximum operating temperature of 250°F (121°C). Above this temperature, even water under low pressure will begin to boil and change to steam, creating a dangerous condition. In most cases, low-pressure boilers operate below 200°F (93°C).

Medium- and high-pressure hot-water boilers are built to operate at pressures above 160 psig (1,103 kPag) and temperatures above 250°F (121°C). Keeping the pressure higher raises the boiling point of the water. In extremely large systems, such high pressures and temperatures allow the water to be circulated great distances (from a central heating plant to other buildings, for example) without losing significant heating capacity before arriving at the intended destination. Operating and working on such systems can also be extremely dangerous.

Very large hydronic heating systems often use high-temperature, high-pressure water, especially in central plant applications for college campuses and military bases. At such extreme temperatures (over 450°F or 232°C in some cases) and under high pressure, escaping water instantly flashes to steam when even a small fracture or tiny leak develops in piping or components. Quite often, evaporation of the superheated steam occurs so rapidly that no steam plume is seen at the site of the leak. This is due to its sudden exposure to atmospheric pressures at which water boils at 212°F (100°C). If the steam condenses at all, it may be some distance from the leak.

> **WARNING!**
> Walking near or moving a hand across the path of escaping steam from a high-pressure system can result in serious injury. Steam at any pressure can produce serious physical injury. Use extreme caution when servicing or in the presence of such systems.

Most boilers in larger commercial applications are made of cast iron or steel. Small commercial facilities are more likely to use boilers of copper-finned tube construction. This design continues to increase in popularity because of heat transfer efficiency, reduced installation cost, and application flexibility.

2.1.1 Copper-Finned Tube Boilers

Boilers using the copper-finned tube design offer some distinct advantages in commercial applications. Because traditional steel and cast-iron units generally store a significant volume of hot water, they are subject to greater heat losses when heating needs are low. They can also require more space for installation and service.

With the copper-finned tube design, heat transfer is greatly enhanced, providing for rapid recovery. Their small sizes also eliminate the need for hot water to be stored within the boiler itself and they produce hot water on demand. Such boilers are gas-fired, generally using natural gas or propane.

Atmospheric models, such as the one shown in *Figure 5*, often offer modulating or staged-heating capacity to further enhance operating efficiency. When compared to traditional cast-iron or steel alternatives, they are also very lightweight, making for easier installation and/or replacement.

One other unique feature these boilers can offer is outdoor installation. The unit pictured in *Figure 5* does not have a top connection for flue venting and can be installed outdoors without any additional protection.

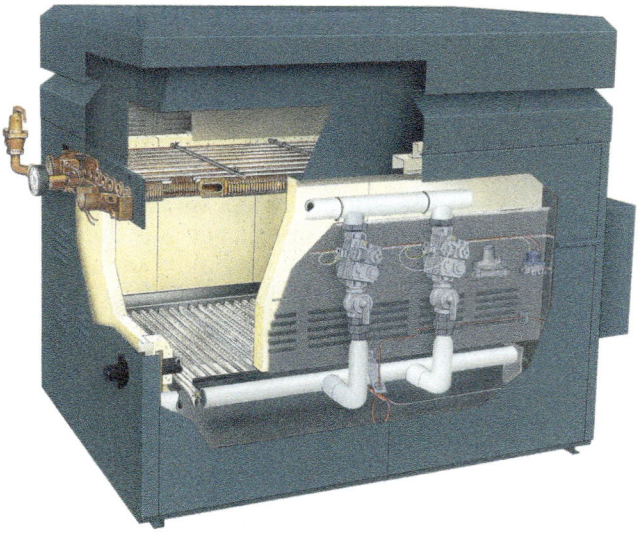

Figure 5 Commercial copper-finned tube boiler cutaway.

Boilers installed outdoors are subject to damage from freezing in the event of failure of the burner or pumping system. For outdoor installations, an antifreeze solution is added to protect the boiler and water circuit from freeze damage. The addition of antifreeze impacts the heat transfer and viscosity of the water. Both issues must be considered in the pump selection process. Generally, to achieve the desired overall heat transfer effect, system flow rates must be increased when antifreeze is used. Galvanized piping is not recommended when glycol-based antifreeze is used. The zinc in the pipe coating may react with the glycol to form sludge in the piping system.

Other copper-finned boiler models feature induced-draft combustion, allowing for additional installation flexibility and higher levels of efficiency. An example of an induced-draft model is shown in *Figure 6*. The induced-draft design allows the combustion chamber to be sealed; air for the combustion process can be ducted in from outdoors.

> **CAUTION**
> Proper water flow through a copper finned-tube boiler is critical to the life of the boiler. Consult the manufacturer's instructions and engineering guides to determine the correct pipe size and flow requirements.

Although very high heating capacities are available in copper-finned tube units, multiple units are often selected and piped in parallel to further enhance efficiency by matching capacity to the heating load. Using multiple boilers to achieve desired maximum capacity also provides redundancy when one or more units are in need of repair or maintenance.

2.1.2 Cast-Iron Sectional Boilers

Cast-iron boilers are formed by assembling together individual cast-iron heat exchanger sections (*Figure 7*). Each section is basically a separate boiler. The number of sections used determines the size of the boiler and its energy rating. Cast-iron boiler capacities range from those required for small residences up to large commercial systems of 13,000 MBh.

> **NOTE**
> The non-metric unit MBh is the combination of the Roman numeral M (which stands for the Latin *mille*—1,000) and Bh (which stands for Btu/hour). Therefore, 1 MBh = 1,000 Btu/hour.

In a cast-iron boiler, the water circulates inside the cast sections with the flue gases on the outside of the sections (*Figure 8*). The cast sections

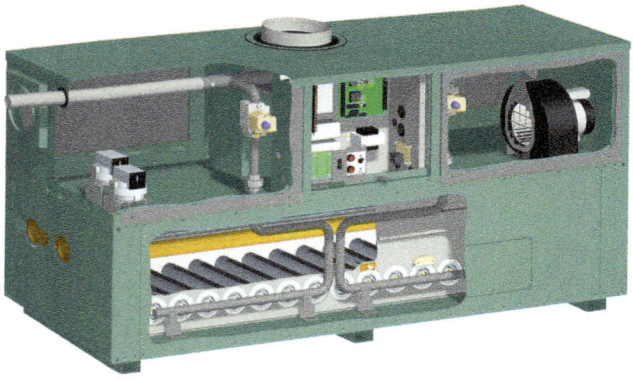

Figure 6 Boiler with sealed combustion chamber.

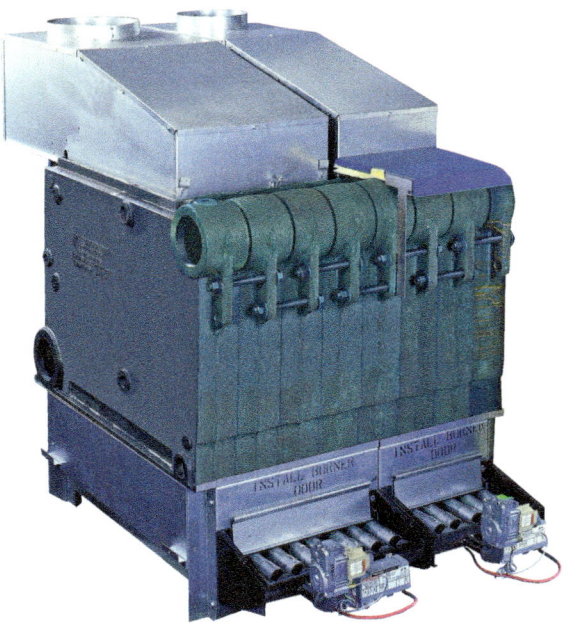

Figure 7 Cast-iron sections in a packaged boiler.

are usually mounted vertically, but they can be mounted horizontally. In both arrangements, the heating surface is large relative to the volume of water. This allows the water to heat up quickly.

The boiler shown in *Figure 8* uses modular construction. This allows different combinations of boiler sections, bases, and flue collectors to be assembled to match heating requirements. Cast-iron boilers used for residential and smaller commercial jobs usually are supplied completely assembled (packaged). A packaged boiler is one that includes the burner, boiler, controls, and auxiliary equipment. Larger boilers can be packaged units, or they can be assembled on the job site.

The terms *dry-base*, *wet-leg*, and *wet-base* may be used by manufacturers when describing cast-iron and other boilers. These terms refer to the location of the boiler's water-filled sections relative to the combustion chamber (firebox). They are defined as follows:

- *Dry-base* – Firebox is located beneath the water vessels.
- *Wet-leg* – Firebox top and sides are enclosed by water vessels.
- *Wet-base* – Firebox is surrounded by water vessels, except for necessary openings.

2.1.3 Steel Firetube and Watertube Boilers

Steel boilers are fabricated into one assembly of a given size and rating. The heat exchanger surface is usually formed by vertical, horizontal, or slanted tubes. Two types of steel boilers are the firetube boiler (*Figure 9* [A]) and the watertube boiler (*Figure 9* [B]). In the firetube boiler, flue gases are contained inside the tubes, with the heated water on the outside. In the watertube boiler, the heated water is contained inside the tubes, with the flue gases on the outside.

Smaller capacity steel boilers are usually vertical firetube units. Medium- and large-capacity steel boilers normally use horizontal or slant-mounted tubes. They can be either firetube or watertube boilers. Steel boilers range in size from small residential units of 150,000 Btuh to large systems of 23,500 MBh and above.

A common steel firetube boiler is the Scotch Marine boiler (*Figure 10*). As with most boilers, Scotch Marine boilers are made to produce either hot water or steam. They are used mainly for heating and industrial applications. As shown, this boiler has a low profile, allowing it to be installed in areas where headroom is limited. Steel boilers, including the Scotch Marine boiler, are usually built as packaged units.

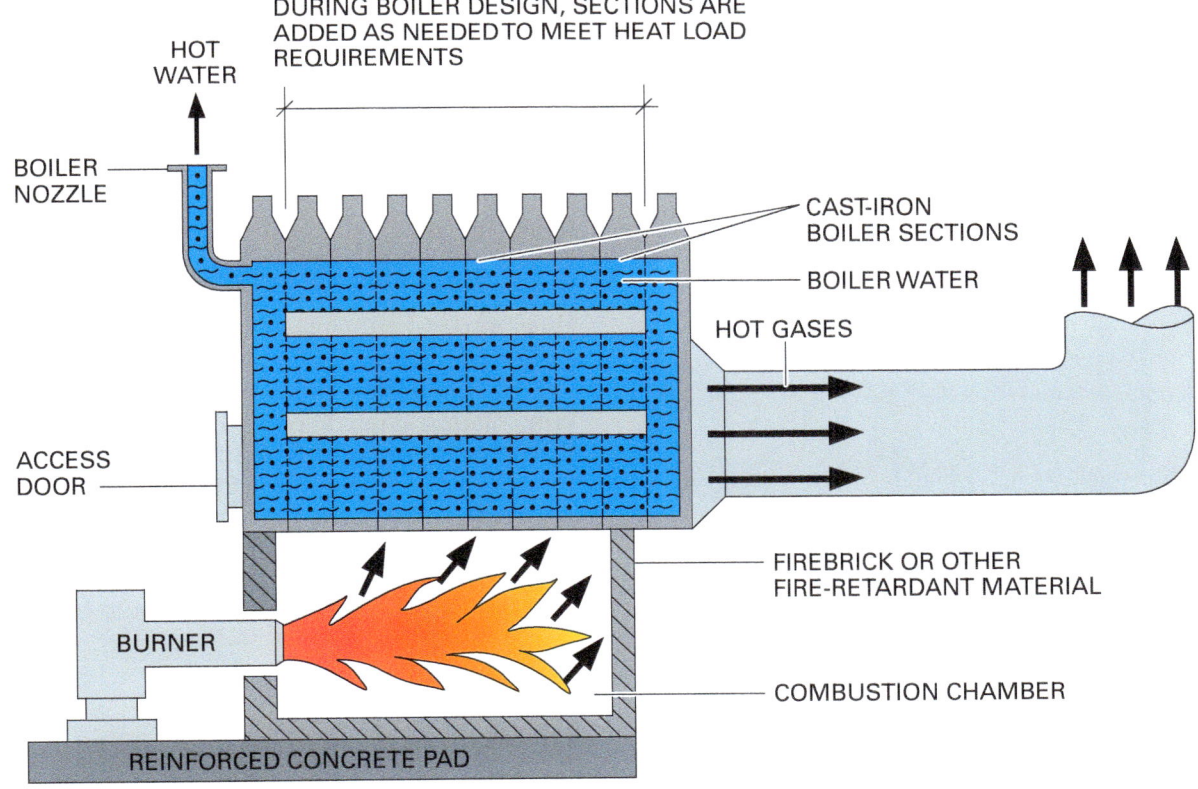

Figure 8 Cast-iron packaged boiler operation.

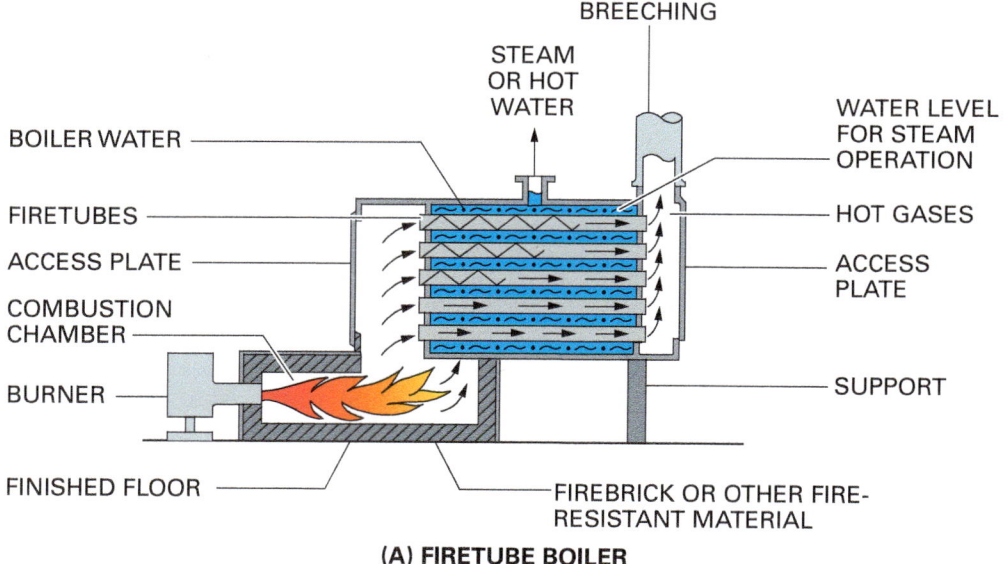

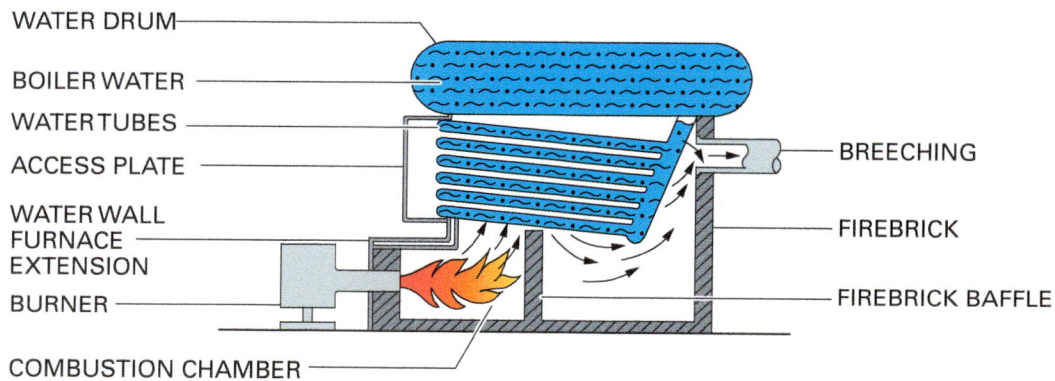

Figure 9 Firetube and watertube boilers.

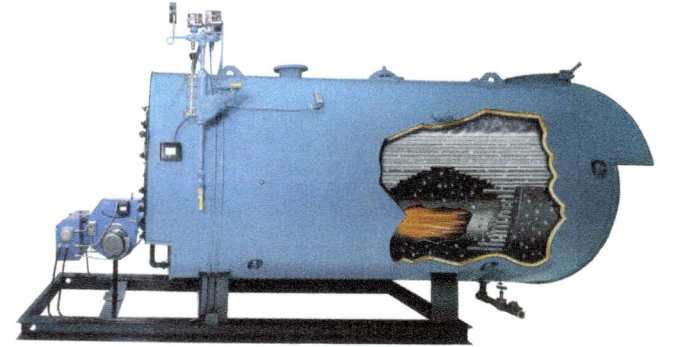

Figure 10 Scotch Marine boiler.

2.1.4 Steel Vertical Tubeless Boilers

Steel vertical tubeless boilers are another type of boiler used to produce hot water or steam (*Figure 11*). In this boiler, the water vessel surrounds the combustion chamber. A top-mounted, fuel-fired burner sends a cyclonic (spinning) flame down the length of the furnace. Flame retarders in the furnace retain the flame for maximum heat transfer.

As the flame swirls downward, the water spirals upward in the water vessel. Hot gases from the flame are carried up the outside of the water

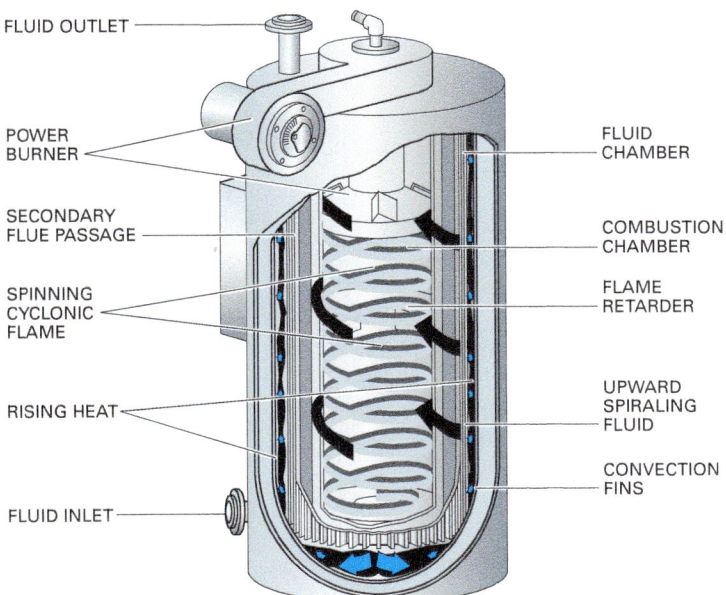

Figure 11 Vertical tubeless boiler.

vessel in a secondary flue passage convection area. Full-length convection fins welded on the outside of the water vessel transfer the remaining heat to the water.

2.1.5 Electric Boilers

Electric boilers use immersion resistance heater elements to heat the water in the boiler (*Figure 12*). Heating results from electrical resistance as the current flows through the heater elements. These elements are totally immersed in the water within the boiler pressure vessel. They are installed in flanges mounted at the top of the boiler assembly. This feature makes them easy to remove for inspection and maintenance.

In smaller boilers, heater elements are often mounted horizontally. Capacities of resistance-type electric boilers can range from those required for small residences and light commercial use up to a maximum capacity of about two megawatts, or 200 **boiler horsepower (BoHP)**. (One BoHP equals 9.803 kilowatts.) Boiler operation is controlled by an electric/electronic controller. This controller activates and deactivates the individual heater elements in response to load demands. It also allows for incremental loading of the electric service to reduce line fluctuations and power surges during startup.

> **NOTE:** Boiler horsepower (BoHP) should not be confused with brake horsepower, which is abbreviated BHP.

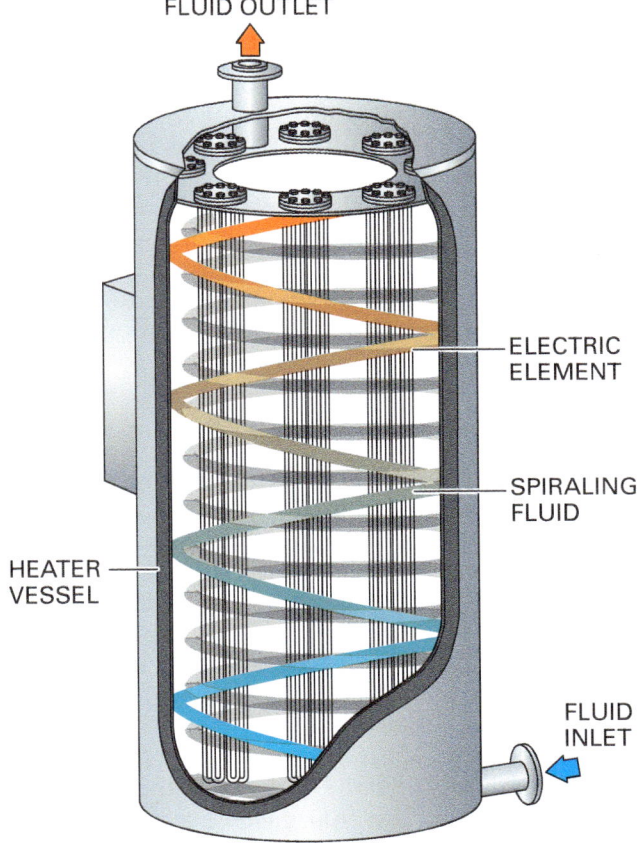

Figure 12 Electric hot-water boiler.

Another type of electric boiler, called an electrode boiler, relies on the electrical resistance of water for heating the water. In this boiler, streams of a saltwater solution flow past electrodes supplied with very high AC voltages

> **Indirect Electric Boilers**
>
> Although most electric boilers directly heat the water flowing through a building for heating purposes, others heat this water stream indirectly. By maintaining a reasonable pressure inside the boiler, the water inside can be heated well above the 212°F (100°C) sea-level boiling point. The building heating water loop is then routed through a heat exchanger in or near the boiler tank and is heated by the surrounding water inside. The water for building heat never makes direct contact with the boiler tank or boiler water.

(7,000VAC–13,000VAC). An electric current passes through the conductive water. Current flowing through the water heats it to steam by electrical resistance. The steam vapor leaves behind the dissolved salts.

The capacity of an electrode boiler is controlled by regulating the amount of water that comes in contact with the electrodes. If there is no contact, there is no current flow, and therefore no heating of the water. Partial contact will cause some heat to be generated. Maximum heat is generated when there is total contact between the electrodes and the water. Hot-water electrode boilers are available with capacities up to 20 megawatts (about 2,000 BoHP).

2.1.6 Commercial Condensing Boilers

In a standard boiler installation using natural gas, LPG, or fuel oil, the combustion gases pass through the heat exchanger to heat the water. Any residual heat in the flue gasses exits the unit and goes up the chimney or stack. The flue-gas heat content consists of sensible heat in the gas itself plus the latent heat in water vapor, which is a normal product of combustion. This thermal energy is wasted in such units.

Newer-design, high-efficiency boilers called *condensing boilers* (*Figure 13*), recover the sensible and latent heat in the flue gases by cooling the gas and condensing the water vapor. This process increases the thermal efficiency of a fossil-fuel boiler by as much as 25 percent. Most condensing boilers can extract more than 95 percent of the thermal energy released from combustion. The boiler design typically preheats the return water before it enters the boiler, reducing the amount of fuel required to heat the water.

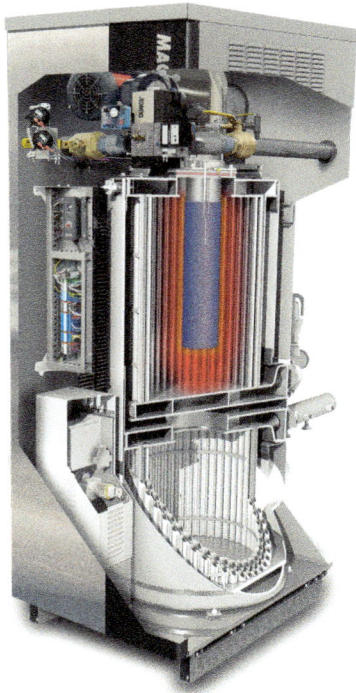

Figure 13 Cutaway of a modern commercial condensing boiler.

Because of the cooler exhaust exiting the boiler, condensing boilers do not require a standard high-temperature breeching or flue. Generally, low-temperature flue gases are vented through PVC piping, often near ground level. The condensed water collects in a sump and is drained away. In some cases, the condensate must be treated first to neutralize acids.

Condensing boilers offer many economic advantages, but there are some disadvantages to the concept. Their construction and operation are more complicated than a standard fuel-fired boiler. They require a variety of safety interlocks to ensure the heat-recovery and exhaust system is working properly before lighting the main burners. In cold climates, the exhaust pipe is subject to freezing the residual water vapor, which can lead to blockage and shutting down the system when it is needed most. Proper design and installation of the exhaust system can avoid this problem. In the long term, well-designed condensing boilers can make the most economical sense for building owners.

2.1.7 Boiler Operating/Safety Controls

A variety of controls are used to manage and maintain safe conditions on hot-water boilers. Most of these devices, such as the pressure/temperature gauge, pressure (safety) relief valve, and thermal/electronic probe safety and operating controls, are described in detail in NCCER Module 03203 (*Introduction to Hydronic Systems*). If you are unfamiliar with these devices, review them before proceeding.

A float-operated, low-water cutoff control (*Figure 14*) is a device used to protect boilers from damage caused by operating with too-low or no water. This control consists of a float that actuates a related electrical switch. The switch contacts are connected in the boiler burner control circuit.

If the water in the boiler drops below the minimum safe level, the float drops and the switch contacts open, turning off fuel supply to the burners. Operating levels for the control are factory-set and nonadjustable. Some low-water fuel cutoff controls have an opening through which a tool can be inserted and positioned to simulate a low-water condition. Other controls have this feature built in. *Figure 14* shows a typical installation using a float-operated, low-water fuel cutoff control. However, several locations and piping arrangements can be used. Because there is no normal water line to be maintained in a hot-water boiler, any location of the control above the lowest permissible water level is satisfactory.

Copper-finned, tube-style units transfer heat effectively with no significant volume of water contained on board. Because of this, they are subject to damage very quickly if the water flow is stopped while combustion continues. Since there is no water-filled vessel or space other than the tubes, a low-water float-type device is of no value for protection. For this reason, copper-finned tube units should be equipped with a water flow switch (*Figure 15*). These simple units have a paddle placed into the water piping. They are adjusted so that a flow volume insufficient for safe boiler operation will not push the paddle forward, allowing the integral switch to remain open. The switch is usually wired directly in series with the combustion controls to shut the unit down when water flow is too low or absent. Flow switches are also used in chilled-water systems.

Figure 15 Flow switch.

Because of the large amount of gas used to fuel large commercial burners, these boilers are equipped with special flame safeguard controls.

2.2.0 Centrifugal Pumps

A pump is a mechanical device designed to increase the energy of a fluid so that it can be moved from one location to another. Hydronic pumps are generally driven by electric motors and are used to pump a variety of liquids.

All pumps operate by the general principle that fluid flows into a low-pressure region at one end (the suction side) after being forced out by higher pressure at the other end (the discharge side).

Pumps are classified as either positive-displacement or centrifugal. Positive-displacement pumps are often used to pump high-viscosity liquids (thick liquids that flow slowly) or liquids containing particulate matter. Positive-displacement pumps are generally high-pressure pumps.

Centrifugal pumps are used to pump liquids that have low viscosities, such as water or other "thin" fluids. They are generally used in applications that do not require extreme pressure differentials, though this is not a limitation. They are by far the most common type used in commercial hydronic systems. To accommodate a variety of design situations and installations, they are available in a number of different configurations.

2.2.1 Centrifugal Pump Construction

A centrifugal pump operates by adding energy to the fluid at a relatively constant overall liquid velocity. Liquid entering the pump is rotated by an impeller. This rotation creates a centrifugal force within a stationary casing that accelerates (flings) the liquid to the outside of the casing. As the liquid leaves the vanes and is no longer accelerated,

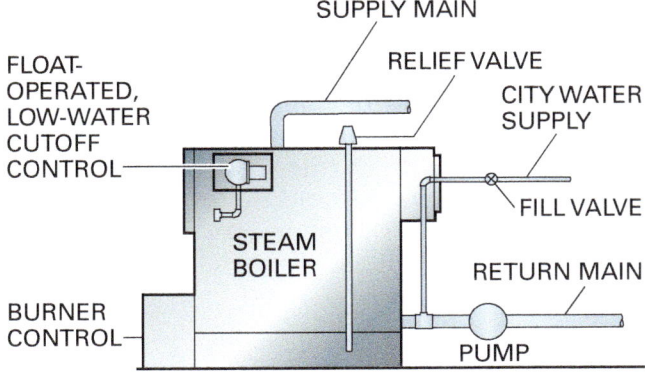

Figure 14 Float-operated, low-water cutoff control.

the liquid exchanges velocity for pressure and discharges at the higher system-supply pressure. It is the difference in pressure between the outlet and inlet of the pump that causes the liquid to circulate through the piping system. When operating normally, the liquid's inlet and outlet velocities should be the same, assuming the same pipe sizes. *Figure 16* shows a centrifugal pump.

The parts of centrifugal pumps may vary in size and shape depending on the manufacturer, but will have the same functions. Centrifugal pumps include the following primary parts, which are shown in *Figure 17*:

- *Bearings* – Parts that support the shaft and impeller in the casing.
- *Discharge port* – Where fluid leaves the pump casing and enters the piping system.
- *Impeller shrouds* – Parts that enclose the blades and keep the flow of fluid in the impeller area.
- *Impeller vanes or blades* – Parts of the impeller that accelerate the fluid.
- *Impeller* – The rotating part that accelerates the fluid and raises its pressure. There are many different types of impellers used for different purposes.
- *Mechanical seal* – Creates a rotating seal between the pump shaft and casing. In some pumps, they are used instead of packing to eliminate leakage.
- *Packing gland* – Another type of seal between the rotating shaft and the stationary pump casing that contains dense, moldable packing material forced into the gland space by a nut or retainer.
- *Pump casing* – The housing surrounding the shaft, bearings, packing gland, and impeller, enclosing the fluid inside the pump. Pump casings can be of split or single-piece castings.
- *Pump shaft* – The bearing-supported part that connects the motor to the impeller and rotates it.
- *Suction port* – Where fluid enters the pump casing.
- *Volute* – Sweeping shape within the pump casing that works together with open-vane impellers to help guide and accelerate the fluid toward the discharge port of the pump.
- *Wearing rings* – Replaceable rings used in some pumps to maintain impeller alignment and make up for the lack of shaft bearings at the far end of the impeller. They permit a small amount of fluid leakage between the impeller and the casing in the suction area. The leakage creates a hydraulic seal and helps the pump operate more efficiently.

A variety of impeller styles are available, depending upon the application and operating characteristics desired. *Figure 18* provides examples of some different styles. The most common designs used in commercial hydronic comfort systems are the single-suction closed impeller and the double-suction impeller. Some impellers, such as the single-suction closed style, can be precisely machined to a given diameter to provide very specific performance characteristics to meet specifications.

Cavitation can occur in a water-circulating pump when the pressure at the inlet of the impeller falls below the vapor pressure of the water, causing the water to vaporize and form bubbles. These bubbles are carried through the pump impeller to an area of higher pressure where they implode (collapse) with extreme force that can pit or erode the impeller vanes or outlet area.

The problem of cavitation can be eliminated by maintaining a minimum suction pressure at the pump inlet to overcome the pump's internal losses. This minimum suction pressure is called the *net positive suction head required* (*NPSHR*). Some symptoms that indicate cavitation is occurring in a pump include the following:

- Snapping and crackling noises at the pump inlet
- Pump vibration
- Variations in pump sounds
- Reduced or no water flow
- Oscillations or drop in discharge pressure

The type of centrifugal pump used depends upon the requirements. Some systems require that large amounts of fluids be pumped; others require high pressures. Two types of centrifugal pump designs that meet one of these requirements or the other are double-suction and multistage pumps.

2.2.2 Double-Suction Centrifugal Pumps

A double-suction centrifugal pump is used to pump larger volumes of fluid. The pump casing is designed to allow inlet fluid flow into the impeller through openings on both sides of the impeller. The fluid passes out through a single discharge opening. Due to the forces generated in moving large volumes of fluid, this design balances the axial force on the shaft and motor, minimizing bearing wear. *Figure 19* shows a double-suction centrifugal pump with a horizontally-split case.

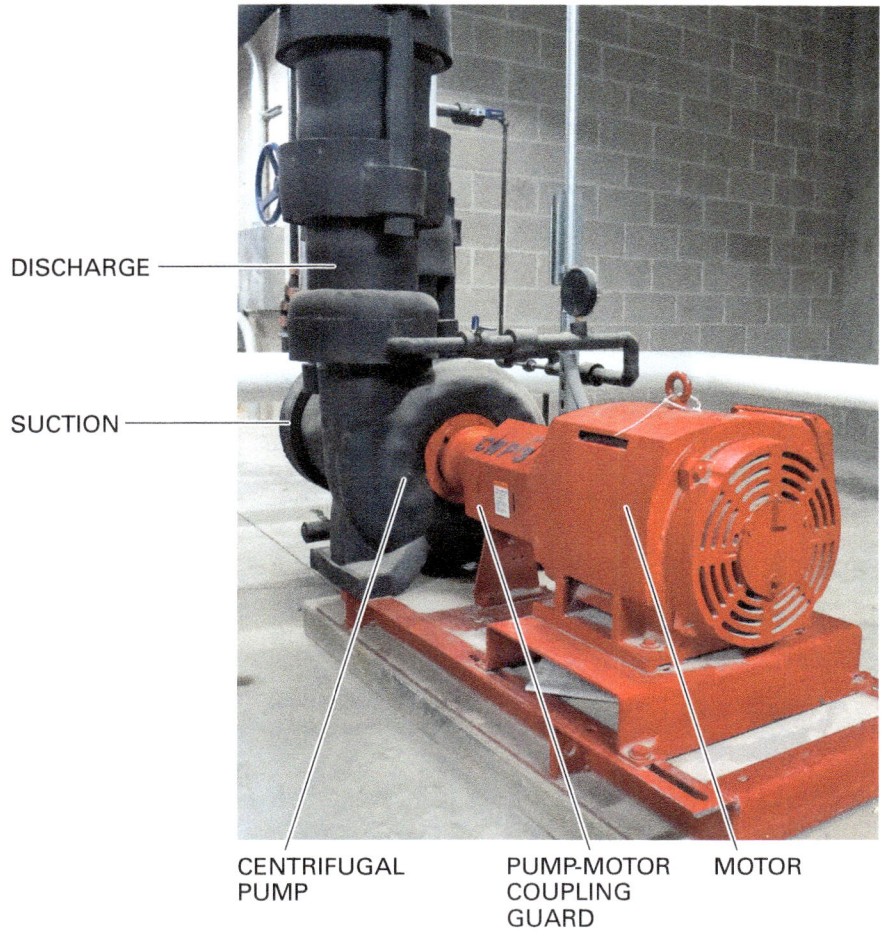
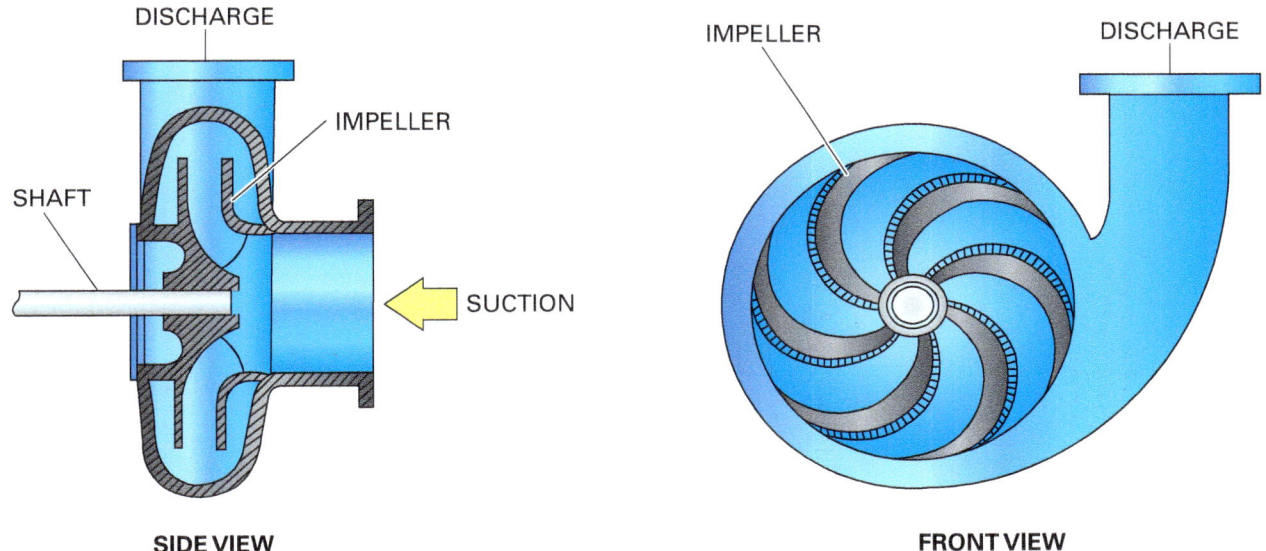

Figure 16 Centrifugal pump.

Pump Selection

The websites for many major pump manufacturers have an online software program that can be used to help select a pump for a particular application. User inputs to this program typically include the required flow rate, total pump head, speed of pump operation, and the model desired.

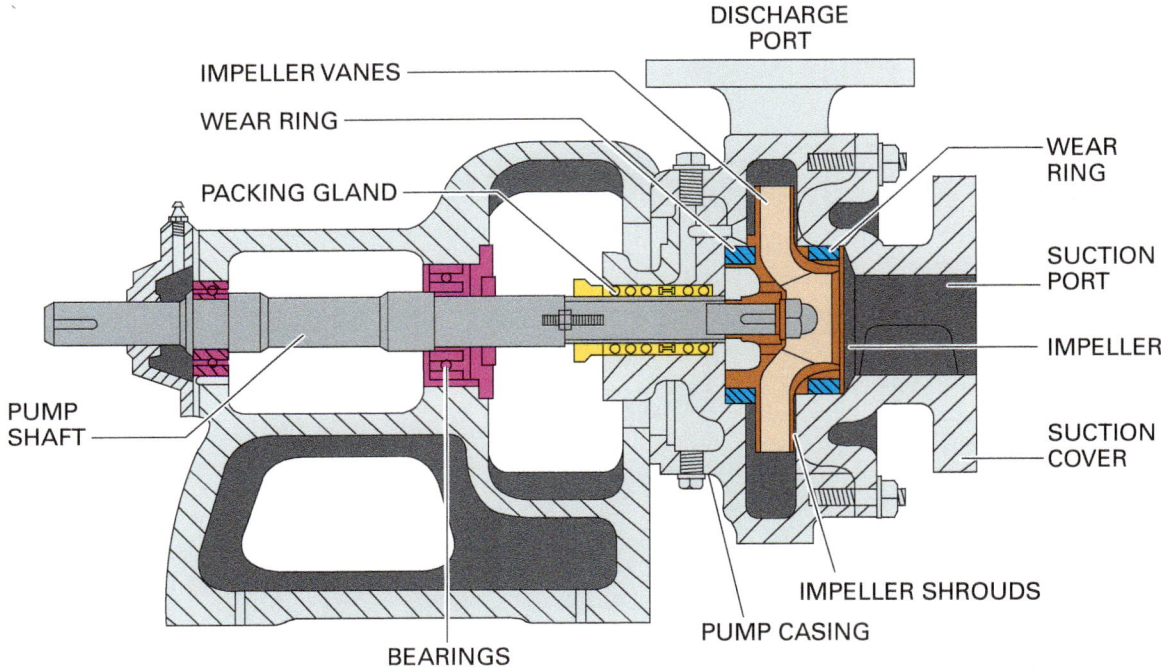

Figure 17 Centrifugal pump components.

Variable-Speed Centrifugal Pumps

Energy-saving variable-speed centrifugal pumps are becoming more common in hydronic systems, especially in zoned systems where loads are constantly changing. These pumps are now being driven by electronically commutated motors (ECMs) that are already popular for indoor blower and condenser fan applications in residential and commercial HVACR systems. An electronic control senses temperature conditions in the hydronic system and its zones. The control then signals the pump to increase or decrease its speed. This, in turn, delivers the correct volume of water at the correct temperature to satisfy the load.

Figure Credit: Armstrong Fluid Technology

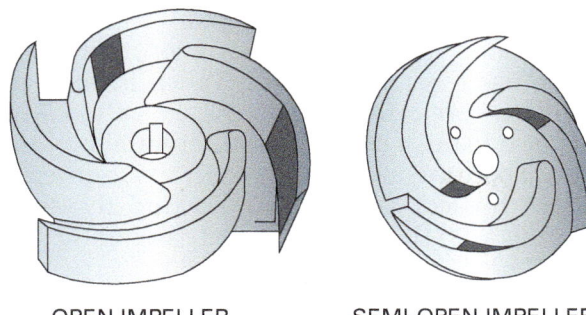
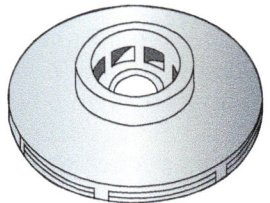
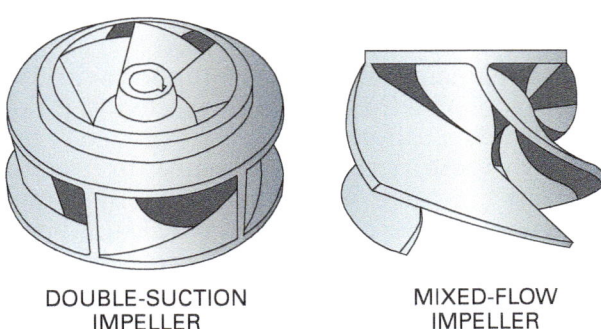

Figure 18 Impeller types.

2.2.3 Multistage Centrifugal Pumps

Multistage centrifugal pumps (*Figure 20*) contain two or more stages and can be either single- or double-suction pumps. Each stage has an impeller that is used to increase the pressure of the fluid being pumped. Fluid is discharged from one stage to the next through passages inside the pump casing until the desired discharge pressure is reached. Multi-stage feedwater pumps are needed for supplying large industrial boilers with operating pressures in the range of hundreds or even thousands of pounds per square inch.

2.3.0 Valves and Components

Common valves and their uses, as well as some specialty valves, have been discussed in earlier modules. Larger commercial and industrial systems generally use these same valves along with additional types for proper and flexible system operation.

2.3.1 Multi-Purpose Valves

In larger systems, a multi-purpose valve is commonly installed in the discharge side of the circulating pump (*Figure 21*). They are also referred to as *triple-duty valves*. This single valve functions as a shutoff valve, a check valve, and a balancing valve. It has a calibrated stem to return the valve to the set position after the valve is used as a shutoff valve. When in the set position, it acts as a balancing valve.

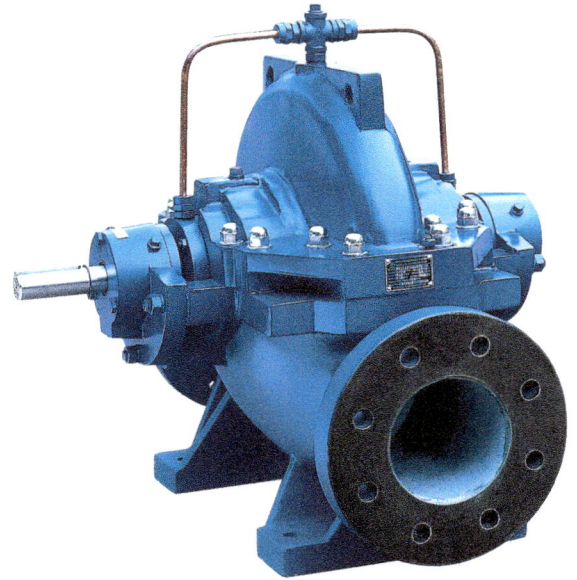

Figure 19 Double-suction centrifugal pump.

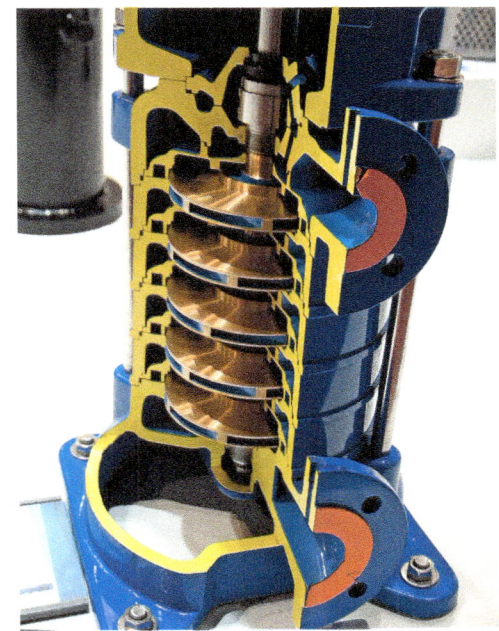

Figure 20 Multistage centrifugal feedwater pump.

The valve is tapped to install two readout valves connected to instrumentation that enable it to be used as a flow meter. When balancing a system, the pressure drop across the valve can be measured using a connected differential pressure gauge. The measured pressure drop can then be used with a conversion chart to find the corresponding total flow rate of water passing through the valve.

2.3.2 Balancing Valves and Flow Meters

Balancing valves (*Figure 22*) and flow meters provide a means of balancing a water system. A balancing valve can function as both a precision balancing device as well as an isolation valve.

As a balancing valve, it is adjusted to a predetermined setting that relates to the degree of resistance needed to achieve the desired water flow rate at a given pressure. The system designer usually determines the initial setting, which is based on a specific design pressure. The installer sets it accordingly. When set to the closed position, the valve is fully closed and serves as an isolation valve.

Most balancing valves are equipped with two ports that enable them to be connected to a flow meter. When a differential pressure gauge is connected to these ports, the pressure drop across the balancing valve can be measured. By using a conversion chart or calculator, the measured pressure drop can be converted to the corresponding total flow rate of water through the valve. Some balancing valves give direct readouts of flow, but they are rare.

Flow measurements can also be made by measuring the pressure drop through other devices. A venturi tube or orifice-plate installed in the system can be used for this purpose (*Figure 23*). Note that these devices are used to simply measure and monitor flow rates; they do not include any mechanism to adjust the flow.

The orifice plate may be installed between a single pair of flanges that have pressure differential taps built in by the manufacturer. These devices insert a specific fixed-area restriction in the path of water flow, thereby creating calibrated pressure drops at corresponding flow rates. The flow rate through the device can be determined by accurate measurement of this pressure drop. Measurement of the pressure drop is typically made using a manometer or differential pressure gauge. A differential pressure gauge used for this purpose typically reads in feet of head. For proper operation, the manufacturer's installation instructions must be followed.

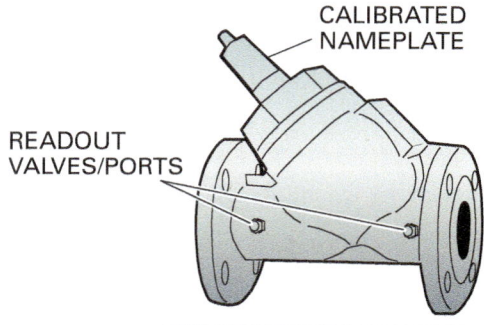

STRAIGHT-THROUGH MODEL

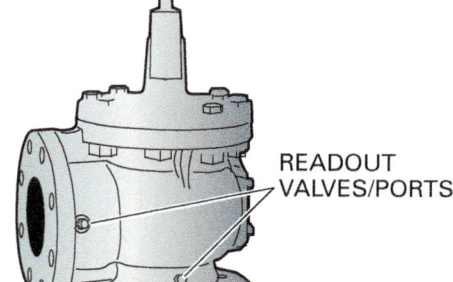

ANGLE MODEL

Figure 21 Multi-purpose valve.

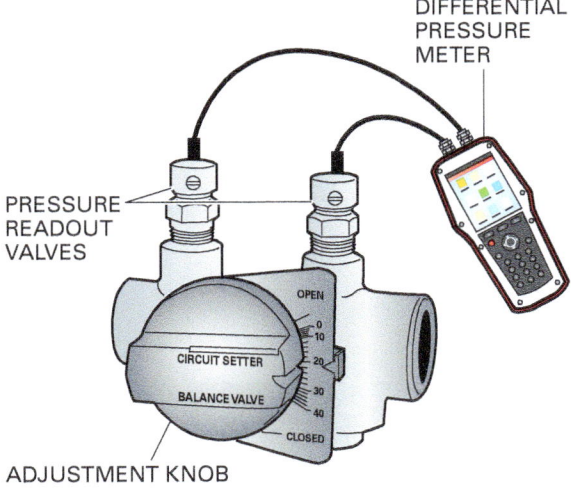

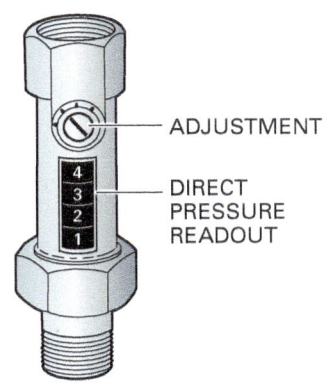

Figure 22 Balancing valve.

Because of its streamlined shape, a Venturi tube is more accurate than an orifice plate. Water passing through the reduced area of the Venturi throat increases in velocity and drops in pressure, creating a pressure differential between the inlet and the throat areas. After passing through the throat, the flow area gradually increases, which lowers the water velocity and increases pressure.

2.3.3 Butterfly Valves

Butterfly valves are popular in large piping systems as a simpler means of controlling flow, when compared to globe, ball, or gate valves. They are typically used in low- to medium-pressure and low- to medium-temperature applications. They generally weigh less than other types of valves because of their narrow body design.

A butterfly valve (*Figure 24*) has a round disc that fits tightly in its mating seat and rotates 90 degrees in one direction to open and allow fluid to pass through the valve. The butterfly valve can be operated quickly by turning the hand wheel or hand lever one-quarter of a turn, or 90 degrees, to open or close the valve. These valves can be used completely open, completely closed, or partially open for non-critical throttling operations. Note that they are not designed for balancing.

When the butterfly valve is equipped with a hand lever, the position of the lever indicates whether the valve is open, closed, or partially open. When the lever is perpendicular to the flow through the valve, the valve is shut; when the handle is in line with the flow, the valve is fully open. Butterfly valves that are 12" (30 cm) in diameter and larger are usually equipped with a handwheel or gear-operated actuator because of the large amount of fluid flowing through the valve and the great amount of pressure pushing against the seat when the valve is fully closed.

Butterfly valves have an arrow stamped on the side, indicating the proper direction of flow through the valve. They must be installed in the proper flow direction, or the seat will not seal and the valve will leak.

There are three common types of butterfly valve bodies in use: wafer, wafer lug, and two-flange. The wafer butterfly valve (*Figure 25*) is designed for quick installation between two flanges. No additional flange gasket is needed between the valve and the flanges because the valve seat is lapped over the edge of the body to make contact with the valve faces. Bolt holes are provided in larger wafer valves to help line up the valve with the flanges.

Wafer lug valves (*Figure 26*) are the same as wafer valves except that they have bolt lugs completely around the valve body. Like the wafer valve, no additional gasket is needed between the valve and the flanges because the valve seat is lapped over the edges of the body. The lugs are normally drilled to match ANSI 150-pound steel drilling templates. The lugs on some wafer

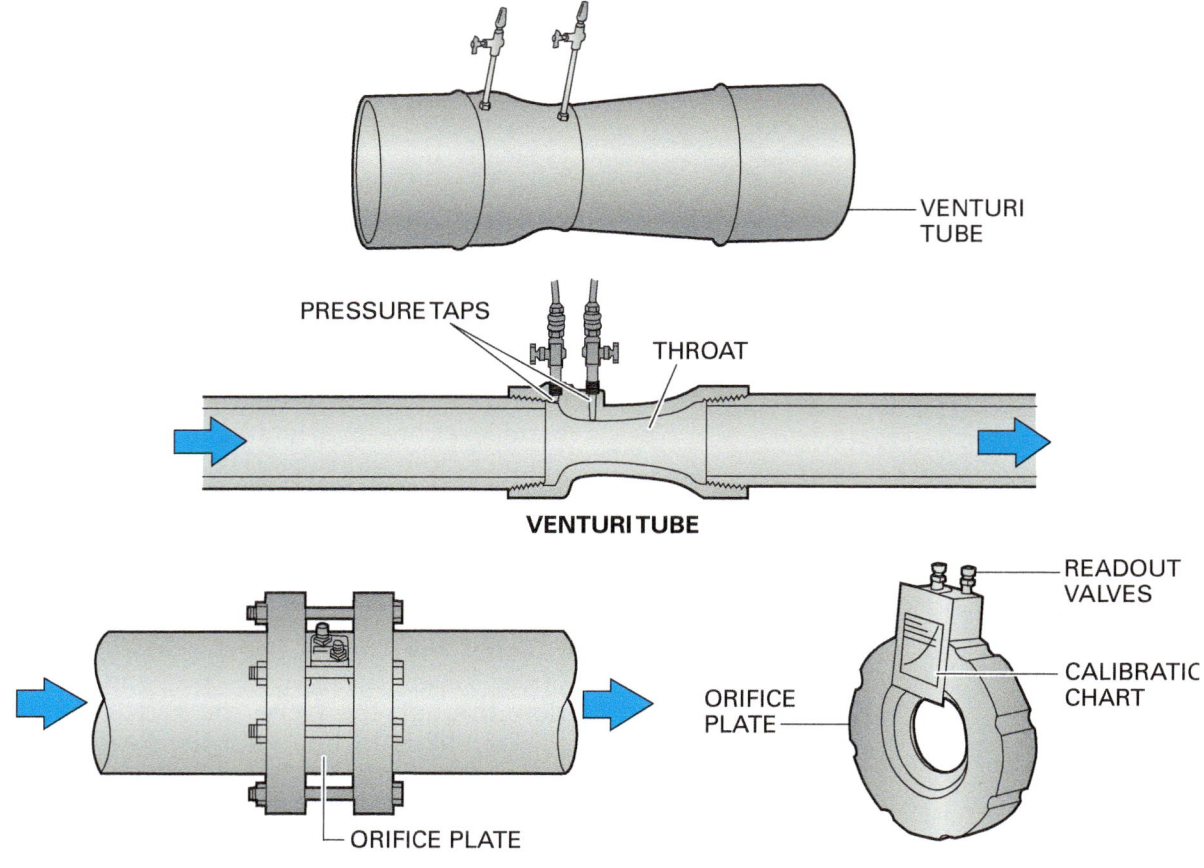

Figure 23 Venturi tube and orifice plate flow-metering device.

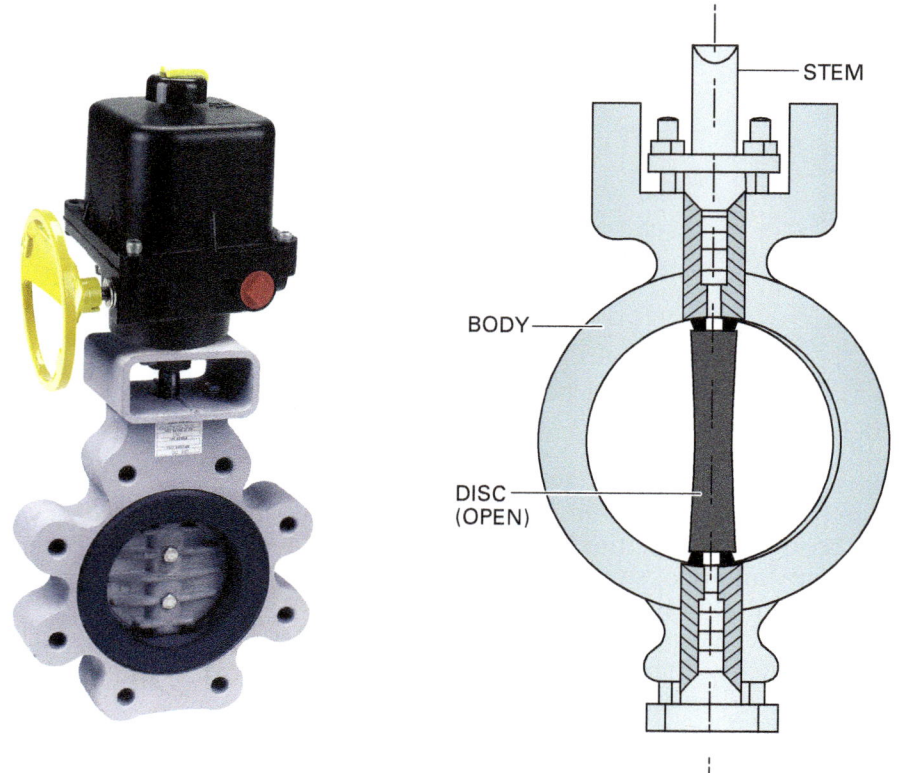

Figure 24 Butterfly valve.

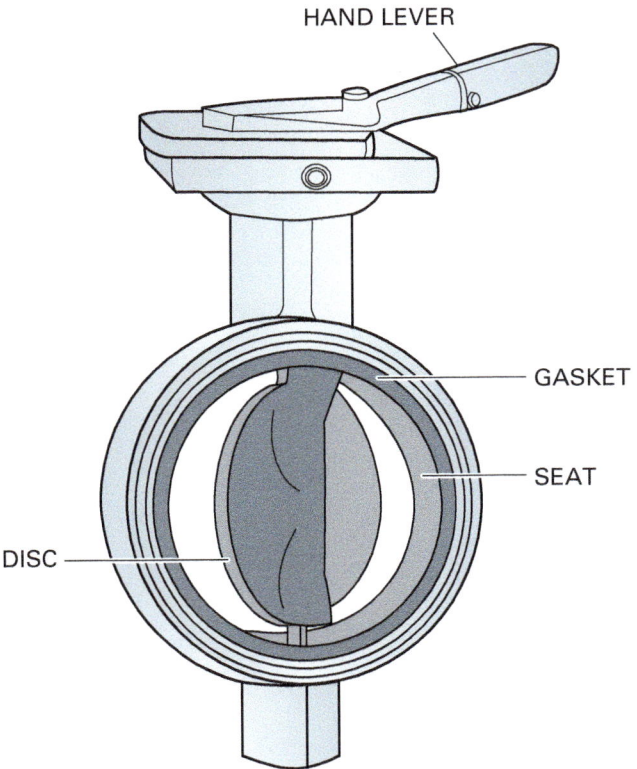

Figure 25 Wafer butterfly valve.

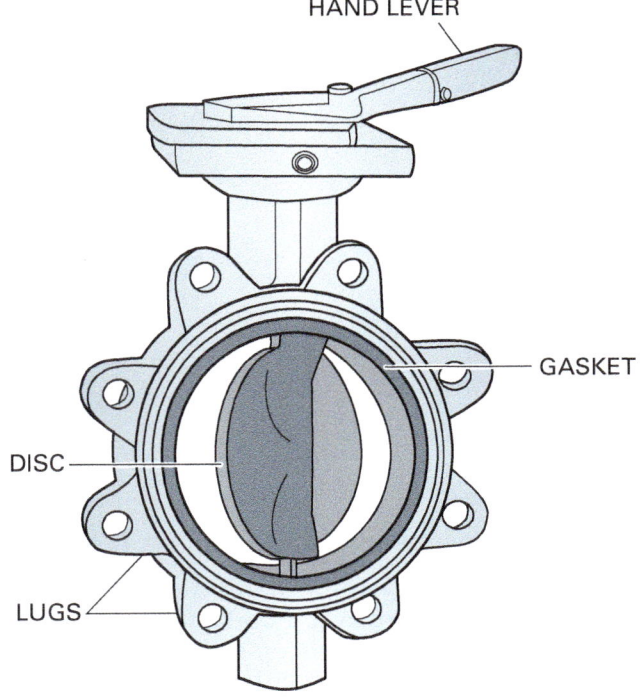

Figure 26 Wafer lug butterfly valve.

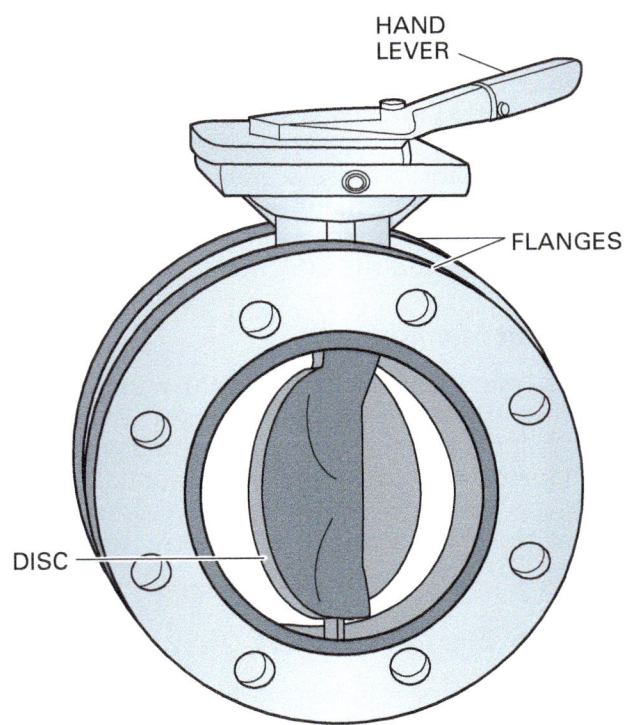

Figure 27 Two-flange butterfly valve.

lug valves are drilled and tapped so that when the valve is closed, downstream piping can be dismantled for cleaning or maintenance while the upstream piping is left intact. When tapped wafer lug valves are used for pipe end applications, only one pipe flange is necessary. During installation, both wafer and wafer lug valves must be in the open position prior to tightening the flange-mating hardware to prevent damage to the valve seat.

The body of a two-flange butterfly valve is made with a flange on each end (*Figure 27*). The valve seat is not lapped over the flange ends of the valve, so flange gaskets are required between the flanged body and the mating flanges. The two-flange valve is made with either flat-faced flanges or raised-face flanges, and the mating flanges must match the valve flanges.

Waterlogged Expansion Tanks

You should suspect a waterlogged expansion tank if the boiler relief valve discharges whenever the boiler water is being heated. With no room for the expanding water to go, the system pressure can rise very quickly. If this happens, check the expansion tank and any related air-control components for leaks.

Figure 28 Pressurized expansion tanks.

2.3.4 Expansion and Compression Tanks

Commercial hot water systems also incorporate expansion or compression tanks (*Figure 28*), as is the case with smaller systems. The tanks are often larger to accommodate the greater volume of water in the system, since the water requires more room for expansion as it heats up.

If a pressurized expansion tank fails to hold the pre-charge of air, you may suspect a leaking air valve or a leaking bladder/diaphragm. Most bladder-type tanks used commercially are larger than those used in smaller applications. The bladder is accessible for replacement through a large flange. With most diaphragm tanks (now used less often than bladder tanks), the diaphragm is not accessible and the tank must be replaced.

2.3.5 Air Management

Air management in commercial systems is extremely important to ensure proper water circulation and prevent noise. The same types of air management devices used in residential applications are generally employed in commercial systems. However, because of the greater volume of water and the potential for much larger volumes of air to collect and create problems with circulation, large commercial systems are often equipped with an air separator (*Figure 29*). This type of air separator is also used in chilled-water systems. They are highly efficient at collecting entrained air from water and enable large systems to operate more effectively.

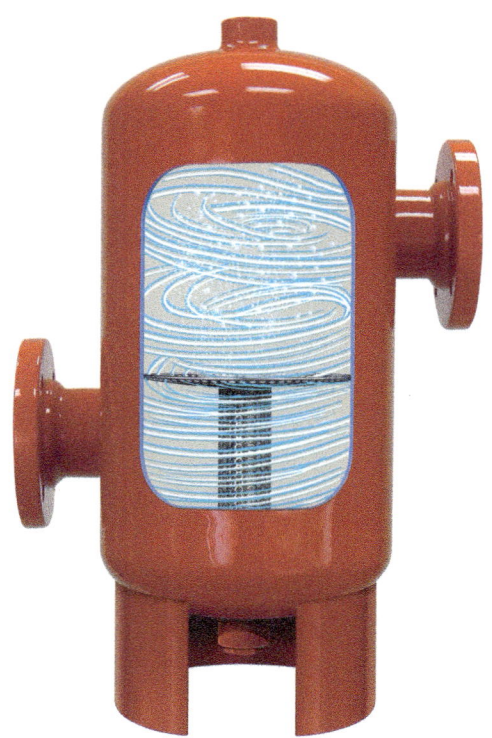

Figure 29 Commercial air separator.

The unit shown is available in sizes large enough to accommodate 36" (DN 750) pipe. It reduces the water velocity to a speed that allows the air bubbles to rise, preventing them from being swept along by the water stream. At the same time, it encourages a centrifugal spin in the water stream. This causes the heavier water to move outward and the lighter air to remain

The Science of Air Separation

The solubility of air in water at different temperatures and pressures is based on a law of physics known as Henry's law. This law describes how both the pressure of the gas and temperature of water affect the gas's solubility. As system pressure / air pressure increase, water is able to keep more air in solution at a given temperature. When system pressure decreases, air dissolved in the system water forms bubble and separates until the dissolved air pressure again equals system pressure.

Higher water temperature decreases its ability to hold air in solution. As temperature falls, water can hold more air in solution. The result is that air separates more readily from water at a higher temperature and lower pressure. This also means that chilled-water systems, which must operate at higher pressures and lower temperatures because of the needs of the water distribution system, represent the most challenging air-management condition.

closer to the center. The air is collected there and discharged at the top of the tank.

Air management in hydronic systems is accomplished through one of two strategies: air control or air elimination. In air control (*Figure 30*), a compression tank is mounted on the air separator discharge and air is discharged to it. The air and water are in contact inside the tank, but the air is separated from the water in circulation. Heated water expands into the compression tank, pushing against the air cushion. In air elimination (*Figure 31*), a bladder-type tank is used on the suction side of the pump, providing the area for heated water to expand into. The air is vented to atmosphere through a high-capacity air vent on top of the separator.

2.4.0 Common Commercial Piping Systems

Water piping systems provide routing and distribution of hot water, chilled water, or both. Piping systems used in residential and light commercial applications tend to be simple. (Those piping systems were covered in NCCER Module 03203, *Introduction to Hydronic Systems*.) Large commercial and industrial piping systems tend to be more complex because the heating and cooling loads in a large building can vary widely. Some areas of a large building might need cooling, while at the same time another area would require heat.

Although there are some two- and three-pipe systems remaining in older buildings, they are rare. For this reason, this module will focus on conventional systems in common commercial use today.

2.4.1 Four-Pipe Systems

The four-pipe system is a dual-temperature piping system. Because of its design simplicity, room control, and operational economy, it is commonly used in large commercial applications. It is constructed of two separate two-pipe circuits that use separate supply and return lines to carry hot and chilled water to the terminals. One supply and return circuit carries the hot water, while the other supply and return circuit carries the chilled water. *Figure 32* shows a simplified four-pipe system. The chilled and hot water streams never mix together in the four-pipe system.

In this system, three-way valves are often used at the input and output of a single coil. The system also uses terminals with separate chilled- and hot-water coils. This allows heating and cooling to potentially occur at the same time, thus providing for humidification and reheating of the circulated air. However, this operating mode is rare because it uses a great deal of energy.

Two-way valves can also be used at the input to each of the coils. However, when two-way valves close, all flow ceases in the portion of the circuit. When many or all the valves close, this causes problems for the pump. To overcome this problem, the pump must be cycled on and off in response to certain levels of demand. A better way is to modulate the speed of the pump, but an even better option is to use three-way valves. Systems with three-way valves typically maintain the same amount of flow at all times. Regardless of whether the water is directed by the valve through a coil or around it, the flow rate in the circuit remains steady.

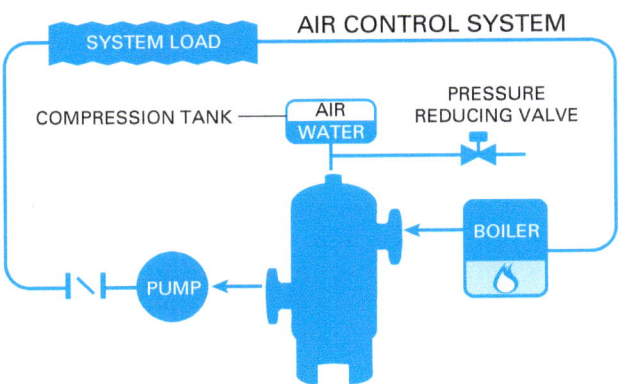

Figure 30 Air-control component arrangement.

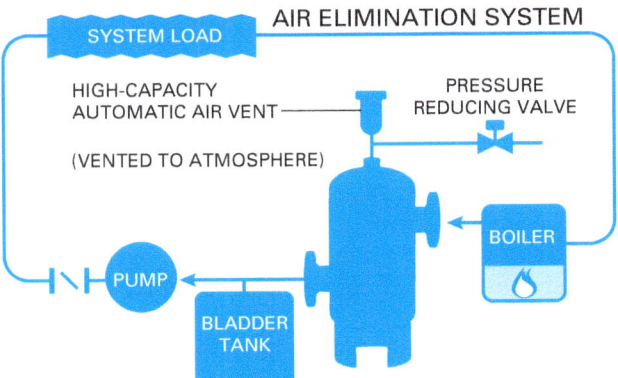

Figure 31 Air-elimination component arrangement.

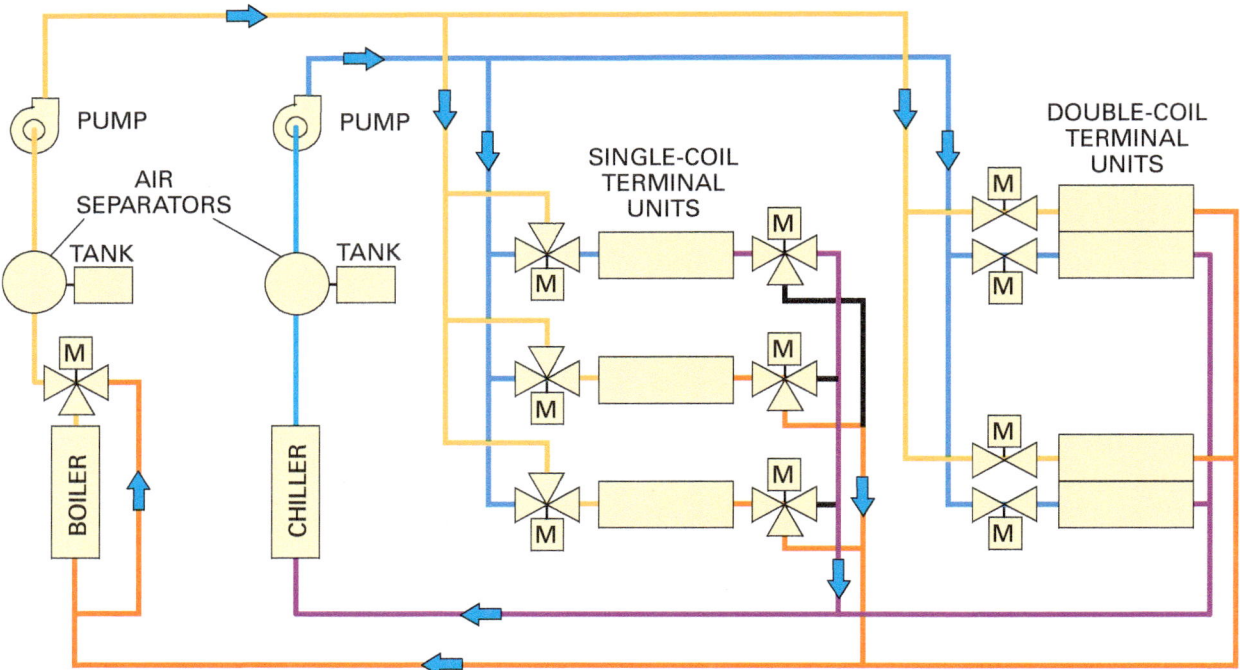

Figure 32 Four-pipe water system with one-coil and two-coil terminals.

The hot- and chilled-water piping systems are separate, so pressure regulation is less critical in the four-pipe system than in older two- and three-pipe systems. Also, much higher hot-water supply temperatures are possible, especially when separate heating coils are used in the terminals. This allows a reduction in the size of the piping for that circuit. Another advantage of the four-pipe system is that various types of heating terminals, such as baseboard and finned-tube units, can be connected to the system in spaces that do not need cooling.

2.4.2 Primary-Secondary Piping Systems

Primary-secondary hot-water or chilled-water systems are used in applications that have peak and varying loads. They supply variable-flow, constant-temperature chilled/hot water using methods that provide for energy conservation. *Figure 33* shows a simplified diagram for a primary-secondary chilled-water system. For a hot-water system, the chillers would be replaced by boilers.

As shown, a primary chiller loop for this system consists of two 500-ton chillers and one 250-ton chiller, each with a related circulating pump. Each of the primary pumps is sized to handle only the low head and steady flow rate for the related chiller. The building load is divided into two secondary loops, each with a secondary circulating pump, terminal devices, and control valves.

Again, each of the pumps in the secondary loop is sized to handle the flow and pressure requirements for this specific loop. In this system, primary loop pumps circulate the water through the primary loop; the secondary loop pumps move water through the secondary loops. The transfer of chilled water between the primary loop and each of the secondary loops is made by a short section of pipe (the bypass line) to both the primary and secondary circuits. Modulating control valves in the bypass loops maintain the required differential pressure across the secondary loops to ensure stable system flows.

During periods of peak load, all three chillers in the primary loop produce 45°F (7°C) water to supply the secondary loops. The pump in each of the secondary loops is running and the maximum amount of flow is in demand.

During periods of reduced load, only the 250-ton chiller and its pump may need to operate to supply 45°F (7°C) water to the terminals. With variable-speed pumps used in the secondary loops, each of the secondary pumps may be operating at different speeds to satisfy the individual loop flow requirements. Different combinations of chillers and pumps are used in the primary and secondary loops to satisfy the cooling load as it changes.

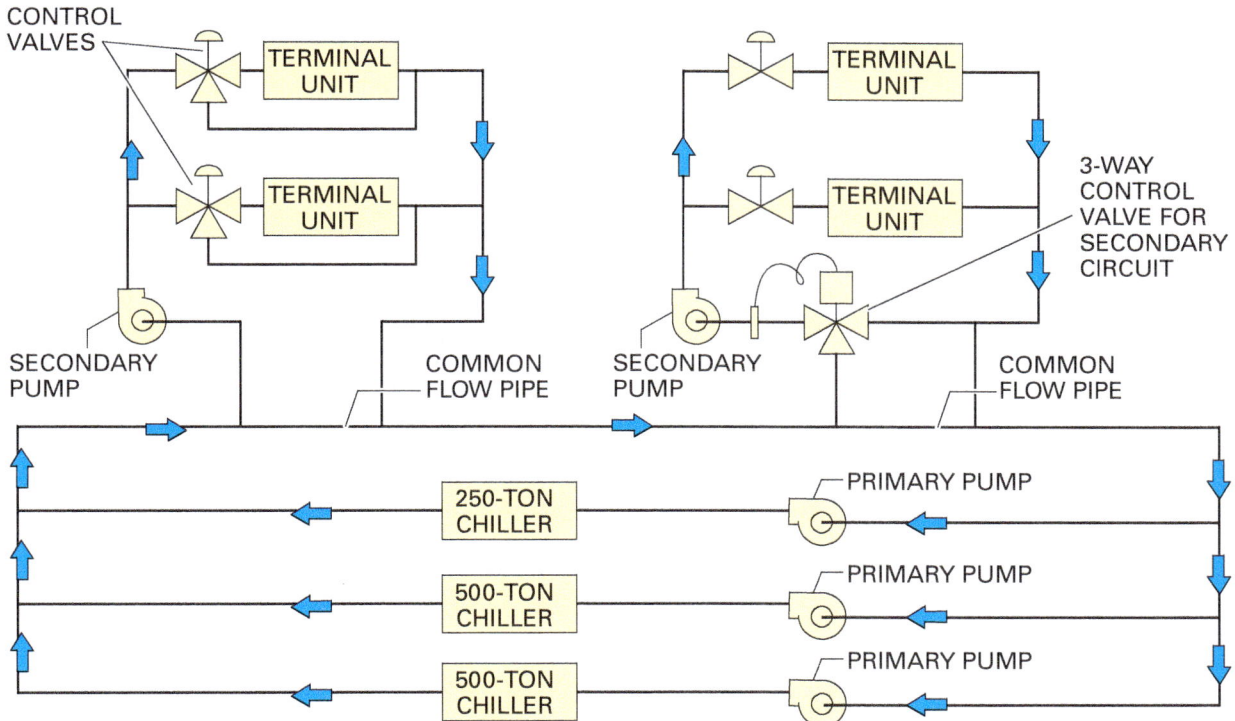

Figure 33 Simplified primary-secondary chilled-water system.

2.5.0 Water System Balancing

The basic concepts regarding water balance, as well as the calculation of needed flow rates and pumping requirements, were presented in NCCER Module 03203. Here, the hydronic system balancing procedures will be discussed. In a large building, the procedure can be a lengthy and complex task. Hydronic system balancing is most often done by firms that specialize in the procedure, especially when a new building or system is first commissioned. There are several reasons for this. The job often requires skills that many technicians do not possess. In addition, especially when a new building is being commissioned, owners and mechanical engineering firms prefer to have the analysis of system performance provided by a company other than the installing contractor.

Testing, balancing, and adjusting hydronic heating or cooling systems can best be done by following a systematic procedure. The entire process should be carefully planned and performed. All activities, including organization, procurement of test instruments and equipment, and execution of the balancing procedure, must be properly scheduled. Coordinate air balancing with water balancing, because many systems function differently on a seasonal basis and temperature performance is a big factor when testing and balancing.

Before attempting to balance a water system, the boiler and/or chiller and related components should be operating in accordance with the manufacturer's instructions. Normally, the system startup procedures include making sure that the chiller is correctly charged with refrigerant and that the boiler and chiller water systems are correctly filled with treated water. They also include ensuring that all chiller/boiler-related safety and operating controls have been tested, adjusted, and set for proper operation.

2.5.1 Pre-Balance Checks

Before beginning the balancing procedure for a selected water system, complete the following pre-balancing procedure:

Step 1 Obtain all technical data, including original plans and specifications or as-built plans pertaining to the hydronic equipment. This includes the following:
- Pump curves and performance data
- Pressure, flow-temperature characteristics, and other performance data for boilers, chillers, heat exchangers, and coils
- Motor nameplate data and ratings
- Performance characteristics of balancing and flow-meter devices

Step 2 Locate and study the following diagrams:
- Temperature-control and flow-control valve characteristics
- Special piping circuits

Step 3 Locate and record the ratings and settings of pressure-reducing valves, pressure-relief valves, and temperature-relief valves.

Step 4 Locate the operating and maintenance instructions for important equipment.

Step 5 Obtain or prepare a schematic layout of each system. Identify connection points of primary and secondary piping circuits, manual valves, automatic valves, flow meters, air vents, coils, and expansion tanks. Note any changes in the design layout that were made during construction (use as-built schematics). Number all water-flow balance points and metering points for rapid identification and reporting.

Step 6 Study the water system piping arrangement to determine the steps and sequence required for balancing and the instruments needed for the task. Note system features that would contribute to an unbalanced condition, such as dirty fins on terminals. Determine which circuits must be balanced by temperature rather than fluid flow because of the absence of flow-regulating devices. Note the location of strainers and other piping configurations that may contribute to the collection of dirt and sludge.

Step 7 Obtain all necessary instruments and calibration data. If more than one instrument of a similar type is used, calibrate each one to the same level of precision.

Step 8 Make a complete visual inspection of the system and make sure that all air distribution equipment and controls are functioning properly. Do the same for hydronic equipment and controls.

2.5.2 Water System Balancing Procedure

Proper water balancing is essential. The following general procedure can be used for testing, balancing, and adjusting a water system:

Step 1 Make sure that all preliminary visual checks and service and maintenance tasks have been performed.

WARNING! When working with water systems, watch out for rotating, pressurized, or hot components. Follow all safety instructions labeled on the equipment and given in the manufacturer's service manual. Air balancing requires that equipment and systems be energized and operational; use extreme caution when working near energized load and circuit controls.

Step 2 Check all electrical wiring in power and control circuits.

Step 3 Check that all manual valves are open or properly preset.

Step 4 Verify that all automatic valves are in the fully open position.

Step 5 Look for broken thermometers and gauges on the equipment that could be misleading while balancing.

Step 6 Read and record the system pressure at the pump with the pump de-energized.

Step 7 Operate the system, bleed the air from the piping, and allow flow conditions to become stabilized.

Step 8 Record the operating voltage and amperage draw of all electrical motors.

Step 9 Record the pump speed (rpm).

Step 10 Slowly close the balancing cock in the pump discharge line with the pump running and record the shutoff discharge and suction pressures at the pump gauge connections.

Step 11 Use the shutoff head to determine the pump operating curve (impeller size) and compare it with the design curve data. Check the pump differential at the pump. Use one gauge to read the differential pressure. If two gauges are used, the gauge readings must be corrected to the center line elevation of the pump. If there is a discrepancy in the pump output compared to design specifications, either correct the velocity head or report the information to the proper authority.

Step 12 Record the pump discharge pressure by opening the discharge balancing cock slowly to the fully open position. Record the suction pressure and total head in feet at the same time.

Step 13 Using the total head and the pump curve established in Step 11, read the water flow. If the total head is greater than design specifications, the water flow will be less than design. If the total head is less than design, water flow will be more than the selected rating. If the water flow is more than the selected rating, the pump discharge pressure should be increased by partially closing the balancing cock until water flow is approximately 110 percent of design specifications.

Step 14 Start any secondary pumps and readjust the balancing cock in the primary circuit pump discharge if necessary. Record the new adjusted readings.

Step 15 Take a complete set of pressure-drop readings through all equipment such as chillers, boilers, hot-water coils, and chilled-water coils before adjusting any balancing cocks at the equipment. Compare the readings with design specifications. Determine which units are above/below design requirements.

Step 16 Make a preliminary adjustment in the balancing device on all units with high water flow. Adjust them to about 10 percent above the design flow rate.

Step 17 Perform another complete set of pressure and amperage draw readings in the pump system. If the pump total flow has fallen below design, open the balancing cock at the pump discharge to raise the pump output, bringing the flow rate within 105–110 percent of design.

Step 18 Decrease the flow a second time on all units that are still receiving more than their design flow rate. This will also serve to increase water flow through units that are not presently receiving sufficient water.

Step 19 Keep repeating the procedure (adjusting unit flow devices then pump output) until all units are within ±10 percent of the design water flow.

Step 20 Make a final check of the water flow rates at the pump and equipment and of the pump motor amperage draw and record the findings.

Step 21 After all the balancing is completed, mark all balancing cocks, gauges, and thermometers at the final settings or range of operation for future reference.

Step 22 Check all water flow safety shutdown controls for proper action.

Step 23 Prepare and submit all necessary reports.

These are the basic steps that can be used for balancing regular water (hydronic) systems. However, special types of hydronic distribution systems, such as condenser water and cooling tower systems, steam and hot-water boilers, and heat exchangers will require additional procedures by certified professionals.

Additional Resources

NCCER Module 03203, *Introduction to Hydronic Systems*.

Air-Cooled Chillers, TDP-622. Current edition. Syracuse, NY: Carrier Corporation.

Water-Cooled Chillers, TDP-623. Current edition. Syracuse NY: Carrier Corporation.

Refrigeration and Air Conditioning: An Introduction to HVAC/R, Larry Jeffus. Current edition. New York, NY: Pearson.

2.0.0 Section Review

1. Although they can be built to withstand higher pressures, the typical maximum working pressure of a low-pressure hot-water boiler is _____.
   a. 250 psig (1,724 kPag)
   b. 160 psig (1,103 kPag)
   c. 30 psig (207 kPag)
   d. 10 psig (69 kPag)

2. A centrifugal pump operates by accelerating a(n) _____.
   a. impeller
   b. piston
   c. fluid
   d. refrigerant

3. A butterfly valve can be fully opened or closed by rotating the handle _____.
   a. two full turns
   b. one full turn
   c. one-half turn
   d. one-quarter turn

4. In a four-pipe hydronic system, what type of valve is used on each single-coil terminal?
   a. Butterfly valve
   b. Three-way valve
   c. Venturi tube
   d. Multi-purpose valve

5. Which information is *not* needed when performing a pre-balance check?
   a. Chiller ref_rigerant type
   b. Motor nameplate data
   c. Pump curves
   d. Boiler performance data

SECTION THREE

3.0.0 CHILLED-WATER SYSTEMS

Objective

Explain the basic concepts of chilled-water systems and the related components.
a. Explain the basic concepts of chilled-water cooling systems.
b. Identify various types of chillers and their common controls.
c. Identify various types of cooling towers and evaporative condensers.

Performance Tasks

1. Identify the major components of commercial hot-water heating and chilled-water cooling hydronic systems.

Trade Terms

Flash economizer: An energy-conservation process used in chillers that accomplishes intercooling of the refrigerant between the condenser and evaporator using a flash chamber or similar device. Flash gas of coolant produced in the flash chamber flows directly to the second stage of the chiller compressor, bypassing the evaporator and the first compressor stage.

Lithium bromide: A caustic salt solution commonly used in absorption chillers.

Secondary coolant: Any liquid, such as water, that is cooled by a system refrigerant and then used to absorb heat from the conditioned space without a change in state. Also known as indirect coolant.

Thermal economizer: An energy-conservation process used in chillers in which warm condensed refrigerant is brought into contact with the coldest (inlet) water tubes in a condenser or heat exchanger. This causes the condensed refrigerant to subcool to a temperature below the condensing temperature.

A chilled-water system eliminates the need to circulate refrigerant through a building to provide comfort cooling. Instead, a piece of equipment called a *chiller* (*Figure 34*) uses refrigerant to cool the water. The chilled water is circulated throughout the building and back to the chiller where the process is repeated. This uses basic water piping throughout the building rather than refrigerant piping, which is more expensive and represents a unique challenge to install in large buildings.

After chilled water is delivered to the building, it may be circulated through a large cooling coil in a central air handler to cool a large area, or it can be delivered to the conditioned space and circulated through a small cooling coil housed in a terminal to cool the room or zone.

3.1.0 Basic Chilled-Water System Concepts

This section focuses on chilled-water systems that operate above 40°F (4°C) and are used for comfort cooling. Other types of chilled-water systems may be used to provide comfort air conditioning or support an industrial process. These systems cool a brine or other type of water and antifreeze solution, operating at water temperatures below 32°F (0°C).

Chilled-water systems are used to cool buildings that have either a separate heating system or no heating system at all. They can also be used as part of a dual-temperature system. *Figure 35* shows a simplified diagram of a basic chilled-water system. It consists of a chiller, cooling terminal, chilled-water pump, and related chilled-water piping. The chiller unit shown is a mechanical refrigeration system. Some chilled-water systems use absorption-type chillers to produce cooling.

The chiller in *Figure 35* is based on the same refrigeration cycle as all other common cooling systems. The chiller evaporator (cooler) is a water-to-refrigerant heat exchanger through which both chilled water and system refrigerant flow. The vaporization of the liquid refrigerant cools the chilled water.

Figure 34 Packaged air-cooled chiller.

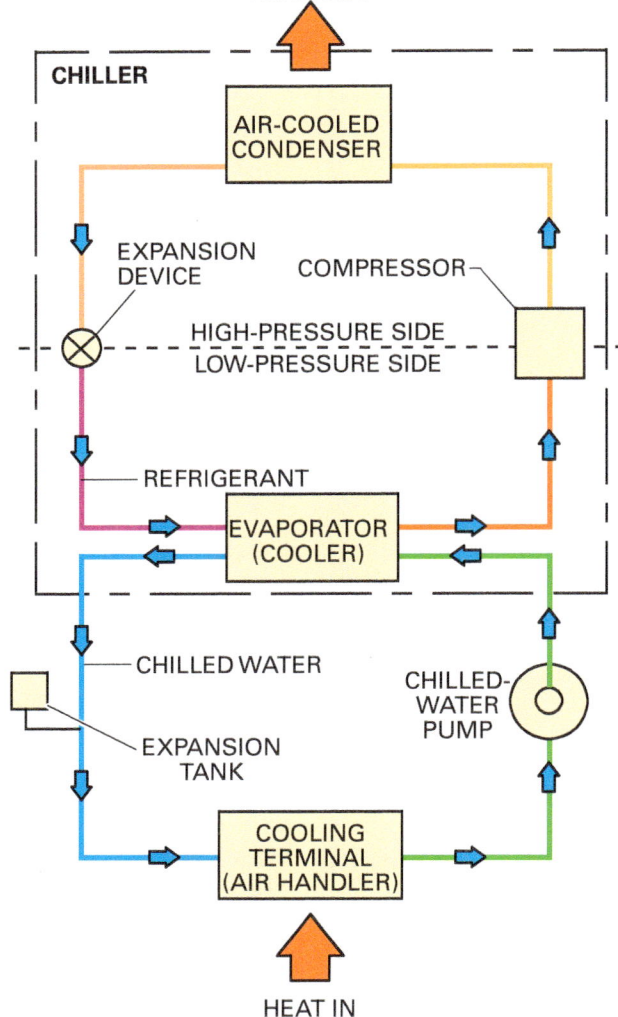

Figure 35 Basic chilled-water system.

For comfort cooling applications, the chilled water temperature supplied to the building is generally around 45°F (7°C), with the returning water temperature expected to be around 10°F (5–6°C) warmer. Of course, the return water temperature can vary with the load. The chilled water produced acts as a secondary coolant, which is any liquid cooled by a system refrigerant, then used to transfer heat from the conditioned space without a change in state.

The refrigerant vapor from the chiller evaporator flows into the chiller compressor. There, the warm, low-pressure refrigerant vapor is converted into a high-temperature, high-pressure vapor so that it can be condensed into a liquid in the condenser. The system shown in the example has an air-cooled condenser.

In some systems, the condenser is a separate unit and not built into the chiller itself. The air-cooled condenser in this type of chiller has a series of tubing coils through which the refrigerant flows. As cooler outside air is forced across the condenser tubing, the hot refrigerant vapor condenses into a liquid, giving up its heat to the outside air. The high-temperature, high-pressure liquid refrigerant then flows through the expansion device.

The expansion device regulates the flow of refrigerant into the evaporator and decreases its pressure and temperature. This converts the high-temperature, high-pressure liquid refrigerant into the low-temperature, low-pressure mixture of refrigerant liquid and vapor needed to absorb heat in the evaporator. There, the refrigerant vaporizes as it absorbs the heat of vaporization from the chilled water flowing through the heat exchanger, completing the refrigeration cycle.

The water cooled by the chiller is circulated through the piping system by the chilled-water pump. When there is a call for cooling from the building controls, the water is recirculated through the chiller evaporator and cooling terminal.

In a typical system, the terminal is an air handler containing chilled-water coils and a fan. Cooling of the room air takes place by blowing the warm room air over the chilled-water coils. This transfers the heat from the warmer room air through the coil tubing to the chilled water.

An isometric view of the system is shown in *Figure 36*. Note that the chiller unit used in this system is a packaged air-cooled reciprocating chiller.

3.2.0 Chillers

Chillers fall into two classes: mechanical and non-mechanical. Mechanical chillers are by far the most common and use various types of compressors to provide cooling. The compressor maintains a pressure difference between the high-pressure and low-pressure sides of the chiller system. This pressure difference causes the refrigerant to flow throughout the refrigerant circuit. The heat exchanges and changes of refrigerant state that occur were just described in the previous section.

Non-mechanical chillers do not have a compressor. Instead, they use different physical properties of the refrigerant mixture to provide cooling. This type of chiller is called an *absorption chiller*.

Mechanical water chillers can be assembled from components in the field or they may be assembled at the factory into a packaged unit. This section will focus on packaged chillers. A packaged chiller is typically one that includes the compressor, condenser, evaporator, expansion device, related controls, and auxiliary equipment. The system water pumps are often not a part of

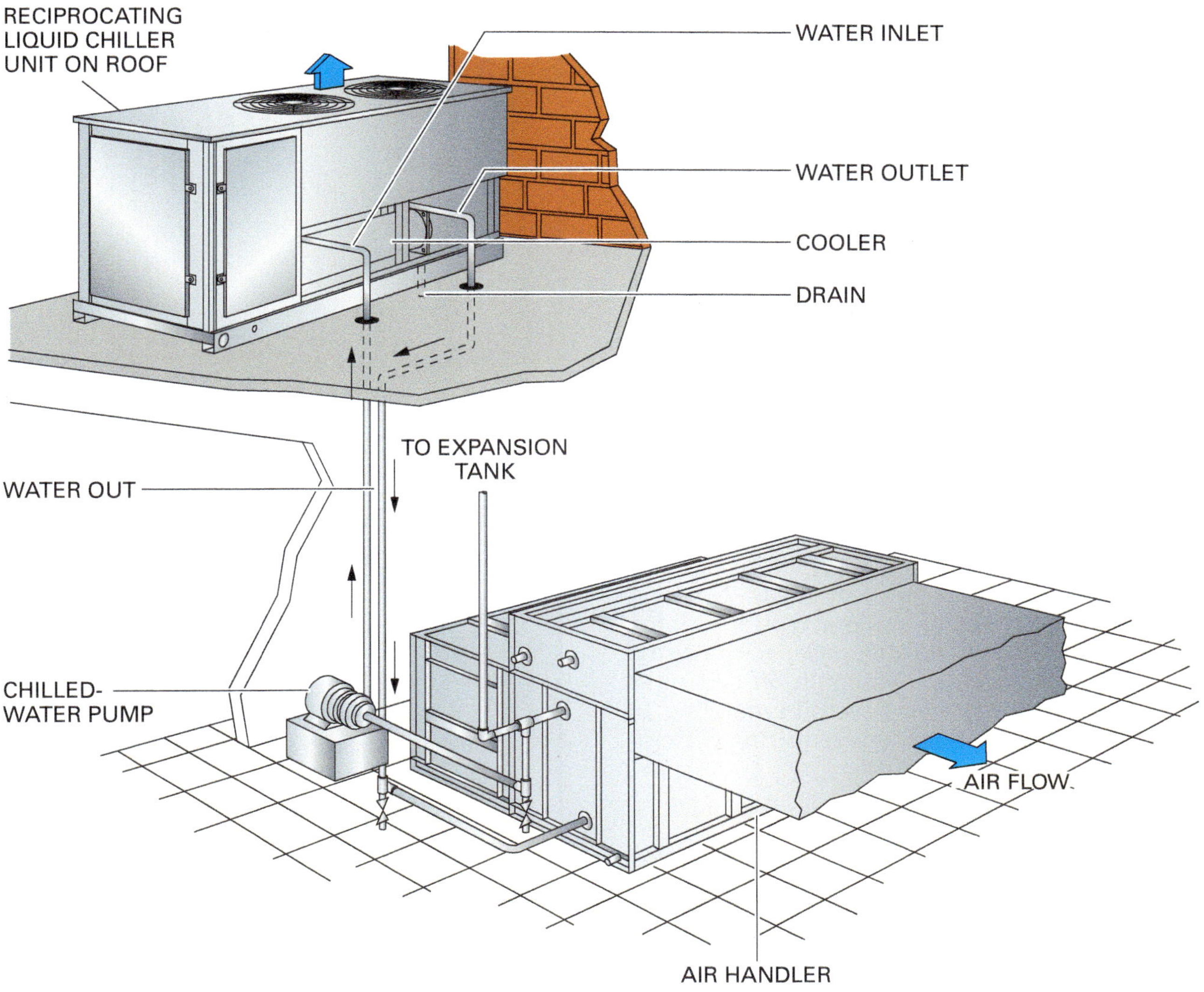

Figure 36 Chilled-water system with a packaged air-cooled chiller.

the package. This is because they are not part of the refrigerant circuit, and every application has different pumping requirements. The basic refrigeration components used in mechanical chillers operate the same way as similar components in other, more basic, cooling systems.

Three common types of mechanical chillers are reciprocating or scroll compressor chillers, centrifugal chillers, and screw chillers.

3.2.1 Reciprocating and Scroll Chillers

Reciprocating or scroll chillers (*Figure 37*) use one or more compressors to provide the necessary cooling capacity. When multiple compressors are used, the compressors are generally staged by temperature controls. As load increases and the water temperature begins to increase from the

Figure 37 Packaged air-cooled scroll chiller.

03305 Commercial Hydronic Systems

setpoint, more compressors are started. In some cases, two compressors may be connected on a single refrigeration circuit.

Although hermetic reciprocating and scroll compressors dominate the small- to medium-sized chiller market today, models using one or more semi-hermetic compressors remain in use and are still available from some manufacturers. The chiller condenser can be air cooled, water cooled, or evaporative. Water-cooled condensers may be tube-in-tube, shell-and-tube, or shell-and-coil.

The expansion device controls refrigerant flow from the condenser to the evaporator to maintain enough suction superheat to prevent liquid refrigerant from flooding the compressor. HCFC-22 refrigerant was widely used in new reciprocating water chillers. Today's models use non-chlorine refrigerants such HFC-410A. Reciprocating and scroll chillers have cooling capacity sizes ranging from about 5 tons to as high as 225 tons. However, other types of chillers typically dominate the market over 100 tons. Capacity control is generally accomplished using on-off cycling of the compressor(s), cylinder or scroll unloading, or hot-gas bypass for applications requiring extreme accuracy in water temperature.

3.2.2 Centrifugal Chillers

The centrifugal chiller (*Figure 38*) uses a centrifugal compressor consisting of one or more stages. Both open and hermetic compressors can be used. Open compressors can be driven by gas or steam turbines, engines, or electric motors. Centrifugal compressors use a high-speed impeller with many blades that rotate in a spiral-shaped housing. The impeller is driven at high speed inside the compressor housing. Refrigerant vapor is fed into the housing at the center of the impeller. The impeller throws the incoming vapor outward in a circular path from between the blades and into the compressor housing. This centrifugal action adds energy to the refrigerant, creating pressure in the high-velocity gas and forcing it out the discharge port.

The condenser in a centrifugal chiller is usually water-cooled, with the refrigerant condensing on the outside of the tubes. Air-cooled condensers can also be used, but they are uncommon because they greatly increase the unit's power consumption. The evaporator (cooler) can be a direct-expansion type in smaller units, but is usually a flooded type in the larger units.

Centrifugal chillers have cooling capacities ranging from about 200 tons to 8,000 tons. Capacity control is normally accomplished using

Figure 38 Centrifugal chiller.

GOING GREEN

Central Plant Facilities

Central plant facilities that provide hot and chilled water to large service areas have long been used for college campuses, airports, high-technology manufacturing centers, and some municipalities. Central plant technology is experiencing renewed interest from energy-conscious system designers. Designers of new buildings and retrofitted high-density complexes are leaning more toward this technology. Many of these facilities combine cogeneration capability that uses the heat byproducts of standalone electricity generation to drive the heating water system or power chillers. Typical cooling tonnages for these facilities begin in the 3,000-ton range and are seen consistently in the 8,000- to 10,000-ton range.

Condenserless Packages

Some chillers are supplied without a condenser assembly attached, saving space in the mechanical room. Due to design issues, this option is generally found only with reciprocating, scroll, or screw chillers. The remote condenser can be air-cooled or water-cooled. As with packaged chillers that are equipped with a condenser on board, a standard direct-expansion refrigerant circuit is employed.

adjustable compressor inlet guide vanes or compressor speed control.

HCFC-123 and HFC-134a are common refrigerants used in centrifugal chillers. HCFC-123 is a low-pressure refrigerant, while HFC-134a is a high-pressure refrigerant. A variety of low-pressure CFC refrigerants such as CFC-11 were used in the past, but those chillers have been replaced with those using new refrigerants.

Subcooling or intercooling devices are often used in a centrifugal chiller to conserve energy. Subcooling is a process by which the condensed refrigerant is subcooled in either a subcooling section of the water-cooled condenser or in a separate heat exchanger. This is done by bringing the warm, condensed liquid refrigerant into contact with the coldest (inlet) water tubes in the condenser or in a heat exchanger. This causes the condensed refrigerant to subcool further, increasing the refrigeration cycle efficiency and lowering power consumption. A device called a thermal economizer accomplishes subcooling in this way.

Intercooling is a process in which the condensed refrigerant leaving the condenser is passed through a flash (surge) chamber before reaching the evaporator. The pressure in this chamber falls between the condenser and evaporator pressures. Because its pressure is lower than the condensing pressure, some of the liquid refrigerant flashes into a vapor. This cools the remaining liquid. The flash gas, having absorbed heat, is drawn off and returned directly to the inlet of the compressor second stage, where it is mixed with discharge gas already compressed by the first-stage impeller. The flash gas must pass through only one stage of compression to reach condenser pressure, saving power.

The liquid refrigerant remaining in the flash chamber is metered through a valve and flows into the evaporator. Because pressure in the evaporator is lower than the flash chamber pressure, some of the liquid flashes and cools the remainder of the refrigerant to evaporator temperature. A device called a flash economizer accomplishes intercooling using a flash chamber or similar device.

Purge units are another type of device used on centrifugal chillers that operate with low-pressure refrigerants, such as HCFC-123. Chillers using these refrigerants operate at below-atmospheric pressures. Because of this, air and water tend to leak into and accumulate in the system, rather than the refrigerant leaking out. A purge unit must be used to separate and remove noncondensable gases and water vapor from the system.

Belly Heaters

Belly heaters are used on low-pressure systems to keep the pressure from falling and remaining below the atmospheric pressure during the off cycle, allowing noncondensable substances to enter from the surrounding environment.

Impellers in Series

Several impellers can be put in series (stages) to create greater pressure difference and to pump a sufficient volume of vapor. A compressor that uses one impeller is called a single-stage machine; one that uses two impellers is called a *double-stage machine*, and so on. When more than one stage is used, the discharge from the first stage is fed into the inlet of the next stage.

Purge units are generally small refrigeration units mounted on the chiller assembly. They draw in refrigerant vapors from the chiller, and then chill the vapor below the point of condensation for the refrigerant itself. Any vapor that remains is considered a noncondensable gas other than refrigerant and is released to the atmosphere. Any moisture is captured by filter-driers. The condensed liquid refrigerant is then returned to the chiller. The purge units operate on high-pressure refrigerants, like most common cooling and refrigeration units.

Older purge systems that managed CFC refrigerants were quite inefficient and tended to release a significant amount of refrigerant along with the air. Because the refrigerants were relatively inexpensive and environmental issues were not being considered, this issue presented little concern. The chiller was simply recharged periodically. To comply with new EPA guidelines, purge units have been redesigned to operate much more effectively.

3.2.3 Screw Chillers

Screw chillers (*Figure 39*) use an oil-injected hermetic or open-screw compressor. Screw compressors use a matched set of screw-shaped rotors, one male and one female, enclosed within a cylinder. The compressor drives the male rotor, which drives the female rotor. Normally, the driven male rotor turns faster than the female rotor because it has fewer lobes. Typically, the male has four lobes, while the female has six. As these rotors turn,

Figure 39 Air-cooled screw chiller.

they mesh with each other and compress the gas between them. The screw threads form boundaries that separate several compression chambers. Gas entering the compressor is moved through a series of progressively smaller compression stages until it exits at the compressor discharge in its fully compressed state.

Screw chillers can have a water-cooled shell-and-tube condenser, with the refrigerant condensing on the outside of the tubes. However, air-cooled condensers are also very popular, since they are generally less expensive to install. Water-cooled chillers are located indoors, in equipment rooms, while air-cooled chillers are located outside the building. When the chiller is used with a remotely-located air-cooled or evaporative condenser, a liquid receiver may replace the condenser on the package assembly. A remote air-cooled condenser allows the chiller to be inside the building. This can be an important consideration, as screw chillers can be quite noisy. This is especially true when they operate at less than 100 percent of their capacity.

The evaporator can be either a direct-expansion or flooded type. When the direct-expansion evaporator is used, refrigerant usually flows through the tubes. With a flooded evaporator, refrigerant is normally outside the tubes. Because the screw compressor is oil-injected, an oil separator is also part of the package. A flash economizer, similar to the one described for the centrifugal chiller, can be used to improve chiller efficiency. In the screw compressor, flash gas produced in the economizer is injected into the compressor at an intermediate point of the compression cycle.

New packaged screw chillers use HFC-134a, HFC-407C, or HFC-410A refrigerant. Screw chillers have cooling capacity sizes ranging from about 40 to 850 tons. Capacity control is normally accomplished with a hydraulically operated slide valve.

3.2.4 Absorption Chillers

Absorption chillers use a solution-concentration cycle to provide cooling. In the absorption cycle, evaporation and condensation of the refrigerant occur at different pressures. The cycles differ in that the absorption cycle uses a heat-operated generator to produce the pressure difference, while the mechanical refrigeration cycle uses a compressor. Both cycles require energy for operation, thermal energy in the absorption cycle, and mechanical energy in the compression cycle.

The refrigerant (water) and absorption solution (a solution of water and lithium bromide) are the working fluids of the absorption cycle. Lithium bromide is a caustic salt solution. The absorption cycle, or lithium bromide cycle, is based on two principles:

Absorption Chiller Heat Sources

Because heat is the primary energy source needed for absorption chillers, they are normally used where a source of steam or high-temperature water is readily available. The heat can also be from direct natural-gas combustion or even hot industrial gases.

- Water can be made to boil in the temperature range of 40°F to 45°F (4°C to 7°C) by maintaining a low absolute pressure (partial vacuum) of about 0.25 to 0.30 inches of mercury (in Hg), or 6.4–7.6 mm Hg. This means that water acts like a volatile refrigerant to absorb heat from the chilled water flowing through tubes in the chiller evaporator.
- Lithium bromide solution has the ability to absorb water vapor. This ability is best demonstrated when common table salt clogs up a salt shaker in humid weather because the salt has absorbed moisture from the air.

> **WARNING!**
> Lithium bromide is can be very harmful. It reacts with the skin's moisture to cause burns, damages the respiratory tract and lungs if inhaled, and is corrosive. Contact with the eyes can cause irritation, blurred vision, and possibly serious eye damage. When handling lithium bromide, always use appropriate personal protective equipment (PPE) as recommended in the SDS. Disposal of lithium bromide is controlled by local, state, and federal agencies. Check local, state, and Federal EPA regulations and comply with all disposal requirements.

Absorption chillers are available in capacities ranging from 50 tons to as high as 1,800 tons. As a rule, they are used in applications where a significant amount of steam is readily available, often where refuse-burning facilities are nearby or onboard sea-going vessels with large steam boilers in operation. They are relatively rare in commercial applications, and the operating cycle is quite complex and detailed. For this reason, an in-depth study of their operation is not included here.

3.2.5 Chiller Operating/Safety Controls

Almost all modern chillers have one or more control or interface panels used by the operator to control or monitor the operating status of the chiller (*Figure 40*). Operation of the chiller from these panels is under the control of a microprocessor programmed for the task. Status inputs are supplied to the microprocessor by remote sensors. The sensors may be

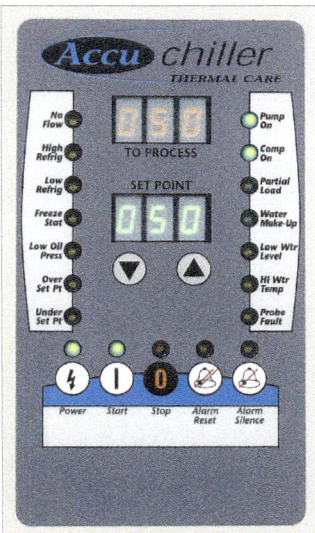

Figure 40 Chiller operator interface.

thermistors for measuring temperature, transducers for measuring pressure, or contacts that supply or interrupt various signals.

Typically, chiller performance status is shown on various types of displays. Status messages can be generated whether or not the chiller is running. Operation of the chiller from the panel(s) is often done by entering commands using a computer-type keyboard in response to menu-driven displays. When troubleshooting, these same panels can be used to receive diagnostic messages or run tests. Many chillers are designed to interface easily with laptops or touch-screen tablets and smartphones.

The following are some common chiller operations performed either automatically or manually by a typical microprocessor control center:

- Startup and shutdown of the chiller
- Adjustment of chiller compressor capacity control to maintain the proper leaving chilled-water temperature
- Control of system operation per programmed occupied/unoccupied schedules
- Registering cooler, condenser, and lubricating system pressures and temperatures
- Generating alarm and alert messages for abnormal conditions and de-energizing the chiller when system safety dictates
- Recording the total machine operating hours

Because chilled-water systems represent a significant investment, use large amounts of energy, and are essential to the comfort of large numbers of occupants, they are often integrated into computer-controlled building automation systems. Through the use of these systems, even the smallest details of the chiller's current operation and operating history are readily available to building managers on site and other personnel in remote locations. When problems occur, service personnel can review detailed system information and the nature of the condition before arriving on site.

3.3.0 Cooling Towers and Evaporative Condensers

Condensers are used to remove heat from the chiller refrigerant system. They take in high-pressure, high-temperature refrigerant gas from the compressor and change it into cool, high-pressure liquid refrigerant. They do this by removing the superheat and latent heat of vaporization from the refrigerant by transferring this heat to air, water, or both.

> **WARNING!**
> All cooling towers provide the opportunity for water stagnation in or around the cooling tower. This stagnation can be from the mist plume, the catch basin of the tower, or the formation of puddles on the surrounding roof or ground-mounted mechanical area. If any of these areas of water collection are within 10 feet of a fresh air intake to the building HVAC systems or equipment, they can promote bacterial contamination, such as *Legionella*. Although this is a portion of the system design and may be an ongoing issue with older buildings, there should be an awareness of the potential health risk. The building owner should be notified if such conditions exist.

For the condenser to operate properly, the condensing medium of air or water must always be at a lower temperature than the refrigerant it is condensing. Water-cooled condensers reject their heat to circulating water, which then circulates through a cooling tower or evaporative condenser. Water-cooled condensers are normally used in conjunction with a cooling tower. This section will focus on the cooling towers and evaporative condensers typically used.

Legionnaires' Disease

Legionnaires' disease is a very serious pneumonia-like lung infection caused by bacteria called *Legionella*. *Legionella* bacteria naturally live in fresh water and rarely causes illness. In situations where water remains above its natural temperature, such as hot tubs and cooling towers, *Legionella* can grow out of control unless proper maintenance is provided. When small droplets of water that contain the bacteria get into the air, people can easily breathe them in.

Cooling towers often produce the water droplets that can carry the bacteria to humans. Occasionally, *Legionella* gets into the lungs when drinking what is thought to be safe water, and mild choking allows a small amount to enter the lungs. Fortunately, Legionnaires' disease is not considered contagious, but research continues. If you develop pneumonia-like symptoms and believe you have been exposed to *Legionella*, seek medical care immediately. Be sure to mention if you have been in a hot tub or have been working around or near cooling towers.

The bacterium was named when, in 1976, many people who attended an American Legion convention in Philadelphia became ill. About 5,000 cases of Legionnaires' disease are diagnosed each year in the United States alone. It proves to be deadly for one out of ten, leading to roughly 50 deaths per year.

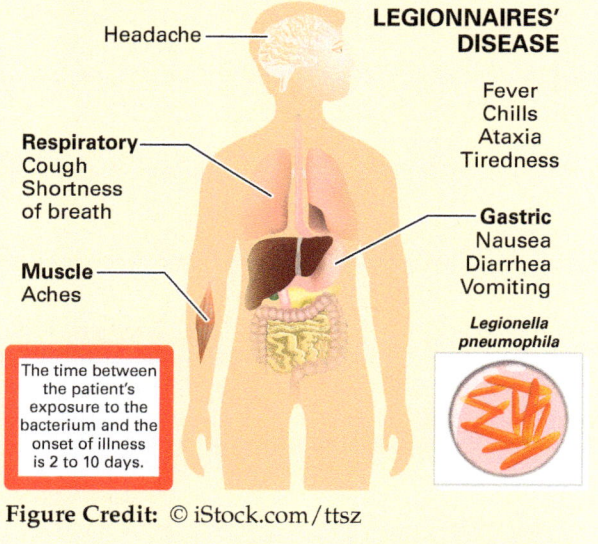

Figure Credit: © iStock.com/ttsz

3.3.1 Cooling Towers

The water portion of a chiller's water-cooled condenser is connected by piping to a cooling tower that is located outdoors. In the cooling tower, the heat absorbed by the water in the condenser is rejected from the system into the atmosphere by evaporation. The cooled water is then returned from the cooling tower to the condenser for reuse. Cooling towers pass outdoor air over the warm water to remove the heat from the water. There are many kinds of cooling towers used with water-cooled condensers. One type is a direct-contact, natural-draft tower (*Figure 41*). It is called a *direct-contact tower* because the heated water comes in direct contact with the air. These towers are mounted outdoors, usually on the roof, to make use of natural air currents and heights for the net positive suction head on the pump. They are made of a metal frame covering several layers or tiers of fiberglass-reinforced plastic decks or splash bars.

Because they use natural draft, no blowers are used to move air through the tower. Water is piped up from the condenser located in the building below to the top of the decks and is discharged in sprays over the decks. Spaces between the boards in the decks permit the water to drip or run from deck to deck, while being spread out and exposed to breezes that enter the tower from

Cooling Towers

Cooling towers are usually designed to reduce water temperature by approximately 10°F (5.6°C) based on the location of the system. A cooling tower can only cool to a point about 3°F to 7°F (2°C to 4°C) above the ambient wet-bulb temperature. Because they depend upon evaporation of water to the atmosphere for a cooling effect, they are more effective and possibly smaller in areas of the country where humidity levels are low.

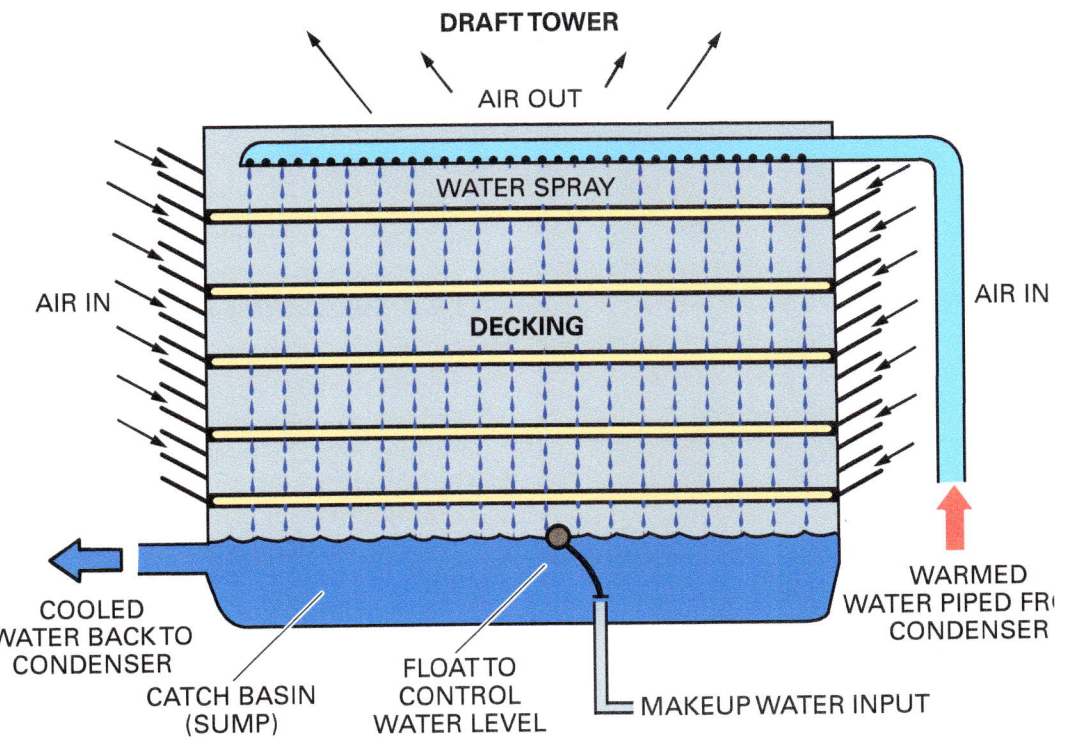

Figure 41 Direct-contact natural-draft cooling tower.

the open sides. The cooled water is collected in a sump at the bottom of the tower, and is pumped back to the condenser for reuse.

Natural-draft towers are becoming increasingly rare. They are generally much larger than newer designs. Packing more capacity into smaller packages is a common goal for almost every HVACR equipment manufacturer.

Because cooling towers work partially on evaporation, any water lost in the evaporation process must be replaced in order to maintain correct system operation. This is done by a float-controlled makeup water system (*Figure 42*) that senses the water level in the sump and adds water as needed. Electronic water level sensing systems are also used.

A variation of the natural-draft tower is an indirect-contact tower, sometimes called a *closed-circuit cooler* (*Figure 43*). The indirect-contact tower has two separate water circuits. One circuit consists of cooling tubes through which the water from the chiller condenser is flowing. This circuit is closed to the air, thereby protecting the chiller condenser water from being contaminated by airborne dirt and impurities.

The second water circuit is open to the air and consists of a pump that delivers water from a sump at the bottom of the tower to the top of the tower. There it is discharged in sprays over the tower decking in the same way as in the direct-contact tower. In this tower, heat in the closed tubes is first transferred to the water, then

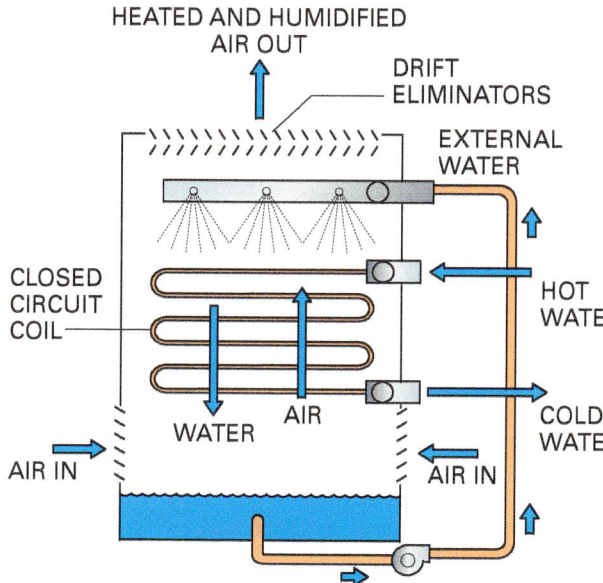

Figure 43 Indirect-contact natural-draft cooling tower.

from the water to outdoor air. Drift eliminators (droplet separators) prevent moisture discharge from the cooling tower, reducing the amount of makeup water needed.

Centrifugal and other types of fans may be used to increase the volume of airflow through cooling towers and indirect-contact towers. When fans are used, the tower is called a *forced-draft tower* or *induced-draft tower*.

A forced-draft tower is one in which the fans push the air so that it flows up and over the wet decking surfaces. In an induced-draft tower, the fans pull the air rather than push it, so that it flows up and over the decking. Because they use fans, forced-draft and induced-draft towers tend to be smaller than natural-draft cooling towers. *Figure 44* shows an example of an indirect-contact, induced-draft cooling tower.

3.3.2 Evaporative Condensers

Like the water-cooled condenser and water tower combination, evaporative condensers first transfer heat to the water, then from the water to the outdoor air. The evaporative condenser combines the functions of a water-cooled condenser and cooling tower in one enclosure or package (*Figure 45*). The condenser water evaporates directly off the tubes of the condenser with each pound of water evaporated, removing about 1,000 Btu from the refrigerant flowing through the tubes.

Air enters the bottom of the unit and flows upward over the refrigerant-filled condensing tube bundle. At the same time, water is being sprayed over the tube bundle. Both the air and sprayed water absorb heat from the refrigerant inside.

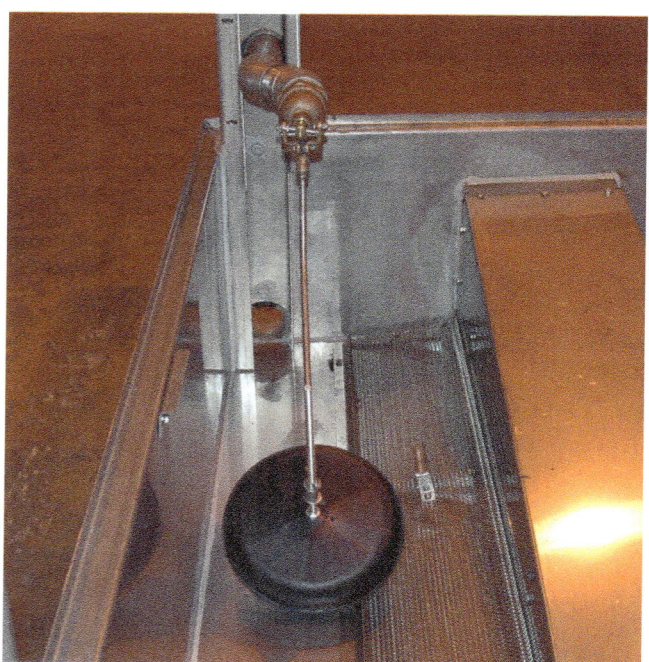

Figure 42 Cooling tower float-controlled makeup water system.

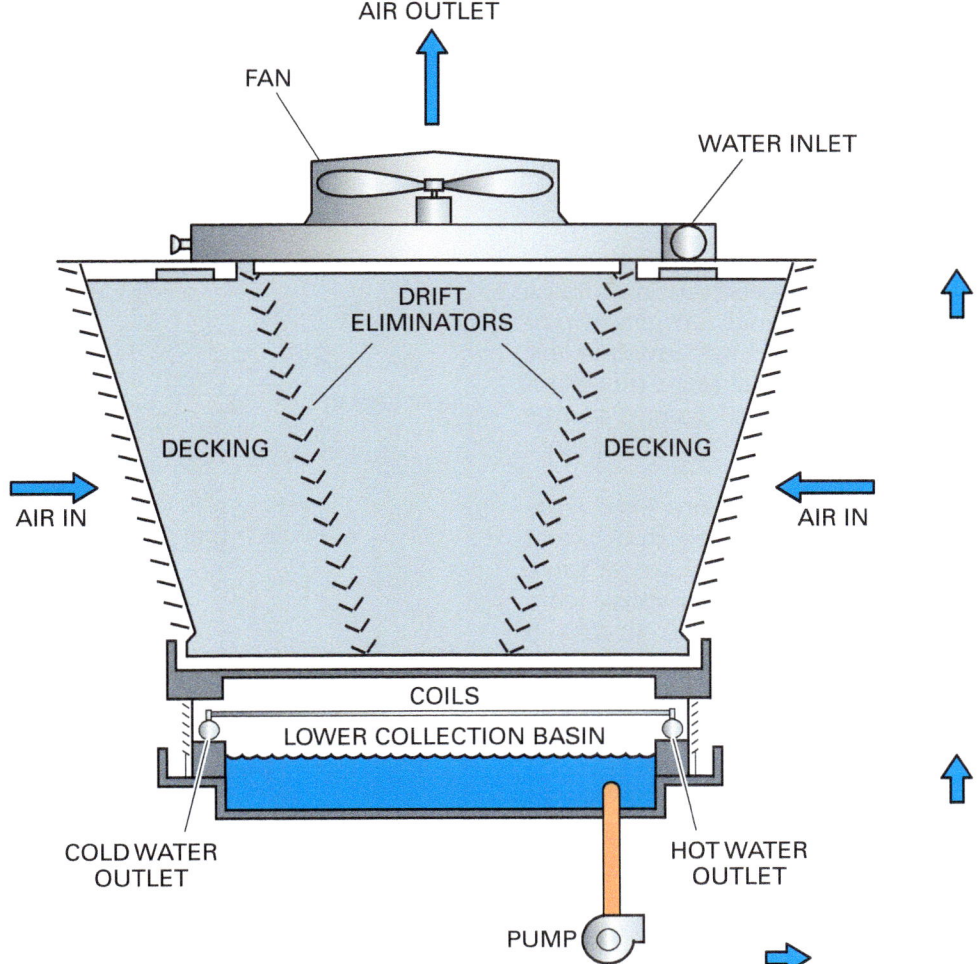

Figure 44 Indirect-contact induced-draft cooling tower.

Cooling Tower Water Treatment

Proper water treatment is essential to prevent and control biological growth and mineral deposits in cooling towers. Without proper treatment, a wide variety of biological growth will occur, sometimes with hazardous results to humans in the vicinity of the evaporating moisture. These same growths also cause a loss of capacity by impeding the heat transfer process.

Mineral deposits also impede heat transfer; as water is evaporated, minerals contained in the water supply remain behind. Although chemical treatment can assist in reducing these deposits, most cooling towers require bleed-off (a consistent minimal draining of the tower water) to dilute and flush away minerals in circulation. The chemical treatment approach should be designed with consideration for local conditions, the anticipated or identified types of biological growth, and the surrounding environment.

Water eliminators located above the water spray keep water entrained in the rising air from escaping the condenser. Water in the airstream strikes the eliminators and falls back into the condenser basin. Air is moved through the condenser assembly by one or more fans. Cooled by both air and water, the refrigerant in the coil condenses into a subcooled liquid at the output of the coil.

Relocating the refrigerant condenser to the base of the evaporative condenser generally requires a considerable amount of additional refrigerant piping. Because many water-cooled systems that use an evaporative condenser are quite large, this factor also significantly increases the refrigerant charge required and exposes the circuit to a greater potential for damage and leakage.

Refrigerant piping is usually more expensive, and its installation is more complex than water piping. For this reason, cooling towers are far more popular for most applications. Evaporative condensers are more likely to be found in refrigeration and food processing applications than in comfort cooling. They are often used in large refrigeration systems using anhydrous ammonia as the refrigerant.

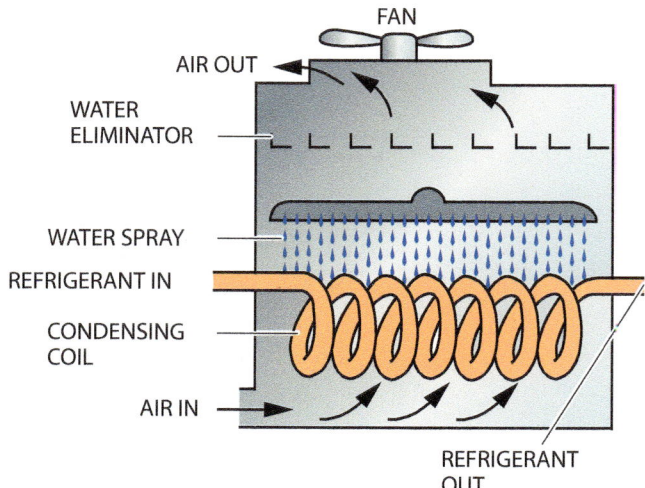

Figure 45 Evaporative condenser.

Additional Resources

Air-Cooled Chillers, TDP-622. Current edition. Syracuse, NY: Carrier Corporation.

Water-Cooled Chillers, TDP-623. Current edition. Syracuse NY: Carrier Corporation.

Refrigeration and Air Conditioning: An Introduction to HVAC/R, Larry Jeffus. Current edition. New York, NY: Pearson.

3.0.0 Section Review

1. Chillers that produce water used for comfort cooling typically operate above _____.
    a. 20°F (–7°C)
    b. 30°F (–1°C)
    c. 40°F (4°C)
    d. 50°F (10°C)

2. What does a mechanical chiller use to provide cooling?
    a. Compressor
    b. Centrifugal pump
    c. Air compressor
    d. Absorption generator

3. If the cooling tower you are working on contains a refrigerant circuit, it is a(n) _____.
    a. natural-draft cooling tower
    b. evaporative condenser
    c. hybrid cooling tower
    d. closed-circuit cooling tower

SUMMARY

Commercial hydronic systems use pipes to transport heated or cooled water from its source to where it is needed. The term *hydronic* refers to all heating and cooling systems using water. An air conditioning system that circulates chilled water to cooling coils is known as a *chilled-water system*. Hot-water heating systems are used mainly to provide comfort heating.

A boiler is a heat exchanger that uses radiant heat and hot flue gases to generate hot water for heating and process loads. The main differences between hot-water and steam boilers are in the operating and safety controls and accessories. Also, hot-water boilers are filled to capacity with liquid water, while steam boilers include a steam space during operation.

Piping in hydronic systems provides for the routing and distribution of hot water, cold water, or both. In hydronic comfort systems, water balancing means the proper delivery of hot water, chilled water, or both, in the correct amounts to each of the areas being conditioned in a structure. The satisfactory distribution of conditioned water depends upon a well-designed piping system and properly chosen water system components. Balancing the water system requires adjusting flow controls so that the right amount of hot or chilled water is circulated in the required spaces at the proper velocity to provide satisfactory heating and cooling.

Chilled-water systems eliminate the need to circulate refrigerant through refrigerant system components installed in a building to provide comfort cooling. Instead, a chiller uses refrigerant to cool the water. Then, the chilled water is circulated throughout the building and back to the chiller where the process is repeated.

Review Questions

1. Liquid water changes density with a change in _____.
   a. temperature
   a. velocity
   b. flow rate
   c. polarity

2. A hydronic piping system was originally designed for a flow rate of 50 gpm (189 L/min) with an initial pressure drop of 6 psi (41.4 kPa). If the flow rate were increased to 100 gpm (379 L/min), the new pressure drop would be approximately _____.
   a. 2 psi (14 kPa)
   b. 12 psi (83 kPa)
   c. 24 psi (166 kPa)
   d. 72 psi (496 kPa)

3. The highest static pressure in a hydronic system is located at the _____.
   a. highest point in the system
   b. lowest point in the system
   c. discharge of the circulating pump
   d. average height of all piping and tanks in the system

4. Low-pressure hot-water boilers are generally limited to a maximum operating temperature of _____.
   a. 180°F (82°C)
   b. 212°F (100°C)
   c. 250°F (121°C)
   d. 500°F (260°C)

5. One characteristic the copper-finned tube atmospheric boiler offers compared to other commercial types is _____.
   a. greater total heating capacity
   b. an extended life
   c. they cannot be installed outdoors
   d. they are very lightweight

6. Which boiler type allows the heating capacity to be increased by adding more sections?
   a. Firetube
   b. Watertube
   c. Cast-iron
   d. Vertical tubeless

7. In steel firetube boilers, _____.
   a. hot flue gases are contained in the tubes, with heated water on the outside
   b. heated water circulates through the tubes, with the flue gases on the outside
   c. cast-iron sections contain the flue gases
   d. the water and the flue gases both travel through separate tubes

8. Which type of boiler generally uses a water flow switch?
   a. Steel firetube
   b. Steel watertube
   c. Copper-finned tube
   d. Cast-iron

9. In some centrifugal pumps, the packing gland _____.
   a. reduces pump shaft friction
   b. attaches the impeller to the shaft
   c. stores lubricating oil
   d. helps control fluid leakage around the shaft

10. Multi-purpose valves are commonly installed _____.
    a. on the inlet side of the a pump
    b. on the discharge side of a pump
    c. at the inlet of each terminal device
    d. on the boiler makeup water line

11. Commercial air separators allow air to separate from the water stream by significantly reducing the water _____.
    a. pressure
    b. velocity
    c. temperature
    d. flow volume

12. The chilled water produced by a chiller is considered to be _____.
    a. a secondary coolant
    b. the primary coolant
    c. a brine solution
    d. more difficult to pump

13. The purpose of a purge unit on a low-pressure chiller is to _____.
    a. provide additional subcooling of the refrigerant
    b. automatically maintain the proper refrigerant charge
    c. remove noncondensable materials from the refrigerant circuit
    d. remove the remaining air from the chilled water

14. For cooling, all indirect-contact cooling towers rely on _____.
    a. only natural air circulation
    b. vapor compression heat transfer
    c. refrigerant vaporization
    d. a separate cooling water spray

15. Evaporative condensers differ from cooling towers in that _____.
    a. they do not use water for cooling
    b. cooling towers do not transfer heat through evaporation
    c. the refrigerant condenser itself is contained inside the evaporative condenser
    d. they cannot be used with most refrigerants

Trade Terms Introduced in This Module

Boiler horsepower (BoHP): The heat necessary to vaporize 34.5 pounds (15.7 kg) of water per hour at 212°F (100°C) and 0 psig (0 kPag). This is equal to a heat output of 33,475 Btus per hour or about 140 square feet (13 m2) of steam radiation. 1 BoHP is also equal to 9.803 kilowatts.

Firetube boiler: A boiler design where the hot flue gases flow inside the tubes, which are surrounded by the water to be heated.

Flash economizer: An energy-conservation process used in chillers that accomplishes intercooling of the refrigerant between the condenser and evaporator using a flash chamber or similar device. Flash gas of coolant produced in the flash chamber flows directly to the second stage of the chiller compressor, bypassing the evaporator and the first compressor stage.

Lithium bromide: A caustic salt solution commonly used in absorption chillers.

Oxygen scavengers: Chemicals added to boiler feed water that chemically combine with dissolved oxygen to prevent it from causing corrosion within the system. Typically creates a sludge that must be purged with boiler blowdowns or skimming.

Secondary coolant: Any liquid, such as water, that is cooled by a system refrigerant and then used to absorb heat from the conditioned space without a change in state. Also known as indirect coolant.

Superheated steam: Steam at a temperature above the saturation temperature for its pressure.

Thermal economizer: An energy-conservation process used in chillers in which warm condensed refrigerant is brought into contact with the coldest (inlet) water tubes in a condenser or heat exchanger. This causes the condensed refrigerant to subcool to a temperature below the condensing temperature.

Venturi tube: A short tube with flaring ends and a constricted throat that is used for measuring flow velocity by comparing the throat pressure to the inlet pressure. (Throat pressure decreases as the fluid's velocity increases.)

Watertube boiler: A boiler design where the water to be heated flows inside the tubes, with the hot flue gases flowing around the outside of the tubes.

Additional Resources

This module presents thorough resources for task training. The following reference material is recommended for further study.

NCCER Module 03203, *Introduction to Hydronic Systems*.
Air-Cooled Chillers, TDP-622. Current edition. Syracuse, NY: Carrier Corporation.
Water-Cooled Chillers, TDP-623. Current edition. Syracuse NY: Carrier Corporation.
Refrigeration and Air Conditioning: An Introduction to HVAC/R, Larry Jeffus. Current edition. New York, NY: Pearson.

Figure Credits

© iStock.com/alacatr, Module opener
Raypak, Inc., Figures 5, 6
Courtesy of ECR International Inc., Figure 7
Burnham Commercial, Figure 10
Laars Heating Systems Company, Figure 13
Photo courtesy of Paragon Pump Co., Figure 19
© iStock.com/Photon-Photos, Figure 20
Armstrong Fluid Technology, SA01
Courtesy of Dwyer Instruments, Inc., Figure 24 (photo)
Xylem Applied Water Systems, Figure 28
Courtesy of Xylem Inc., Figures 29–31
Daikin Industries, Ltd., Figure 34, 37, 39
© Lee Torrens/Dreamstime.com, Figure 38
Courtesy of John C. Schaub, Inc., Figure 40
© iStock.com/ttsz, SA02
Courtesy of EVAPCO, Inc., Figure 42

Section Review Answer Key

Answer	Section Reference	Objective
Section One		
1. a	1.1.0	1a
2. b	1.2.1	1b
Section Two		
1. c	2.1.0	2a
2. c	2.2.1	2b
3. d	2.3.3	2c
4. b	2.4.1	2d
5. a	2.5.1	2e
Section Three		
1. c	3.1.0	3a
2. a	3.2.0	3b
3. b	3.3.2	3c

NCCER CURRICULA — USER UPDATE

NCCER makes every effort to keep its textbooks up-to-date and free of technical errors. We appreciate your help in this process. If you find an error, a typographical mistake, or an inaccuracy in NCCER's curricula, please fill out this form (or a photocopy), or complete the online form at **www.nccer.org/olf**. Be sure to include the exact module ID number, page number, a detailed description, and your recommended correction. Your input will be brought to the attention of the Authoring Team. Thank you for your assistance.

Instructors – If you have an idea for improving this textbook, or have found that additional materials were necessary to teach this module effectively, please let us know so that we may present your suggestions to the Authoring Team.

NCCER Product Development and Revision
13614 Progress Blvd., Alachua, FL 32615

Email: curriculum@nccer.org
Online: www.nccer.org/olf

❏ Trainee Guide   ❏ Lesson Plans   ❏ Exam   ❏ PowerPoints   Other _____

Craft / Level: _____   Copyright Date: _____

Module ID Number / Title: _____

Section Number(s): _____

Description: _____

_____

_____

_____

_____

Recommended Correction: _____

_____

_____

_____

_____

Your Name: _____

Address: _____

_____

Email: _____   Phone: _____

Steam Systems

Overview

Steam has a wide variety of uses in the commercial and industrial sectors. With the ability to carry large quantities of stored energy, steam can be applied in comfort heating, process heating, mechanical drive, pressure control, and a variety of other applications. The majority of steam's heat content is stored in its latent form, so it can be transported efficiently at low cost. It is generally a safer alternative to high-temperature water for carrying heat in large quantities.

Module 03306

Trainees with successful module completions may be eligible for credentialing through NCCER's National Registry. To learn more, go to **www.nccer.org** or contact us at **1.888.622.3720**. Our website has information on the latest product releases and training, as well as online versions of our *Cornerstone* magazine and Pearson's product catalog.

Your feedback is welcome. You may email your comments to **curriculum@nccer.org**, send general comments and inquiries to **info@nccer.org**, or fill in the User Update form at the back of this module.

This information is general in nature and intended for training purposes only. Actual performance of activities described in this manual requires compliance with all applicable operating, service, maintenance, and safety procedures under the direction of qualified personnel. References in this manual to patented or proprietary devices do not constitute a recommendation of their use.

Copyright © 2018 by NCCER, Alachua, FL 32615, and published by Pearson, New York, NY 10013. All rights reserved. Printed in the United States of America. This publication is protected by Copyright, and permission should be obtained from NCCER prior to any prohibited reproduction, storage in a retrieval system, or transmission in any form or by any means, electronic, mechanical, photocopying, recording, or likewise. To obtain permission(s) to use material from this work, please submit a written request to NCCER Product Development, 13614 Progress Blvd., Alachua, FL 32615.

03306 V5

From *HVAC Level Three, Trainee Guide*. NCCER.
Copyright © 2018 by NCCER. Published by Pearson. All rights reserved.

03306
Steam Systems

Objectives

When you have completed this module, you will be able to do the following:

1. Describe the properties of water as they are related to steam systems.
   a. Describe the basic properties of water.
   b. Describe the pressure-temperature relationship of water.
2. Describe the basic steam cycle and the primary components related to its operation.
   a. Describe the steam cycle principles of operation.
   b. Identify steam boilers and their common accessories.
   c. Identify common steam heat exchangers and terminal devices.
3. Identify common steam system piping arrangements and valves.
   a. Identify common steam system piping arrangements.
   b. Identify common condensate return systems and describe waterside care.
   c. Explain how steam and condensate system piping is sized.
   d. Identify steam-based pressure-reducing and thermostatic control valves.
4. Identify, install, and maintain various types of steam traps.
   a. Identify various types of steam traps.
   b. Explain the basic concepts of installing steam traps.
   c. Explain how to maintain various steam traps.
   d. Explain how to troubleshoot various steam traps.

Performance Task

Under the supervision of your instructor, you should be able to do the following:

Perform two of the following:
1. Perform selected operating procedures on low-pressure steam boilers and systems.
2. Maintain selected steam traps.
3. Identify common piping configurations used with steam systems.

Trade Terms

Drip leg
Dry steam
Flash steam
Saturated steam

Trim
Turndown ratio
Water hammer

Industry-Recognized Credentials

If you are training through an NCCER-accredited sponsor, you may be eligible for credentials from NCCER's Registry. The ID number for this module is 03306. Note that this module may have been used in other NCCER curricula and may apply to other level completions. Contact NCCER's Registry at 888.622.3720 or go to **www.nccer.org** for more information.

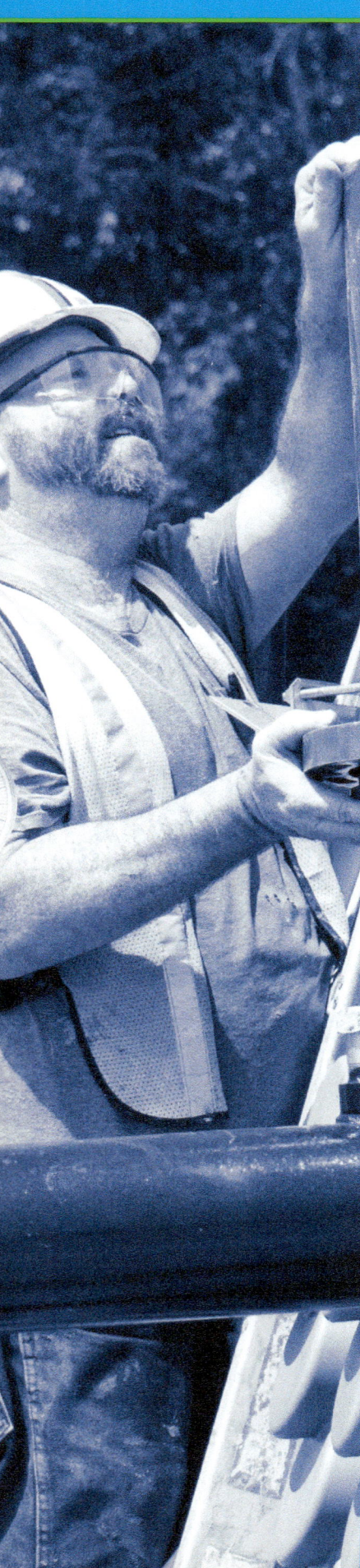

Contents

1.0.0 Steam-Related Fundamentals of Water 1
  1.1.0 Basic Properties of Water 1
  1.2.0 Pressure-Temperature Relationship of Water 3
2.0.0 Basic Steam System Cycle and Operational Components 6
  2.1.0 Steam Cycle Principles of Operation 6
  2.2.0 Steam Boilers, Controls, and Accessories 7
    2.2.1 Steam Boilers 7
    2.2.2 Pressure Gauges 9
    2.2.3 Water Level Indicators 10
    2.2.4 Low- and High-Water Cutoff and Water Feeder Controls 11
    2.2.5 High-Pressure Limit and Other Controls 13
    2.2.6 Pressure-relief valve 14
  2.3.0 Steam System Loads 14
    2.3.1 Heat Exchangers/ Converters 14
    2.3.2 Terminals 15
3.0.0 Steam System Valves and Piping 19
  3.1.0 Steam System Piping 19
    3.1.1 One-Pipe Systems 19
    3.1.2 Two-Pipe Systems 20
  3.2.0 Condensate Return/ Feedwater System Components 22
    3.2.1 Condensate Pump and Receiver Tank 22
    3.2.2 Vacuum Pumps 24
    3.2.3 Deaerating Feedwater Heater 25
    3.2.4 Flash Tanks 25
    3.2.5 Boiler Blowdown and Skimming 26
    3.2.6 Boiler Water Treatment 27
  3.3.0 Steam and Condensate Pipe Sizing 27
    3.3.1 Velocity Sizing 28
    3.3.2 Steam Line Pressure Drop 28
    3.3.3 Condensate Line Sizing 30
  3.4.0 Valves 31
    3.4.1 Pressure-Reducing Valves 31
    3.4.2 Thermostatic Valves 33
4.0.0 Steam Traps and Strainers 36
  4.1.0 Steam Trap Types 36
    4.1.1 Mechanical Steam Traps 36
    4.1.2 Thermostatic Steam Traps 37
    4.1.3 Thermodynamic Steam Traps 38
    4.1.4 Fixed-Orifice Steam Traps 39
    4.1.5 Strainers 39
  4.2.0 Installing Steam Traps 39

4.3.0 Maintaining Steam Traps 40
   4.3.1 Inverted Bucket Trap 41
   4.3.2 Liquid Expansion Thermostatic Trap 41
   4.3.3 Balanced Pressure Thermostatic Trap 42
   4.3.4 Bimetal Thermostatic Trap 42
   4.3.5 Float and Thermostatic Trap 42
   4.3.6 Thermodynamic Disc Trap 42
4.4.0 Troubleshooting Steam Traps 42
   4.4.1 Diagnostic Methods 43

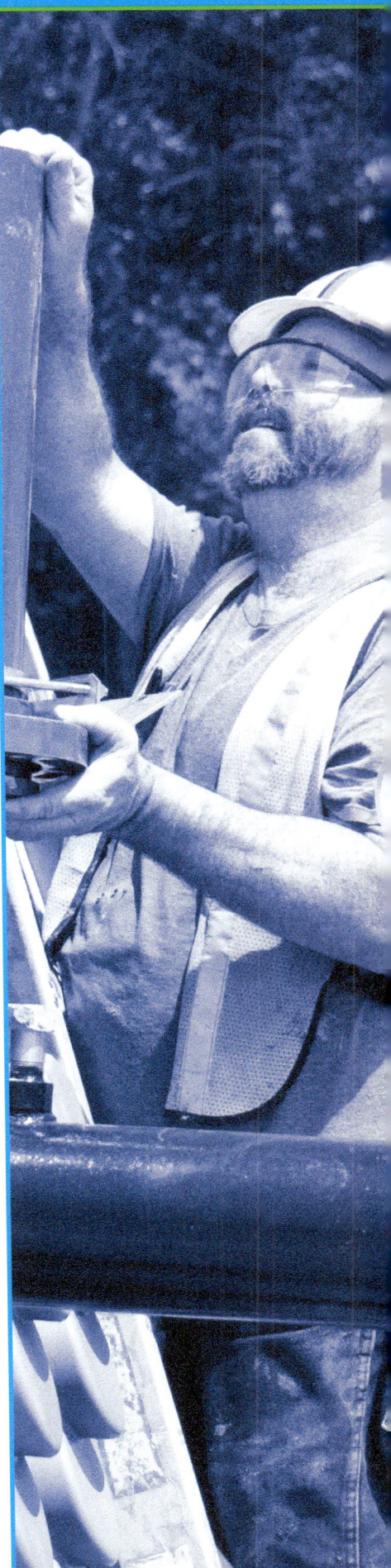

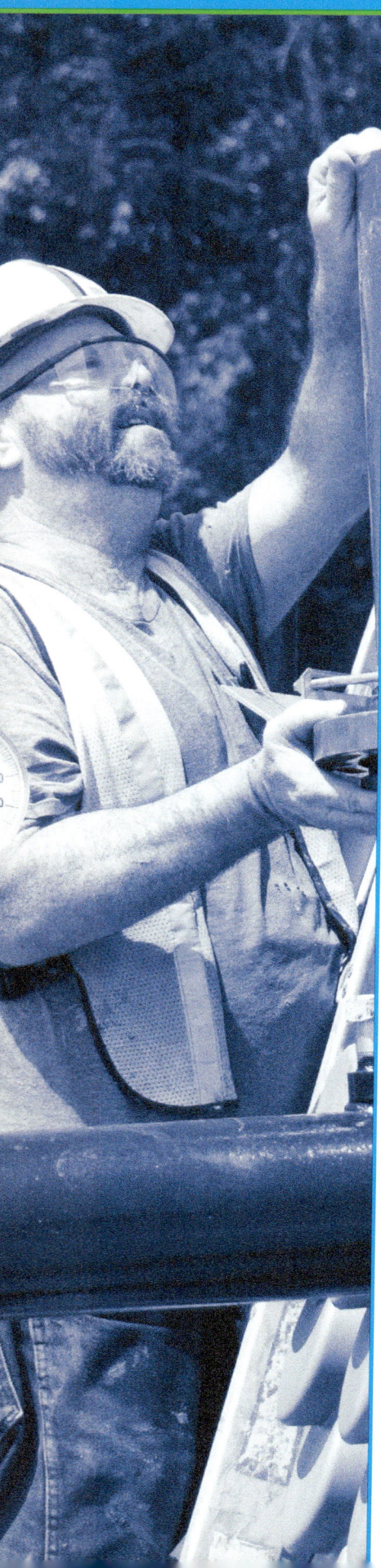

Figures and Tables

Figure 1   Changing states of water 2
Figure 2   Change of state terminology 4
Figure 3   Example table entries for properties of saturated steam 5
Figure 4   Basic steam cycle 6
Figure 5   Typical cast-iron steam boiler controls and accessories 8
Figure 6   Commercial and industrial steam boilers 9
Figure 7   Efficient residential cast-iron steam boilers 10
Figure 8   Pigtail siphon 10
Figure 9   Water gauge glass 11
Figure 10 Water column 11
Figure 11 Water levels 11
Figure 12 Low-water cutoff 12
Figure 13 Electronic low-water cutoff 13
Figure 14 Electric water feeder 13
Figure 15 High-pressure limit control 13
Figure 16 Pressure-relief valve 14
Figure 17 Horizontal shell and tube heat exchanger 15
Figure 18 Typical installation of a heat exchanger when used as
           a converter 16
Figure 19 Unit heater 16
Figure 20 Single-row steam coil 17
Figure 21 Steam distribution coil 17
Figure 22 Basic one-pipe upfeed system 20
Figure 23 Two-pipe gravity system 21
Figure 24 Two-pipe system with mechanical condensate return 22
Figure 25 Vacuum return system 23
Figure 26 Duplex condensate receiver 23
Figure 27 Typical condensate pump and receiver system 24
Figure 28 Vacuum heating units 24
Figure 29 Flash tank connections 25
Figure 30 Steam pipe sizing for steam velocity 29
Figure 31 Steam pipe sizing for pressure drop 30
Figure 32 Pressure-reducing valve 32
Figure 33 Single-stage pressure-reducing station 33
Figure 34 Use of pressure-reducing valves connected in series 33
Figure 35 Use of pressure-reducing valves connected in parallel 34
Figure 36 Typical individual thermostatic control 34
Figure 37 Temperature-sensing element attached to a valve 34
Figure 38 Installed steam traps 37
Figure 39 Inverted bucket trap 37
Figure 40 Thermostatic steam traps 38
Figure 41 Thermodynamic stainless-steel steam trap 39
Figure 42 Strainers 40
Figure 43 Basic trap installation 41
Figure 44 Ultrasonic tester 44
Figure 45 Pyrometer 44
Figure 46 Thermowells 44

SECTION ONE

1.0.0 STEAM-RELATED FUNDAMENTALS OF WATER

Objectives

Describe the properties of water as they are related to steam systems.
- a. Describe the basic properties of water.
- b. Describe the pressure-temperature relationship of water.

Trade Terms

Dry steam: The steam that exists after all the water has been vaporized and is moving beyond its saturation temperature (being superheated).

Saturated steam: The steam produced and existing at the temperature that corresponds to the boiling temperature of water at the existing pressure (saturation).

Steam is the most widely used form of energy for industrial use in the world. Water itself is generally plentiful, as well as inert, making it a desirable compound to work with in many applications. The energy stored within water, once it is has been sufficiently heated to the point of vaporization and beyond, can be five to six times greater than liquid water can generally carry. Coupled with the fact that pumps are not required to move steam from the point of creation to the intended load, it easily becomes the best means of storing and moving large amounts of energy in a variety of applications.

1.1.0 Basic Properties of Water

Depending on its heat content and temperature, water can exist in three states: ice (solid), water (liquid), and water vapor (gas, such as steam). When water changes from one state to another, some interesting things occur. *Figure 1* shows this graphically for water at sea level. If heat is added to one pound of ice at 0°F, a thermometer will show a rise in temperature until the reading reaches 32°F. This is the point at which the ice starts changing into water. If heat is continually added, the thermometer reading remains fixed at 32°F instead of rising. It continues to read 32°F until the entire pound of ice is melted. The increase in temperature from 0°F to 32°F registered by the thermometer is sensible heat. Sensible heat is heat that can be sensed by a thermometer or by touch. The heat that was added to the ice and caused its change in state from a solid to a liquid, but did not register on the thermometer, is latent heat. Latent heat is the heat energy absorbed or rejected when a substance is changing state (solid to liquid, liquid to gas, or vice versa) without a change in the measured temperature.

If you continue to add heat to the pound of water, the thermometer will once again show an increase in temperature until the temperature reaches 212°F. At this point, the water starts boiling and changes state from water into steam (water vapor). As even more heat is added, the thermometer reading remains at 212°F until all the water has turned into steam. If you continue to add heat, the thermometer will once again register sensible heat.

As shown in *Figure 1*, it takes a great deal more heat to cause a change of state in water than is needed for a degree change in its temperature. It requires 144 Btus of latent heat to melt the pound of ice before the temperature began to rise. This is 144 times as much heat as is needed to raise the temperature of water one degree (specific heat). The change from water to steam requires an even greater amount of latent heat. It takes 970 Btus to change the water to steam. None of the latent heat consumed during the change of state of water registers on the thermometer. Once water reaches the threshold of changing states and begins the change process, the thermometer becomes stable and unchanged until the phase change is complete. This property of latent heat helps to explain why large amounts of energy can be stored and transported through the use of steam.

Figure 2 shows the common terms used in steam systems when defining the latent heat added or removed in changing water to and from the solid, liquid, or vapor states. It also shows the terms *subcooling* and *superheat* as they relate to sensible heat changes in water. Remember that the processes of subcooling and superheating are sensible processes. As a substance such as water is changing state, a latent process is at work and there is no change in temperature during this period.

Did You Know?
Steam Energy

An astounding 80 percent of the energy needed to produce pulp and paper comes from steam, while roughly 35 percent of the total energy required for all product output from industry in the United States comes from this vapor phase of water.

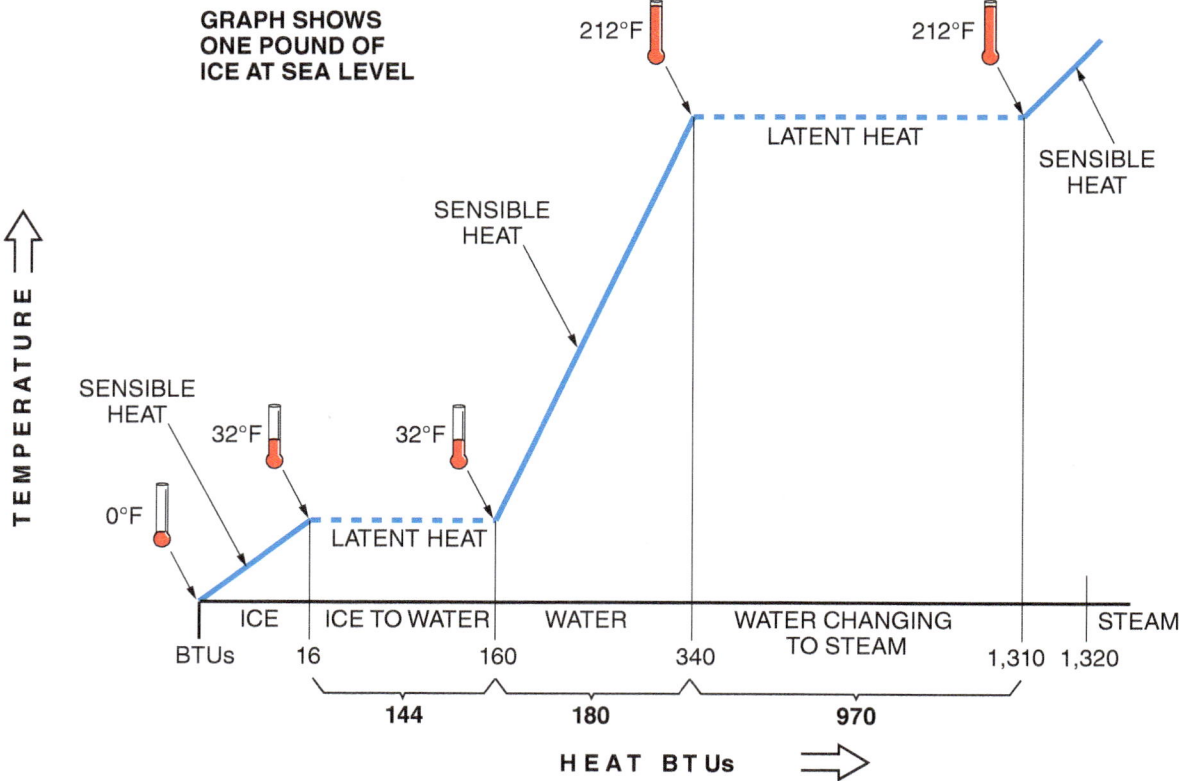

Figure 1 Changing states of water.

- *Latent heat of fusion* – The heat gained or lost in changing ice to water or water to ice.
- *Latent heat of vaporization* – The heat gained in changing water to steam at a given pressure. The temperature at which this occurs is known as the boiling point or the saturation temperature. As water absorbs latent heat and the change of state begins, steam can be seen rising from the water surface. This steam is commonly referred to as **saturated steam**. After all the water has been vaporized to steam at the saturation temperature and begins to become superheated, the steam is referred to as **dry steam**.
- *Latent heat of condensation* – The heat given up or removed from steam in changing back to water.
- *Superheat* – The measurable heat added to the steam after all the water has reached its boiling point and completely changed into steam. For example, the saturation temperature of water at atmospheric pressure is 212°F. If the resulting steam is superheated 10°F, the steam temperature is 222°F. Steam that has been heated above the saturation temperature is called superheated steam.
- *Subcooling* – The measurable heat that is removed after all the water has condensed from the vapor state to the liquid state. It is the reverse of superheat, and the subcooling process cannot begin until the change of state to liquid is complete. For example, the condensing temperature of water at atmospheric pressure is 212°F. If the water is subcooled 10°F, the temperature is cooled to 202°F.

Figures 1 and *2* show that most of the heat content of steam is stored as latent heat. It is this latent heat (970 Btu/lb at sea level) that makes steam so efficient as a carrier of heat. As it condenses, it will release this stored heat. Large quantities can be transmitted efficiently with little change in temperature. Remember that steam is pressure- and temperature-dependent. This means that the system can be controlled by varying either the steam pressure or temperature.

1.2.0 Pressure-Temperature Relationship of Water

Steam tables illustrate the pressure-temperature relationships and other properties of saturated steam over a broad range of temperatures. *Figure 3* shows entries from such a table. One entry is for the properties of steam at sea level as just described. The other entries are given for the purpose of providing comparisons.

Columns A through C show the pressure-temperature relationships. Columns A and B show the corresponding gauge and absolute pressures. Absolute pressure is the pressure in pounds per square inch (psia) above a perfect vacuum. Gauge pressure (psig), which starts at zero psig, is pressure in pounds per square inch above atmospheric pressure. Gauge pressure plus 14.7 equals the related absolute pressure.

For each pressure, there is a corresponding temperature. Column C shows that the saturation temperature of steam at sea level (0 psig/14.7 psia) is 212°F. As steam cools down to these conditions, the latent process of condensation begins. If water is heated up to these conditions, the latent process of vaporization is initiated.

Column D shows the heat of saturated liquid. This is the amount of sensible heat required to raise the temperature of a pound of water from 32°F to the boiling point at the pressure and temperature shown. It is expressed in Btu/lb. In some tables, this value is called the enthalpy of the liquid (hf).

Column E shows the latent heat. This is the amount of heat required to change a pound of boiling water to steam, expressed in Btu/lb. This same amount of heat is released when a pound of

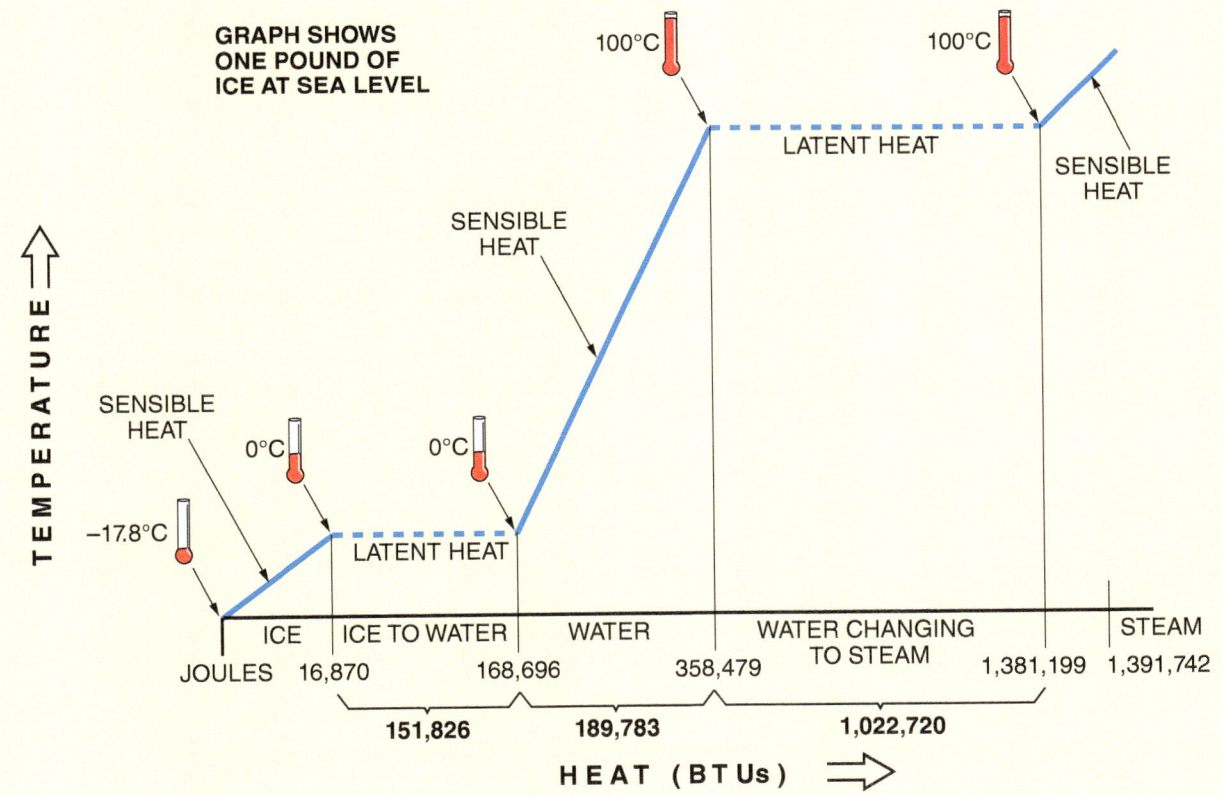

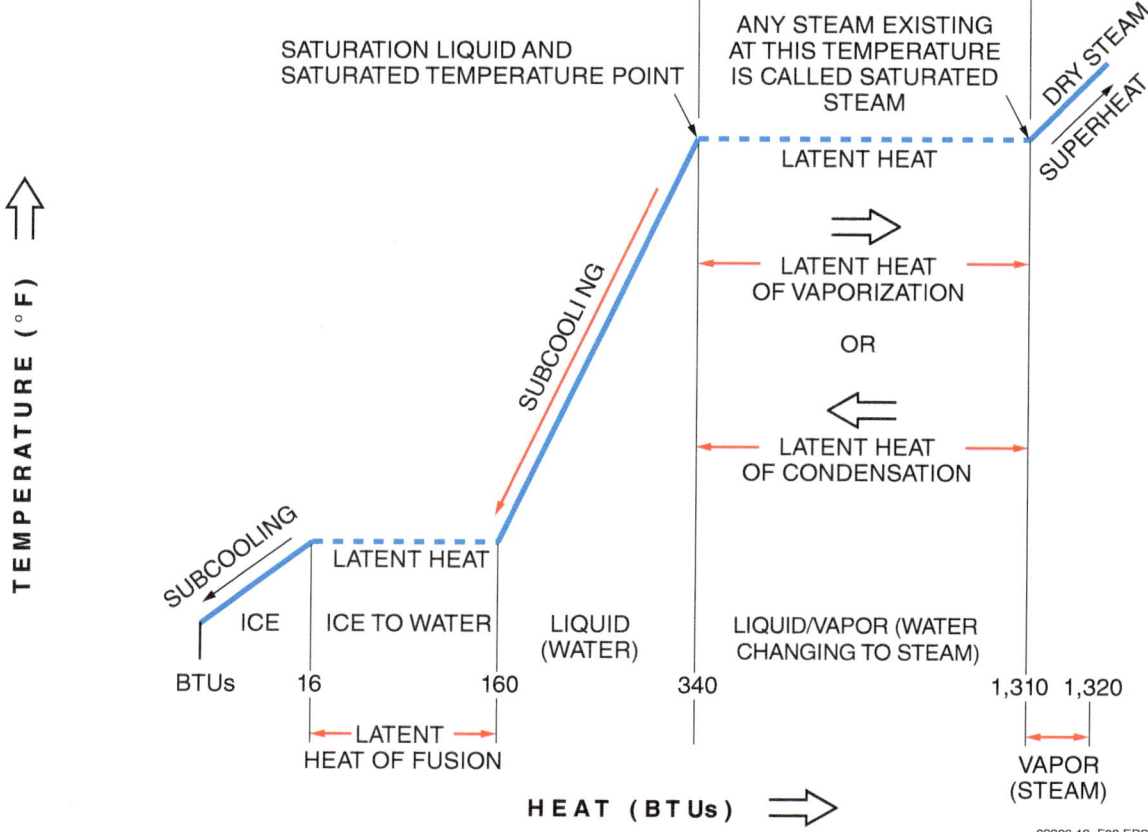

Figure 2 Change of state terminology.

steam is condensed back into a pound of water. This heat quantity is different for every pressure-temperature combination. In some tables, this value is called the enthalpy of the evaporation (hfg).

Column F shows the total heat of the steam. This is the sum of the sensible heat and the latent heat given in Columns D and E, expressed in Btu/lb. It is the total heat in the steam above 32°F.

In some tables, this value is called the enthalpy of the steam (hg).

Column G shows the specific volume of steam. This is the volume per unit of mass for the steam. It is the space that one pound of steam will occupy at the pressure-temperature shown. It is expressed in cubic feet per pound. Note that, as pressure and the total heat contained within the steam increases on the chart, the volume of the steam decreases.

You Know This...

The use of the terms *sensible* and *latent heat*, and the terms *superheat* and *subcooling*, have exactly the same meaning in a discussion of steam as they do in a discussion of refrigerants.

Properties of Saturated Steam Tables

Tables showing the various properties of saturated steam are often found in manufacturers' literature, as well as reference texts for steam systems. In addition to their use for determining pressure-temperature relationships, the various data about the other properties of steam given in such tables are commonly used in calculations made when designing and sizing components for a steam system.

| | Column A
Gauge
Pressure | Column B
Absolute
Pressure
PSIA | Column C
Steam
Temp.
°F | Column D
Heat of
Sat. Liquid
btu/lb | Column E
Latent
Heat
btu/lb | Column F
Total Heat
of Steam
btu/lb | Column G
Specific
Volume
cu ft/lb |
|---|---|---|---|---|---|---|---|
| Inches of Vacuum | 29.743 | 0.08854 | 32.00 | 0.00 | 1075.8 | 1075.8 | 3306.00 |
| | 29.515 | 0.2 | 53.14 | 21.21 | 1063.8 | 1085.0 | 1526.00 |
| | 27.886 | 1.0 | 101.74 | 69.70 | 1036.3 | 1106.0 | 333.60 |
| | 19.742 | 5.0 | 162.24 | 130.13 | 1001.0 | 1131.1 | 73.52 |
| | 9.562 | 10.0 | 193.21 | 161.17 | 982.1 | 1143.3 | 38.42 |
| | 7.536 | 11.0 | 197.75 | 165.73 | 979.3 | 1145.0 | 35.14 |
| | 5.490 | 12.0 | 201.96 | 169.96 | 976.6 | 1146.6 | 32.40 |
| | 3.454 | 13.0 | 205.88 | 173.91 | 974.2 | 1148.1 | 30.06 |
| | 1.418 | 14.0 | 209.56 | 177.61 | 971.9 | 1149.5 | 28.04 |
| PSIG | 0.0 | 14.696 | 212.00 | 180.07 | 970.0 | 1150.4 | 26.80 |
| | 1.3 | 16.0 | 216.32 | 184.42 | 967.6 | 1152.0 | 24.75 |
| | 2.3 | 17.0 | 219.44 | 187.56 | 965.5 | 1153.1 | 23.39 |
| | 5.3 | 20.0 | 227.96 | 196.16 | 960.1 | 1156.3 | 20.09 |
| | 10.3 | 25.0 | 240.07 | 208.42 | 952.1 | 1160.6 | 16.30 |
| | 15.3 | 30.0 | 250.33 | 218.82 | 945.3 | 1164.1 | 13.75 |
| | 20.3 | 35.0 | 259.28 | 227.91 | 939.7 | 1167.1 | 11.90 |
| | 25.3 | 40.0 | 267.25 | 236.03 | 933.7 | 1169.7 | 10.50 |
| | 30.3 | 45.0 | 274.44 | 243.36 | 928.6 | 1172.0 | 9.40 |
| | 40.3 | 55.0 | 287.07 | 256.30 | 919.6 | 1175.9 | 7.79 |
| | 50.3 | | 297.97 | 267.50 | 911.6 | 1179.1 | 6.66 |
| | 60.3 | | | 277.43 | 904.5 | 1181.9 | 5.82 |
| | | | | 286.39 | 897.8 | 1184.0 | |

Figure 3 Example table entries for properties of saturated steam.

Additional Resources

Low Pressure Boilers, 3rd Edition. Fredrick M. Steingress, Daryl R. Walker. Orland Park, Illinois: American Technical Publishers.

High Pressure Boilers, 4th Edition. Fredrick M. Steingress, Harold J. Frost, Daryl R. Walker. Orland Park, Illinois: American Technical Publishers.

1.0.0 Section Review

1. Steam at a temperature of 222°F is referred to as _____.

   a. saturated
   b. superheated
   c. subcooled
   d. condensed

2. Steam tables provide a comparison of pressure-temperature relationships based on steam at sea level that is _____.

   a. saturated
   b. superheated
   c. subcooled
   d. vaporized

SECTION TWO

2.0.0 BASIC STEAM SYSTEM CYCLE AND OPERATIONAL COMPONENTS

Objectives

Describe the basic steam cycle and the primary components related to its operation.
  a. Describe the steam cycle principles of operation.
  b. Identify steam boilers and their common accessories.
  c. Identify common steam heat exchangers and terminal devices.

Performance Task

1. Perform selected operating procedures on low-pressure steam boilers and systems.

Trade Terms

Trim: External controls and accessories attached to the boiler itself, such as sight glasses and water feeder controls.

The cycle for steam systems is essentially the same for all systems. But the type of equipment used with the steam system depends on the heat content and steam volume required. The heat content and volume of steam required usually dictates the steam pressure and temperature along with the type of boilers, their primary system equipment, and various types of ancillary equipment needed. Some of the types of boilers and their primary equipment along with some of ancillary equipment used with them are described in the following sections.

2.1.0 Steam Cycle Principles of Operation

Figure 4 shows the basic components of a simple steam cycle. They include the following:

- Boiler
- Control valve
- Heat exchanger
- Trap
- Deaerator and condensate receiver
- Condensate pump
- Vacuum breaker

The steam cycle begins at the boiler. In the boiler, water is vaporized to make steam. The initial source of energy used for this purpose is typically from the combustion of a fuel such as oil, gas, or coal. Once the water has reached the boiling point and vaporization begins, the pressure begins to rise. Remember that the water temperature required for boiling also increases due to this change in pressure. As shown in *Figure 3*, once the pressure reaches 1.3 psig, the boiling point increases to over 216°F and continues to increase as pressure rises. Steam is found at the highest temperature and pressure in the boiler itself. Steam flows through the supply piping and components as the result of natural movement from an area of higher pressure to an area of lower pressure, much like air through an air distribution system.

The control valve is an automatic, temperature-operated control device. Its purpose is to regulate or meter the flow of steam into the heat exchanger at a rate comparable to the load. This ensures that only the proper amount of steam needed to take full advantage of its stored latent heat is admitted. The steam system is designed so the control valve has the needed residual pressure to assure proper operation of the valve. Pressure drop across steam valves must be used to calculate the entering steam pressure to the final user of the steam.

The heat exchanger may be a steam coil in a system airstream, a steam-to-water heat exchanger, or any other type of terminal used with steam systems. Its purpose is to transfer the heat carried in the steam to the conditioned space or other

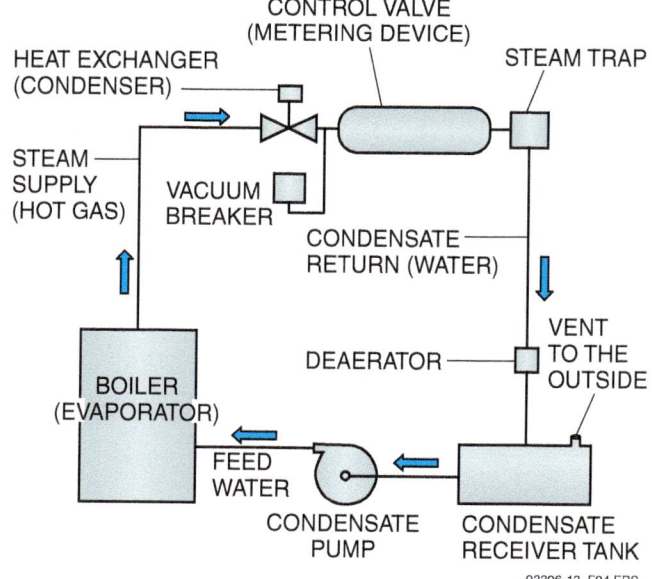

Figure 4 Basic steam cycle.

medium. The heat exchanger is like the condenser in a refrigeration system, where the steam (gas) is converted back into water (liquid). There, the latent heat and any sensible superheat gained in the boiler by vaporization is rejected to the cooler conditioned space, causing the steam to cool and condense back into water. The condensing process in the heat exchanger occurs at a lower pressure than in the boiler. The amount of pressure depends on the temperature of the medium being heated, the pressure losses in the supply piping, and the pressure drop across the control valve. It is this pressure difference that causes the steam to flow without a pump.

Steam traps are among the most important devices used in the steam cycle. The trap in the piping at the output of the heat exchanger holds the steam in the heat exchanger until it all condenses. Otherwise, the live steam would pass through the heat exchanger into the return line, wasting the useful heat released as a result of condensation. The trap will allow only liquid condensate and/or air, never steam, to flow away from the heat exchanger. There are a variety of styles and types suited to different applications. Traps also require the most attention and maintenance to ensure proper operation, because failure of a steam trap can create a number of problems.

Where elevations permit, steam systems can use a gravity return to route the flow of condensate water directly back to the boiler for re-evaporation into steam. However, most systems use gravity flow to route the condensate into a receiver tank, where it collects and is stored. When the boiler controls call for the water to replenish the amount boiled off, a condensate pump is turned on and produces the necessary pressure differential to pump the condensate (feed water) from the tank to the boiler. This method allows gravity to be used to the extent possible, with the condensate pump doing only the remaining work. Return of the condensate to the boiler for re-evaporation completes the steam cycle.

Steam systems that use a condensate receiver tank and pump in the condensate return line operate at atmospheric pressure, with the return opened to the atmosphere in one or more places. This has an advantage. When steam is applied to heat exchangers that have coils subjected to outside airflow, or cold water such as in domestic hot-water heaters, the pressure in the heating coil can suddenly drop below atmospheric pressure (vacuum range). In extreme cases, this sudden drop in pressure can cause the tubes in the coil to collapse. This can be prevented by allowing air to enter the heat exchanger by way of a vacuum breaker device or through a thermostatic trap that senses the drop in pressure. A small amount of the air that enters the heat exchanger will be absorbed into the condensate, but most of it flows in the return line to the condensate receiver tank. From there it is discharged by vents back into the atmosphere.

2.2.0 Steam Boilers, Controls, and Accessories

Unlike hot-water boilers, steam boilers are not completely filled with water. A sufficient amount of space must be left at the top for steam to form. Steam boilers are generally categorized by their working pressure.

Many of the boiler controls and accessories used with steam systems are the same as those used with hot-water boilers and described earlier. Typically, the main difference is that these controls and accessories have different temperature and pressure ratings. For this reason, this section focuses on the controls and accessories that are unique to steam boilers and systems.

In a well-designed steam system, the boiler is equipped with several operating and safety controls needed to guarantee safe and proper operation of the boiler (*Figure 5*). Operating and safety controls normally used with all steam boilers include the following:

- Pressure gauge
- Water gauge glass/water column
- Low-water cutoff/water feeder
- High-pressure limit control
- Pressure-relief (safety) valve

2.2.1 Steam Boilers

Steam boilers are divided into three broad categories: those that produce low-pressure steam; those that produce medium-pressure steam; and

Piping Friction

The pressure of the steam leaving the boiler gradually falls as it flows through the supply piping because of the piping friction. The pressure differential at the boiler supply and return outlets allows steam to flow, just like a hydronic system. With steam however, no pump is required.

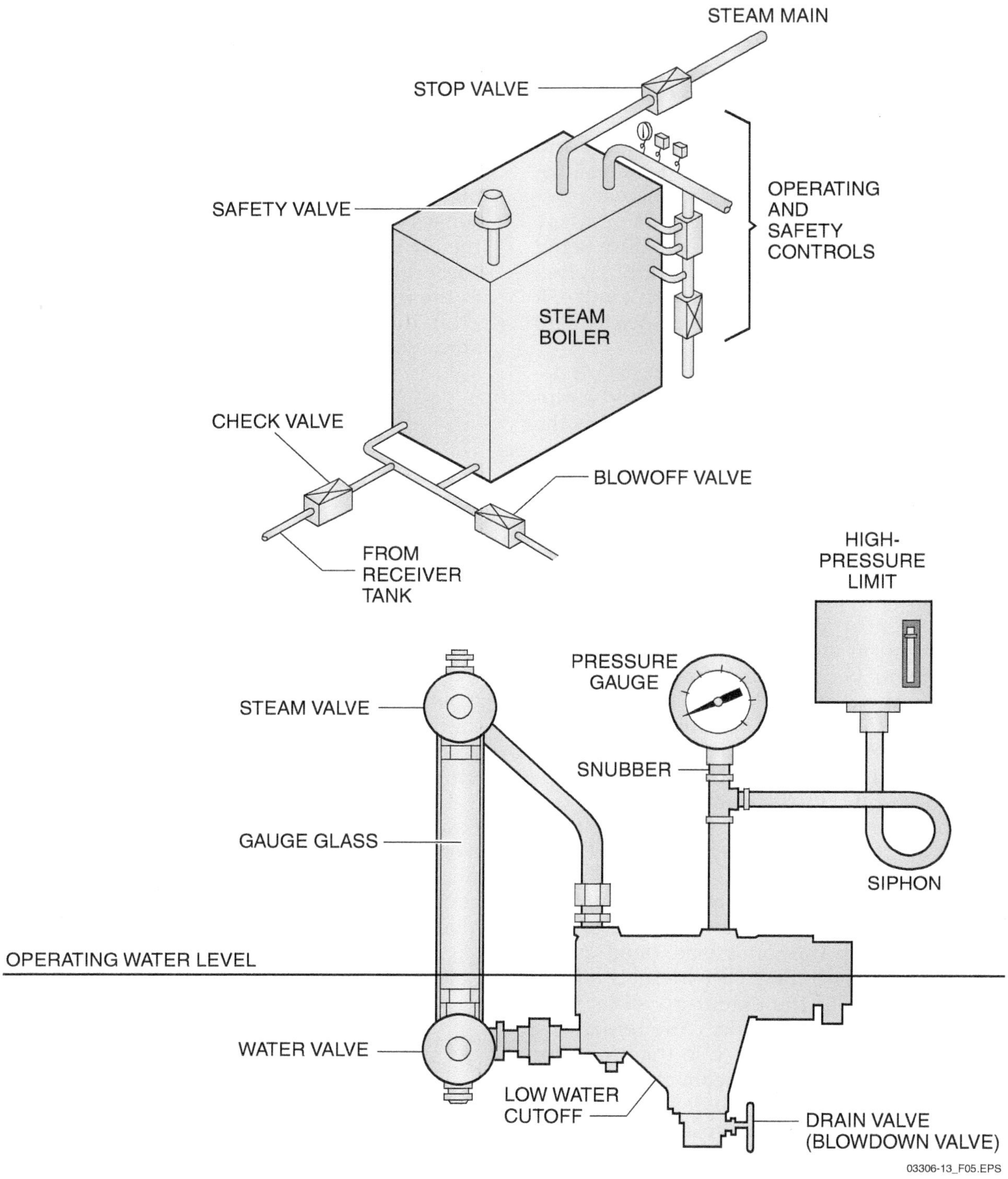

Figure 5 Typical cast-iron steam boiler controls and accessories.

those that produce high-pressure steam. Low-pressure steam boilers have working pressures of up to 15 psi (250°F). Medium-pressure boilers operate from 15 psi to 60 psi. High-pressure boilers operate at pressures from 60 psi and up. It is important to note that these ratings indicate the maximum working pressure of the boiler. For safety purposes, the actual operating pressure of the system would be somewhat lower than the maximum. Low-pressure steam boilers (*Figure 6*) are normally used in commercial applications, apartment houses, and single- or multiple-unit industrial facilities. High-pressure industrial boilers are used mainly to produce process steam for facilities such as college campuses, hospitals, or industrial plants.

Although larger boilers and higher-pressure boilers are generally designed specifically for steam applications, many low-pressure boilers are identical in construction to those used for hot water applications. For added safety, a number of such boilers are rated at a maximum working pressure of 30 psig for hot water, but have a reduced maximum rating of 15 psig for steam applications. Low-pressure steam boilers often differ from their hot water components only in the trim; that is, the external operating and safety controls or accessories attached to the boiler itself.

Overall, steam boiler designs vary more widely than hot-water designs. Typical designs include the following:

- Cast-iron, often built in matching sections for field assembly
- Steel firetube and watertube boilers
- Steel vertical tubeless boilers
- Electric boilers

Boilers used in residential (*Figure 7*) and some light commercial low-pressure applications are generally of cast-iron sectional or one-piece designs. Although many older units were quite inefficient, federal minimum efficiency standards, which took effect in 1992, require boilers to have a minimum Annual Fuel Utilization Efficiency (AFUE) rating of 80 percent. Boilers earning the EPA's ENERGY STAR designation must have a minimum AFUE rating of 85 percent. These boilers generally incorporate a number of energy-saving options, such as electronic ignition and sealed combustion systems using outside air for the combustion process.

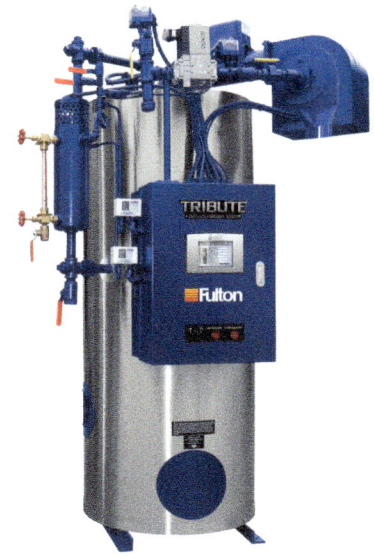

COMMERCIAL GAS-FIRED VERTICAL TUBELESS BOILER

INDUSTRIAL OIL-FIRED HORIZONTAL FIRETUBE BOILER

03306-13_F06.EPS

Figure 6 Commercial and industrial steam boilers.

With the exceptions that follow, steam boilers operate in the same manner as hot-water boilers:

- Steam boilers are partially filled with water to allow room for steam creation and expansion.
- The pressure-temperature levels at which steam and hot-water boilers operate are different.
- The operating and safety controls are different.

2.2.2 Pressure Gauges

The pressure (steam) gauge shows the pressure within the boiler in pounds per square inch gauge (psig). This gauge is installed at the highest point of the boiler's steam space. A siphon assembly is generally installed between the boiler and the gauge. Two different types of siphons are used: pigtail siphons and gauge siphons. Both provide separation for the gauge or other control from the raw steam, which may cause damage to the gauge, as well as erroneous readings. *Figure 8* provides an example of a pigtail siphon. Pigtail

Using the Condensate Receiver and Pump

Another advantage of using the condensate receiver and pump is that they permit condensate return lines to be located below the water level in the boiler. This makes it possible to locate heat exchanger terminals at levels lower than the boiler. In large systems, more than one may be required.

GAS-FIRED STEAM BOILER

OIL-FIRED STEAM BOILER

Figure 7 Efficient residential cast-iron steam boilers.

siphons are always installed vertically and are filled with water before installing the gauge.

2.2.3 Water Level Indicators

The water gauge glass is used to check the water or steam level in the boiler (*Figure 9*). The steam level can be checked by opening the steam valve at the top of the water gauge. To check the water level, the water valve at the bottom of the water gauge is opened. Normally, the water gauge glass is connected directly to the boiler. A drain valve (blowdown valve) is usually connected to the bot-

Figure 8 Pigtail siphon.

tom of the water gauge. When opened, the valve is used to purge the gauge passages of trapped water and any buildup of sludge and sediment.

> **WARNING!**
> Blowdown or drain valves opened while a steam boiler is in operation can release water at a temperature above the atmospheric boiling point, causing flash steam. Use extreme caution when using the blowdown valve. The blowdown valve discharge must always be piped to an appropriate drain, and in a manner that will prevent hot water or flash steam from contacting personnel.

> **NOTE**
> The terms *bleed-off* and *blowdown* are often used interchangeably, but they apply to different systems. Bleed-off is generally used to define the process of bleeding water from a cooling tower or other hydronic loop to reduce the concentration of dissolved solids in the remaining water. Blowdown is associated with the process of removing dissolved solids and/or sludge from the surface or bottom of a steam boiler. Use of the proper term will ensure clear communication.

On larger boilers, the water gauge may be attached to a water column (*Figure 10*). It diverts water or steam from the boiler for application to the water gauge. This helps to stabilize water levels for more accurate water gauge readings.

Three globe valves on the water column provide a backup method for checking the boiler steam and water levels if the water gauge is damaged. These globe valves are often called gauge cocks or try cocks. Opening the top, middle, or bottom valve allows either steam, a mixture of steam and water, or water to escape. A blowdown valve connected to the bottom of the water col-

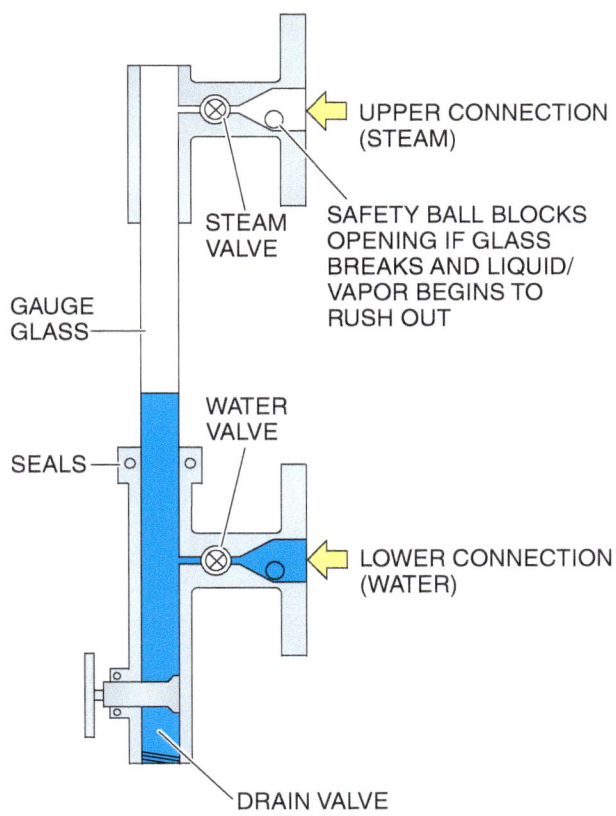

Figure 9 Water gauge glass.

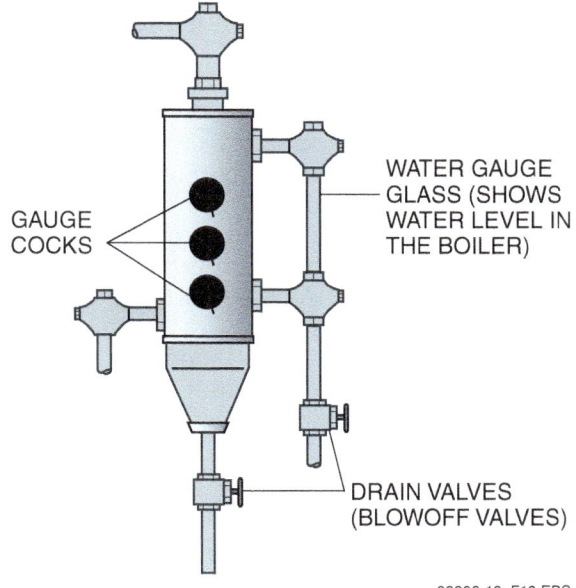

Figure 10 Water column.

umn is used to clear the column of trapped water and any buildup of sludge and sediment.

In a steaming boiler, the gauge glass will generally indicate a lower water level than is accurate, as shown in *Figure 11*. This is because the water in the gauge glass is not subject to currents or agitation, it is cooler than the water in the boiler, and it does not contain steam bubbles.

2.2.4 Low- and High-Water Cutoff and Water Feeder Controls

Low-water cutoff controls provide valuable protection for steam boilers by interrupting power to the heat source when the water falls below the minimum safe level. Water feeders are used to control the flow of makeup water to the boiler. In systems that effectively collect and return steam condensate to the boiler for reuse, the water feeder uses the condensate reservoir and associated pumps as the primary water source. Fresh water, generally from a municipal water supply, becomes the backup source when condensate levels are insufficient. These boilers are generally classified as steam heating boilers. However, steam boilers are also used in many applications where little or no condensate is collected for reuse. In these cases, the water feeder may be quite

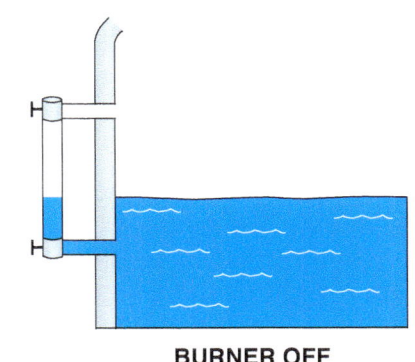

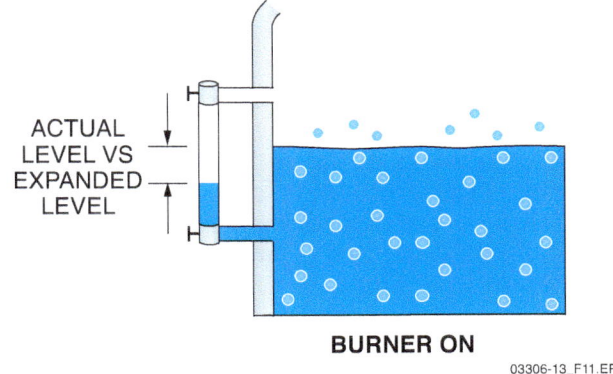

Figure 11 Water levels.

active, using only the fresh water supply to make up for the lost steam. Boilers used in this manner are classified as steam-process boilers.

Typically, low-water cutoff control for steam boilers is done using the same methods as those for hot-water boilers, employing float-operated controls (*Figure 12*). However, steam boiler operation is adversely affected by a high water level as well. While a water boiler must be completely full of water, a steam boiler cannot operate in a flooded condition. As a result, both low- and high-water cutoff controls are needed for a steam boiler. Some steam boilers may be equipped with electronic probe-type controls (*Figure 13*) to provide these functions, much like hot-water boilers. A high-water cutoff is not mandatory in most cases. A low-water cutoff is required for safety purposes. A boiler firing with little or no water inside is a serious hazard.

Both the float-operated and electronic controls can be used to provide low-water and high-water cutoff duty. The control circuit switches have both normally open and normally closed contacts. The switch is then wired depending on the desired action. For example, a low-water float control would be wired normally open, and the contacts close when the water level is high enough for operation. Conversely, for high-water cutoff, the switch would be wired normally closed, and the contacts will open if the water level rises too high.

A marker on the float bowl of the float-operated control used with steam boilers shows the cutoff level for the device. The control is normally installed so that the marker on the float bowl is about ½ inch higher than the lowest visible point in the water gauge glass. For some float-operated controls, the operating level is established automatically by fittings on the device that permit its installation right in the gauge glass tappings of the boiler. If a low-water condition occurs, float-actuated switch contacts in the cutoff control open the electrical circuit to the boiler burner, preventing the burner from firing. Note that water will still be visible in the gauge glass. However, the boiler is shut down and cannot be operated until makeup water is manually added to the boiler. If this situation persists, it can cause a loss of heating until the water in the boiler is returned to a safe level. This problem is usually eliminated by installing a water feeder control between the boiler's supply water line input and its condensate return main line. This control functions to automatically maintain a constant water level in the boiler.

In some boiler installations, the water feeder control is a separate unit (*Figure 14*). There are many kinds of feeder controls. Typically, they use a solenoid-operated valve controlled by the related low-water cutoff control. The low-water cutoffs have a second switch that is actuated by the cutoff float. This switch is used to control the related solenoid-operated water valve (electric water feeder). This second switch operates at a higher water level than the cutoff level switch. If the boiler water drops to this first operating level, the float closes the switch contacts, completing the electrical circuit to the water feeder. This energizes the water feeder solenoid-operated valve, allowing makeup water to flow through the valve and into the boiler. When the boiler water level is restored, the float level actuates the switch to open the electrical circuit to the water feeder. This shuts off the supply of makeup water to the boiler.

In other boiler installations, the water feeder control is combined with the low-water cutoff control into one assembly. These controls are called feeder-cutoff combination controls. Usually, the water feeder part of the control is mechanically actuated by the float. It is inactive as long as the condensate water returning to the boiler keeps the water in the boiler at the normal operating level. If for some reason the boiler water drops below the normal operating level, the first thing the feeder cutoff combination control does is to add the amount of water required to keep the boiler in operation. It mechanically opens its feed valve, adds the small amount of makeup water necessary, and when the float reaches the normal water

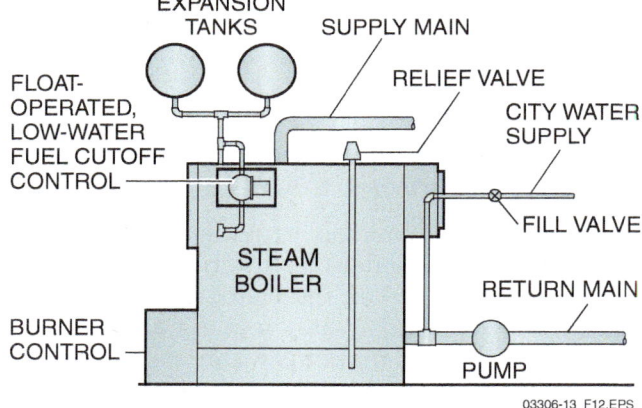

Figure 12 Low-water cutoff.

Water Columns and ASME

Water gauges must be mounted on a boiler at a location that meets ASME code. ASME tests and approves gauges for mounting in specific locations. If the system fails and the gauge is not installed at the correct location, the installer may be liable for any damages that occur.

Figure 13 Electronic low-water cutoff.

level, it closes again. A marker on the float bowl casting shows the feeder closing level. Feeders are normally installed so that the marker on the float bowl is about 2 to 2½ inches below the normal water line.

If the feeder part of a combined feeder cutoff control cannot keep the water in the boiler at a safe level, the low-water fuel cutoff portion of the control will activate. When the water in the boiler drops below the minimum safe level, the float drops and opens the switch contacts, resulting in the fuel supply to the burners being turned off.

The selection of low-water cutoff and water feeder devices generally depends on design system pressure and desired mounting style. It is recommended that steam boilers over 400,000 Btuh input and/or operating above 15 psig be fitted with two low water cutoff devices for redundancy. The first cutoff can be a type that resets itself automatically, while the second must be a manual-reset model. Water feeder and low water cutoff controls provide valuable protection and are essential for proper system operation. They should be disassembled, cleaned, calibrated, and reinstalled annually, at a minimum.

2.2.5 High-Pressure Limit and Other Controls

The high-pressure limit control (*Figure 15*), also known as a pressuretrol, is activated by a rise in boiler pressure. The high-pressure setpoint should always be lower than the setting of the pressure-relief valve, and a pressuretrol style that requires a manual reset should always be used for high-limit duty. Although a few models use snap switches, most use a mercury bulb much

Figure 14 Electric water feeder.

Figure 15 High-pressure limit control.

like older wall thermostats. Control circuit power is routed through the mercury bulb. Should the boiler pressure rise to the point where it exceeds the pressuretrol setpoint, the switch contacts open and break the burner control circuit. This shuts down the boiler.

Pressuretrols are also commonly used to provide on-off and sequential control of boilers. They can be adjusted in the field so that the boilers all start at high-fire (unison), or in sequence where the boilers start at some firing rate other than high-fire. The control is activated when the steam pressure drops below a preset level. This causes automatic feeding of an air/fuel mixture to the furnace, ignition of the burners, and opening of the main air damper(s). As the boiler steam pressure rises, it crosses fixed control points. At each control point, a predetermined mixture of fuel and airflow is selected and applied to the burners. When the boiler pressure reaches the operational pressure limit, the control shuts off the boiler.

2.2.6 Pressure-Relief Valve

A pressure-relief valve is used to protect the boiler and the system from steam pressures exceeding the pressure rating of the boiler or the system (*Figure 16*). It does not operate unless an overpressure condition exists. The low-pressure steam boiler is built for a maximum working pressure of 15 psi. The pressure-relief valve is set to open at the same pressure and discharge the steam into the atmosphere. More than one pressure-relief valve must be used on boilers having over 500 square feet of heating surface. This is a precaution to guard against the failure of a single valve. Safety relief valves should always be installed as directed by the manufacturer. They are usually installed in an upright position at the top of the boiler and at the highest point on the steam side of the boiler.

> **WARNING!**
> Unless stated otherwise, the pressure setting of a safety relief valve is not field-adjustable and must not be altered. Misadjustment may cause an explosion, which can result in death or personal injury. The relief valve must also be full-sized, without reductions on its outlet or inlet. A pressure-relief valve must be immediately replaced if it does not open when its pressure rating is exceeded, or if it leaks when the boiler is operating at normal pressures. Operation of the boiler with the relief valve leaking can cause a dangerous buildup of scale in the relief valve. This can cause the relief valve not to function should an overpressurization condition occur.

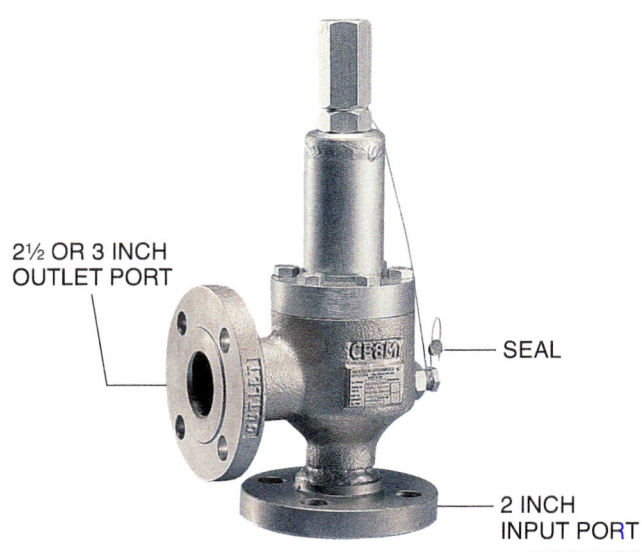

Figure 16 Pressure-relief valve.

Pressure-relief valves should be checked per the manufacturer's instructions intermittently during operation and after long periods of inactivity.

2.3.0 Steam System Loads

Steam system loads can be manufacturing processes, steam-to-water heat exchangers used for heating potable water, or for supplying hot water used as a medium for space heating. They can also be loads comprised of various types of heating terminals that use steam directly as the medium for space heating.

2.3.1 Heat Exchangers/ Converters

A heat exchanger is used to transfer heat from one medium to another. A heat exchanger is used in steam systems mainly to generate hot water for hot-water heating systems, as a steam-to-water heat exchanger. *Figure 17* shows a horizontal shell-and-tube heat exchanger (converter) used in this manner. It consists of tubes encased in a shell. The tubes provide a closed water circuit that allows redirection of the water through the converter more than once. The hot water is circulated through the exchanger tubes, while the steam circulates in the shell that surrounds the tubes. There, the heat is transferred from the steam to the water. The result is that hot water at the output of the heat exchanger is available to heat all or part of a building.

Figure 18 shows a typical piping diagram for a heat exchanger used as a hot-water converter in a hot-water heating system. The modulating steam control valve shown is generally thermostatic,

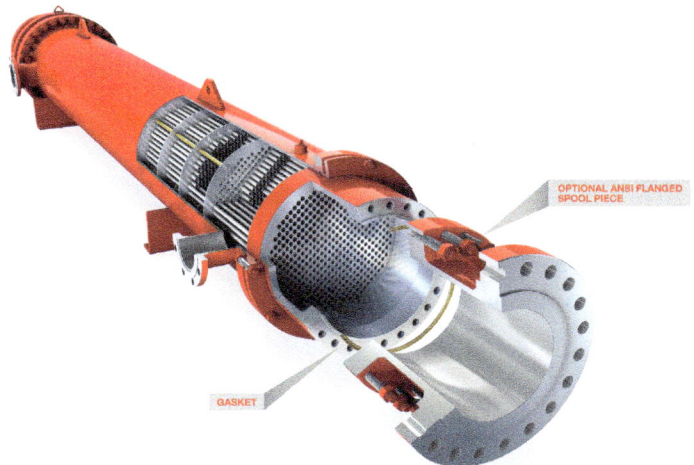

Figure 17 Horizontal shell and tube heat exchanger.

regulating the flow of steam based on the leaving water temperature.

A steam-to-water heat exchanger can also be used to heat cold water to produce hot water at the output of the heat exchanger for domestic use. When used in this way, the heat exchanger is sometimes called an instantaneous heater.

For added safety against contamination, most heat exchangers used to heat potable water are of double-walled construction; that is, two walls separate the steam from the potable water. This is specifically required if conditioning chemicals have been added to the steam system.

2.3.2 Terminals

Many kinds of heating terminals can be used in steam-heating systems. Natural convection units include radiators, convectors, and baseboard convectors. Forced convection units include unit heaters (*Figure 19*), unit ventilators, cabinet heaters, induction units, fan coil units, and central air handling units.

With the exceptions covered here, the terminals used in steam-heating systems are built and operate in the same way as the terminals used in hot-water heating and dual-temperature systems. Fan coils, induction units, and unit ventilators can be used for both heating and cooling. Dual-coil units have a steam coil used for heating and a separate coil for cooling.

There are two types of steam heating coils, single-row steam coils (*Figure 20*) and steam distribution coils. It can be difficult to distinguish the two types of coils from the outside. Single-row coils are used for general heating applications where the following conditions exist:

- The coil will not be subjected to freezing temperatures
- The flow of steam will likely not be modulated
- Precise control of the air temperature across the entire face of the coil is not necessary

Steam distribution coils (*Figure 21*) are generally specified where one or more of these issues must be considered. These coils are built with perforated inner tubes that allow the steam to be distributed evenly throughout the coil. They are generally considered freeze-resistant. Many HVAC applications require this type of coil. Systems with outside air intakes that use steam to preheat the air are often exposed to temperatures below freezing. Most HVAC applications also require the flow of steam to be modulated,

Controlling Mercury Hazards

Mercury is a major environmental pollutant that is known to cause cancer. Every effort is being made to remove mercury from the environment and to regulate its disposal. The pressure switches in older boiler systems contain a large amount of mercury, and are therefore a target of this effort. As mercury switches reach the end of their useful lives, or as boilers are overhauled, the mercury switches are being replaced with other devices such as snap-action bimetal and electric switches. Disposal of switches and other devices containing mercury is carefully regulated. Most states have enacted laws governing the storage and disposal of mercury, so it is important to know the applicable regulations in your state.

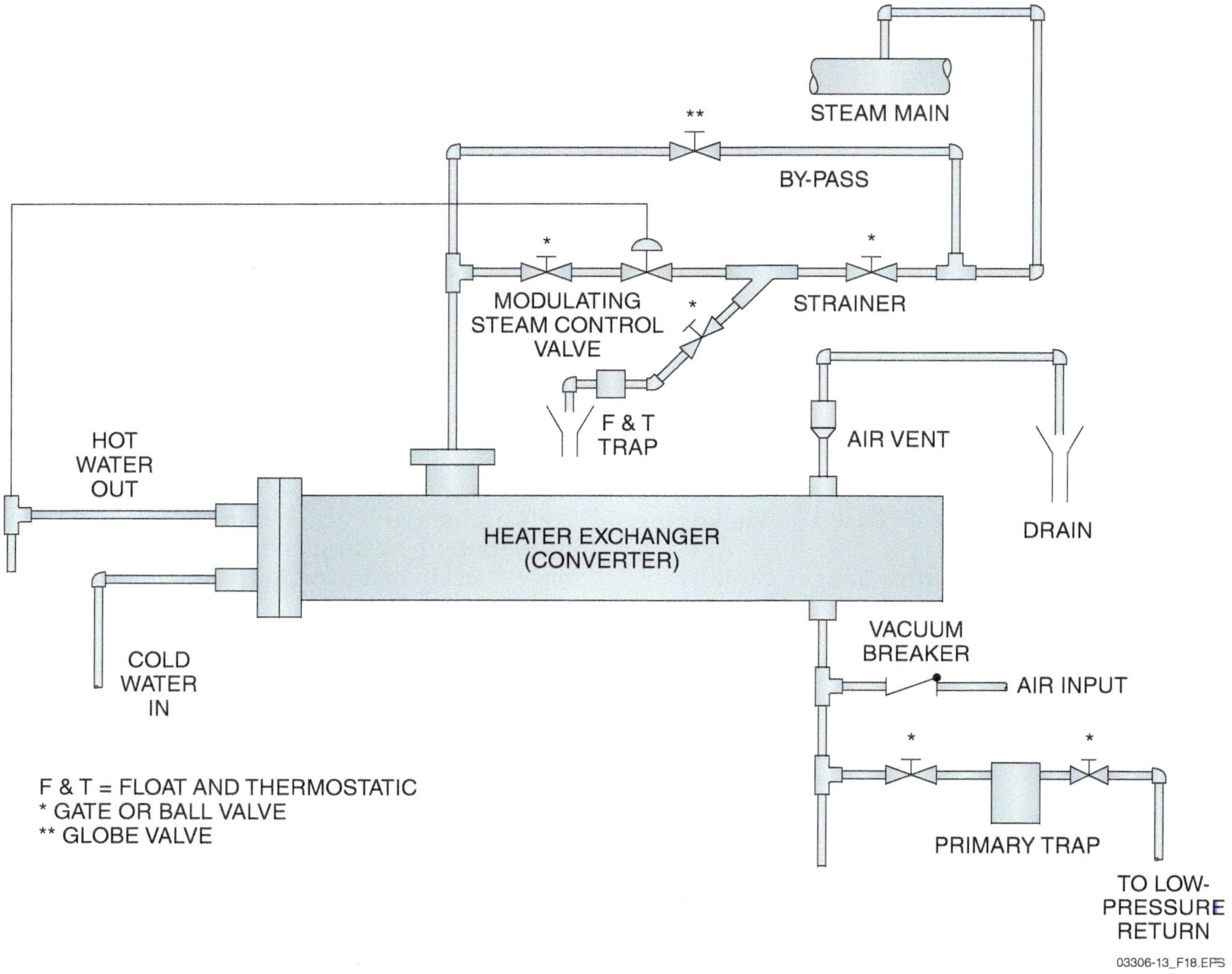

Figure 18 Typical installation of a heat exchanger when used as a converter.

Figure 19 Unit heater.

Figure 20 Single-row steam coil.

Figure 21 Steam distribution coil.

making the steam distribution coil the best overall choice. Single-row coils are used for general heating where no attempt is made to obtain equal air temperatures over the entire length of the coil. They are also used in airstreams where the temperature does not drop below 32°F. Steam distribution coils are made with perforated inner tubes that are used to distribute the steam along the full length of the inner surface of the tubes. They are typically used where the steam must be throttled to obtain control. They are also used where even temperatures are required along the entire length of the tube, or where freezing temperatures are frequently encountered in the air moving over the coil.

Single-Coil Terminals

When a single-coil terminal is used, a steam-to-water heat exchanger must be used in the steam system to transfer the heat from the steam to a secondary hot-water source for use in the terminal.

Drainable Steam Coils

Draining the condensate from a steam coil helps prevent it from freezing. Some manufacturers offer drainable non-freeze coils for this reason. However, if the coil is to be out of service for an extended period, it should be blown out with low-pressure air. If it will be exposed to freezing temperatures, antifreeze may be added to the coil.

Additional Resources

Low Pressure Boilers, 3rd Edition. Fredrick M. Steingress, Daryl R. Walker. Orland Park, Illinois: American Technical Publishers.

High Pressure Boilers, 4th Edition. Fredrick M. Steingress, Harold J. Frost, Daryl R. Walker. Orland Park, Illinois: American Technical Publishers.

The Control of Boilers, 2nd Edition. Sam G. Dukelow. Research Triangle Park, North Carolina: International Society of Automation (ISA).

2.0.0 Section Review

1. The steam cycle begins at the _____.
    a. load
    b. control valve
    c. boiler
    d. condensate tank

2. Boilers with the EPA's ENERGY STAR logo must have a minimum AFUE rating of _____.
    a. 80 percent
    b. 85 percent
    c. 90 percent
    d. 95 percent

3. When used as a domestic hot water heater, a steam-to-water heat exchanger is sometimes called a(n) _____.
    a. flash steam heater
    b. instantaneous heater
    c. momentary demand heater
    d. water desuperheater

Section Three

3.0.0 Steam System Valves and Piping

Objectives

Identify common steam system piping arrangements and valves.
  a. Identify common steam system piping arrangements.
  b. Identify common condensate return systems and describe waterside care.
  c. Explain how steam and condensate system piping is sized.
  d. Identify steam-based pressure-reducing and thermostatic control valves.

Performance Task

1. Perform selected operating procedures on low-pressure steam boilers and systems.
3. Identify common piping configurations used with steam systems.

Trade Terms

Drip leg: A drain for condensate in a steam line placed at a low point or change of direction in the line and used with a steam trap.

Flash steam: Formed when hot condensate is released to a lower pressure and re-evaporated.

Turndown ratio: In steam systems, the ratio of downstream pressure to upstream pressure, usually applied to steam pressure-reducing valves. More than a 50 percent turndown ratio is generally considered a large pressure reduction.

Water hammer: A condition that occurs when hot steam comes into contact with cooled condensate, builds pressure, and pushes the water through the line at high speeds, slamming into valves and other devices. Water hammer also occurs in domestic water systems. If the cause is not corrected, water hammer can damage the system.

Various steam and auxiliary steam systems are used based on the size of the boiler system and the design steam temperature. A number of steam piping systems and valves, as well as auxiliary steam system piping that include condensate/feedwater piping and components, are covered in the following sections.

3.1.0 Steam System Piping

Steam piping systems are classified as either one-pipe or two-pipe systems. The type of system is determined by the piping method used to supply steam to and return condensate from the system terminals. Each of these piping systems is further classified by the method used to return the condensate (gravity or mechanical flow), and the directions of steam and condensate flow in the system.

3.1.1 One-Pipe Systems

In a one-pipe system, steam flows to and condensate is returned from the terminal unit through the same pipe. There is only one connection to each heating terminal unit, which functions both as the supply and the return. Normally, one-pipe systems use a gravity return, but they may use a condensate pump. This is done when there is insufficient height above the boiler water level to develop enough pressure to return the condensate directly to the boiler. One-pipe gravity systems do not use steam traps. Air vents are used at each terminal unit and at the ends of all supply mains to vent air so the system can fill with steam. If a one-pipe system uses a condensate pump, air vents are used at each terminal unit, but steam traps are used at the end of each supply main. One-pipe systems are used mostly in residential and small commercial systems. They are best applied in homes or small buildings with basements, where space and height exist for the needed drainage pitch. Though limited in potential applications, their advantage rests in their simplicity. Very few moving or functional components are required, increasing reliability and minimizing service on properly installed systems. However, they are more susceptible to water hammer because the condensate and steam frequently make contact and the condensate is not quickly removed from the system. This is especially true during system startup, when piping and components are relatively cold.

There are several configurations of one-pipe steam systems. *Figure 22* shows a basic one-pipe up-feed gravity system. The term *upfeed* means that the steam supply main is below the level of the heating terminals that they serve. Steam flow is then fed upward in the riser for input to the terminal units.

As shown in *Figure 22*, the steam supply main rises from the boiler to a high point, then

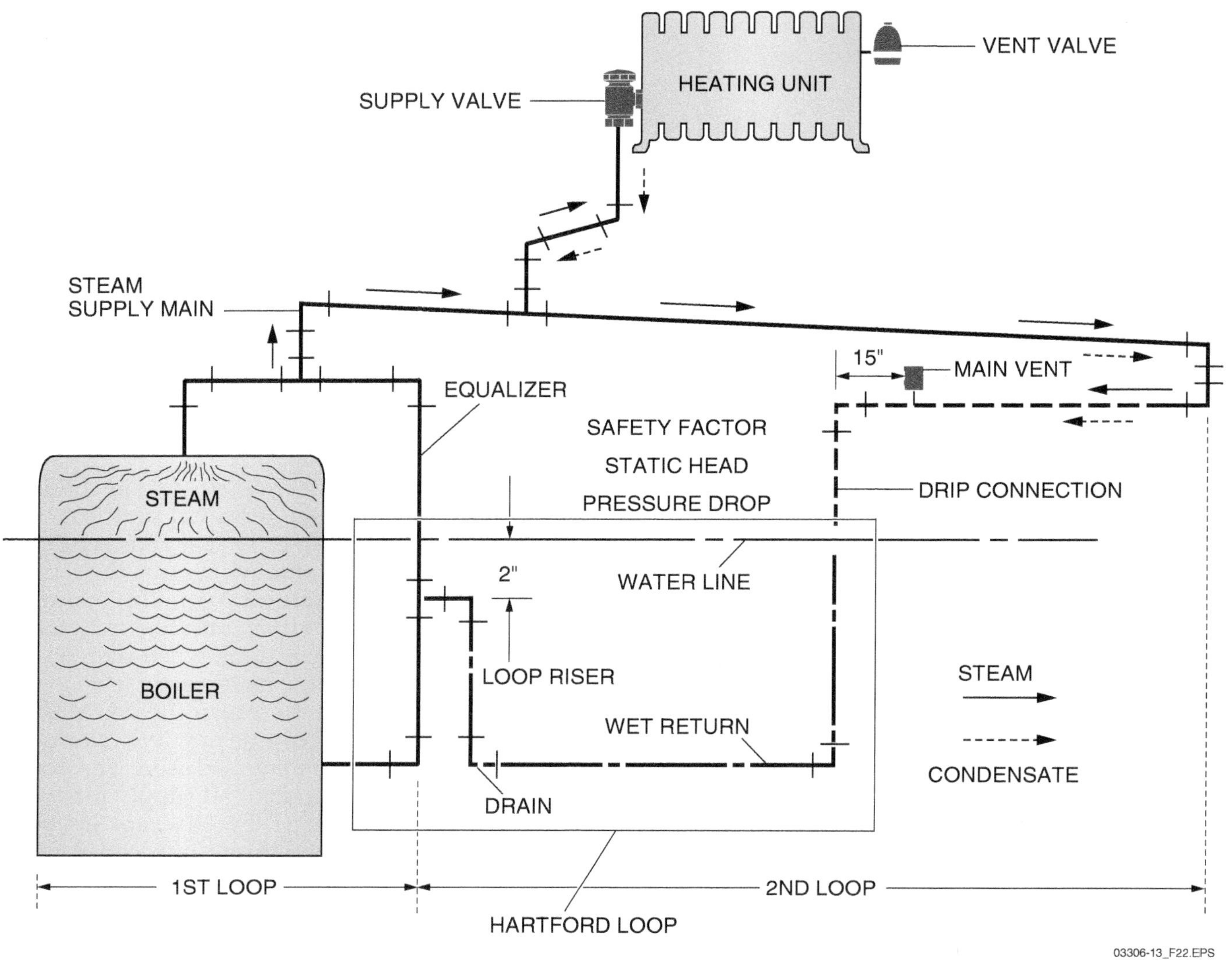

Figure 22 Basic one-pipe upfeed system.

is pitched downward from this point. The pitch should be no less than one inch down per 20 feet and should slope in the direction of the gravity flow of the condensate. The piping normally runs full size to the last terminal unit takeoff and is then reduced in size after it drops down below the boiler water line. When the return is above the boiler water line, it is called a dry return. In this case, it is only considered to be dry as it is not flooded or filled with water—it carries condensate, along with some steam and air. The section of return piping installed below the level of the water in the boiler is called the wet return. It is completely filled with water and does not carry steam or air. When the system is in operation, the steam and condensate flow in the same direction in the horizontal mains and in opposite directions in the branches and risers. Because the steam and condensate flow in the same direction in the horizontal mains, this type of system is also called a parallel-flow system.

A Hartford loop is a pressure-balancing loop that introduces full boiler pressure on the return side of the boiler. This pressure prevents reverse circulation or water leaving the boiler by the return line. The Hartford loop is actually two loops; the first is an equalizer line and the second is an extension from the steam main through the return and back to a connection at the equalizer loop.

3.1.2 Two-Pipe Systems

The two-pipe system uses separate pipes to supply the steam and return the condensate from the terminal units.

Thermostatic traps are used at the outlet of each terminal. They keep the steam in each terminal unit until it gives up its latent heat, then the trap opens to pass the condensate and permit more steam to enter the terminal. Float and thermostatic steam traps should be used at the end of each steam main, at the bottom of each riser, and along the horizontal mains. *Figure 23* shows a basic two-pipe gravity system.

Two-pipe systems can have either gravity or mechanical condensate returns. However, most systems use a mechanical return (*Figure 24*) provided either by a condensate pump or a vacuum pump. When steam systems increase in size, higher steam pressures are needed to obtain steam circulation, and some means other than gravity must be used to return the condensate to the boiler. A condensate pump and receiver tank are commonly used for this purpose. Normally, the condensate from the return piping system flows by gravity into a receiver tank. A condensate pump mounted near the receiver pumps the accumulated condensate into the boiler feedwater return piping. A typical condensate pump and receiver tank used for this purpose were covered earlier in this module. Note that in *Figure 24* air vents have been eliminated from the design. With a mechanical condensate return, air is released from the system through an open vent pipe on the receiver tank. This vent must open at a higher level than the boiler water level.

A vacuum-return system (*Figure 25*) is similar to a condensate return system, except that a vacuum pump is installed to provide a low vacuum in the return line to return the condensate to the boiler. A vacuum-return system is one that operates with a pressure in the condensate-return piping that is below atmospheric pressure. The steam supply lines and terminal units are maintained at a positive pressure. The vacuum pump produces a vacuum in the system return lines by removing the air and vapor and venting it to the atmosphere in a related receiver tank.

It is important to note that the vacuum pump does not draw or pump the condensate itself. By quickly exhausting air from the system, the vacuum pump causes the steam to flow more rapidly, reducing warmup times, as well as noise from water hammer. The vacuum pump cycles as needed to maintain a predetermined vacuum level in the return lines. Operating the return line in a vacuum, rather than at or slightly above atmospheric pressure, increases the pressure differential in the overall system. As a result, steam pressure settings at the boiler can sometimes be lowered for significant fuel savings and improved efficiency. Much of this efficiency can easily be lost through air leaks. It is essential that return piping be leak-free to maximize the benefits associated with vacuum return and prevent excessive pump operation.

It is commonly understood that water boils, creating steam, at a lower temperature when exposed to pressures less than atmospheric. For example, at 15 in Hg vacuum, water boils at roughly 180°F. Some vacuum systems, known as variable vacuum systems, rely on this advantage by placing the supply side of the system, as well as the return piping, into a vacuum. This obviously requires greater capacity (expressed in cfm) from

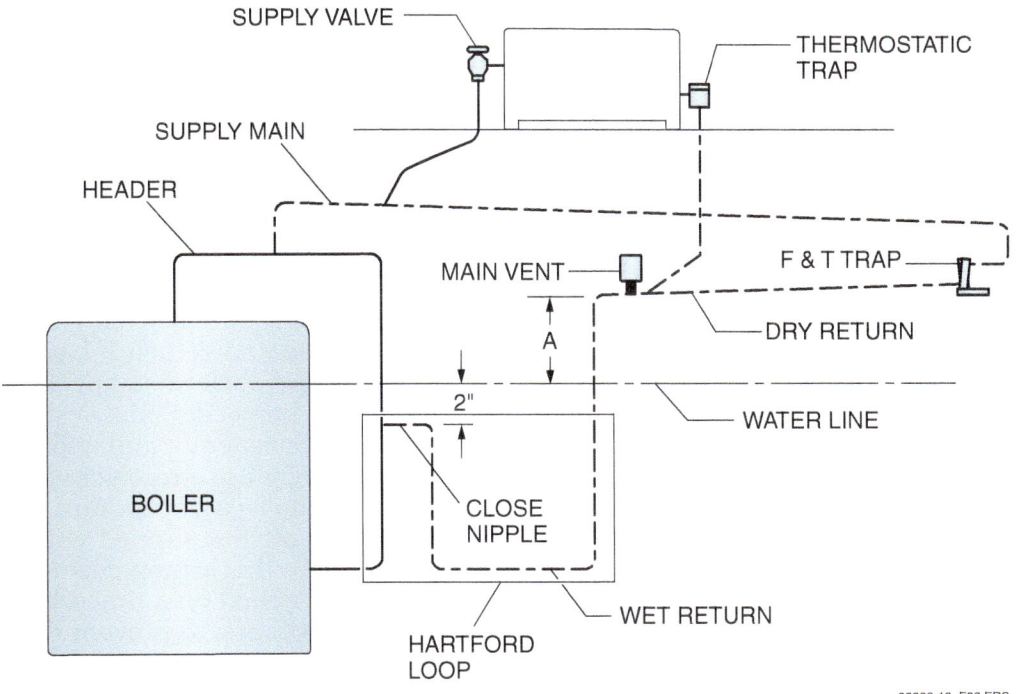

Figure 23 Two-pipe gravity system.

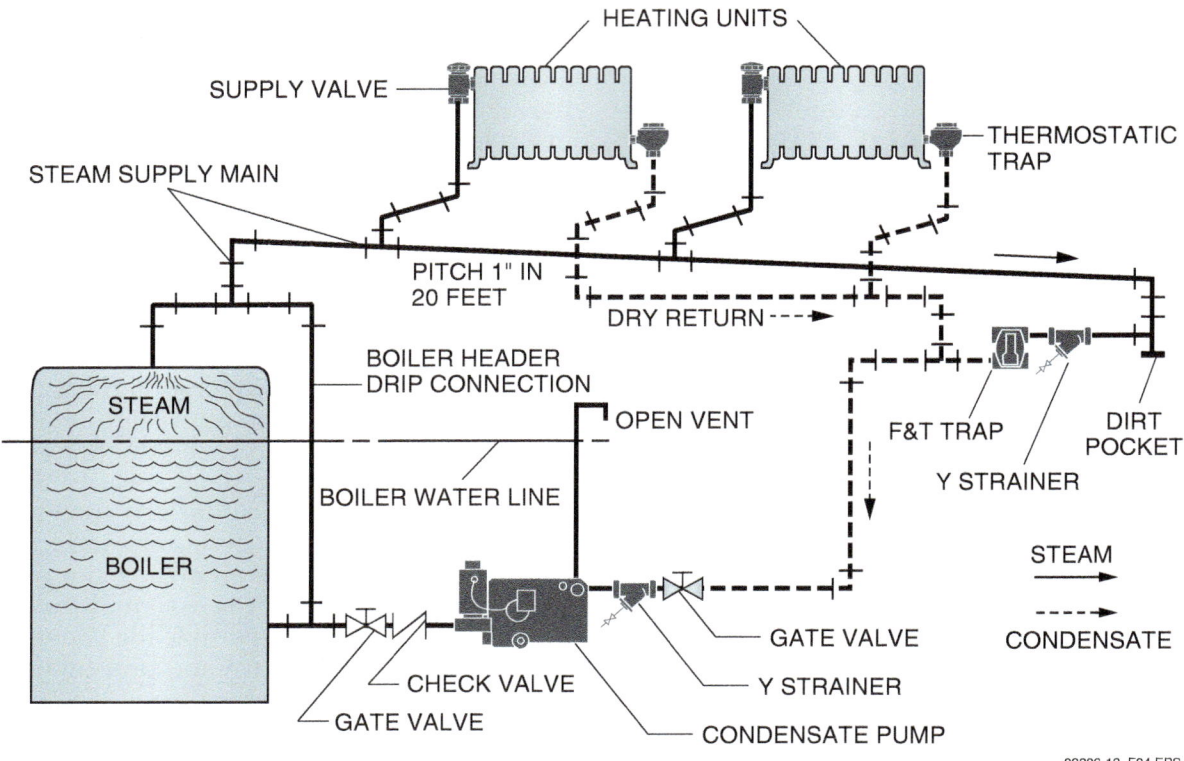

Figure 24 Two-pipe system with mechanical condensate return.

the vacuum pump, and air tight piping systems become even more important. By reducing the boiling point of the water, even greater operating efficiencies can be obtained.

The vacuum pump controller, which determines the vacuum level that the pump must maintain, is generally designed to respond to outdoor temperatures. During mild weather, vacuum levels can be maintained at a lower value (18 to 25 in Hg), resulting in lower temperature steam and reduced fuel usage. As outdoor temperatures fall, creating the need for additional capacity, vacuum levels are maintained at a higher value (2 to 15 in Hg) to increase the temperature of the steam provided to the heating system. The controller is programmed to monitor outdoor temperatures and reset the vacuum level across the range of heating season outdoor temperatures expected for the specific geographic region.

Operation at lower pressures will increase the specific volume of the steam. Because of this increased volume, it takes less steam to fill the system, resulting in more efficient system operation.

3.2.0 Condensate Return/ Feedwater System Components

Condensate is the byproduct of a steam system. It forms in the distribution system because of the heat transfer from the steam to the area or substance being heated. Some condensation is also formed as a result of unavoidable radiation. Once the steam has condensed, giving up its latent heat, the hot condensate must be removed immediately and returned to the boiler. This is because the amount of heat in a pound of condensate is negligible when compared to the heat in a pound of steam.

The components that may be used in a condensate return/feedwater system are a condensate pump and receiver tank, vacuum pump, and de-aerating feedwater heater.

3.2.1 Condensate Pump and Receiver Tank

On larger steam-heating systems, a condensate pump and receiver are normally used to accumulate condensate and return it back to the boiler. There are many variations and types of equipment that can be used. Some systems may use a condensate receiver equipped with a single pump, while others may use a receiver with two pumps, often referred to as a duplex unit (*Figure 26*). Duplex units are often equipped with an alternator arrangement. In this arrangement, pumps are automatically switched on a timed basis, or when a pump failure occurs, to prevent the system from flooding and shutting down. Alternating their operation also balances pump wear.

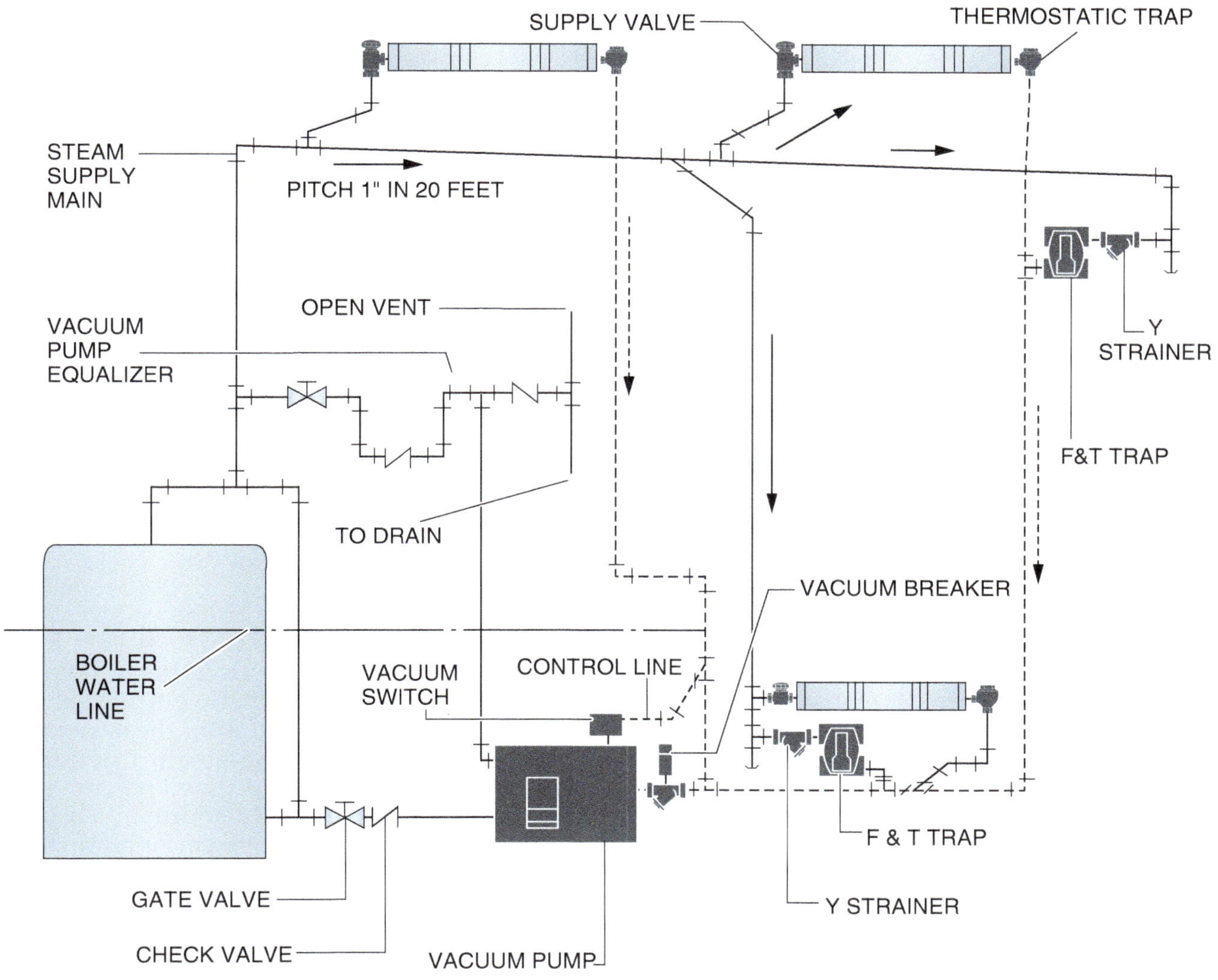

Figure 25 Vacuum return system.

Figure 26 Duplex condensate receiver.

Figure 27 shows one type of condensate and receiver tank system. The receiver tank is an open reservoir into which the return piping discharges its condensate. This system uses a pump controller installed at the normal boiler water line to control the condensate (boiler feed) pump. Whenever the pump controller senses a drop in the normal boiler water level, it turns on the condensate pump. This causes water from the receiver tank to be pumped into the boiler as needed to replenish the boiler water level. A check valve in the condensate pump discharge line prevents the boiler water from backing up into the receiver when the condensate pump is turned off. A gate valve in the line allows the pump to be serviced without the need to drain the boiler.

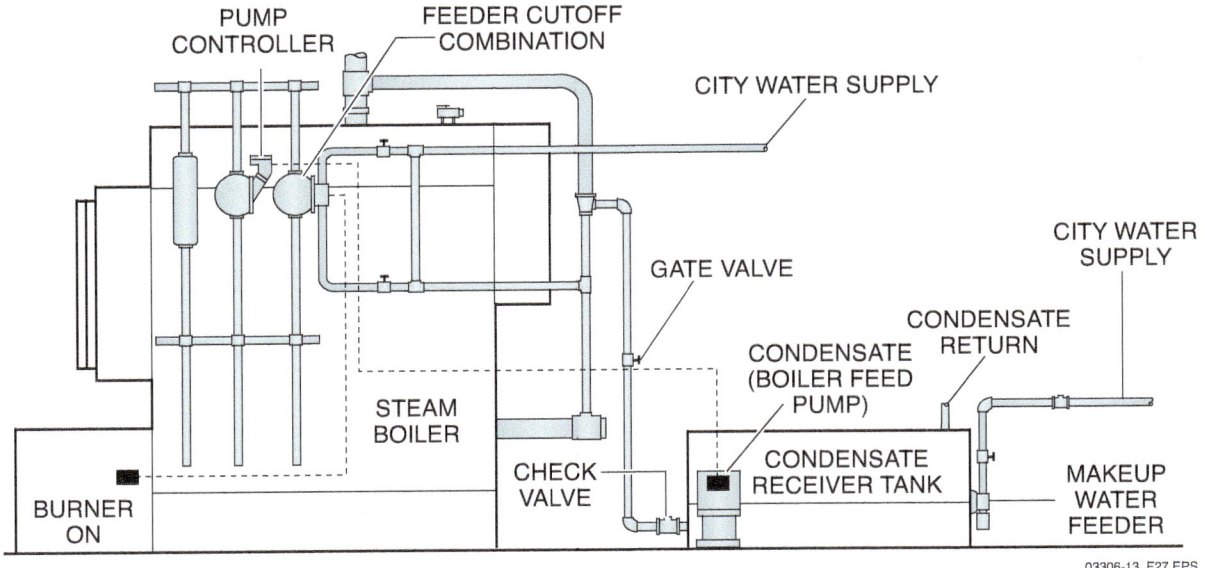

Figure 27 Typical condensate pump and receiver system.

The receiver tank is also equipped with a float-actuated makeup water feeder control. If the level of condensate water in the receiver tank drops below the normal operating level, the control float drops and opens its valve to allow makeup water to enter the receiver tank from the city water supply. Additional boiler protection is provided by a combined feeder cutoff control. In the event of a power failure, condensate pump failure, or other problem, this control will first attempt to supply the boiler with sufficient water from the city water supply to prevent it from overheating. If the boiler water drops below a safe level, the control then acts to turn off the boiler burners.

3.2.2 Vacuum Pumps

Some systems use a vacuum pump in the condensate return line. Vacuum pump operation is similar to condensate pump operation in that the vacuum pump causes condensate from the condensate return line to flow into a receiver tank. Normally, a feed pump is then used to pump the condensate water from the receiver tank into the boiler. The vacuum unit and condensate receiver and pump are often combined into a single unit, such as the unit pictured in *Figure 28*.

At system startup, the condensate vacuum pump produces a vacuum in the system by removing the air and vapor from the system and venting it to the atmosphere. By quickly exhausting the air and condensate from the system, the vacuum pump causes the steam to circulate rapidly, reducing warm-up time and providing quieter system operation. In addition to causing condensate flow during system operation, the

Figure 28 Vacuum heating units.

Returning Condensate to the Boiler

In small systems, the condensate may be returned to the boiler by gravity. In most systems, however, pumping is required. Usually, the condensate flows by gravity to a collecting tank (receiver tank) from which it is pumped directly to the boiler or to a boiler feedwater system.

vacuum pump also turns on whenever needed to maintain the vacuum level in the system.

3.2.3 Deaerating Feedwater Heater

In some larger steam systems, the condensate water pumped out of the receiver tank does not get pumped into the boiler. Instead, it is pumped into a deaerating feedwater heater, also called a deaerator. This device uses system steam to preheat the water before it is supplied to the boiler in order to remove most of the dissolved oxygen. This is especially important on systems that lose 25 percent or more of the steam condensate from the circuit during normal operation, or are drained and refilled regularly. The reason is that fresh makeup water carries large quantities of dissolved oxygen. Deaeration is also beneficial in systems that operate at pressures above 75 psig.

3.2.4 Flash Tanks

When hot condensate under pressure is released to a lower pressure, part of it is re-evaporated and becomes flash steam. This normally happens when hot condensate is discharged into the condensate return line at a pressure lower than its saturation pressure. Some of the condensate flashes into steam and flows along with the liquid condensate through the return lines back to the boiler. This tends to cause an undesirable pressure increase in the condensate return lines. Flash tanks are used in medium-pressure and high-pressure systems to remove flash steam from the condensate lines by venting it either to the atmosphere or into a low-pressure steam main for reuse.

When flash steam is used as an energy source, high-pressure steam returns are usually piped to a flash tank (*Figure 29*). Flash tanks can be mounted vertically or horizontally. However, ver-

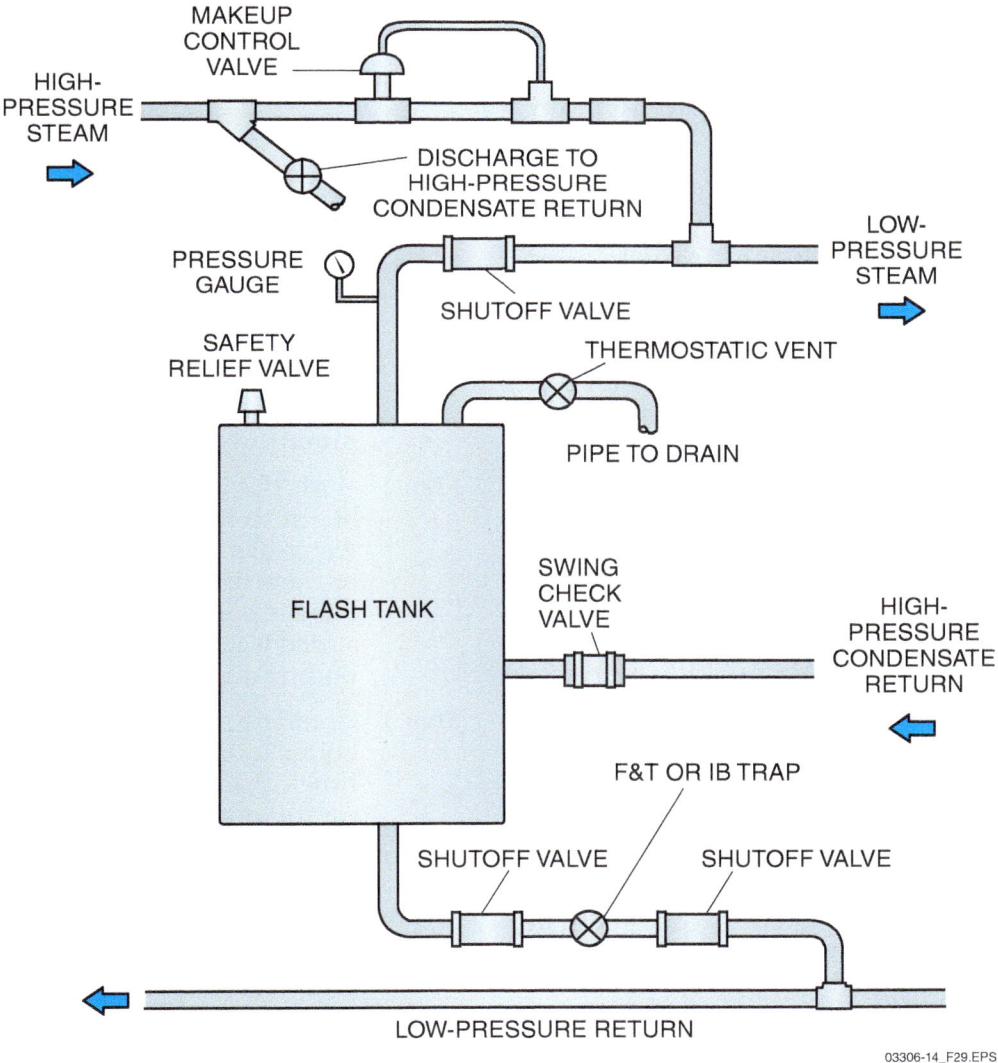

Figure 29 Flash tank connections.

tical mounting provides for better separation of steam and water.

When the condensate and any flash steam in the return line reach the tank, the high-pressure condensate is released into the lower pressure of the tank, causing some of it to flash into steam. This flash steam is discharged into the low-pressure steam main. The remaining condensate is pumped back to the boiler or discharged to a waste drain.

For proper flash tank operation, the condensate lines should pitch toward the tank. If more than one condensate line feeds the tank, a check valve should be installed in each line to prevent backflow. The top of the tank should have a thermostatic air vent to vent any air that accumulates in the tank. The bottom of the tank should have an inverted bucket or float and thermostatic-type steam trap.

The demand steam load on the flash tank should be greater than the amount of flash steam available from the tank. If it is not, the low-pressure system can be over-pressurized. A safety relief valve must be installed at the top of the flash tank to protect the low-pressure line from any over-pressurization. Since the flash steam produced in the flash tank should always be less than the amount of low-pressure steam that is needed, a makeup valve in the high-pressure steam line is used to supply the additional steam needed to maintain the correct pressure in the low-pressure line. This steam makeup valve is actually a pressure-reducing valve. It is much like those discussed earlier, but has an external pilot line to monitor the low pressure downstream of the valve itself.

3.2.5 Boiler Blowdown and Skimming

> **WARNING!** Boilers operate under high pressure and are potentially dangerous. When working with or around boilers, follow all applicable manufacturer's instructions, OSHA regulations, and job-site requirements relating to safety and the use of safety equipment.

During operation, makeup water is added to boilers to maintain the proper water level. As a result of adding this water, sediment, dissolved salts, or similar organic matter gradually accumulate in the boiler. Some of these impurities cause foam to form on the surface of the boiler water. This is usually indicated by drops of water appearing with the steam. If all the foreign matter is not purged from the boiler on a regular basis, boiler operation will degrade. The process of purging the boiler is called blowdown. Every boiler is equipped with one or more valves for this purpose. Two kinds of blowdown valves may be found on a boiler: surface blowdown, also known as skimming valves, and/or bottom blowdown valves.

Surface blowdown valves are used to skim off impurities that cause foaming on the surface of the water inside the boiler. This foaming is caused by high surface tension due to a scum buildup from oil, grease, and/or sediment on the surface of the water. Foaming prevents steam bubbles from breaking through the water surface and hinders steam production. Surface blowdown valves are located on the boiler at the normal operating water level. Not all boilers are equipped with a surface blowdown valve. Use of the surface blowdown valve to skim off the impurities requires some experience.

Skimming should always be done according to the instructions given in the boiler manufacturer's service literature. The following represents common skimming procedures.

> **WARNING!** Never add cold water to a completely empty boiler when the burners are ignited. Doing so will cause serious damage and may create an explosion hazard.

Step 1 If not permanently connected to a drain, run a temporary connection from the boiler's skimming valve to a suitable drain.

Step 2 With the boiler empty and cool, slowly begin to add water. After a quantity of water has entered the boiler, turn on the burners and adjust the firing rate and/or burner controls so that the water being added is kept just below the boiling point. Boiling and turbulence must be avoided.

Step 3 Gradually raise the hot water level in the boiler to the point where the water just flows from the skimming valve, being careful not to raise it above this point.

Step 4 Continue to skim the boiler water until there is no trace of impurities. Water may be checked to make sure that it is free from oil by drawing off a sample. If the sample is reasonably free of oil, it will not froth when heated to the boiling point.

> **NOTE:** Skimming will not clean the boiler of sediment that has accumulated at the bottom. After skimming, the boiler should be cleaned further by performing a blowdown procedure.

> **NOTE:** Only attempt blowdown at a light load. Ideally, temporarily suspend the boiler's heating process to halt water turbulence and allow the solids to settle out.

Blowdown is another method used to purge the boiler of foreign matter. This method is used to drain off some of the water from the bottom of the boiler. The drained water is then replaced it with an equal amount of fresh, clean water. The bottom blowdown valve is used to drain off the sediment, scale, and other impurities that settle out of the water as it is heated and accumulate at the lowest point in the boiler. If the boiler is extremely dirty, blowdown may require that the boiler be completely drained and refilled with water one or more times. This process must be continued until the water discharged from the boiler runs clear. On smaller systems, the hot boiler water discharged through the bottom blowdown valve is drained directly into a nearby sewer. In larger systems, it is piped through a line to a tank where it is allowed to cool before going into the sewer. Codes require that wastewater from a steam or hot-water boiler not be hotter than 140°F. Boiler blowdown should always be performed as directed in the boiler manufacturer's service instructions. The following represents common blowdown procedures:

Step 1 Check the water level in the boiler.

Step 2 Partially open the bottom blowdown valve. Once the water starts draining, fully open the valve.

Step 3 Remain at the blowdown valve. Monitor the gauge glass during blowdown to make sure that the water level is not lowered to a dangerously low point.

Step 4 When the desired amount of water has been drained from the boiler, close the blowdown valve.

3.2.6 Boiler Water Treatment

The water used in a boiler must be kept clean for efficient operation. Never add dirty or rusty water to a boiler. Hard water may eventually interfere with the efficient operation of a boiler. For this reason, hard water should be chemically treated with water softeners before being used. Scale, corrosion, fouling, and foaming can all cause boiler problems. Scale deposits on heating surfaces increase boiler temperatures and lower operating efficiency. Corrosion can damage metal surfaces, resulting in metal fatigue and failure. Fouling clogs nozzles and pipes with solid materials, thereby restricting circulation and reducing the heat transfer efficiency. Foaming results in overheating and can cause water impurities to be carried along with the steam into the system.

There are various chemical and mechanical methods of water treatment used to help prevent problems caused by water impurities. At a minimum, proper water treatment includes blowdown, pH control, and the addition of chemicals to neutralize other contaminants. The specific methods and devices used for water treatment are covered in a separate module.

3.3.0 Steam and Condensate Pipe Sizing

Proper steam and condensate line sizing is essential for any steam system to operate efficiently and provide the needed capacity for each load. To size the piping system, information regarding the volume of steam needed, as well as the steam pressure in the system must be known.

As a general rule, it is best that steam boilers operate at a pressure near their design pressure. In some cases, especially in larger facilities with a significant number of loads, this may require that the steam system pressure be considerably higher than actually required at any one of the operating loads. Operating steam boilers at pressures significantly lower than the boiler is designed for reduces the volume of space in the boiler that normally is allowed for steam. This places the water level somewhat closer to the steam outlet, and the flow velocity of the steam to the outlet becomes faster due to the smaller area of confinement.

Using Flash Steam

The heat content of flash steam is the same as that of live steam at the same pressure. Use of the flash steam in a low-pressure steam main allows this heat to be used rather than wasted. It is commonly used for space heating, as well as heating or preheating of water, oil, and other liquids.

When operated in this manner, the likelihood of water droplet carryover is greatly increased.

There are additional advantages to controlling steam pressure higher than is needed downstream. The size of steam distribution mains becomes smaller to carry the same amount of energy, thus saving on installation costs. Due to smaller surface areas with smaller piping, heat loss is also reduced. Steam pressure can then be reduced using pressure-reducing valves near the inlet to each of the individual loads. These valves can be physically smaller, further reducing installation costs.

Appropriate pipe sizing is important to ensure a sufficient steam supply to the loads. Undersized piping causes high pressure drops and greater velocities, increasing noise and erosion of internal piping surfaces. It is also quite common for a new load to be added to a steam system long after the initial installation is complete. A piping system that is already undersized or marginal may prevent even a relatively small load from being added without re-piping, or an insufficient supply of steam to all loads could result. On the other hand, piping that is too large creates unnecessary heat losses and higher installation costs.

Two primary methods are used to size steam piping: sizing by desired velocity or sizing to maintain an acceptable pressure drop. Systems with shorter mains and branch lines are often sized by velocity. Longer piping runs may be sized first by velocity, then rechecked to ensure that acceptable pressure drops have not been exceeded. There are a number of factors that impact steam pipe sizing in various parts of the system.

3.3.1 Velocity Sizing

When sizing based on velocity, steam lines should be designed for speeds of 80 to 120 feet per second (fps), or 4,800 to 7,200 feet per minute (fpm). In very limited cases, where noise may not be an issue and pressure drops are not as important, velocities up to 200 fps or 12,000 fpm have been used. Again, these values are not recommended for the reasons stated earlier concerning undersized piping.

Figure 30 provides a steam velocity chart for pipe sizing. As mentioned earlier, the volume of steam to be carried, stated in lb/hr (pounds per hour) must be known. For new pipe sizing, the system design steam pressure must also be known. Note that this chart is for Schedule 80 pipe. The appropriate chart for the type of pipe being used in the installation must be used.

Example 1

A steam system is being designed that must carry a minimum of 1,500 lb/hr of steam through the steam main. Maximum design velocity must not exceed 100 fps, and the designated steam pressure will be 100 psig. Select an appropriate pipe size for the steam main.

Step 1 Enter the left side of the chart at 100 psig. Since the velocity cannot exceed 100 fps in this example, read across the chart horizontally at the line of values for 80 fps (120 fps is obviously too fast).

Step 2 Continue across the chart until you reach a value equal to, or greater than, 1,500 lb/hr. The closest value that is not less than our minimum of 1,500 lb/hr is 1,660 lb/hr.

Step 3 Read up the column to the top line of steam pipe sizes. Pipe size will be 2 inches.

When troubleshooting steam systems, the chart can also be used to check existing conditions. For example, if the system seems too noisy in operation, check the pipe size against the volume of steam being transported to see if velocities exceed the prescribed maximum of 120 fps for quiet operation. This can happen when new loads are added to an existing piping system, and steam pressures are increased to provide the necessary additional volume of steam through the existing piping.

3.3.2 Steam Line Pressure Drop

Steam lines are generally sized based on velocity first, then checked against the appropriate chart for pressure drop in the system. The allowable pressure drop varies widely, based on the system itself. Systems operating at 100 psig, with one or more loads that require a minimum of 80 psig at the inlet, certainly cannot be sized at a 25 psig allowable loss. Perhaps a more appropriate choice here would be a 5 to 10 psig loss. However, a system operating as low as 5 psig should probably be sized for a pressure drop no greater than 0.3 psig to maintain the needed pressure downstream. As a general rule, the higher the steam pressure, the greater the pressure drop that can likely be tolerated.

Figure 31 is one example of a steam pressure drop chart. Note that this chart is also designated for Schedule 80 pipe. Pressure drops for a given volume of flow will differ in other classes of pipe. Also note that the chart is based on a steam pressure of 100 psig. For pressures other than 100 psig, a correction factor must be applied to the value read from the chart. The correction factors for various pipe sizes are listed at the top of the

Capacity of Sch. 80 Pipe in lb/hr steam															
Pressure psi	Velocity ft/sec	½"	¾"	1"	1¼"	1½"	2"	2½"	3"	4"	5"	6"	8"	10"	12"
5	50	12	26	45	70	100	190	280	410	760	1250	1770	3100	5000	7100
	80	19	45	75	115	170	300	490	710	1250	1800	2700	5200	7600	11000
	120	29	60	110	175	245	460	700	1000	1800	2900	4000	7500	12000	16500
10	50	15	35	55	88	130	240	365	550	950	1500	2200	3770	6160	8500
	80	24	52	95	150	210	380	600	900	1500	2400	3300	5900	9700	13000
	120	35	72	135	210	330	590	850	1250	2200	3400	4800	9000	14400	20500
20	50	21	47	82	123	185	320	520	740	1340	1980	2900	5300	8000	11500
	80	32	70	120	190	260	520	810	1100	1900	3100	4500	8400	13200	18300
	120	50	105	190	300	440	840	1250	1720	3100	4850	6750	13000	19800	28000
30	50	26	56	100	160	230	420	650	950	1650	2600	3650	6500	10500	14500
	80	42	94	155	250	360	655	950	1460	2700	3900	5600	10700	16500	23500
	120	62	130	240	370	570	990	1550	2100	3950	6100	8700	16000	25000	35000
40	50	32	75	120	190	260	505	790	1100	1900	3100	4200	8200	12800	18000
	80	51	110	195	300	445	840	1250	1800	3120	4900	6800	13400	20300	28300
	120	75	160	290	460	660	1100	1900	2700	4700	7500	11000	19400	30500	42500
60	50	43	95	160	250	360	650	1000	1470	2700	3900	5700	10700	16500	24000
	80	65	140	250	400	600	1000	1650	2400	4400	6500	9400	17500	27200	38500
	120	102	240	410	610	950	1660	2600	3800	6500	10300	14700	26400	41000	58000
80	50	53	120	215	315	460	870	1300	1900	3200	5200	7000	13700	21200	29500
	80	85	190	320	500	730	1300	2100	3000	5000	8400	12200	21000	33800	47500
	120	130	290	500	750	1100	1900	3000	4200	7800	12000	17500	30600	51600	71700
100	50	63	130	240	360	570	980	1550	2100	4000	6100	8800	16300	26500	35500
	80	102	240	400	610	950	1660	2550	3700	6400	10200	14600	26000	41000	57300
	120	150	350	600	900	1370	2400	3700	5000	9100	15000	21600	38000	61500	86300
120	50	74	160	290	440	660	1100	1850	2600	4600	7000	10500	18600	29200	41000
	80	120	270	450	710	1030	1800	2800	4150	7200	11600	16500	29200	48000	73800
	120	175	400	680	1060	1520	2850	4300	6500	10700	17500	26000	44300	70200	97700
150	50	90	208	340	550	820	1380	2230	3220	5500	8800	12900	22000	35600	50000
	80	145	320	570	900	1250	2200	3400	4900	8500	14000	20000	35500	57500	79800
	120	215	450	850	1280	1890	3400	5300	7500	13400	20600	30000	55500	85500	120000
200	50	110	265	450	680	1020	1780	2800	4120	7100	11500	16300	28500	45300	64000
	80	180	410	700	1100	1560	2910	4400	6600	11000	18000	26600	46000	72300	100000
	120	250	600	1100	1630	2400	4350	6800	9400	16900	25900	37000	70600	109000	152000

Figure 30 Steam pipe sizing for steam velocity.

Blowdown Separators

Boilers may be equipped with blowdown separators. These devices help reduce corrosion and scaling by separating steam from water and removing dissolved solids. Blowdown separators also lower the temperature of the drain water, which may be necessary to comply with environmental regulations and municipal plumbing codes.

chart. For example, if a steam system is to operate at 5 psig, then the value read from the chart at 100 psig must be multiplied by 5.2 to achieve the correct pressure drop value.

The values in the chart provide the pressure drop that exists in 100 feet of pipe. Your final answer must also be adjusted for the actual length of the pipe in your installation.

Example 2

Using the values and resulting pipe size in *Example 1*, calculate the pressure drop in 75 feet of pipe.

Step 1 Enter the chart at the bottom, locating the vertical line representing 1,500 lb/hr of steam flow.

Step 2 Follow the line vertically until it intersects with the diagonal line representing 2-inch pipe.

Step 3 Follow the horizontal line left from this point of intersection and read the resulting pressure drop for 100 feet of Schedule 80 pipe. The result would be 2.0 psig of pressure loss per 100 feet of pipe.

Step 4 To adjust for the actual length of pipe (75 feet in our example), multiply 2.0 psig/100 feet by 0.75. The resulting pressure drop will then total 1.5 psig in this particular steam line.

3.3.3 Condensate Line Sizing

The steam condensate can carry as much as 15 percent of the energy originally used in the process of creating steam. With ever-rising energy costs, it is imperative that steam systems capture and return the condensate back to the process to preserve this expended energy.

Condensate line sizing can be more difficult than sizing of the steam lines, due to the variety of differing conditions found in the system. A single set of recommendations does not apply. The lines carrying condensate from the load (such as a steam coil in ductwork for a comfort heating application) and into the trap represent a different set of conditions than those found in condensate lines transporting condensate away from a trap.

Condensate piping systems are classified based on the design pressures. Gravity returns describe those systems where the condensate line is at or very near atmospheric pressure, and the condensate generally returns to a vessel vented to the atmosphere. Low-pressure returns are those operating at a pressure below 15 psig, while medium-pressure returns are those operating between 15 and 100 psig. Systems that operate with the condensate line experiencing pressures above 100 psig are classified as high-pressure returns.

In all cases, the sizing of condensate lines is based on several primary factors:

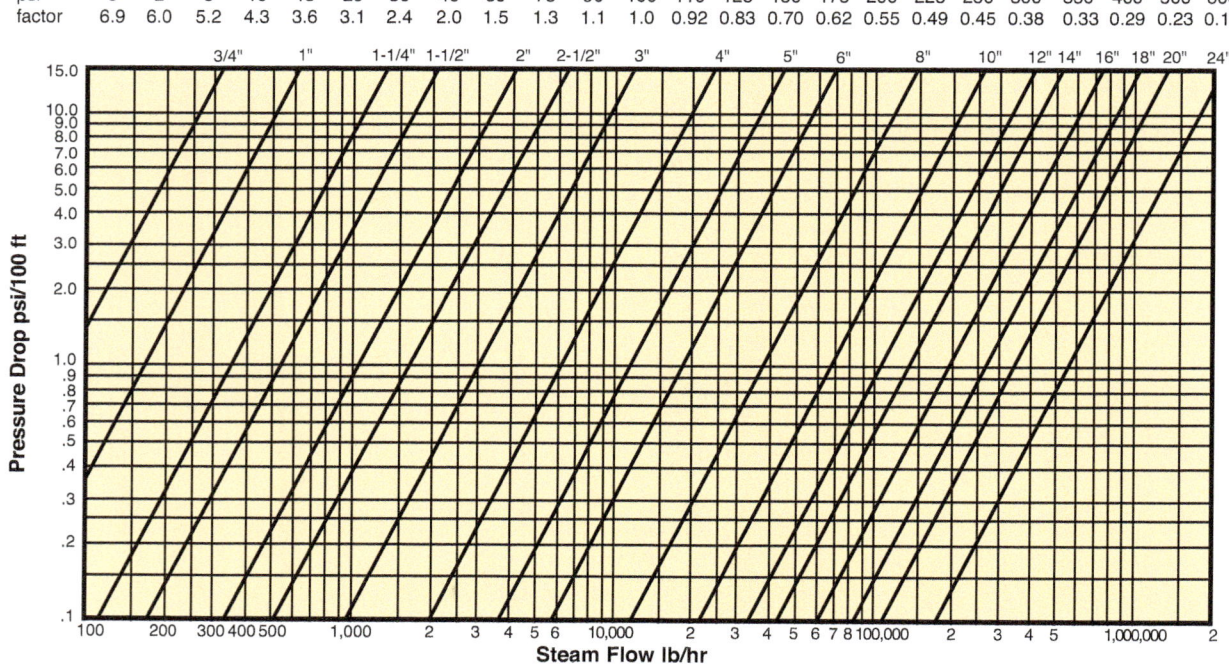

Figure 31 Steam pipe sizing for pressure drop.

- *Pressure* – The difference in pressure from one end of the line to the other is certainly a factor. This pressure difference may cause some flash steam to be generated and may help promote flow in the line.
- *Condition of the condensate* – For a proper selection, the designer must know whether the condensate is primarily liquid, or whether a significant amount of flash steam can be expected.
- *Volume* – The anticipated volume of condensate flow is also an important factor. This condition changes dramatically between startup of a system and full, consistent operation. During startup, the temperature difference at the load and throughout the piping system, relative to the temperature of the steam supply, is greater. Far more condensate will be generated during this period in all locations, including steam mains, branch lines, and especially at the system loads.

Condensate must be drained away from load apparatus by drain line(s) and through trap(s) to prevent excessive water hammer and potential damage to piping and the internal components of the trap(s). As a general rule, there is no significant pressure drop in this line to consider, so flash steam resulting from changes in pressure is negligible or non-existent. There is also no pressure differential that would assist flow because condensate movement is by gravity. The condition and volume of the condensate will, however, change dramatically from startup to operation at full load. It is not uncommon for the volume of condensate to be three times greater during startup than at full load. The flow of noncondensables (air) through the system must be accommodated as well.

It cannot be assumed that the pipe outlet size of the load or apparatus is the proper size to promote good drainage. Once the steam trap has been selected based on design criteria, however, it is quite common for the trap inlet size to be the correct pipe size to use between the two points. In most cases, sizing the line for a volume twice that of the expected flow during a running load is sufficient to accommodate startup loads without being oversized for normal operating conditions.

Wherever possible, discharge lines from traps should incorporate gravity as the primary mover. A pressure drop does exist at the trap. Flash steam may result from this pressure difference, and will be more significant at steam pressures above 30 psig. As condensate is discharged from the trap, some small amounts of live steam may also make it past the trap. In the trap discharge line then, liquid and vapor are moving together. This is known as two-phase flow, much the same as the flow of liquid and vapor in sewer systems.

The mixture leaving the trap will exhibit more properties of steam than properties of liquid water, especially at steam pressures above 30 psig. Sizing is therefore evaluated much like steam lines. The same maximum velocities are sometimes used, although lower velocities generally work best to minimize water hammer and the creation of flash steam. Undersized drain lines from the trap, unable to handle the flow as needed, can dramatically slow down or flood the process.

As a result of heat loss through piping walls, condensate will also gather in steam mains, especially during system startup. Steam mains should be pitched, usually ½ inch down per 10 feet of run, to allow the condensate to gravity-drain toward a gathering point. Collection legs, also known as drip legs, are installed on the bottom of the main along the horizontal run at 150 to 300-foot intervals. The collection leg should be the same pipe diameter as the main itself. In cases where the pitch of the steam main causes it to be too low, a riser may be needed to route the line up higher. An additional collection leg should be installed at the base of all risers in stem mains. The inlet of a steam trap is then connected to each collection leg to drain the condensate away.

In most cases, it is necessary to drain condensate from more than one load into a common condensate return line. The line should never become completely filled, or flooded, with condensate. It should slope toward the vented receiver tank or flash tank, when one is included. In most cases, sizing for liquid-only (no steam, flash or otherwise) condensate drainage should be for velocities at or below 7 fps. The common line is generally sized based on the combined volume of flow used to size the individual trap discharge lines routed to the common line.

3.4.0 Valves

Many valves used to regulate and direct the flow of steam within a steam piping system are the same as those used in hot-water piping systems. These valves include gate valves, ball valves, globe valves, angle valves, and check valves. The focus of this section is on pressure-reducing valves and thermostatic valves used in steam systems.

3.4.1 Pressure-Reducing Valves

Pressure-reducing valves are used to reduce a higher steam pressure to a controlled lower pressure. Such a condition often exists when a process steam boiler supplies steam at a higher pressure than can be used in a related secondary steam-heating system. The valve output pressure level must be held constant, regardless of any pressure fluctuations in the input steam supply or variations in the downstream load. There are several kinds of pressure-reducing valves. One simple type of pressure reducing valve (*Figure 32*) controls the output pressure by achieving a balance between an internal spring pressure acting on a diaphragm and the pressure of the delivered steam. The spring pressure is usually adjusted manually to set the desired pressure downstream.

Larger and more complex pressure-reducing valves use external tubes to sense output pressure further downstream. Other valves are positioned using pneumatic or electric actuators for greater accuracy. This allows them to be interfaced with building or process automation systems using a variety of sensor inputs.

Figure 33 shows a typical single-stage pressure regulator. Notice that the pressure regulator in the figure can be bypassed by opening a globe valve. This allows steam to be supplied to the downstream load should the pressure regulator valve malfunction. Also, the gate valve installed on each side of the pressure-relief valve allows the pressure-reducing valve to be isolated for repair without shutting down the system.

Suitable drains and steam traps must be used to prevent accumulation of condensate on the input side of the pressure-reducing valve. The assembly, which consists of the pressure regulator and related valves and traps, is often called a pressure-reducing station.

Pressure reduction can be accomplished with one or more valves (stages). Multistage pressure reduction is commonly used where an intermediate pressure is required or where the total pressure reduction is large. A large pressure reduction, also referred to as the turndown ratio, is one in which the ratio of downstream pressure to upstream pressure is greater than 50 percent. *Figure 34* demonstrates the installation of pressure-reducing valves in series.

When there is a wide variation in the steam load, such as can occur between summer and winter load conditions, pressure-reducing valves are often installed in parallel. *Figure 35* shows a typical pressure-reducing station with valves installed in parallel. Normally, these valves have different operating pressures that provide for sequenced valve operation. A smaller regulator valve sized to handle about one-third of the maximum load is connected in parallel with a larger valve sized to handle the remaining load. The smaller regulator valve is set for a slightly higher pressure than the larger one. During times when the load is low, the smaller regulator handles the load. As the load increases, the reduced pressure causes the larger regulator to operate and provide any additional required flow.

There are several common factors that should be considered for a successful pressure-reducing valve installation:

- Follow the manufacturer's guidelines for proper sizing. Oversized valves tend to exceed the desired pressure output and experience greater wear, especially at light loads. Undersized valves are unable to pass a sufficient amount of steam to satisfy downstream needs.
- Pressure-reducing valves operate properly in one direction only. Make sure the valve is properly oriented in the line. Most have an arrow cast into the valve body to indicate the proper flow direction. Pressure-reducing valves also must be installed horizontally, and should always have a steam trap installed immediately upstream.

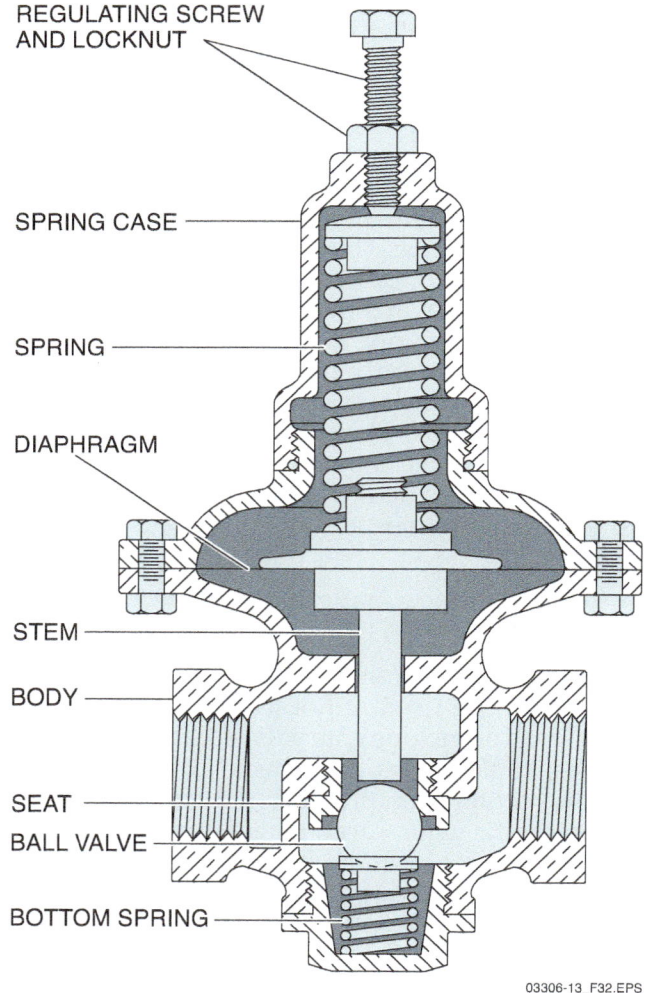

Figure 32 Pressure-reducing valve.

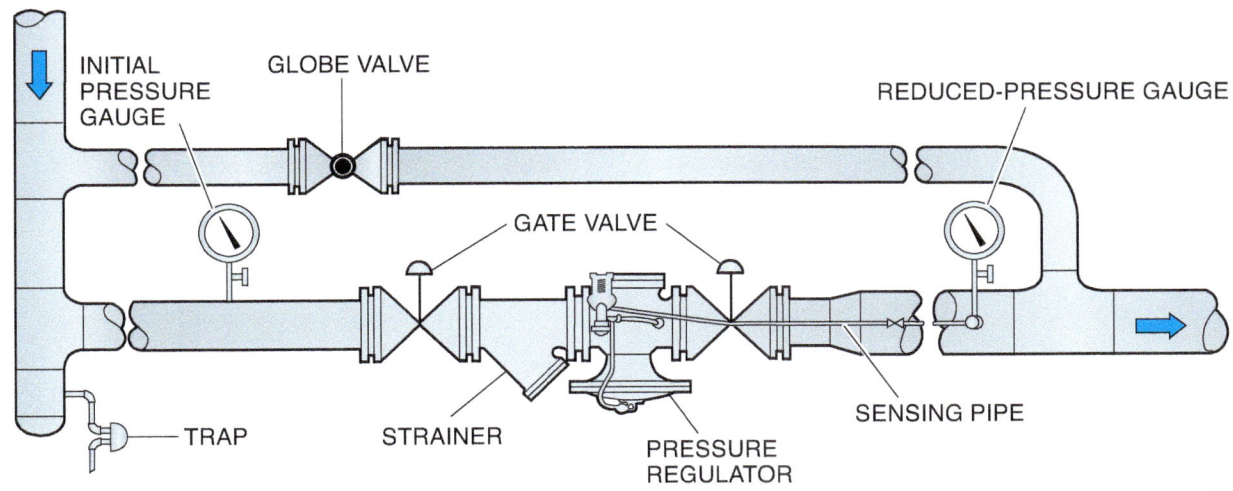

Figure 33 Single-stage pressure-reducing station.

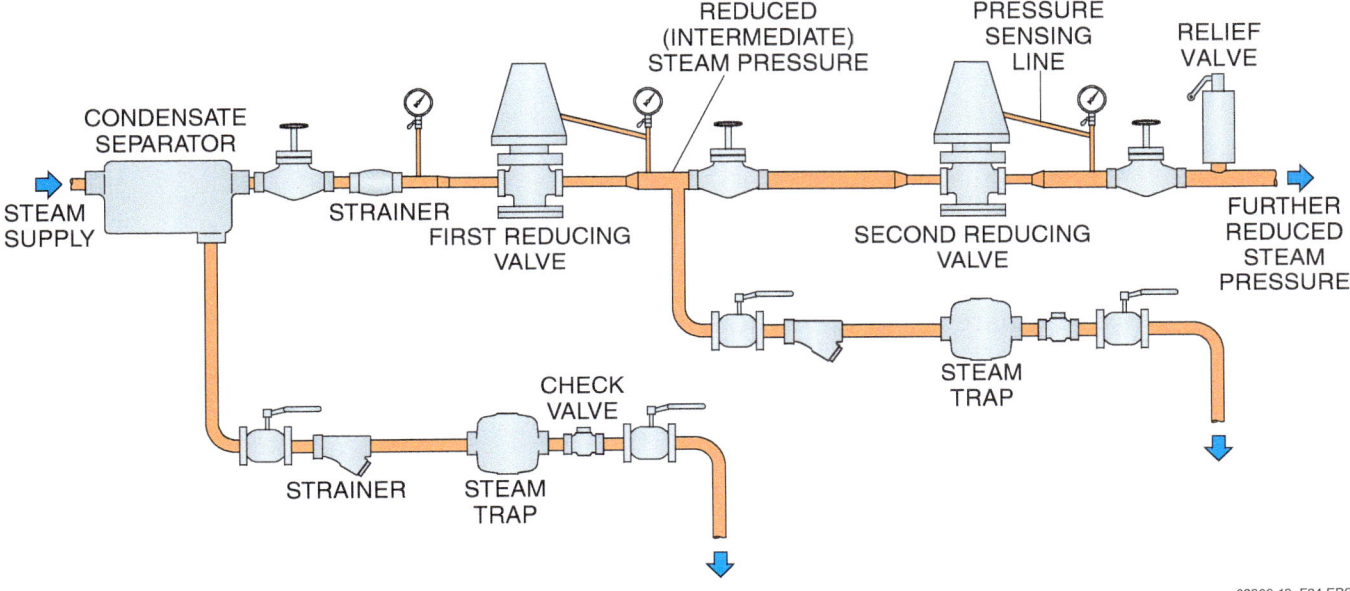

Figure 34 Use of pressure-reducing valves connected in series.

- Pressure gauges should be installed near both the inlet and outlet. Valves that use a sensing tube for remote pressure input should also have a gauge installed at the same location as the sensing port.
- Dirt and other foreign matter are the most likely causes of valve failure. The use of strainers upstream of the valve is recommended. Strainers should be removed and cleaned soon after a new installation has been placed in operation.

3.4.2 Thermostatic Valves

Thermostatic valves control the flow of steam based on temperature inputs. They can be applied many different ways. Some possible applications include comfort heating radiation, steam water heaters, steam-jacketed vessels, various types of component molds, and tank heating coils. In the simplest applications, for example, a thermostatic valve can be used to control the volume of steam flow to a radiator or other type of steam terminal (*Figure 36*) to provide zone temperature control. A dial provides for operator setpoint adjustment. Much like pressure-reducing valves, many thermostatic valves are self-contained, with a liquid-filled temperature-sensing element attached directly to the valve such as the one in *Figure 37*. A number of thermostatic elements are available, with selection dependent upon the desired temperature range. As temperature at the sensing element increases, the liquid expands and applies

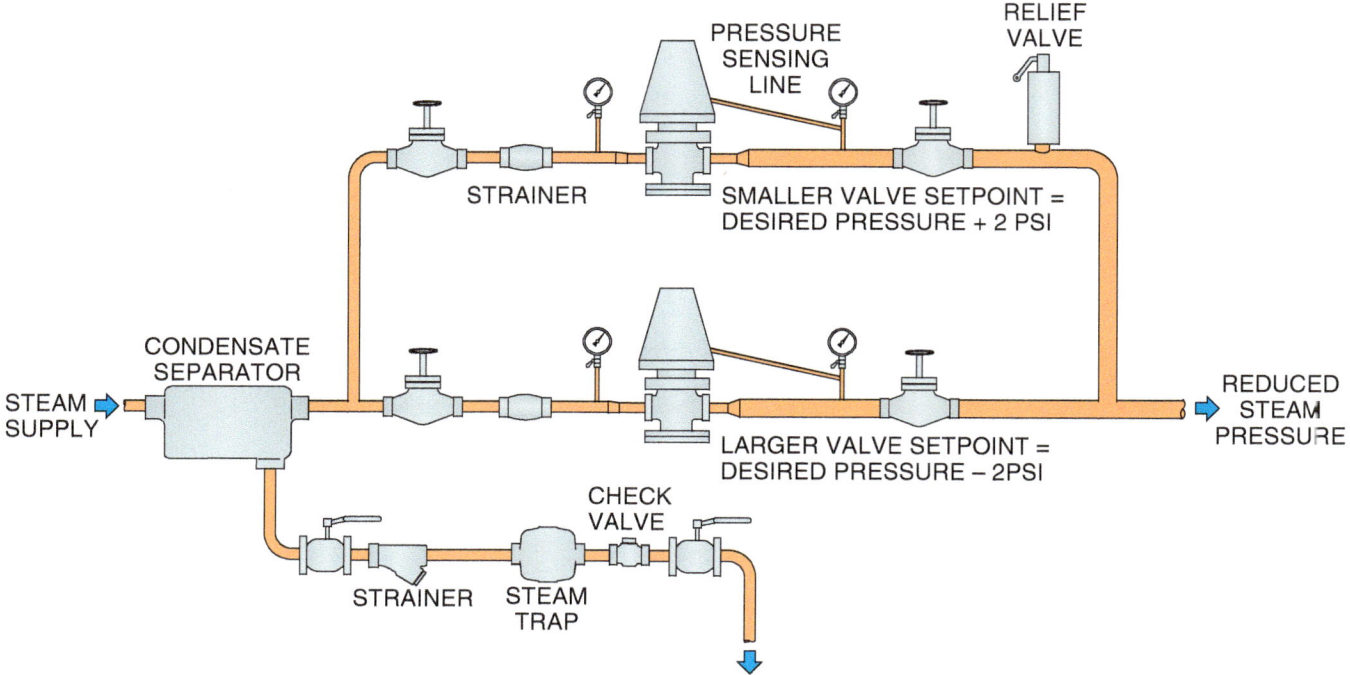

Figure 35 Use of pressure-reducing valves connected in parallel.

an opening force against the closing force of the valve spring. Setpoint adjustments change the spring tension, thereby changing the force required to open or close the valve.

Thermostatic valves share most of the same common installation considerations with pressure-reducing valves. Although a pressure gauge upstream of the valve is generally recommended, downstream pressure may not need to be monitored. Instead, thermometers should be placed in the controlled medium to provide a quick visual indication of the temperature.

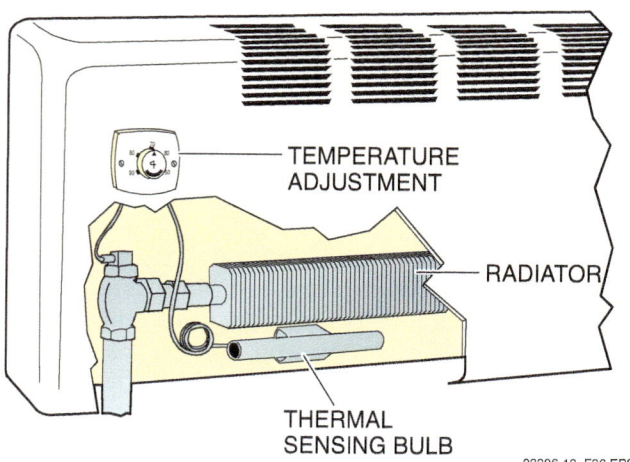

Figure 36 Typical individual thermostatic control.

Figure 37 Temperature-sensing element attached to a valve.

Additional Resources

Low Pressure Boilers, 3rd Edition. Fredrick M. Steingress, Daryl R. Walker. Orland Park, Illinois: American Technical Publishers.

High Pressure Boilers, 4th Edition. Fredrick M. Steingress, Harold J. Frost, Daryl R. Walker. Orland Park, Illinois: American Technical Publishers.

Piping for High Pressure Boilers: The Installation and Inspection of High-Pressure Boiler Piping for Code Compliance with the ASME and National Board Code Requirements. Steve Kalmbach. Tucson, Arizona: Wheatmark®.

3.0.0 Section Review

1. A condensate system vacuum pump is used to pump condensate.
   a. True
   b. False

2. The load demand on the low-pressure steam line where a flash tank discharges its collected steam should always be _____.
   a. greater than the flash tank output
   b. the same as the flash tank output
   c. half as much as the flash tank outputs
   d. one-third as much as the flash tank output

3. To properly size the piping of a steam system, primary consideration must be given to _____.
   a. flash tank capacity
   b. condensate line size and lengths
   c. steam volume and pressure
   d. the number of steam traps to be used

4. The turndown ratio that normally requires the use of multiple pressure-reducing valves is a ratio of more than _____.
   a. 40 percent
   b. 50 percent
   c. 60 percent
   d. 70 percent

Section Four

4.0.0 Steam Traps and Strainers

Objectives

Identify, install, and maintain various types of steam traps.
   a. Identify various types of steam traps.
   b. Explain the basic concepts of installing steam traps.
   c. Explain how to maintain various steam traps.
   d. Explain how to troubleshoot various steam traps.

Performance Task

2. Maintain selected steam traps.

The steam trap is one of the most important components used in steam systems. It is likely that they will demand a great deal of the total time invested in steam maintenance and repairs. Steam traps provide the point of separation between live steam and water in the system and must provide this service effectively for the system to function properly.

As steam is allowed to cool, it condenses to its liquid form. Most of this condensing process takes place at the load, such as a heat exchanger where steam gives up its stored energy to heat water (*Figure 38*). However, it can also take place in the distribution and piping system and collect in low spots. The condensate can then be picked up by steam as it passes, propelling it at a high rate of speed into bends and fittings. Known as water hammer, this phenomenon can cause serious damage and, at a minimum, create noise. For this reason, steam traps are not only used on the leaving side of loads such as heat exchangers and coils, but also at strategic low points throughout the piping system.

For steam to remain efficient, it must be dry and at the proper temperature and pressure. Essentially, a steam trap is a self-actuating drain valve that maintains effective heat transfer in the steam system by consistently purging it of collected condensate and air.

Air is always present during system startup and in the boiler feed water. In addition, the feed water may contain carbonates that dissolve and release carbon dioxide gas (CO_2). Air and carbon dioxide can cause problems in a steam line. Air is an excellent insulator and can reduce the efficiency of the steam at the terminal heat transfer surfaces. Carbon dioxide can dissolve in the condensate and form carbonic acid, which is extremely corrosive. Scale and debris are also transferred with the condensate to areas that are difficult to repair when clogged. Use of properly installed steam traps and strainers throughout the system can prevent these problems.

4.1.0 Steam Trap Types

There are three general categories of steam traps: mechanical, thermostatic, and thermodynamic. Among these many types, there are also various materials used, such as cast iron, stainless steel, forged steel, and cast steel. Steam traps are fitted to the pipe with screwed, flanged, socket-welded, or compression fittings. All types of steam traps automatically open an orifice, drain the condensate, then close before steam is lost.

4.1.1 Mechanical Steam Traps

Mechanical steam traps respond to the difference in density between steam and steam condensate. The trap opens to condensate and closes to steam. The typical mechanical trap is called an inverted bucket trap and has only two moving parts, the valve lever assembly and the bucket (*Figure 39*). The trap is normally installed in the drain line between the steam-heated unit and the condensate return header.

The cycle starts with the bucket down and the valve all the way open. The condensate fills the body and the bucket, causing the bucket to sink and open the discharge valve. Gases and air flow through first, then steam raises the bucket and closes the valve. The gas slowly bleeds out through a small vent in the top of the bucket. The valve remains open until enough steam has collected to float the bucket again. The inverted bucket trap operates intermittently, responding to the accumulation of condensate by opening, then reclosing to allow steam to pass.

Because the inverted bucket trap operates on a pressure differential, it must be primed before being put into service.

The inverted trap can handle high-pressure steam. It resists damage from pressure surges and water hammer, and tolerates freezing if the body is made of a ductile material. Its disadvantages are its limited air discharge capacity and a tendency to be noisy.

Figure 38 Installed steam traps.

> **CAUTION**
> Water hammer can occur when hot steam comes into contact with cooled condensate, builds pressure, and pushes the water through the line at high speeds, slamming into valves and other devices, causing audible noise. In addition, water hammering creates vibrations throughout the system components that can eventually weaken joints and loosen connections, causing them to rupture. Water hammer is most likely to occur in single-pipe systems.

4.1.2 Thermostatic Steam Traps

Thermostatic steam traps respond to temperature changes in the steam line. They open when the cooler condensate is present and close to the higher steam temperatures. The following descriptions, individually illustrated in *Figure 40*, represent four different types of thermostatic steam traps:

- *Liquid expansion thermostatic trap* – Remains closed until the condensate cools below 212°F, opening the thermostat downstream of the

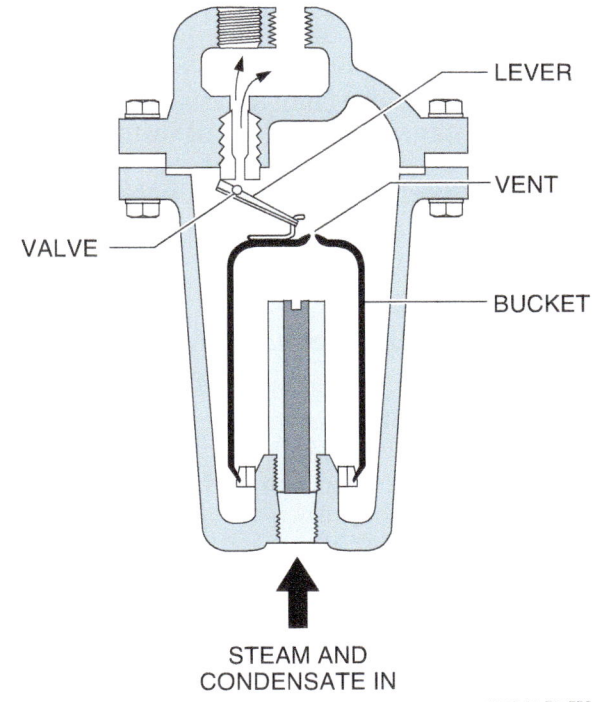

Figure 39 Inverted bucket trap.

03306 Steam Systems   Module Ten 37

valve. This discharges the cool condensate at a constant temperature.

- *Balanced pressure thermostatic trap* – Opens when condensate cools slightly below the steam saturation temperature at any pressure within the trap's range. Controlled by a liquid-filled bellows, the discharge valve closes when hot condensate vaporizes the liquid in the bellows. It opens when the condensate cools enough to lower the pressure inside the element. Balanced pressure traps are smaller than mechanical traps, open wide when cold, readily purge gases, and are unlikely to freeze.
- *Bimetal thermostatic trap* – Has bimetal strips that respond to cool condensate by bending to open the condensate valve. When steam hits the strips, they expand and close the valve. They require considerable cooling to reopen, and the pressure-compensating characteristics vary among the models. The closing force of some designs varies with the steam pressure, approximating the response of a balanced pressure trap. This trap has a large capacity, but responds slowly to process load changes. However, it vents gases well. It is not easily damaged by freezing, water hammer, or corrosion and it can handle the high temperatures of superheated steam.
- *Float and thermostatic (F and T) trap* – Has a float that rises when condensate enters and opens the main valve that lies below the water level. At the same time, another discharge valve at the top of the trap, operated by a thermostatic bellows, opens to release cooler gases. This trap design vents gases well and responds to wide and sudden pressure changes, discharging condensate continuously. In the failure mode, the main valve is closed but the air vent is usually open. Float thermostatic traps should always be protected from freezing because they usually contain water.

4.1.3 Thermodynamic Steam Traps

Thermodynamic steam traps use the heat energy in hot condensate and steam to control the opening and closing of the trap. They operate well at higher temperatures and can be installed in any position because their function does not depend on gravity.

The thermodynamic disc steam trap is the most widely used. In this unit, the only moving part is the disc (*Figure 41*). Flash steam pressure from hot condensate keeps the disc closed. Flash steam is steam that is formed when hot conden-

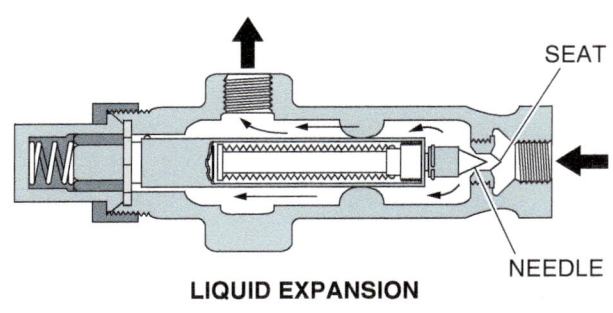
LIQUID EXPANSION

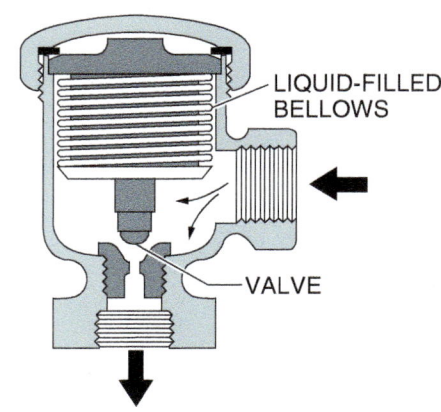
BALANCED PRESSURE

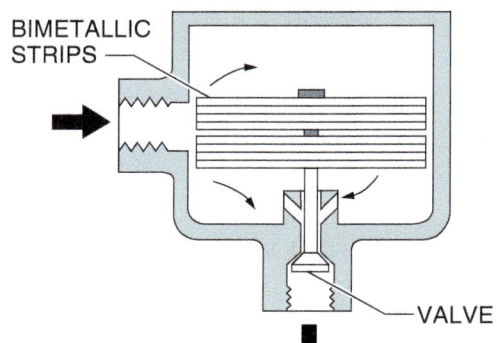
BIMETALLIC

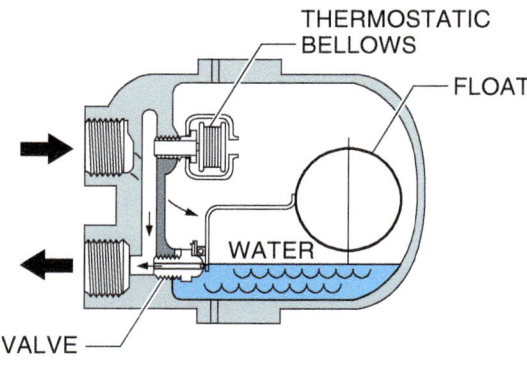
FLOAT AND THERMOSTATIC

Figure 40 Thermostatic steam traps.

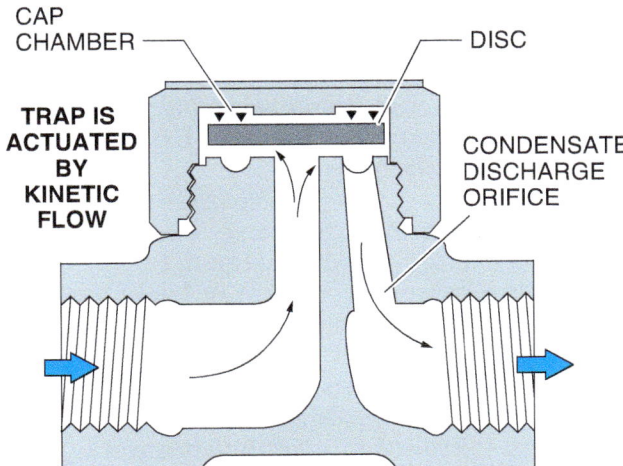

Figure 41 Thermodynamic stainless-steel steam trap.

sate is released to a lower pressure and flashes back to steam as a result. When the flash steam above the disc is condensed by cooler condensate, the disc is pushed up and remains open until hot condensate flashes again to build up pressure in the cap chamber. At the same time, flashing condensate on the underside of the disc is discharged at a high velocity, lowering the pressure in this area and closing the disc before steam can escape.

The thermodynamic steam trap can withstand freezing, high pressure, superheating, and water hammer without damage. The audible clicking of the disc indicates when the discharge cycle occurs. These traps last longer if they are not oversized.

4.1.4 Fixed-Orifice Steam Traps

As the name implies, the fixed-orifice trap depends on a single opening of a predetermined size to pass a calculated amount of condensate under specific pressure and load conditions. In actual practice, it is rare that steam and condensate flow rates remain precise, rendering the fixed-orifice trap impractical under most conditions. They have a tendency to waterlog (hold back too much condensate) on startup when condensate loads are high, yet pass live steam once the piping system is hot if the load begins to drop. In cases where it can be successfully applied, the fixed-orifice steam trap has the distinct advantage of having no moving parts.

4.1.5 Strainers

In all systems, a strainer should be installed ahead of the steam trap. The scale and corrosion in any steam system has to be stopped before it enters the trap, or it will cause clogging and damage. Strainers need to be cleaned and inspected on a regular basis. Strainers are available in many styles and are sometimes incorporated into the steam trap. *Figure 42* shows three different types of strainers.

4.2.0 Installing Steam Traps

To mount a steam trap anywhere other than the correct position would render it useless. Two general rules for steam-trap mountings are that they should be lower than any line in the system, and that a strainer has to be upstream of the trap. Isolation drains, couplings, check valves, and drip leg placement are also important to each application. Their use depends on the needs and contents of the system. A drip leg is a drain for condensate in a steam line placed at a low point in the line and used with a steam trap.

Each steam trap manufacturer has resource books that detail how to install steam-trap systems. The following is a list of general guidelines for proper steam-trap installation:

- Provide a separate trap for each piece of equipment or apparatus. Short-circuiting may occur if more than one piece of equipment is connected to a single trap.
- When piping from the steam header or main to the load, always tap the steam supply off the top to prevent any collected condensate from draining into the load apparatus.
- Install a supply valve ahead of the steam trap and close to the steam header to allow maintenance or modifications to be performed without shutting down the steam header.
- Install a steam supply valve close to the equipment entrance to allow equipment maintenance work to be performed without shutting down the supply line.

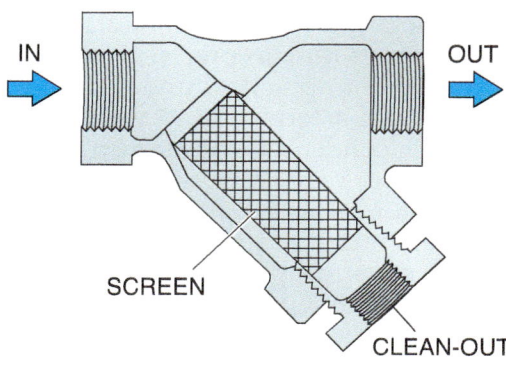

- Install a strainer and strainer flush valve ahead of each trap to keep rust, dirt, and scale out of working parts and to allow blowdown removal of foreign material from the strainer basket.
- Provide unions on both sides of the trap for its removal or replacement.
- Install a test valve downstream of the trap to allow observation of discharge when testing.
- Install a check valve downstream or upstream of the trap to prevent condensate backflow during shutdown or in the event of unusual conditions.
- Install a downstream shutoff valve to cut off equipment condensate piping from the main condensate system for maintenance work.
- Do not install a bypass unless there is some urgent need for it. Bypasses are an additional expense to install and are frequently left open, resulting in the loss of steam and inefficient equipment operation. When a bypass is necessary, a globe valve should be used to allow manual balancing of condensate flow, as the steam process continues during maintenance of the primary trap.

Figure 43 shows a basic trap installation. A steam trap should have a provision for inspection without interrupting the process flow.

4.3.0 Maintaining Steam Traps

Most steam traps fail in the open position, which is difficult to pinpoint because it does not affect equipment operation. When the trap fails open, the steam flows freely through the system, keeping the heat transfer high but failing to hold back the condensate. Traps that fail in the closed position are easy to identify because the backup of condensate cools the system. Clogged strainers, which protect the trap from debris, also have the same effect.

An effective planned maintenance (PM) program includes scheduled checks of the entire system. Clogged strainers, leaking joints, leaking valve packing, or missing insulation are some of the items to check. When the internal parts of a steam trap wear out, the water seal deteriorates and steam flows through the valve assembly, further worsening the condition. Common causes of steam-trap failure include the following:

- Scale, rust, or corrosion buildup preventing the valve from closing
- Valve assembly wear
- Defective or damaged valve seat
- Physical damage from severe water hammer
- Foreign material lodged between seat and valve

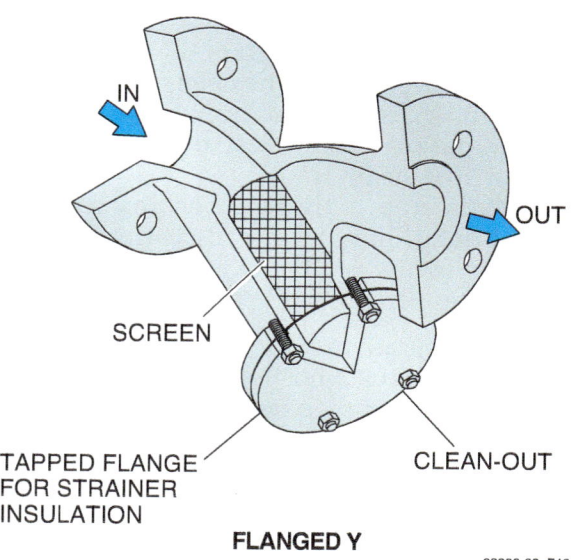

Figure 42 Strainers.

- Connect the condensate discharge line to the lowest point in the equipment to avoid water pockets and water hammer.
- Install a shutoff valve upstream of the condensate discharge piping to cut off discharge of condensate from equipment and allow service work to be performed.

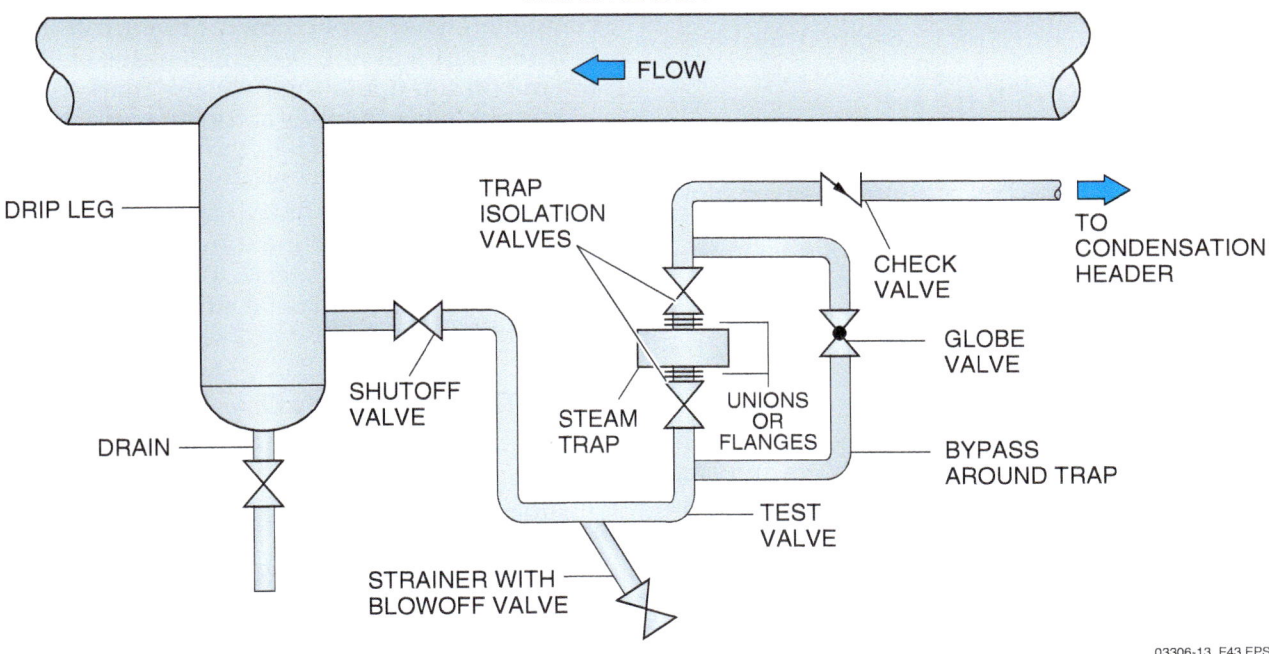

Figure 43 Basic trap installation.

- Blocked, clogged, or damaged strainers
- Increased back pressure

Other failures are specific to the type of steam trap. The two primary indications of failure in a steam trap are allowing live steam to pass freely (will not close), or failure to allow condensate to pass through (will not open). The following sections explain the causes and solutions for failures on each type of steam trap.

4.3.1 Inverted Bucket Trap

If an inverted bucket trap blows steam, check for loss of the water seal. Isolate the trap, wait for the condensate to accumulate, and start the steam flow again. If this solves the problem, try to discover the cause of the water seal loss. It could be caused by superheat, sudden pressure fluctuations, or the trap being installed so that the water seal runs out by gravity.

Another solution is to try installing a check valve ahead of the trap. If the steam blow persists, check for dirt or wear on the valve and linkage. Replace the valve, the seat, and the lever. If this fails, check the bucket. If the bucket or lever is distorted, the trap failure was probably caused by water hammer. Trace the source of the problem and try to eliminate it.

If the trap will not pass condensate, check that the maximum operating pressure marked on the trap is not lower than the actual pressure to which it is subjected. If it is, the valve cannot open. Install a valve and seat assembly with the correct pressure rating. Make sure that it has sufficient capacity to handle the maximum load. Check the internal parts and be sure that the air vent hole in the bucket is not obstructed. This could cause pockets of air to be trapped in the vent. This condition is called air binding.

4.3.2 Liquid Expansion Thermostatic Trap

If a liquid expansion thermostatic trap blows steam, check for dirt or wear on the valve and seat. If wear has occurred, replace all of the internal parts. Remember that this type of trap is not self-adjusting to changes in pressure. If it has been set to close at a high pressure, it may not close off at a lower pressure. Try adjusting the trap to a lower setting, making sure that it does not waterlog excessively. If it does not appear to react to this change, replace all of the internal parts. If a trap will not pass condensate, check that it has not been adjusted to a setting that is too cold.

> **Steam Trap Maintenance**
>
> Nonfunctioning steam traps are a significant cause of lost energy. Steam traps must be inspected and cleaned in accordance with the manufacturer's instructions at regular operating intervals.

4.3.3 Balanced Pressure Thermostatic Trap

If a balanced pressure thermostatic trap blows steam, isolate the trap and allow it to cool before inspecting it for dirt. If the seat is wire-drawn, replace all of the internal parts, including the thermostatic element. The original parts have probably been strained by continuous steam blow.

If the valve and seat seem to be in good order, check the element. You should not be able to compress it when it is cool. Any flabbiness of the element indicates failure. Flattening of the convolutions indicates water hammer damage. If the water hammer cannot be eliminated at its source, a stronger type of trap must be used.

If the trap will not pass condensate, the element may be overextended due to excessive internal pressure, making it impossible for the valve to lift off its seat. An overextended element could be caused by superheat or by someone opening the trap while it is still very hot and before the vapor inside has had time to condense.

4.3.4 Bimetal Thermostatic Trap

If a bimetal thermostatic trap blows steam, check for dirt and wear on the valve. A bimetal trap has limited power to close because of its method of operation, and the valve may be held off its seat by the accumulation of soft deposits.

This type of trap is usually supplied preset. Check that any locking device on the manual adjustment is still secure. If it is not secure, see if the trap responds to adjustment. If cleaning has no effect, replace all of the internal parts.

The valve on a bimetal trap is on the downstream side of the valve orifice, so these traps tend to fail in the open position. Failure to pass cold condensate indicates either gross maladjustment or complete blockage of the valve orifice or built-in strainer.

4.3.5 Float and Thermostatic Trap

If an F and T trap blows steam, check the trap for dirt at the main valve and the air vent valve. If the trap has a steam lock release, check to make sure that it is not opened too far. It should be open no more than one quarter of a full turn.

Make sure that the valve mechanism has not been knocked out of line either by rough handling or water hammer, preventing the valve from seating. Check that the ball float is free to fall without fouling the casing, which would cause the mechanism to hang up.

Test the air vent the same way as the element of a balanced pressure trap is tested. If the internal parts of a float trap must be replaced, install a complete set as supplied by the manufacturer.

If the trap will not pass condensate, check that the maximum operating pressure marked on the trap is not lower than the actual pressure to which the trap is subjected. If it is, the valve cannot open. Install a valve and seat assembly with the correct pressure rating. Make sure that this has sufficient capacity to handle the maximum load. If the ball float is leaking or damaged, it is probably caused by water hammer. Also, be sure that the air vent or steam lock release is working correctly.

4.3.6 Thermodynamic Disc Trap

If a thermodynamic disc trap blows steam, the trap will probably give a continuous series of abrupt discharges. Check the trap and strainer for dirt and wipe the disc and seat clean. If this does not correct the problem, it is possible that the seat faces and disc have become worn. The extent of wear is evident by the amount of shiny surface that replaces the normal crosshatching of machining.

If records show that thermodynamic traps on one particular installation suffer repeatedly from rapid wear, the trap may be oversized, associated pipework may be undersized, or there may be excessive back-pressure.

If the trap will not pass condensate, the discharge orifices may be plugged with dirt. This is most likely caused by air binding, particularly if it occurs regularly during startup. Check the air venting arrangements of the equipment. In extreme cases, it may be necessary to install an air vent parallel with the trap, or to use a float trap with a built-in air vent instead of a thermodynamic trap.

4.4.0 Troubleshooting Steam Traps

To diagnose and solve problems with steam traps requires listening to the noises they make and measuring temperature and pressure. The temperature of steam can be measured with handheld or remote-sensing pyrometers that read in degrees centigrade or Fahrenheit. Steam is produced at 212°F when at atmospheric pressure. When the steam systems are functioning, the temperature will be substantially higher, as it is confined inside the system. The pressure will be higher as well. Pressure readings are in two scales, absolute (psi) and gauge (psig). Absolute pressure is read from zero up, which is to say that normal atmospheric pressure is 14.7 psi at sea level. Gauge pressure is relative to atmospheric pressure at sea level; therefore, zero psig is equal to 14.7 psi absolute. Pressures below zero gauge are expressed in inches of mercury. Steam trap manufacturers have excellent books and web documents that detail how to troubleshoot and repair steam traps.

4.4.1 Diagnostic Methods

Three basic diagnostic methods of reading a steam system are sight, sound, and temperature. Another process, less used to date, is the use of a conductance probe, which compares the difference in electrical conductivity between condensate and live steam.

Each of the three main methods has advantages and limitations. Sight is fairly immediate; it is easy to tell whether there is condensate coming from a trap. If the condensate line is open, or if a test valve has been added on the condensate line, the condensate can be vented to the atmosphere. It will be obvious that the trap is removing condensate from the line. If live steam is going into the condensate line, the trap has failed open, and it is necessary to resolve the problem. If no condensate is coming from the line, the trap may have failed closed. The limitation of sight diagnosis is that the worker must be able to tell the difference between flash steam and live steam, and it is necessary for the worker to check by activating a test valve. It is also possible for the trap to be overloaded or loaded from the return system, producing an apparent open failure.

> **WARNING!**
> In a high-pressure steam system, the condensate will flash back to steam when the pressure drops as the result of exposure to the atmosphere through a leak. This exposure causes a rapid change in volume, which can result in the release of high-pressure steam.

The two most commonly used diagnostic tools are temperature sensing and sound, either at human hearing ranges or ultrasound. Note that temperature and sound together offer more information than one or the other alone.

From a diagnostic point of view, there are two general types of traps. Traps either flow or dribble continuously or intermittently (open and close). F and T traps discharge fairly continuously if they are working properly. Inverted bucket, bellows thermostatic, thermodynamic, and most thermostatic traps flow intermittently.

A few common characteristic symptoms of traps when tested might include the following:

- A thermal analysis will show the inlet as the area at the highest temperature for any of the traps. In most functioning traps, the condensate outlet, whether it is a closed return or an open return, should be cooler than the inlet. Remember that the condensate will most frequently be at or near steam temperature. However, it should still be at a lower temperature than the steam inlet. The live steam inlet should be hotter than the condensate exiting the trap. If this is not the case, the trap is probably malfunctioning.

- Normal failure for an inverted bucket trap is open failure. The trap loses its prime, and steam and condensate blow through steadily. If this happens, the continuous rushing sound of the blow-through may be accompanied by the bucket banging against the inside of the trap. The temperature of the trap and of the condensate line will rise because it is no longer holding condensate. The sound of the bucket linkage rattling inside the trap may indicate that the trap is beginning to loosen up and needs checking. With ultrasound equipment, the inverted bucket or thermodynamic traps would show a cyclic curve on the screen in normal function.

- Normal failure for a thermodynamic trap is open; that is, passing steam. Usually, the disk clicks shut audibly once for each cycle. When the disk is cycling normally, it will shut about 4 to 10 times a minute. When the trap fails, the disk no longer clicks as it rises and falls, and steam blows through. If the disc produces a continuous, rapid, rattling sound, the disc is worn, and more problems will probably develop. If a continuous hissing sound is produced, the trap is not cycling. Again, the condensate line will heat up due to escaping steam and the absence of cooler condensate.

- Bellows or bimetallic thermostatic traps operate on a difference in temperature between condensate and steam. Both types are intermittent in operation. When there is little condensate, they remain closed most of the time. When there is a lot of condensate, for example at startup, they may run continuously for a long time. In the case of bimetallic traps, misalignment may allow steam leaks. With sound testing equipment, a constant rushing sound in the closed part of the cycle would indicate a leak.

- Float and thermostatic traps tend to fail closed. A leaky float will not float on the condensate, or the float may have been crushed by water hammer. In either case, the F and T trap will remain closed and will not cycle. The trap will remain cool, and a temperature sensor will reveal that. With listening equipment, a failed float will be silent in the absence of discharging condensate.

- The thermostatic element in an F and T trap is normally quiet; a rushing sound would indicate that the element had failed open. A rattling or metallic clanging noise might mean that the mechanical linkage had sustained some damage.

The sight method uses a test valve that vents the process steam to the atmosphere for visual inspection. This test is subjective and depends on skill and experience. Most traps are dealing with condensate that is at steam temperatures; that is, somewhere above the boiling point of water. When condensate is released from the pressure of the steam system in the trap or in a condensate release, a certain amount of the condensate will immediately turn into flash steam. The difference is fairly visible once you have seen it a few times; the live steam comes out at first as a hard, blue-white straight flow, often with a clear bluish area at the outlet. Flash steam usually billows and spreads more quickly, as much of the pressure is dissipated in the initial release. The sound probes are either audible-range or ultrasonic-range probes that pick up noise from the steam system and send an audible signal to headphones. The technician performing this test must also have skill and experience at interpreting the noise as leaks, discharges, or other steam trap problems. *Figure 44* shows a portable ultrasonic tester.

Temperature-sensing tests are done with infrared thermometers (*Figure 45*), also known as pyrometers, that are handheld for external readings, or use mounted thermowells (*Figure 46*) that house devices to read the steam temperature. A thermowell is a permanently installed well or cavity in a process pipe or tank into which a glass thermometer or thermocouple can be inserted. The thermometer or thermocouple can be temporarily or permanently installed.

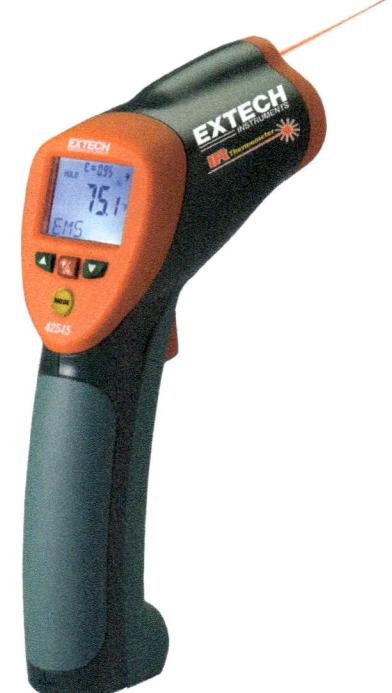

Figure 45 Pyrometer.

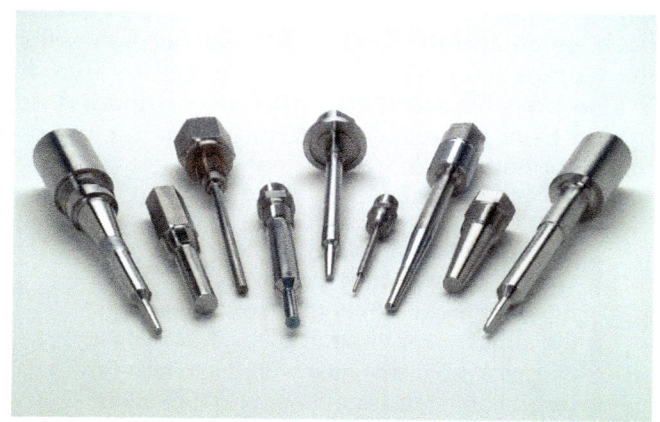

Figure 46 Thermowells.

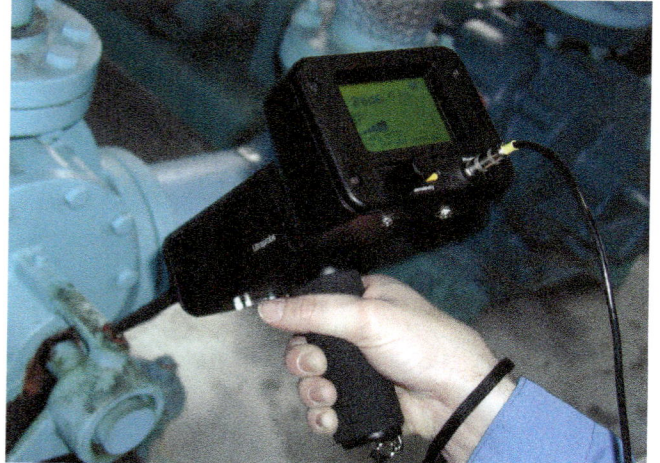

Figure 44 Ultrasonic tester.

Additional Resources

Mechanical Products Manual, 14th Edition, available at **www.spiraxsarco.com**.

4.0.0 Section Review

1. Installation of an inverted bucket trap requires it to be _____.
   a. primed before use
   b. mounted with the outlet up
   c. mounted with the inlet horizontal
   d. mounted with the inlet vertical

2. Steam traps must be installed with a strainer downstream from the trap.
   a. True
   b. False

3. A problem that can be isolated by installing a check valve upstream of an inverted bucket trap is a trap that is _____.
   a. blowing steam
   b. not passing condensate
   c. causing water hammer
   d. not venting

4. In a malfunctioning trap passing steam, the outlet temperature is usually _____.
   a. the same as the inlet temperature
   b. lower than the inlet temperature
   c. 10°F higher than the inlet temperature
   d. at least 25°F higher than the inlet temperature

Summary

Steam systems generate and distribute steam used for comfort heating and commercial and industrial processes. Steam heating systems can be separate systems used only to heat buildings, or they can provide energy in a combined system for both comfort and process use.

Boilers used in steam systems are divided into the two broad categories of low-pressure and high-pressure boilers. Low-pressure steam boilers are normally used in residential and commercial applications for both heating and process use. High-pressure boilers are typically used in larger systems for college campuses, hospitals, and industrial applications.

Steam piping systems are either one pipe or two pipe in design. The piping approach is determined by the requirements to supply steam to the terminals and by the destination of the condensate. Condensate can be returned by gravity when appropriate, but many systems will require mechanical assistance when the condensate is to be returned to the boiler for re-use.

Steam traps provide a means to separate live steam from condensate, trapping the steam inside the terminal until it has condensed, thereby giving up the vast majority of the energy it contains before being allowed to pass through the trap. Steam traps automatically open an orifice, drain away accumulated condensate, and then re-close once the condensate is removed. They can be classified based on their operating characteristics as mechanical, thermostatic, or thermodynamic. Steam traps that fail can be quite costly in terms of energy loss, so the proper maintenance of steam traps is crucial to control energy costs and ensure good system performance.

Review Questions

1. The amount of latent heat required to vaporize or evaporate, 5 pounds of water into steam is _____.
   a. 970 Btus
   b. 1,114 Btus
   c. 1,294 Btus
   d. 4,850 Btus

2. The latent heat of condensation is defined as _____.
   a. the heat removed during the change of state from steam to water
   b. the heat gained during the change of state from water to steam
   c. the heat lost in changing water to ice
   d. any heat added after ice changes to water

3. The majority of maintenance attention in a steam system is generally devoted to the _____.
   a. piping system
   b. pressure regulating valves
   c. steam traps
   d. heat exchangers

4. Steam boilers _____.
   a. have exactly the same type of controls as hot water boilers
   b. are entirely filled with water
   c. must operate at extreme pressures only
   d. are partially filled with water

5. A boiler with a maximum operating pressure of 75 psi would be classified as a _____.
   a. high-pressure boiler
   b. low-pressure boiler
   c. process boiler
   d. medium-pressure boiler

6. An assembly generally installed between the boiler and an external pressure gauge to protect the gauge from contacting raw steam is a _____.
   a. strainer
   b. siphon
   c. heat exchanger
   d. trap

7. Steam-to-water heat exchangers designed for potable hot water use are of _____.
   a. stainless steel construction
   b. aluminum construction
   c. hermetically sealed construction
   d. double-walled construction

8. When the condensate return line is above the boiler water line, it is called a _____.
   a. steam line
   b. Hartford loop line
   c. dry return line
   d. wet return line

9. Most two-pipe steam systems use a _____.
   a. mechanical condensate return
   b. gravity condensate return
   c. fixed-orifice trap
   d. very long startup cycle

10. In a vacuum return system, excessive vacuum pump operation can be caused by _____.
    a. too much condensate
    b. low steam pressure
    c. air leaks
    d. heat exchangers

11. To maintain and ensure a sufficient water supply is available for the boiler, the condensate receiver tank is equipped with a _____.
    a. flash reservoir
    b. makeup water valve
    c. thermodynamic trap
    d. vacuum pump

12. When more than one condensate line is connected to a flash tank, each line should be equipped with its own _____.
    a. check valve
    b. pressure reducing valve
    c. condensate pump
    d. sight glass

13. Collected flash steam can be discharged into a _____.
    a. floor drain
    b. vacuum pump
    c. boiler
    d. low-pressure main

 03306 Steam Systems

14. Operating steam boilers at pressures significantly lower than their design operating pressure increases the likelihood of _____.
    a. excessive leaks
    b. thermal shock and hammering
    c. water droplet carryover
    d. trap damage

15. Steam system piping that is too large results in _____.
    a. excessive pressure drop and increased system noise
    b. increased trap failures
    c. high installation costs and greater heat loss
    d. condensate pump overload

16. Steam supply piping for a system designed to operate at 5 psi should generally be designed for a pressure drop no greater than _____.
    a. 0.3 psi
    b. 0.5 psi
    c. 1.0 psi
    d. 2.0 psi

17. Mechanical steam traps take advantage of the difference in density between steam and _____.
    a. air
    b. steam condensate
    c. gases
    d. debris

18. Fixed orifice steam traps are advantageous because they _____.
    a. are practical under most conditions
    b. have very precise flow rates at all times
    c. have no moving parts
    d. always pass condensate at startup

19. Bimetallic steam traps function based on the difference between condensate and steam _____.
    a. air volume
    b. temperature
    c. weight
    d. pressure

20. The three basic diagnostic methods for steam systems, especially traps, are _____.
    a. pressure, temperature, and time
    b. temperature, flow, and sound
    c. current, pressure, and sound
    d. sight, sound, and temperature

Trade Terms Introduced in This Module

Drip leg: A drain for condensate in a steam line placed at a low point or change of direction in the line and used with a steam trap.

Dry steam: The steam that exists after all the water has been vaporized into steam at its saturation temperature (being superheated).

Flash steam: Formed when hot condensate is released to a lower pressure and re-evaporated.

Saturated steam: The pure or dry steam produced at the temperature that corresponds to the boiling temperature of water at the existing pressure (saturation).

Trim: External controls and accessories attached to the boiler itself, such as sight glasses and water feeder controls.

Turndown ratio: In steam systems, the ratio of downstream pressure to upstream pressure, usually applied to steam pressure-reducing valves. More than a 50 percent turndown ratio is generally considered a large pressure reduction.

Water hammer: A condition that occurs when hot steam comes into contact with cooled condensate, builds pressure, and pushes the water through the line at high speeds, slamming into valves and other devices. Water hammer also occurs in domestic water systems. If the cause is not corrected, water hammer can damage the system.

Additional Resources

This module presents thorough resources for task training. The following resource material is suggested for further study.

The Control of Boilers, 2nd Edition. Sam G. Dukelow. Research Triangle Park, North Carolina: International Society of Automation (ISA).

High Pressure Boilers, 4th Edition. Fredrick M. Steingress, Harold J. Frost, Daryl R. Walker. Orland Park, Illinois: American Technical Publishers.

Low Pressure Boilers, 3rd Edition. Fredrick M. Steingress, Daryl R. Walker. Orland Park, Illinois: American Technical Publishers.

Mechanical Products Manual, 14th Edition, available at **www.spiraxsarco.com**.

Piping for High Pressure Boilers: The Installation and Inspection of High-Pressure Boiler Piping for Code Compliance with the ASME and National Board Code Requirements. Steve Kalmbach. Tucson, Arizona: Wheatmark®.

Figure Credits

Courtesy of Fulton Boiler Works, Inc., Figure 6, SA02

Courtesy of ECR International, Inc., Figure 7

Kele, Inc., Figure 8

Xylem Applied Water Systems, Figure 13

Courtesy of PexSupply.com, Figure 14

Courtesy of Honeywell International, Figure 15

Anderson Greenwood, Figure 16

API Heat Transfer Inc., Figure 17

Courtesy of Reznor/Thomas & Betts, Figure 19

Daikin McQuay, Figures 20, 21

Courtesy of Xylem Inc., Figure 22, 24, 25, 32

Courtesy of NES Company Inc., Figure 28

Courtesy of Danfoss, Figure 37

Watson McDaniel Company, Figure 41 (photo)

Courtesy of UE Systems Inc., Figure 44

Courtesy of Extech Instruments, a FLIR Company, Figure 45

Ultra Electronics Nuclear Sensors & Process Instrumentation, Figure 46

Section Review Answer Key

Answer	Section Reference	Objective
Section One		
1.b	1.1.0	1a
2.a	1.2.0	1b
Section Two		
1.c	2.1.0	2a
2.b	2.2.1	2b
3.b	2.3.1	2c
Section Three		
1.b	3.1.2	3a
2.a	3.2.4	3b
3.c	3.3.0	3c
4.b	3.4.1	3d
Section Four		
1.a	4.1.1	4a
2.b	4.2.0	4b
3.a	4.3.1	4c
4.a	4.4.1	4d

NCCER CURRICULA — USER UPDATE

NCCER makes every effort to keep its textbooks up-to-date and free of technical errors. We appreciate your help in this process. If you find an error, a typographical mistake, or an inaccuracy in NCCER's curricula, please fill out this form (or a photocopy), or complete the online form at **www.nccer.org/olf**. Be sure to include the exact module ID number, page number, a detailed description, and your recommended correction. Your input will be brought to the attention of the Authoring Team. Thank you for your assistance.

Instructors – If you have an idea for improving this textbook, or have found that additional materials were necessary to teach this module effectively, please let us know so that we may present your suggestions to the Authoring Team.

NCCER Product Development and Revision
13614 Progress Blvd., Alachua, FL 32615

Email: curriculum@nccer.org
Online: www.nccer.org/olf

❏ Trainee Guide   ❏ Lesson Plans   ❏ Exam   ❏ PowerPoints   Other _____

Craft / Level: _____   Copyright Date: _____

Module ID Number / Title: _____

Section Number(s): _____

Description: _____

_____

_____

_____

_____

Recommended Correction: _____

_____

_____

_____

_____

Your Name: _____

Address: _____

_____

Email: _____   Phone: _____

Retail Refrigeration Systems

OVERVIEW

Retail refrigeration systems are everywhere. Refrigerated coolers, freezers, cases, and dispensers of all kinds are widely used in retail establishments that sell or serve food and beverages, such as grocery stores and convenience stores. The specialized field of retail refrigeration offers many career opportunities for HVACR service technicians.

Module 03304

Trainees with successful module completions may be eligible for credentialing through the NCCER Registry. To learn more, go to **www.nccer.org** or contact us at 1.888.622.3720. Our website has information on the latest product releases and training, as well as online versions of our *Cornerstone* magazine and Pearson's product catalog.

Your feedback is welcome. You may email your comments to **curriculum@nccer.org**, send general comments and inquiries to **info@nccer.org**, or fill in the User Update form at the back of this module.

This information is general in nature and intended for training purposes only. Actual performance of activities described in this manual requires compliance with all applicable operating, service, maintenance, and safety procedures under the direction of qualified personnel. References in this manual to patented or proprietary devices do not constitute a recommendation of their use.

Copyright © 2018 by NCCER, Alachua, FL 32615, and published by Pearson, New York, NY 10013. All rights reserved. Printed in the United States of America. This publication is protected by Copyright, and permission should be obtained from NCCER prior to any prohibited reproduction, storage in a retrieval system, or transmission in any form or by any means, electronic, mechanical, photocopying, recording, or likewise. To obtain permission(s) to use material from this work, please submit a written request to NCCER Product Development, 13614 Progress Blvd., Alachua, FL 32615.

03304 V5

From *HVAC Level Three, Trainee Guide*. NCCER.
Copyright © 2018 by NCCER. Published by Pearson. All rights reserved.

03304
RETAIL REFRIGERATION SYSTEMS

Objectives

When you have completed this module, you will be able to do the following:

1. Describe retail refrigeration applications and the related refrigeration cycle.
   a. Describe the refrigeration cycle for medium-temperature systems.
   b. Describe the refrigeration cycle for low-temperature systems.
   c. Describe various approaches to defrosting.
2. Identify various types of refrigeration equipment and their application in retail refrigeration.
   a. Identify and describe reach-in coolers and freezers.
   b. Identify and describe various walk-ins and merchandisers.
   c. Identify various types of commercial ice machines and their applications.
3. Identify and describe components related to the operation of retail refrigeration systems.
   a. Identify and describe the primary components used in retail refrigeration applications.
   b. Identify and describe secondary components and accessories used in retail refrigeration applications.
   c. Identify and describe common refrigeration system controls.
4. Explain the basic principles of maintaining and troubleshooting various retail refrigeration systems.
   a. Explain how to maintain and troubleshoot a typical reach-in freezer.
   b. Explain how to maintain and troubleshoot a typical cubed-ice machine.

Performance Tasks

Under the supervision of your instructor, you should be able to do the following:

1. Set up a defrost timer for a freezer.
2. Isolate a fault in refrigeration equipment or an ice machine.
3. Clean an ice machine.

Trade Terms

Phase change
Slinger ring
Unit cooler

Industry Recognized Credentials

If you are training through an NCCER-accredited sponsor, you may be eligible for credentials from NCCER's Registry. The ID number for this module is 03304. Note that this module may have been used in other NCCER curricula and may apply to other level completions. Contact NCCER's Registry at 888.622.3720 or go to www.nccer.org for more information.

Contents

1.0.0 Refrigeration System Operation 1
   1.1.0 Medium-Temperature Refrigeration Cycle 3
   1.2.0 Low-Temperature Refrigeration Cycle 7
   1.3.0 Defrost Systems 8
      1.3.1 Off-Cycle Defrost 8
      1.3.2 Timed Defrost 9
      1.3.3 Electric Defrost 9
      1.3.4 Hot-Gas Defrost 11
2.0.0 Retail Refrigeration Equipment 13
   2.1.0 Reach-In Coolers and Freezers 13
      2.1.1 Ice Merchandisers 13
   2.2.0 Merchandising Walk-In Systems 15
   2.3.0 Commercial Ice Makers 15
      2.3.1 Cubed-Ice Machines 16
      2.3.2 Tube-Ice Machines 17
      2.3.3 Flaked-Ice Machines 19
3.0.0 Retail Refrigeration System Components 21
   3.1.0 Primary Refrigeration Cycle Components 21
      3.1.1 Compressors 23
      3.1.2 Condensers 23
      3.1.3 Evaporators 24
      3.1.4 Expansion (Metering) Devices 26
   3.2.0 Secondary Refrigeration Cycle Components
           and Accessories 26
      3.2.1 Receivers 26
      3.2.2 Accumulators 26
      3.2.3 Crankcase Pressure Regulators 27
      3.2.4 Evaporator Pressure Regulating Valves 27
      3.2.5 Refrigerant-Side Head Pressure Control 28
   3.3.0 Common Refrigeration System Controls 29
      3.3.1 Thermostats 29
      3.3.2 Pressure Controls 30
      3.3.3 Time-Delay Relays 30
      3.3.4 Solenoid Valves 31
4.0.0 Troubleshooting Retail Refrigeration Systems 34
   4.1.0 Troubleshooting and Maintenance of a Reach-In Freezer 34
   4.2.0 Troubleshooting and Maintenance of Cubed-Ice Machines 38
      4.2.1 Cleaning and Maintenance 38
      4.2.2 Operation and Troubleshooting 39

Figures

Figure 1   Basic refrigeration system. 2
Figure 2   Typical medium-temperature refrigeration cycle for HFC-134a. 4
Figure 3   Typical low-temperature refrigeration cycle for HFC-407C. 6
Figure 4   24-hour defrost timers. 9
Figure 5   Evaporator electric-heat arrangement. 9
Figure 6   Infrared quartz heater assembly. 10
Figure 7   Heated condensate pan. 10
Figure 8   Typical defrost timer. 10
Figure 9   The hot-gas defrost process. 11
Figure 10 Two-door upright reach-in merchandiser. 14
Figure 11 Under-counter reach-in unit. 14
Figure 12 Indoor ice merchandiser. 14
Figure 13 Outdoor ice merchandiser. 14
Figure 14 Walk-in merchandiser. 15
Figure 15 A supermarket meat-display case. 15
Figure 16 Cubed-ice machine with bin. 16
Figure 17 Cubed-ice maker with an electrical cutting grid. 17
Figure 18 Cubed-ice maker with cube-shaped chambers
           on the evaporator plate. 18
Figure 19 Tube ice. 18
Figure 20 Flaked-ice machine with bin. 19
Figure 21 Flaked-ice machine operation. 19
Figure 22 Refrigeration condensing unit. 21
Figure 23 Semi-hermetic and hermetic compressors. 22
Figure 24 Digital scroll compressor. 22
Figure 25 Air-cooled fin-and-tube condenser. 23
Figure 26 Walk-in refrigeration unit cooler. 24
Figure 27 Commercial cubed-ice machine evaporator. 24
Figure 28 Capillary-tube metering device. 25
Figure 29 Thermostatic expansion valve. 25
Figure 30 Electronically controlled expansion valves (EEVs). 25
Figure 31 Suction accumulator. 26
Figure 32 Crankcase pressure regulating valve. 27
Figure 33 Evaporator pressure regulator. 27
Figure 34 Refrigerant-side, head-pressure control valve. 28
Figure 35 Head-pressure control valve installation position. 28
Figure 36 Typical refrigeration pressure switch. 30
Figure 37 Time-delay relay. 31
Figure 38 Bypass timer wiring. 31
Figure 39 Solenoid valve. 31
Figure 40 Solenoid valve with manual override. 32
Figure 41 Reach-in freezer. 34
Figure 42 Freezer wiring diagram. 36
Figure 43A Troubleshooting chart example (1 of 2). 38
Figure 43B Troubleshooting chart example (2 of 2). 40
Figure 44 Cubed-ice machine wiring diagram. 41
Figure 45 Energized-parts chart. 41
Figure 46 Ice thickness check. 43

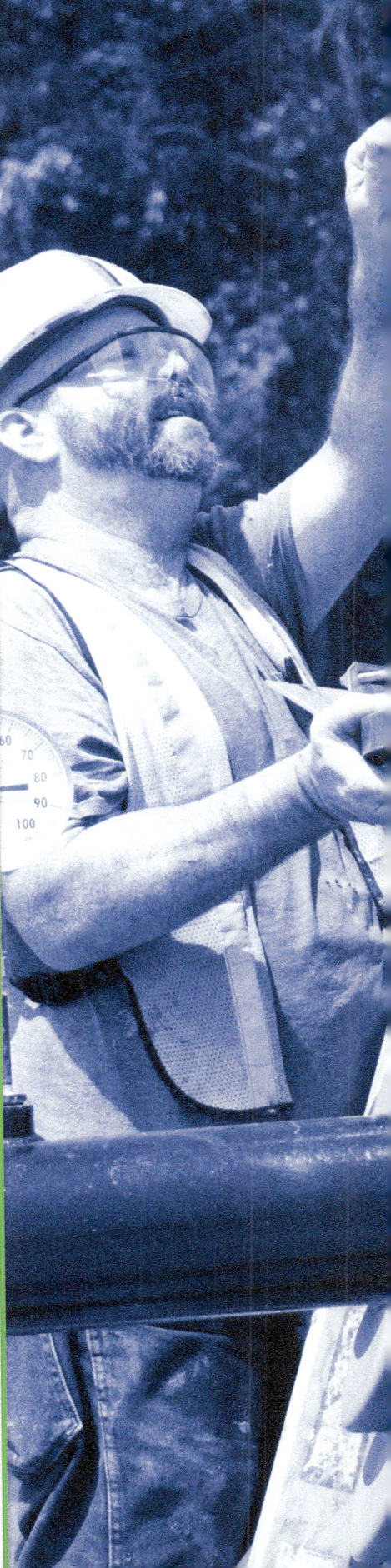

SECTION ONE

1.0.0 BASIC REFRIGERATION SYSTEM OPERATION

Objective

Describe retail refrigeration applications and the related refrigeration cycle.
a. Describe the refrigeration cycle for medium-temperature systems.
b. Describe the refrigeration cycle for low-temperature systems.
c. Describe various approaches to defrosting.

Performance Tasks

1. Set up a defrost timer for a freezer.

Trade Terms

Phase change: The transition of a refrigerant's physical state (or phase) to another, such as from liquid to vapor.

Refrigeration has a significant impact on out daily lives. Historically, ways of preserving foods were limited to curing, dehydrating, or using natural resources such as snow and winter ice to avoid spoilage.

Bacteria are the reason many foods cannot remain edible for long periods of time. By the mid-1800s, scientists understood that bacteria are responsible for most food spoilage. After they discovered that bacteria are unable to grow at lower temperatures, development of modern approaches to refrigeration began in earnest. Throughout the 1800s, advances continued in methods to protect and preserve foods. In 1882, the British sailing ship *Dunedin* carrying meat from New Zealand to England was fitted with a coal-fired apparatus to freeze the meat for the voyage. The successful delivery of frozen meat products by *Dunedin* established the practice of maritime refrigerated shipment.

Retail refrigeration systems and applications are now evident at every turn. Although the preservation of a variety of products remains the primary function, style and convenience built into refrigerated fixtures play a significant role in marketing as well. Quiet, visually appealing equipment attracts the consumer, increases sales, and helps to reinforce confidence that contents are safe and properly preserved.

Many types and styles of mechanical refrigeration systems are used to cool or freeze foods and other perishables. In most cases, consumable products reach the retail level already cooled or frozen.

The retail fixture then is used to maintain those conditions while providing easy access to the products. Operation of systems used for these applications is basically the same as that of comfort cooling systems, with the addition of components and controls necessary to achieve the desired refrigerated effect in the fixture's environment. *Figure 1* shows a basic air-cooled, refrigeration-system flow diagram.

Primary refrigerant circuit components are the compressor, evaporator, condenser, and expansion (metering) device. However, when the circuit is applied to refrigeration, a number of other factors are different. The installed location of primary components and interconnecting lines, evaporator temperatures, condensing environment, and types of refrigerants used are all significant variables.

Refrigeration applications are generally classified by required space temperature for a given

> **NOTE**
> NCCER Module 03101 (*Introduction to HVAC*) and NCCER Module 03107 (*Introduction to Cooling*) describe the basic refrigeration cycle and its components. Reviewing that material may assist in understanding differences in the operating cycle when applied to lower-temperature applications.

application. These classifications are important to remember because some components, such as compressors and expansion devices, are classified for duty according to the following temperature ranges:

- *Above 60°F (16°C)* – Air conditioning/high-temperature refrigeration
- *30°F to 60°F (–1°C to 16°C)* – Medium-temperature refrigeration
- *–40°F to 30°F (–40°C to –1°C)* – Low-temperature refrigeration

The basic refrigeration cycle for most applications is similar to that for air conditioning. However, many technicians are uneasy about servicing refrigeration systems because they have spent much of their career working with comfort cooling systems, where a 40°F (4°C) evaporator temperature is the established standard. In refrigeration, different evaporator temperatures must be reached to achieve the desired storage temperature. In reality, while simple changes in a few components yield far different results, the theory of operation remains the same.

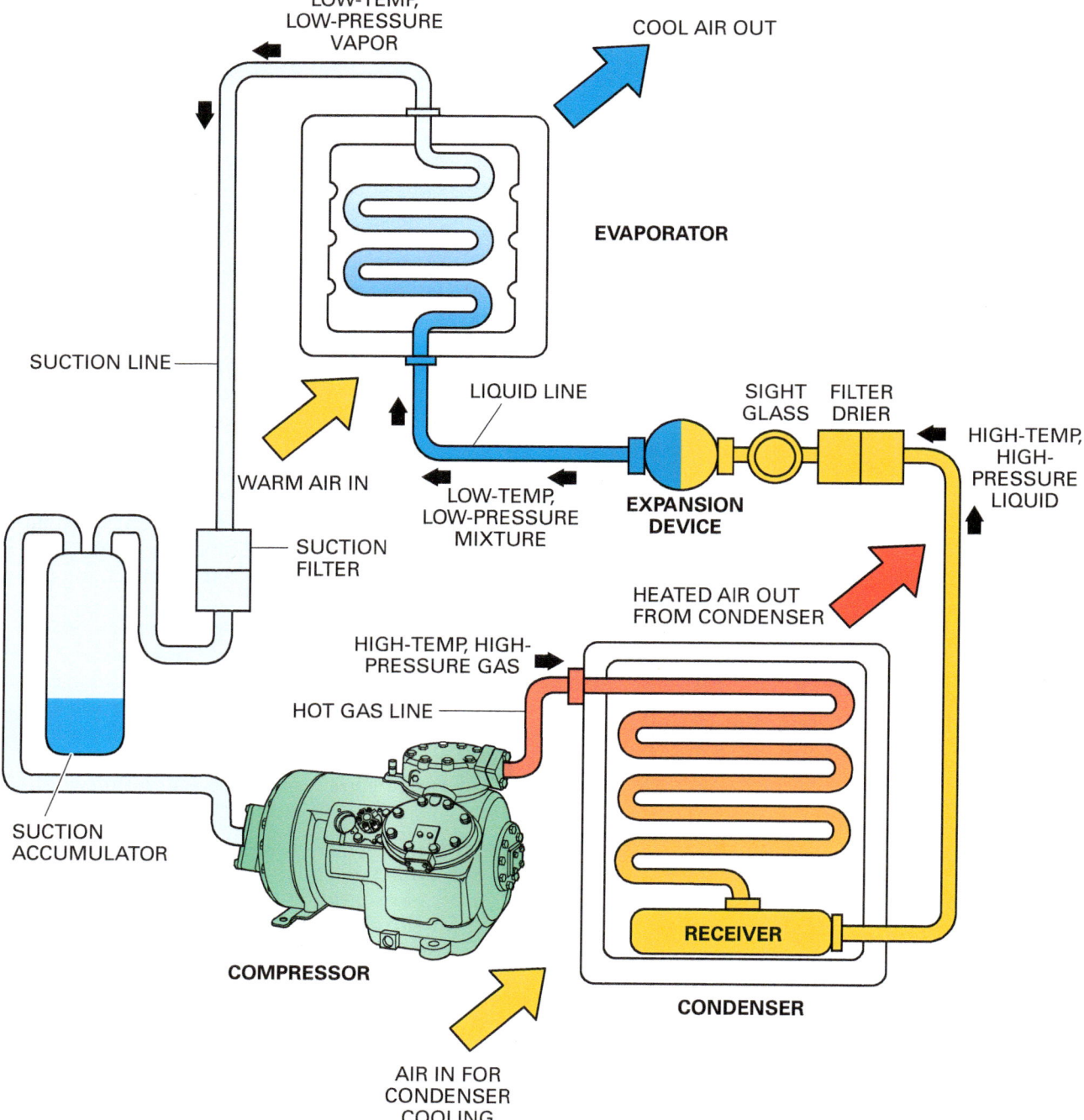

Figure 1 Basic refrigeration system.

Did You Know?

Ice

Before the development of mechanical refrigeration, people preserved food by using blocks of ice. The ice would be harvested from nearby lakes in the winter and stored in icehouses. The ice blocks were covered with hay and sawdust to insulate them from the heat. Throughout the year, ice was removed from storage and delivered to customers. The delivery was often done using horse-drawn ice wagons. In areas where the temperature seldom dropped below freezing, large masses of sawdust-insulated blocks of ice had to be delivered from colder locations in the north by rail or riverboat.

Figure Credit: Cape Pond Ice Company

1.1.0 Medium-Temperature Refrigeration Cycle

An example of a typical medium-temperature refrigeration cycle, using HFC-134a as the refrigerant, is shown in *Figure 2*.

Assume that the system is for a retail fixture inside a conditioned space, such as a convenience store. The air temperature in the vicinity of the fixture and unit condenser is 75°F (24°C), and the desired box temperature is 38°F (3°C). In actual practice, these conditions can fluctuate dramatically throughout the course of a day.

The following numbers correspond to numbered points shown in *Figure 2*. Note that this example is theoretical and does not account for normal pressure drops through piping or components in a normally operating system.

1. When the refrigerant leaves the expansion device, it is a mixture of liquid and vapor. The vapor results from some liquid flashing at the expansion device, absorbing heat to cool the remaining liquid to the desired evaporator temperature. A 75-percent liquid to 25-percent vapor volume can be considered reasonably normal.

   To achieve the desired box temperature of 38°F (3°C), an evaporator temperature lower than 38°F is needed. For this example, use an evaporator temperature of 24°F (−4°C) at an HFC-134A pressure of 21.4 psig (147 kPag). This creates a temperature difference (ΔT) of 14°F (7.8°C) between the refrigerant-saturated suction temperature in the evaporator coil and the desired box air temperature.

2. As the refrigerant mixture flows through the evaporator, it absorbs heat from the warmer air in the box and continues to change phase from liquid to vapor. The temperature of the refrigerant remains unchanged at its saturation temperature until it completes its phase change. The mixture of liquid and vapor will remain at 24°F (−4°C) until all liquid has entered its vapor phase.

   To ensure the maximum amount of evaporator area is providing usable refrigeration, the point where the phase change is complete is ideally located in the last 3 to 7 percent of the coil. Once phase change is complete, the vapor can begin to increase in temperature and its superheat can be measured.

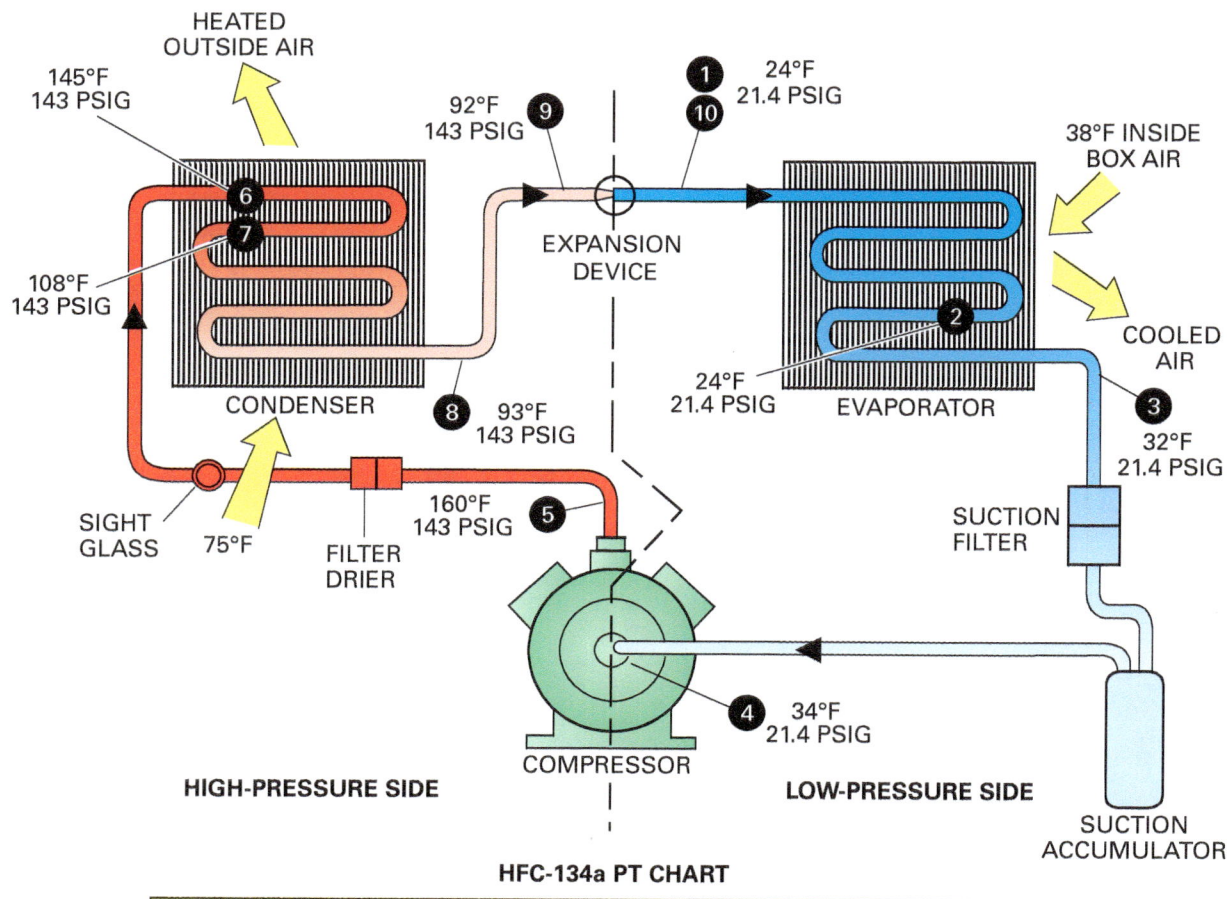

Figure 2 Typical medium-temperature refrigeration cycle for HFC-134a.

3. Refrigerant exiting the evaporator coil and entering the suction line should now be in a single state: vapor. The completion of the phase change process can be proven by the existence of measurable superheat. Temperature and pressure measurements taken at this point should reveal that the refrigerant has increased in temperature above that shown for the refrigerant's saturation on the pressure-temperature (P-T) chart. This provides evidence that the phase change is complete and ensures that damaging liquid refrigerant is not on its way to the compressor.

   In this example, the refrigerant has gained 8°F (4.4°C) of superheat for a suction line temperature of 32°F (0°C). It is important to note that the suction line will often form frost, which should not be considered an indication of the presence of liquid refrigerant.

4. The vapor returning to the compressor received only a very small amount of heat (2°F or 1.1°C) as it traveled through the suction line. In actual practice, this value can vary considerably, depending on the length of the suction line, quality of insulation, and the ambient temperature in the area of operation. In a small reach-in retail case, the suction line may be as short as a few inches or centimeters, while other applications with remote condensing units may require much longer lines.

5. After the compression process, highly superheated vapor now exits the compressor. In this example, the discharge vapor temperature has reached 160°F (71°C) at a pressure of 143 psig (986 kPag) as the direct result of compression. A quick check of the P-T chart reveals that the superheat is now at 52°F (28.8°C). All of this superheat must be removed from the vapor before condensation back to the liquid state can begin. A small amount of this superheat is often lost to the surrounding air as the vapor travels to the condenser. The hot gas line is rarely insulated, and a large ΔT between the line and the surrounding air is common.

6. As the refrigerant vapor enters the condenser coil, superheat is quickly removed. As the refrigerant temperature continues to fall toward the temperature of the condenser air, the ΔT between condenser air and refrigerant is reduced and the process slows somewhat. All superheat must be removed and the vapor temperature must fall to the saturation temperature for HFC-134a at the given pressure before the condensing process can begin. For this example, the appropriate condensing temperature would be 108°F (42°C).

7. When the vapor has cooled to 108°F in the condenser, condensation begins and the refrigerant temperature will remain unchanged throughout the process. As refrigerant flows through the condenser, the ratio of liquid to vapor increases until only liquid remains.

8. When the phase change is complete, refrigerant liquid can begin to experience measurable cooling into the subcooled region of the P-T chart. In this example, the refrigerant temperature has fallen to 93°F (34°C) at the exit of the condenser. Because the saturation temperature for HFC-134a is 108°F (42°C) at the pressure of 143 psig (986 kPag), the subcooling is calculated to be 15°F (8.3°C).

   Remember that without measurable subcooling as an indicator, the condensation process has not been completed and some amount of vapor remains in the liquid. This causes a serious loss of refrigerating capacity in the system, because any vapor that travels through the expansion device and into the evaporator is virtually useless in absorbing any significant amount of heat.

9. As is the case with other refrigerant lines connecting primary components, the heat loss or gain in the liquid line can be large or small. In this example, the liquid line is very short and has had a temperature reduction of only 1°F (0.6°C). If the liquid line travels through an area at a significantly different temperature, a greater increase or decrease in temperature can occur.

10. At this point, the liquid refrigerant has passed through the expansion device, having a pressure drop of over 121 psig (834 kPag) in the process. Although the refrigerant was subcooled and entirely in its liquid state as it entered the expansion device, some of the refrigerant quickly flashed into the vapor state, absorbing heat and cooling the remaining liquid to the proper saturation temperature for the new pressure condition of 21.4 psig (147 kPag). The cycle is complete and ready to begin again.

 03304 Retail Refrigeration Systems

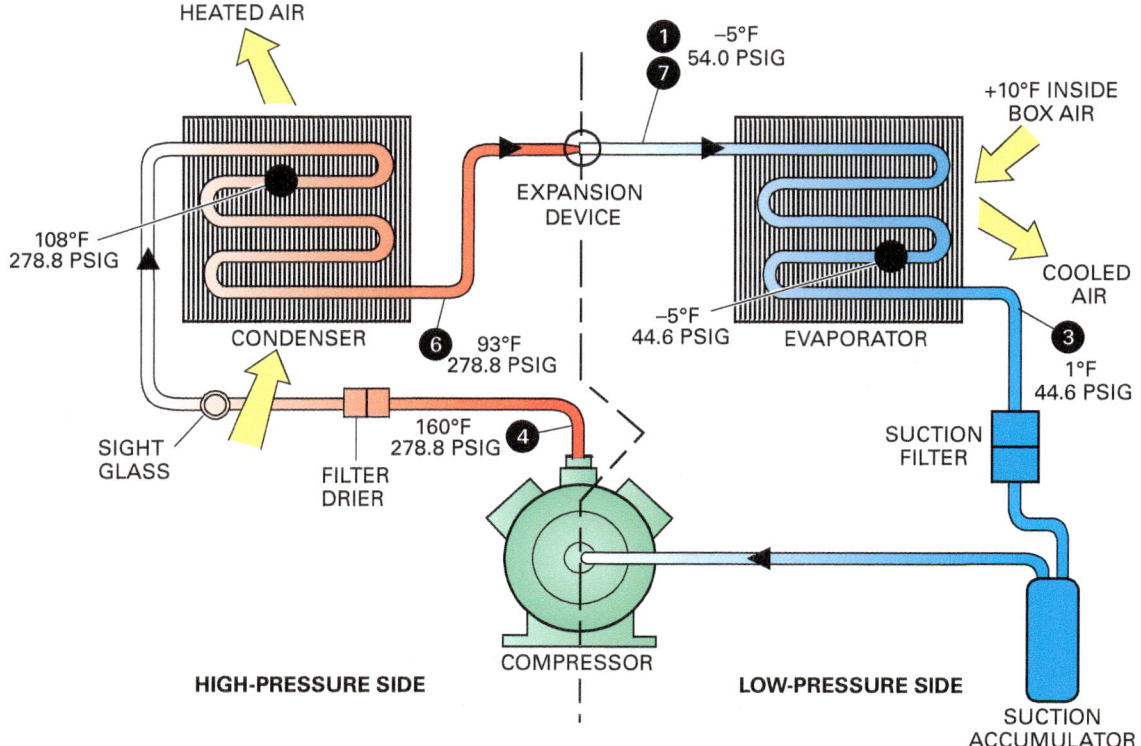

R-407C PRESSURES/TEMPERATURES

TEMP. °F	PRESS. (psig) P_L	PRESS. (psig) P_V	TEMP. °F	PRESS. (psig) P_L	PRESS. (psig) P_V	TEMP. °F	PRESS. (psig) P_L	PRESS. (psig) P_V	TEMP. °F	PRESS. (psig) P_L	PRESS. (psig) P_V
−10	50.0	41.3	24	84.9	70.7	58	141.4	120.4	92	226.6	198.6
−8	51.5	42.6	26	87.5	73.0	60	145.5	124.2	94	232.7	204.3
−6	53.2	43.9	28	90.3	75.3	62	149.8	128.0	96	238.9	210.1
−4	54.8	45.3	30	93.1	77.8	64	154.1	131.9	98	245.2	216.1
−2	56.6	46.8	32	96.0	80.3	66	158.6	135.9	100	251.7	222.1
0	58.4	48.3	34	98.9	82.9	68	163.1	140.1	102	258.3	228.4
2	60.2	49.8	36	102.0	85.5	70	167.8	144.3	104	265.0	234.7
4	62.1	51.4	38	105.1	88.3	72	172.6	148.7	106	271.8	241.2
6	64.1	53.0	40	108.4	91.1	74	177.5	153.1	108	278.8	247.9
8	66.2	54.8	42	111.7	94.0	76	182.5	157.7	110	286.0	254.7
10	68.3	56.5	44	115.1	97.0	78	187.6	162.4	112	293.2	261.7
12	70.4	58.3	46	118.1	100.1	80	192.8	167.2	114	300.6	268.8
14	72.7	60.2	48	122.1	103.2	82	198.1	172.1	116	308.2	276.1
16	75.0	62.2	50	125.8	106.5	84	203.6	177.2	118	315.9	283.5
18	77.3	64.2	52	129.5	109.8	86	209.2	182.3	120	323.7	291.1
20	79.8	66.3	54	133.4	113.3	88	214.9	187.6	122	331.7	298.7
22	82.3	68.4	56	137.3	116.8	90	220.7	193.1	124	339.9	306.8

Note: R-407C is zeotropic. P_L is the saturation pressure for 100% liquid; P_V is the saturation pressure for 100% vapor.

Figure 3 Typical low-temperature refrigeration cycle for HFC-407C.

GOING GREEN

Infiltration

Badly worn door gaskets and hinges can seriously affect the performance and energy efficiency of a refrigerated fixture. The constant infiltration of warm and moist outside air through poorly fitted doors and gaskets can prevent even the best fixture from reaching its setpoint. It can also increase frost accumulation, further reducing coil heat transfer and increasing refrigeration equipment runtime and operating costs. Maintaining and replacing door gaskets and hinges with high wear and abuse is as important to a fixture's efficient function as having proper refrigerant charge.

1.2.0 Low-Temperature Refrigeration Cycle

The cycle used to produce freezing temperatures is not significantly different from the medium-temperature cycle. This section provides an example using HFC-407C refrigerant. Remember that HFC-407C is a zeotropic blend and has a relatively significant glide. The associated P-T chart includes columns for all-vapor and all-liquid pressures. However, this characteristic does not significantly alter the refrigeration process.

Figure 3 represents a typical low-temperature HFC-407C refrigeration circuit, such as that used in a reach-in freezer. For this example, assume that the condenser entering air temperature is 75°F (24°C), since the condenser is located indoors. Because this is a freezer application, the desired box temperature will now be 10°F (–12°C). The following numbers correspond to the numbered points in *Figure 3*:

> **NOTE**
> The metric HFC-407C Pressures/Temperatures table for this discussion is provided in the *Appendix*.

1. To achieve the desired box temperature of 10°F (–12.2°C), a much lower saturated refrigerant temperature must be created in the evaporator. For this example, use an evaporator temperature of –5°F (–20.6°C) at an HFC-407C pressure of 54.0 psig (372 kPag). This creates a ΔT of 15°F (8.3°C) between the refrigerant in the evaporator coil and the desired box air temperature.

2. The mixture of liquid and vapor in the evaporator will not rise above –5°F (–21°C) until the last liquid fraction has returned to the vapor phase. To ensure that the maximum amount of evaporator area is providing usable refrigeration, the point where the phase change is complete is ideally located in the last 3 to 7 percent of the coil. When the phase change is complete, the vapor can begin to gain heat and superheat can be measured.

3. Refrigerant exiting the evaporator coil and entering the suction line should now be in a single state: vapor. The completion of the phase change process can be proven by the existence of measurable superheat. Temperature and pressure measurements taken at this point should reveal that the refrigerant has increased in temperature above that shown for saturation on the P-T chart, providing evidence that the phase change is complete. This ensures that damaging liquid refrigerant is not on its way to the compressor. In this example, the refrigerant vapor has increased in temperature 6°F (3.3°C) (superheat) following the phase change.

4. After compression, highly superheated vapor exits the compressor. In this example, the discharge vapor temperature has reached 160°F (71°C) at a pressure of 278.8 psig (1,922 kPag).

5. All superheat must be removed and the vapor temperature must fall to the saturation temperature for the most volatile vapor fraction of HFC-407C at the given pressure before the condensing process can begin. Since the refrigerant is zeotropic, the more volatile fraction will begin condensing at a higher temperature. As refrigerant moves through the condenser, the ratio of liquid to vapor constantly changes until only liquid remains. For this example, the lowest condensing temperature would be 108°F (42°C), which is the saturation temperature for the liquid phase of the least-volatile fraction phase at the existing pressure.

6. In this example, the refrigerant temperature has fallen to 93°F (34°C) at the exit of the condenser. Because the saturation temperature for liquid HFC-407C is 108°F (42°C) at a pressure of 278.8 psig (1,922 kPag), the subcooling is 15°F (8.3°C).

7. At this point, the liquid refrigerant has passed through the expansion device, having a pressure drop of roughly 225 psig (1,550 kPag) in the process. The cycle is complete and ready to begin again.

Refrigeration cycles are generally the same. Differing temperatures and pressures in various systems result from characteristics of the different refrigerants and metering device used, but the process remains unchanged. With correct information and an understanding of the cycle, a technician should have no trouble understanding system operating characteristics and identifying when the refrigeration cycle is not operating as designed.

One important note about superheat values for refrigeration applications when compared to comfort cooling. Although typical desired superheat values for comfort cooling may be 10°F to 20°F (5.6°C to 11.1°C), refrigeration systems generally operate at lower superheat values. As saturated suction temperatures are reduced, compression ratios become higher and compressor energy efficiency is significantly reduced. Energy efficiency ratios (EERs) for medium-temperature applications are typically around 7.0 to 8.0, while low-temperature systems may operate with EERs as low as 4.0.

As a result, gaining as much capacity per watt from the system as possible is essential for low-temperature applications. One way to achieve this is through reduced superheats, exposing as much evaporator internal volume as possible to boiling liquid refrigerant. All significant heat exchange takes place in this phase-change process.

Once the liquid has boiled into vapor, little heat is exchanged as it superheats. Superheat values of 6°F to 12°F (3.3°C to 6.7°C), measured at the evaporator, are considered acceptable for high- and medium-temperature applications. Superheat values for low-temperature applications may range from 3°F to 8°F (1.7°C to 4.4°C) to maximize the refrigerating effect in the evaporator. Great care must be taken to ensure adequate compressor protection from liquid floodback exists when low superheat values are chosen. However, there is much to gain in energy efficiency.

1.3.0 Defrost Systems

As long as moisture is inside the refrigerated space, it will condense on the surface of the evaporator, as it does in comfort cooling systems. However, when the coil surface is below freezing, frost and ice will form, slowing and eventually blocking airflow through the coil. The coil must be defrosted to resume proper operation.

In retail refrigeration applications, there are four primary defrost methods:

- Off-cycle defrost
- Timed defrost
- Electric defrost
- Hot-gas defrost

Regardless of the defrost approach used, several factors regarding frost accumulation on the evaporator coils are common. Fixtures that were selected and designed to operate at large air-to-refrigerant ΔTs will generally build frost faster, since their coil temperature will likely be well below the dew point for the box. This high air-to-refrigerant ΔT can be used when the product is hermetically sealed and can't dry out, such as canned or bottled drinks. This encourages greater moisture condensation on the coil, but is more energy efficient. Fixtures with small air-to-refrigerant ΔTs will not accumulate as much condensate or frost as quickly. Foods that can be damaged through dehydration must be refrigerated with fixtures having small air-to-refrigerant ΔTs.

Another factor in frost accumulation is related to fixture usage and infiltration of outside air. Most retail refrigeration fixtures experience high customer usage. In hot, humid weather, a fixture that is opened many times each day will constantly be exposed to warm, humid air. Not only does this significantly increase the volume of airborne moisture in the enclosure, but the infiltration of hot air replacing refrigerated air greatly increases the refrigeration load as well.

Conditions like this can overwhelm and interfere with fixture performance, in spite of defrost cycles. Consider the potential difference in performance of an identical reach-in refrigerated fixture installed in two different locations, one with little consumer traffic in Arizona, and the other in a busy, 24-hour convenience store in coastal Florida.

All heat added to the refrigerated enclosure during the defrost cycle, regardless of the source, must again be removed when the process is complete. This fact has a large effect on the refrigeration load. In addition, the time a system is projected to spend in defrost mode must be subtracted from the available time for refrigeration. For example, if the total refrigeration load for a fixture is calculated as 48,000 Btus over a 24-hour period (2,000 Btuh), but the unit is anticipated to operate in defrost for 2 hours per day, then the refrigeration equipment must be able generate this much capacity in 22 hours instead of 24 hours. That would require 2,182 Btuh of capacity instead of 2,000.

1.3.1 Off-Cycle Defrost

Off-cycle defrost is a simple passive defrost method. In fixtures that maintain temperatures at or above 36°F (2°C), the coil is simply allowed to defrost during the normal off cycle. Fixtures in this category usually operate at a coil/saturated suction temperature of 16°F to 31°F (–9 to –1°C), depending on the desired humidity levels for the stored product and box design.

Because the fixture temperature is above freezing, the coil will build frost, but should defrost naturally when the unit has cycled off. The evaporator fan continues to run during the off cycle.

One assist to this natural defrost approach is to ensure that the fixture temperature is not set lower than needed. Maintaining the box even a few degrees colder than necessary can slow or defeat the natural defrost process, resulting in frozen coils and rising box temperatures. When this occurs, the fixture requires shut down for a period long enough to rid the coil of ice.

1.3.2 Timed Defrost

Systems operating at slightly lower temperatures than those using simple off-cycle defrost require a longer period to clear accumulated ice. Fixtures that operate at 32°F to 36°F (0°C to 2°C) often benefit from this approach. Typically, a 24-hour timer (*Figure 4*) with normally closed contacts, placed in series with the compressor controls, is set for a reasonable period, yet longer than the typical off cycle. The evaporator fan continues to run during the off cycle.

Fixture temperature may rise a few degrees above normal during the extended period, but will quickly recover. Like all defrost scenarios, setting the timer for a defrost period that coincides with the fixture's lightest period of usage (late at night, for example) helps to prevent the fixture temperature from rising beyond an acceptable level. Several cycles per day may be required.

1.3.3 Electric Defrost

Electric defrost systems are common for medium- and low-temperature fixtures that operate at temperatures too cold for off-cycle or timed defrost to be effective, and for freezers in the retail refrigeration class. Electric defrost components are inexpensive, reliable, and simple to troubleshoot and repair.

Figure 5 shows a typical arrangement for electric defrost heaters installed on both the face and back of an evaporator coil. For proper operation, the heater must be in good thermal contact with the coil fins to effectively distribute heat and minimize the required defrost period.

A timer stops the refrigeration equipment and evaporator fans during the defrost, then energizes the electric heater. In many cases, the shape and size of the defrost heater are unique to a given evaporator coil and manufacturer. This requires original parts when replacement is necessary. Some coil designs require that the heater be placed deep inside the coil assembly, making repair and replacement harder, but resulting in an effective and rapid defrost.

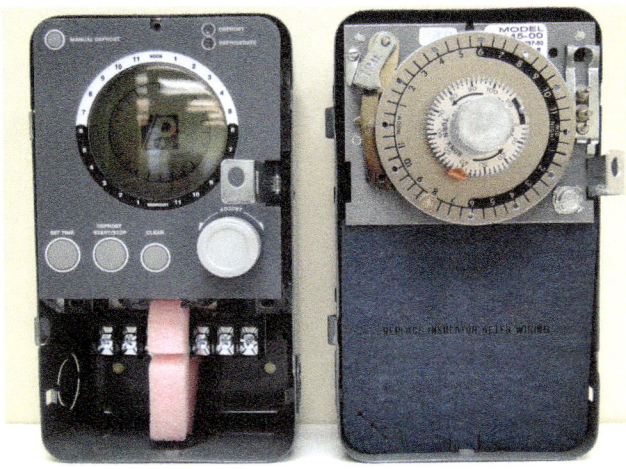

Figure 4 24-hour defrost timers.

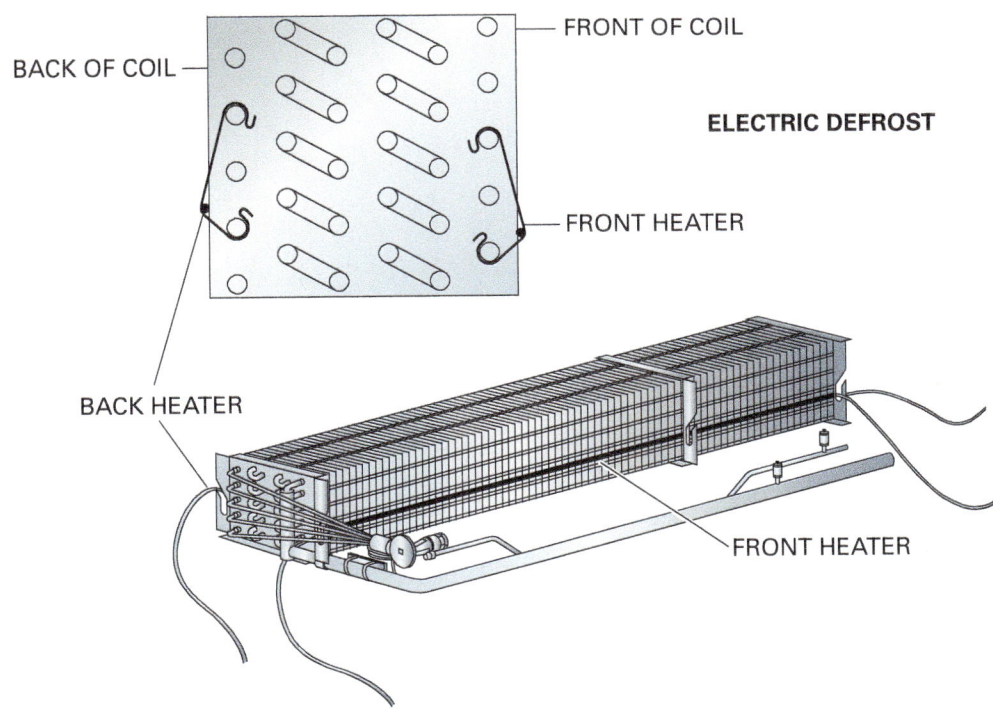

Figure 5 Evaporator electric-heat arrangement.

Some units are equipped with radiant electric heaters (*Figure 6*). The resistance heater wire is spirally wound and enclosed inside a quartz tube, with the wires passed through the ends of the sealed tube cap. Using infrared radiant heat, defrost can be accomplished without the heater being in contact with the coil. These heaters can reach a high temperature very quickly and are corrosion resistant. However, the quartz-glass enclosure is fragile.

When electric defrost is required, condensate removed from the coil must also be maintained above freezing until it has completely left the freezing environment. Many condensate collection pans, either as part of the evaporator coil assembly or as a separate collection point inside the fixture, may also be equipped with electric heaters such as the one shown in *Figure 7*. The heater simply re-evaporates the water to the surrounding air.

Some installations with the drain line routed through a freezing area may require the condensate drain line to be kept above freezing as well. For this, electric pipe heating cable (heat tape) is attached to the exterior of the line.

Electric defrost systems also use a 24-hour timer for defrost control, except for large and/or complex systems. Electro-mechanical timers have a dial assembly, as shown in *Figure 8*, allowing the timer to be configured for the time, duration,

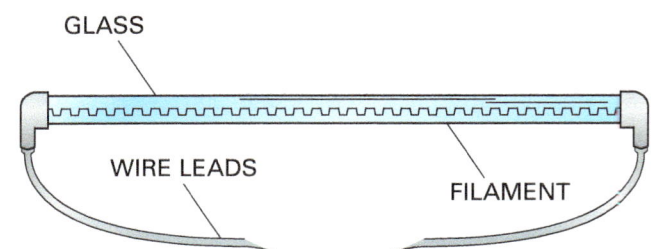

Figure 6 Infrared quartz heater assembly.

Figure 7 Heated condensate pan.

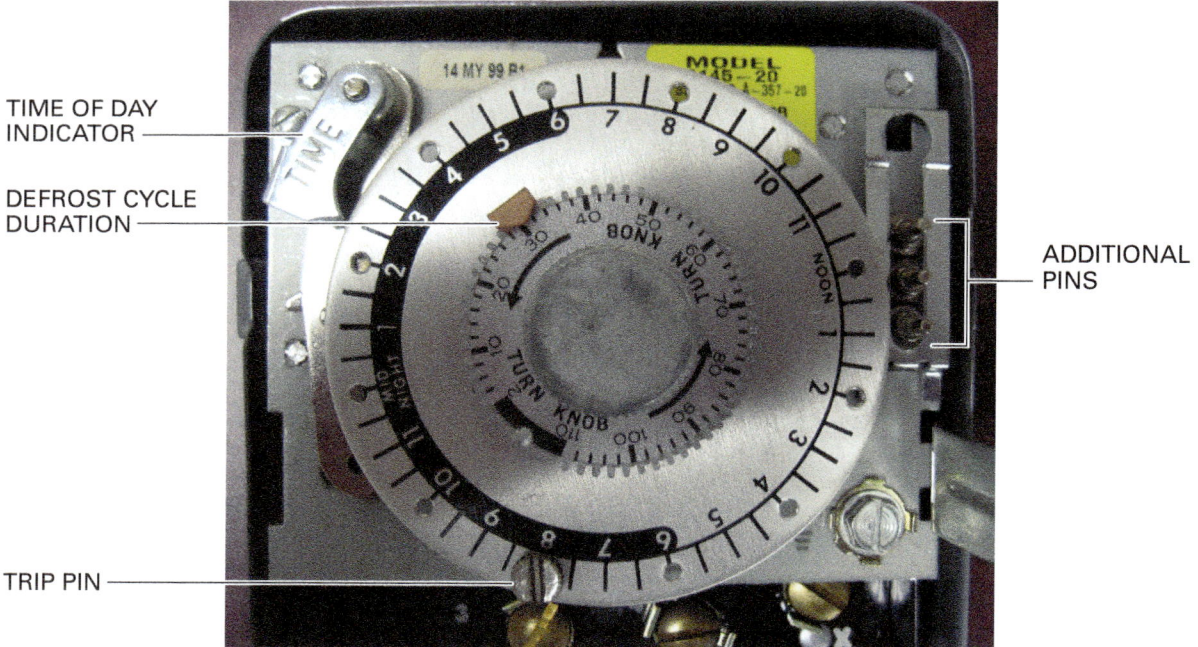

Figure 8 Typical defrost timer.

and number of defrost periods during a 24-hour period. The inner defrost dial sets the duration of the defrost period; defrost is terminated after the period has elapsed. The timer contacts then return to normal position and the refrigeration cycle resumes. The contacts can generally handle the electrical load of the defrost heater(s) directly; no additional relay or contactor is needed in retail fixtures.

Electric defrost is often terminated by a temperature sensor or line-mounted thermostat. By monitoring the temperature of the coil, or refrigerant suction line close to the coil, the thermostat indicates that the coil has reached the correct temperature. Defrost can be terminated before the timer has reached its allowed maximum time. The proper setpoint for temperature termination may differ among products, since this is based on the manufacturer's chosen location and laboratory testing. Line-mounted thermostats are generally non-adjustable, so using the correct replacement is important.

The final choice of defrost control and termination is often an option provided to the end user, either as a standard feature or at an added cost. Defrost termination can therefore be time-only, temperature-only, or a combination of the two.

1.3.4 Hot-Gas Defrost

Although not as common as the other defrost methods, hot-gas defrost is widely considered to be the fastest and most efficient. Because of the additional refrigerant circuit components required, this method generally costs more to install, making it less popular. The refrigerant circuit is significantly more complex. It is typically used only on large systems.

Hot-gas defrost uses a solenoid valve(s) to send discharge gas directly from the compressor to the evaporator, bypassing the metering device (*Figure 9*). The hot gas condenses in the cold evaporator as the frost melts. The latent heat of vaporization released during condensation raises the coil surface temperature above freezing. The pressure regulator shown in *Figure 9* is set to maintain an evaporator pressure (upstream of the valve) that is well above the saturated suction temperature for the refrigerant in use; it is usually operated using a solenoid coil, as it must be opened completely during the refrigeration cycle.

Care must also be taken in these systems to avoid compressor overload and the potential for liquid refrigerant to return to the compressor. The figure shown here is a very simple example of such a system. It is essential to understand the details of operation for any hot-gas defrost system you encounter to properly service and maintain it. This information should be acquired directly from the manufacturer.

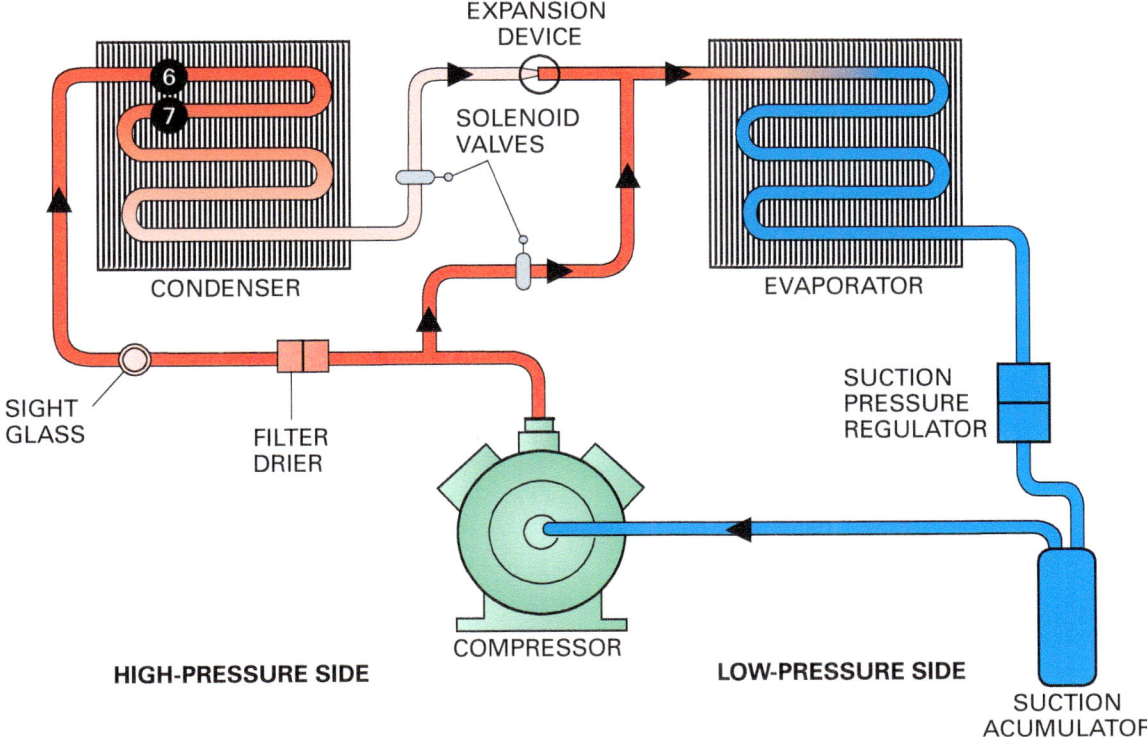

Figure 9 The hot-gas defrost process.

Additional Resources

NCCER Module 03101, *Introduction to HVAC*.

NCCER Module 03107, *Introduction to Cooling*.

Refrigeration and Air Conditioning: An Introduction to HVAC/R, Larry Jeffus. Current Edition. New York, NY: Pearson.

1.0.0 Section Review

1. In a medium-temperature refrigeration unit, the refrigerant phase change from liquid to vapor takes place in the _____.

   a. suction line
   b. evaporator
   c. liquid line
   d. condenser

2. A typical superheat range in low-temperature refrigeration applications is _____.

   a. 3°F to 8°F (1.7°C to 4.4°C)
   b. 6°F to 12°F (3.3°C to 6.7°C)
   c. °8°F to 16°F (4.4°C to 8.9°C)
   d. 10°F to 20°F (5.6°C to 11.1°C)

3. Simple off-cycle defrost can usually be applied to refrigerated fixtures that maintain a minimum temperature of _____.

   a. 28°F (–2°C)
   b. 32°F (0°C)
   c. 36°F (2°C)
   d. 40°F (4°C)

SECTION TWO

2.0.0 RETAIL REFRIGERATION EQUIPMENT

Objective

Identify various types of refrigeration equipment and their application in retail refrigeration.
 a. Identify and describe reach-in coolers and freezers.
 b. Identify and describe various walk-ins and merchandisers.
 c. Identify various types of commercial ice machines and their applications.

Trade Terms

Unit cooler: A packaged unit containing evaporator coil, metering device, and evaporator fan(s) that serves as the evaporator section for a refrigeration appliance.

Retail refrigeration equipment and systems are relatively mobile and can be installed in a variety of locations with a minimal amount of external utilities. Most retail systems, for our purposes, would be considered packaged or unitary. Although the refrigerated cases in supermarkets may be considered under the retail heading, they represent some of the most complex systems in the industry and are rarely self-contained. As a result, they are not reviewed in great detail here.

One important consideration for many packaged retail refrigeration units is their environment. Many manufacturers specify that indoor units are designed for indoor spaces maintained at 75°F (24°C) and 55% relative humidity. Although many units are placed in areas that conform to these requirements, many others are not. Technicians and end-users should not expect manufacturer-specified operation and capacity from units placed in areas that are excessively hot and/or humid.

This section, reviews the following types of equipment and their features:

- Reach-in coolers and freezers
- Merchandising walk-in systems
- Commercial ice makers

2.1.0 Reach-In Coolers and Freezers

Reach-in coolers and freezers are available in styles and types that suit almost any merchandising need. Despite the large variety available, refrigeration cycles and operating sequences are similar. The refrigerated enclosure is insulated with urethane or polystyrene materials, either in board-form or sprayed.

Figure 10 represents a typical two-door reach-in unit. Some common features and options for these units include the following:

- Interior lighting
- Self-closing doors that can be locked open for stocking
- A variety of shelving options
- Shelf moldings to allow for price tags
- Digital thermometers
- A unitary condensing unit that can be removed for service or replacement
- Adjustable leveling legs

Figure 11 shows a typical under-counter unit used to store food or beverages. Because units of this type are not directly accessible by consumers, appearance is less important than a durable and maintainable design.

Typical options and features for under-counter units include the following:

- Stainless steel cabinet construction
- A unitary condensing unit that can be removed for service or replacement
- Casters for easy movement, or adjustable legs
- Epoxy-coated evaporator coils to protect the aluminum from corrosion
- Exterior thermometers

Condensing units in packaged systems are mounted either above or below the refrigerated case, depending on manufacturer preference and fixture style. Medium-temperature units may incorporate either off-cycle or timed defrost, with electric defrost available as a factory-installed option. Low-temperature freezer units are equipped with electric defrost almost exclusively. The metering device is usually a capillary tube or thermostatic expansion valve (TEV).

2.1.1 Ice Merchandisers

Ice merchandisers for bagged retail ice have a simple design and function. Models are designed for indoor or outdoor use in sizes ranging from roughly 25 ft.^{3} to over 100 ft.^{3} (0.7 to 3 m^3) of internal volume. Indoor models (*Figure 12*) are gen-

Figure 10 Two-door upright reach-in merchandiser.

Figure 12 Indoor ice merchandiser.

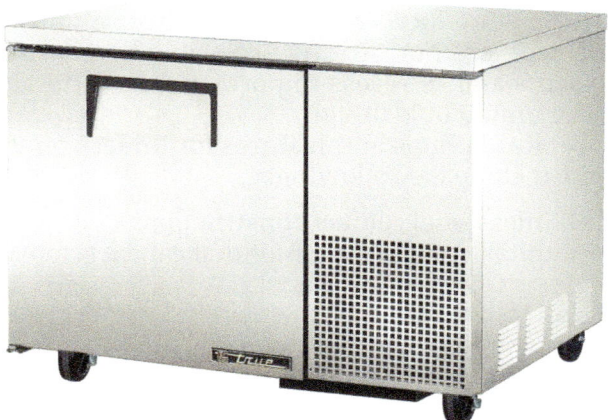

Figure 11 Under-counter reach-in unit.

erally equipped with a glass door with internal lighting. Outdoor units (*Figure 13*) have metal doors that can be locked and are built to withstand the elements and possible physical abuse. Both styles are usually controlled by a simple thermostat at box temperatures of 20°F to 25°F (–7°C to –4°C).

Ice merchandisers are generally offered with two major options: cold-wall construction (with no automatic defrost) and unitary evaporator assemblies (with electric defrost). In cold-wall models, the cabinet walls are actually plate

Figure 13 Outdoor ice merchandiser.

evaporators, with the serpentine evaporator tubing bonded to the metal sheets. No evaporator fan is used and defrost is accomplished when the merchandiser is empty. Simply shut down the unit and allow it to warm to ambient temperature. An example of this type is shown in *Figure 13*. Those that use unitary evaporator coil assemblies have a defrost timer and use electric defrost elements on the evaporator coil. *Figure 12* is an example of this type; the coil assembly is visible at the top of the box, where the evaporator fan is visible.

2.2.0 Merchandising Walk-In Systems

Many of today's convenience stores feature merchandising walk-in coolers and freezers (*Figure 14*). These units are simply walk-in coolers or freezers with one or more walls replaced by doors, allowing the consumer direct access to products.

Specialized shelves placed directly behind the doors allow for rapid stocking from within the walk-in unit itself. Products are received and stored inside the same enclosure, never leaving the refrigerated area except in the consumer's hands. Well-stocked shelving at the door opening help to block the consumer's view of perishables in storage behind the merchandise. The merchandising-shelf system often has product replacement feeding, allowing new product to fall or slide into place as one is removed. The doors of walk-in units are often equipped with a means to quickly defog the glass after the door has been opened and closed. Freezers have low-wattage electric heaters around the doorframe to prevent the door from freezing closed.

Supermarket open-face coolers (*Figure 15*) are non-enclosed, medium-temperature coolers generally used for displaying perishable items such

Figure 15 A supermarket meat-display case.

as vegetables, meats, and dairy items. They provide the customer with more convenient access to products than walk-in type cases. Since they are open to ambient air, the warmth of the store environment places a significant demand on the refrigeration system.

Walk-in and open-faced coolers generally use a unit cooler or remote condensing unit, located either outdoors or indoors on top of the refrigerated enclosure. Because the refrigeration capacity need is often greater than smaller self-contained fixtures, semi-hermetic compressors are often used. TEVs are most often used as the metering device.

2.3.0 Commercial Ice Makers

Packaged ice-making equipment is mainly of three types: cubed-ice, tube-ice, and flaked-ice machines. These are widely used in restaurants, convenience stores, hotels, and other point-of-use locations. Typically, ice is formed on a surface at roughly 10°F (–12°C). This is the target surface temperature for many ice machines.

These ice machines have some common features. They are designed to operate automatically, control their own water supply, freeze water into the desired shape and size, collect the ice in a storage bin, and shut down when the bin is full. In most cases, the bin is simply an insulated enclosure with a tough plastic or vinyl liner, refrigerated only by the ice it contains. A drain removes melted ice water. Smaller ice makers may also be placed on top of a small bin. Such units are equipped with a mechanical drive to dispense small amounts of ice on demand. These small units are often coupled with beverage dispensing systems accessible to the consumer.

Figure 14 Walk-in merchandiser.

Refrigeration Rack Systems

Although many refrigerated retail merchandising systems are self-contained and operate independently, supermarkets generally use centralized refrigeration equipment to serve all the open and closed display cases. Rack systems use multiple compressors connected to one or two common refrigerant circuits, staged as necessary to accommodate the changing loads. For efficiency, the medium- and high-temperature load circuit is separated from the low-temperature load circuit. These systems allow refrigeration to continue in some areas while other areas are in defrost. Each refrigerated fixture that requires a higher temperature than the coldest fixture on its circuit is fitted with an evaporator pressure regulator. Advanced controllers stage compressors and monitor system performance continuously.

Figure Credit: Courtesy of Zero Zone, Inc.

Most retail-sized ice machines feature air-cooled condensers, but water-cooled models are available. Self-contained units can usually be placed in ambient temperatures between 40°F and 115°F (4°C to 46°C). Ambient temperature can significantly affect the speed at which ice is made. The temperature of the supply water can also impact production speed because warmer water requires more cooling before ice formation can begin.

Care must be taken when locating air-cooled models to prevent recirculation of condenser air. Larger ice makers use remote condensing units to remove rejected heat from the conditioned space. For obvious reasons, most ice makers are not designed to operate in environments that approach or drop below freezing. If they are shut down during the winter but left in an area prone to freezing, the water supply must be turned off and water drained from all internal components.

2.3.1 Cubed-Ice Machines

Cubed-ice machines (*Figure 16*) generally range in capacity from 65–2,000 pounds (30–910 kg) of ice per 24-hour period. Storage bins mated with ice machines commonly store between 120 pounds and 1,500 pounds (55–680 kg) of ice. The selection of cuber and bin combination is based on customer requirements. Applications in which large quantities of ice are used at once versus those with steady consumption throughout the day would lead a buyer to different cuber/bin combinations.

Cube sizes and shapes vary significantly, depending on the customer's needs. Although the machines themselves operate in the same way, the grid where ice is formed is changed to produce the desired shape and size. Although sizes vary by manufacturer, a half-dice cube may be $\frac{3}{8}" \times \frac{7}{8}" \times \frac{7}{8}"$ deep; a dice cube would be $\frac{7}{8}" \times \frac{7}{8}" \times \frac{7}{8}"$ deep; and a full cube would

Figure 16 Cubed-ice machine with bin.

High-Quality Ice

Water quality is important to the quality of ice. Aerated ice and ice with a high mineral content may appear cloudy and unappealing. The entering water supply should be filtered and potentially conditioned to help avoid these problems.

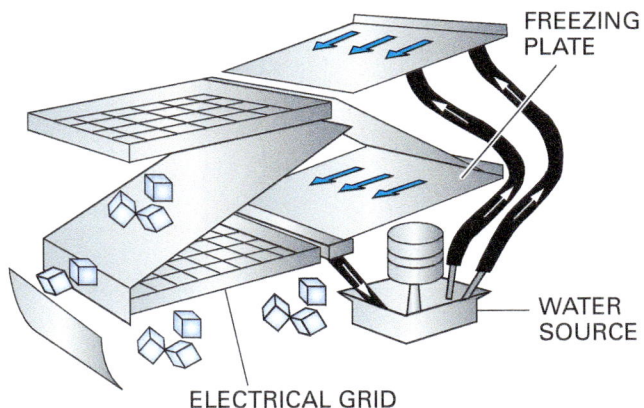

Figure 17 Cubed-ice maker with an electrical cutting grid.

be $1\frac{1}{8}" \times 1\frac{1}{8}" \times \frac{7}{8}"$. Other sizes and shapes are also available, such as nuggets, tubular nuggets, and even octagons. Metric cube-ice machines, such as those sold in the United Kingdom, can produce various sizes of dice cubes between 26 mm × 30 mm × 29 mm deep and 39 mm × 42 mm × 43 mm deep.

Cubed-ice machines have four basic modes of operation: fill, freeze, harvest, and drain-down. In the fill mode, new water is supplied for the next batch of cubes. In the freeze mode, water applied to the evaporator freezes. When the ice has reached a predetermined size, the evaporator is defrosted to quickly loosen the ice from the evaporator, melting only a small portion of the ice in the process. The harvest mode is like a typical defrost mode in other refrigeration applications and heat pumps. Drain-down, or purge, rids the machine of minerals left behind when the water freezes.

Cubed ice should also be very hard, a result of being sufficiently subcooled before harvesting. Ice that is harvested too early will be smaller and melt quicker.

Some units harvest the ice in a single solid sheet, which falls from the evaporator onto an electrical wire grid. The grid is heated and the wires melt through the ice in knife-like fashion, separating the ice sheet into cubes of various sizes, such as 1" × 1" (25 mm × 25 mm). *Figure 17* provides an example of this design. After the cubes are cut by the wire grid, they fall into the storage bin.

The other popular design (*Figure 18*) either sprays water into cube-shaped chambers permanently attached to a horizontal plate evaporator, or water is pumped to the top of a vertically positioned evaporator of the same design and allowed to cascade down by gravity. When the harvest mode begins, hot refrigerant gas (or possibly electric heat) is directed to the plate evaporator to loosen the cubes, which fall into the bin. The hot gas condensing in the coil quickly adds heat to the back of the evaporator, loosening the cubes. This is the same approach used to defrost a heat pump. With either style, ice cube size is an option when ordering.

An ice thickness sensor is typically used on a cubed-ice evaporator. It is set at a precise distance from the evaporator. As ice builds on the plate, the cascading water will eventually make contact with the sensor. When the sensor has been wetted continuously for 10 seconds, indicating the ice has reached proper thickness, the harvest mode begins. Water flow then ceases.

Cubed-ice makers often have a complex sequence of operation to ensure that a consistent ice cube is produced under a variety of operating conditions. Shutdown of the process is done by either a temperature sensor located near the top of the bin or by using a photoelectric eye that is blocked by the presence of ice. Both approaches can indicate a full bin.

2.3.2 Tube-Ice Machines

Tube ice is a popular form of bulk ice all over the world. This form of ice (*Figure 19*) is delivered as short sections of clear, cylindrical ice. The process for creating tube ice is similar to cubes, except the ice freezes inside long cylinders. When the freeze period ends, controlled by a timer, the cylinders are warmed briefly by a burst of hot refrigerant gas during the thaw cycle. The ice in contact with the cylinder wall surface melts, releasing the ice, which falls into the cutter mechanism.

For smaller applications, a series of rotary saw blades mounted on a common shaft cut the rods of ice into uniform lengths. The spacing of the blades determines the length of tube ice. One prominent tube-ice maker can be set make tube ice in three sizes between 1" and 1.5" (25–38 mm) long.

In industrial applications, the ice from multiple machines is collected by a screw conveyor and delivered to a large bin.

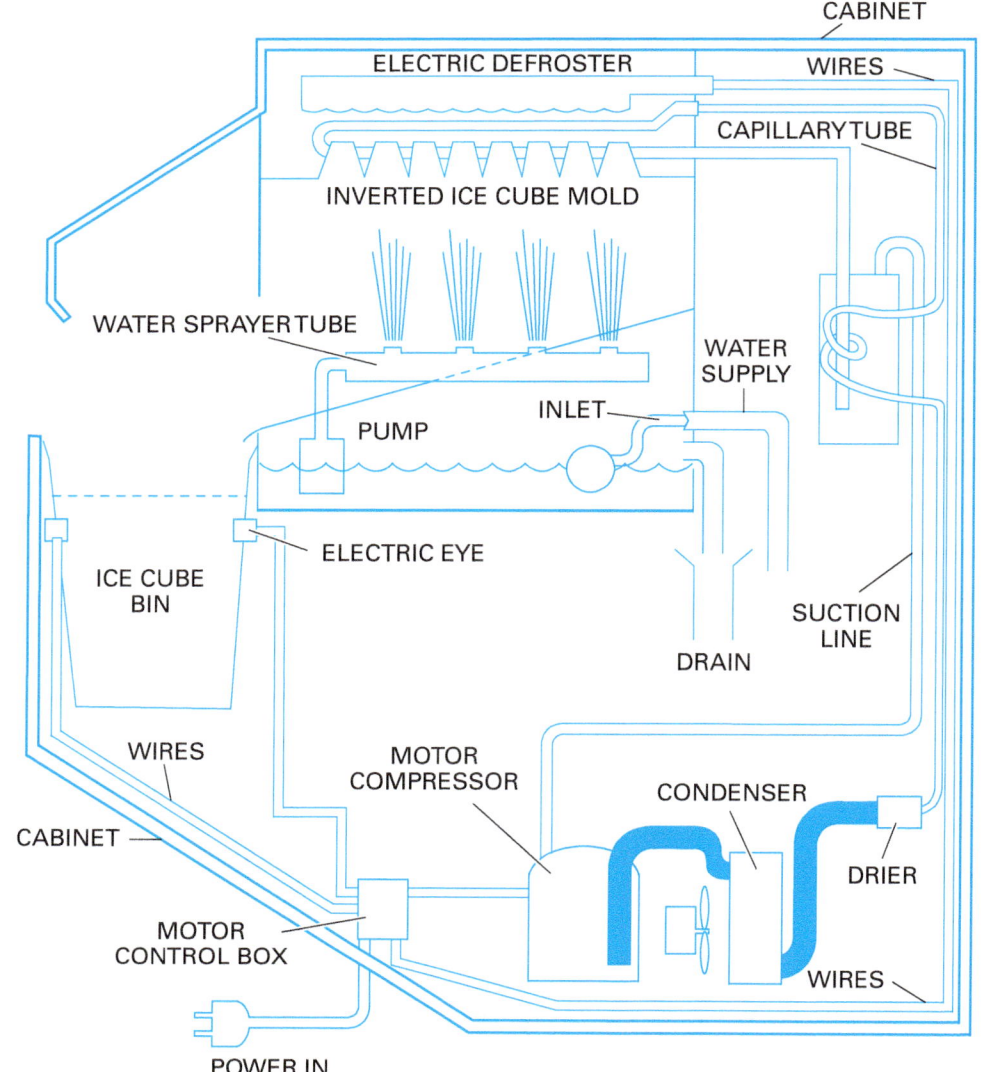

Figure 18 Cubed-ice maker with cube-shaped chambers on the evaporator plate.

Figure 19 Tube ice.

2.3.3 Flaked-Ice Machines

Ice flakers (*Figure 20*) operate in a continuous ice-making mode. Although they may look similar to cube machines, they are different in their operation. When the unit is started and the evaporator assembly reaches the proper temperature for freezing, it continues to operate until the bin is full.

The freezing assembly for flaked-ice makers (*Figure 21*) consists of an insulated vertical cylinder.

The outer walls of the cylinder form the evaporator, which is filled with boiling, low-temperature refrigerant. Ice forms on the cylinder's inner walls, which are continuously bathed with a constant supply of water. An auger, driven by a constantly and slow turning gear motor drive, scrapes the ice from the walls of the cylinder. The flaked ice is pushed out of the top of the cylinder by the auger and falls into the bin.

As is the case with cubed-ice machines, flaked-ice units are now generally designed around HFC-134a and HFC-404A refrigerants.

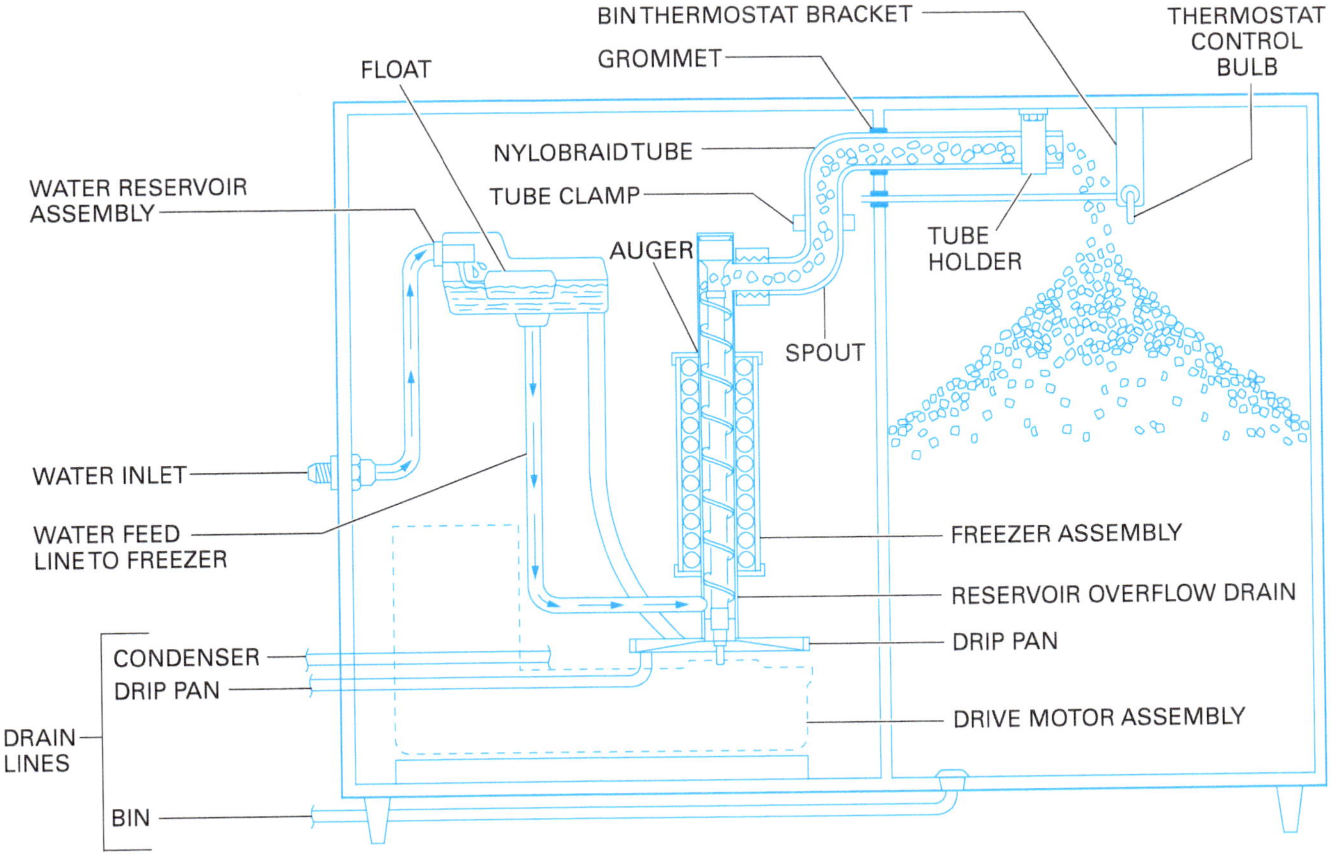

Figure 20 Flaked-ice machine with bin.

Figure 21 Flaked-ice machine operation.

Additional Resources

Refrigeration and Air Conditioning: An Introduction to HVAC/R, Larry Jeffus. Current Edition. New York, NY: Pearson.

2.0.0 Section Review

1. The box-temperature range for ice merchandisers is typically _____.
    a. –10°F to 5°F (–23°C to –15°C)
    b. 5°F to 15°F (–15°C to –9°C)
    c. 10°F to 20°F (–12°C to –7°C)
    d. 20°F to 25°F (–7°C to –4°C)

2. What type of metering device is typically used with walk-in coolers?
    a. Thermostatic expansion valve (TEV)
    b. Capillary tube
    c. Metering piston
    d. Bypass piston

3. The harvest mode of some cubed-ice machines resembles the _____.
    a. heating mode of a heat pump
    b. purge cycle of a flaked-ice machine
    c. cooling mode of a heat pump
    d. defrost mode of a heat pump

SECTION THREE

3.0.0 RETAIL REFRIGERATION SYSTEM COMPONENTS

Objective

Identify and describe components related to the operation of retail refrigeration systems.
a. Identify and describe the primary components used in retail refrigeration applications.
b. Identify and describe secondary components and accessories used in retail refrigeration applications.
c. Identify and describe common refrigeration system controls.

Trade Terms

Slinger ring: A ring attached to the outer edge of the fan that picks up water as the fan rotates and slings it onto the condenser coil.

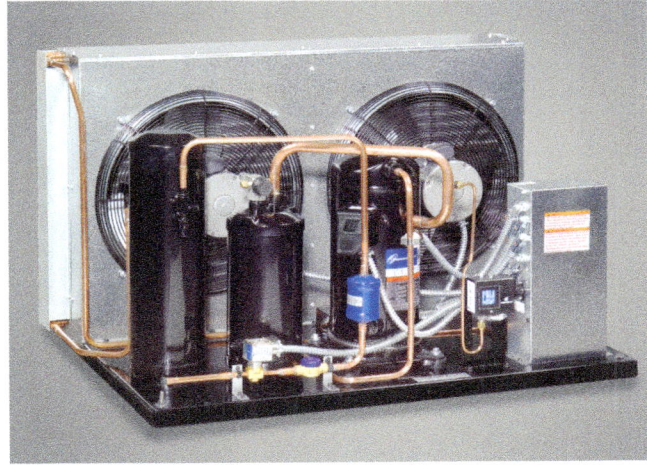

Figure 22 Refrigeration condensing unit.

Many of the components found in retail refrigeration systems are similar or identical to components found in comfort cooling systems. Because retail refrigeration equipment operates at lower temperatures than comfort cooling systems, the components must accommodate that difference. For example, compressors must be rated for refrigeration duty, which means they must be able to operate reliably at higher compression ratios.

Different refrigerants are often used. Refrigerants that are ideal for comfort cooling may not be suited for refrigeration applications. In comfort cooling, fixed-orifice metering devices such as capillary tubes can provide adequate refrigerant flow to the evaporator. In refrigeration systems, capillary tubes cannot provide good refrigerant flow control over the wider temperature ranges in which refrigeration units must operate. For this reason, expansion valves, with their modulating control of refrigerant flow, are more commonly used for retail fixtures.

3.1.0 Primary Refrigeration Cycle Components

Many retail refrigeration applications use air-cooled condensing units (*Figure 22*). This type of unit is a package assembled on a single chassis containing the compressor, compressor-start components, condenser and condenser fan(s), refrigerant receiver (if equipped), and a variety of controls and valves.

These units can be used in self-contained refrigeration fixtures, or they can be remotely installed and piped to the fixture. Used indoors, self-contained fixtures reject heat from the box, as well as heat of compression, to the space in which they are installed. This additional heat can be a significant comfort problem if not taken into consideration when loads for the comfort cooling system are calculated.

Although more costly to install, remote condensing units installed outdoors have the advantage of rejecting heat outside of the conditioned space. Otherwise, all heat rejected from the cabinet, including the heat of compression, must be transferred outdoors by the air conditioning system.

On the other hand, the heat from indoor units can benefit comfort in the winter. In most cases, remote installation is reserved for larger applications such as walk-in coolers or freezers, as well as larger ice machines. In some installations, a single condensing unit serves multiple evaporators or fixtures.

3.1.1 Compressors

For retail refrigeration use, hermetic and small semi-hermetic reciprocating compressors (*Figure 23*) have long dominated the market. However, scroll compressors (*Figure 24*) have also made their way onto the scene, with design changes that improve their suitability for reduced-temperature applications.

Early attempts to use scroll compressors in refrigeration applications resulted in overheated and failed compressors because of high

(A) SEMI-HERMETIC COMPRESSOR

> ### Improving Energy Efficiency
> **GOING GREEN**
>
> New models of retail refrigeration equipment use scroll compressors to improve energy efficiency. Scroll compressors have no pistons to compress gas, so they achieve 100-percent volumetric efficiency, which provides reduced energy costs in many applications. Re-expansion losses that typically occur in reciprocating models are eliminated. Similarly, losses at valves are eliminated, because suction valves are not required.

(B) HERMETIC SCROLL COMPRESSOR

Figure 23 Semi-hermetic and hermetic compressors.

Figure 24 Digital scroll compressor.

compression ratios and the low density of the returning refrigerant vapor. However, several manufacturers have now developed scroll compressors for the refrigeration market.

For low-temperature applications, many scroll compressors are fitted with a liquid or vapor injection device, designed to provide for improved internal cooling and reduced refrigerant discharge temperatures. Many scroll compressors used in refrigeration are not equipped with internal pressure relief valves. Therefore, providing reliable high-pressure safety controls is a must. Control systems often use a discharge-line thermostat to help prevent compressor damage or failure from overheating. They are generally set to open at 250°F to 260°F (121°C to 127°C).

Systems that use hermetic reciprocating compressors can be fitted with additional refrigerant flow-control components for capacity control. Semi-hermetic compressors with multiple cylinders can be used with integral capacity control to relieve compression in one or more cylinders. Scroll compressors are now available with internal capacity controls that modulate compressor capacity in small increments between 10 and 100 percent. This reduces energy consumption and provides very stable temperature control.

3.1.2 Condensers

Most retail refrigeration systems use air-cooled condensers because of the ease of installation, mobility, and simplicity of the design. The condenser can be mounted on the condensing unit, within inches of the compressor, or remotely installed. Aluminum fin or copper tube coils are typical (*Figure 25*).

Proper condensing pressures and temperatures must be maintained for the refrigerant circuit to operate effectively. As condensing temperatures rise, capacity and efficiency are reduced. Some retail systems operate in a conditioned environment, but many do not. This must be considered in final equipment selection. Even when placed in a conditioned space such as a floral showroom or convenience store, it should not be assumed that the condensers will remain clean. For example, when used near cooking apparatus, air-cooled condensers can foul rapidly from airborne grease depositing on the coil fins. Regular condenser cleaning must be an integral part of retail refrigeration system maintenance.

Reduced condensing temperatures can provide increased energy efficiency. However, low condensing pressures can result in a loss of performance because of insufficient refrigerant flow through the expansion device. This condition often occurs in winter for condensers or condensing units that are exposed to outdoor conditions.

Either air-side or refrigerant-side controls can be used to maintain minimum required condensing pressures. On the air side, the condenser fan(s) can be cycled on and off by pressure, or the condenser fan speed can be modulated for a more stable result. On the refrigerant side, controls installed in the refrigeration circuit can be used to control the active surface of the condenser by flooding a portion of its internal volume with liquid refrigerant.

Some commercial refrigeration installations may rely on water-cooled condensers. Manufacturers offer condensers suitable for high-, medium-, and low-temperature applications. The condenser may be integral to the refrigeration package, placed in another space, or even located outdoors. The more distant the location, the greater amount of refrigerant piping required. The cooling water circuit may connect to a separate air-cooled heat exchanger (dry cooler) or draw from another water source.

3.1.3 Evaporators

Many types and styles of evaporators are necessary in retail refrigeration systems. Although most are very typical air-cooling models, built much like comfort cooling coils, evaporators used for liquid cooling and ice making can be unique in their construction and design.

Forced-draft, finned-tube evaporators are most commonly used in smaller retail refrigeration fixtures and systems. For larger applications, such as walk-in freezers or coolers, a unit like the one in *Figure 26* is used. The evaporator coil is generally a packaged assembly consisting of the coil, evaporator fan, metering device, and defrost heaters (when needed). These are often referred to as *unitary coolers*, or *unit coolers*. Unlike most comfort cooling applications that require blower wheels to overcome resistance to airflow in duct the system, these units generally use free-blow (no ductwork) propeller-type fans made of plastic or aluminum.

As in comfort cooling, condensation occurs on the evaporator coil surface when coil temperatures are below the dew point of the enclosure. Condensate must be collected and drained out. For portability of small refrigerated fixtures, condensate is often collected in a small heated pan installed in the fixture cabinet. An electric

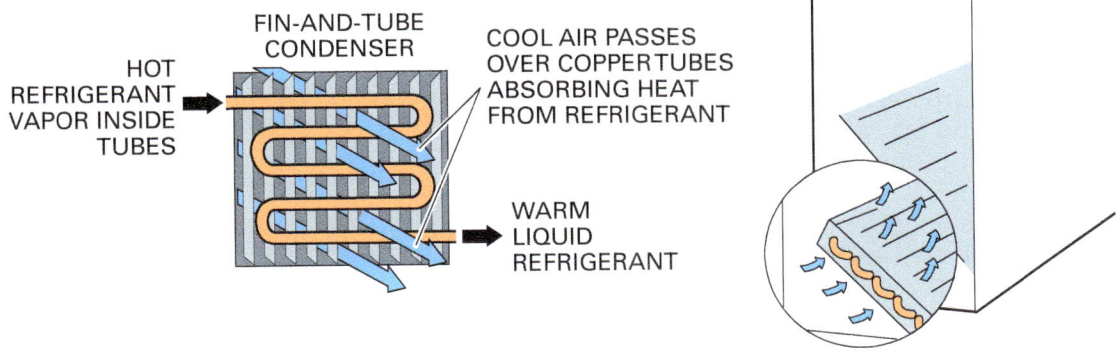

Figure 25 Air-cooled fin-and-tube condenser.

Figure 26 Walk-in refrigeration unit cooler.

heating element embedded in the pan re-evaporates the condensate back to the air. For coils that operate below freezing, the condensate collection pan under the coil assembly generally uses an electric heater to prevent freezing and blockage of condensate flow. The condensate line itself, where it passes through the refrigerated space, also requires freeze protection. Heat tape wrapped around the outside of the line prevents freezing.

Some units may also use this condensate to advantage. When the condensate is sent to a pan under the condenser coil, it can be picked up by the condenser fan blades fitted with a **slinger ring**. This slings water droplets onto the surface of the condenser coil to enhance the condensing process. Moisture is evaporated back to the surrounding air as a result. However, this feature has become uncommon.

Maintaining humidity in the refrigerated space is important, depending on the intended use of a refrigerated fixture. If storing fresh produce, flowers, and similar products, humidity levels must remain high to prevent dehydration of the product. In other cases, especially when the stored product is sealed, low humidity levels may be needed to prevent fog from forming. Proper selection of the evaporator is essential to maintain the desired condition.

To maintain higher humidity levels, evaporator coils must operate at a narrow ΔT between the desired storage temperature and the refrigerant (4–8°F or 2–4°C, for example). To meet this requirement, coil assemblies generally have large surface areas with high airflow rates. Conversely, when low humidity levels are desirable, the evaporator coil is often smaller and operates with low airflow values. In these applications, the air-to-refrigerant ΔT may be as high as 20–30°F (11–17°C). Unitary evaporator coils that operate at 30°F (–1°C) and below generally form frost and will require some method of frost removal.

Many commercial cubed-ice machines use evaporator coils similar to simple plate evaporators. A length of tubing is firmly attached to a plate on one side, while the opposite side is used to produce the desired cube size and shape. A variety of cube shapes and sizes are available.

Figure 27 shows one such evaporator assembly in which the water flows over the vertical plates to form individual cubes. In other units, the evaporator assembly is a single-sided vertical grid cell plate. Water cascades down the face of the grid, building ice cubes in the cells.

3.1.4 Expansion (Metering) Devices

Put briefly, an expansion (metering) device controls the flow of refrigerant into the evaporator, which provides the pressure drop that lowers the boiling point of the liquid to the desired temperature. It also meters the flow of liquid refrigerant into the evaporator to match the rate of refrigerant boiling into vapor. Typically, only refrigerant va-

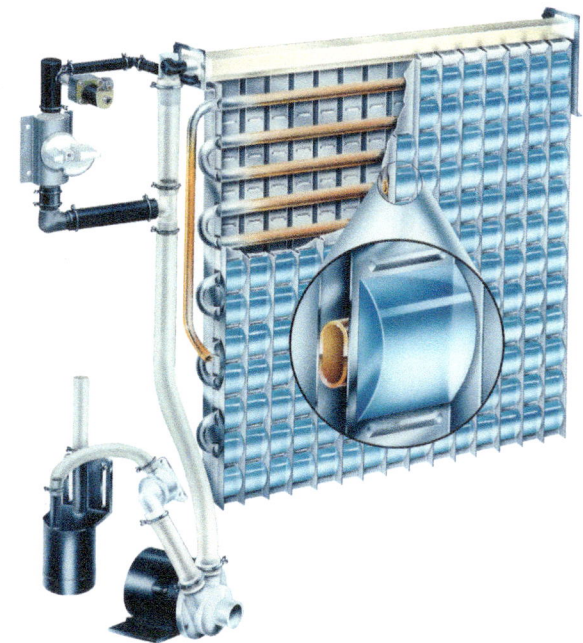

Figure 27 Commercial cubed-ice machine evaporator.

por should leave the evaporator. At the same time, most coil internal surfaces should be in contact with boiling liquid for maximum capacity.

> **NOTE:** Metering devices are covered in greater detail in NCCER Module 03303, *Metering Devices*.

Fixed metering devices such as capillary tubes (*Figure 28*) are available, but they are not as widely used in retail refrigeration because they cannot effectively meter refrigerant in many load conditions and may overfeed or underfeed refrigerant. They are also very sensitive to the amount of refrigerant charge in the system. Very small changes in the refrigerant charge have a significant effect on the operating superheat. For this reason, TEVs are widely used in retail refrigeration applications.

The TEV in *Figure 29* controls the amount of refrigerant flow using the superheat temperature sensed at the evaporator outlet. The device provides a flow rate to match the evaporator load. As the evaporator load changes, the TEV responds by opening or closing. TEVs are not nearly as charge-sensitive as fixed devices. Using TEVs ensures that a specific superheat value can be maintained effectively under a wide variety of conditions, even if the refrigerant charge is not precise.

Electronic expansion valves (EEVs), shown in *Figure 30*, are gaining popularity in the industry because they are not refrigerant-specific. By installing different orifice sizes, each valve body can be used to cover many capacity ranges. They function through a wide range of loads and are remotely adjustable. EEVs offer precise operation and flexibility at extremely low superheat values, allowing maximum use of the evaporator's heat transfer surface. However, they are presently used in more critical applications than most common retail units represent.

These valves can be very compact, offering thousands of possible positions, called *steps*. This term is derived from using a stepper motor to open the valve in very small, constant distances. While older EEVs were equipped with motors offering a few hundred steps or less, over 6,000 steps are now possible. This provides very precise refrigerant metering. Valve positioning is controlled by inputs from one or more temperature sensors and microprocessor-based controls. Many valves can change positions at a rate of 200 steps per second.

Figure 29 Thermostatic expansion valve.

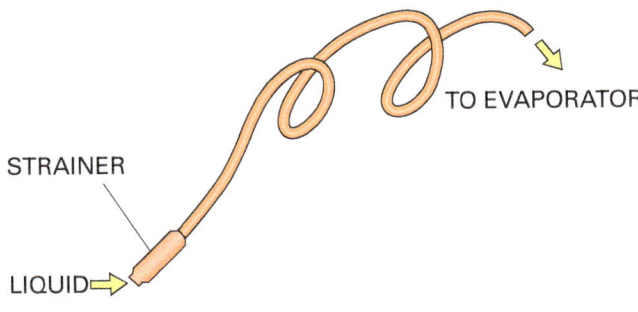

Figure 28 Capillary-tube metering device.

Figure 30 Electronically controlled expansion valves (EEVs).

EEV Superheat Control

There are two basic approaches to controlling superheat with EEVs. Superheat is all about pressure-temperature relationships. These relationships are specific to each refrigerant, so programming the valve requires a pressure transducer at the coil outlet, a temperature sensor, and the input of the proper P-T chart into the software of the microprocessor control. This is considered the most accurate method, since it is directly related to the thermodynamics involved.

Another simpler approach uses two less-expensive temperature sensors. This method discounts the pressure-temperature relationship completely. One sensor is located at the evaporator inlet, and the other at the evaporator outlet. Software compares the two and assumes any rise in temperature through the coil is superheat. This method is not as precise and may require that a slightly higher superheat value be programmed in as a safety net. One problem that arises relates to the position of the evaporator inlet sensor: to remain accurate, it must be located in a position that is always exposed to refrigerant in its liquid state. However, this two-temperature method can be used with any refrigerant without reprogramming the valve for a specific compound or blend.

Systems with traditional TEVs are often equipped with a solenoid valve installed in the liquid line upstream of the TEV. During the compressor off cycle, the solenoid valve closes to prevent refrigerant migration to the evaporator. Once in the evaporator and exposed to the cold temperatures of the enclosure, the refrigerant condenses. It can then return to the compressor as floodback and cause damage. Since EEVs offer the advantage of tight closure during the off cycle, they eliminate the need for a solenoid valve. Expect to see more use of EEVs as the industry strives to increase the energy efficiency of retail refrigeration products.

3.2.0 Secondary Refrigeration Cycle Components and Accessories

A variety of supporting components are used in refrigeration systems. Although some of the components discussed here may be used in comfort cooling applications, they are more common in refrigeration applications.

3.2.1 Receivers

Receivers provide a place where liquid refrigerant can be stored in refrigeration units. Because variations in load conditions can cause rapid changes in refrigeration load, it is important to have an adequate supply of refrigerant to satisfy the needs of the evaporator under all load conditions.

Many small receivers use a back-seating shutoff valve on the outlet. Receivers are also equipped with a dip tube, drawing liquid refrigerant from the bottom to ensure that pure liquid is sent to the metering device. The liquid refrigerant level inside will change in relation to the system flow rate at any given time.

Receivers are used only on systems with a TEV or other modulating metering device. They are of no real value in systems with a fixed metering device.

3.2.2 Accumulators

Because refrigeration systems tend toward lower superheat values, the chance of liquid refrigerant leaving the evaporator becomes higher. The accumulator (*Figure 31*) is essentially a trap designed to hold liquid refrigerant in the system suction line.

The accumulator uses a U-shaped dip tube to separate liquid from vapor. The compressor draws refrigerant vapor from one end of the U-tube. The other end is contained within and terminates near the top of the vessel to ensure

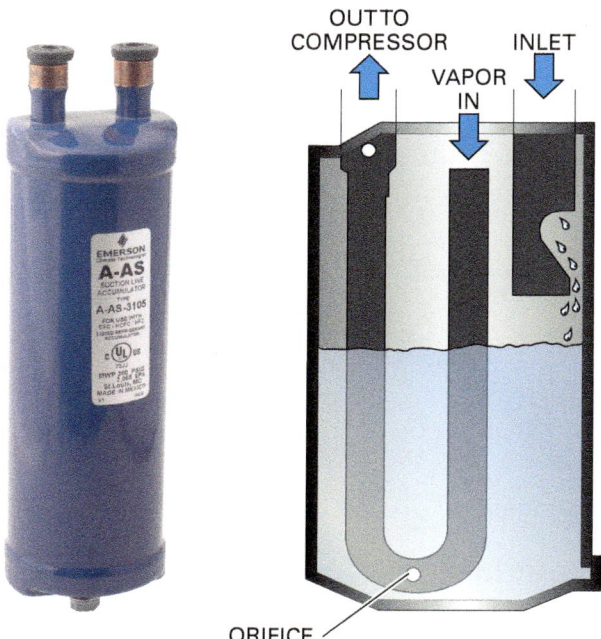

Figure 31 Suction accumulator.

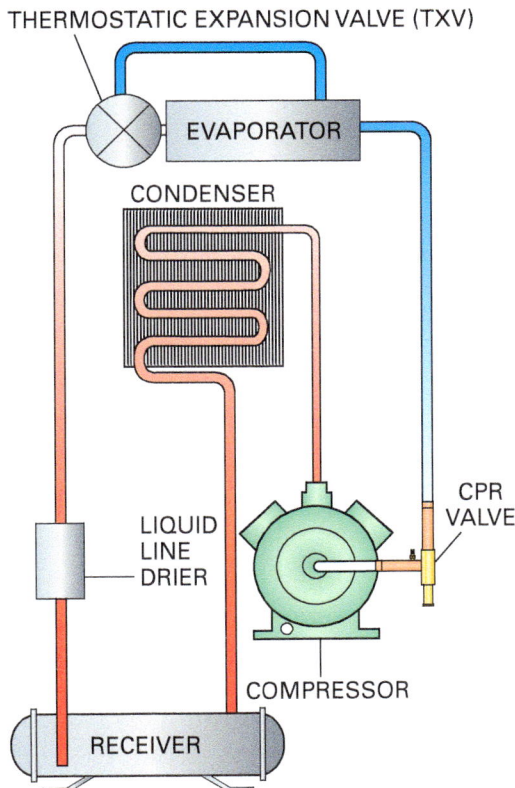

Figure 32 Crankcase pressure regulating valve.

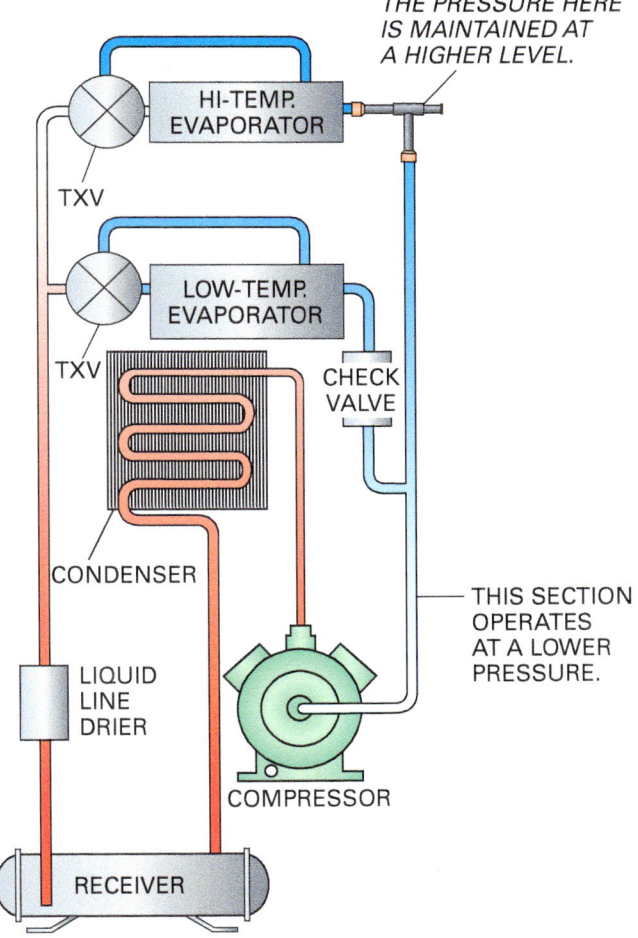

that only vapor is drawn from the accumulator. A small orifice at the lowest point in the U-tube allows small amounts of trapped compressor oil and liquid refrigerant to return to the compressor without damaging the compressor. Accumulators are also used in some comfort cooling applications and heat pumps.

3.2.3 Crankcase Pressure Regulators

To prevent overload of the compressor motor, crankcase pressure regulating (CPR) valves are installed in the suction line just upstream of the compressor, as shown in *Figure 32*. The valve controls the maximum pressure at the compressor suction inlet.

Refrigeration equipment must often be started up when the entire fixture is at room or ambient temperature, even though the fixture may be designed to maintain –10°F (–23°C). This load is well outside of the compressor's normal operating range for the application. This same extreme load can occur following a defrost cycle, especially in systems that use hot-gas defrost. The CPR valve ensures that refrigerant vapor is fed to the compressor in a reasonable volume that prevents it from being overloaded.

The valve setpoint is adjusted by changing the internal spring pressure so that the valve moves

Figure 33 Evaporator pressure regulator.

towards its closed position on a rise of suction pressure. CPR valves are very simple to set by starting the system in a known-overload condition while checking compressor current with an ammeter. The valve spring can then be adjusted to a lower pressure setting while watching the actual current draw of the compressor. The valve is properly set when the current falls to, or drops very slightly below, the rated load amperage for the compressor.

3.2.4 Evaporator Pressure Regulating Valves

Some larger retail refrigeration systems use a single compressor to maintain temperature in multiple compartments or fixtures that have individual temperature settings. The evaporator pressure regulating (EPR) valve allows evaporators serving higher temperature areas to maintain proper refrigerant pressures for their applications, therefore maintaining the saturated suction temperature of the refrigerant. One example of such an installation is shown in *Figure 33*.

The pressure in the common suction line serving all of the fixtures must be low enough to properly refrigerate the coldest fixture. When the pressure rises above the set point, EPRs mounted on each of the evaporators operating at higher temperatures respond to the inlet pressure of the evaporator to open wider, allowing more refrigerant to flow through the coil. When evaporator pressure falls, the valve moves toward its closed position. The evaporator will maintain the proper pressure that corresponds to the designated saturated suction temperature.

3.2.5 Refrigerant-Side Head Pressure Control

Refrigeration units installed in locations subject to low ambient temperatures generally require a method of maintaining head pressure. Cold air entering the condenser causes a reduced condensing pressure. If the pressure falls too low, the pressure drop through the metering device may not be sufficient to ensure adequate refrigerant is fed to the evaporator.

A head-pressure control valve (*Figure 34*) is the most accurate and stable means of ensuring that an adequate head pressure is maintained. Although condenser fan cycling controls are often used for this function, condensing pressures rise and fall through the control's range. As a result, the constantly changing condensing pressure causes the flow rate of the metering device to be unstable. The refrigerant-side head-pressure control maintains condensing pressures in a far more stable manner.

Figure 35 demonstrates the typical installation of this valve. As the air temperature across the condenser falls, a corresponding decrease in the condensing pressure occurs. When the pressure falls to the valve's set point, a portion of the discharge gas is routed through port B of the valve

Figure 34 Refrigerant-side, head-pressure control valve.

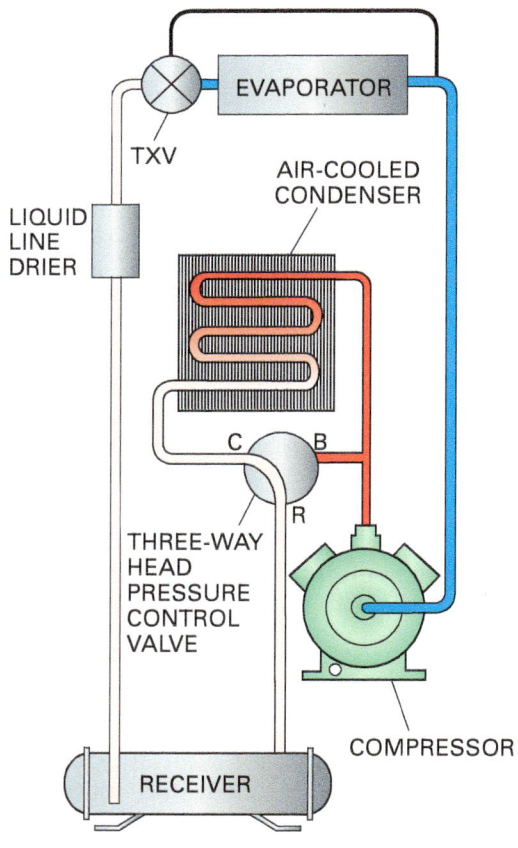

(A) NORMAL OPERATION

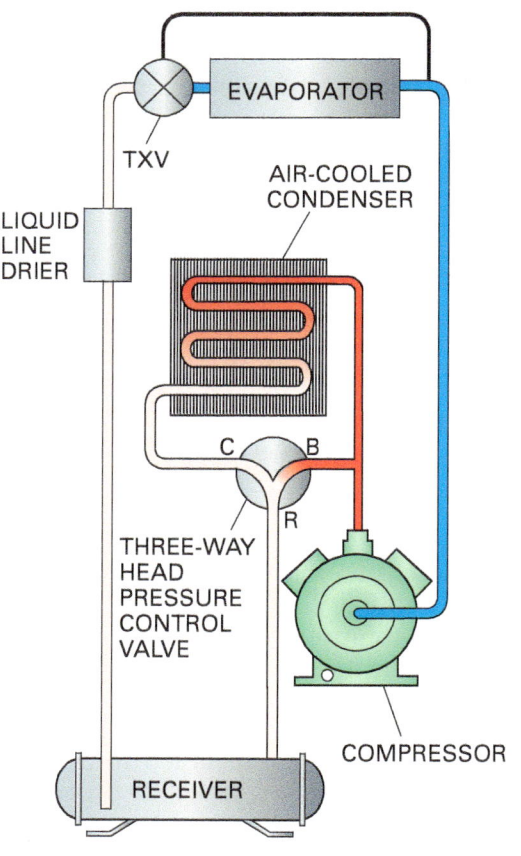

(B) LOW HEAD PRESSURE

Figure 35 Head-pressure control valve installation position.

Pilot-Operated EPRs

Normally, EPRs are installed as close to the evaporator as possible. If they must be installed away from the evaporator or downstream from a riser, a pilot-operated EPR like this one is typically used. The pilot line is connected to the evaporator to monitor the pressure and position the valve. Pilot-operated valves are also used to operate larger valves that require extra power to move the internal components against pressure differences.

Figure Credit: Courtesy of Emerson Climate Technologies

and into the liquid line to the receiver through port R. Liquid from the condenser mixes with the vapor at the valve. The action causes a rise in pressure at the condenser outlet, which impedes the flow of refrigerant liquid out of the condenser. Condensed liquid begins to back up, effectively reducing the available condensing surface and raising the head pressure.

Head-pressure control valves are often non-adjustable. To effectively operate a head pressure control valve, the system must include a receiver, since this device requires a larger volume of refrigerant during low-ambient conditions than is needed during periods of high load and high condensing pressures. Additional refrigerant, held in reserve in the receiver, is needed to partially flood the condenser coil as the valve modulates to maintain appropriate pressure.

This specific type of valve is usually refrigerant-specific and is factory preset at an appropriate value for the application and refrigerant used. It has a sensing dome, as shown in *Figure 34*, which must be exposed to the ambient temperature. Other models and styles that can be adjusted in the field are available. Head pressure control valves are generally not a single valve, but two separate valves that operate independently. Since retail refrigeration systems are generally compact, the type shown here is far more popular.

3.3.0 Common Refrigeration System Controls

Some typical controls for retail refrigeration systems include thermostats, pressure controls, time delay relays, and solenoid valves. Again, while these are similar to comfort cooling controls, they are often applied in different ways in retail refrigeration situations and may not have the same appearance as commercial models.

3.3.1 Thermostats

Thermostats sense temperature at a given point and respond with an action. While many different styles are available, some of them are designed to be very equipment-specific. Electronic temperature controls are popular in the retail refrigeration industry. They offer reliable operation and a wide temperature range, allowing identical controls to be installed on high-, medium-, or low-temperature systems. Installing a small temperature sensor and routing flexible wiring back to the control is much simpler than routing a long capillary tube through complex cabinet components, as required by older remote-bulb thermostats.

A typical thermostat features a setpoint range of –30°F to 130°F (–34°C to 54°C) and an adjustable differential of 1°F to 30°F (0.6°C to 17°C). The range can be modified by using different temperature sensors. An LED indicator is often used to provide status. Normally open/normally closed operation is field selectable.

Some applications for thermostats in refrigeration include the following:

- Controlling the refrigeration circuit and compressor to maintain desired fixture temperature
- Providing a low-temperature safety control for the refrigeration circuit
- Delaying the restart of evaporator fans following a defrost cycle. This allows refrigeration equipment to operate long enough to refreeze any remaining moisture on the evaporator coil before the fans start, preventing water from being blown off the coil.
- Cycling of the condenser fan(s) based on ambient temperature
- Terminating the defrost cycle because of evaporator coil or suction line temperature
- Shutting down an ice-making process in response to bin temperature

Thermostats for these applications are available with a wide variety of ranges and differentials. They can be adjustable or have a fixed setting chosen for specific applications.

Single-stage thermostats provide an action for a single temperature setting, while multi-stage thermostats respond to two or more temperature points. Those used as safety controls often must be reset manually. Manual-reset controls prevent equipment damage by keeping the equipment off until a service technician repairs the condition that caused the safety control to open.

3.3.2 Pressure Controls

Refrigeration systems require a variety of pressure controls. You should recall from other modules that pressure sensors and switches detect fluid pressures using a variety of methods. Older technology may use Bourdon tubes, which are curved, helical, or spiral-shaped, flattened tubes to detect pressure. They tend to straighten when pressure inside them increases. The end of the tube is mechanically connected to an indicator pointer or switch.

Most often, pressure-sensing devices use bellows or diaphragm switches (*Figure 36*). These adjustable switches provide high- and low-pressure service for cycling and safety limits. They can use a single pressure input, or accept two separate pressure inputs and react to the differential between the two. An oil-pressure safety switch is a good example of a differential-pressure switch.

Like thermostats, pressure switches perform a variety of functions and are available in different pressure ranges to suit the application.

Figure 36 Typical refrigeration pressure switch.

Some of the functions that pressure switches provide include the following:

- Controlling refrigeration equipment by the pressure sensed in the suction line. Since suction pressure is related to saturated suction temperature, pressure switches can provide start-stop control of refrigeration circuits (primarily those using fixed metering devices)
- Controlling one or more condenser fans by condensing pressure
- Shutting down the compressor because of excessively low pressures caused by problems such as loss of charge or evaporator fan failure
- Shutting down the compressor caused by a high-pressure condition resulting from condenser fan failure, refrigerant overcharge, or other problems
- Controlling pump-down cycles on systems equipped with a liquid-line solenoid valve. When the valve closes, the compressor continues to run until most of the refrigerant has been pumped out of the evaporator and suction pressure falls to the setting of the pressure switch.
- Monitoring the differential between the suction pressure and oil-pump discharge pressure to determine compressor net oil pressure

3.3.3 Time-Delay Relays

Time-delay relays are used to delay a switching action or ensure that a sequence of actions takes place at the proper time. Many are very compact, such as the one shown in *Figure 37*, but still have an adjustable time-delay range.

Most time-delay relays used in refrigeration applications can be specified as follows:

Figure 37 Time-delay relay.

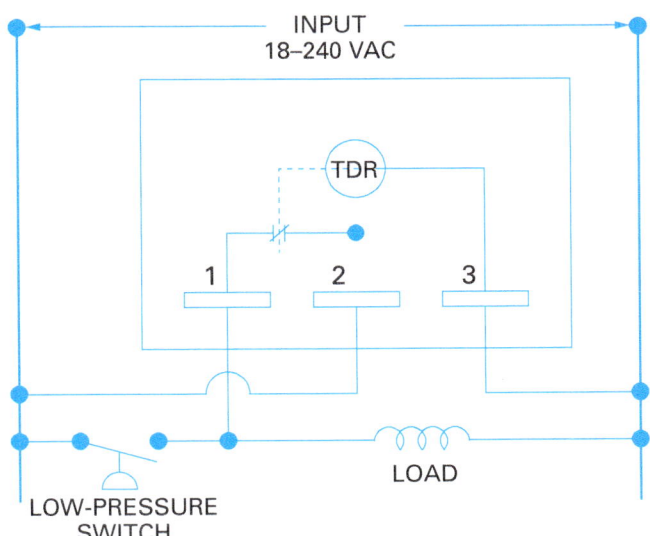

Figure 38 Bypass timer wiring.

- *Delay-on-make* – When energized, contacts remain in their existing position (usually closed) until the specified time period elapses, then they change position. These relays are often used to delay a step in a sequence of operation or to prove that a specific condition has existed for a predetermined period before allowing the system to respond.
- *Delay-on-break* – When de-energized, the relay contacts immediately change position (usually from closed to open) until the time period has elapsed. Then they return to the previous position (usually closed). Delay-on-break timers are often used as compressor short-cycle protectors by opening their contacts when de-energized, preventing the compressor from starting until the time-delay period (usually two to five minutes) has elapsed.
- *Bypass duty* – One frequent application for the bypass timer is to bypass a low-pressure switch for a short period of time after a compressor starts. Several factors (all normal) can cause a short-term period of low suction pressure when a compressor first starts. Briefly bypassing the low-pressure switch allows suction pressure to build to normal levels and prevents a nuisance shutdown of the compressor. Bypass-duty relays are usually of the delay-on-make style, but not in all cases.

The normally-closed switch contacts of the bypass timer (*Figure 38*) are placed in parallel with the low-pressure switch, providing an electrical path around the switch until the selected time period has elapsed. Timer contacts then open to allow the low-pressure switch to regain control when the system is in stable operation.

More sophisticated timers can provide multiple timing events or a repeated sequence of timed events for special applications, such as ice makers. Timed functions can also be built into printed circuit boards and microprocessor controls.

3.3.4 Solenoid Valves

Solenoid valves (*Figure 39*) are used to deny or permit the flow of either liquids or gases in refrigeration systems. The valve shown has extended fittings that allow the valve to be brazed into the line with less chance of overheating and damaging the valve body.

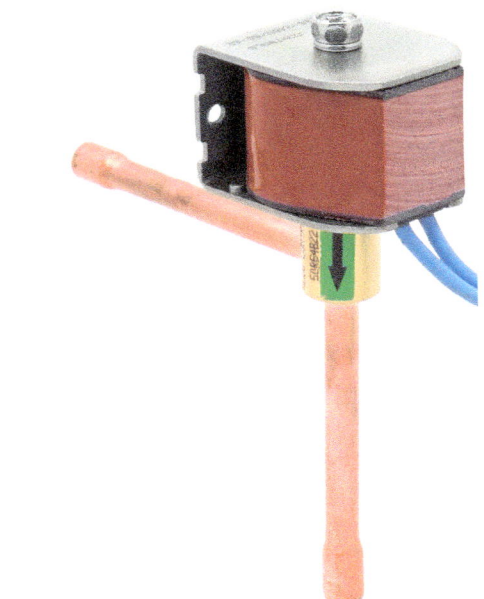

Figure 39 Solenoid valve.

Solenoid valves are selected by refrigerant flow rate and pressure drop through the valve. The resulting choice may also fit the system line size, but line size is not the primary factor.

Solenoid valves can be provided with a manual override stem, allowing the technician to temporarily open the valve manually when a coil fails or the valve sticks closed (*Figure 40*). Although most solenoid valves are normally closed and are energized to open, types that are normally open and close when energized are also available. The solenoid coil is typically replaceable and available in all commonly used voltages.

Applications for solenoid valves include the following:

- Provide flow control of liquid refrigerant to the metering device that is controlled by the fixture thermostat. This prevents refrigerant migration during the off cycle and enables a pump-down circuit on systems so equipped.
- Providing independent temperature control on systems with a single compressor serving multiple evaporators. The valves are used in the liquid line to each evaporator.
- Controlling liquid refrigerant flow to injection-cooling TEVs.
- Controlling the flow of hot gas refrigerant for defrost, capacity control, or low-ambient condenser circuiting.

Figure 40 Solenoid valve with manual override.

Solenoid-Valve Service Tool

One solenoid-valve manufacturer has developed a multi-function service tool for use with its products. The body of the tool is a powerful permanent magnet that simulates the coil of a solenoid valve. When placed over the plunger's enclosure tube, the magnet can open a valve even with a 300 psi (2,070 kPa) pressure differential across it. The body of the tool acts as a spanner for taking apart the valve body. It also has a manual stem adjustment tool for solenoid valves that can be manually opened.

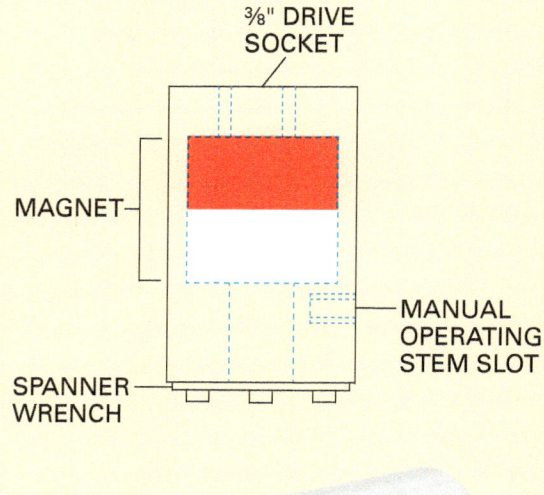

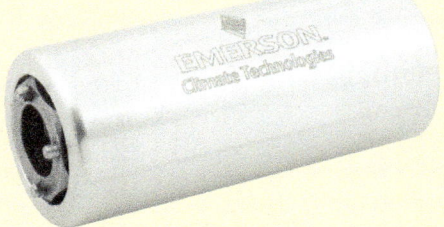

Figure Credit: Courtesy of Emerson Climate Technologies (Photo)

3.0.0 Section Review

1. What device is often installed with scroll compressors used in low-temperature applications to improve compressor reliability?

    a. Suction unloader
    b. Hot-gas bypass valve
    c. Liquid- or vapor-injection device
    d. Cylinder unloader

2. What is the purpose of the orifice in the U-tube of an accumulator?

    a. The return of oil or liquid refrigerant
    b. Pressure equalization
    c. Refrigerant separation
    d. Pressure relief

3. Delay-on-break time delays are often used to prevent _____.

    a. nuisance cycling of the low-pressure switch
    b. nuisance cycling of the high-pressure switch
    c. compressor short-cycling
    d. evaporator fan short-cycling after shutdown

SECTION FOUR

4.0.0 TROUBLESHOOTING RETAIL REFRIGERATION SYSTEMS

Objective

Explain the basic principles of maintaining and troubleshooting various retail refrigeration systems.
  a. Explain how to maintain and troubleshoot a typical reach-in freezer.
  b. Explain how to maintain and troubleshoot a typical cubed-ice machine.

Performance Tasks

2. Isolate a fault in refrigeration equipment or an ice machine.
3. Clean an ice machine.

Other than a difference in work environments, all basic troubleshooting skills previously presented for troubleshooting comfort systems apply. Troubleshooting and maintenance procedures for two different retail-refrigeration systems are examined: a reach-in freezer and a cubed-ice machine. Note that systems can differ widely, so specific sequences of operation and service information for a given system should always be consulted before beginning the troubleshooting process.

4.1.0 Troubleshooting and Maintenance of a Reach-In Freezer

For a typical reach-in freezer (*Figure 41*) with electric defrost, a lack of simple maintenance is one of the primary causes of performance problems. Although the periodic maintenance required for a reach-in freezer is not challenging, it is critical to proper operation. The most important of these tasks is to inspect and maintain the evaporator and condenser coils.

Depending on where the unit is located, the condenser coil can become fouled very quickly and cause high condensing pressures. In the early stages, this condition may only raise the unit's operating cost without affecting performance. But as fouling worsens, performance will begin to suffer.

Evaporator coils generally do not foul as quickly, but are subject to becoming clogged. The aluminum fins have the potential to rapidly corrode or deteriorate duew to the off-gassing of certain foods. The fins then become coated with aluminum oxide or a similar residue, reducing heat transfer.

Generally, refrigeration systems of this size have a relatively small refrigerant charge. Refrigerant operating pressures should not be checked during every maintenance visit, unless you suspect problems. Each time pressure gauges are used, a small loss of refrigerant occurs, so constant checking will eventually cause performance problems. Most motors on units of this size do not require lubrication, but they should be checked to ensure this is the case.

Before troubleshooting, review the manufacturer's data and normal operating conditions. For example, suppose the features of a typical freezer include the following:

- The unit power switch is located on the electrical box behind the lower louvered access panel. The condensing unit is also located in this area.
- The unit is equipped with an electronic temperature control that has no Off position. The electro-mechanical defrost timer and temperature control are located inside the same enclosure where the power switch is mounted.

Figure 41 Reach-in freezer.

- The defrost timer has been factory-preset for two defrost cycles per day at 6:00 a.m. and 10:00 p.m., with a 40-minute fail-safe defrost termination period. The defrost cycle is normally terminated by defrost thermostat located on the evaporator suction line. The three-wire thermostat will terminate defrost when a line temperature of 58°F (14°C) is reached. As refrigeration resumes, it will not allow the evaporator fans to operate until the line temperature has dropped below 32°F (0°C). This prevents condensate on the coil from being blown onto the product. Should the defrost thermostat fail to terminate defrost, the timer will terminate the cycle after 40 minutes.
- The electric defrost heaters are attached to the fins of the evaporator coil using spring clips.
- Fluorescent lamps provide interior lighting and can be turned off with a switch located inside the cabinet above the door.
- A fan switch is located at the top of the doors. The evaporator fans stop when the door is opened to prevent cold air from being blown out of the box.
- Low-wattage electric heaters are installed in the structure framework and in the perimeter of the doors to prevent freezing. The heaters are not energized until the box has reached operating temperature. The thermostat controlling their operation is also referred to as the alarm thermostat, because it can be wired to generate a visual or audible alarm when the temperature rises above 18°F (–8°C). The thermostat switches to its normal position, energizing the door heaters, when the box temperature has reached 0°F (–18°C).
- The refrigeration circuit is equipped with a TEV metering device and a refrigerant receiver is installed in the liquid line.
- A semi-hermetic compressor is used, and the system refrigerant is HFC-404A. To avoid overloading the compressor, a crankcase-pressure regulating valve is installed and factory set at 10 psig (69 kPag).
- The manufacturer indicates the standard ambient temperature should be 75°F (24°C) and lists the expected head pressure as 230 psig to 240 psig (1,586–1,655 kPag). The expected suction pressure is 5 psig to 6 psig (34–41 kPag) with the box temperature stable at –10°F (–23°C). The design air-to-refrigerant ΔT is 15°F (8.3°C), resulting in an expected evaporator air discharge temperature of –25°F (–32°C).

This critical information allows technicians to recognize proper and improper operation. Before beginning any troubleshooting procedure, understand the sequence of operation and indicators of proper and improper operation. Only then will you be able to determine when system performance is not as it should be. The wiring diagram in *Figure 42* and the troubleshooting chart examples in *Figure 43A* and *Figure 43B* are helpful troubleshooting tools.

Although most troubleshooting charts do not cover every potential situation, it will help you quickly diagnose many problems you are likely to encounter. Your ability to locate and repair unusual problems will increase with hands-on experience.

Silver and Biological Films

Silver is being used in products at an increasing rate to assist in the prevention of biological growth. Even clothing items can now be impregnated with silver for this purpose. Many ice machine manufacturers are adding elemental silver in the form of silver ions to plastic components such as ice bin walls to delay biological growth. One such product is known as AlphaSan®. Such additives, compounded directly into the part during manufacturing, release silver ions slowly to the surface of the plastic. Not only do they reduce or prevent biological growth, but the additives also help avoid discoloration of the part over time. They do not impact the taste and flavor of the ice, but the presence of these products is absolutely not a replacement for proper periodic cleaning.

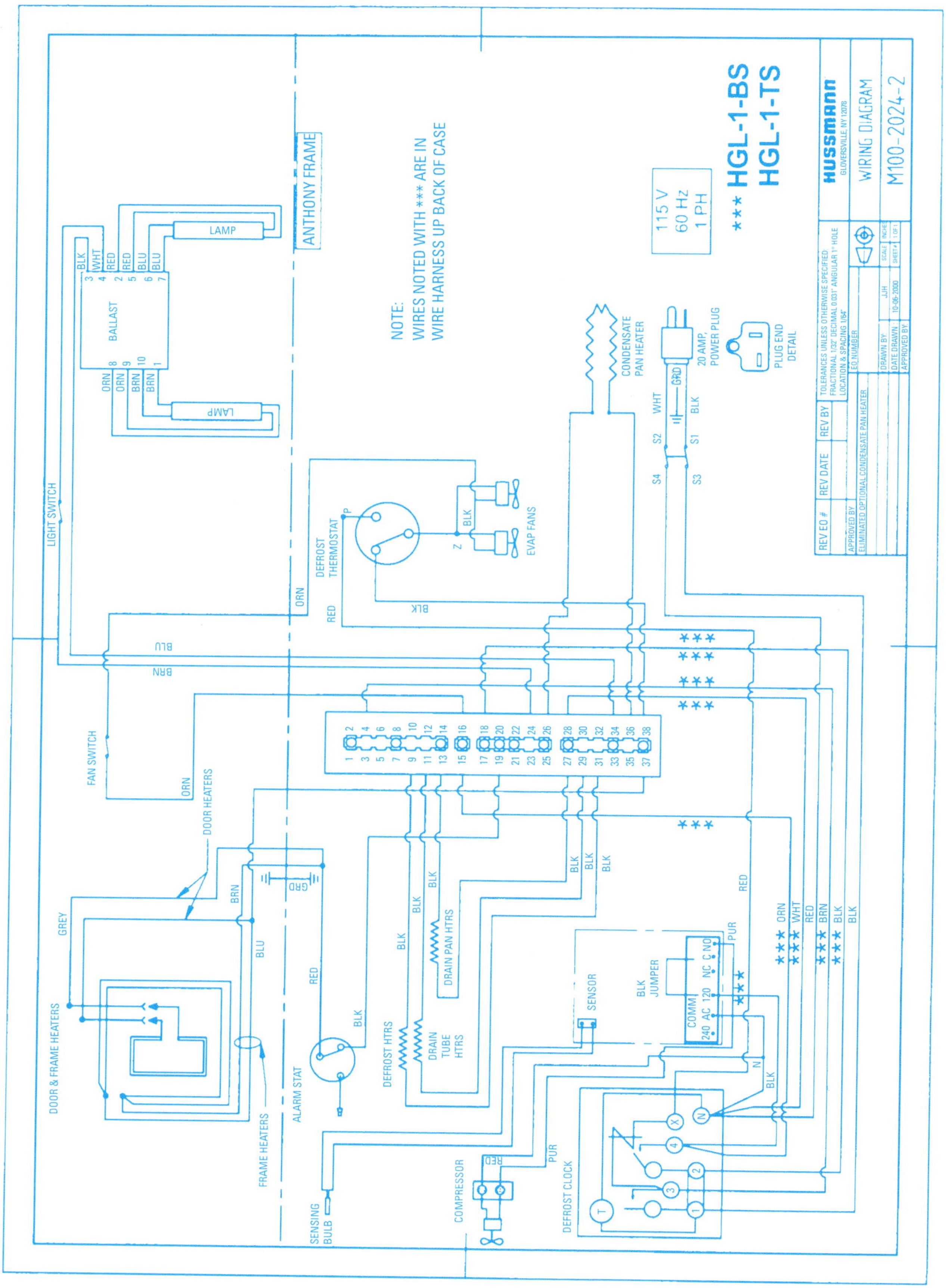

Figure 42 Freezer wiring diagram.

TROUBLE	PROBABLE CAUSE	SOLUTION
Compressor runs continuously, product too warm	1. Short of refrigerant 2. Inefficient compressor 3. Dirty condenser	1. Leak check. Change drier. Evacuate and recharge 2. Replace 3. Clean
High head pressure	1. Cabinet location too warm 2. Restricted condenser air flow 3. Defective condenser fan motor 4. Air or non-condensable gases in system	1. Relocate cabinet 2. Clean condenser to remove air flow restriction 3. Replace 4. Leak check, change drier, evacuate, and recharge
Warm storage temperatures	1. Temperature control not set properly 2. Short of refrigerant 3. Cabinet location too warm 4. Too much refrigerant 5. Low voltage. Compressor cycling on overload 6. Condenser dirty	1. Reset control 2. Leak check, change drier, evacuate, and recharge. 3. Relocate 4. Change drier, evacuate, and recharge 5. Check power 6. Clean
Compressor runs continuously, product too cold	1. Defective control 2. Control feeder tube not in positive contract 3. Short on refrigerant	1. Replace 2. Assure proper contract 3. Leak check, change drier, evacuate, and recharge
Compressor will not start, no noise	1. Blown fuse or breaker 2. Defective or broken wiring 3. Defective overload 4. Defective temperature control 5. Power disconnected	1. Replace fuse or reset breaker 2. Repair or replace 3. Replace 4. Replace 5. Check service cord or wiring connections
Compressor will not start, cuts out on overload	1. Low voltage 2. Defective compressor 3. Defective relay 4. Restriction or moisture 5. Inadequate air over condenser 6. Defective condenser fan motor 7. CRO not set properly	1. Contact electrician 2. Replace 3. Replace 4. Leak check, replace drier, evacuate and recharge 5. Clean condenser 6. Replace 7. Reset to 10 psi
Icing condition in drain pan	1. Low voltage 2. Cabinet not level 3. Defective drain tube heater 4. Defective drain pan heater	1. Check voltage at compressor 2. Check front to rear leveling, adjust legs accordingly 3. Replace 4. Replace

Figure 43A Troubleshooting chart example (1 of 2).

> **WARNING!**
> It is important to remember that ice machines produce a consumable product, requiring that the unit be properly maintained and kept sanitary. Failure to properly maintain and clean these units could lead to widespread illness from contamination.

4.2.0 Troubleshooting and Maintenance of Cubed-Ice Machines

Most cubed-ice machines have a more complex sequence of operation than a freezer. Troubleshooting and properly testing an ice machine can be time consuming because it takes time to complete a full cycle. After any adjustments or changes are made, the unit should be put through several complete cycles to ensure that it is operating properly. An air-cooled cubed-ice machine is being used here to review troubleshooting and maintenance procedures.

> **WARNING!**
> Ice machine cleaning and sanitizing products can be harmful. It is a violation of federal law to use them in any manner not consistent with their stated purpose. Always wear rubber gloves and goggles/face shield when handling and using these products.

4.2.1 Cleaning and Maintenance

Properly performing a maintenance inspection and cleaning of an ice machine can be a significant benefit to the technician as well as to the equipment. The technician can become familiar with the location of components and observe normal operations without the pressure of trying to correct a problem.

> **NOTE**
> The following process contains general instructions. Cleaning or maintenance should not be performed without access to unit-specific manufacturer's instructions.

Primary maintenance activities for cubed-ice machines are outlined as follows:
- *Interior cleaning and sanitizing* – This maintenance should be conducted at least every 6 months for most applications. Environmental factors can require increasing the frequency of these inspections and checks to as often as every 30 days. Poor ventilation system performance or maintenance can allow higher temperatures and dust levels in the building air. The water supply may have a high mineral content. If problems appear before this

TROUBLESHOOTING LIGHTING SYSTEM	
Lights won't start	1. Check light switch 2. Check continuity to ballast 3. Check to see if bulbs are inserted properly into sockets 4. Check voltage
Lights flicker	1. Allow lamps to warm up 2. Check lamp sleeve for cracks 3. Check sockets for moisture and proper contact 4. Bulb replacement may be necessary 5. Check voltage 6. New bulbs tend to flicker until used
Ballast hums	1. Check voltage 2. Replace ballast

Figure 43B Troubleshooting chart example (2 of 2).

interval occurs, other preventive actions may be needed, such as the installing improved water filtration or relocating the unit to a conditioned area (from an excessively hot area) to help impede bacterial growth.

– Ice machine cleaners help remove lime and other mineral deposits left on surfaces from the water supply. Ice-machine sanitizers disinfect and remove algae and slime. Always ensure that cleaning and sanitizing agents are used according to the manufacturer's recommendations and are used in their proper concentrations.

– For this process, the ice machine should be shut down following a harvest operation so that the evaporator is clear of ice. All ice should be removed from the bin. Thorough cleaning requires significant disassembly of the unit and a lot of time to complete. Schedule these activities to ensure that an adequate supply of ice will be on hand.

– Ice-machine cleaners should be added and circulated throughout the water distribution system using the water pump. Most manufacturers provide a cleaning control-switch position for this operation. The cleaner should remain in circulation for at least 30 minutes unless directed otherwise.

–Disassemble and remove all components possible that are in contact with water—water curtains, distribution tubes, water sumps, water level probe, and pump assembly. Clean these components manually using fresh solution, allowing them to soak 15 to 20 minutes for heavily concentrated mineral deposits. Use a soft-bristle brush or sponge for cleaning. With the components removed, use fresh solution to wipe down all accessible surfaces of both the internal ice machine and the bin. Then, rinse or wipe down again with clean water.

– Prepare sanitizer solutions and again clean removed components and wipe down all accessible internal surfaces of the unit and bin. Reassemble the unit and add sanitizing solution to the water sump. Circulate the sanitizing solution through the water distribution system for approximately 30 minutes. Disassemble the unit a second time and wipe down removed components and accessible areas of the bin and unit. Rinsing with clear water is generally not needed.

– Reassemble the unit, restore power and water, and return the unit to service. Observe the unit's operation through several cycles to ensure that all components have been properly reinstalled and are working as expected.

- *Ice machine cabinet and bin exterior* – External surfaces should be cleaned for an attractive appearance. Customers will be turned off by a poorly maintained exterior.
- *Air-cooled condenser coils* – As with all condensers, clean these using a soft brush and vacuum. A coil cleaning solution may be needed if grease or other deposits are evident. Some ice machines are equipped with washable filters to help protect the coil; these can be washed using a mild soap and water solution. Straighten bent fins with a fin comb.
- *Water filters and conditioning systems* – Many systems are equipped with water filters to remove odor, taste, and particulates from the water supply. Filter elements should be replaced during the cleaning/sanitizing process, or when fouled.

Specialized water treatment systems are often installed to help prevent biological growth by adding controlled amounts of chlorine dioxide to the water. Another type of installation generates ozone gas (O_3) and disperses it into the ice maker's water supply line. Ozone is a very effective antibacterial chemical and is also effective as a demineralizing agent. However, the suspended mineral oxides must be filtered out to prevent them from depositing on internal surfaces. Icemaker filters can also remove the metal ions that form hard scale on evaporator plates and chlorine that affects the taste of ice. These water-treatment products should not be considered a replacement for proper manual cleaning and sanitizing. However, they may reduce the amount of time required for a cleaning operation.

4.2.2 Operation and Troubleshooting

Before beginning a troubleshooting process, ensure that the condenser coil and other heat transfer surfaces are clean. Fouled surfaces and clogged coils will cause abnormal operation and high refrigerant pressures. Pressure gauges should be connected only when necessary. Since the refrigerant charge is often very small, even slight losses that occur when gauges are connected and the hoses are filled may have a significant effect. The proper refrigerant charge is vital to their proper operation.

The electrical system of the ice machine (*Figure 44*) includes circuits for the compressor, water pump, condenser fan(s), water and refrigerant solenoid valves, control transformer, control relays, and microprocessor (if used). Basic electrical troubleshooting principles apply to these items when a problem exists.

Knowing and understanding the precise timing and sequence of operation is essential. Although units may appear to be identical, the sequence of events may differ. Always ensure that proper literature is on hand for the equipment being serviced.

The following sequence of operation applies to the cubed-ice machine shown in *Figure 44*. Also refer to the energized-parts chart (*Figure 45*) for details of the operating sequence of that same cubed-ice machine. A watch or digital timer is helpful when timing each event.

- The toggle switch must be in the Ice position and the water curtain covering the evaporator plate must be in position before the unit will start.
- The water pump and water dump solenoid are energized for 45 seconds before the compressor starts to purge the unit of old water. The harvest valve is also energized during the water purge and remains energized for an additional 5 seconds after the purge process stops.
- The compressor starts after the water purge process and remains running throughout the ice-making and harvest cycles. The water fill valve is also energized. The condenser fan motor circuit receives power, but the fan cycling control may cause it to cycle on and off.

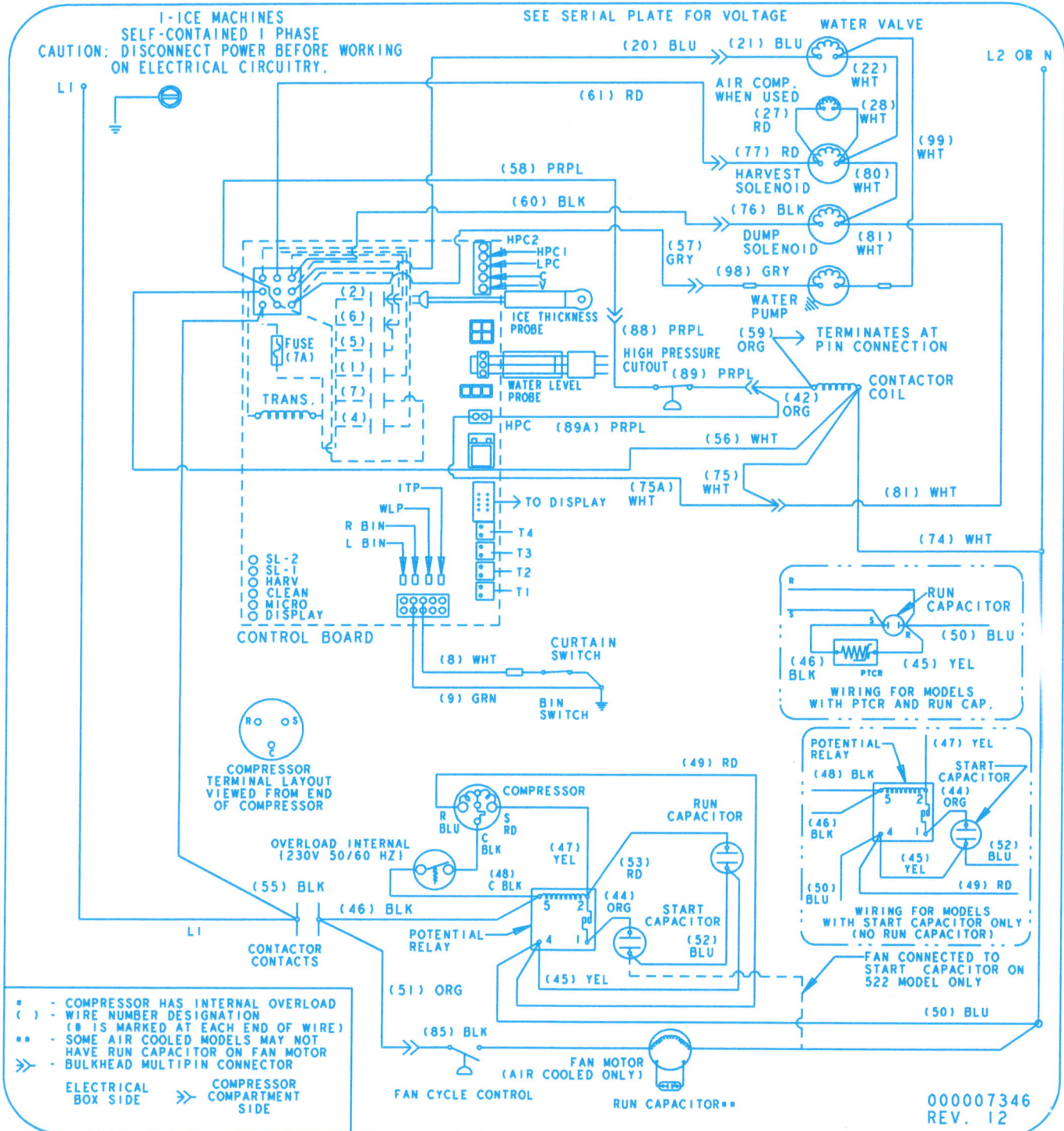

Figure 44 Cubed-ice machine wiring diagram.

Indigo Sequence of Operation
Air & Water-Cooled and Traditional Remotes
Energized Parts Chart

 = energized part

Ice Making Sequence of Operation	Control Board Relays					Contactor		LENGTH of "ON" Time
	WATER PUMP	HARVEST VALVE(S) **HPR Solenoid** AirPump(s)*	WATER INLET VALVE	WATER DUMP VALVE	CONTACTOR COIL **Liquid Line Solenoid**	Compressor	Condenser Fan Motor†	
Start-Up 1. Water Purge	⚡			⚡				45 Seconds
2. Refrigeration System Start-up a. equalization		⚡						5 Seconds
b. compressor start		⚡			⚡	⚡	⚡	5 Seconds
Freeze Sequence 3. Prechill			May Cycle On/Off during pre-chill		⚡	⚡	⚡	Initial Start-Up is 120 Seconds / 30 Seconds thereafter
4. Freeze Six minute Lock in on every cycle (Max Freeze 35 Min)	⚡		Cycles off then On two more times		⚡	⚡	⚡	Until Ice Contact with Ice Thickness Probe
Harvest Sequence 5. Water Purge	⚡	⚡		⚡	⚡	⚡	⚡	Factory Set at 45 Seconds
6. Harvest a. Normal (0-3.5 Minutes)		⚡			⚡	⚡	⚡	Bin Switch Activation
b. Water Assit (3.5-6.5 Minutes)	⚡	⚡	Energizes and fills to the High WLP		⚡	⚡	⚡	
c. Water Dump (6.5-7 Minutes) Max Harvest 7 Minutes	⚡	⚡		⚡	⚡	⚡		
Bin Switch Open 30 Seconds								
7. Automatic Shut-Off								Until Bin Switch Re-closes & 3 min. delay

*NOT USED ON ALL MODELS
** REMOTES ONLY

† CONDENSER FAN MOTOR: Self-contained Air-cooled - The fan motor is wired through a fan cycle pressure control; therefore, it may cycle on and off. Remotes do not have a fan cycle pressure control with the exception of the JC995, which does have Condenser Fan Cycling control.
♦ 3 MINUTE DELAY FOR RESTART

Figure 45 Energized-parts chart.

- The compressor runs to chill the evaporator plate for 30 seconds before the water pump starts. The water fill valve remains on until the probe indicates that a sufficient water level exists in the sump.
- The water pump starts after the 30-second period. Water is evenly distributed across the top of the plate and trickles down into the individual cube cells. Unfrozen water collects in the evaporator trough. Some water remains behind as ice in the cells, continuing to build in thickness as the cycle continues. The water fill valve may cycle open one more time to replenish the water level.
- As ice builds, water flow is pushed out slightly away from the evaporator plate. Eventually, the flowing water makes contact with the ice thickness probe. After roughly 10 seconds of constant contact proving that ice has formed, the harvest cycle can begin. The unit allows at least 6 minutes to elapse in the ice-making cycle before initiating a harvest cycle.
- The harvest valve is opened, allowing hot compressor discharge gas to circulate through the evaporator. At the same time, the water pump continues to run as it did during the ice-making cycle, but the water dump valve is energized for 45 seconds and the water is purged from the sump. The water fill valve is energized during the last 15 seconds of this period to help flush out the existing water. After the 45-second period, the water pump, fill valve, and dump valve are all de-energized.
- The harvest valve remains open, warming the evaporator plate and breaking the bond between the ice and the plate. Cubes are formed as a single sheet with a thin bridge between the individual cubes. The sheet slides out and breaks apart into individual cubes as it falls into the storage bin. The water curtain swings out as the sheet falls, actuating the bin switch in the process. If the curtain and the bin switch remain open for more than 30 seconds, this is a signal that the ice has not completely fallen out and the unit stops, assuming the storage bin is now full. When the bin is not full, the actuation of the bin switch is momentary and is a signal to the unit to return to the ice-making cycle.
- If the bin switch is open for more than 30 seconds and the unit shuts down, it must remain off for at least 3 minutes before it will restart. When enough ice is removed from the bin to allow the curtain to swing closed, ice making can again begin.

- The maximum amount of time allowed to complete the ice-making cycle is 60 minutes, after which the unit will automatically initiate a harvest cycle.
- The maximum amount of time allowed for a harvest is $3\frac{1}{2}$ minutes after which the unit will return to the ice-making cycle.

Over time, water vapor will collect and freeze on the back of the evaporator plate. To rid the unit of this accumulation, after 200 freeze/harvest cycles have been completed, the microprocessor will initiate a warm-water rinse in the following sequence:

- Both the Clean and Harvest LEDs come on, indicating that the unit has entered a warm-water rinse cycle.
- The compressor and harvest valve are energized.
- The water pump is energized.
- The water inlet valve is energized until it makes contact with the water level probe.
- This process warms the water for 5 minutes, then the compressor and harvest valve are de-energized.
- The water pump remains on for an additional 5 minutes, then is de-energized.
- The process can be terminated by moving the control switch to Off, then back to Ice.

Most ice machines are test-operated at the factory and adjusted as necessary. The following tests and adjustments should be conducted, after cleaning and sanitizing at initial startup, and after a prolonged shutdown:

- The water level sensor is designed to maintain proper water level in the sump and is not adjustable. If the water level appears incorrect, check the probe for physical damage and clean it if necessary.
- The ice thickness probe is factory set to maintain an ice bridge thickness of $\frac{1}{8}$" (3 mm) (*Figure 46*). This amount is sufficient to keep the ice cubes in a sheet until the mass falls from the evaporator, yet thin enough to break apart when the sheet falls to the bin. If the ice bridge appears to be too thick or too thin, set the ice thickness probe distance using the adjusting screw. Begin with a $\frac{1}{4}$" (6 mm) gap between the thickness probe and the evaporator as a starting point.

This unit is controlled by the microprocessor, which precisely handles the large number of timed events. Microprocessor controls are very

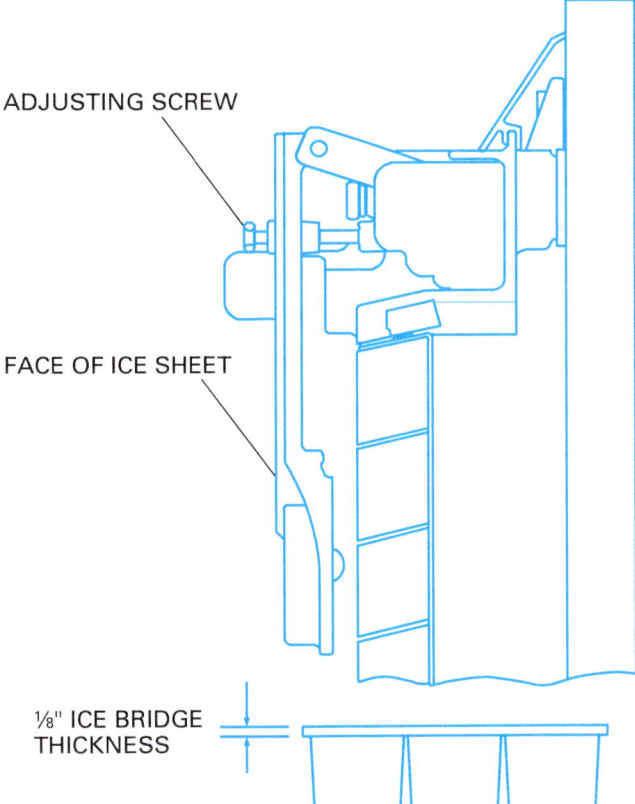

Figure 46 Ice thickness check.

reliable, so if one fails, technicians are not expected to troubleshoot the control to the component level. If a microprocessor chip is bad, its entire board needs to be replaced. However, because this type of control can be costly, be sure the control board has failed before replacing it.

Before condemning the microprocessor, make sure the control is receiving correct input signals and delivering the correct output signals. If it is doing both, and the equipment is not responding, the problem lies somewhere other than the microprocessor control.

One of the best ways to evaluate a problem is by using the proper sequence of operation and a watch/timer to time each step in the cycle. Look for events that do not occur when expected. Use the LEDs on the microprocessor board to help determine what it is attempting to do, then see if controls external to the board respond properly. When they do not, test the affected components for proper operation.

To prevent loss of refrigerant, service gauges should only be connected when necessary. When refrigerant must be added, or when refrigeration cycle components must be replaced, use the proper service practices. Always weigh in the refrigerant when possible to ensure accuracy. Also, always remember to charge zeotropic refrigerants in their liquid state. Remember that charging by subcooling is used for TEVs, while superheat is used when charging units with fixed restrictors. Superheat and subcooling should be checked even when the charge has been weighed in to ensure proper system operation. Indeed, subcooling and superheat values should be checked any time gauges have been connected to the system.

Here are some other troubleshooting hints:

- Always listen carefully to all the customer's complaints. While they may not understand details of the operating sequence, valuable information can be gained from their observations. In addition, technicians who develop good listening skills are often the first to advance.
- Pay close attention to the unit when it is operating properly. If you are familiar with normal unit operation, use sight, sound, and touch (carefully) to detect abnormalities before using tools and instruments.
- Develop a logical approach to troubleshooting all systems. Patience and logical thinking are important troubleshooting tools at your disposal, especially when working with ice makers and other complex systems. Each step depends on the proper execution of previous steps. Without patience and the use of available data to confirm suspicions, technicians can be easily fooled into an incorrect diagnosis.

Additional Resources

Refrigeration and Air Conditioning: An Introduction to HVAC/R, Larry Jeffus. Current Edition. New York, NY: Pearson.

4.0.0 Section Review

1. A very important step in maintaining a reach-in freezer is _____.
   a. changing compressor oil
   b. lubricating fan motors
   c. checking refrigerant charge
   d. keeping coils clean

2. Ice machine cleaner should be circulated through the water distribution system for at least _____.
   a. 30 seconds
   b. 10 minutes
   c. 30 minutes
   d. 60 minutes

SUMMARY

Retail refrigeration equipment provides an essential service—the preservation of consumable and perishable goods. This segment of the HVACR trade includes a variety of equipment and systems designed for specific needs of the products to be maintained. Although there are similarities in the refrigeration circuit to comfort cooling systems, a much wider variety of refrigerants and operating parameters are found in retail refrigeration. Retail refrigeration equipment is often placed in direct view of the consumer, providing direct access to food products. As a result, the equipment should look presentable and operate reliably.

Systems that operate under the classification of high-temperature refrigeration are often very simple in design and require limited troubleshooting skills. However, creating storage temperatures below freezing requires equipment capable of eliminating frost buildup through a variety of approaches. Components that enable equipment operation in severe environments also increase the overall complexity for the service technician.

Commercial ice-making systems generally represent the most complex equipment to be serviced, with many individual and well-timed steps required for the equipment to operate properly and produce the required amount of ice. Troubleshooting these systems requires equipment knowledge and a logical troubleshooting process. Reviewing product-specific manufacturer guidelines is critical so that the technician can provide high-quality service and repairs.

Review Questions

1. Which temperature application operates at an evaporator temperature below −30°F (−34°C)?
   a. High-temperature
   b. Medium-temperature
   c. Moderate-temperature
   d. Low-temperature

2. After vapor has cooled to the condensing temperature in the condenser, what happens until all vapor has condensed to a liquid?
   a. It begins to vaporize
   b. It remains at same temperature
   c. The mixture freezes
   d. The pressure decreases

3. A technician can determine if all refrigerant in a condenser has changed to its liquid state by _____.
   a. the presence of some superheat
   b. a sudden increase in pressure
   c. a sudden decrease in pressure
   d. the presence of some subcooling

4. What process happens immediately after all superheat is removed from a refrigerant?
   a. Subcooling
   b. Vaporization
   c. Condensation
   d. Oil separation

5. When low superheat values are chosen, particular care must be taken to ensure adequate protection from _____.
   a. liquid floodback
   b. contamination
   c. overheating
   d. voltage imbalance

6. The timer for a defrost period should be set _____.
   a. to coincide with the fixture's lightest period of usage
   b. to coincide with the fixture's heaviest period of usage
   c. at least 6 times per day
   d. for long times in dry climates

7. Which defrost method does not require physical contact with the evaporator coil to melt the frost?
   a. Resistive electric heaters
   b. Hot-gas defrost coils
   c. Infrared electric heaters
   d. Nichrome electric heaters

8. Ice merchandisers using cold-wall construction are defrosted _____.
   a. with hot gas
   b. manually
   c. electrically
   d. twice a day

9. In commercial ice makers, the target temperature for forming ice on a surface is near _____.
   a. 32°F (0°C)
   b. 0°F (−18°C)
   c. 10°F (−12°C)
   d. −40°F (−40°C)

10. Ice thickness sensors on cubed-ice machines function _____.
    a. by instant contact with the ice
    b. based on water temperature
    c. based on refrigerant pressures
    d. by contact with water for a specified time

11. Condensate from a refrigerated fixture can be _____.
    a. thrown on the condenser coil for re-evaporation
    b. recycled to form new ice cubes
    c. placed back into the refrigerated area
    d. used to cool the compressor

12. Which device has the refrigerant feed rate determined by inputs from temperature sensors?
    a. Electronic expansion valve
    b. Capillary tube
    c. Thermostatic expansion valve
    d. Refrigerant injector

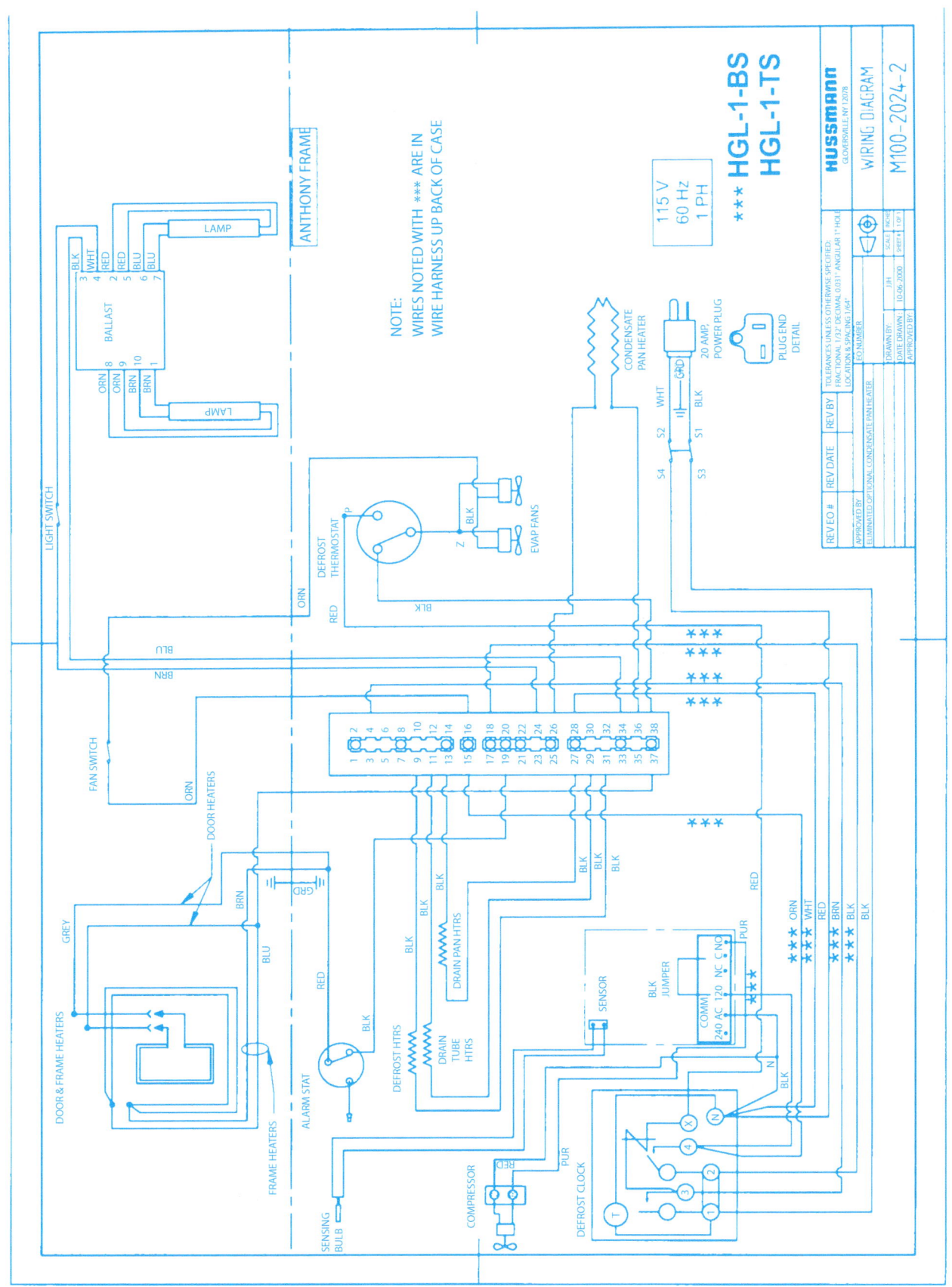

Figure Credit: Hussmann Corporation

Figure RQ01

Indigo Sequence of Operation
Air & Water-Cooled and Traditional Remotes
Energized Parts Chart

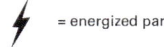 = energized part

Ice Making Sequence of Operation	Control Board Relays					Contactor		LENGTH of "ON" Time
	WATER PUMP	HARVEST VALVE(S) **HPR Solenoid** AirPump(s)*	WATER INLET VALVE	WATER DUMP VALVE	CONTACTOR COIL **Liquid Line Solenoid**	Compressor	Condenser Fan Motor†	
Start-Up 1. Water Purge	⚡			⚡				45 Seconds
2. Refrigeration System Start-up a. equalization		⚡						5 Seconds
b. compressor start		⚡			⚡	⚡	⚡	5 Seconds
Freeze Sequence 3. Prechill			May Cycle On/Off during pre-chill		⚡	⚡	⚡	Initial Start-Up is 120 Seconds / 30 Seconds thereafter
4. Freeze Six minute Lock in on every cycle (Max Freeze 35 Min)	⚡		Cycles off then On two more times		⚡	⚡	⚡	Until Ice Contact with Ice Thickness Probe
Harvest Sequence 5. Water Purge	⚡	⚡		⚡	⚡	⚡	⚡	Factory Set at 45 Seconds
6. Harvest a. Normal (0-3.5 Minutes)		⚡			⚡	⚡	⚡	Bin Switch Activation
b. Water Assit (3.5-6.5 Minutes)	⚡	⚡	Energizes and fills to the High WLP		⚡	⚡	⚡	✕✕✕
c. Water Dump (6.5-7 Minutes) Max Harvest 7 Minutes	⚡	⚡		⚡	⚡	⚡	⚡	
Bin Switch Open 30 Seconds								✕✕
7. Automatic Shut-Off								Until Bin Switch Re-closes & 3 min. delay

*NOT USED ON ALL MODELS
** REMOTES ONLY

† CONDENSER FAN MOTOR: Self-contained Air-cooled - The fan motor is wired through a fan cycle pressure control; therefore, it may cycle on and off. Remotes do not have a fan cycle pressure control with the exception of the JC995, which does have Condenser Fan Cycling control.
♦ 3 MINUTE DELAY FOR RESTART

Figure Credit: Manitowoc Ice

Figure RQ02

13. The primary function of the suction accumulator is to _____.
    a. store refrigerant for system operation at heavy load
    b. remove oil from refrigerant in the suction line
    c. store liquid refrigerant to prevent compressor damage
    d. prevent refrigerant vapor formation during defrost

14. Allowing the head pressure to get too low as a result of low ambient conditions can cause _____.
    a. low flow through the metering device
    b. the compressor to overheat
    c. damage to the air-cooled condenser
    d. the metering device to freeze

15. Electronic temperature controls generally _____.
    a. have a very narrow operating range
    b. are applied to low-temperature systems only
    c. use capillary tubes filled with refrigerant
    d. are easier to install than remote bulb thermostats

16. One use of a bypass timer is to bypass which of the following components when the compressor starts?
    a. Fixture thermostat
    b. Compressor contactor
    c. Low-pressure switch
    d. Defrost thermostat

17. Refer to *Figure RQ01*. When the fan switch is closed, the other device that controls power to the evaporator fans is the _____.
    a. defrost thermostat
    b. defrost timer motor
    c. alarm thermostat
    d. defrost heaters

18. Which of the following statements regarding maintenance of a cubed-ice machine is *not* true?
    a. Proper cleaning requires significant disassembly of all components in direct contact with water.
    b. Sanitizers must be used according to the chemical's and manufacturer's instructions.
    c. Cleaning and inspection should be performed at least annually.
    d. Components should be manually cleaned and soaked in fresh cleaning solution for 15 to 20 minutes.

19. Refer to *Figure RQ02*. The ice-machine compressor is de-energized during which operating event(s)?
    a. The harvest water purge cycle.
    b. The initial startup water purge cycle and the automatic shut-off cycle.
    c. The freeze prechill cycle.
    d. The initial startup water purge cycle and the harvest water purge cycle.

20. When charging a refrigeration unit with a zeotropic refrigerant, you should _____.
    a. replace the compressor oil
    b. charge in the vapor state
    c. charge in the liquid state
    d. use any blended refrigerant

Trade Terms Introduced in This Module

Phase change: The transition of a refrigerant's physical state (or phase) to another, such as from liquid to vapor.

Slinger ring: A ring attached to the outer edge of the fan that picks up water as the fan rotates and slings it onto the condenser coil.

Unit cooler: A packaged unit containing evaporator coil, metering device, and evaporator fan(s) that serves as the evaporator section for a refrigeration appliance.

Appendix A

HFC-407C Pressure-Temperature Chart (Metric)

R-407C PRESSURES-TEMPERATURES CHART (METRIC)

TEMP.	PRESS. (kPag)		TEMP.	PRESS. (kPag)		TEMP.	PRESS. (kPag)		TEMP.	PRESS. (kPag)	
°C	P_L	P_V	°C	P_L	P_V	°C	P_L	P_V	°C	P_L	P_V
−23.3	345	285	−3.3	603	503	14.4	975	830	33.3	1,562	1,369
−22.2	355	294	−2.2	623	519	15.6	1,003	856	34.4	1,604	1,409
−21.1	367	303	−1.1	643	536	16.7	1,033	883	35.6	1,647	1,449
−20.0	378	312	0.0	662	554	17.8	1,062	909	36.7	1,691	1,490
−18.9	390	323	1.1	682	572	18.9	1,094	937	37.8	1,735	1,531
−17.8	403	333	2.2	703	590	20.0	1,125	966	38.9	1,781	1,575
−16.7	415	343	3.3	725	609	21.1	1,157	995	40.0	1,827	1,618
−15.6	428	354	4.4	747	628	22.2	1,190	1,025	41.1	1,874	1,663
−14.4	442	365	5.6	770	648	23.3	1,224	1,056	42.2	1,922	1,709
−13.3	456	378	6.7	794	669	24.4	1,258	1,087	43.3	1,972	1,756
−12.2	471	390	7.8	814	690	25.6	1,293	1,120	44.4	2,022	1,804
−11.1	485	402	8.9	842	712	26.7	1,329	1,153	45.6	2,073	1,853
−10.0	501	415	10.0	867	734	27.8	1,366	1,187	46.7	2,125	1,904
−8.9	517	429	11.1	893	757	28.9	1,404	1,222	47.8	2,178	1,955
−7.8	533	443	12.2	920	781	30.0	1,442	1,257	48.9	2,232,	2,007
−6.7	550	457	13.3	947	805	31.1	1,482	1,293	50.0	2,287	2,059
−5.6	567	472	−3.3	603	503	32.2	1,522	1,331	51.1	2,344	2,115

Notes:
1. R-407C is zeotropic. P_L is the saturation pressure for 100% liquid; P_V is the saturation pressure for 100% vapor.
2. Celsius temperatures correspond to Farenheit temperatures listed in Figure 3.

Additional Resources

This module presents thorough resources for task training. The following reference material is recommended for further study.

NCCER Module 03101, *Introduction to HVAC*.
NCCER Module 03107, *Introduction to Cooling*.
Refrigeration and Air Conditioning: An Introduction to HVAC/R, Larry Jeffus. Current Edition. New York, NY: Pearson.

Figure Credits

©Gigisomplak/Dreamstime, Module opener
Cape Pond Ice Company, SA01
Hussmann Corporation, Figures 5, 26, 42, 43, Review Questions Figure 1
Courtesy of Sealed Unit Parts Co., Inc., Figure 7
Master-Bilt Products, an unincorporated division of Standex International Corporation, Figures 10, 12, 41
Courtesy of True Manufacturing, Figure 11
© Anna Krasnopeeva/Dreamstime, Figure 13
© Anna Krivitskaia/Dreamstime, Figure 15
Photo courtesy of Zero Zone, Inc., SA02
Manitowoc Ice, Figures 16, 20, 44–46, Review Questions Figure 2
© Shahizal Rizwan Ahmat Raslan/Dreamstime, Figure 19
Courtesy of Emerson Climate Technologies, Figures 31A, 39, 40, SA03, SA04B
Courtesy of Emerson, Figures 22, 24
Hoshizaki America, Inc., Figure 27
Courtesy of Parker Hannifin, Sporlan Division, Figures 30, 34
Courtesy of ICM Controls, Figure 37

Section Review Answer Key

Answer	Section Reference	Objective
Section One		
1. b	1.1.0	1a
2. a	1.2.0	1b
3. c	1.3.1	1c
Section Two		
1. d	2.1.1	2a
2. a	2.2.0	2b
3. d	2.3.1	2c
Section Three		
1. c	3.1.1	3a
2. a	3.2.2	3b
3. c	3.3.3	3c
Section Four		
1. d	4.1.0	4a
2. c	4.2.1	4b

NCCER CURRICULA — USER UPDATE

NCCER makes every effort to keep its textbooks up-to-date and free of technical errors. We appreciate your help in this process. If you find an error, a typographical mistake, or an inaccuracy in NCCER's curricula, please fill out this form (or a photocopy), or complete the online form at **www.nccer.org/olf**. Be sure to include the exact module ID number, page number, a detailed description, and your recommended correction. Your input will be brought to the attention of the Authoring Team. Thank you for your assistance.

Instructors – If you have an idea for improving this textbook, or have found that additional materials were necessary to teach this module effectively, please let us know so that we may present your suggestions to the Authoring Team.

NCCER Product Development and Revision
13614 Progress Blvd., Alachua, FL 32615

Email: curriculum@nccer.org
Online: www.nccer.org/olf

❏ Trainee Guide   ❏ Lesson Plans   ❏ Exam   ❏ PowerPoints   Other _____

Craft / Level: _____ Copyright Date: _____

Module ID Number / Title: _____

Section Number(s): _____

Description:

_____

_____

_____

_____

Recommended Correction:

_____

_____

_____

_____

Your Name: _____

Address: _____

_____

Email: _____ Phone: _____

Customer Relations

OVERVIEW

To a customer, the HVACR service technician is the face of a company. The customer's attitude toward the company, and the likelihood of that customer doing future business with the company, are determined largely by the customer's impressions of, and interactions with, that technician. For those reasons, a service technician must learn how to deal effectively with customers.

Module 03316

Trainees with successful module completions may be eligible for credentialing through NCCER's National Registry. To learn more, go to **www.nccer.org** or contact us at **1.888.622.3720**. Our website has information on the latest product releases and training, as well as online versions of our *Cornerstone* magazine and Pearson's product catalog.

Your feedback is welcome. You may email your comments to **curriculum@nccer.org**, send general comments and inquiries to **info@nccer.org**, or fill in the User Update form at the back of this module.

This information is general in nature and intended for training purposes only. Actual performance of activities described in this manual requires compliance with all applicable operating, service, maintenance, and safety procedures under the direction of qualified personnel. References in this manual to patented or proprietary devices do not constitute a recommendation of their use.

Copyright © 2018 by NCCER, Alachua, FL 32615, and published by Pearson, New York, NY 10013. All rights reserved. Printed in the United States of America. This publication is protected by Copyright, and permission should be obtained from NCCER prior to any prohibited reproduction, storage in a retrieval system, or transmission in any form or by any means, electronic, mechanical, photocopying, recording, or likewise. To obtain permission(s) to use material from this work, please submit a written request to NCCER Product Development, 13614 Progress Blvd., Alachua, FL 32615.

03316 V5

From *HVAC Level Three, Trainee Guide*. NCCER.
Copyright © 2018 by NCCER. Published by Pearson. All rights reserved.

03316
CUSTOMER RELATIONS

Objectives

When you have completed this module, you will be able to do the following:

1. Explain the service technician's role in customer relations.
   a. Explain how personal habits, behaviors, and attitudes affect customer relations.
   b. Explain how to properly communicate with customers.
2. Describe basic conduct required for a service call.
   a. Describe how to conduct the three phases of a service call.
   b. Describe ways to handle challenging customer situations.

Performance Task

Under the supervision of your instructor, you should be able to do the following:

1. Participate in at least three different role-playing scenarios related to challenging customer service situations.

Trade Terms

Empathy
Rapport

Industry-Recognized Credentials

If you are training through an NCCER-accredited sponsor, you may be eligible for credentials from NCCER's Registry. The ID number for this module is 03316. Note that this module may have been used in other NCCER curricula and may apply to other level completions. Contact NCCER's Registry at 888.622.3720 or go to **www.nccer.org** for more information.

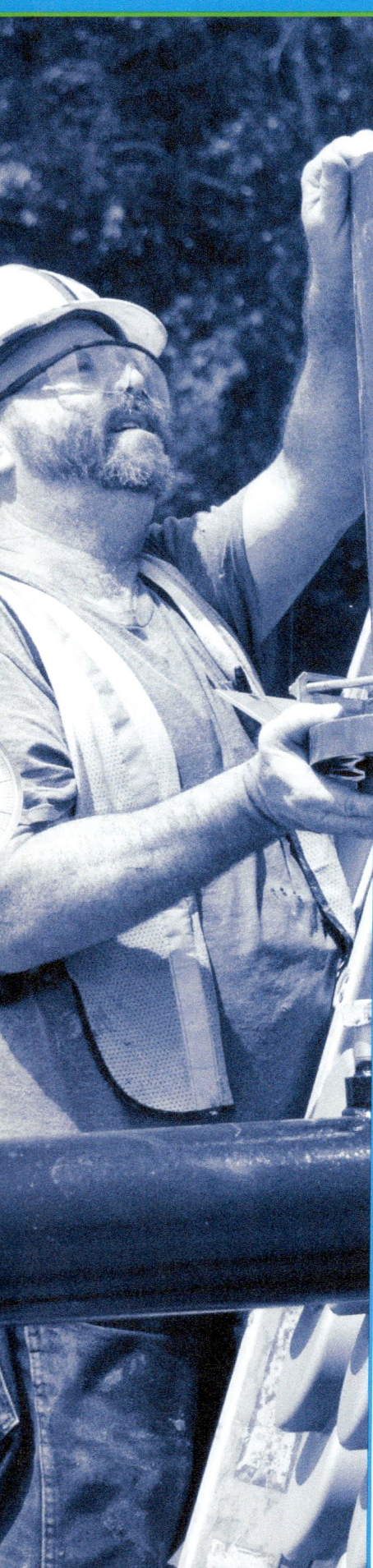

Contents

1.0.0 The Technician's Role in Customer Relations 1
   1.1.0 How Personal Habits, Behaviors, and Attitudes Affect Customer Relations 1
   1.2.0 Customer Communication 3
      1.2.1 Keeping Communications Positive 4
      1.2.2 The Positive Approach 4
      1.2.3 Showing Concern for Customers 5
      1.2.4 Selling Yourself and Your Company 6
2.0.0 Handling Service Calls 8
   2.1.0 Conducting the Service Call 8
      2.1.1 The Opening 8
      2.1.2 Performing the Service 9
      2.1.3 The Closing 10
   2.2.0 Handling Challenging Customers 10
      2.2.1 Fearful Customers 10
      2.2.2 Opinionated Customers 11
      2.2.3 Argumentative Customers 11
      2.2.4 Sloppy Customers 12
      2.2.5 Angry Customers 12
      2.2.6 Critical Customers 13
      2.2.7 Customers with Unresolvable Problems 13
      2.2.8 Customers Who Request Help With Odd Jobs 14

Figures and Tables

Figure 1 Keep your service van clean 2
Figure 2 Construction site PPE 3
Figure 3 Social media has increased an individual's network
of friends and acquaintances to levels never before possible 3
Figure 4 It can be difficult to communicate with unhappy customers 4
Figure 5 Changing filters can increase equipment life and efficiency 6
Figure 6 Show the customer how to avoid frustration and adjust the
thermostat for greater comfort and efficiency 6
Figure 7 A pleasant greeting is a good way to start a service call 8
Figure 8 Disposable shoe covers 9

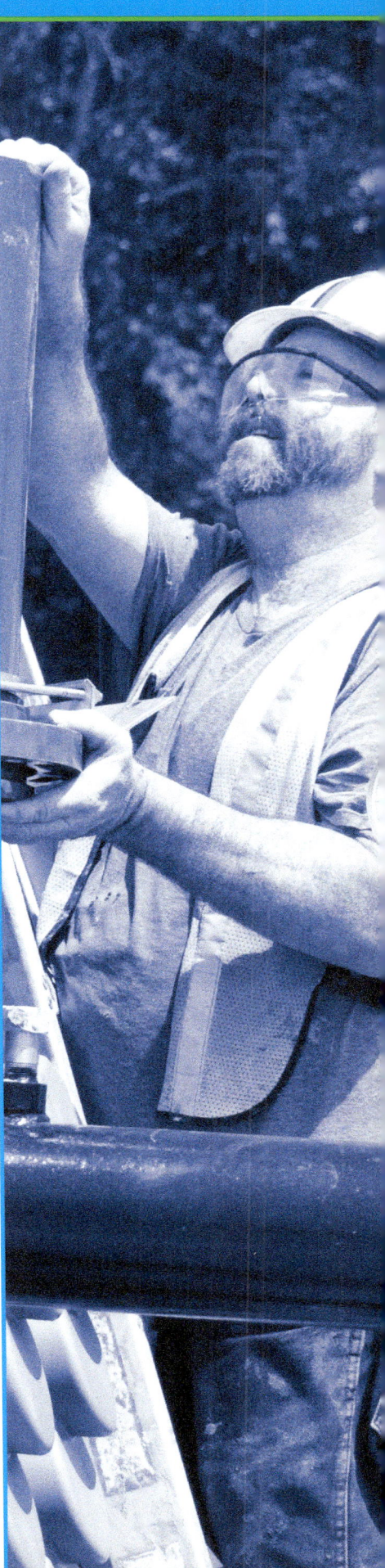

SECTION ONE

1.0.0 THE TECHNICIAN'S ROLE IN CUSTOMER RELATIONS

Objectives

Explain the technician's role in customer relations.
  a. Explain how personal habits, behaviors, and attitude affect customer relations.
  b. Explain how to properly communicate with customers.

Performance Task

1. Participate in at least three different role-playing scenarios related to challenging customer service situations.

Trade Terms

Empathy: The ability to identify with another person's situation or feelings.

Rapport: Establishing a relationship of mutual trust.

Good customer relations are essential to the success of most businesses, including those in the HVAC industry, where each service technician represents the company and influences what customers think about the business. Keeping these impressions positive requires that all employees work toward generating customer good will. Providing good service is a part of this, but so are good personal habits, good work practices, and customer-pleasing attitudes. Customer goodwill often results in more business through referrals and repeat servicing of existing customers.

This module provides guidelines for dealing with customers during service calls. Although the examples used in the module lean toward working with residential customers, the basic principles apply to all types of service calls.

1.1.0 How Personal Habits, Behaviors, and Attitudes Affect Customer Relations

Positive habits, behaviors, and attitudes help to ensure that a customer's first impression of you and your company is a good one; that the customer's needs are clearly understood; and that appropriate customer service is provided to meet these needs. When you provide this level of effective service to your customers, you can increase your company's share of business.

Your ability to relate well to your customers is critical. As a service technician, you are likely to be the only person from your company that the customer ever sees. In the customer's eyes, you are the company, so appearances do count. First impressions are made only once, and it generally happens in the first 60 seconds. There is an old saying: "You don't get a second chance to make a first impression." Mastering the art of making a good impression will not only make your job easier, it will also make it more likely that a customer will call your company the next time HVAC service is needed. Positive responses and feedback from customers that reach your place of business also reinforce your value to the employer and provide support for pay increases.

Because customers lack technical training and test equipment, they have to rely on a service company for help with heating and cooling problems. Although they may not be able to judge the service technician's professional competence, they can, and do, form an impression based on the technician's appearance and attitudes. The customer's first glimpse of a technician can determine that customer's opinion of the worker and

---

It's All About Perception

Customers and potential customers base their buying decisions on their personal attitudes and perceptions. If someone sees you speeding or driving aggressively in your company vehicle, it can affect that person's willingness to do business with your company. If your truck is dirty on the outside, or cluttered and disorganized on the inside, many customers will see that as a reflection of you. The same applies to your personal appearance. If you need a shave or a haircut, if your clothes are dirty or wrinkled, or even if it is as simple as not having your shirt tucked in, some customers will be put off by it. Many customers, especially older people, could be put off by heavy tattooing. It is usually best to cover them up, rather than displaying them.

the company the technician represents. Examine your first impression. Do you:

- Practice good personal hygiene?
- Get enough sleep and look alert?
- Wear a neat, clean uniform?
- Wear shoe covers or remove shoes when you will be walking on carpeting or finished floors?
- Carry a pencil and pad to take notes?
- Park on the street, rather than the customer's driveway? While this is not always feasible, it is a good idea. Vehicles often have oil or fluid leaks that will stain a driveway.

Once the customer has formed a first impression of you and your company, your on-site work habits will confirm or change that impression. Here are some questions to ask yourself on a regular basis, rather than just once:

- Are you on time? Keep in mind that customers must often take time away from work or from other activities in order to be available for the service call. Make every possible effort to be on time for the appointment. Respect the time of your clients. If you are going to be unavoidably late, have the courtesy to call and let them know. When you arrive, apologize for your tardiness.
- Is your tool set complete and neatly packed? Check your tool kit and your truck every day before setting out on your service calls to make sure you have everything you need.
- Do you show concern, courtesy, and respect to the customer by listening carefully while the customer identifies the problem? Listening attentively may provide you with useful information about the problem. It also helps you establish rapport with the customer.
- Do you tackle the problem promptly and quietly?
- Do you remember to protect the work area? Do you carry rags for cleaning up and drop cloths to protect floors and carpets?
- Do you respect the job site by not tracking in dirt and by cleaning up after yourself?
- Do you take the time to explain carefully to the customer what's wrong and how the unit will perform satisfactorily after you've repaired it?
- Do you refrain from smoking on the premises, avoid smelling of smoke, or spitting?

- Do you respect company equipment, tools, and vehicles? Trucks and vehicles displaying the company name are rolling advertisements (*Figure 1*). Is your vehicle clean and in good repair? Are your driving habits courteous?
- Do you avoid using alcohol and drugs while on the job or around vehicles?
- Do you avoid profanity and horseplay? Customers, especially those with children on the premises, are likely to be offended by inappropriate language.
- Do you avoid wearing political or sports-related articles that might cause a confrontation?

If you are working at a commercial construction site, you must be aware of and respect the site safety requirements. This includes wearing the proper personal protection equipment (PPE) (*Figure 2*), which can include:

- Hard hat
- Ear protection
- Safety glasses
- Reflective vest
- Work gloves
- Steel-toed boots

Site managers are perfectly within their rights to refuse entry to someone who is not properly attired. The right thing to do is to learn in advance what PPE is required. Don't assume anything, because requirements vary from site to site. You will make a bad first impression if the site manager has to refuse you entry to the site. If and when that does happen, your attitude and response to

Figure 1 Keep your service van clean.

Customer Technical Knowledge and Skills

When talking with customers, never assume that they do not have technical knowledge or skills. Although many do not, some customers have a broad technical background and can readily understand technical concepts.

Figure 2 Construction site PPE.

Figure 3 Social media has increased an individual's network of friends and acquaintances to levels never before possible.

it will be remembered by others long term. Every hour of every working day, you are an advertisement for your company. Your appearance and behavior must be consistent with the positive image your company wants you to reflect.

Off the job, you can increase word-of-mouth advertising (free advertising created by satisfied customers who recommend your company to friends) by having a positive attitude about your employers, co-workers, and customers. Positive statements about your job are free advertising for your company.

The internet and social media have changed the face of customer relations. A person who is dissatisfied with a service company can go to a website and lodge a complaint that hundreds of other potential customers can read. Many people these days will not even do business with a company without looking for opinions from others on websites. People living in a residential subdivision are likely to be part of an internet social group; they can instantly share a positive or negative opinion of a contractor to hundreds of their neighbors. Both good and bad news travel at lightning speed through social media (*Figure 3*). Unfortunately, negative information typically spreads farther and faster than positive information. One reason for this is the power of anger and/or resentment as motivators. Both are powerful motivators for unhappy customers to spread their dissatisfaction. Happy, pleased clients are not as likely to access social media to share this information, but will share it when it is the context of a conversation.

1.2.0 Customer Communication

It is important to know the technical requirements of your job. Knowing the steps in completing a typical call is important, too. Confidence is an important trait to convey to a customer. However, you can gain a little extra leverage by learning how to interact well with customers. This may make all the difference for your company and may determine whether you or your competitor is called the next time. When you practice good customer relations throughout the service call, you give the customer confidence in your service and your company and stimulate that customer to call you again. One of the most important things to remember about communication is that it is a two-way street. Be sure to let customers communicate their feelings and ideas. Part of this process is allowing them to vent their frustrations about the equipment or about the inconvenience or discomfort that the equipment failure caused. For example, a customer who has who has been without heat on a cold day is likely to be very uncomfort-

---

Think About It
Personal Habits, Behaviors, and Attitudes

If you were asked to hire a new service technician for your company, what qualities would you look for?

able (*Figure 4*). The customer will want to let you know about it, and listening with sincerity and a concern for the situation will be required.

It is sometimes necessary to contact the customer by phone. When that happens, it's important to ensure that the communication is clear and precise. Here are some tips to follow when making a phone call to a customer:

- Before you make the call, think about what you want to say and the questions you want to ask. It's always a good idea to write them down.
- When the person answers, identify yourself and your company. Speak clearly and (if practical) take notes to be sure you are getting the correct information. Repeat it back to make sure.
- Prepare in advance in case you have to leave a message on an answering machine. Be ready to give your name and company, the purpose of your call, and a way to contact you. Speak slowly and clearly, especially when you are giving a phone number or e-mail address. Speak the numbers slowly, and then repeat the number.

1.2.1 Keeping Communications Positive

On every service call, you should work to keep the communications positive. This means paying close attention to verbal and nonverbal elements such as:

- The technical elements of the job
- What the customer needs
- Your problem-solving expertise
- What you've done to make the customer happy

Here are some simple but effective tips:

- Do the best technical job you can, since actions speak louder than words.
- Always treat the customer with courtesy, concern, and respect. Treating others as you'd like to be treated yourself is particularly important, especially in unpleasant situations. Remember, it is always easy to be courteous when things are going well. A truly mature person is courteous even when things aren't going well.
- Treat each service call as if it were an emergency. To the owner, it is. Usually, customers do not call a service technician unless they are without heat in cold weather or without cooling in hot weather. When comfort is at stake, repair becomes a top priority in the customer's mind.
- Show empathy for the customer. This means putting yourself in the customer's shoes. Think about how you and your family would feel on a cold day if your furnace had not been working for many hours. Be sincere in your empathy; the customer is likely to detect insincerity and resent it.
- Leave your personal life at home. Don't let personal frustrations affect your attitude and your relations with customers. If you are having personal problems, avoid using customers as sounding boards.
- If your schedule is running late, report this to your supervisor or the customer. If you're running late, the customer must often make other arrangements to fulfill later commitments. The main thing is: don't leave the customer wondering about what is going on. Make sure they know you are running late and see if they need to reschedule.

1.2.2 The Positive Approach

The following are some important tips on keeping things positive:

- *Smile* – Be genuinely interested in the customer. You may think you were called to repair a furnace. You were really called to make the customer's home comfortable again.

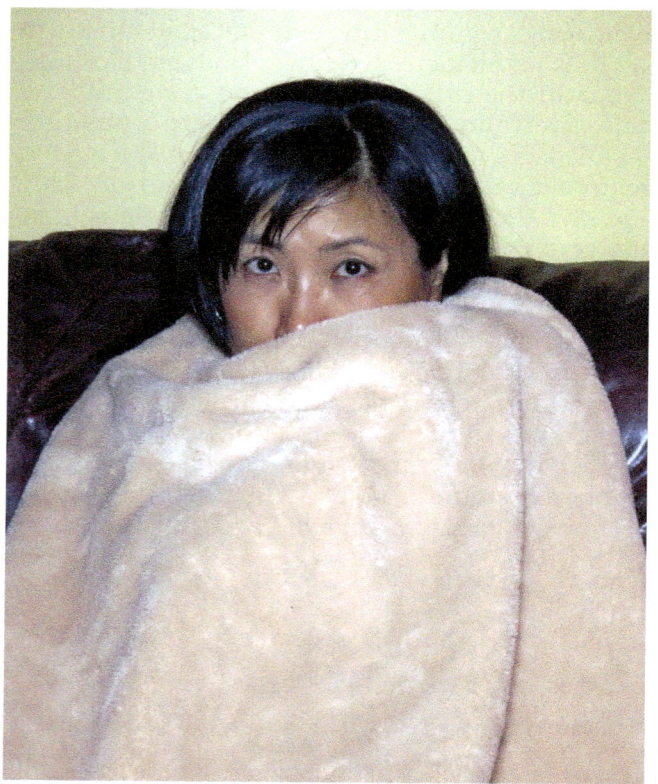

Figure 4 It can be difficult to communicate with unhappy customers.

- *Use positive statements with the customer* – Compliment the equipment if you know it is good. Compliment the cleanliness of the furnace area if that contributes to the operation of the equipment. Speak of your employer in positive terms, because that builds customer confidence in your company.
- *Do not criticize the client or the system* – Whenever possible, avoid negative statements. Be careful of what you say about the brand of equipment, the design of the system, or the quality of the workmanship, because you may offend the customer. Homeowners do not want to hear that you think they own inferior equipment. Even if it is inferior, fix what is possible or explain respectfully to the customer what must be replaced. Also, this is not the time to remind the customer that the problem might not have happened if they had taken your advice about signing a maintenance agreement. It's better to bring that up later, in a positive way, after you have corrected their problem.
- *Do not criticize your own organization* – In some cases, you may be called upon to service a system recently repaired by one of your co-workers. Rather than criticizing his or her work, apologize for the inconvenience and assure the customer you will get to the bottom of the problem and report your findings to them directly. Then follow through, and take responsibility for any errors on behalf of your employer.

1.2.3 Showing Concern for Customers

Another part of good, positive, on-the-job communications is showing concern for your customers. Do this by:

- *Being a good listener* – Listen carefully when the customer explains what went wrong, what was observed, etc. If you yawn, slouch, or let your eyes wander, it is a sign of inattention and is likely to aggravate the customer. Ask questions as appropriate in order to gather the information you need to solve the problem. Remember that although the customer may not have the technical vocabulary you do, the customer's own words can supply you with important clues. After you've been in the business for a while, you will probably have heard most of the equipment problems and customer complaints. The client's description may be long and rambling. Yet, it's important to stay engaged and treat each situation like it is new to you. While it may be old hat to you, it is unique and often stressful for the customer.
- *Talking in terms of the customer's interests* – Always try to look at the problem from the customer's point of view. For example, Mrs. Jones may be upset that you are late. Maybe the emergency you fixed for your last customer caused your tardiness. However, to Mrs. Jones, the important thing may be that she's running late in picking up her child from school. If this understandable concern for her child is added to concern over a furnace that isn't heating, Mrs. Jones might be short-tempered. If she speaks to you abruptly, try turning the conversation around by saying, "I know you must be concerned about my being late. I apologize. Please show me your furnace so we can get you back on schedule quickly."
- *Keeping your personal problems to yourself* – Remember, you are on the job to solve the customer's problem. Don't let your problems spill over into your dealings with customers. If you are in a bad mood because of personal problems, don't take it out on the customer.
- *Answering questions honestly, but positively* – If Mr. Brown asks if his equipment is worn out, you may reply, "It needs to be replaced," but you should avoid negative comments such as, "This pile of junk needs to be replaced."
- *Respecting your customer's opinions* – Suggest alternatives that fall short of telling a customer, "You're wrong!" Show that the customer's opinion may have value. For example, if Mr. Smith announces that the thermostat is no good and you suspect there is a wiring problem, don't say, "No, you're wrong. It's in the wiring." Instead, say, "I think it's the wiring, but I'll check out the thermostat too."
- *Not socializing while on the call* – Generally, service technicians are paid by the customer by the hour. If you explain pleasantly that you appreciate the offer of a cup of coffee, but that you're sure they'll understand if you decline, the customer will not be offended by your trying to keep the charges to a minimum. Unless it is absolutely essential, leave the mobile phone in your pocket. Better yet, leave it in the service vehicle. Customers hearing a technician engaging in any conversation that is unrelated to the job at hand will often become unhappy very quickly.

Further, you continue to show concern for your customers when you make it easy for them to contact you for repeat business. When possible, do the following:

- Leave a business card.

- Put a sticker with your company's phone number on the equipment itself or leave one with the customer for placement near the phone.
- Show the customer some simple maintenance techniques that will prolong equipment life, such as changing filters frequently (*Figure 5*). Tell the customer why this maintenance is helpful.
- Teach the owner steps for more efficient use of the equipment, if applicable. For example, show the owner how to program the thermostat (*Figure 6*) so less energy is used during times when the structure is not occupied.
- Discuss billing accurately and honestly.
- If the opportunity arises, discuss additional equipment to improve comfort and efficiency that the customer can purchase from your company. Ask for the customer's permission to notify the sales staff. Follow up by providing this lead to the sales staff.
- Perform customer follow-up, if possible.

1.2.4 Selling Yourself and Your Company

You may not consider yourself a salesperson, but selling is a key element of your job even though it may not always be a conscious effort. In reality, you are selling whenever you go on a service call. First, you are selling yourself because your appearance, professionalism, courtesy, and self-confidence will sell the customer on you.

Second, you are selling your company. The impression you make, along with your successful completion of the job, will determine whether the customer is likely to call your company again. Third, every service call you make is a selling opportunity. If you handle a trouble call for a customer who has no annual service contract, there is an opportunity to convince the customer that periodic maintenance will reduce or eliminate emergency calls and the discomfort that goes along with them.

You also have the opportunity to point out the need to replace aging equipment. You might be surprised how practical your customers can be. Many of them can readily understand the long-term benefit of replacing aging equipment with

Figure 5 Changing filters can increase equipment life and efficiency.

Figure 6 Show the customer how to avoid frustration and adjust the thermostat for greater comfort and efficiency.

new energy-efficient equipment and will see beyond the initial cost. It never hurts to ask. Besides, how would you feel if your company got a $100 service call and a competitor got a $5,000 contract to replace a worn out system for the same customer a year later? One way to avoid such embarrassment is to note the age of the customer's heating and cooling equipment and pass that information along to your company's sales department.

Think About It
Showing Concern for Customers

What would you do in these situations?

1. Mr. Gray owns a brand of furnace that is no longer being manufactured. You don't think parts are available to fix the burner. What would you say?
2. Mrs. Brown called for a technician to fix her furnace. As you enter the job site, she says, "It's cold in here." What questions would you ask her?
3. Mr. Green says his blower and burner are not operating in sequence and may need to be replaced. You think the problem is a faulty limit switch. How do you build his confidence in your skills without hurting his feelings?
4. Can you suggest some ways of advertising your company that are simple and free?

Additional Resources

Tools for Success: Critical Skills for the Construction Industry. Upper Saddle River, NJ: Prentice Hall.

Applied Communication Skills for the Construction Trades. Upper Saddle River, NJ: Prentice Hall.

1.0.0 Section Review

1. First impressions are usually made in the first 60 seconds of contact.
    a. True
    b. False

2. If you are going to be late for an appointment, the best thing to do is _____.
    a. just get there as quickly as you can
    b. call the dispatcher and have them reschedule for another day and time
    c. report it to your supervisor or let the customer know you will be late
    d. leave the first job unfinished in order to be on time for the next appointment

 03316 Customer Relations

SECTION TWO

2.0.0 HANDLING SERVICE CALLS

Objectives

Describe basic conduct required for a service call.
a. Describe how to conduct the three phases of a service call.
b. Describe ways to handle challenging customer situations.

Performance Task

1. Participate in at least three different role-playing situations related to challenging customer situations.

There is more to a repair call than simply showing up ready to work. In a typical case, the technician starts by interviewing the customer to find out information about the problem and, in many cases, soothing ruffled feathers. Once the repair is complete, there are wrap-up tasks that must be completed.

2.1.0 Conducting the Service Call

During each service call, you can do several things to enhance the customer's image of you and your company.

The typical service call has the following three parts:

- *The opening* – Meeting the customer and identifying the problem.
- *Performing the service* – Solving the problem.
- *The closing* – Leaving the customer with a positive impression.

Remember that your objective in servicing your customer's equipment goes beyond technical competence. Your objective is to win the customer's repeat business.

2.1.1 The Opening

On most service calls, the HVACR technician is going to face an unhappy and uncomfortable customer. The customer may have lost heating on a cold day or cooling on a hot day. In most cases, the problem has been going on for hours, possibly for more than a day. In many cases, the customer is looking for someone to blame. If it is equipment your company installed, the customer will be happy to blame you, even if you weren't personally involved. Always remember that your real responsibility is to solve a client's problem, and they write the checks that pay your wage. An air conditioning or heating unit has never paid a bill; only satisfied customers do that.

The first order of business is to defuse the situation. Listen attentively and sincerely to the customer's complaints and try to get him or her to describe how the problem manifested itself. Engage the customer by asking questions: Is the system operating but not providing enough heating or cooling? If it is not working at all, was there any noticeable problem before it stopped working? Were there unusual noises or odors? If this is a new customer, ask about previous problems. Engaging the customer in this way not only provides you with information about the problem, it helps to make a customer feel like an ally in the battle to restore normalcy. However, be careful about using industry jargon. Using technical terms may make you seem clever, but is more likely to further annoy the customer.

As you open the service call, you are already making your first impression. Your personal appearance is good. Your vehicle looks good. You show respect for the customer by appearing promptly and politely at the door. Now what? You should do the following:

- Smile and display confidence and polite respect for the customer (*Figure 7*). Make eye contact with the customer.
- Promptly identify yourself. Many customers waiting for a service worker are naturally cautious about admitting a stranger. Identify yourself clearly and show an appropriate ID,

Figure 7 A pleasant greeting is a good way to start a service call.

if you have one. Give the customer a company business card with your name on it. If your company does not provide business cards with the names of individual service technicians, encourage the company to at least have company business cards available so that you can give one to the customer in case they need to contact the company for follow up questions or concerns.

- Be respectful. If a customer introduces himself including his first name, don't assume you can call him by his first name. Some customers resent such familiarity. Above all, concentrate on remembering the customer's name in case you have to get their attention later on.
- Understand your customer's needs. Listen to your customer to learn what the problem is. Ask questions about what the customer has seen, heard, smelled, and felt. Ask when things occurred and how often they happened. Often, the customer can help speed your diagnosis by providing clues to the equipment problem. When you listen attentively, everyone wins.

It's important to ask open-ended questions, the kind that result in specific, information-gathering answers. You don't want to be led to the equipment without any information from the customer. Ask as many questions as you need to get the exact symptoms. Don't say, "I'm here to fix your furnace. Where is it?" Instead, say, "I came to repair your furnace. What seems to be the problem?"

- In a residential situation, do not enter the home until you are invited to do so, either verbally or by gesture. Before entering, don some disposable shoe covers (*Figure 8*). They should not be worn until you are at the entrance. Otherwise, they will get soiled as you walk from the truck to the door.

2.1.2 Performing the Service

While servicing the equipment, you should be adding to your customer's positive impression of you and your company. The customer has an equipment problem. You are there to solve it. Once you have listened to the customer explain the problem, you can build customer confidence by getting right to work to determine the probable cause of the malfunction. Practice good work habits. Avoid general conversation while working. Remember, many companies charge customers by the hour for service. The customer must feel that you are filling the time with productive work, not idle chatter.

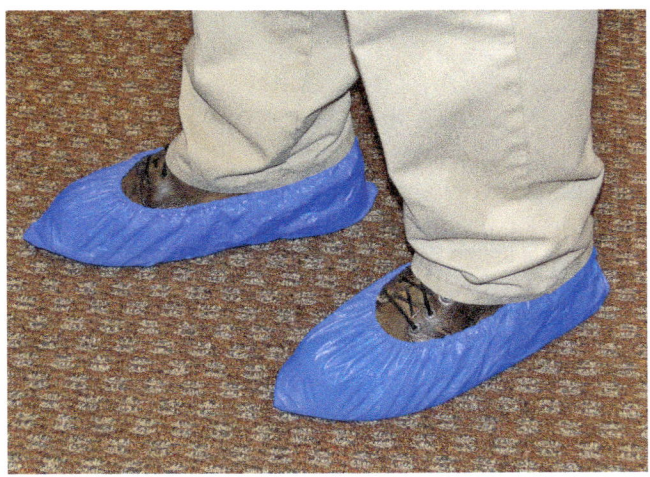

Figure 8 Disposable shoe covers.

If you make a mistake that is going to lengthen the duration of the service call, do not try to cover it up by telling the customer that the problem is worse than you thought and bill them for the extra time. Be honest. Own up to your mistakes and learn from them. Although constant errors will not be tolerated by employers or clients, a great deal of respect and admiration can be gained by stepping up when an error is made. Assure your client that any additional repair costs that resulted from your negligence or error will not be included on the invoice.

It is important to arrive prepared, with the proper tools and, to the extent feasible, replacement parts appropriate to the situation. The company dispatcher should have obtained information about the nature of the customer's problem and the type of equipment. If this is an existing customer of your company, you should have access to the make and model of the equipment. This will make it easier to be prepared and avoid wasting time. If you have to run back to the shop for what the customer might think is a standard part, they might think they are getting billed for unnecessary time.

During this part of the service call, you will need your professional skills to:

- Analyze the symptoms.
- Perform necessary testing to isolate the problem.

Think About It
Customer Relations

What questions can you ask Mrs. Jones when responding to an air conditioning service call?

- Determine the solution.
- Explain the solution to the customer, including your best estimate of what it will cost.
- If the customer approves, remedy the situation.

While the work is ongoing, maintain a clean, organized work area. Not only does this look far more impressive to a customer, it also helps to ensure your work activities will be completed safely. When the work is complete, make sure that all trash is picked up and inventory your tools. Take the extra step to ensure that the area is cleaner than it was when you arrived by picking up any other trash that is nearby. Successful technicians carry simple cleaning materials to clean the cabinets of furnaces and condensing units. Make sure that the equipment serviced looks its best before departing. Many companies also provide labels with company contact information to their technicians that are applied to the equipment before departing.

2.1.3 The Closing

This very important step helps set the tone for your customer's image of you and your company. When you finish the work:

- Neatly pack your tools.
- Return the premises to its original condition. (Replace covers, wipe off dirty fingerprints, clean up drop cloths, etc.)
- Explain to the customer that the problem has been solved. Explain what parts you replaced and offer to show the customer the defective parts.
- Demonstrate that the equipment works.
- Wrap up the call by relating your service to the customer, not the equipment. Don't say: "I fixed the thermostat; your furnace will turn on now." Rather, say: "I repaired your thermostat so you can be warm and comfortable again." A positive closing like this builds confidence in you and your company, and shows that your mind is in the right place.

Following up after the repair is important, but it is often overlooked. If time permits, call the customer a day or two later to ask if the repairs were satisfactory. If you made adjustments, stop by to see if they were acceptable. Often, a simple phone call from you can be the touch that creates a repeat customer. In your company, if the service technicians don't have the time to follow up on service visits, perhaps someone from the office staff can handle these calls.

2.2.0 Handling Challenging Customers

When you deal with customers, remember: the customer wants safety, comfort, and convenience. If your habits, behaviors, and attitudes inspire that customer to believe your company will provide for their safety and comfort in a convenient manner, the customer will return to your company rather than seeking out a competitor next time.

Sometimes, the service technician encounters problems that go beyond failed machinery. For example:

- A customer may be unduly worried about safety.
- A customer may be upset about something totally unrelated to the problem, but is taking out that irritation on the service technician.
- A customer may be upset over equipment performance and your company's bill.

In cases like these, the service technician's actions and attitude will often determine whether the service visit is a successful one or not.

The following sections cover several situations that require more advanced customer relations skills. Decide how you would handle these customers and compare your ideas with those of your fellow trainees.

Remember, there may be no single right way out of an angry confrontation, but your attitudes and reactions can go a long way to help resolve even the most difficult situations.

2.2.1 Fearful Customers

Sometimes customer fears are based on experience, rumors, or misunderstandings. As the technician, you have the technical information that can calm your fearful customer. There is always a

Going the Extra Mile

People often neglect to change or clean the air filters for their equipment. Sometimes they are not physically able to do it. If you are on a service call for a new customer that does not have an annual service contract, offer to check the condition of their filters. This takes very little time and could increase the value of the service call. Make a note of the filter sizes so you can put them on the truck the next time the customer calls. If the filters have been neglected, it is a good opportunity to mention the value of a service contract.

reason why your customer has a fear. Find it and work from the facts to ease the customer's mind. If you listen with concern and take customers' fears seriously, they will feel reassured and be more willing to listen to your explanation.

Exercise Question:

Mr. Fearful thinks his furnace is leaking gas because he smelled some when the pilot light blew out. You have re-ignited the pilot and are closing the call. What can you say to calm his fear that the house will explode from a leak at the pilot light?

Exercise Solution:

Fear is real. Often you can discover the reason for it by asking the following questions:

- When does this happen? When did it start?
- When it happens, what does it look like?
- What does it sound like?
- What does it smell like?
- Before it happens, are you doing something different from what you used to do?

After you have found the answers, don't just say, "Yes, when the pilot light goes out there is a smell. It is fixed." This has not addressed Mr. Fearful's concerns for the next time it happens. Instead, explain: "When a pilot goes out, there is a smell of gas. A minute quantity escapes but will not ignite. It is also mixed with a definite odor that alerts you to re-light the burner. The small amount of gas will not cause a fire. The vent pipe will exhaust it. Your family is not in any danger." Possibly suggest that a gas sniffer alarm and carbon monoxide sensor be installed to give the customer peace of mind.

Since this explanation not only addresses the technical problem, but also the fear associated with it, the customer will be more satisfied.

2.2.2 Opinionated Customers

Opinions are beliefs based not entirely on facts, but also on what seems probable in a person's mind. Everyone has opinions. Everyone also likes to get approval for opinions held. If your opinions are ridiculed, you feel hurt and offended.

It is important that customers do not feel their opinions have been challenged or belittled. With a customer, it's best to turn the conversation to the technical problem at hand (even when you agree with the opinion).

Exercise Question:

Mrs. Opinionated has had a bad week, including the failure of her air conditioner. She's trying to draw you into her earlier problems and her opinions about them. You're trying to finish quickly because you have two other service calls before lunch. What do you say?

Exercise Solution:

Taking what seems to be a negative approach might offend her. So don't say: "Yeah, we all have problems. My problem is fixing air conditioners. Where is it?" Instead, say: "I've had times like that myself. I can help you with one of your problems, though. Please show me your air conditioner so I can help you solve that problem." This offers sympathy while at the same time diverting the customer to the task at hand.

2.2.3 Argumentative Customers

Some customers are just the argumentative type. They want you to challenge their opinions, just to get an argument going. You're sensitive to the customer, and you don't want to seem aloof. You don't want to express negative opinions either. You also know that some topics are really argument-prone, such as politics, religion, children, and in-laws. Here, too, the safest approach is to avoid controversy by diverting the conversation to the task at hand.

Exercise Question:

Mr. Argumentative has no airflow from his furnace. You've discovered a broken fan belt. He wants to pull you into a discussion of local politics. He thinks the mayor is a crook, but he senses you disagree and wants to argue. You do not want

---

Think About It

Closing

After you've replaced the refrigerant charge in Mrs. Jones' air conditioner, what closing statement can you make?

Verify That the System Works

One of the most important things you can do at the end of a service call is to demonstrate to the customer that the system works. Don't leave them wondering about it; show them.

03316 Customer Relations

Module Twelve 11

to be drawn into this discussion. How would you handle this?

Exercise Solution:

The best way to handle this is to state a general fact that will not offend, and to avoid prolonging or expanding the political discussion while you gently turn the conversation to the problem you're there to solve. (Even if you disagreed with the customer, the technique would still be same: shifting attention to the fan belt problem.) Say, "Some people feel that way. But my feeling is that you'll be a lot more comfortable after I replace this fan belt."

2.2.4 Sloppy Customers

As you make service calls, you'll see all kinds of housekeeping: some spotless; some cluttered; some downright dirty. Always, it's best to avoid offending your customer. When admitted to the job site, avoid being judgmental in your words and gestures. Since you're a visitor, the owner will often offer an excuse or reason if things are messy. In some homes, the owner's sloppy habits can adversely affect the operation of the equipment.

You can handle such sensitive situations in the following ways:

- *Focus on the task at hand* – Don't look like you're judging the job site. Avoid comments like "What a mess!" and raised eyebrows or rolled eyes that say the same thing. Assume that the owner is the unfortunate victim of several small domestic catastrophes that will be remedied soon.
- *Put the owner at ease* – If you must reply, avoid degrading and comparing them unfavorably. Say: "This happens at our home, too."
- *Be diplomatic* – If it's cluttered in front of the furnace, move only what's necessary and try to restore things when you're finished. Don't be obvious in moving things. As you clear space, avoid grunts, groans, and comments. If obstructions are blocking the flow of combustion air, let the customer know that this could be a safety hazard. Also, if you see exposed detergents, bleach, or chemicals near a furnace, point out the hazards.
- *Don't sound like an accuser* – Avoid "you" statements. Talk about what's best for the health of the furnace or make up a similar situation that you solved, using your own family as an example.

2.2.5 Angry Customers

In the last section, you were advised to avoid "you" statements to a customer when it might suggest a negative judgment; for example, accusing them of bad housekeeping. There is, however, a good use of a "you" statement, and that's when the customer has a right to be angry.

When equipment breaks down twice in three days, or when the customer has tried to meet the service technician two days in a row without success, the customer may feel justified in venting anger at the first available target: the technician who appears at the door. What should you do?

First, let the angry customer vent his or her emotions. Next, acknowledge the customer's feelings. Sometimes, this means saying something sympathetic like, "I'd feel that way, too." Finally, assure the customer that you'll do your best to solve the problem.

Think About It

Mr. Sloppy

Mr. Sloppy takes you into his home to fix the air conditioner. The musty air smells of cat litter; dusty newspapers fill all the chairs; the kitchen looks like a scene from a disaster movie. Mr. Sloppy says, "Isn't this place a mess?" What would you say?

Exercise Question:

Mrs. Smith's furnace was repaired just last week. It has broken down again, leaving the family without heat for two days. The children have colds. The Smiths have already paid a sizable sum for the repair that did not last. As you walk in the door, Mrs. Smith snarls at you. What do you say?

Exercise Solution:

Here is an appropriate answer: "Hello, Mrs. Smith. I hear your furnace isn't working again. I'm sure you're very concerned about your children's comfort. I know just how you feel. My air conditioner quit twice on me last summer during a heat wave. I'll get to the problem as quickly as I can. Let's see what the matter is." You have shown concern for her feelings and worries; you have acknowledged that she is justified in her concern; and you have assured her that you will address the problem quickly and to her satisfaction.

2.2.6 Critical Customers

Sometimes you'll encounter a customer who can't be satisfied, no matter what you say or do. Often this happens because the customer is under a lot of stress or has had a bad experience that hasn't been dealt with properly. When this customer explodes, you may be the nearest target. How can you handle these situations with sensitivity and tact?

You've heard of cases where a parent, angry about a bad day at work, comes home and yells at the kids. This displacement of anger can also happen when a person has an unresolved conflict that is upsetting.

- A man who refused to pay extra for a weekend call now has a mess on his hands because the pipes in his poorly insulated laundry room are frozen.
- A woman having trouble with her sales accounts at work doesn't really want to be home waiting for a service technician when she should be dealing with her office problems.

The first step in handling these situations is to remember that everyone has bad days. Approach the customer with the attitude that someday you will feel out of sorts. On these days, you hope those around you will make allowances and be tolerant. Give your customer the same understanding you'd like to have at these times.

If the customer explodes, don't take it personally. After all, you may have just been a convenient target. You may not be responsible for the problem. Review what you have done. If your work was correct, shrug off the comments silently.

If your customer is justifiably upset, let the customer vent some anger. Empathize with the feelings, explaining that you would probably feel the same under the same circumstances. Reassure the customer that you will fix the HVAC technical problem as quickly as you can, so at least part of the day will be a positive experience.

Get the job done as quickly as you can. Get out of the situation as soon as possible without offending the customer. Stick to the facts. Focus on the mechanical problem as you solve it.

2.2.7 Customers with Unresolvable Problems

Once in a while, you will have a customer whose problem you cannot resolve. For example, the wiring may be too outdated to support the air conditioner the customer wants installed, or the existing furnace may be inadequate to heat a newly expanded space. What should you do?

Here is a poor response: "Sorry, sir. Your electrical service is not adequate to handle this air conditioner. I can't do electrical work."

Think About It
Mr. Negative

Mr. Negative is unhappy. The builders who just completed his home left several jobs unfinished. He begins a tirade about builders and service workers in general, including you, in his speech, even though you've only just arrived. The tirade begins anew after you have already started work on his furnace. How do you handle this difficult customer?

A better response might be: "I found that your electrical service was too small to handle the new air conditioner. We can't do this kind of work, but here is the name of an electrical company we often deal with. Or, you can check the phone book for the names of other electricians. If you prefer, we can subcontract a licensed electrical contractor to do the job, then after the electrical service is fixed, we'll be glad to come back to install the air conditioner."

A typical homeowner might prefer to contact a company that is recommended, rather than searching out several competing companies and taking a chance on the quality of their work. Your positive, helpful attitude can leave a favorable impression, even if the customer already has an electrician. You have identified the problem and proposed a good solution, then offered to complete the job when the customer is ready.

2.2.8 Customers Who Request Help With Odd Jobs

Sometimes a customer asks a service technician to perform a task that is outside company regulations. In these situations, explain why you can't help. Then suggest another way to solve the problem.

Exercise Question:
While working on her furnace, Mrs. Oddjob asks you to move her freezer. What do you say?

Exercise Solution:
A good answer is: "I'm sorry, Mrs. Oddjob, but company regulations won't let me handle appliances we don't service. I know you're anxious to move that freezer. Perhaps a neighbor can help you. It wouldn't be fair to my later customers if I spent too much time at any one place. As you know, you're billed by the hour for my time. I know you want to keep your costs down." This response recognizes that the customer has a problem, but explains why you can't help her solve it.

Think About It
Mr. Knowitall

Mr. Knowitall wants to use his existing furnace to heat the home addition he just proudly completed himself. He has asked you to check out the ducts and connect the furnace, since he has already made up his mind it can handle the additional load. He's an expert at everything; a legend in his own mind. The addition is 950 square feet on the north side, with two outside walls. You determine that the existing furnace is inadequate to service the additional load. How do you approach this client with the news?

Additional Resources

Tools for Success: Critical Skills for the Construction Industry. Upper Saddle River, NJ: Prentice Hall.

2.0.0 Section Review

1. It is a good idea to establish rapport with a customer by chatting about general topics while you are working on their equipment.
    a. True
    b. False

2. If an angry customer starts berating you because equipment your company installed has stopped working, you should _____.
    a. immediately turn around and leave
    b. shift the blame to others in your company
    c. empathize with the customer
    d. tell the customer you don't want to hear it

Summary

Knowing how to diagnose and repair failures in HVACR equipment is only half the battle. The HVACR technician, as the company's representative, must know how to properly interact with customers. When customers call-in for repair service, it usually means that they are suddenly without heat, air conditioning, or refrigeration, so to them it is a crisis. They are likely to be uncomfortable and may be angry as well. In order to make a good first impression, the technician needs to be careful to be on time, well-groomed, and have a positive, friendly attitude. It is very important to communicate with the customers in positive ways that reflect concern for their problems. Listening patiently and attentively to customers is a good way to earn their confidence.

Learning how to deal with difficult customers is an important aspect of a technician's job. Technicians are likely to encounter customers who are angry, critical, demanding, or opinionated. It is important to learn how to deal with each type.

Review Questions

1. A customer begins forming a first impression about you and your company _____.
   a. after you have fixed their equipment
   b. after you have submitted your bill
   c. as soon as you appear at their premises
   d. as soon as you speak

2. You arrive at a service call wearing a NY Yankees ball cap. Unfortunately, the customer is a Boston Red Sox fan who hates the Yankees, so he immediately starts insulting the Yankees. Looking back on this incident, what do you think you would do in the future?
   a. Turn around and walk away because you won't do work for a Red Sox fan.
   b. Leave the ball cap in the truck to avoid the possibility of confrontation.
   c. Argue the relative merits of the two teams.
   d. Brush past the customer and find the equipment that needs to be repaired.

3. When speaking with a customer, always _____.
   a. start by telling them about your training and experience
   b. break the ice by chatting about local politics
   c. tell the customer if you think their equipment is low-quality
   d. show courtesy, concern, and respect

4. One way to make a good first impression is to use a lot of technical terms when talking to the customer.
   a. True
   b. False

5. A customer is giving a long, boring description of a problem. Because it is obvious that the customer has no technical knowledge, you should _____.
   a. tune the customer out and start thinking about how you will approach the problem
   b. nod appreciatively and pretend you are interested even if you are thinking about something else
   c. listen carefully because it is part of establishing rapport with the customer
   d. talk about similar problems you have encountered and how you solved them

6. During the introduction at the door, your customer says: "Hi, I'm Tom Collins." That means you can call him by his first name.
   a. True
   b. False

7. After telling the customer how long it will take you to complete a repair, you accidently tear a length of flex duct in the system. Replacing it will lengthen the service call considerably and the customer is paying by the hour. What should you do?
   a. Tell the customer about the mistake and explain that the repair costs will not be included in their bill.
   b. Tell the customer that you discovered an additional problem with the system that will lengthen the service call and increase their cost.
   c. Tell the customer the problem was caused by another technician who previously serviced the system.
   d. Ignore the damage because it will eventually result in another service call and increase your company's billing.

8. Mrs. Smith is worried because she has read about families dying of carbon monoxide poisoning from a defective furnace. How should you respond?
    a. tell her to relax and stop worrying
    b. explain that it hardly ever happens
    c. calm her and use factual information to ease her mind
    d. ask your supervisor to contact her

9. In order to access a furnace, a lot of clutter must be moved from in front of it. The way to deal with this problem is _____.
    a. refuse to continue working until the customer moves it
    b. move it out of the way and leave it there when you are finished work
    c. call your supervisor and explain the problem
    d. move it out of the way and return it when the work is finished

10. A customer was supposed to have had an electrical circuit installed to serve a new humidifier you have come to install. The electrical work is not done and the customer asks you to do it, even though you are not an electrician. What should you do?
    a. Go ahead and do the wiring because it's an easy job.
    b. Advise the customer to have the work done by a licensed electrician and provide a reference.
    c. Temporarily run an extension cord from the nearest electrical outlet so you can do the humidifier installation.
    d. Contact your supervisor and see if it is okay to do the electrical work.

Trade Terms Introduced in This Module

Empathy: The ability to identify with another person's situation or feelings.

Rapport: Establishing a relationship of mutual trust.

Additional Resources

This module presents thorough resources for task training. The following resource material is suggested for further study.

Tools for Success: Critical Skills for the Construction Industry. Upper Saddle River, NJ: Prentice Hall.
Applied Communication Skills for the Construction Trades. Upper Saddle River, NJ: Prentice Hall.

Figure Credits

iStock/lisafx, Module opener, SA03
Topaz Publications, Inc., Figures 1, 2, 4, 5, 7, 8
iStock/piranka, Figure 6
iStock/tlnors, SA01
iStock/theprint, SA02

Section Review Answer Key

Answer	Section Reference	Objective
Section One		
1. a	1.1.0	1a
2. c	1.2.1	1b
Section Two		
1. b	2.1.2	2a
2. c	2.2.6	2b

NCCER CURRICULA — USER UPDATE

NCCER makes every effort to keep its textbooks up-to-date and free of technical errors. We appreciate your help in this process. If you find an error, a typographical mistake, or an inaccuracy in NCCER's curricula, please fill out this form (or a photocopy), or complete the online form at **www.nccer.org/olf**. Be sure to include the exact module ID number, page number, a detailed description, and your recommended correction. Your input will be brought to the attention of the Authoring Team. Thank you for your assistance.

Instructors – If you have an idea for improving this textbook, or have found that additional materials were necessary to teach this module effectively, please let us know so that we may present your suggestions to the Authoring Team.

NCCER Product Development and Revision
13614 Progress Blvd., Alachua, FL 32615

Email: curriculum@nccer.org
Online: www.nccer.org/olf

❏ Trainee Guide ❏ Lesson Plans ❏ Exam ❏ PowerPoints Other _____

Craft / Level: _____ Copyright Date: _____

Module ID Number / Title: _____

Section Number(s): _____

Description: _____

_____

_____

_____

_____

Recommended Correction: _____

_____

_____

_____

_____

Your Name: _____

Address: _____

_____

Email: _____ Phone: _____

Glossary

AL-CU: An abbreviation for aluminum and copper that is commonly marked on electrical connectors to indicate the device is suitable for use with conductors made with either metal.

Annual Fuel Utilization Efficiency (AFUE): An industry standard used to define furnace efficiency.

Atomize: Reduce to a fine spray.

Base speed: The rating given on a motor's nameplate at which the motor will develop rated horsepower at its rated load and voltage.

Boiler horsepower (BoHP): The heat necessary to vaporize 34.5 pounds (15.7 kg) of water per hour at 212°F (100°C) and 0 psig (0 kPag). This is equal to a heat output of 33,475 Btus per hour or about 140 square feet (13 m^2) of steam radiation. 1 BoHP is also equal to 9.803 kilowatts.

Break-away torque: The torque required to loosen a fastener. This is usually lower than the torque to which the fastener has been tightened.

Burnout: The condition in which the breakdown of the motor winding insulation causes the motor to short out or ground electrically. The breakdown of the insulating coating typically results in the formation of strong acids.

Cad cell: A light-sensitive device (photocell) in which the resistance reacts to changes in the amount of light. In an oil furnace, it acts as a flame detector.

Capacitance boost: A procedure used to start a stuck PSC compressor. It involves momentarily connecting a start (boost) capacitor across the run capacitor of the stuck compressor to increase the starting torque in an attempt to start the stuck compressor.

Connector: A device used to physically and electrically connect two or more conductors.

Cooling compensator: A fixed resistor installed in a thermostat to act as a cooling anticipator.

Coping: The upper masonry course of a chimney designed to keep water out.

Deadband: A temperature band, usually 3°F (1.6°C), that separates heating and cooling in an automatic changeover thermostat.

Demand defrost: A defrost control strategy that typically does not depend on time to determine when a defrost cycle is needed. Instead, the system responds directly to current conditions, such as temperature or pressure drop through the outdoor coil, to determine when frost is present.

Differential: The difference between the cut-in and cut-out points of a control.

Direct-acting valve: A valve in which the solenoid coil operates the valve stem directly to open or close.

Downdraft: A situation in which outdoor air is forced down a chimney, preventing proper draft. Downdraft can be caused by not extending the top of the chimney above the roof line.

Drip leg: A drain for condensate in a steam line placed at a low point or change of direction in the line and used with a steam trap.

Droop: A mechanical condition caused by heat that affects the accuracy of a bimetal thermostat.

Dry steam: The steam that exists after all the water has been vaporized into steam at its saturation temperature (being superheated).

Dump zone: An area used as a point to discharge unneeded supply air from a zoning system. The dump zone is typically in the conditioned space, but not an area where maintaining a specific space temperature is essential. A dump zone generally has no supply air grilles, but a return grille is often installed to relieve room pressure and return mixed air back to the system.

Electronically commutated motor (ECM): A variable-speed electric motor based primarily on a DC-driven permanent magnet design that responds to microprocessor inputs to achieve the desired speed.

Empathy: The ability to identify with another person's situation or feelings.

Fault isolation diagram: A troubleshooting aid usually contained in the manufacturer's installation, startup, and service instructions for a particular product. Fault isolation diagrams are also called *troubleshooting trees*. Normally, fault isolation diagrams begin with a failure symptom, then guide the technician through a logical decision-action process to isolate the cause of failure.

Field wiring: Wiring that is installed during installation of the system, such as a thermostat hookup to a unit.

Firetube boiler: A boiler design where the hot flue gases flow inside the tubes, which are surrounded by the water to be heated.

Flame rectification: A process in which exposure of a sensing rod to a flame produces a tiny direct current which can be used to control the gas valve.

Flame-retention burner: A high-efficiency oil burner.

Flash economizer: An energy-conservation process used in chillers that accomplishes intercooling of the refrigerant between the condenser and evaporator using a flash chamber or similar device. Flash gas of coolant produced in the flash chamber flows directly to the second stage of the chiller compressor, bypassing the evaporator and the first compressor stage.

Flash steam: Formed when hot condensate is released to a lower pressure and re-evaporated.

Floodback: A condition in which there is liquid refrigerant in the suction vapor being returned to an operating compressor.

Flooded starts: A condition in which slugging, foaming, and inadequate lubrication occur at compressor startup due to oil in the compressor having absorbed refrigerant during the off cycle. When the compressor starts and the pressure in the crankcase drops quickly, the refrigerant boils rapidly, foaming the oil.

Free cooling: A mode of economizer operation. During cooling operation, the system compressor is turned off and the indoor fan is used to bring outside air into the building to provide cooling.

Insulated gate bipolar transistor (IGBT): A type of transistor that has low losses and low gate-drive requirements. This allows it to be operated at higher switching frequencies.

Invar®: An alloy of steel containing 36-percent nickel. It is one of the two metals in a bimetal device used in temperature controls.

Joist: Framing member used to support floors and ceilings.

Label diagram: A troubleshooting aid usually placed in a convenient location inside the equipment. It normally depicts a wiring diagram, a component arrangement diagram, a legend, and notes pertaining to the equipment.

Ladder diagram: An electrical diagram that provides a schematic diagram of circuits, with the load lines arranged like the rungs of a ladder between vertical lines representing the power source.

Lithium bromide: A caustic salt solution commonly used in absorption chillers.

Lockout: A safety feature that prevents a furnace from automatically cycling if the flame fails.

Lug: A wiring terminal used on wire sizes No. 8 and larger. Also a term used to describe a screw-type terminal that simply applies pressure to a stripped wire, without the need for a crimp-on terminal.

Mechanical cooling: The cooling provided in the conventional manner by the compressor and mechanical refrigeration system.

Modulating gas valve: A gas valve that precisely controls burner capacity by adjusting the flow of gas to burners in small increments.

Net stack temperature: A value needed to perform a combustion efficiency analysis, calculated by subtracting room temperature from the temperature measured in a furnace's vent or vent connector.

Nominal size: A means of expressing the size of a bolt or screw. It is the approximate diameter of a bolt or screw.

Over-firing: A situation caused by an oversized nozzle in which excess oil is pumped into the combustion chamber.

Oxygen scavengers: Chemicals added to boiler feed water that chemically combine with dissolved oxygen to prevent it from causing corrosion within the system. Typically creates a sludge that must be purged with boiler blowdowns or skimming.

Phase change: The transition of a refrigerant's physical state (or phase) to another, such as from liquid to vapor.

Piezoelectric: A property of many natural and synthetic materials where a voltage is produced when the material is mechanically stressed (placed under pressure, bent, or twisted). Used in strain gauges and pressure sensors.

Pilot-operated valve: A valve in which a solenoid operates a smaller valve, which then ports more pressure and/or volume to a larger valve to open or close it. Large valves use this strategy due to the limitations in the magnetic field of solenoid coils.

Polling: A network-communications arrangement where a VVT system seeks input from individual zones, determines what demands have priority, and then responds accordingly. Electronic participants in a polling system are referred to as voters.

Pressure-dependent operation: A VVT-system approach where dampers are positioned based on demand only, or operate in two positions. The result is that the design volume of air may or may not be flowing through the damper at any time. The volume of air flow through each damper is dependent on the duct pressure; it fluctuates as the pressure fluctuates. A minimum damper position is set, but air volume is not a controlling factor, and so the minimum damper position is a fixed point.

Pressure-independent operation: A VVT-system approach where dampers are positioned by demand and by the actual volume of air flowing through the damper. To achieve a specific amount of airflow, the damper positions itself as necessary. Maximum and minimum airflow setpoints are in effect, and the damper will modulate as necessary to remain within these boundaries.

Primary control: A combination switching and safety control that turns the oil burner on and off in response to signals from the thermostat and safety controls.

Rapport: Establishing a relationship of mutual trust.

Refnet joints: Engineered joints used as connection points in a VRF-system refrigerant network. Refnet joints connect the main refrigerant lines to branch runs or to indoor units.

Refractory: Materials such as ceramics designed to withstand extremely high temperatures.

Reverberator: A wire or corrugated-metal screen used to cover the ceramic grids in a gas-fired radiant heater, which boosts the radiant-heat output of the heater by converting convective heat produced by the grid into radiant heat.

Run-down resistance: The torque required to overcome the resistance of associated hardware such as locknuts and lock washers, when tightening a fastener.

Saturated steam: The pure or dry steam produced at the temperature that corresponds to the boiling temperature of water at the existing pressure (saturation).

Secondary coolant: Any liquid, such as water, that is cooled by a system refrigerant and then used to absorb heat from the conditioned space without a change in state. Also known as *indirect coolant*.

Seismic activity: Movement of the earth such as during an earthquake.

Seismic restraint: A device or method used to prevent or restrict the movement of equipment during a seismic event.

Set or seizure: In the last stages of rotation in reaching a final torque, the fastener may lock up; this is known as setting or seizing. This is usually accompanied by a noticeable popping effect.

Slinger ring: A ring attached to the outer edge of the fan that picks up water as the fan rotates and slings it onto the condenser coil.

Slugging: The entrance of liquid refrigerant and/or a significant amount of oil into a compressor cylinder. Slugging can occur at startup due to condensed refrigerant and/or collected oil near the compressor inlet.

Smoke spot test: A test to determine the level of smoke in the flue gases.

Soot: Fine carbon particles resulting from incomplete combustion. An accumulation of soot on heat exchangers will reduce combustion efficiency and result in increased carbon monoxide production.

Submerged flame: A method of gas-fired radiant heater operation that uses burning gas within two-layer, porous-ceramic grids.

Superheated steam: Steam at a temperature above the saturation temperature for its pressure.

Temperature differential: The difference in temperature between the flue gases and the gases above the combustion chamber.

Tensile strength: The maximum stress a material can endure before being pulled apart.

Terminal: A device used for connecting wires to an object.

Termination: The connection of a wire to its destination.

Thermal economizer: An energy-conservation process used in chillers in which warm condensed refrigerant is brought into contact with the coldest (inlet) water tubes in a condenser or heat exchanger. This causes the condensed refrigerant to subcool to a temperature below the condensing temperature.

Thread classes: Threads are distinguished by three classifications according to the amount of tolerance the threads provide between the bolt and nut, and are designated by a number and letter combination that indicates the degree of fit for a thread. Higher numbers indicate a tighter fit.

Thread series: The combinations of diameter and pitch in threads that are applied to specific thread diameters to identify the coarseness or fitness of a thread.

Thread standards: An established set of standards for machining threads.

Torque: Applying a specific amount of twisting force to a threaded fastener.

Trim: External controls and accessories attached to the boiler itself, such as sight glasses and water feeder controls.

Turndown ratio: In steam systems, the ratio of downstream pressure to upstream pressure, usually applied to steam pressure-reducing valves. More than a 50 percent turndown ratio is generally considered a large pressure reduction.

Unit cooler: A packaged unit containing evaporator coil, metering device, and evaporator fan(s) that serves as the evaporator section for a refrigeration appliance.

Venturi tube: A short tube with flaring ends and a constricted throat that is used for measuring flow velocity by comparing the throat pressure to the inlet pressure. (Throat pressure decreases as the fluid's velocity increases.)

Vibration isolator: A device that prevents the transmission of vibrations from operating machinery to surrounding objects.

Water hammer: A banging noise in water pipes that occurs when the water supply solenoid valve on a humidifier abruptly shuts off water to the humidifier.

Water hammer: A condition that occurs when hot steam comes into contact with cooled condensate, builds pressure, and pushes the water through the line at high speeds, slamming into valves and other devices. Water hammer also occurs in domestic water systems. If the cause is not corrected, water hammer can damage the system.

Watertube boiler: A boiler design where the water to be heated flows inside the tubes, with the hot flue gases flowing around the outside of the tubes.

Weatherized furnace: A furnace that has been well insulated and has a weather-resistant case as specified by government standards.

Wiring diagram: A troubleshooting aid, sometimes called a schematic, that provides a picture of what the unit does electrically and shows the actual external and internal wiring of the unit.

Index

A

Absorption chillers, (03305):34–35
Accessory devices, connecting, (03312):13
Accessory troubleshooting
   basics, (03312):1
   electrical problems, (03312):1–2
   electronic air cleaners (EACs), (03312):8–12
   humidifiers, (03312):2–8
   operating sequence in, (03312):2, 3, 4
   outside air accessories
      economizers, (03312):18–19
      recovery ventilators, (03312):21–22
   service/product literature, (03312):2, 5
   ultraviolet (UV) lamps, (03312):11–12
AC-choke VFD, (03314):64
Accumulators, refrigeration systems, (03304):26–27
Acid contamination, (03210):17–18
Acid/moisture test kits, (03210):18, 30
Acorn nut, (03313):5
Adjustable-frequency drive (AFD), (03314):63
AFD. *See* Adjustable-frequency drive (AFD)
AFUE. *See* Annual Fuel Utilization Efficiency (AFUE)
Air
   for combustion, (03209):24
   combustion, ventilators for, (03312):19, 23
   discharge air sensor, (03312):15
   outside air, accessory troubleshooting
      economizers, (03312):18–19
      recovery ventilators, (03312):21–22
Air cleaners, electronic (EACs)
   airflow rates, (03312):11
   cleaning, (03312):10–11
   collector section, (03312):8
   filter, (03312):8
   function of, (03312):8
   humidifiers, (03312):2
   operation, typical, (03312):8–9
   overload conditions, (03312):11
   power supply, (03312):8
   prefilter, (03312):8
   schematic diagram, (03312):8–10
   troubleshooting, (03312):8–12
Air conditioning, humidistats, (03312):7
Air conditioning systems
   basic, (03210):3–4
   high-pressure side, (03210):3–4
   low-pressure side, (03210):3–4
Air contamination, (03210):16–17
Air distribution systems, troubleshooting
   blower motor, (03209):16–17
   ductwork, (03209):16–17
   guide, (03209):33–34
   induced-draft systems, (03209):17–18
   obstructed airflow, (03209):16
Airflow
   checking, (03311):21
   function of, (03315):4
   obstructed, (03209):16
   problems
      condensers, (03210):11–12
      evaporators, (03210):10–11

Air handlers
   ductless split systems, (03315):26, 27–29
   seismic restraints, (03313):26
   vibration insulators, (03313):23–25
Air management, boilers, (03305):1, 22–23
Air purification, ultraviolet (UV) lamps for, (03312):11–12
Air separation, (03305):22
Air tightness
   average, (03312):8
   building, (03312):7–8
   loose, (03312):8
Air-to-air heat pumps, (03311):2, 25
Air-to-water heat pumps, (03311):2, 4
AL-CU, (03313):31, 33, 38
AL-CU connectors, (03313):33
AlphaSan®, (03304):35
Aluminum connectors, (03313):33
Aluminum-to-copper connections, (03311):22
American Petroleum Institute (API) gravity, (03310):8
American Society of Mechanical Engineers (ASME), (03306):12
Anchor bolts
   concrete inserts, (03313):14–15
   expanding concrete anchors, (03313):17, 18
   non-expanding concrete anchors, (03313):16–17
   Tapcon® screws, (03313):17–18
Angry customers, (03316):12–13
Annual Fuel Utilization Efficiency (AFUE)
   defined, (03310):1, 36
   furnace efficiency stated as, (03209):24
   gas furnaces, (03310):19
   oil furnace, (03310):19
   ratings, mapped, (03209):6, 9
Annual Fuel Utilization Efficiency (AFUE), boilers, (03306):9
API. *See* American Petroleum Institute (API) gravity
Aquastat, (03314):58
Argumentative customers, (03316):11–12
ASME. *See* American Society of Mechanical Engineers (ASME)
ASTM International, steel bolts and screws grade markings, (03313):2, 3
Atomize, (03310):1, 2, 36
Attitudes effect on customer relations, (03316):1–3
Automatic-changeover thermostats, (03314):16–17
Average air tightness, (03312):8

B

Bacteria, (03304):1
Balanced pressure thermostatic steam traps, (03306):38, 42
Balancing valves, (03305):17, 18–19
Barometric dampers, (03315):4, 5, 6
Base speed, (03314):62, 66, 85
Beam-type torque wrench, (03313):11
Bearings, centrifugal pumps, (03305):14
Behaviors effect on customer relations, (03316):1–3
Bellows thermostatic steam traps, (03306):43
Belly-band heaters, (03210):25
Belly heaters, (03305):33
Bimetal thermostatic steam traps, (03306):38, 42, 43
Bimetal thermostats, (03314):14, 15–16

Biodiesel heating, (03310):3
Blind rivets, (03313):20
Blowdown valves, (03306):10–11, 26–27
Blower motor vibration insulators, (03313):22–23
Blowers
   airflow problems, (03210):10
   humidifiers, (03312):2, 7
Blower wheels, dirty, (03209):16–17
BoHP. *See* Boiler horsepower (BoHP)
Boiler horsepower (BoHP), (03305):6, 11, 44
Boilers
   air management, (03305):1, 22–23
   commercial hot-water
      cast-iron sectional, (03305):8
      combustion systems, (03305):6–7
      condensing, (03305):12
      construction, (03305):7
      copper-finned tube, (03305):7–8, 13
      electric, (03305):11–12
      electrode, (03305):11–12
      firetube, (03305):6, 9–10, 44
      heat sources, (03305):6
      indirect electric, (03305):12
      low-pressure, (03305):7
      medium-pressure, (03305):7
      operations, (03305):12–13
      safety controls, (03305):12–13
      Scotch Marine, (03305):9–10
      steel vertical tubeless, (03305):10–11
      watertube, (03305):6, 9–10, 44
   gas-fired. *See also* heating systems, gas
      combustion air, (03209):25
      installation, (03209):25
      troubleshooting control circuits, (03209):11–13
   high-temperature, high-pressure water, (03305):7
   natural gas, (03305):6, 7
   outdoor installation, (03305):7–8
   steam, (03305):7
   steam leaks, (03305):7
Boilers, steam
   blowdown procedure, (03306):27
   blowdown valve, (03306):10–11, 26–27
   hot-water boilers vs., (03306):9
   operating/safety controls
      high-pressure limit control, (03306):13–14
      pressure-balancing, (03306):20
      pressure gauges, (03306):9–10
      pressure-relief valve, (03306):14
      water cutoff/feeder controls, (03306):11–13
      water level indicators, (03306):10–11
   siphon assembly, (03306):9–10
   skimming procedure, (03306):26–27
   water treatment, (03306):27
Bolts
   extractions, (03313):8
   grade markings, (03313):2
   grades of, (03313):2, 3
   hold-down, (03313):22
   machine bolts, (03313):2–4
   stud bolts, (03313):2–4
Bolt threads, (03313):2
Bottoming taps, (03313):7
Braking, dynamic, VFD
   DC braking of an AC motor, (03314):70–71
   with an AC drive, (03314):71–72
Break-away torque, (03313):1, 12, 38
Building air tightness, (03312):7–8
Bumpers, (03313):27

Burners
   direct-fired radiant, (03209):19
   flame-retention, (03310):1, 6, 36
   high-pressure gun-type, (03310):2–3
   pressure-type oil
      biodiesel, (03310):3
      combustion air, (03310):3
      components, (03310):3
      flame-retention burner, (03310):6
      high-pressure gun-type, (03310):2–3
      ignition system, (03310):3, 5–6
      lockouts, (03310):3
      low-pressure gun-type, (03310):2
      nozzle assembly, (03310):4–5, 6
      nozzle flow rate, (03310):6
      nozzle spray patterns, (03310):5
      oil pumps, (03310):4
      power assembly, (03310):3–4
      pressure-regulating valves, (03310):3
      solenoid oil flow valves, (03310):4
Burnouts, (03210):1, 17–19, 39
Butterfly valves, (03305):19–21
Bypass dampers, (03315):4–6
Bypass duty relays, (03304):31
Bypass humidifiers, (03312):9
BZD Barometric Zone Damper™, (03315):6

C
Cad cell, (03310):1, 3, 7, 36
Cad cell flame detector, (03310):26–27
Cadmium sulfide cell, (03310):7
Cage nuts, (03313):5, 7
Capacitance boost, (03210):21, 26, 39
Capacitance checkers, (03210):27
Capacitors
   start and run checks, (03210):27, (03314):56
   VFD, (03314):64
Capacitor start, capacitor run (CSR) motors, (03314):52–56
Capillary tubes, (03210):15, (03304):25
Cap screws, (03313):2–4
Carbon dioxide (CO_2)
   buildup, mitigating, (03310):19
   byproduct, (03209):24
   in flue gas, (03209):25
   levels, checking, (03310):20
Carbon emissions, (03209):21
Carbon monoxide (CO), (03209):24–25, (03310):19
Carborundum, (03209):14
Carrier Infinity™ system, (03311):23
Castellated nuts, (03313):5
Cast-iron sectional boilers, (03305):8
Castle nuts, (03313):5
Cavitation, (03305):14
Centrifugal chillers, (03305):32–33
Centrifugal compressors, (03305):32
Centrifugal pumps
   construction, (03305):13–14
   double-suction, (03305):14, 17
   illustrated, (03305):15
   impeller types, (03305):17
   multistage, (03305):17
   operations, (03305):13
   parts, (03305):14, 16
   variable-speed, (03305):16
CFC-11 refrigerant, (03305):33
CFC refrigerants, (03305):33
Check valves, (03311):2, 29–31

Chilled-water systems
　basics, (03305):29–30
　chillers
　　absorption, (03305):34–35
　　centrifugal, (03305):32–33
　　compression, (03305):30
　　mechanical, (03305):30
　　nonmechanical, (03305):30
　　operations, (03305):35–36
　　packaged, (03305):29, 30–31
　　pumps, (03305):30–31
　　reciprocating, (03305):31–32
　　safety controls, (03305):35–36
　　screw, (03305):33–34
　　scroll, (03305):31–32
　condensers, (03305):36
　cooling towers, (03305):36–38, 39
　evaporative condensers, (03305):36, 38, 40
　function, (03305):29
　piping systems
　　four-pipe systems, (03305):23–24
　　primary-secondary systems, (03305):24
　water system balancing
　　pre-balance checks, (03305):25–26
　　procedure, (03305):25, 26–27
Chillers
　absorption, (03305):34–35
　centrifugal, (03305):32–33
　compression, (03305):30
　condenserless, (03305):32
　freezestats, (03314):8–9
　mechanical, (03305):30
　nonmechanical, (03305):30
　operations, (03305):35–36
　packaged, (03305):29, 30–31
　pumps, (03305):30–31
　reciprocating, (03305):31–32
　safety controls, (03305):35–36
　screw, (03305):33–34
　scroll, (03305):31–32
　vibration insulators, (03313):24
Chimney liners, (03310):32
Circuit breaker checks, (03314):45–48
Circulator pump, (03314):59, 60
Click-type torque wrench, (03313):11, 12
CO. *See* Carbon monoxide (CO)
CO_2. *See* Carbon dioxide (CO_2)
Coal, carbon emissions from, (03209):21
Coal furnaces, (03314):15
Coarse pitch threads, (03313):2
Color coding terminations, (03313):31
Combustion air
　pressure-type oil burners, (03310):3
　ventilators for, (03312):19, 23
Combustion analysis, gas heating systems
　benefits, (03209):24
　combustion efficiency
　　determining, (03209):24
　　field analysis, (03209):24
　conducting
　　combustion efficiency test analysis, (03209):28
　　flue gas, sampling, (03209):28
　　flue gas temperature, measuring, (03209):27–28
　　other gases present, identifying, (03209):29
　　paperwork, (03209):29
　　preliminary inspection, (03209):26
　　sampling holes, (03209):26–27
　　smoke spot testing, (03209):28
　　vent draft, (03209):28
　equipment, (03209):26
　gas combustion byproducts, (03209):24–26
Combustion analyzer, (03209):26
Combustion chambers, (03310):1–2
Combustion efficiency, (03209):24, (03310):19–22
Combustion noise, troubleshooting, (03310):43
Communication
　in customer relations, (03310):15, (03316):3–5, 8–9
　positive, (03316):4–5
　VRF systems, (03315):40–41
　VVT systems, (03315):19–20, 21
Complaints, common causes of primary, (03310):28
Compression chillers, (03305):30
Compression tanks, (03305):22
Compression-type connectors, (03313):31
Compressors
　centrifugal, (03305):32
　failure
　　electrical, replacing following, (03210):30–31
　　mechanical, replacing following, (03210):28–29
　function, refrigeration cycle, (03210):1, 2
　hermetic, (03304):21, (03305):32
　hermetic reciprocating, (03304):22–23
　hold-down bolts, (03313):22
　ice on, (03210):12
　impellers, (03305):32–33
　mechanical problems
　　causes and results of, (03210):25
　　no pressure change, (03210):27–28
　　replacing following failure, (03210):28–29
　　rings, (03210):25–26
　　seizing, (03210):26–27
　　valves, (03210):25–26
　oil
　　changing following an electrical failure, (03210):33–35
　　color, information from, (03210):25
　　volume, determining, (03210):24
　open, (03305):32
　problems with
　　floodback, (03210):22
　　flooded starts, (03210):21, 23–24, 39
　　overheating, (03210):24–25
　　slugging, (03210):21–22, 39
　refrigeration systems, (03304):21–23
　replacing following a mechanical failure, (03210):28–29
　replacing following an electrical failure
　　changing oil, (03210):33–35
　　determining burnout type, (03210):30–31
　　mild burnout procedures, (03210):31
　　severe burnout procedures, (03210):31–33
　rotary type, (03210):28
　scroll, (03304):21–23
　scroll type, (03210):26, 28
　semi-hermetic, (03304):21
　startup/shutdown, (03210):26
　troubleshooting
　　rotary compressors, (03210):28
　　scroll compressors, (03210):28
　variable-speed, (03311):17
　vibration insulators, (03313):22–23
　VRF systems, (03315):38, 40
Compressor short-cycle timers, (03314):4–5
Concrete anchors
　expanding, (03313):17, 18
　non-expanding, (03313):16–17
Concrete inserts, (03313):14–15
Condensate line sizing, (03306):30–31

Condensate pumps, (03306):7
Condensate receiver tank, (03306):7
Condensation, latent heat of, (03306):2
Condenserless packaged chillers, (03305):32
Condensers
　airflow problems, (03210):11–12
　function, (03210):1, 2, (03305):36
　refrigeration cycle, (03210):1–2
　refrigeration systems, (03304):13, 23–24
Condensing commercial boilers, (03305):12
Condensing furnaces
　combustion air, (03209):26
　troubleshooting control circuits, (03209):10–11
Condensing units
　ductless split systems, (03315):27
　VRF systems, (03315):38, 40
Conductors, (03313):33–34
Connectors
　AL-CU, (03313):33
　aluminum, (03313):33
　compression-type, (03313):31
　copper, (03313):33
　crimp type, (03313):31, 32, 34
　defined, (03313):31, 38
Consumer relations, communication in, (03310):15
Contactors , (03314):6, 7–8
Contamination
　acid, (03210):17–18
　air, (03210):16–17
　moisture, (03210):16–17
　preventing, (03210):17
　testing for, (03210):18–19
Control circuits
　heat pumps, (03311):1–2
　HVACR, (03314):5
　isolating a faulty circuit component, (03314):48–50
　isolating a faulty circuit via the process of elimination, (03314):48
　operating sequences
　　basic, (03314):35–36
　　　cooling, (03314):36–38
　　　heating, (03314):36–38
　　pneumatic, (03314):40
　troubleshooting, oil-fired heating systems, (03310):23–28
　VFD, (03314):65
Control devices, operation of HVACR
　contactors, (03314):6, 7–8
　control circuits, (03314):5
　fan controls, furnace, (03314):10–11
　limit controls, furnace, (03314):11–12
　motor starters, (03314):6–7
　relays
　　basic, (03314):2–4
　　contact checks, (03314):7–8
　　lockout, (03314):5
　　solid-state, (03314):5, 6
　　time-delay, (03314):4
　switches
　　control circuits, (03314):8
　　fan/limit, (03314):12
　　freezestats, (03314):8–9
　　pressure, (03314):8, 9
　　testing, (03314):7–8
　thermostats, outdoor, (03314):9–10
　timers, compressor short-cycle, (03314):4–5
Controllers
　VRF systems, (03315):41
　VVT unit controllers, (03315):19

VVT zone controllers, (03315):18–19
Control panels
　infrared gas-fired heaters, (03209):20–21
　zone, (03315):6–9
Controls, hydronic systems
　aquastat, (03314):58
　circulator pump, (03314):59, 60
　hot-water systems, (03314):57–58
　low-water cutoff, (03314):58, 60
　reset controller, (03314):58, 59
　zoned operating sequence, (03314):60–61
　zone valves, (03314):59–60
Control systems, HVACR
　automated, characteristics of, (03314):1
　electronic, (03314):1
　function, (03314):1
　pneumatic, (03314):1
Coolant, secondary, (03305):29, 30, 44
Cooling
　free cooling, (03312):15, 27
　mechanical cooling, (03312):15, 27
Cooling compensator, (03314):14, 16, 85
Cooling-only systems, heat pumps vs., (03311):1–2
Cooling-only thermostats, (03314):16
Cooling systems, corrosion in, (03305):1
Coping, (03310):23, 30, 36
Copper connectors, (03313):33
Copper-finned tube boilers, (03305):7–8, 13
Corrosive properties of water, (03305):1
Cotter pins, (03313):19–20
Countersunk lock washers, (03313):5
CPR. *See* Crankcase pressure regulators (CPR)
Crankcase heaters, (03210):25, 26
Crankcase pressure regulators (CPR), (03304):27
Crimp connectors, (03313):34
Crimp-on terminations, (03313):31, 32
Crimp type connectors, (03313):31, 32
Critical customers, (03316):13
CSR. *See* Capacitor start, capacitor run (CSR) motors
Cubed-ice machines, (03304):16–17, 18, 24, 38–41
Cup point set screw, (03313):5
Current imbalance, (03314):45
Customer interview, (03314):39
Customer relations
　attitudes effect on, (03316):1–3
　behaviors effect on, (03316):1–3
　communication's role in, (03316):3–5, 8–9
　concern, demonstrating, (03316):5–6, 8–9
　at construction sites, (03316):2–3
　criticism and, (03316):5
　customers, types of
　　angry, (03316):12–13
　　argumentative, (03316):11–12
　　critical, (03316):13
　　fearful, (03316):10–11
　　Mr. Knowitall, (03316):14
　　Mr. Negative, (03316):13
　　opinionated, (03316):11
　　sloppy, (03316):12
　　who request help with odd jobs, (03316):14
　　with unresolvable problems, (03316):13–14
　the Internet and, (03316):3
　listening in, (03310):15, (03316):5, 8–9
　off the job, (03316):3
　personal habits effect on, (03316):1–3
　on the phone, (03316):4, 10
　selling as element of, (03316):6

on service calls
    cleanliness element, (03316):9, 10
    the closing, (03316):10
    communication, (03310):15, (03316):3–5, 8
    concern, demonstrating, (03316):5–6, 8–9
    the extra mile, (03316):10
    first impressions, (03316):1–3, 8–9
    the opening, (03316):8
    performing the service, (03316):9–10
    repeat business, generating, (03316):5–6, 10
    residential access, (03316):9
    socializing, (03316):5, 9
social media and, (03316):3
Customers, types of
  angry, (03316):12–13
  argumentative, (03316):11–12
  critical, (03316):13
  fearful, (03316):10–11
  Mr. Knowitall, (03316):14
  Mr. Negative, (03316):13
  opinionated, (03316):11
  sloppy, (03316):12
  who request help with odd jobs, (03316):14
  with unresolvable problems, (03316):13–14

D

Daikin, (03315):41
Damper motors, zone, (03315):10
Dampers
  barometric, (03315):4, 5, 6
  bypass, (03315):4–6
  motorized, (03315):4, 5
  troubleshooting, (03315):15–16
  VVT bypass dampers, (03315):17–18
  zone
    typical zoning system, (03315):2–4, 7, 10
    VVT systems, (03315):18
    wireless, (03315):9
Data-interface, VRF systems, (03315):41
DB. *See* Defrost board (DB)
DC-capacitor VFD, (03314):64
DCV. *See* Demand control ventilation (DCV)
Deadband, (03314):14, 17, 85
Defrost board (DB), (03311):14–16, 33
Defrost methods
  electric, (03304):9–11
  hot-gas, (03304):11
  off-cycle, (03304):8
  timed, (03304):9
Delay-on-break relays, (03304):31
Delay-on-make relays, (03304):31
Demand control ventilation (DCV), (03312):16
Demand defrost, (03311):19, 32, 43
Density of water, (03305):1
Department of Energy (DOE), AFUE regulations, (03209):9
Diagrams. *See also* Schematic symbols
  fault isolation, (03314):32, 41, 85
  label, (03314):32, 41, 85
  ladder, (03314):1, 4, 85
  wiring, (03314):1, 2, 85
Die nuts, (03313):8
Dies, (03313):8
Differential, (03314):14, 17, 85
Differential pressure switches, (03314):8
Digital torque wrench, (03313):12
Direct-acting valve, (03311):19, 26, 43
Direct-fired radiant burners, (03209):19
Direct ignition devices, troubleshooting, (03209):13–14

Discharge air sensor, (03312):15
Discharge port, centrifugal pumps, (03305):14
DOE. *See* Department of Energy (DOE)
Door interlock switch, (03209):11
Double-pole, double-throw (DPDT) relays, (03314):3–4
Double-suction centrifugal pumps, (03305):14, 17
Dowel pins, (03313):19–20
Downdraft, (03310):23, 30, 36
DPDT. *See* Double-pole, double-throw (DPDT) relays
Draft-inducer proving switch, (03314):8
Draft inducers, (03209):17–18, 19
Draft regulator, troubleshooting, (03310):31
Drain valves, (03306):10–11, 26–27
Drier cores, (03210):32, 33
Drip leg, (03306):19, 31, 49
Droop, (03314):14, 18, 85
Dry-base, (03305):9
Dry-bulb economizers, (03312):16–17, 18
Dry steam, (03306):1, 2, 49
Ductless split systems
  air handlers, (03315):26
  applications, (03315):25
  condensing units, (03315):27
  hydronic, (03315):25
  indoor units
    air handlers, (03315):27–29
    components, (03315):27
    multiple, (03315):29–30
  installation, (03315):26, 30–34
  outdoor units, (03315):27
  power, (03315):27
  refrigerant tubing, (03315):26
  remote controls, (03315):26–27, 32–34
  thermostats, (03315):26
  troubleshooting, (03315):34, 36–38, 39
  wiring, (03315):26
Ductwork, vibration insulators for, (03313):24–25
Dump zone, (03315):1, 5, 59

E

EACs. *See* Electronic air cleaners (EACs)
Earthquake-resistant buildings, (03313):29
ECM. *See* Electronically commutated motor (ECM)
Economizers
  basic system, (03312):16
  benefits of, (03312):18
  control modules, (03312):18
  control signals, (03312):15
  discharge air sensor, (03312):15
  dry-bulb, (03312):16–17, 18
  enthalpy, (03312):17–18
  enthalpy sensor, (03312):15, 16
  flash, (03305):29, 33, 44
  function of, (03312):15
  integrated, (03312):18
  operation, typical, (03312):15–16
  outside air accessory troubleshooting, (03312):18–19
  schematic, (03312):18
  setpoint, (03312):15
  thermal, (03305):29, 33, 44
  thermostat, (03312):15–16
EEV. *See* Electronic expansion valves (EEV)
Electrical problems in accessories, troubleshooting, (03312):1–2
Electric boilers, (03305):11–12
Electric defrost, (03304):9–11

Electronic air cleaners (EACs)
  airflow rates, (03312):11
  cleaning, (03312):10–11
  collector section, (03312):8
  filter, (03312):8
  function of, (03312):8
  humidifiers, (03312):2
  operation, typical, (03312):8–9
  overload conditions, (03312):11
  power supply, (03312):8
  prefilter, (03312):8
  schematic diagram, (03312):8–10
  troubleshooting, (03312):8–12
Electronically commutated motor (ECM), (03311):7, 16, 17, 43
Electronically commutated motors (ECMs)
  benefits of, (03314):73
  characteristics, (03314):73
  ICM vs., (03314):74
  identification, (03314):73–74
  installation, (03314):75–76
  operation, (03314):73
  setup, (03314):75–76
  speed control, (03314):74–75
  troubleshooting, (03314):76–79
Electronic expansion valves (EEV), (03304):25–26
Electronic gas/oil combustion analyzer, (03310):19–20
Electronic humidistats, (03312):6–7
Electronic time delay, troubleshooting, (03310):32
Empathy, (03316):1, 4, 18
Energy recovery ventilators (ERVs), (03312):19–21
Enthalpy economizers, (03312):17–18
Enthalpy sensor, economizers, (03312):15
Environmental Protection Agency (EPA), (03310):9
EPA. See Environmental Protection Agency (EPA)
Epoxy, (03313):16
EPR. See Evaporator pressure regulator (EPR)
ERVs. See Energy recovery ventilators (ERVs)
Evaporative humidifiers, problem prevention, (03312):2–3
Evaporator coils, dirty, (03210):10
Evaporator pressure regulator (EPR), (03304):27–28, 29
Evaporators
  airflow problems, (03210):10–11
  forced-draft, finned-tube, (03304):23
  function, (03210):1, 2
  refrigeration cycle, (03210):1
  refrigeration systems, (03304):23–24
Expanding concrete anchors, (03313):17, 18
Expansion devices, (03304):24–26
Expansion tanks, (03305):21, 22
External lock washers, (03313):5

F

Factory wiring, (03311):21
Fan controls, furnace, (03314):10–11
F and T. See Float and thermostatic (F and T) steam traps
Fan/limit switches, (03314):12
Fans, zone, (03315):10
Fan vibration insulators, (03313):22
Fasteners
  anchor bolts
    concrete inserts, (03313):14–15
    expanding concrete anchors, (03313):17, 18
    non-expanding concrete anchors, (03313):16–17
    Tapcon® screws, (03313):17–18
  dies, (03313):8, 10
  extractions, (03313):8–9, 10
  non-threaded
    keys, (03313):19, 20
    lock rings, (03313):18
    pins, (03313):19, 20
    retainer rings, (03313):18
    rivets, (03313):19–21
  taps, (03313):7–8, 10
  threaded
    cap screws, (03313):2–4
    flat washers, (03313):5, 7
    grade designations, (03313):2
    lock washers, (03313):5, 7
    machine bolts, (03313):2–4
    machine screws, (03313):2–4
    nuts, (03313):4–5, 6
    reusing, (03313):13
    screws, (03313):6–7, 9
    set screws, (03313):5, 8, 9
    stud bolts, (03313):2–4
    thread designations, (03313):1–2
    thread repair inserts, (03313):5, 7
    tightening sequence, (03313):12–13
    torquing, (03313):9, 11–13
  toggle bolts, (03313):14, 15
Fastener threads
  grade designations, (03313):2
  thread designations, (03313):1–2
  thread repair inserts, (03313):5
Fault isolation, equipment problem area, (03314):43
Fault isolation diagrams, (03314):41, 85
Fearful customers, (03316):10–11
Federal Emergency Management Agency (FEMA), *Installing Seismic Restraints for Mechanical Equipment*, (03313):26, 29
FEMA. See Federal Emergency Management Agency (FEMA)
Field wiring, (03311):19, 21, 43
Films, silver and biological, (03304):35
Filter driers, (03210):32
Filter driers, VRF systems, (03315):42
Filter Reset button, (03314):16
Filters, dirty, (03209):16
Fine pitch threads, (03313):2
Firetube boiler, (03305):6, 9–10, 44
First impressions, (03316):1–3, 8–9
Fixed metering devices, troubleshooting, (03210):12–13
Fixed-orifice steam traps, (03306):39
Flaked-ice machines, (03304):19
Flame rectification, (03209):1, 32
Flame-retention burner, (03310):1, 6, 36
Flame sensors, troubleshooting, (03209):3, 14–16
Flare joints, HFC-410A, (03315):46
Flash economizer, (03305):29, 33, 44
Flash steam, (03306):19, 25, 27, 49
Flat washers, (03313):5, 7
Float and thermostatic (F and T) steam traps, (03306):20–21, 38, 42, 43–44
Floodback
  causes and results of, (03210):9–10, 23
  defined, (03210):1, 39
  indicators of, (03210):6
Flooded starts
  causes and results of, (03210):23–24
  defined, (03210):21, 39
Flow meters, (03305):18–19, 20
Flue and chimney exhaust, troubleshooting, (03310):30–31
Flue gases
  characteristics, (03209):25

sampling, (03209):28
temperature, measuring, (03209):27–28
testing, (03310):19–22
Food spoilage, (03304):1
Four-pipe piping systems, (03305):23–24
Free cooling, (03312):15, 27
Freezestats, (03314):8–9
Frost accumulation, (03304):8
Fuel oil
   grades of, (03310):8
   odors from, (03310):29–30
   piping systems, (03310):8–12
Fuel supply, troubleshooting, (03310):29
Furnaces
   biodiesel, (03310):3
   bumpers, (03313):27
   coal, (03314):15
   fan controls, (03314):10–11
   gas, pressure switches, (03314):8
   gravity-type, (03314):15
   limit controls, (03314):11–12
   multi-fuel system, (03310):20
   natural gas, (03314):15
   seismic restraints, (03313):26
Furnaces, gas, (03209):25. *See also* Heating systems, gas
   AFUE efficiency, (03209):24
   AFUE regulations, (03209):9
   condensing
      combustion air, (03209):26
      troubleshooting control circuits, (03209):10–11
   energy efficiency requirements, (03209):6
   flame sensors, (03209):3
   ignitions, electronic, (03209):6
   induced-draft
      airflow-related problems, (03209):17–18
      combustion air, (03209):25
      installation, (03209):25
      troubleshooting control circuits, (03209):6, 9–10
   low-NO_x, (03209):25
   natural-draft
      combustion air, (03209):25
      installation, (03209):25
      obsolescence, (03209):6
      troubleshooting control circuits, (03209):6, 7–8
   outdoor units, electronic ignitions, (03209):6
   pressure switches, (03314):8
Furnaces, natural gas, (03305):7
Fuse checks, (03314):45–46
Fusion, latent heat of, (03306):2

G

Gas combustion, byproducts of, (03209):24–26
Gauge siphon, (03306):9
Geothermal heat pumps, (03311):4
Grade designations, threaded fasteners, (03313):2
Gravity-type furnaces, (03314):15

H

H_2O. *See* Water (H_2O)
Hammer drills, (03313):18
Hangers, vibration-isolating, (03313):24
Hartford loop, (03306):20
Hazardous-waste disposal, (03314):25
HCFC-123 refrigerant, (03305):33
Head pressure, (03305):3
Head pressure, causes of high, (03314):8
Heat anticipators, (03314):15
Heat converters, (03306):14–15, 16

Heaters. *See also* Heating systems
   belly, (03305):33
   crankcase, (03210):25, 26
   indirect radiant, (03209):19
   infrared gas-fired
      applications, (03209):19
      control panels, (03209):20–21
      draft inducers, (03209):19
      high-intensity, (03209):19–20
      line-of-sight principle, (03209):19
      low-intensity, (03209):19, 22
      maintenance safety, (03209):21
      operations, (03209):20–21
      purpose, (03209):19
      venting, (03209):19
   swimming pools, (03311):1–2
Heat exchangers
   dirty, (03209):16
   oil-fired heating systems, (03310):2
   steam systems, (03306):6–7, 14–15, 16
   ventilators, (03312):19
Heating-cooling thermostats, (03314):16
Heating-only thermostats, (03314):15
Heating systems, gas. *See also* Furnaces, gas; Heaters
   boilers, gas-fired
      combustion air, (03209):25
      installation, (03209):25
      troubleshooting control circuits, (03209):11–13
   combustion analysis, conducting
      combustion efficiency test analysis, (03209):28
      flue gas, sampling, (03209):28
      flue gas temperature, measuring, (03209):27–28
      other gases present, identifying, (03209):29
      paperwork, (03209):29
      preliminary inspection, (03209):26
      sampling holes, (03209):26–27
      smoke spot testing, (03209):28
      vent draft, (03209):28
   combustion efficiency
      determining, (03209):24
      field analysis, (03209):24
   equipment, (03209):26
   gas combustion byproducts, (03209):24–26
Heating systems, gas, troubleshooting
   air distribution systems
      blower motor, (03209):16–17
      ductwork, (03209):16–17
      induced-draft systems, (03209):17–18
      obstructed airflow, (03209):16
   combustion systems, (03209):2
   common problems, (03209):33–34
   control circuits
      condensing furnaces, (03209):10–11
      digital/electronic controls, (03209):2–5
      gas-fired boilers, (03209):11–13
      induced-draft furnaces, (03209):6, 9–10
      natural-draft furnaces, (03209):6, 7–8
      variation in, (03209):1
   flame sensors
      combined HSI/flame sensor, (03209):15–16
      method, (03209):14–15
   flow chart, (03209):35–36
   gas valves, (03209):1
   guide, (03209):33–34
   ignition devices, (03209):13–14
      direct ignition devices, (03209):13–14
      hot-surface igniters (HSIs), (03209):14

Heating systems, gas, troubleshooting
ignition devices (*continued*)
pilot re-ignition devices, (03209):13–14
spark igniters, (03209):13–14
ignition systems, (03209):1–2
older furnaces, (03209):1
packaged units, (03209):1
skills required, (03209):1
subsystems, (03209):1
Heating systems, hot-water
air management, (03305):22–23
boilers
cast-iron sectional, (03305):8
combustion systems, (03305):6–7
condensing, (03305):12
construction, (03305):7
copper-finned tube, (03305):7–8, 13
electric, (03305):11–12
electrode, (03305):11–12
firetube, (03305):6, 9–10, 44
heat sources, (03305):6
indirect electric, (03305):12
low-pressure, (03305):7
medium-pressure, (03305):7
operations, (03305):12–13
safety controls, (03305):12–13
Scotch Marine, (03305):9–10
steel vertical tubeless, (03305):10–11
watertube, (03305):6, 9–10, 44
centrifugal pumps
construction, (03305):13–14
double-suction, (03305):14, 17
illustrated, (03305):15
impeller types, (03305):17
multistage, (03305):17
operations, (03305):13
parts, (03305):14, 16
variable-speed, (03305):16
flow meters, (03305):18–19, 20
piping systems
four-pipe systems, (03305):23–24
primary-secondary systems, (03305):24
tanks
compression, (03305):22
expansion, (03305):22
valves
balancing, (03305):17, 18–19
butterfly, (03305):19–21
multi-purpose, (03305):17–18
triple-duty, (03305):17
water system balancing
pre-balance checks, (03305):25–26
procedure, (03305):25, 26–27
Heating systems, millivolt, (03314):15
Heating systems, oil-fired
basics, (03310):1–2
bypass plug, (03310):11
cad cells
safety controls, (03310):7
servicing, (03310):18
troubleshooting, (03310):26–27
combustion chambers, (03310):1–2
combustion efficiency testing, (03310):19–22
flue gas testing, (03310):19–22
heat exchangers, (03310):2
nozzle assembly
erosion, (03310):30
pressure-type oil burners, (03310):4–5, 6
servicing, (03310):17–18
troubleshooting, (03310):28–29
oil burner assembly
electrode tips, resetting the, (03310):17–18
pressure-type, (03310):3
servicing, (03310):17–19
troubleshooting, (03310):28–29
pressure-type oil burners
components, (03310):3
flame-retention burner, (03310):6
high-pressure gun-type, (03310):2–3
ignition system, (03310):3, 5–6
lockouts, (03310):3
low-pressure gun-type, (03310):2
nozzle assembly, (03310):4–5, 6
oil pumps, (03310):4
power assembly, (03310):3–4
pressure-regulating valves, (03310):3
solenoid oil flow valves, (03310):4
primary components, (03310):2
primary control
safety function, (03310):6–7
troubleshooting, (03310):23–26
safety controls
cad cells, (03310):7
limit switch, (03310):7–8, 16
primary control, (03310):6–7
service call checks, (03310):16
stack switch, (03310):8
servicing
customer interaction, (03310):15
oil burner assembly, (03310):17–19
oil filter change, (03310):16
procedure, (03310):15
safety controls check, (03310):16
safety practices, (03310):14–15
temperature rise check, (03310):16–17
smoke spot testing, (03310):20, 21
solenoid oil flow valves
pressure-type oil burners, (03310):4
troubleshooting, (03310):31
storage tank disposal, (03310):29
supply system
fuel oil, (03310):8
fuel supply troubleshooting, (03310):29
piping systems, (03310):8–12
thermostats, (03310):6–7
troubleshooting
burner motors, (03310):38
cad cell flame detector, (03310):26–27
combustion noise, (03310):43
common causes of primary complaints, (03310):28
control circuit, (03310):23–28
cycle problems, (03310):40
draft regulator, (03310):31
electronic time delay, (03310):32
flue and chimney exhaust, (03310):30–31
fuel supply, (03310):29
ignition system, (03310):27–28
no heat, (03310):42
noise, (03310):40, 43
nozzle assembly, (03310):28–29
odors, (03310):29–30, 43
oil burner, (03310):28–29
operating costs, (03310):43
overheating, (03310):39, 43
primary control, (03310):23–26
solenoid oil flow valves, (03310):31

summary overview, (03310):37–43
underheating, (03310):42
underhouse heating, (03310):39–40
Heat pumps
advantages, (03311):1
air-to-air, (03311):2, 14, 25
air-to-air split system, (03311):13
air-to-water, (03311):2, 4
balance point, (03311):13–14
check valves, (03311):2
classifications, (03311):2, 4
components, (03311):1
control circuits, (03311):1–2
cooling-only systems vs., (03311):1–2
dual-fuel, (03314):10
duel-fuel, (03311):13
electromechanical, (03311):14
electronic controls, (03311):14
functions, (03311):1
geothermal, (03311):4
heat sources, (03311):4
metering devices, (03311):2
microprocessor controls
defrost board (DB), (03311):14–16
integrated control motor (ICM) board, (03311):14–16
schematic, (03311):15
timing circuits, (03311):14
variable-speed heat pumps, (03311):16–17
operating modes
cooling mode, (03311):3, 5
defrost mode, (03311):5–6
heating mode, (03311):3, 5
operating sequence
cooling mode, (03311):7, 8
defrost mode, (03311):10, 12–13
heating mode, (03311):7, 9–10, 11
operation, (03311):1–2
reversing valve, (03311):1
SEER ratios, (03311):16
split-systems, (03311):13–14
swimming pool heaters, (03311):1–2
thermostats, (03314):10, 17
thermostats and thermistors, (03311):13–14, 21–26
troubleshooting
check valves, (03311):29–31
cooling mode, (03311):19–20
defrost controls, (03311):31–35
defrost mode, (03311):19
restricted airflow, (03311):21
reversing valves, (03311):27–29
solenoid valves, (03311):26–27
thermistors, (03311):26
thermostats, (03311):21–25
variable-speed heat pumps, (03311):17
wiring, (03311):21
without equipment, (03311):19
variable-speed, (03311):16–17
Heat pumps, airflow, (03315):4
Heat recovery ventilators (HRVs), (03312):19–22
Heat sinks, (03311):30
Heli-coils, (03313):5, 7
Hermetic compressors, (03304):21, (03305):32, (03313):22–23
Hermetic reciprocating compressors, (03304):22–23
HFC-134a refrigerant, (03305):33
HFC-407C, (03304):6, 7, 50
HFC-410A, (03210):10, (03315):46
High-pressure limit control, steam boilers, (03306):13–14
High-pressure safety switches, (03314):8

High-pressure steam boilers, (03306):8
High-voltage power sources, testing
circuit breaker checks, (03314):45–48
current imbalance, (03314):45
fuse checks, (03314):45–46
high voltage effects, (03314):43, 45
low voltage effects, (03314):43, 45
voltage imbalance, (03314):45
Hot-gas defrost, (03304):11
Hot gas line, function, (03210):2
Hot-surface igniters (HSIs), (03209):1–2, 13, 14
Hot water, humidifiers and, (03312):9
Hot-water boilers, steam boilers vs., (03306):9
Hot-water heating systems
air management, (03305):22–23
boilers
cast-iron sectional, (03305):8
combustion systems, (03305):6–7
condensing, (03305):12
construction, (03305):7
copper-finned tube, (03305):7–8, 13
electric, (03305):11–12
electrode, (03305):11–12
firetube, (03305):6, 9–10, 44
heat sources, (03305):6
high-temperature, high-pressure water, (03305):7
indirect electric, (03305):12
low-pressure, (03305):7
medium-pressure, (03305):7
operations, (03305):12–13
safety controls, (03305):12–13
Scotch Marine, (03305):9–10
steam leaks, (03305):7
steel vertical tubeless, (03305):10–11
watertube, (03305):6, 9–10, 44
centrifugal pumps
construction, (03305):13–14
double-suction, (03305):14, 17
illustrated, (03305):15
impeller types, (03305):17
multistage, (03305):17
operations, (03305):13
parts, (03305):14, 16
variable-speed, (03305):16
flow meters, (03305):18–19, 20
piping systems
four-pipe systems, (03305):23–24
primary-secondary systems, (03305):24
tanks
compression, (03305):22
expansion, (03305):22
valves
balancing, (03305):17, 18–19
butterfly, (03305):19–21
multi-purpose, (03305):17–18
triple-duty, (03305):17
water system balancing
pre-balance checks, (03305):25–26
procedure, (03305):25, 26–27
HRVs. *See* Heat recovery ventilators (HRVs)
HSI/flame sensors, troubleshooting, (03209):15–16
Humidifiers
blower operation, (03312):2, 7
bypass type, (03312):9
capacity chart, (03312):8
common problems
low/high levels of humidity, (03312):6–8
noise, (03312):6

Humidifiers
common problems (continued)
not providing moisture, (03312):4
not running, (03312):4
water overflow, (03312):6
control circuits, (03312):4–5
function of, (03312):2
hot water and, (03312):9
problem prevention, (03312):2–3
ratings, (03312):7
saddle valves, (03312):11
selecting, (03312):7
troubleshooting, (03312):2–8
Humidistats, (03312):6–7
Humidistat setpoint, (03312):6
Hunting TXVx, (03210):13–14
Hydronic controls
aquastat, (03314):58
circulator pump, (03314):59, 60
hot-water systems, (03314):57–58
low-water cutoff, (03314):58, 60
reset controller, (03314):58, 59
zoned operating sequence, (03314):60–61
zone valves, (03314):59–60
Hydronic systems. *See also* Chilled-water systems; Hot-water heating systems
ductless, (03315):25
leaks, checking for, (03305):1
pressure drop, (03305):2–3
seismic restraints, (03313):26, 28

I
Ice, high quality, (03304):17
Ice flakers, (03304):19
Ice for preservation, (03304):3
Ice makers
biological growth, prevention of, (03304):35
cleaning, (03304):38–39
commercial, (03304):15–16
cubed-ice machines, (03304):16–17, 18, 24, 38–41
flaked-ice machines, (03304):19
troubleshooting and maintenance, (03304):38–41
tube-ice machines, (03304):17–18
Ice merchandisers, (03304):13–15
ICM. *See* Integrated control module (ICM) motors; Integrated control motor (ICM) board
IGBT. *See* Insulated gate bipolar transistor (IGBT)
Ignition devices, gas heating systems
hot-surface igniters (HSIs), (03209):1–2, 13–14
intermittent-pilot spark igniters, (03209):1
troubleshooting
direct ignition devices, (03209):13–14
hot-surface igniters (HSIs), (03209):13–14
pilot orifice, (03209):13–14
pilot re-ignition devices, (03209):13–14
spark igniters, (03209):13–14
Ignition systems
oil-fired heating systems, (03310):27–28
pressure-type oil burners, (03310):3, 5–6
Impeller blades, (03305):14
Impellers
centrifugal pumps, (03305):14, 17
compressors, (03305):32–33
in series, (03305):33
Impeller shrouds, (03305):14
Impeller vanes, (03305):14
Indirect radiant heaters, (03209):19
Indoor comfort, defined, (03209):1

Induced-draft furnaces
airflow-related problems, (03209):17–18
combustion air, (03209):25
installation, (03209):25
troubleshooting control circuits, (03209):6, 9–10
Inductive loads, (03314):51–52
Infiltration, (03304):5
Infrared gas-fired heaters
applications, (03209):19
control panels, (03209):20–21
draft inducers, (03209):19
high-intensity, (03209):19–20
line-of-sight principle, (03209):19
low-intensity, (03209):19, 22
maintenance safety, (03209):21
operations, (03209):20–21
purpose, (03209):19
venting, (03209):19
Infrared humidifiers, (03312):4–5
Insulated gate bipolar transistor (IGBT), (03314):62, 65, 85
Integrated control module (ICM) motors, (03311):17, (03314):74
Integrated control motor (ICM) board, (03311):14–16
Integrated economizers, (03312):18
Intercooling, (03305):33
Intermittent-pilot spark igniters, (03209):1
Internal-external lock washers, (03313):5
Internal lock washers, (03313):5
Internet, customer relations and the, (03316):3
Invar®, (03314):14, 85
Inverted bucket trap steam traps, (03306):36, 41, 43

J
Jam nuts, (03313):5
J-nuts, (03313):5
Joist, (03310):23, 30, 36

K
Key fasteners, (03313):19, 20
Knurled cup point set screw, (03313):5

L
Label diagrams, (03314):1, 32, 41, 85
Ladder diagram, (03314):4, 85
Lag screws, (03313):6, 9
Latent heat
of condensation, (03306):2
defined, (03306):1
of fusion, (03306):2
of vaporization, (03306):2
Left-hand thread symbol, (03313):2
Legionnaires' disease, (03305):36
Limit controls , (03314):11–12
Limit switch, (03310):7–8
Line-voltage thermostats, (03314):18
Liquid expansion thermostatic steam traps, (03306):37–38, 41
Liquid line, function, (03210):2
Listening in customer relations, (03310):15, (03316):5, 8–9
Lithium bromide, (03305):29, 34–35, 44
Locked rotor amperage (LRA), (03210):26
Lockout, (03310):1, 3, 36
Lockout relays , (03314):5
Lockout/tagout procedure, (03314):34–35
Lock ring pliers, (03313):19
Lock rings, (03313):18
Lock washers, (03313):5, 7
Loose air tightness, (03312):8
Loss-of-charge switch, (03314):8

Low-intensity infrared gas-fired heaters, (03209):19, 22
Low-NO$_x$ furnaces, (03209):25
Low-pressure gun-type burner, (03310):2
Low-pressure steam boilers, (03306):8, 9
Low-pressure switches, (03314):8
Low-voltage power source testing, (03314):50–51
Low-water cutoff control, (03305):13
Low-water cutoff device, (03314):58, 60
LP gas, (03209):3
LRA. *See* Locked rotor amperage (LRA)
Lubricating oil, opening/closing cans of, (03210):17
Lug, (03313):31, 38

M
Machine bolts, (03313):2–4
Machine screws, (03313):2–4
Mechanical chillers, (03305):30
Mechanical cooling, (03312):15, 27
Mechanical seal, centrifugal pumps, (03305):14
Mechanical steam traps, (03306):36–37
Medium-pressure steam boilers, (03306):8
Mercury, recycling, (03314):15, 25
Mercury-bulb thermostats, (03314):15, 25
Mercury switches, (03306):15
Metering devices
　function, (03210):1, 2
　heat pumps, (03311):2
　refrigeration cycle, (03210):1–2
　refrigeration systems, (03304):13, 24–26
　VRF systems, (03315):35
Meters, flow, (03305):18–19, 20
Metric MJ-profile threaded screws, (03313):2
Metric M-profile threaded screws, (03313):2
Metric screw threads, (03313):2
Microchannel coils, (03210):11–12
Microprocessor control circuits, (03209):2–5
Modulating gas valves, (03209):1, 9, 32
Moisture contamination, (03210):16–17
Motors
　basic motor testing, (03314):52–54
　capacitor start, capacitor run (CSR), (03314):52–53
　dynamic DC braking of an AC motor, (03314):70–71
　ECMs
　　benefits of, (03314):73
　　characteristics, (03314):73
　　ICM vs., (03314):74
　　identification, (03314):73–74
　　installation, (03314):75–76
　　operation, (03314):73
　　setup, (03314):75–76
　　speed control, (03314):74–75
　　troubleshooting, (03314):76–79
　electronically commutated (ECM), (03311):7, 16, 17, 43
　grounded conditions, (03210):27
　ICM, (03314):74
　integrated control module (ICM), (03311):17
　multi-speed, (03314):54
　open, shorted, or grounded windings checks, (03314):54–55
　permanent split capacitor (PSC), (03314):52–53
　three-phase, (03314):54
　unmarked terminals of a PSC/CSR motor, identifying, (03314):55–56
　VFD, (03314):64, 67
Motor starters, (03314):6–7
Mr. Knowitall, (03316):14
Mr. Negative, (03316):13
Multi-purpose valves, (03305):17–18

Multistage centrifugal pumps, (03305):17

N
National Electrical Code® (NEC®), (03314):33
Natural-draft furnaces
　combustion air, (03209):25
　installation, (03209):25
　obsolescence, (03209):6
　troubleshooting control circuits, (03209):6, 7–8
Natural gas, carbon emissions from, (03209):21, (03305):7
Natural gas boilers and furnaces, (03305):6, 7, (03314):15
Navigation Remote Controller™, (03315):41
Negative temperature coefficient (NTC) thermistors, (03314):19, 21
NEMA-rated contactors, (03314):6
Net stack temperature, (03209):24, 27, 32
Nitrogen dioxide (NO$_2$), (03209):25
Nitrogen oxides (NO$_x$), (03209):24–25
Nitrous oxide (NO$_x$) emissions, (03310):20
NO$_2$. *See* Nitrogen dioxide (NO$_2$)
Noise, troubleshooting, (03310):40, 43
Nominal size, (03313):1, 2, 38
Non-expanding concrete anchors, (03313):16–17
Nonmechanical chillers, (03305):30
Non-threaded fasteners
　keys, (03313):19, 20
　lock rings, (03313):18
　pins, (03313):19, 20
　retainer rings, (03313):18
　rivets, (03313):19–21
Normal operation, restoring, (03314):35
NO$_x$. *See* Nitrogen oxides (NO$_x$); Nitrous oxide (NO$_x$) emissions
Nozzle assembly
　oil-fired heating systems, (03310):28–29
　pressure-type oil burners, (03310):4–5, 6
NTC. *See* Negative temperature coefficient (NTC) thermistors
Nuts, (03313):4–5, 6, 12

O
O$_2$. *See* Oxygen (O$_2$)
Odors
　flue pipe and chimney exhaust, (03310):30
　refrigerant smell to indicate burnout, (03210):19
　troubleshooting, oil-fired systems, (03310):29–30, 43
Off-cycle defrost, (03304):8
Oil
　carbon emissions from, (03209):21
　in compressors
　　changing following an electrical failure, (03210):33–35
　　color, information from, (03210):25
　　determining volume, (03210):25
Oil burner assembly, servicing, (03310):17–19
Oil burners, troubleshooting, (03310):28–29
Oil filter change, (03310):16
Oil-fired heating systems
　basics, (03310):1–2
　bypass plug, (03310):11
　cad cells
　　safety controls, (03310):7
　　servicing, (03310):18
　　troubleshooting, (03310):26–27
　combustion chambers, (03310):1–2
　combustion efficiency testing, (03310):19–22
　flue gas testing, (03310):19–22
　heat exchangers, (03310):2

Oil-fired heating systems (*continued*)
    nozzle assembly
        erosion, (03310):30
        pressure-type oil burners, (03310):4–5, 6
        servicing, (03310):17–18
        troubleshooting, (03310):28–29
    oil burner assembly
        electrode tips, resetting the, (03310):17–18
        pressure-type, (03310):3
        servicing, (03310):17–19
        troubleshooting, (03310):28–29
    pressure-type oil burners
        components, (03310):3
        flame-retention burner, (03310):6
        high-pressure gun-type, (03310):2–3
        ignition system, (03310):3, 5–6
        lockouts, (03310):3
        low-pressure gun-type, (03310):2
        nozzle assembly, (03310):4–5, 6
        oil pumps, (03310):4
        power assembly, (03310):3–4
        pressure-regulating valves, (03310):3
        solenoid oil flow valves, (03310):4
    primary components, (03310):2
    primary control
        safety function, (03310):6–7
        troubleshooting, (03310):23–26
    safety controls
        cad cells, (03310):7
        limit switch, (03310):7–8, 16
        primary control, (03310):6–7
        service call checks, (03310):16
        stack switch, (03310):8
    servicing
        customer interaction, (03310):15
        oil burner assembly, (03310):17–19
        oil filter change, (03310):16
        procedure, (03310):15
        safety controls check, (03310):16
        safety practices, (03310):14–15
        temperature rise check, (03310):16–17
    smoke spot testing, (03310):20, 21
    solenoid oil flow valves
        pressure-type oil burners, (03310):4
        troubleshooting, (03310):31
    storage tank disposal, (03310):29
    supply system
        fuel oil, (03310):8
        fuel supply troubleshooting, (03310):29
        piping systems, (03310):8–12
    thermostats, (03310):6–7
    troubleshooting
        burner motors, (03310):38
        cad cell flame detector, (03310):26–27
        combustion noise, (03310):43
        common causes of primary complaints, (03310):28
        control circuit, (03310):23–28
        cycle problems, (03310):40
        draft regulator, (03310):31
        electronic time delay, (03310):32
        flue and chimney exhaust, (03310):30–31
        fuel supply, (03310):29
        ignition system, (03310):27–28
        no heat, (03310):42
        noise, (03310):40, 43
        nozzle assembly, (03310):28–29
        odors, (03310):29–30, 43
        oil burner, (03310):28–29
        operating costs, (03310):43
        overheating, (03310):39, 43
        primary control, (03310):23–26
        solenoid oil flow valves, (03310):31
        summary overview, (03310):37–43
        underheating, (03310):42
        underhouse heating, (03310):39–40
Oil-pressure safety switches, (03314):8
Oil pumps, (03310):4
Oil sample traps, (03210):31
Oil separators, VRF systems, (03315):40
Oil storage tank disposal, (03310):29
Oil test kits, (03210):18–19, 30
One-pipe piping systems, (03306):19–20
Open adjusting dies, (03313):8
Open compressors, (03305):32
Operations chillers, (03305):35–36
Opinionated customers, (03316):11
OSHA, *Lockout and Tagging of Circuits*, (03314):34
Outside air, accessory troubleshooting
    economizers, (03312):18–19
    recovery ventilators, (03312):21–22
Overfeeding TXVs, (03210):13–14
Over-firing, (03310):14, 16, 36
Oxygen (O_2)
    in air, (03209):24
    corrosive properties of, (03305):1
    in flue gas, (03209):25
Oxygen scavengers, (03305):1, 44

P

Packaged chillers, (03305):29, 30–31
Packing gland, centrifugal pumps, (03305):14
Panels, troubleshooting, (03315):13–15
Paperwork, (03209):29
Parallel-flow piping systems, (03306):20
Permanent split capacitor (PSC) motors, (03314):52–56
Personal habits effect on customer relations, (03316):1–3
Personal protective equipment on construction sites, (03316):2–3
Phase change, (03304):1, 3, 49
Photocell flame detector, (03310):26
Pigtail siphon, (03306):9–10
Pilot-operated valve, (03311):19, 27, 43
Pilot orifice, troubleshooting, (03209):13–14
Pilot re-ignition devices, troubleshooting, (03209):13–14
Pin fasteners, (03313):19, 20
Pipe dies, (03313):8
Pipe taps, (03313):8
Piping
    primary-secondary systems, (03305):24
    refrigerant
        burying, (03210):23
        ductless split systems, (03315):26
        installation, (03305):40
        overheating, (03311):30
        VRF systems, (03315):42–44
    scale in, (03305):1
Piping systems
    four-pipe, (03305):23–24
    sizing
        condensate line sizing, (03306):30–31
        pressure drop sizing, (03306):28, 30
        velocity sizing, (03306):28, 29
    steam heating systems
        condensate line sizing, (03306):30–31
        one-pipe systems, (03306):19–20
        parallel-flow systems, (03306):20

pressure drop sizing, (03306):28, 30
two-pipe systems, (03306):20–22
valves, (03306):31–34
velocity sizing, (03306):28, 29
Piping systems, vibration insulators, (03313):25
Piston rings, (03210):25
Plug-in relays, (03314):2
Plug taps, (03313):7
Polling, (03315):1, 20, 59
Pollution control
   biodiesel heating, (03310):3
   gas-fired heating equipment testing for, (03209):25
   nitrous oxide (NO_x) emissions, (03310):20
Pop rivet tool, (03313):20, 21
Positive communication, (03316):4–5
Positive-displacement pumps, (03305):13
Positive temperature coefficient (PTC) thermistors, (03314):19, 20, 21
Power assembly, pressure-type oil burners, (03310):3–4
Pressure-dependent operation, (03315):1, 20–22, 59
Pressure drop sizing, (03306):28, 30
Pressure gauge, steam boilers, (03306):9–10
Pressure-independent operation, (03315):1, 20–21, 59
Pressure-reducing valves, (03306):31–33
Pressure-relief valve, steam boilers, (03306):14
Pressures, troubleshooting abnormal, (03210):7–9
Pressure switches, (03314):8, 9
Pressure-temperature chart, HFC-407C, (03304):6, 50
Pressuretrol, (03306):13–14
Pressure-type oil burners
   biodiesel, (03310):3
   combustion air, (03310):3
   components, (03310):3
   flame-retention burner, (03310):6
   high-pressure gun-type, (03310):2–3
   ignition system, (03310):3, 5–6
   lockouts, (03310):3
   low-pressure gun-type, (03310):2
   nozzle assembly, (03310):4–5, 6
   nozzle flow rate, (03310):6
   nozzle spray patterns, (03310):5
   oil pumps, (03310):4
   power assembly, (03310):3–4
   valves
      pressure-regulating valves, (03310):3
      solenoid oil flow valves, (03310):4
Primary control
   defined, (03310):1, 3, 36
   safety function, (03310):6–7
   troubleshooting, (03310):23–26
Primary-secondary piping systems, (03305):24
Propane, (03209):3
PSC. *See* Permanent split capacitor (PSC) motors
PTC. *See* Positive temperature coefficient (PTC) thermistors
Pulse-width modulated (PWM) VFD, (03314):63–64
Pump casing, centrifugal pumps, (03305):14
Pumps
   centrifugal
      construction, (03305):13–14
      double-suction, (03305):14, 17
      function, (03305):13
      illustrated, (03305):15
      impeller types, (03305):17
      multistage, (03305):17
      parts, (03305):14, 16
      variable-speed, (03305):16
   chillers, (03305):30–31
   circulator, (03314):59
   condensate, (03306):7
   operations, (03305):13
   positive-displacement, (03305):13
   selecting, (03305):15
Pump shaft, centrifugal pumps, (03305):14
Purge units, (03305):33
PWM. *See* Pulse-width modulated (PWM) VFD
Pyrometer, (03306):44

R

Rapport, (03316):1, 2, 18
Receivers, (03304):26
Reciprocating chillers, (03305):31–32
Recovery ventilators, troubleshooting, (03312):21–23
Recycling mercury, (03314):15, 25
Refnet joints, (03315):25, 42, 59
Refractory, (03310):1, 2, 36
Refrigerant
   charge, low/excessive, (03210):7–9
   charging
      HFC-410A, (03210):10
      liquid, (03210):30
      chiller, (03305):33–34
      condensers and, (03305):36, 38–39
      smell to indicate burnout, (03210):19
      vapors, (03305):33
Refrigerant piping
   burying, (03210):23
   ductless split systems, (03315):26
   installation, (03305):40
   overheating, (03311):30
   VRF systems, (03315):42–44
Refrigerant-side head pressure control, (03304):28–29
Refrigeration cycle
   circuit operation, (03210):2
   component functions
      compressor, (03210):1, 2
      condenser, (03210):1, 2
      evaporator, (03210):1, 2
      hot gas line, (03210):2
      liquid line, (03210):2
      metering device, (03210):1, 2
      suction line, (03210):2
   illustrated, (03210):2
   lines and accessories, (03210):15
   low-temperature, (03304):6–8
   medium-temperature, (03304):3–5
   piping, (03210):1–2
   typical, (03210):3–5
Refrigeration cycle, troubleshooting
   accessory problems, (03210):14–16
   condenser airflow problems, (03210):11–12
   contamination
      acid, (03210):17–18
      air, (03210):16–17
      moisture, (03210):16–17
      preventing, (03210):17
   evaporator airflow problems, (03210):10–11
   fixed metering devices, (03210):12–13
   operating conditions
      preliminary inspection, (03210):5–6
      system operating conditions, (03210):6–7
   pressures, abnormal, (03210):7–9
   refrigerant charge, low/excessive, (03210):7–9
   refrigerant line problems, (03210):14–16
   temperatures, abnormal, (03210):7–9
   thermostatic expansion valves (TXVs), (03210):13–14
   troubleshooting aids, (03210):7

Refrigeration equipment, retail
  ice makers
    commercial, (03304):15–16
    cubed-ice machines, (03304):16–17, 18, 24, 38–41
    flaked-ice machines, (03304):19
    troubleshooting and maintenance, (03304):38–41
    tube-ice machines, (03304):17–18
  ice merchandisers, (03304):13–15
  reach-in coolers and freezers
    condensing units, (03304):13
    features and options, (03304):13
    metering devices, (03304):13
    troubleshooting and maintenance, (03304):34–37
    under-counter units, (03304):13–14
  troubleshooting and maintenance
    cubed-ice machines, (03304):38–41
    reach-in freezer, (03304):34–37
  walk-in merchandisers, (03304):15
Refrigeration equipment, seismic restraints, (03313):26
Refrigeration systems
  applications, (03304):1
  basic, (03304):2
  circuit components, (03304):1
  historically, (03304):1, 3
  humidity, (03304):24
  importance of, (03304):1
  retail, (03304):1
Refrigeration systems, components
  controls
    pressure controls, (03304):30
    solenoid valves, (03304):31–32
    thermostats, (03304):29–30
    time-delay relays, (03304):30–31
  primary
    compressors, (03304):21–23
    evaporators, (03304):23–24
    expansion (metering) devices, (03304):13, 24–26
  secondary
    accumulators, (03304):26–27
    crankcase pressure regulators, (03304):27
    evaporator pressure regulating valves, (03304):27–28
    receivers, (03304):26
    refrigerant-side head pressure control, (03304):28–29
Refrigeration systems, operation
  defrost methods
    electric, (03304):9–11
    hot-gas, (03304):11
    off-cycle, (03304):8
    timed, (03304):9
  frost accumulation, (03304):8
  low-temperature cycle, (03304):6–8
  medium-temperature cycle, (03304):3–5
Relative humidity (RH), (03312):6
Relay noise, troubleshooting, (03310):40
Relay packs, VVT systems, (03315):19
Relays
  basic, (03314):2–4
  coil element, (03314):2–3
  contact checks, (03314):7–8
  contacts, (03314):2
  double-pole, double-throw (DPDT), (03314):3–4
  function, (03314):2
  invention of, (03314):2
  lockout, (03314):5
  plug-in, (03314):2
  schematic, (03314):2, 3
  single-pole, double-throw (SPDT), (03314):3
  single-pole, single-throw (SPST), (03314):2, 3

  solid-state, (03314):5, 6
  start relay checks, (03314):56–57
  time-delay, (03304):30–31, (03314):4
Remote controls
  ductless split systems, (03315):26–27, 32–34
  VRF systems, (03315):41, 45, 49
Reset controller, (03314):58, 59
Resistance, thermistors, (03314):19, 21
Resistive loads, (03314):51–52
Resistors, heat anticipators, (03314):15
Retainer rings, (03313):18
Rethreading dies, (03313):8
Reverberator, (03209):19, 20, 32
Reversing valves, (03311):1, 27–29
RH. *See* Relative humidity (RH)
Rings, centrifugal pumps, (03305):14
Rings in compressors, testing, repairing, and replacing, (03210):25–26
Rivets, (03313):19–21
Rotary compressors, (03210):28
Run-down resistance, (03313):1, 12, 38
Running burn, (03210):18

S

Saddle valves, (03312):11
SAE. *See* Society of Automotive Engineers (SAE)
Safety
  commercial hot-water boilers, (03305):12–13
  electrical, (03314):32–33
  infrared gas-fired heaters maintenance, (03209):21
  lockout/tagout procedure, (03314):34–35
  LP gas, working with, (03209):3
  OSHA lockout/tagout rule, (03314):34
  restoring normal operation, (03314):35
  servicing oil-fired heating systems, (03310):14–15
  solid-state controls, (03314):33–34
Safety controls
  chillers, (03305):35–36
  oil-fired heating systems check, (03310):16
  steam boilers
    high-pressure limit control, (03306):13–14
    pressure-balancing, (03306):20
    pressure gauges, (03306):9–10
    pressure-relief valve, (03306):14
    water cutoff/feeder controls, (03306):11–13
    water level indicators, (03306):10–11
Sampling holes, (03209):26–27
San Francisco earthquake, (03313):26
Saturated steam, (03306):1, 2, 4–5, 49
Scale, (03305):1
Schematic symbols. *See also* Diagrams
  common, (03314):86–87
  relays, (03314):2–4
Scotch Marine boiler, (03305):9–10
Screw adjusting dies, (03313):8
Screw chillers, (03305):33–34
Screw drives, (03313):9
Screw extractors, (03313):8–9, 10
Screws
  cap screws, (03313):2–4
  function of, (03313):6
  head types, (03313):9
  lag screws, (03313):6, 9
  machine screws, (03313):2–4
  Metric MJ-profile threaded screws, (03313):2
  Metric M-profile threaded screws, (03313):2
  self-drilling sheet metal screws, (03313):7
  set screws, (03313):5, 8, 9

sizes, (03313):6
Tapcon® screws, (03313):17–18
thread-forming screws, (03313):6, 9
Screw threads, (03313):2
Scroll chillers, (03305):31–32
Scroll compressors, (03210):26, 28, (03304):21–23
Seals, mechanical, (03305):14
Seasonal energy-efficiency ratio (SEER), (03311):16
Secondary coolant, (03305):29, 30, 44
SEER. *See* Seasonal energy-efficiency ratio (SEER)
Seismic activity, (03313):22, 26, 38
Seismic building codes, (03313):30
Seismic curb, (03313):26, 27
Seismic gas shut-off valve, (03313):27
Seismic restraints, (03313):22, 26–29, 38
Self-drilling screws, (03313):7
Self-drilling sheet metal screws, (03313):7
Self-locking nuts, (03313):5
Semi-hermetic compressors, (03304):21
Semi-hermetic compressors, vibration insulators, (03313):23
Sensible heat, (03306):1, 2
Sensors, zone, (03315):9–10
Service calls
    customer relations
        cleanliness element, (03316):9, 10
        the closing, (03316):10
        communication, (03310):15, (03316):3–5, 8
        concern, demonstrating, (03316):5–6, 8–9
        the extra mile, (03316):10
        first impressions, (03316):1–3, 8–9
        the opening, (03316):8
        performing the service, (03316):9–10
        repeat business, generating, (03316):5–6, 10
        residential access, (03316):9
        socializing and, (03316):5, 9
    customers, types of
        angry, (03316):12–13
        argumentative, (03316):11–12
        critical, (03316):13
        fearful, (03316):10–11
        Mr. Knowitall, (03316):14
        Mr. Negative, (03316):13
        opinionated, (03316):11
        sloppy, (03316):12
        who request help with odd jobs, (03316):14
        with unresolvable problems, (03316):13–14
    following up after, (03316):10
    mistakes, handling, (03316):9
    preparing for, (03316):9
    procedure prior to beginning work, (03310):15
Set or seizure, (03313):1, 12, 38
Set screws, (03313):5, 8, 9
Sheet metal screws, (03313):7
Shell-and-tube heat exchanger, (03306):14
Sight glass, (03210):24
Silicon carbide HSIs, (03209):14
Silicon nitride HSIs, (03209):14
Silver film, (03304):35
Single-pipe oil supply systems, (03310):8–9, 11
Single-pole, double-throw (SPDT) relays, (03314):3
Single-pole, single-throw (SPST) relays, (03314):2, 3
Single-stage oil pump, (03310):4
Siphon assembly, steam boilers, (03306):9
Slinger ring, (03304):21, 24, 49
Sloppy customers, (03316):12
Slotted nuts, (03313):5
Slugging, (03210):21–22, 39
Smog, (03209):25

Smoke spot test, (03310):14, 20, 21, 36
Smoke spot testing, (03209):28
Social media, customer relations and, (03316):3
Society of Automotive Engineers (SAE), steel bolts and screws grade markings, (03313):2, 3
Solenoid oil flow valves
    pressure-type oil burners, (03310):4
    troubleshooting, (03310):31
Solenoid valves, (03304):31–32, (03311):26–27
Solenoid-valve service tool, (03304):32
Solid-state controls safety, (03314):33–34
Solid-state relays, (03314):5, 6
Soot, (03310):14, 19, 36
Soot buildup, (03209):6
Soot removal, (03310):15
Spark igniters, troubleshooting, (03209):13–14
SPDT. *See* Single-pole, double-throw (SPDT) relays
Split ring lock washers, (03313):5
Split-system compressors, vibration insulators, (03313):22
SPST. *See* Single-pole, single-throw (SPST) relays
Stack switch, (03310):8
Starting burn, (03210):18
Static pressure, (03305):3–4
Static-pressure sensors, (03315):9, 10
Steam
    dry, (03306):1, 2, 49
    flash, (03306):19, 25, 49
    saturated, (03306):1, 2, 49
    superheated, (03305):6, 7, 44
Steam boilers
    blowdown procedure, (03306):27
    blowdown valve, (03306):10–11, 26–27
    hot-water boilers vs., (03306):9
    operating/safety controls
        high-pressure limit control, (03306):13–14
        pressure-balancing, (03306):20
        pressure gauges, (03306):9–10
        pressure-relief valve, (03306):14
        water cutoff/feeder controls, (03306):11–13
        water level indicators, (03306):10–11
    siphon assembly, (03306):9–10
    skimming procedure, (03306):26–27
    water treatment, (03306):27
Steam coils, (03306):15, 17
Steam cycle
    basic components, (03306):6
    principles of operation, (03306):6–7
Steam gauge, (03306):9–10
Steam piping systems
    condensate line sizing, (03306):30–31
    one-pipe systems, (03306):19–20
    parallel-flow systems, (03306):20
    pressure drop sizing, (03306):28, 30
    two-pipe systems, (03306):20–22
    valves
        pressure-reducing, (03306):31–33
        thermostatic, (03306):33–34
    velocity sizing, (03306):28, 29
Steam system loads
    heat exchangers/converters, (03306):14–15, 16
    terminals, (03306):15, 17
Steam systems
    condensate pumps, (03306):19–24
    condensate receiver tank, (03306):22–24
    condensate return, (03306):19–24
    deaerating feedwater heater, (03306):25
    flash tanks, (03306):25–26
    vacuum-return system, (03306):21–22, 24–25

Steam traps
    diagnostic methods, (03306):43–44
    diagnostic tools, (03306):44
    failure, causes of, (03306):40–41
    function of, (03306):7, 36
    installing, (03306):39–40
    maintenance, (03306):40, 41
    strainers, (03306):39, 40
    troubleshooting, (03306):42–44
    types of
        balanced pressure thermostatic, (03306):38, 42
        bellows thermostatic, (03306):43
        bimetal thermostatic, (03306):38, 42, 43
        fixed-orifice, (03306):39
        float and thermostatic (F and T), (03306):20–21, 38, 42, 43–44
        inverted bucket trap, (03306):36, 41, 43
        liquid expansion thermostatic, (03306):37–38, 41
        mechanical, (03306):36–37
        thermodynamic, (03306):38–39, 43
        thermodynamic disc, (03306):42
        thermostatic, (03306):37–38
Steel vertical tubeless boilers, (03305):10–11
Stranded wire terminations, (03313):31, 32
Stud bolts, (03313):2–4
Subcooling, (03210):2, (03305):33, (03306):1, 2, 4
Submerged flame, (03209):19, 32
Suction line
    function, refrigeration cycle, (03210):2
    ice on, (03210):12
Suction port, centrifugal pumps, (03305):14
Suction pressure, causes of low, (03314):8
Superheat, (03306):1, 2, 4
Superheated steam, (03305):6, 7, 44
Swimming pool heaters, (03311):1–2
Switches
    control circuits, (03314):8
    differential pressure, (03314):8
    door interlock, (03209):11
    draft-inducer proving, (03314):8
    fan, (03315):11
    fan/limit, (03314):12
    freezestats, (03314):8–9
    high- and low-limit, (03315):9–10, 12–13
    high-pressure safety, (03314):8
    loss-of-charge, (03314):8
    low-pressure, (03314):8
    oil-pressure safety s, (03314):8
    pressure, (03314):8, 9
    system mode, (03315):11
    temperature-sensitive, (03314):15
    testing, (03314):7–8

T

Tanks
    compression, (03305):22
    expansion, (03305):21, 22
Tap and die set, (03313):10
Tapcon® screws, (03313):17–18
Taper pins, (03313):19–20
Taper taps, (03313):7
Taps, (03313):7–8
Tattoos, (03316):1
TE. *See* Thermal-electric (TE) expansion valve testing
The telephone, customer relations on, (03316):4, 10
Temperature
    abnormal, troubleshooting, (03210):7–9
    net stack, (03209):24, 27, 32

Temperature differential, (03310):14, 16, 36
Temperature rise check, (03310):16–17
Temperature sensors, infrared gas-fired heaters, (03209):20
Tensile strength, (03313):1, 2, 38
Terminal, (03313):31, 38
Terminals, steam system loads, (03306):15, 17
Terminations
    basic requirements, (03313):31
    color coding, (03313):31
    crimp-on, (03313):31, 32
    defined, (03313):31, 38
    installing, (03313):33–34
    stranded wire, (03313):31, 32
TEV. *See* Thermostatic expansion valve (TXV/TEV)
Thermal circulation, (03305):1
Thermal economizer, (03305):29, 33, 44
Thermal-electric (TE) expansion valve testing, (03314):19, 20
Thermidistat™, (03312):6–7
Thermistors
    application, (03314):19
    failure, (03314):20
    function, (03314):8, 18–19
    heat pumps, (03311):13–14
    negative temperature coefficient (NTC), (03314):19, 21
    positive temperature coefficient (PTC), (03314):19, 20, 21
    resistance, (03314):19, 21
    start thermistor checks, (03314):57
    thermal-electric (TE) expansion valve testing, (03314):19, 20
    values, (03314):21
Thermocouples, (03209):1
Thermocouples in thermowells, (03306):44
Thermodynamic disc steam traps, (03306):42
Thermodynamic steam traps, (03306):38–39, 43
Thermometers, infrared, (03306):44
Thermostatic expansion valve (TXV/TEV), (03210):13–14, (03304):25, (03311):2
Thermostatic steam traps, (03306):37–38
Thermostatic valves, (03306):33–34
Thermostat Recycling Corporation (TRC), (03314):15
Thermostats
    automatic-changeover, (03314):16–17
    bimetal, (03314):14, 15–16
    communicating type, (03311):23–24
    cooling-only, (03314):16
    ductless split systems, (03315):26
    economizers, (03312):15–16
    electromechanical, (03314):8
    Filter Reset button, (03314):16
    heat anticipators, (03314):15
    heating-cooling, (03314):16
    heating-only, (03314):15
    heat pumps, (03311):13–14
    infrared gas-fired heaters, (03209):20
    installation
        cycle rate adjustments, (03314):24
        final check, (03314):24
        guidelines, (03314):21–22
        mercury-thermostat disposal, (03314):25
        wiring, (03314):22–24
    line-voltage, (03314):18
    mercury-bulb, (03314):15, 25
    oil-fired heating systems, (03310):6–7
    outdoor, (03311):25, (03314):9–10
    programmable, (03314):17–18
    programmable electronic, (03314):14, 28–29
    refrigeration systems, (03304):29–30
    room type, (03311):25

set-points, (03314):15
troubleshooting
    cooling operation, (03314):27
    electronic thermostats, (03314):28–29
    fan-switch operation, (03314):26
    heating operation, (03314):27
    heat pumps, (03311):21–25
    jumper wires when, (03314):25–26, 27
    visual checks, (03314):25
    zoned systems, (03315):15–16
two-stage, (03312):16, (03314):17
voltage, (03314):15
VRF systems, (03315):45
VRV systems, (03314):30
zone, (03315):2–3, 9, 10–11
Thermowells, (03306):44
Thread classes, (03313):1, 2, 38
Thread-cutting screws, (03313):7, 9
Thread designations, threaded fasteners, (03313):1–2
Threaded fasteners
    cap screws, (03313):2–4
    flat washers, (03313):5, 7
    grade designations, (03313):2
    lock washers, (03313):5, 7
    machine bolts, (03313):2–4
    machine screws, (03313):2–4
    nuts, (03313):4–5, 6
    reusing, (03313):13
    screws, (03313):6–7
    set screws, (03313):8
    stud bolts, (03313):2–4
    thread designations, (03313):1–2
    thread repair inserts, (03313):5, 7
    tightening sequence, (03313):12–13
    torquing, (03313):11–13
Thread-forming screws, (03313):6, 9
Thread repair inserts, (03313):5, 7
Thread series, (03313):1, 2, 38
Thread standards, (03313):1, 38
Tight buildings, (03312):8
Timed defrost, (03304):9
Time-delay relays, (03304):30–31, (03314):4
Timers, compressor short-cycle, (03314):4–5
Tinner's rivets, (03313):20
Toggle bolts, (03313):14, 15
Torque, (03313):1, 9, 38
Torque wrenches, (03313):11–12
Torquing threaded fasteners, (03313):11–13
TRC. See Thermostat Recycling Corporation (TRC)
Trim, (03306):6, 9, 49
Triple-duty valves, (03305):17
Troubleshooting
    basics
        customer interview, (03314):39
        fault isolation, equipment problem area, (03314):43
        manufacturers' troubleshooting aids, (03314):41–43
        system analysis, (03314):40–41
        system physical examination, (03314):40
    defined, (03314):39
Troubleshooting aids, (03210):7
Troubleshooting aids, manufacturer, (03314):41–43
Troubleshooting chart, sample, (03311):44–50
Troubleshooting guide, sample, (03311):51
Tube-ice machines, (03304):17–18
Turndown ratio, (03306):19, 32, 49
Two-piece rectangular dies, (03313):8
Two-pipe oil supply systems, (03310):8, 9–10
Two-pipe piping systems, (03306):20–22

Two-stage oil pump, (03310):4
Two-stage thermostats, (03314):17
TXVs. See Thermostatic expansion valve (TXV/TEV)

U
Ultrasonic testers, (03306):44
Ultraviolet (UV) lamps, (03312):11–12
UNC. See Unified National Coarse (UNC) thread
Underfeeding TXVx, (03210):13–14
UNEF. See Unified National Extra Fine (UNEF) thread
UNF. See Unified National Fine (UNF) thread
Unified National Coarse (UNC) thread, (03313):1
Unified National Extra Fine (UNEF) thread, (03313):2
Unified National Fine (UNF) thread, (03313):2
Unit cooler, (03304):13, 15, 23–24, 49
U-nuts, (03313):5
UV. See Ultraviolet (UV) lamps

V
Vacuum breaker device, (03306):7
Valves
    balancing, (03305):17, 18–19
    blowdown, (03306):10–11, 26–27
    butterfly, (03305):19–21
    check, (03311):29–31
    in compressors, testing, repairing, and replacing, (03210):25–26
    crankcase pressure regulators (CPR), (03304):27
    direct-acting, (03311):19, 26, 43
    electronic expansion valves (EEV), (03304):25–26
    evaporator pressure regulator (EPR), (03304):27–28, 29
    gas, modulating, (03209):1, 9, 32
    gas shut-off, (03313):27
    head-pressure control, (03304):28–29
    modulating gas, (03209):1, 9, 32
    multi-purpose, (03305):17–18
    pilot-operated, (03311):19, 27, 43
    pressure-reducing, (03306):31–33
    pressure-regulating valves, (03310):3
    pressure-relief, (03306):14
    reversing, (03311):1, 27–29
    solenoid, (03304):31–32, (03310):4, 31, (03311):26–27
    thermal-electric (TE) expansion valve testing, (03314):19, 20
    thermostatic, (03306):33–34
    thermostatic expansion (TXV/TEV), (03210):13–14, (03304):25, (03311):2
    triple-duty, (03305):17
    zone valves hydronic controls, (03314):59–60
Vaporization, latent heat of, (03306):2
Variable frequency drive (VFD)
    AC-choke, (03314):64
    benefits of, (03314):63
    control circuits, (03314):65
    controller, (03314):64
    DC-capacitor, (03314):64
    dynamic braking
        DC braking of an AC motor, (03314):70–71
        operation, (03314):65–66
        with an AC drive, (03314):71–72
    function, (03314):63
    inverter, (03314):65
    motors, (03314):64
    operation basics, (03314):63–66
    programmable parameters, (03314):66
    pulse-width modulated (PWM), (03314):63–64
    selection of
        drives, (03314):67, 70

Variable frequency drive (VFD)
  selection of (*continued*)
    motors, (03314):67
    persons qualified for, (03314):66–67
  technology, (03314):63
  typical, (03314):63–64
Variable frequency drive (VFD) Application Checklist, (03314):68–69
Variable refrigerant flow (VRF) systems
  basics, (03315):34–35, 38
  ductless, (03315):40
  equipment
    communication, (03315):40–41
    compressors, (03315):38, 40
    condensing units, (03315):38, 40
    controllers, (03315):41
    data-interface, (03315):41
    oil separators, (03315):40
    remote controls, (03315):41, 45, 49
  indoor units, (03315):40
  installation, (03315):41–46
  metering devices, (03315):35
  outdoor units, (03315):40–41
  troubleshooting, (03315):46–56
  twinned units, (03315):41
Variable refrigerant volume (VRV) systems, (03314):30
Variable refrigerant volume (VRV™), (03315):35
Variable-speed centrifugal pumps, (03305):16
Variable-speed heat pumps, (03311):16–17
Variable vacuum systems, (03306):21–22
Variable volume and temperature (VVT) systems
  applications, (03315):17
  components
    bypass dampers, (03315):17–18
    relay packs, (03315):19
    typical system, (03315):17
    unit controllers, (03315):19
    zone controllers, (03315):18–19
    zone dampers, (03315):18
  operation
    communication, (03315):19–20, 21
    occupied cooling, (03315):22–23
    occupied heating, (03315):23
    pressure-dependent, (03315):20–22
    pressure-independent, (03315):20–21
    simultaneous cooling and heating, (03315):23–24
    unoccupied period, (03315):23
    wiring, (03315):19–20
  programming, (03315):16–17
Velocity sizing, (03306):28, 29
Vent draft, analyzing, (03209):28
Ventilation, demand control (DCV), (03312):16
Ventilators
  for combustion air, (03312):19, 23
  energy recovery (ERVs), (03312):19–21
  function of, (03312):19
  heat exchangers in, (03312):19
  heat recovery (HRVs), (03312):19–22
  make-up air devices, (03312):23
  recovery ventilators, troubleshooting, (03312):21–23
Venting infrared gas-fired heaters, (03209):19
Venturi tube, (03305):6, 18, 20, 44
VFD. *See* Variable frequency drive (VFD)
Vibration insulators
  between objects, (03313):23
  defined, (03313):22, 38
  ductwork, (03313):24–25
  field-fabricated, (03313):29

  flexible metal, (03313):25
  load range, (03313):23
  piping systems, (03313):25
  seismic restraints, (03313):26
  selecting, (03313):26, 29
  types of
    blower motors, (03313):22–23
    compressor, (03313):22
    fan, (03313):22
Vibration-isolating hangers, (03313):24
Voltages
  high voltage effects, (03314):43, 45
  imbalance, (03314):45
  low voltage effects, (03314):43, 45
  low-voltage power source testing, (03314):50–51
  safety factor, (03314):43
Volute, (03305):14
VRF. *See* Variable volume and temperature (VVT) systems
VRV. *See* Variable refrigerant volume (VRV) systems;
  Variable refrigerant volume (VRV™)
VVT. *See* Variable volume and temperature (VVT) systems

W

Washers
  flat, (03313):5, 7
  lock, (03313):5, 7
Water
  basic properties, (03306):1
  changing states of, (03306):1–4
  pressure-temperature relationship, (03306):3–4, 21
Water (H_2O)
  movement through piping systems, factors affecting
    head pressure, (03305):3
    pressure drop, (03305):2–3
    static pressure, (03305):3–4
  properties of, (03305):1–2
Water column, (03306):10–11, 12
Water cutoff controls, steam boilers, (03306):11–13
Water feeder controls, steam boilers, (03306):11–13
Water gauge glass, (03306):10–11
Water hammer, (03306):19, 36–37, 49, (03312):1, 6, 27
Water level indicators, steam boilers, (03306):10–11
Water pressure, calculating changes in, (03305):2
Water pumps, seismic restraints, (03313):26, 28
Water treatment, cooling towers, (03305):39
Watertube boiler, (03305):6, 9–10, 44
Wearing rings, centrifugal pumps, (03305):14
Weatherized furnace, (03310):1, 36
Wet-base, (03305):9
Wet-leg, (03305):9
Wet-pack, (03310):2
Wing nuts, (03313):5
Wire crimper, (03313):33
Wire strippers, (03313):33
Wiring
  aluminum-to-copper connections, (03311):22
  diagram abbreviations, (03311):52
  ductless split systems, (03315):26
  factory, (03311):21
  field, (03311):19, 21, 43
  VRF systems, (03315):42–46
  VVT systems, (03315):19–20
  zone, (03315):10
Wiring diagram, (03314):1, 2, 85

Y

YELLOW JACKET®, (03210):30

Z

Zone control, function of, (03315):1
Zoned hydronic operating sequence, (03314):60–61
Zoned system
  operation
    cooling, (03315):11–12
    fans, (03315):10
    heating, (03315):11–12
    high-limit, (03315):12–13
    low-limit, (03315):12–13
    minimum-on/-off times, (03315):12
    multistage systems, (03315):12
    principles of, (03315):10–11
    ventilation modes, (03315):10
  troubleshooting
    dampers, (03315):15–16
    first steps, (03315):13
    non-zoned systems, (03315):13
    panels, (03315):13–15
    thermostats, (03315):15–16
  typical, (03315):2
Zoned system, components
  control panels, (03315):6–9
  damper motors, (03315):10
  dampers
    bypass dampers, (03315):4–6
    zone dampers, (03315):2–4, 7, 9, 10
  sensors, (03315):9–10
  static-pressure sensors, (03315):9, 10
  switches, high- and low-limit, (03315):9–10
  thermostats, (03315):2–3, 9, 10–11
  typical system, (03315):1–2
  wiring, (03315):10
Zone valves, hydronic systems, (03314):59–60